Results and Problems in Cell Differentiation

Volume 68

More information about this series at http://www.springer.com/series/400

Waclaw Tworzydlo • Szczepan M. Bilinski
Editors

Evo-Devo: Non-model Species in Cell and Developmental Biology

Springer

Editors
Waclaw Tworzydlo
Institute of Zoology and Biomedical Research
Jagiellonian University
Krakow, Poland

Szczepan M. Bilinski
Institute of Zoology and Biomedical Research
Jagiellonian University
Krakow, Poland

ISSN 0080-1844 ISSN 1861-0412 (electronic)
Results and Problems in Cell Differentiation
ISBN 978-3-030-23458-4 ISBN 978-3-030-23459-1 (eBook)
https://doi.org/10.1007/978-3-030-23459-1

This Springer imprint is published by the registered company Springer Nature Switzerland AG.
The registered company address is: Gewerbestrasse 11, 6330 Cham, Switzerland

Preface

The evo-devo, short for evolutionary developmental biology, is a field of biological research that addresses questions about the developmental grounds of evolutionary changes and evolution of developmental processes. It compares mechanisms operating during development of various taxa to infer the ancestral condition of these processes and elucidate how they had evolved. In the chapters contained in the first part of the book, theoretical background of evo-devo is discussed. Alan Love and Yoshinari Yoshida review different categories of model organisms (systems) and illustrate how evo-devo model organisms are (and should be) selected. Peter Dearden discusses the phenomenon of the developmental hourglass and suggests a new "twisted ribbon" model that links variability in subsequent generations. Finally, Laurent Formery and colleagues provide an overview of hypotheses explaining the origin and gradual evolution of deuterostome nervous system. The second part combines chapters presenting new and emerging models in the evo-devo field. Among others, cnidarian *Hydra* is proposed to be a good model in analyses of the processes underlying regeneration; tunicates *Oikopleura* and *Ciona* seem to be ideal for inquiring into gene loss during evolution and cell signaling. Two non-*Drosophila* insect species, the two-spotted cricket, *Gryllus bimaculatus*, and the hairy rove beetle, *Creophilus maxillosus*, emerge as models in analyses of early embryonic development of holometabolous insects and asymmetric divisions, respectively. Slipper snails (calyptraeids), *Crepidula*, have been successfully used in studies of sex determination and sex change. Finally, phylogenic position of an enigmatic taxon Xenoturbellida makes all six described species of *Xenoturbella* ideal for inferring information on the evolution of bilaterians and their bauplans. The last part of the volume contains chapters with evo-devo approach in a broad comparative context. Various aspects of developmental processes of reptiles, marsupial frogs, mollusks, and viviparous earwigs are discussed in a phylogenetic context. Chapters focusing on the oogenesis of tardigrades, annelids, and chelicerates as well as limb regeneration and evolution of the nervous system are also included in this part.

Krakow, Poland

Waclaw Tworzydlo
Szczepan M. Bilinski

Abstract

Evolutionary developmental biology or evo-devo is a field of biological research that compares mechanisms underlying developmental processes of different organisms to infer the ancestral condition of these processes and elucidate how they had evolved. It addresses questions about developmental bases of evolutionary changes and evolution of developmental processes. The content of the book is divided into three parts. In the chapters contained in the first one, the theoretical background of evo-devo is discussed. The second part provides chapters presenting new and emerging model organisms in the evo-devo field. The last part of the volume contains chapters with evo-devo approach in a broad comparative context. To the best of our knowledge, there is no other book combining those three different evo-devo aspects: theoretical considerations, a comprehensive list of emerging model species, and comparative analyses of developmental processes. This will give the readers a new perspective on naturally existing diversity of processes operating in cells and during development of various animal groups and enrich the minds of current and newcomer researchers.

Contents

Part I
Theoretical Background

Chapter 1
Reflections on Model Organisms in Evolutionary Developmental Biology

Alan C. Love and Yoshinari Yoshida

Abstract This chapter reflects on and makes explicit the distinctiveness of reasoning practices associated with model organisms in the context of evolutionary developmental research. Model organisms in evo-devo instantiate a unique synthesis of model systems strategies from developmental biology and comparative strategies from evolutionary biology that negotiate a tension between developmental conservation and evolutionary change to address scientific questions about the evolution of development and the developmental basis of evolutionary change. We review different categories of model systems that have been advanced to understand practices found in the life sciences in order to comprehend how evo-devo model organisms instantiate this synthesis in the context of three examples: the starlet sea anemone and the evolution of bilateral symmetry, leeches and the origins of segmentation in bilaterians, and the corn snake to understand major evolutionary change in axial and appendicular morphology.

1.1 Introduction

In a prescient methodological paper of the early 1990s, James Hanken argued that evolutionary developmental biology (evo-devo) must combine a model-systems (or model organisms) strategy, exemplified in developmental biology, with a comparative strategy, exemplified in areas of evolutionary biology, to study cranial development and evolution (Hanken 1993). This insight had historical precedent (de Beer 1985 [1937]) and was straightforward conceptually—evo-devo should combine aspects of developmental models and aspects of evolutionary models—but subtle to implement, in large part because these approaches to studying organisms and their traits are in tension with each other (Collins et al. 2007; Minelli and Baedke 2014). Developmental model organisms are used to establish core

A. C. Love (✉) · Y. Yoshida
Department of Philosophy and Minnesota Center for Philosophy of Science, University of Minnesota - Twin Cities, Minneapolis, MN, USA
e-mail: aclove@umn.edu

W. Tworzydlo, S. M. Bilinski (eds.), *Evo-Devo: Non-model Species in Cell and Developmental Biology*, Results and Problems in Cell Differentiation 68,
https://doi.org/10.1007/978-3-030-23459-1_1

similarities exemplified by many taxa (i.e., conservation), especially with an eye toward medical application (Bier and McGinnis 2003), whereas a comparative strategy is utilized to establish significant patterns of difference (i.e., evolutionary change) manifested across a clade (Harvey and Pagel 1991). Given this tension between investigating similarities and differences of model organisms in evo-devo, it is unsurprising that researchers often discuss their status and distinctiveness (Jenner 2006; Collins et al. 2007; Jenner and Wills 2007; Milinkovitch and Tzika 2007; Russell et al. 2017), especially in calls to introduce and standardize new ones (Crotty and Gann 2009; Lapraz et al. 2013; Braasch et al. 2015; Medina Jiménez et al. 2017).

Although it is commonly recognized that evo-devo model organisms are associated with distinctive reasoning practices, much attention has been given to the lingering effects of standard model organisms from developmental biology, such as *Drosophila*, because these played a key role in touchstone studies that helped establish contemporary evo-devo (Carroll 1995; McGinnis et al. 1984). "Model organisms, in particular the big six, are conceptual carry-over from developmental biology, but their study was crucial in establishing evo–devo as a new discipline" (Jenner and Wills 2007, p. 311). The recommendations that follow from this type of concern are laudable (e.g., "judicious choice of new model organisms is necessary to provide a more balanced picture") and build on analyses showing how numerous features of standard developmental model organisms are not representative, such as rapid rates of development and high degrees of canalization (Bolker 1995). *C. elegans* embryogenesis is not representative of nematodes for pattern formation and cell specification (Schulze and Schierenberg 2011), mice do not exemplify key features of early mammalian development (Berg et al. 2011), and zebrafish fin formation is not a good proxy for the ontogeny of tetrapod appendages more generally (Metscher and Ahlberg 1999). However, these recommendations conceal an underlying logic to the combining of different modeling approaches that is suggested in Hanken's earlier discussion and was glimpsed by others at the same time (Kellogg and Shaffer 1993).

The purpose of this chapter is to reflect on and make explicit the distinctiveness of model organisms in the reasoning of evolutionary developmental research. We argue that model organisms in evo-devo instantiate a distinctive synthesis of model systems strategies from developmental biology and comparative strategies from evolutionary biology that can negotiate the essential tension between developmental conservation and evolutionary change (Collins et al. 2007; Minelli and Baedke 2014). Evo-devo model organisms are typically selected and evaluated given their potential for solving problems concerning both the evolution of development and the developmental basis of evolutionary change, as well for their experimental tractability (Sommer 2009). In condensed form, the results acquired by studying experimentally tractable model organisms in evo-devo are generalized or extrapolated to particular taxa (developmental modeling emphasizing similarities) to which other taxa are then compared phylogenetically (evolutionary modeling emphasizing differences). They fulfill Hanken's original aim of integrating both strategies to study complex phenomena like cranial development and evolution: "Solution of many outstanding problems ... will require a combined approach that incorporates the technical and conceptual strengths of each" (Hanken 1993, p. 455).

We begin our discussion with a review of different categories of model systems that have been advanced to understand practices found in the life sciences. This sets the stage to comprehend how evo-devo model organisms instantiate a synthesis of different modeling strategies from both development and evolution. We illustrate this type of synthesis with three examples: the starlet sea anemone used to address questions related to the evolution of bilateral symmetry in metazoans, species of leeches used to study processes of segmentation and represent lophotrochozoans for comparison with better studied chordates and ecdysozoans, and the corn snake used to investigate the origin of increased numbers of vertebrae and the reduction or absence of limbs. In conclusion, we propose that the "judicious choice of new model organisms is necessary" in evo-devo not only to provide a more balanced picture of developmental processes but also as an ongoing methodological strategy for answering questions about the evolution of development and the developmental basis of evolutionary change.

1.2 Different Categories of Biological Model Systems

The reasoning and material practices surrounding model organisms and model systems in the life sciences are diverse. What one disciplinary approach refers to as a model organism can be quite distinct from what another disciplinary approach designates a model organism. This is exacerbated by the fact that biologists also routinely discuss "non-model" organisms, which have different contrast classes for what are supposed or assumed to be "model" organisms (e.g., Russell et al. 2017). We have no delusions of policing this semantic diversity, nor would we presume to attempt to because these ambiguities of meaning can be productive in ongoing research. Instead, to the end of characterizing the distinctiveness of evo-devo model organisms, we rehearse a variety of different categories that have been introduced to capture different facets and distinctions latent in these diverse meanings associated with "model organisms."

1.2.1 Representation and Manipulation

Biological model systems are evaluated in terms of two major criteria: representation and manipulation (Love and Travisano 2013). The former concerns what biological systems a model can represent and to what extent. The latter concerns ease of empirical examination of a model. Ankeny and Leonelli (2011) emphasize two dimensions underlying the representational role of biological model systems: scope and target. *Representational scope* describes how widely and to what biological systems the results and lessons learned by studying a model system are projected. If a research group uses zebrafish to learn about vertebrates in general, then zebrafish as a model organism is supposed to have vertebrates as its representational scope. *Representational target* indicates what biological phenomena are explored by studying the model system. If the research group uses zebrafish to explore nervous system development, then in this situation the development of the nervous system is the representational target. In many cases, the choice of representational target is connected to the representational scope. For example, the study of genetic and cellular *mechanisms* in development (representational target) and the discovery that they are widely conserved evolutionarily (representational scope) is a significant motivator for the continued use of model systems (Gerhart and Kirschner 2007; Ankeny and Leonelli 2011). Additionally, the representational target might be a higher-level *phenomenon* instead of a molecular mechanism, and either of these can be scrutinized narrowly (*specificity*) or with respect to a range of variation on the theme (*variety*) (Love and Travisano 2013).

Many different factors are relevant to the criterion of manipulation for model systems. Examples include the availability of the biological system, the cost of initiating inquiry on it, possible experimental techniques that can be applied, and how quickly one can produce data and results (Love and Travisano 2013). Although there are differences in the degree to which these factors must be present and how they are fulfilled (e.g., regular availability might be achieved through chemically preserving and storing specimens), it is widely accepted that manipulation is a crucial criterion for biological model systems. Furthermore, the criteria of representation and manipulation are interrelated. In some cases, they exhibit a trade-off relationship, such as an organism that faithfully represents organisms of interest might be difficult to experimentally manipulate or an organism that is easy to manipulate might represent a group or phenomenon of interest poorly. Biologists choose and evaluate model systems by considering these criteria jointly, giving weight to different factors in terms of what they aim to accomplish (i.e., their research purposes).

1.2.2 Model Organisms and Experimental Organisms

In addition to those already described, previous philosophical studies have introduced a number of significant distinctions about biological model systems. Ankeny and Leonelli (2011) distinguish *model* organisms from what they call *experimental* organisms. These differ in their representational scope, representational target, manipulative requirements, and purposes of research. "Model organisms" correspond to a limited number of species that have been widely used in recent biological research, such as mouse, zebrafish, fruit fly, and *C. elegans*. Studies of model organisms have various genetic, developmental, physiological, ecological, and evolutionary phenomena as their target, especially those that occur in organisms generally. Results of these studies (e.g., identified mechanisms) are generalized to a wide range of other species. Important manipulative requirements for model organisms (in this sense) are the ability to undertake genetic analysis and successful standardization to minimize confounding variation (e.g., pure lines and standard strains). They are studied for the purpose of developing "an integrative understanding of intact organisms in terms of their genetics, development, and physiology, and in the longer run of evolution and ecology, among other processes" (Ankeny and Leonelli 2011, p. 319).

In contrast, "experimental organisms" are studied to explore particular biological phenomena. They exemplify the principle proposed by August Krogh: "For a large number of problems there will be some animal of choice or a few such animals on which it can be most conveniently studied" (Krogh 1929, p. 202). Historical examples include sea urchins for studying fertilization or early embryogenesis and frogs for studying the role of electricity in muscle contraction. Experimental organisms are chosen to answer specific questions, which means that each one has specific phenomena as its representational targets (Burian 1993). In the case of experimental organisms, the results of the research are often (though not necessarily) generalized or extrapolated more narrowly than results derived from model organisms. They also are not as standardized and less suitable for many forms of genetic analysis. However, this is not an inherent drawback because manipulative requirements vary depending on what questions are being addressed. The giant squid axon was strategic for electrophysiological experiments to ascertain neuronal function because its size (~0.5 mm in diameter) made it possible to reliably apply voltage clamps (Hodgkin and Huxley 1952).

1.2.3 Exemplars Versus Surrogates

Another helpful distinction has been made between *exemplary* and *surrogate* models (Bolker 2009). This distinction consists in what kind of inference predominates in the research using the model system. Exemplary models are studied as examples of larger groups of organisms, such as zebrafish as a model of vertebrates or *Drosophila* as a model of animals generally. The results acquired investigating an exemplary model are generalized to a larger group of which the model species is a member. Standard developmental biology model organisms are exemplary in this sense: "The motivation for their study is not simply to understand how that particular animal develops, but to use it as an example of how all animals develop" (Slack 2006, p. 61). On the other hand, surrogate models serve as substitutes for particular biological systems. The results derived from research on a surrogate model are extrapolated to a specific target. A predominant example is the use of mouse as a model of *Homo sapiens* (e.g., Cheon and Orsulic 2011). The inference is from a proxy to a target instead of from an exemplar to other taxa more generally, as is the case for exemplary models. The different roles played by exemplary models and surrogate models arise from different investigative purposes. For exemplary models, the goal is to identify widely shared biological mechanisms or better understand evolutionary processes. As a result, exemplary models are associated with basic research. In contrast, surrogate models are used in more applied fields, such as biomedicine or conservation research, to better understand medical and ecological problems and develop potential solutions to them.

1.2.4 Model Taxa and Model Life Histories

Although most philosophical accounts have been concerned primarily with organisms of specific *species* that are used as models, there also are perspectives that focus on differently arrayed biological systems. The notion of a model *taxon* refers to a clade that is used to investigate diverse questions about genetic, developmental, physiological, ecological, and evolutionary phenomena both within the clade and with the aim of identifying generalizations applicable to other clades (Griesemer 2015). Thus, the representational scope of a model taxon is understood as taxa within and beyond the clade, with results ascertained through investigation applying differentially to individual taxa of smaller and larger sizes. Inquiry is organized around interrelated "packages of phenomena" as the representational target—rather than individual phenomena—and constitutes central features of the evolution of the taxon. For example, lungless salamanders have been studied as a model taxon and have provided explanatory insights about the evolution of anatomical features (e.g., the tongue or body size) and associated functional systems (e.g., feeding, locomotion), especially with respect to mechanisms that contribute to these patterns (e.g., miniaturization derived from changes in developmental timing or large cell size resulting from constraints of genome size) (Wake and Larson 1987; Wake 2009). Another prominent example of a model taxon used to investigate ecological and evolutionary questions is the squamate genus *Anolis*, which is composed of a large number of lizard species with good phylogenetic resolution, while also being regionally delimited and relatively accessible (Sanger 2012; Stroud and Losos 2016).

Griesemer uses the term "export" to refer to the predominant inference made from model taxa, where exportation is distinguished from inferences involving simple generalization. The latter is common in molecular biology and operates by assuming that species are instances of the same type. Exportation is based on the idea that taxa are historical individuals and in lineal relationships with one another. Although discussions of model organisms tend to focus on how particular results acquired by studying them are extrapolated, generalized, or exported, Griesemer also emphasizes additional kinds of payoffs that can be acquired by studying model taxa. These include methodological lessons about how to investigate different taxa.

A final category of biological model system is a model *life history* (Love and Strathmann 2018). These are temporal sequences that occur within the ontogeny of organisms, characterized in terms of functional and morphological properties, which are used as models of sequences within development found in other species. Marine invertebrate larvae are a primary example of a model life history (Love 2009). Unlike a model taxon, these stages of life history are not unified within a single monophyletic group; often, studies of marine larvae involve cross-clade comparisons of functional requirements for specific ecological settings that exhibit a broad but disjoint taxonomic distribution (i.e., their representational scope). The representational targets of investigations of model life histories are functional or morphological traits in ontogenetic sequences relevant to questions about developmental, ecological, and evolutionary phenomena. Some resulting generalizations revolve around particular instantiations of larval forms that exemplify a broader type (e.g., pluteus or trochophore), while others pertain to behavioral and ecological patterns, such as feeding versus nonfeeding or planktonic versus benthic. Model life histories concentrate attention on problems related to ecology, evolution, adaptation, and phylogeny and serve to coordinate research by scientists from different disciplines, often at marine stations where the availability, cost, and infrastructural prerequisites for manipulation are in place to support the relevant configuration of approaches simultaneously.

1.3 Examples of Evo-Devo Model Organisms

The reasoning strategies associated with evo-devo model organisms have distinctive features and are not sufficiently characterized by any one of the accounts of biological model systems reviewed above. One central feature is the importance of phylogenetically informed comparison. Although results and lessons acquired by studying model organisms in evo-devo are generalized or extrapolated, similar to what is seen for the use of many other model systems, this is followed by a process of comparison with taxa related by specific patterns of common descent. These processes of comparison are crucial for elucidating the origin of novel traits in a lineage, the evolution of properties of ontogeny, and dissecting the relevant influence of developmental processes on evolutionary mechanisms, all of which comprise research purposes governing the use of model organisms in evolutionary developmental research (Collins et al. 2007).

*1.3.1 The Starlet Sea Anemone (*Nematostella vectensis*)*

A good example of an evo-devo model organism is the starlet sea anemone, *Nematostella vectensis* (Darling et al. 2005; Genikhovich and Technau 2009; Layden et al. 2016). *Nematostella* has many practical advantages for experimental studies: it can be maintained easily in little space and at a low cost, adults reproduce under laboratory conditions about once a week and throughout the year, eggs are large enough for manipulation, and the generation time is relatively short (Darling et al. 2005). Resources and experimental techniques available include an annotated genome, visualization techniques such as in situ hybridization and immunohistochemical analysis, and knockdown/knockout techniques from molecular genetic analysis (e.g., Morpholinos, CRISPR/Cas9, and TALEN), as well as overexpression through mRNA injection (Layden et al. 2016).

An important problem to which *Nematostella* can contribute answers is the evolution of bilaterality. Bilateral symmetry was one of the major evolutionary novelties that originated in animals more than 600 million years ago, and evo-devo researchers have tried to reveal when and explain how the two major body axes (anterior–posterior [A–P] and dorsal–ventral [D–V]) emerged. To accomplish this requires comparing bilaterian axis formation (e.g., in *Drosophila* or mouse) with the ancestral pattern of development. *Nematostella* belongs to the phylum Cnidaria, which is an outgroup of Bilateria that includes corals, jellyfish, hydras, and sea anemones. It is expected to serve as a model of the ancestral pattern of development that basal metazoans exhibited in the past on the assumption that extant members of this outgroup have retained significant features of this pattern. Although it appears to be radially symmetrical, *Nematostella* has two body axes. The oral–aboral (O–A) axis runs from the mouth to the other end of the body; the directive axis runs across the pharynx and is orthogonal to the O–A axis. Molecular developmental studies have revealed relations between these body axes of *Nematostella* and the A–P and D–V axes of bilateria, which give clues as to how the primary and secondary axes of symmetry in animals originated.

For example, Wnt/β-catenin signaling is known to play a crucial role in A–P axis specification across bilaterians (Petersen and Reddien 2009). Thus, the function of Wnt/β-catenin signaling in O–A axis specification in *Nematostella* has been examined. In many bilaterian species, Wnt/β-catenin signaling is differentially activated at certain stages of embryonic development and in different locations of the embryo. During early embryogenesis, the side of the embryo with a high Wnt/β-catenin signaling activity develops into the posterior end, while the side with lower signaling activity becomes the anterior end (Petersen and Reddien 2009). In *Nematostella*, from the mid-blastula stage, Wnt gene expression exhibits a demarcated, staggered expression within the oral half of the embryo (Kusserow et al. 2005). Furthermore, overactivation of Wnt/β-catenin signaling promotes oral identity, while inhibition of Wnt/β-catenin signaling leads to expanded expression of aboral markers and reduction of oral marker expression (Röttinger et al. 2012). These results point toward a role for Wnt/β-catenin signaling in primary axis specification that existed before the separation of Bilateria and Cnidaria (Petersen and Reddien 2009). They also suggest a potential correspondence between bilaterian anterior and cnidarian aboral sides, on the one hand, and between bilaterian posterior and cnidarian oral sides, on the other (Layden et al. 2016), though this remains unclear (see below).

Relationships between the bilaterian D–V axis and the directive axis of *Nematostella* have also been examined. In bilaterian embryogenesis, BMP signaling plays a crucial role in the specification of the D–V axis. For example, BMP and its antagonists (e.g., Chordin) are expressed at opposite sides of the body in vertebrates and insects. Their interactions lead to a gradient of BMP signaling that helps to specify the D–V axis (De Robertis 2008). In *Nematostella*, homologs of BMP pathway genes and Chordin are asymmetrically expressed along the directive axis around and after gastrulation (Finnerty et al. 2004; Matus et al. 2006), and BMP signaling is required for directive axis formation (Saina et al. 2009). However, unlike vertebrates and insects, these genes are expressed on the same side of the body (Rentzsch et al. 2006). Additionally, they seem to constitute, along with other genes, a signaling network of different topology than is found in many bilaterians (Saina et al. 2009). Therefore, although bilaterians and *Nematostella* use common molecular signaling mechanisms in axis specification, it is unclear whether the directive axis of *Nematostella* is homologous to the D–V axis of Bilateria (Rentzsch et al. 2006; Layden et al. 2016). More recent work demonstrates that the directive axis of *Nematostella* is under the control of an axial Hox gene code—a developmental characteristic of the metazoan A–P axis—which indicates that molecular signaling pathways from both primary and secondary axis specification mechanisms in bilaterians are present in directive axis specification (He et al. 2018). This suggests that there is no straightforward homology relationship between the O–A and directive axis of *Nematostella* and the A–P and D–V axes of bilaterians.

*1.3.2 Leeches (*Helobdella*) and Corn Snake (*Pantherophis guttatus*)*

Let us briefly consider two more examples of evo-devo model organisms. Leeches of the genus *Helobdella* have played an important role in research into the evolutionary history of segmentation (Martindale and Shankland 1990; Weisblat and Kuo 2014; Kuo and Lai 2019). Each of the three superphyla of Bilateria (Deuterostomia, Ecdysozoa, and Lophotrochozoa) includes both segmented and unsegmented taxa, which prompts a question about how many times segmentation has evolved in bilaterian lineages (Davis and Patel 1999; Minelli and Fusco 2004). Comparing segmentation mechanisms of different superphyla is crucial to answering this question. However, unlike the other two superphyla that include standard model organisms, our knowledge about developmental mechanisms within Lophotrochozoa has been limited. A major motivation to study *Helobdella* as a model organism is to provide information about segmentation mechanisms of Lophotrochozoa for comparison with the other superphyla (Weisblat and Kuo 2009, 2014; Kutschera and Weisblat 2015).

Another example of an evo-devo model organism is the corn snake *Pantherophis guttatus*, which has been studied as a model for morphological evolution in modified body plans (Guerreiro and Duboule 2014). An increased number of vertebrae and the loss of limbs are characteristic in snakes. Hence, evo-devo researchers have been interested in how such an "extreme body plan" evolved (Woltering 2012), especially because a similar body plan has evolved independently in other lineages, such as in limbless anguids (slow worms or glass lizards) and caecilians (limbless amphibians). Again, as with other evo-devo model organisms, comparison of developmental mechanisms between taxa is an important step in elucidating how such radical morphological changes have occurred. The corn snake together with several other species have contributed to this research by providing details of vertebral formation and the reduction or elimination of appendage formation in snake embryogenesis (Castoe et al. 2013; Cohn and Tickle 1999; Head and Polly 2015; Woltering et al. 2009). This makes it possible to execute informative, phylogenetic comparisons with the development of other tetrapods to answer questions about the evolution of developmental features contributing to the extreme axial and appendicular morphology of snakes (Gomez et al. 2008; Guerreiro et al. 2016; Kvon et al. 2016).

1.4 Evo-Devo Model Organisms as a "Synthesis"

The examples of evo-devo model organisms in Sect. 1.3 illustrate unique features of experimentally tractable model systems that represent a synthesis of the generalization or extrapolation of developmental mechanisms and phylogenetic comparison to answer questions about the evolution of development and the developmental basis of evolutionary change. In this section, we characterize reasoning strategies associated with evo-devo model organisms using the conceptual tools provided by previous philosophical accounts (detailed in Sect. 1.2) and contrast the features of model organisms in evo-devo with other categories of biological model systems to demonstrate their distinctive status.

1.4.1 Not Model Taxa or Model Life Histories

Evo-devo model organisms are typically one species rather than other collectives or units, which makes them distinct from model taxa and model life histories. To use a model taxon, one studies multiple species in the clade with the aim of exporting the lessons learned to other members of the same clade or to different clades. In contrast, an evo-devo model organism is a specific species that is used to produce results that can be generalized or extrapolated to another species rather than across the entire clades. For example, increased Wnt/β-catenin signaling in *Nematostella* promotes oral identity in the establishment of the O–A axis (Röttinger et al. 2012), which can be generalized to primary axis specification in a representative ancestral metazoan species extant prior to the split between Bilateria and Cnidaria hundreds of millions of years ago (Petersen and Reddien 2009). Additionally, leeches from the genus *Helobdella* qualify as model *organisms* rather than a model *taxon* because researchers select some species from the genus and study them as, for example, models of segmentation in other lophotrochozoan species (Kutschera and Weisblat 2015). The entire genus is not studied thoroughly in terms of packages of phenomena; only a few selected species of the genus are studied in terms of individual phenomena (e.g., segmentation). By parallel reasoning, evo-devo model organisms are distinguished from model life histories because the latter category is applied to specific temporal sequences within development. This does not mean that model life histories cannot help to address evolutionary developmental questions. The crucial point is that they do so in a different fashion, such as by generalizing particular functional requirements of dispersal or feeding for larval forms during ontogeny in specific ecological settings.

1.4.2 Sometimes Exemplar, Sometimes Surrogate

An evo-devo model organism can serve as an exemplar of a larger class of species or as a surrogate of a particular species. The former applies when one or a few species belonging to a particular clade are studied to elucidate how traits characteristic of that clade have evolved. Inquiry into the evolution of axial and appendicular morphology in corn snake falls under this category (Guerreiro and Duboule 2014). Evo-devo model organisms also are studied as exemplars of species in a taxon that has been underrepresented in developmental research; information about developmental mechanisms of species in this taxon is required to elucidate the evolution of widely distributed (though not necessarily homologous) traits. *Helobdella* leeches in the investigation of segmentation are exemplary models in this way. Researchers expect them to provide information about segmentation processes in Lophotrochozoa to facilitate answering questions about the origins and evolution of segmentation in Bilateria (Weisblat and Kuo 2009).

On the other hand, the role that *Nematostella* plays as a model organism in evo-devo appears slightly more complicated. In some cases, *Nematostella* is regarded as an exemplar of anthozoan cnidarians or cnidarians in general or even as an exemplar of animals that exhibit developmental processes such as asexual fission and regeneration (e.g., Burton and Finnerty 2009). In other cases, such as studies related to the evolution of bilaterality, *Nematostella* serves as a surrogate model of extinct basal metazoans. *Nematostella* is likely to have retained ancestral features of basal metazoan axis specification and thus has potential to be a surrogate of the last common ancestor of Bilateria and Cnidaria to study how bilaterality evolved. This is indicated in the ways that biologists discuss the representational roles of *Nematostella* in the context of research on the origins of bilateral symmetry: "many ancestral traits have been preserved in *Nematostella* … this makes *Nematostella* a very attractive model system among the representatives of *basal* metazoan lineages" (Genikhovich and Technau 2009, emphasis added). Importantly, *Nematostella* does not have to represent basal metazoans with respect to all traits in order to serve as a surrogate model. What is required in this context is that it represents the last common ancestor of Bilateria and Cnidaria *with respect to body axis specification*. Usefulness as a surrogate can vary depending on which trait is of interest.

Although the distinction between exemplary and surrogate models is useful to capture some of the representational roles played by evo-devo model organisms, other details of how this distinction is characterized are not readily applicable. For example, Bolker (2009) states that a major aim of using exemplary models is to elucidate widely conserved mechanisms. This is not primarily the case for evo-devo model organisms. They are studied to elucidate developmental mechanisms characteristic of species in a taxon to compare with corresponding developmental mechanisms of species in other taxa. Discovering that these mechanisms are widely conserved is not the purpose of investigation; instead, the aim is to uncover how these developmental mechanisms changed (i.e., evolved). Bolker also holds that the central aims of using surrogate models are to understand disease etiology, identify possible therapies, and conserve threatened species. These features of surrogate models are not typically applicable to evo-devo model organisms, though in some cases this is relevant, such as comparative studies of axis specification established by *Hox* gene expression bearing on patterns of human vertebral pathology (ten Broek et al. 2012). The use of *Nematostella* as a surrogate model of basal metazoans is more typical and motivated by an interest in evolutionary history, especially the origin of bilateral symmetry.

1.4.3 Neither Model Organism, Nor Experimental Organism

Evo-devo model organisms are not model organisms in the sense of Ankeny and Leonelli (2011) because they are not intended to have a wide range of species as the representational scope and diverse phenomena as the representational target. The category of experimental organism appears better suited to evo-devo model organisms. As is expected for experimental organisms, an evo-devo model organism represents a limited range of taxa (e.g., Lophotrochozoa in the case of *Helobdella*) and extrapolation or generalization focuses on specific biological phenomenon (e.g., segmentation). However, the characterization of experimental organisms is not adequate to understand evo-devo model organisms either because the latter involve several unique features. One is that the relationship between the choice of an evo-devo model organism and the research questions being asked is more complex than what is found in cases of experimental organisms. The role of an experimental organism is to be a convenient system for studying specific developmental, physiological, genetic, or behavioral phenomenon. Thus, if a species exhibits the phenomenon of interest and satisfies relevant manipulation criteria, it can be a satisfactory experimental organism. On the other hand, exhibiting particular phenomena is only part of the representational requirement for a species to be an evo-devo model organism; its phylogenetic location in the evolutionary tree is also critical (see below).

1.4.4 The Distinctive Synthesis of Evo-Devo Model Organisms

The generalization or extrapolation of experimental results from an evo-devo model organism is followed routinely by a comparison between the taxa in which the generalization or extrapolation applies and other taxa where it does not. Such a comparison is crucial to elucidate the evolution of the traits under scrutiny. Therefore, the goal of using an evo-devo model organism is not to examine mechanisms underlying a particular developmental, physiological, genetic, or behavioral phenomenon. Rather, the examination of these mechanisms is a means to the end of answering broader evolutionary questions. Consider again *Nematostella*. A major motivation for studying it is to address the problem of how bilateral symmetry evolved in the lineage of Bilateria (Darling et al. 2005; Layden et al. 2016). This evolutionary problem is composed of many lower-level questions, such as how the different mechanisms that establish primary and secondary axes in bilaterians operate. Scientific problems constitute hierarchical structures; broad problem domains are composed of many different but related questions (Brigandt and Love 2012; Love 2008, 2014).

To account for the origin and evolution of bilateral symmetry, we have to examine many different but related questions, such as how and in what different ways the A–P axis is determined during bilaterian embryogenesis and how and in what different ways the D–V axis is determined during bilaterian embryogenesis. If *Nematostella* serves successfully as a surrogate model of the last common ancestor of Bilateria and Cnidaria, then mechanisms of its axis specification can be extrapolated to the ancestral metazoan. Researchers can then compare the (hypothetical, extinct) ancestral patterns of axis specification with those in extant bilaterian models. This comparison is a crucial step to help account for the evolution of bilaterality. The same kind of hierarchical organization of questions operates in the other two cases. For example, segmentation is a complex developmental process that requires answers to many distinct questions about the initiation, spatial arrangement, and number of segments. These questions are addressed by investigating segmentation mechanisms in *Helobdella* species and then generalized to Lophotrochozoa for broader comparison with segmentation mechanisms in Arthropoda and Chordata. This broader comparison is a means to the end of addressing the problem of how segmentation has evolved within Bilateria.

Evo-devo model organisms are chosen and evaluated on the basis of their potential contributions to answering research questions about the evolution of development and developmental basis of evolutionary change. Answering such questions involves comparisons of developmental patterns and mechanisms found in different lineages. Therefore, the precise location of a species within the evolutionary tree is a critical factor for an evo-devo model organism. As a consequence, evo-devo model organisms and experimental organisms (sensu Ankeny and Leonelli 2011) are distinct in an important respect. Unlike experimental organisms, evo-devo model organisms are not chosen simply because they are experimentally tractable and exhibit interesting biological phenomena. They have to occupy appropriate phylogenetic positions, as well as exhibit particular phenomena, so that effective comparisons can be made to answer important questions that comprise the research problems of evo-devo, such as the properties underlying evolvability or the origin of novel traits (Jenner 2006; Collins et al. 2007; Jenner and Wills 2007; Milinkovitch and Tzika 2007; Sommer 2009; Minelli and Baedke 2014).

Evo-devo model organisms instantiate a distinctive synthesis of model systems strategies from developmental biology and comparative strategies from evolutionary biology. Regarding the former, they are experimentally tractable species that act as exemplars or surrogates. The results acquired by studying them can be extrapolated to a specific species or generalized to a larger group of species. Regarding the latter, evo-devo model organisms depend heavily on what phylogenetic comparisons they make possible. The purpose of studying them is to answer different questions about the evolution of development and the developmental basis of evolutionary change. Strategic comparisons between taxa is an essential step in this methodology, which means the "judicious choice of new model organisms is necessary" as an ongoing strategy. Overall, evo-devo model organisms, illustrated in the cases of *Nematostella*, *Helobdella* leeches, and corn snake, navigate the essential tension between developmental conservation and evolutionary change by uniquely integrating model systems and comparative approaches to study complex phenomena at the intersection of development and evolution.

1.5 Conclusion

Our analysis herein has concentrated on the core reasoning strategies that underlie the selection and evaluation of most evo-devo model organisms. A number of important conceptual questions have been neglected as a consequence. For example, material practices associated with the laboratory maintenance and manipulation of model organisms for reproducible research involves the idealization of variation exhibited by developmental mechanisms (Love 2010; Minelli and Baedke 2014). These practices make it difficult, if not impossible, to study particular kinds of phenomena, such as phenotypic plasticity, which are relevant to developmental evolution (e.g., Moczek et al. 2011). The nature of these practices and various strategies for negotiating among their inherent trade-offs to address questions about the evolution of development and the developmental basis of evolutionary change requires a separate discussion (Love 2010; Minelli and Baedke 2014). Additionally, we have focused exclusively on animal models in evolutionary developmental research. Although many of our claims about core reasoning strategies are transferable, an explicit treatment of models in other domains is warranted, especially for plant evo-devo (Plackett et al. 2015; Vandenbussche et al. 2016; Yuan 2019) and for anatomical units like the vertebrate limb (Collins et al. 2007; Zuniga 2015).

Answering central questions in evo-devo requires both intensive experimental examination of developmental mechanisms in selected species—a common strategy in developmental biology—and phylogenetic comparison of different species within and across taxa—a common strategy in evolutionary biology. We have argued that model organisms in evo-devo instantiate a distinctive synthesis of these two strategies. Evo-devo model organisms are experimentally tractable species that serve as either exemplars or surrogates. The results acquired through their study are generalized or extrapolated to a larger group or a particular species, to which other species are compared in order to solve evolutionary problems. *Nematostella vectensis* is studied as a surrogate of extinct basal metazoans to which Bilateria is compared to answer questions about the evolution of bilateral symmetry, leeches of the genus *Helobdella* serve as an exemplar of Lophotrochozoa to which Deuterostomia and Ecdysozoa are compared to elucidate the origins and evolution of segmentation, and *Pantherophis guttatus* is an exemplar of snakes to better understand the developmental mechanisms behind their extreme changes in axial and appendicular morphology. The involvement of phylogenetic comparison as an essential part of evo-devo research makes these model organisms a distinctive category that deserves special methodological consideration.

Acknowledgments We are grateful to Max Dresow, Nathan Lackey, Katherine Liu, and Aaron Vesey for feedback on an earlier version of the manuscript.

References

Ankeny R, Leonelli S (2011) What's so special about model organisms? Stud Hist Phil Sci 42:313–323

Berg DK, Smith CS, Pearton DJ, Wells DN, Broadhurst R, Donnison M, Pfeffer PL (2011) Trophectoderm lineage determination in cattle. Dev Cell 20:244–255

Bier E, McGinnis W (2003) Model organisms in the study of development and disease. In: Epstein CJ, Erickson RP, Wynshaw-Boris A (eds) Molecular basis of inborn errors of development. Oxford University Press, New York, pp 25–45

Bolker JA (1995) Model systems in developmental biology. BioEssays 17:451–455

Bolker JA (2009) Exemplary and surrogate models: two modes of representation in biology. Perspect Biol Med 52:485–499

Braasch I, Peterson SM, Desvignes T, McCluskey BM, Batzel P, Postlethwait JH (2015) A new model army: emerging fish models to study the genomics of vertebrate evo-devo. J Exp Zool (Mol Dev Evol) 324:316–341

Brigandt I, Love AC (2012) Conceptualizing evolutionary novelty: moving beyond definitional debates. J Exp Zool (Mol Dev Evol) 318B:417–427

Burian RM (1993) How the choice of experimental organism matters: epistemological reflections on an aspect of biological practice. J Hist Biol 26:351–367

Burton PM, Finnerty JR (2009) Conserved and novel gene expression between regeneration and asexual fission in *Nematostella vectensis*. Dev Genes Evol 219:79–87

Carroll SB (1995) Homeotic genes and the evolution of arthropods and chordates. Nature 376:479–485

Castoe TA et al (2013) The Burmese python genome reveals the molecular basis for extreme adaptation in snakes. Proc Natl Acad Sci USA 110:20645–20650

Cheon D-J, Orsulic S (2011) Mouse models of cancer. Annu Rev Pathol Mech Dis 6:95–119
Cohn MJ, Tickle C (1999) Developmental basis of limblessness and axial patterning in snakes. Nature 399:474–479
Collins JP, Gilbert SF, Laubichler MD, Müller GB (2007) Modeling in EvoDevo: how to integrate development, evolution, and ecology. In: Laubichler MD, Müller GB (eds) Modeling biology. MIT Press, Cambridge, MA, pp 355–378
Crotty DA, Gann A (eds) (2009) Emerging model organisms: a laboratory manual, vol 1. Cold Spring Harbor Laboratory Press, Cold Spring Harbor, NY
Darling JA, Reitzel AR, Burton PM, Mazza ME, Ryan JF, Sullivan JC, Finnerty JR (2005) Rising starlet: the starlet sea anemone, *Nematostella vectensis*. BioEssays 27:211–221
Davis GK, Patel NH (1999) The origin and evolution of segmentation. Trends Genet 9:M68–M72
de Beer GR (1985 [1937) The development of the vertebrate skull. University of Chicago Press, Chicago
De Robertis EM (2008) Evo-devo: variations on ancestral themes. Cell 132:185–195
Finnerty JR, Pang K, Burton P, Paulson D, Martindale MQ (2004) Origins of bilateral symmetry: *Hox* and *Dpp* expression in a sea anemone. Science 304:1335–1337
Genikhovich G, Technau U (2009) The starlet sea anemone *Nematostella vectensis*: an anthozoan model organism for studies in comparative genomics and functional evolutionary developmental biology. Cold Spring Harb Protoc. https://doi.org/10.1101/pdb.emo129
Gerhart J, Kirschner M (2007) The theory of facilitated variation. Proc Natl Acad Sci USA 104:8582–8589
Gomez C, Özbudak EM, Wunderlich J, Baumann D, Lewis J, Pourquié O (2008) Control of segment number in vertebrate embryos. Nature 454:335–339
Griesemer JR (2015) What salamander biologists have taught us about evo-devo. In: Love AC (ed) Conceptual change in biology: scientific and philosophical perspectives on evolution and development. Springer, Dordrecht, pp 271–301
Guerreiro I, Duboule D (2014) Snakes: hatching of a model system for evo-devo? Int J Dev Biol 58:727–732
Guerreiro I et al (2016) Reorganisation of *Hoxd* regulatory landscapes during the evolution of a snake-like body plan. eLife 5:e16087
Hanken J (1993) Model systems versus outgroups: alternative approaches to the study of head development and evolution. Am Zool 33:448–456
Harvey PH, Pagel MD (1991) The comparative method in evolutionary biology. Oxford University Press, Oxford
He S, del Viso F, Chen C-Y, Ikmi A, Kroesen AE, Gibson MC (2018) An axial Hox code controls tissue segmentation and body patterning in *Nematostella vectensis*. Science 361:1377–1380
Head JJ, Polly PD (2015) Evolution of the snake body form reveals homoplasy in amniote *Hox* gene function. Nature 520:86–89
Hodgkin AL, Huxley AF (1952) A quantitative description of membrane current and its application to conduction and excitation in nerve. J Physiol 117:500–544
Jenner RA (2006) Unburdening evo-devo: ancestral attractions, model organisms, and basal baloney. Dev Genes Evol 216:385–394
Jenner RA, Wills MA (2007) The choice of model organisms in evo-devo. Nat Rev Genet 8:311–319
Kellogg EA, Shaffer HB (1993) Model organisms in evolutionary studies. Syst Biol 42:409–414
Krogh A (1929) The progress of physiology. Science 70:200–204
Kuo D-H, Lai Y-T (2019) On the origin of leeches by evolution of development. Devel Growth Differ 61:43–57
Kusserow A et al (2005) Unexpected complexity of the *Wnt* gene family in a sea anemone. Nature 433:156–160
Kutschera U, Weisblat DA (2015) Leeches of the genus *Helobdella* as model organisms for evo-devo studies. Theory Biosci 134:93–104

Kvon EZ et al (2016) Progressive loss of function in a limb enhancer during snake evolution. Cell 167:633–642

Lapraz F et al (2013) Put a tiger in your tank: the polyclad flatworm *Maritigrella crozieri* as a proposed model for evo-devo. EvoDevo 4:29

Layden MJ, Rentzsch F, Röttinger E (2016) The rise of the starlet sea anemone *Nematostella vectensis* as a model system to investigate development and regeneration. Wiley Interdiscip Rev Dev Biol 5:408–428

Love AC (2008) Explaining evolutionary innovation and novelty: criteria of explanatory adequacy and epistemological prerequisites. Philos Sci 75:874–886

Love AC (2009) Marine invertebrates, model organisms, and the modern synthesis: epistemic values, evo-devo, and exclusion. Theory Biosci 128:19–42

Love AC (2010) Idealization in evolutionary developmental investigation: a tension between phenotypic plasticity and normal stages. Philos Trans R Soc B Biol Sci 365:679–690

Love AC (2014) The erotetic organization of developmental biology. In: Minelli A, Pradeu T (eds) Towards a theory of development. Oxford University Press, Oxford, pp 33–55

Love AC, Strathmann RR (2018) Marine invertebrate larvae: model life histories for development, ecology, and evolution. In: Carrier TJ, Reitzel AM, Heyland A (eds) Evolutionary ecology of marine invertebrate larvae. Oxford University Press, Oxford, pp 302–317

Love AC, Travisano M (2013) Microbes modeling ontogeny. Biol Philos 28:161–188

Martindale MQ, Shankland M (1990) Intrinsic segmental identity of segmental founder cells of the leech embryo. Nature 347:672–674

Matus DQ, Thomsen GH, Martindale MQ (2006) Dorso/ventral genes are asymmetrically expressed and involved in germ-layer demarcation during cnidarian gastrulation. Curr Biol 16:499–505

McGinnis W, Garber RL, Wirz J, Kuroiwa A, Gehring WJ (1984) A homologous protein-coding sequence in *Drosophila* homeotic genes and its conservation in other metazoans. Cell 37:403–408

Medina Jiménez BI, Kwak H-J, Park J-S, Kim J-W, Cho S-J (2017) Developmental biology and potential use of *Alboglossiphonia lata* (Annelida: Hirudinea) as an "evo-devo" model organism. Front Zool 14:60

Metscher BD, Ahlberg PE (1999) Zebrafish in context: uses of a laboratory model in comparative studies. Dev Biol 210:1–14

Milinkovitch M, Tzika A (2007) Escaping the mouse trap: the selection of new evo-devo species. J Exp Zool (Mol Dev Evol) 308B:337–346

Minelli A, Baedke J (2014) Model organisms in evo-devo: promises and pitfalls of the comparative approach. Hist Philos Life Sci 36:42–59

Minelli A, Fusco G (2004) Evo-devo perspectives on segmentation: model organisms, and beyond. Trends Ecol Evol 19:423–429

Moczek AP et al (2011) The role of developmental plasticity in evolutionary innovation. Proc R Soc Lond B Biol Sci 278:2705–2713

Petersen CP, Reddien PW (2009) Wnt signaling and the polarity of the primary body axis. Cell 139:1056–1068

Plackett ARG, Di Stilio VS, Langdale JA (2015) Ferns: the missing link in shoot evolution and development. Front Plant Sci 6:972

Rentzsch F, Anton R, Saina M, Hammerschmidt M, Holstein TW, Technau U (2006) Asymmetric expression of the BMP antagonists chordin and gremlin in the sea anemone *Nematostella vectensis*: implications for the evolution of axial patterning. Dev Biol 296:375–387

Röttinger E, Dahlin P, Martindale MQ (2012) A framework for the establishment of a cnidarian gene regulatory network for "endomesoderm" specification: the inputs of ß-catenin/TCF signaling. PLoS Genet 8:e1003164

Russell JJ et al (2017) Non-model model organisms. BMC Biol 15:55

Saina M, Genikhovich G, Renfer E, Technau U (2009) BMPs and chordin regulate patterning of the directive axis in a sea anemone. Proc Natl Acad Sci USA 106:18592–18597

Sanger TJ (2012) The emergence of squamates as model systems for integrative biology. Evol Dev 14:231–233
Schulze J, Schierenberg E (2011) Evolution of embryonic development in nematodes. EvoDevo 2:18
Slack JMW (2006) Essential developmental biology, 2nd edn. Blackwell, Malden, MA
Sommer RJ (2009) The future of evo-devo: model systems and evolutionary theory. Nat Rev Genet 10:416–422
Stroud JT, Losos JB (2016) Ecological opportunity and adaptive radiation. Annu Rev Ecol Evol Syst 47:507–532
ten Broek CMA, Bakker AJ, Varela-Lasheras I, Bugiani M, Van Dongen S, Galis F (2012) Evo-devo of the human vertebral column: on homeotic transformations, pathologies and prenatal selection. Evol Biol 39:456–471
Vandenbussche M, Chambrier P, Rodrigues Bento S, Morel P (2016) Petunia, your next supermodel? Front Plant Sci 7:72
Wake DB (2009) What salamanders have taught us about evolution. Annu Rev Ecol Evol Syst 40:333–352
Wake DB, Larson A (1987) Multidimensional analyses of an evolving lineage. Science 238:42–48
Weisblat DA, Kuo D-H (2009) *Helobdella* (Leech): a model for developmental studies. Cold Spring Harb Protoc. https://doi.org/10.1101/pdb.emo121
Weisblat DA, Kuo D-H (2014) Developmental biology of the leech *Helobdella*. Int J Dev Biol 58:429–443
Woltering JM (2012) From lizard to snake: behind the evolution of an extreme body plan. Curr Genomics 13:289–299
Woltering JM et al (2009) Axial patterning in snakes and caecilians: evidence for an alternative interpretation of the Hox code. Dev Biol 332:82–89
Yuan Y-W (2019) Monkeyflowers (*Mimulus*): new model for plant developmental genetics and evo-devo. New Phytol 222:694–700
Zuniga A (2015) Next generation limb development and evolution: old questions, new perspectives. Development 142:3810–3820

Chapter 2
Hourglass or Twisted Ribbon?

Peter K. Dearden

Abstract Studies of the evolution of axis formation have shown that, consistent with the developmental hourglass model, a great deal of variation occurs at the earliest stages of embryogenesis. This variation has come to be thought of as due to development systems drift, rather than adaptive change. By linking variation in the biology of reproduction in the parental generation with variation in early development, turning the hourglass into a twisted ribbon, I aim to show that early embryo variation is adaptive change shaped by natural selection in the parental generation.

2.1 Evolution of Axis Formation

The formation of the embryonic axis is a critical event in the development of an animal. Formation of the anterior–posterior and dorsoventral axes of an embryo usually happens early in embryogenesis and is foundational to the developmental of the rest of the embryo (Jürgens 1995; Riechmann and Ephrussi 2001; Schier and Talbot 2005).

Remarkably, however, axis formation mechanisms appear to evolve relatively rapidly, at least with respect to the processes that occur later in development (Buchta et al. 2013, Goldstein et al. 1998; Goltsev et al. 2007; da Fonseca et al. 2009; Yanai et al. 2011). In insects, for example, *Bicoid*, a textbook example of an anterior acting axis formation regulator from *Drosophila*, is not present in the genomes of even non-drosophilid flies (Stauber et al. 1999, 2002; Lemke et al. 2008), and axes are determined by different regulators (Lynch et al. 2006; Lynch 2014; Wilson and Dearden 2011). Other components of axis formation, including *oskar*, which helps to define the posterior in *Drosophila*, are also patchily distributed in insect genomes and have different roles in development (Ewen-Campen et al. 2012).

P. K. Dearden (✉)
Genomics Aotearoa and Biochemistry Department, University of Otago, Dunedin, New Zealand
e-mail: peter.dearden@otago.ac.nz

W. Tworzydlo, S. M. Bilinski (eds.), *Evo-Devo: Non-model Species in Cell and Developmental Biology*, Results and Problems in Cell Differentiation 68,
https://doi.org/10.1007/978-3-030-23459-1_2

In mammals, the timing and nature of axis formation is disputed, and evidence points to both early and late origins of the anterior–posterior axis (reviewed in Solter 2016), suggesting variations in mechanisms. In nematodes, molecular mechanisms specifying the anterior–posterior axis vary between species, with evidence existing for novel mechanisms evolving to replace previous ones (Goldstein et al. 1998).

Axis formation variation is difficult to understand, as the formation of axes is required for the development of the rest of the embryo. If these processes are fundamental to embryo patterning, should they not be conserved? Later-acting patterning process are more conserved within a phylum, for example, with limb development in vertebrates (Capdevila and Belmonte 2001), segmentation in insects (Peel et al. 2005), and others conserved across Metazoa, for example, Hox gene expression and function (Pearson et al. 2005). As axis formation is critical to later conserved developmental processes, it seems likely that embryonic development after axis formation must involve buffering evolutionary changes in axis formation into the conserved later-acting processes.

One possible explanation for the rapid evolution of axis formation mechanisms is that such mechanisms are easy to evolve. All that is required is that there is some asymmetry associated with the egg. These asymmetries could be fundamental in any cell, or unequal provisioning of areas of the egg, asymmetry setup by the sperm entry point, gravity, or a myriad of other possibilities. Perhaps so many such asymmetries exist in eggs that it really does not matter which one is chosen, and so axis formation mechanisms change over time, effectively due to developmental drift (Palmer 2004; True and Haag 2001; Cridge et al. 2016).

Understanding how axis formation evolved is stymied somewhat by the difficulty of finding such mechanisms in non-model organisms. Even using modern genomic, potentially unbiased, techniques (Wilson and Dearden 2013), researchers tend to identify mechanisms they know and, given the difficulty of working with non-model organisms, find single explicable mechanisms for axis formation. Thus, we do not know the range of molecules involved in axis formation in any group of species well enough to determine if the molecules involved are all of the same types—suggesting certain solutions to axis formation are better than others or if indeed almost anything asymmetric can serve to trigger axis formation.

Current thought tends to assume that axis formation mechanisms evolve because of internal mechanisms, for example, drift (True and Haag 2001), because they are effectively nonadaptive. Here I hope to show that that is unlikely to be the case.

2.2 The Developmental Hourglass

One way of describing the divergence of the embryonic process along the development of an animal has been the developmental hourglass (Fig. 2.1a). This idea is a way of visualizing the divergence of embryos between species/taxa and produces a pleasing figure that shows that the most divergence is at the start of embryogenesis (when axis formation is likely occurring) and at the end, with a conserved stage,

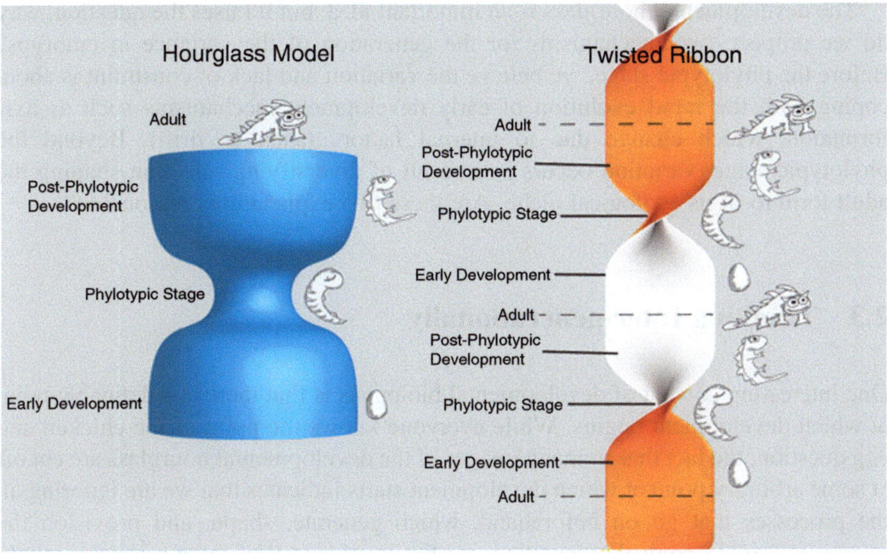

Fig. 2.1 (**a**) Traditional view of a developmental hourglass, with variation at both the start and end of development, and a constraint period at the phylotypic stage. (**b**) Proposed twisted ribbon model, with the generations of organisms linked, indicating the relationship between post- and pre-phylotypic variation

named the phylotypic stage, in mid-embryogenesis (Fig. 2.1a). This idea was first proposed independently by Denis Duboule (1994) and Rudolf Raff (1996) on the basis of the morphology of embryos, but in later years, it was somewhat confirmed, by the finding of similar patterns in gene expression data. These studies have shown that the most conserved genes and the most conserved gene expression patterns occur at a point in mid-embryogenesis, congruent with the phylotypic stage, suggesting that this is the most constrained period in the development of an animal. Hourglass-like descriptions of development based on molecular characteristics have been published for vertebrates, invertebrates, plants and fungi (Hazkani-Covo et al. 2005; Irie and Sehara-Fujisawa 2007; Irie and Kuratani 2011; Quint et al. 2012; Cheng et al. 2015; Drost et al. 2015; Levin et al. 2016; Cutter et al. 2019).

The developmental hourglass then describes well the variation we see in axis formation, early embryogenesis is less conserved and constrained, but that constraint increases as the embryo approaches the phylotypic stage, and variation is reduced. More conserved genes and processes become active, and ancient conserved processes come to the fore. Presumably again, the reduction in variance is related to buffering the extreme variation in early development such that it can easily link to the conserved processes in the phylotypic stage.

Beyond the phylotypic stage, it is suggested that the divergence is a reflection of evolutionary processes that have to drive the divergence of each species to fit the niche it inhabits. In effect, this is "real" divergence, reflecting the evolution of the diversity of adult forms.

The developmental hourglass is an important idea, but it raises the question, why do we propose two mechanisms for the generation of the variance in embryos? Before the phylotypic stage, we believe the variation and lack of constraint is about coping with the rapid evolution of early development mechanisms such as axis formation, which change due to internal factors (such as drift). Beyond the phylotypic stage, variation occurs as a result of diversifying selection shaping the adult form to fit its ecological niche. Are these two explanations reasonable?

2.3 Thinking Inter-generationally

One interesting conceit of developmental biologists is that there is a definable point at which development begins. While everyone knows the point of the chicken and egg question, the fact that so many images of the developmental hourglass are cut off at some arbitrary point at which development starts indicates that we are ignoring all the processes that go on beforehand, which generate, shape, and provision the oocyte, as well as control oviposition and fertilization. All of these processes surely influence the form of the egg and thus influence the "starting point" of development. The role of maternal-effect genes in the development of many species shows that the interface between ovary and egg is much more fluid than described by the cut bottom end of the hourglass. Indeed, many axis formation mechanisms rely on maternal, or maternal and zygotic factors, placing the asymmetries on which axis formation is built back in oocyte development and maturation.

Given this, instead of thinking of hourglasses that reflect variation within the developmental lifetime of one organism, we might better understand variation by thinking across generations.

Given that the barrier between mother and offspring is, at early stages, rather arbitrary, might the variation in axis formation mechanisms in the early embryo be related to maternal variation associated with oocyte development, rather than embryonic variation? Perhaps variation at the start of embryogenesis is actually a reflection and a consequence of variation at the end of embryogenesis and adult life—the pre-phylotypic variation a consequence of the post-phylotypic variation. Perhaps we should not think of developmental hourglasses, but of a twisted ribbon (Fig. 2.1b), where each constrained twist corresponds to the phylotypic stage, where there is less variation. In this model, the pre- and post-phylotypic variation is a consequence of the same thing, the natural selection of that organism to better fit its ecological niche.

The post-phylotypic development of an organism will also affect the development of ovaries and the cell types within. Ovary structure and function vary between species (Jalabert et al. 1991; Büning 1994; Tyler and Sumpter 1996; Van Den Hurk and Zhao 2005; Bird and Bird 2012), as do the reproductive traits of fertilization (Yanagimachi 1994; Poiani 2006), oviposition (Minkenberg et al. 1992), egg shape and size (Christians 2002), and indeed almost all reproductive physiology (Tyler and Sumpter 1996; Kramarenko 2013). These differences are shaped by natural selection as the reproductive traits of an organism are critical to its fitness. All of these adaptive changes in reproductive traits must lead on to variation in early development in the next generation.

Divergence in embryogenesis is thus not buffering divergence in axis formation or early developmental events, but buffering divergence in the fast-evolving reproductive processes that produce and fertilize oocytes. Rather than the hourglass, cut off at the start and end of embryogenesis, the twisted ribbon better explains the origins of variation in the embryo and the link between variation in reproductive traits in the parental generation and variation in the embryo. That natural selection is constantly modifying the key reproductive traits of an organism may explain why hourglasses structures are detectable even between (Kalinka et al. 2010) and within species (Cruickshank and Wade 2008), when developmental drift might not be expected to be rapid enough.

2.4 Is the Divergence After the Phylotypic Stage Buffered by Divergence Before the Phylotypic Stage?

Evolution acts on species to better fit them to their environment. This applies to embryos as much as adults. The selective forces that drive the evolution of novelty in ovary structure, oocyte development, form, and provisioning, in turn, affect the next generation of oocytes and embryos. The twisted ribbon model would imply that the variation we see in early development is part of a buffering process whereby the ecologically driven divergence in reproductive traits in adults must be brought back to a conserved phylotypic stage.

The first part of the developmental hourglass is thus a response to the variation in the adult driven by its niche, which leads to divergence in the development, structure, and environmental responsiveness of the ovary, the form and the provisioning of the oocyte, the mode of fertilization, the environment the egg is in, and the available signals for axis formation.

The consequence of this is that the variation we see in axis formation and early development is not due to internal drift processes, but are consequent on the reproductive traits of adults, particularly the female. The variation in early embryonic processes is adaptive variation reflecting the same selective pressures that form variation after the phylotypic stage.

If this model is correct, how might we test it? How can we distinguish between drift-like processes driving variation in early embryos, from adaptive ones related to reproductive traits? I would propose examining the genes involved in, and methods of, early developmental processes in a wide range of organisms, picking examples where particular reproductive approaches have convergently evolved. By determining if a correlation exists between particular reproductive approaches, and particular genes and processes being used in early embryonic development, one might be able to say that early development varies with the reproductive mode, rather than randomly as might be expected in a drift process.

This calls for more understanding of the kinds of genes and process involved in early embryo development and in axis formation. Discovering such genes must, however, be done in a non-biased way, as our assumptions about what genes and processes are involved will bias the outcome of these experiments. In non-model organisms, this is challenging, but it reflects the current approach of many practitioners of evolution and development who are generating a more detailed understanding of embryogenesis in many species. Only by understanding the diversity of early development can we test why it might exist.

2.5 What Then Is the Phylotypic Stage?

These ideas might better help us understand where variation in development comes from, but they lead us to an important and unanswered question: What is the phylotypic stage? Why is development constrained here? Why would all the variation in adults need to be buffered to produce this constrained stage? Authors (Irie 2017) have suggested that this is when conserved genes are expressed (e.g., Duboule 1994), that this stage is constrained due to pleiotropic interactions between developmental modules (Sander 1983; Galis and Metz 2001), or that this is where integration of the embryo is at its highest (Schmidt and Starck 2004), that it is a reflection of a clade's ancestral developmental mechanisms (Levin et al. 2016), even to the point of it being phylum-specific and a fundamental building block of a phylum's morphology (Irie and Kuratani 2014; Levin et al. 2016). Hourglasses have also been detected in modeled developmental processes, arising spontaneously because of the nature of these genetic networks (Akhshabi et al. 2014; Friedlander et al. 2015). The idea that this is the only constrained point in the lineage of an organism, and in each generation of that lineage, this stage must be generated, gives more, not less, importance to the phylotypic stage. Understanding what this stage represents and why it is so tightly constrained may help us unlock an understanding of the embryogenesis of early metazoans.

2.6 Conclusions

Examining the phenomenon of the developmental hourglass in an intergenerational way may help us explain better the origin of variation in early development. The hourglass is better described as a twisted ribbon linking variability in subsequent generations. If this model is correct, it would imply that axis formation evolution is not drifting—but is a consequence of the evolution of the ovary, oocyte, and reproductive mode from the female parent (and perhaps the male).

By thinking of the developmental hourglass not as something describing the development of one individual, but of a continuous lineage, the phylotypic stage comes more into focus. By understanding why variation before and after the phylotypic stage exists, we inevitably have to ask why there is a conserved constraint at the phylotypic stage.

Acknowledgements The author would like to thank A. Cridge, L. Brownfield, and G. Petersen for discussions that led to this manuscript, P. Dearden for critical reading of drafts, and B. Dearden for editorial input.

References

Akhshabi S, Sarda S, Dovrolis C, Yi S (2014) An explanatory evo-devo model for the developmental hourglass. F1000Research 3:156

Bird AF, Bird J (2012) The structure of nematodes. Academic Press, New York

Buchta T, Özüak O, Stappert D, Roth S, Lynch JA (2013) Patterning the dorsal–ventral axis of the wasp *Nasonia vitripennis*. Dev Biol 381:189–202

Büning J (1994) The insect ovary: ultrastructure, previtellogenic growth and evolution. Springer, New York

Capdevila J, Belmonte JCI (2001) Patterning mechanisms controlling vertebrate limb development. Annu Rev Cell Dev Biol 17:87–132

Cheng X, Hui JH, Lee YY, Wan Law PT, Kwan HS (2015) A "developmental hourglass" in fungi. Mol Biol Evol 32:1556–1566

Christians JK (2002) Avian egg size: variation within species and inflexibility within individuals. Biol Rev 77:1–26

Cridge AG, Dearden PK, Brownfield LR (2016) Convergent occurrence of the developmental hourglass in plant and animal embryogenesis? Ann Bot 117:833–843

Cruickshank T, Wade MJ (2008) Microevolutionary support for a developmental hourglass: gene expression patterns shape sequence variation and divergence in Drosophila. Evol Dev 10:583–590

Cutter AD, Garrett RH, Mark S, Wang W, Sun L (2019) Molecular evolution across developmental time reveals rapid divergence in early embryogenesis. bioRxiv. https://doi.org/10.1101/518621

da Fonseca RN, Lynch JA, Roth S (2009) Evolution of axis formation: mRNA localization, regulatory circuits and posterior specification in non-model arthropods. Curr Opin Genet Dev 19:404–411

Drost HG, Gabel A, Grosse I, Quint M (2015) Evidence for active maintenance of phylotranscriptomic hourglass patterns in animal and plant embryogenesis. Mol Biol Evol 32:1221–1231

Duboule D (1994) Temporal colinearity and the phylotypic progression: a basis for the stability of a vertebrate bauplan and the evolution of morphologies through heterochrony. Development 1994:135–142

Ewen-Campen B, Srouji JR, Schwager EE, Extavour CG (2012) Oskar predates the evolution of germ plasm in insects. Curr Biol 22:2278–2283

Friedlander T, Mayo AE, Tlusty T, Alon U (2015) Evolution of bow-tie architectures in biology. PLoS Comput Biol 11:e1004055

Galis F, Metz JA (2001) Testing the vulnerability of the phylotypic stage: on modularity and evolutionary conservation. J Exp Zool B Mol Dev Evol 291:195–204

Goldstein B, Frisse LM, Thomas WK (1998) Embryonic axis specification in nematodes: evolution of the first step in development. Curr Biol 8:157–160

Goltsev Y, Fuse N, Frasch M, Zinzen RP, Lanzaro G, Levine M (2007) Evolution of the dorsal-ventral patterning network in the mosquito, *Anopheles gambiae*. Development 134:2415–2424

Hazkani-Covo E, Wool D, Graur D (2005) In search of the vertebrate phylotypic stage: a molecular examination of the developmental hourglass model and von Baer's third law. J Exp Zool B Mol Dev Evol 304:150–158
Irie N (2017) Remaining questions related to the hourglass model in vertebrate evolution. Curr Opin Genet Dev 45:103–107
Irie N, Kuratani S (2011) Comparative transcriptome analysis reveals vertebrate phylotypic period during organogenesis. Nat Commun 2:248
Irie N, Kuratani S (2014) The developmental hourglass model: a predictor of the basic body plan? Development 141:4649–4655
Irie N, Sehara-Fujisawa A (2007) The vertebrate phylotypic stage and an early bilaterian-related stage in mouse embryogenesis defined by genomic information. BMC Biol 5:1
Jalabert B, Fostier A, Breton B, Weil C (1991) Oocyte maturation in vertebrates. In: Pang PKT, Schreibman MP (eds) Vertebrate endocrinology: fundamentals and biomedical implications, vol 4. Academic Press, New York, pp 23–90
Jürgens G (1995) Axis formation in plant embryogenesis: cues and clues. Cell 81:467–470
Kalinka AT, Varga KM, Gerrard DK, Preibisch S, Corcoran DL, Jarrells J, Ohler U, Bergman CM, Tomancak P (2010) Gene expression divergence recapitulates the developmental hourglass model. Nature 468:811–814
Kramarenko S (2013) The analysis of the reproductive traits of the pulmonate molluscs: a mini-review. Ruthenica 23(2):115–125
Lemke S, Stauber M, Shaw PJ, Rafiqi AM, Prell A, Schmidt-Ott U (2008) Bicoid occurrence and Bicoid-dependent hunchback regulation in lower cyclorrhaphan flies. Evol Dev 10:413–420
Levin M, Anavy L, Cole AG, Winter E, Mostov N, Khair S, Senderovich N, Kovalev E, Silver DH, Feder M, Fernandez-Valverde SL, Nakanishi N, Simmons D, Simakov O, Larsson T, Liu S-Y, Jerafi-Vider A, Yaniv K, Ryan JF, Martindale MQ, Rink JC, Arendt D, Degnan SM, Degnan BM, Hashimshony T, Yanai I (2016) The mid-developmental transition and the evolution of animal body plans. Nature 531:637
Lynch JA (2014) Diversity of molecules and mechanisms in establishing insect anterior–posterior polarity. Curr Opin Insect Sci 1:39–44
Lynch JA, Brent AE, Leaf DS, Pultz MA, Desplan C (2006) Localized maternal orthodenticle patterns anterior and posterior in the long germ wasp *Nasonia*. Nature 439:728–732
Minkenberg OP, Tatar M, Rosenheim JA (1992) Egg load as a major source of variability in insect foraging and oviposition behavior. Oikos 65:134–142
Palmer AR (2004) Symmetry breaking and the evolution of development. Science 306:828–833
Pearson JC, Lemons D, McGinnis W (2005) Modulating Hox gene functions during animal body patterning. Nat Rev Genet 6:893
Peel AD, Chipman AD, Akam M (2005) Arthropod segmentation: beyond the *Drosophila* paradigm. Nat Rev Genet 6:905–916
Poiani A (2006) Complexity of seminal fluid: a review. Behav Ecol Sociobiol 60:289–310
Quint M, Drost HG, Gabel A, Ullrich KK, Bonn M, Grosse I (2012) A transcriptomic hourglass in plant embryogenesis. Nature 490:98–101
Raff RA (1996) The shape of life: genes, development and the evolution of animal form. University of Chicago Press, Chicago
Riechmann V, Ephrussi A (2001) Axis formation during *Drosophila* oogenesis. Curr Opin Gen Dev 11:374–383
Sander K (1983) The evolution of patterning mechanisms: gleanings from insect embryogenesis. In: Goodwin B (ed) Development and evolution. Cambridge University Press, Cambridge, UK
Schier AF, Talbot WS (2005) Molecular genetics of axis formation in zebrafish. Annu Rev Genet 39:561–613
Schmidt K, Starck JM (2004) Developmental variability during early embryonic development of zebra fish, *Danio rerio*. J Exp Zool B Mol Dev Evol 302:446–457
Solter D (2016) Preformation versus epigenesis in early mammalian development. Curr Top Dev Biol 117:377–391

Stauber M, Jäckle H, Schmidt-Ott U (1999) The anterior determinant bicoid of *Drosophila* is a derived Hox class 3 gene. PNAS 96:3786–3789

Stauber M, Prell A, Schmidt-Ott U (2002) A single Hox3 gene with composite bicoid and zerknüllt expression characteristics in non-cyclorrhaphan flies. PNAS 99:274–279

True JR, Haag ES (2001) Developmental system drift and flexibility in evolutionary trajectories. Evol Dev 3:109–119

Tyler C, Sumpter J (1996) Oocyte growth and development in teleosts. Rev Fish Biol Fisher 6:287–318

Van Den Hurk R, Zhao J (2005) Formation of mammalian oocytes and their growth, differentiation and maturation within ovarian follicles. Theriogenology 63:1717–1751

Wilson MJ, Dearden PK (2011) Diversity in insect axis formation: two orthodenticle genes and hunchback act in anterior patterning and influence dorsoventral organization in the honeybee (*Apis mellifera*). Development 138:3497–3507

Wilson MJ, Dearden PK (2013) RNA localization in the honeybee (*Apis mellifera*) oocyte reveals insights about the evolution of RNA localization mechanisms. Dev Biol 375:193–201

Yanagimachi R (1994) Mammalian fertilization. In: Knobil E, Neill J (eds) The physiology of reproduction. Raven Press, New York

Yanai I, Peshkin L, Jorgensen P, Kirschner MW (2011) Mapping gene expression in two *Xenopus* species: evolutionary constraints and developmental flexibility. Dev Cell 20:483–496

Chapter 3
Ambulacrarians and the Ancestry of Deuterostome Nervous Systems

Laurent Formery, Michael Schubert, and Jenifer C. Croce

Abstract The evolutionary origin and history of metazoan nervous systems has been at the heart of numerous scientific debates for well over a century. This has been a particularly difficult issue to resolve within the deuterostomes, chiefly due to the distinct neural architectures observed within this group of animals. Indeed, deuterosomes feature central nervous systems, apical organs, nerve cords, and basiepidermal nerve nets. Comparative analyses investigating the anatomy and molecular composition of deuterostome nervous systems have nonetheless succeeded in identifying a number of shared and derived features. These analyses have led to the elaboration of diverse theories about the origin and evolutionary history of deuterostome nervous systems. Here, we provide an overview of these distinct theories. Further, we argue that deciphering the adult nervous systems of representatives of all deuterostome phyla, including echinoderms, which have long been neglected in this type of surveys, will ultimately provide answers to the questions concerning the ancestry and evolution of deuterostome nervous systems.

3.1 Introduction

Among metazoans, the deuterostomes constitute one of the two major infrakingdoms of bilaterian animals (Satoh et al. 2014). Members of the deuterostomes include the vertebrates, the tunicates, the cephalochordates, the hemichordates, and the echinoderms, with the former three establishing the chordate superphylum and the latter two constituting the ambulacrarian superphylum (Fig. 3.1). Deuterostomes present very diverse body plans and lifestyles, which together have significantly hampered the establishment of robust phylogenetic relationships within this part of the animal tree of life (e.g. Blair and Hedges 2005; Lowe et al. 2015). Yet, with the advent of

---

L. Formery · M. Schubert · J. C. Croce (✉)
Sorbonne Université, CNRS, Laboratoire de Biologie du Développement de Villefranche-sur-Mer (LBDV), Evolution of Intercellular Signaling in Development (EvoInSiDe) Team, Villefranche-sur-Mer, France
e-mail: jenifer.croce@obs-vlfr.fr

W. Tworzydlo, S. M. Bilinski (eds.), *Evo-Devo: Non-model Species in Cell and Developmental Biology*, Results and Problems in Cell Differentiation 68,
https://doi.org/10.1007/978-3-030-23459-1_3

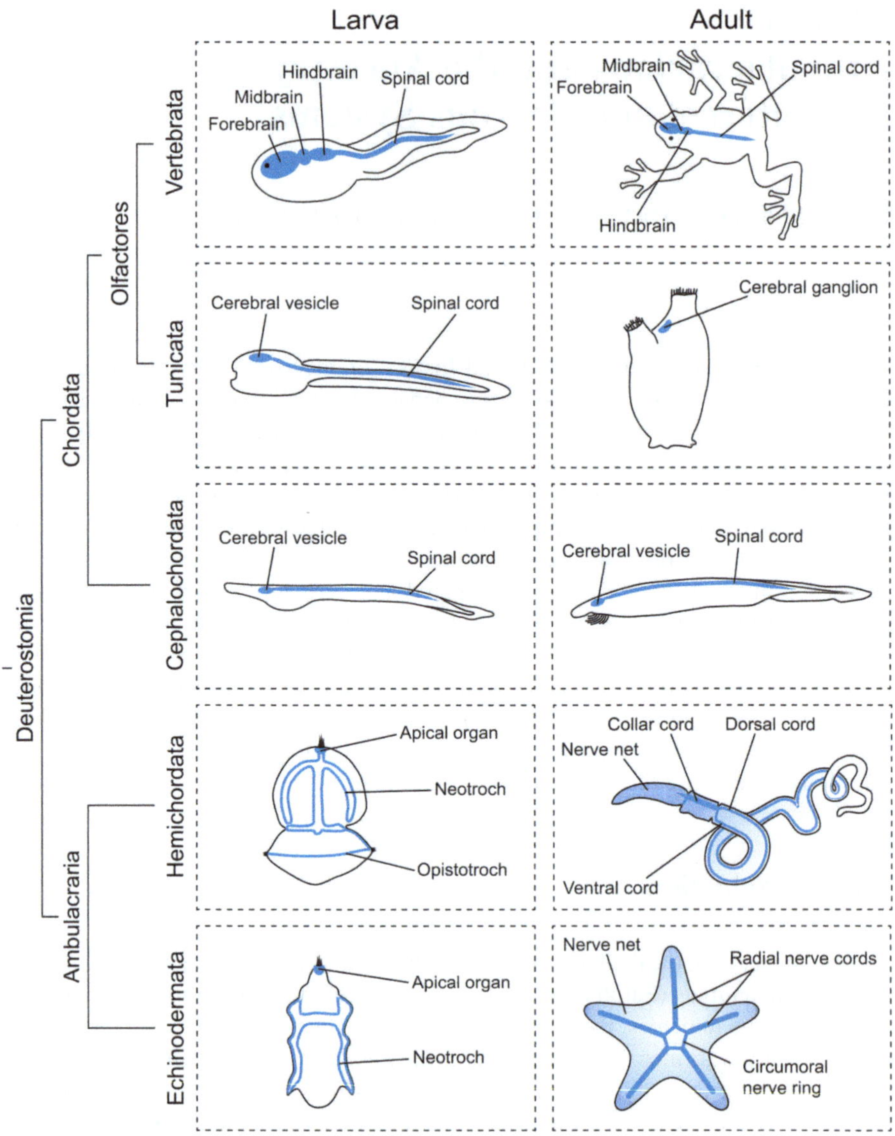

Fig. 3.1 Phylogeny of deuterostomes and schematics of their main neural structures. The tree is based on the current consensus of the phylogenetic relationships between the five extant deuterostome phyla. Vertebrates are represented by frogs, tunicates by ascidians, cephalochordates by amphioxus, hemichordates by enteropneusts, and echinoderms by asteroids. The neural features are indicated in blue for both larva (left panel) and adult (right panel). For chordates, only the central nervous system is highlighted

whole genome sequences and molecular phylogenies, the monophyly of the deuterostomes as well as the major phylogenetic branching patterns within this infrakingdom have by now been firmly established (Delsuc et al. 2006; Dunn et al. 2008; Telford et al. 2015).

Of the many characters investigated to understand the ancestry and evolution of the deuterostomes, particular attention has been paid over the years to the nervous system. It turns out, however, that the two major superphyla of deuterostomes (i.e., chordates and ambulacrarians) are characterized by remarkably different nervous systems (Fig. 3.1). Chordate larvae commonly form a neural tube that, in the adult, gives rise to a central nervous system with an anterior brain and a posterior spinal cord (Holland 2009). In contrast, ambulacrarian larvae possess a nervous system consisting of an apical organ and of neurons underlying the larval ciliary bands. These neural structures are subsequently lost during metamorphosis and hence do not contribute to the adult nervous system of ambulacrarians, which is created de novo (Nakajima et al. 2004; Byrne et al. 2007; Miyamoto et al. 2010). The adult nervous system of ambulacrarians, in turn, can also be subdivided into two main structures: a basiepidermal nerve net (or nerve plexus) and nerve cords (i.e., condensed bundles of neurons) (Chia and Burke 1978; Nakano et al. 2006; Miyamoto et al. 2010). Importantly, although it is largely accepted that the last common chordate ancestor possessed a centralized nervous system, the phylogenetic relationships of the chordate central nervous system with the ambulacrarian larval neural structures, adult nerve nets, or adult nerve cords are still a matter of debate (Lowe et al. 2003; Haag 2005; Nielsen 2006; Nomaksteinsky et al. 2009; Kaul and Stach 2010; Burke 2011).

Interestingly, comparative analyses carried out in cnidarians, arthropods, annelids, and vertebrates have revealed a certain degree of conservation between neuronal cell types and gene expression patterns in the neurogenic territories of these animals (Marlow et al. 2014; Arendt 2018). These conserved features have hence been interpreted as evidence for a common origin of all eumetazoan (cnidarians + bilaterians) nervous systems and for the presence of a condensed, centralized nervous system in the last common ancestor of bilaterians (i.e., the urbilateria) (Holland et al. 2013; Nielsen 2015; Arendt et al. 2016). This hypothesis thus implies the presence of an urbilaterian-like central nervous system in the deuterostome ancestor. However, other surveys conducted on a different selection of bilaterian species have revealed striking differences in the neuroanatomy, neuronal cell types, and gene expression patterns of these animals (Hejnol and Lowe 2015; Martín-Durán et al. 2018). These results have thus led to the contradictory hypothesis that nervous system centralization may have taken place independently in distinct bilaterian phyla (Holland 2003; Northcutt 2012). Consequently, to date, the morphology of the nervous system of the last common ancestor of deuterostomes remains elusive (Holland 2015a; Lowe et al. 2015).

To assess the ancestry of deuterostome nervous systems, careful investigations and interpretations of shared and derived traits between chordate and ambulacrarian nervous systems thus take center stage. The developing and adult nervous systems of chordates have already been extensively studied and characterized both at the anatomical and molecular levels (Baker and Bronner-Fraser 1997; Holland 2009; Candiani et al. 2012; Hudson 2016; Osugi et al. 2017; Zieger et al. 2017). Similarly, our understanding of both the anatomy and molecular organization of the larval and adult nervous systems of hemichordates has improved significantly in the course of the last two decades (Lowe et al. 2003; Nomaksteinsky et al. 2009; Miyamoto et al. 2010; Kaul-Strehlow et al. 2015). In echinoderms, by contrast, most of our knowledge on their nervous system is based on studies on developing embryos and larvae, while the adult nervous system has so far received much less attention (Burke et al. 2006; Hinman and Burke 2018). The main reason for this focus on early development is the derived, pentaradial body plan of the echinoderm adult, which complicates any kind of comparative studies (Hyman 1955; Holland 2015a). However, as argued here, data from echinoderms, and in particular from adult echinoderms, can and should be included in comparative analyses aimed at reconstructing the evolutionary origin and history of deuterostome nervous systems. In this chapter, we thus review our current understanding of the anatomy and molecular organization of the larval and adult nervous systems of ambulacrarians and assess their possible homologies with chordate nervous systems. We further propose that detailed investigations of adult nervous system development in ambulacrarians, including in adult echinoderms, have the potential to resolve the long-standing debate about the ancestry of deuterostome nervous systems and might even provide crucial evidence for clarifying the origin of centralized nervous systems in bilaterians.

3.2 Deuterostomes

The deuterostomes represent a large monophyletic group of animals that is composed of two major superphyla: the chordates and the ambulacrarians (Satoh et al. 2014). It should be mentioned that, in addition to these superphyla, the xenoturbellids, a group of marine worms, have also been placed by some molecular phylogenies as either sister group of the ambulacrarians (Bourlat et al. 2006; Dunn et al. 2008) or basal to all deuterostomes (Mwinyi et al. 2010). However, other molecular analyses have grouped the xenoturbellids together with the acoels and the nemertodermatids in a separate phylum, the Xenacoelomorpha, and positioned this phylum as either sister group of the ambulacrarians (Philippe et al. 2011) or basal to all bilaterians (Hejnol et al. 2009; Cannon et al. 2016; Rouse et al. 2016). Given these current discordant views on the phylogenetic position of the xenoturbellids, representatives of this animal group have thus not been considered in this review.

Historically, animals have been grouped together as deuterostomes on the fundamental basis of specific morphological traits (Grobben 1908). The most important of these traits was the fate of the blastopore that, in deuterostomes, was said to become the anus, while the mouth will form secondarily on the opposite side of the archenteron. Other traits included radial cleavage (early cell divisions are parallel or perpendicular to the original body axis), blastula formation (development of a hollow sphere of cells surrounding a fluid-filled cavity), and enterocoely (emergence of the mesoderm by folding from the archenteron). Even though all these developmental features can indeed be found in all deuterostomes, they have also been reported by now in a number of protostome phyla (Hejnol and Martindale 2009; Martín-Durán et al. 2012). These discoveries have thus significantly reduced the number of characters exclusively shared by deuterostomes, so-called deuterostome synapomorphies.

As of today, the only unambiguous deuterostome synapomorphy is the presence of a pharynx penetrated by several slits (Dominguez et al. 2002; Gillis et al. 2012; Lowe et al. 2015). In extant hemichordates and chordates, pharyngeal gill slits have been described. They are located posterior to the mouth and contribute to gas exchange and feeding. The homology of these structures, in between these animals, has further been supported by several anatomical and molecular characters (Rychel and Swalla 2007; Gonzalez and Cameron 2009; Gillis et al. 2012). In extant echinoderms, by contrast, pharyngeal gill slits are absent, but the fossil record has revealed compelling evidence for the presence of a pharynx pierced by slits in stem echinoderms (Dominguez et al. 2002). Moreover, a perforated pharynx has also been identified in fossils described as stem deuterostomes (Shu et al. 2001, 2003; Ou et al. 2012), thereby reinforcing the notion that a pharynx penetrated by slits is indeed a true deuterostome synapomorphy.

Some authors have further proposed that an endostyle-associated function, so far reported in chordates and hemichordates, may represent an additional deuterostome synapomorphy (Nielsen 2012; Lowe et al. 2015; Satoh 2016). All extant chordates indeed possess a pharynx-associated organ, called the endostyle, which participates in food particle trapping and contributes to endocrine functions. In hemichordates, even though a well-defined endostyle does not seem to be present, the pharynx as a whole has been shown to serve these two functions (Ruppert 2005; Gonzalez and Cameron 2009). Thus, despite the fact that no endostyle equivalent has so far been reported in echinoderms, an endostyle-like structure or at least function might have already been present in the last common ancestor of all deuterostomes (Nielsen 2012; Lowe et al. 2015; Satoh 2016).

3.2.1 Chordates

The chordates unite three distinct phyla of which phylogenetic relationship is by now very well established. The cephalochordates (or amphioxus) represent the sister group to a clade established by the tunicates (or urochordates) and the vertebrates, which is often referred to as the olfactores (Delsuc et al. 2006; Satoh et al. 2014). Chordates display very distinctive adult body plans and anatomies (Fig. 3.1). Cephalochordates are thin and fusiform (Kardong 2012). Tunicates have their entire adult body embedded in a thick tunic made of cellulose (Holland 2016).

Vertebrates exhibit a large variety of adult morphologies, marked, for example, by size differences ranging from 1 cm in frogs to 30 m in whales (Webster and Webster 2013). Despite this diversity, all chordates are characterized by three specific synapomorphies observed at least at some stage in their life cycle: a muscular postanal tail, a notochord, and a dorsal neural tube.

The muscular postanal tail is an elongation of the trunk extending beyond the anus (Stach and Kaul 2011). This postanal tail contains skeletal elements and paired muscles, which are used, for example, for undulatory swimming movements in aquatic species and for balancing or grasping in terrestrial ones. In some chordates, such as cephalochordates and fish, the tail is maintained throughout the whole life cycle, while in others, such as ascidian tunicates and humans, it is only present during embryonic development and subsequently disappears.

The second chordate synapomorphy, the notochord, is a flexible rod-shaped structure derived from the mesoderm (Zhang 2009). It extends all along the anterior–posterior body axis and is located between the digestive tract ventrally and the central nervous system dorsally. In cephalochordates, the notochord acts as the primary axial support of both the larval and the adult body, while, in most tunicates, it is present only in the embryo and is subsequently lost during metamorphosis. In vertebrates, the notochord is found in the embryo and later contributes significantly to the formation of the adult intervertebral disks in the vertebral column.

Finally, the third chordate synapomorphy is the dorsal neural tube. The neural tube is a tubular neuroepithelium with a central lumen. It is derived from the ectoderm and forms in the embryo through a characteristic developmental process called neurulation (Colas and Schoenwolf 2001; Lowery and Sive 2004). During neurulation, the embryonic neuroectoderm, which is located dorsal to the notochord, bends inward, rounds up, and fuses, thereby detaching itself from the remaining epidermal ectoderm and coming to lie just underneath it. Eventually, in the adult, the dorsal neural tube will form the chordate central nervous system, with its anterior part generating a dense neural condensation, the brain, while its more posterior portions becoming the spinal cord (Fig. 3.1). In addition to the central nervous system, chordate nervous systems further include an extensive peripheral nervous system that includes all nerves, ganglia, and sensory cells of the body outside the central nervous system (Buchanan and Tranel 2009; Shepherd 2017). The peripheral nervous system thus serves as a relay between the body and the central nervous system, connecting all neural receptors and effectors in the body to the central nervous system.

3.2.2 Ambulacrarians

The ambulacrarians include two phyla, the hemichordates and the echinoderms, which form a monophyletic clade within the deuterostomes (Satoh et al. 2014; Telford et al. 2015). The hemichordates can further be subdivided into two classes (i.e., the enteropneusts and the pterobranchs) (Osborn et al. 2012; Cannon et al. 2014), while the echinoderms comprise five classes (i.e., the crinoids (sea lilies and feather stars), the echinoids (sea urchins), the holothuroids (sea cucumbers), the asteroids (sea stars), and the ophiuroids (brittle stars)). Among echinoderms, the crinoids are generally placed basal to two sister groups, the Echinozoa (echinoids plus holothuroids) and the Asterozoa (asteroids plus ophiuroids), a phylogenetic relationship that has recently been validated by two large-scale molecular analyses (Telford et al. 2014; Reich et al. 2015).

As adults, the morphology of hemichordates and echinoderms differ greatly (Fig. 3.1). Hemichordates are fusiform worms with a bilateral body plan composed of three regions, which from anterior to posterior are the prosome, the mesosome, and the metasome. The prosome is also commonly referred to as the proboscis in enteropneusts and the cephalic shield in pterobranchs, the mesosome corresponds to the collar in enteropneusts and the neck in pterobranchs, and the metasome is called the trunk in both enteropneusts and pterobranchs (Kaul-Strehlow and Röttinger 2015; Miyamoto and Wada 2018). In echinoderms, by contrast, the adult bodies do not display bilateral symmetry. Instead, they are characterized by a pentaradial body plan. Adult echinoderms are organized along an oral–aboral axis relative to which five axes (or radii) are arranged perpendicularly (Minsuk et al. 2009). Of note, the number of perpendicular axes may vary and can even be higher than five in certain asteroid and ophiuroid species. Each axis is further defined by the presence of an ambulacrum, which is a specific area of the shell that holds the internal water vascular system and that is characterized by a perforated skeleton through which the tube feet emerge. The areas between the ambulacra are referred to as the interambulacra (Serafy and Fell 1985; Mooi and David 2008).

Despite the considerable differences in adult morphology, hemichordates and echinoderms nonetheless share some key features that strongly support their phylogenetic relatedness. First, their representatives chiefly adopt one of two developmental modes. Their adult forms develop either indirectly by drastic metamorphosis from a planktotrophic larva or directly by gradual metamorphosis from a lecithotrophic larva (Raff 1987; Kaul-Strehlow and Röttinger 2015). Second, the larvae of hemichordates (i.e., the tornaria observed in spengelid and ptychoderid enteropneusts) and that of echinoderms (i.e., the vitellaria in crinoids, the echinopluteus in echinoids, the auricularia in holothuroids, the ophiopluteus in ophiuroids, and the bipinnaria and the brachiolaria in asteroids) are very similar. They share a similar body plan with bilateral symmetry (Smith 1997) and are characterized by an ectodermal ciliary band, called the neotroch, which surrounds the mouth and is used for locomotion and food collection (Fig. 3.1) (Nielsen 1987). Given these similarities, it has hence been proposed by some authors that the last common ancestor of hemichordates and echinoderms developed from a common ancestral larva, called the dipleurula larva, from which all extant ambulacrarian larvae have emerged (Nielsen 1998; Swalla 2006).

Another key feature shared by hemichordate and echinoderm larvae is the presence, along the digestive tract, of three bilaterally paired coelomic compartments, from which most of the adult structures form (Fig. 3.2) (Peterson et al. 2000). From the anterior (or animal) to the posterior (or vegetal) pole of the larva, these

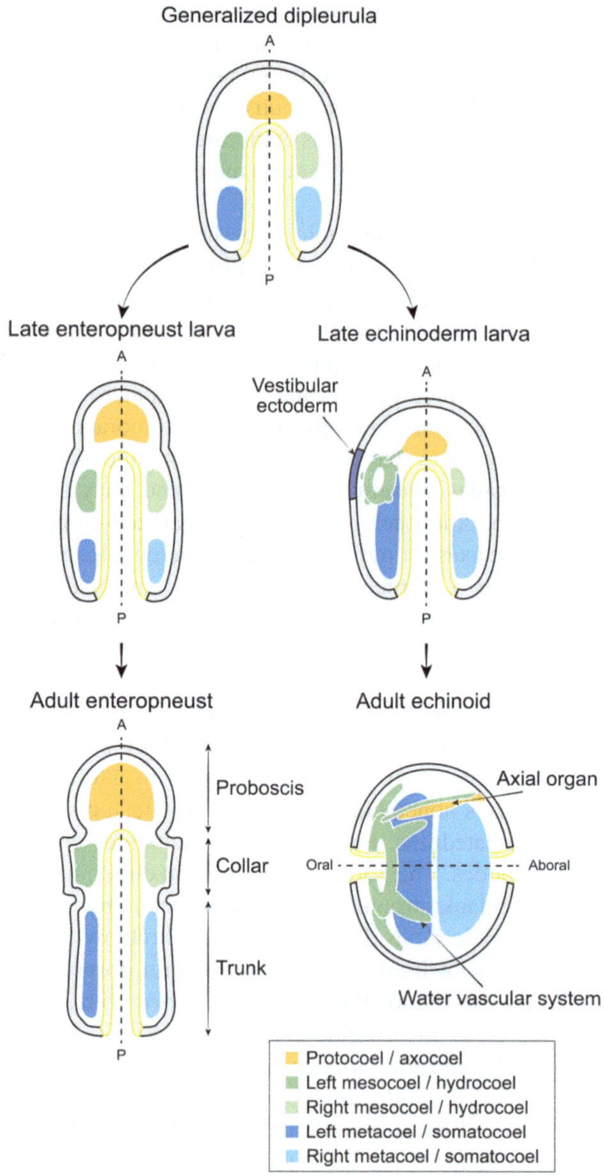

Fig. 3.2 Schematic representation of the coelomic architecture of hemichordate (using enteropneusts as example) and echinoderm (using echinoids as example) larvae and adults (adapted from Peterson et al. 2000 and Tsuchimoto and Yamaguchi 2014). The generalized dipleurula larva of ambulacrarians has five coeloms that will give rise to different adult structures. In hemichordates, the adult inherits the body plan of the larva. In echinoderms, the bilateral larva is reorganized into an adult with pentaradial symmetry. In echinoids, the adult mouth opens on the left side of the larva and the oral–aboral axis of the adult is thus perpendicular to the anterior–posterior axis of the larva. However, note that the relative positioning of the larval anterior–posterior and adult oral–aboral axes is not conserved between the five echinoderm classes (for a review see Peterson et al. 2000; Smiley 1986). In all schematics, the gut is shown in light yellow. Abbreviations: *A* anterior, *P* posterior

three paired coeloms are the protocoel, mesocoel, and metacoel in hemichordates or the axocoel, hydrocoel, and somatocoel in echinoderms (Peterson et al. 2000). In hemichordates, the larval protocoel, mesocoel, and metacoel will, respectively, give rise to the adult prosome, mesosome, and metasome (Kaul-Strehlow and Röttinger 2015). In echinoderms, by contrast, the situation is more complex, as the coeloms undergo a dramatic reorganization during formation of the adult body (Peterson et al. 2000;

Ezhova et al. 2013, 2014). The axocoel thus contributes to the formation of several adult structures, including most notably the excretory axial organ. The right hydrocoel remains in a rudimentary state, while the left hydrocoel generates the most prominent coelom of the adult body, driving both the pentaradial symmetry and the development of the water vascular system. The right and left somatocoels also grow considerably in size and relocate to the aboral side of the hydrocoel, where they create the adult perivisceral coeloms, which surround the gut, the gonads, and all other inner organs. Albeit evident ontogenetic differences between hemichordates and echinoderms, the shared origin of their coelomic compartments nonetheless allows comparisons of specific ambulacrarian body parts. For example, adult hemichordates and echinoderms possess a pulsatile vesicle, filtrating podocytes, and an excretory hydropore, whose developmental origin can be traced back, respectively, to the larval protocoel and axocoel (Willmer 1990; Kaul-Strehlow and Röttinger 2015).

Hemichordates and echinoderms further share significant similarities in the organization of their larval and adult nervous systems (Fig. 3.1). The nervous systems of hemichordate and echinoderm larvae are for instance commonly composed of two domains: an apical organ, located above the mouth and containing serotonergic neurons (Byrne et al. 2007), and a system of neurons and neurite bundles associated with the ciliary bands (Nakajima et al. 2004; Miyamoto et al. 2010). In both hemichordates and echinoderms, the larval nervous system is subsequently lost at metamorphosis, and the adult nervous system is created de novo (Chia and Burke 1978; Nielsen and Hay-Schmidt 2007; Gonzalez et al. 2018). Similarly, the adult nervous system of both hemichordates and echinoderms is made up of two main comparable components: a diffuse nerve net and nerve cords (Knight-Jones 1952; Smith 1965; Stach et al. 2012). In hemichordates, the nerve net is basiepidermal (located within the epidermis or arranged in one or more layers associated with the epidermis) and is spread throughout the adult, although it is denser in the proboscis in enteropneusts and in the trunk in pterobranchs (Stach et al. 2012; Miyamoto and Wada 2018). In echinoderms, the nerve net is also basiepidermal, and it is uniformly distributed throughout the adult.

With respect to the nerve cords, enteropneust hemichordates possess two nerve cords: a dorsal nerve cord that runs from the proboscis anteriorly to the tip of the trunk posteriorly and a ventral nerve cord that is restricted to the trunk. The two nerve cords are further linked, at the boundary between the collar and the trunk, by a circumferential nerve ring. Although referred to as nerve cords and nerve ring, these structures are actually mostly basiepidermal condensations of the basiepidermal nerve net, with the exception of the collar portion of the dorsal cord, also referred to as the collar cord, which is subepidermal (located below the epidermis and separated from it by a basement membrane) (Bullock 1945; Kaul and Stach 2010). In pterobranch hemichordates, several basiepidermal condensations of neurons have also been described. Although not explicitly called nerve cords, these include a ventral stalk nerve, branchial and tentacle nerves, and a dorsal cephalic ganglion that is located in the neck. Even though this ganglion is not subepidermal, anatomical evidence has suggested that it might actually be homologous to the enteropneust collar cord (Stach et al. 2012). By comparison, echinoderms generally have five nerve cords, one associated with each ambulacrum. These nerve cords can either be basiepidermal, such as in crinoids and asteroids, or subepidermal, like in echinoids, holothuroids, and ophiuroids. In any event, these nerve cords are always connected by a set of commissures at their proximal ends on the oral side of the adult, and these commissures always form a circumoral nerve ring.

3.3 On the Significance of Ambulacrarian Larval Neural Structures for Understanding Deuterostome Nervous System Evolution

In all eumetazoans, the first component of the nervous system to arise during embryogenesis is a pro-neural ectodermal domain called the apical or anterior neuroectoderm (ANE) (also referred to sometimes as the animal pole domain) (Arendt et al. 2016; Range and Wei 2016). The ANE domain commonly forms at the animal pole at the onset of gastrulation. Furthermore, its specification commonly requires the suppression of Wnt signaling activity, which is achieved by the expression of a conserved set of Wnt antagonists (Houart et al. 2002; Lagutin et al. 2003; Range et al. 2013). Thus, given the conservation of its position and specification mechanism, the ANE has been suggested to be homologous among eumetazoans and to share a common evolutionary origin dating back to their last common ancestor (Range 2014; Holland 2015a; Arendt et al. 2016), thereby assuming that the last common ancestor of deuterostomes also possessed an ANE.

Later, in chordates, the ANE gives rise to the most anterior region of the dorsal neural tube, which will form the most anterior portion of the adult central nervous system. In vertebrates, for example, the ANE territory will constitute parts of the future telencephalon (Wilson and Houart 2004; Plouhinec et al. 2017). In ambulacrarians, by comparison, the ANE gives rise to the apical organ of the larva, which is positioned above the mouth (Nielsen and Hay-Schmidt 2007; Hinman and Burke 2018) and which will subsequently be lost during metamorphosis, along with all neural structures associated with the ciliary band ectoderm (Chia and Burke 1978; Nakano et al. 2006; Kaul-Strehlow et al. 2015; Gonzalez et al. 2018). Despite these different ontogenetic fates of the chordate and ambulacrarian ANE, a number of molecular and anatomical similarities have been reported between the anterior neural tube of chordates and the apical organ of ambulacrarians. Indeed, following the specification of the ANE, the subsequent differentiation of this neuroectodermal territory into a brain or an apical organ has been shown to be mediated by a similar core set of neurogenic transcription factors that include *Six3*, *FoxQ2*, *Rx*, and *FoxG* (Yankura et al. 2010; Holland et al. 2013; Range 2014). Moreover, at the cellular level, both chordate brains and ambulacrarian apical organs contain clusters of serotonergic neurons, which are characterized by neurites projecting posteriorly (Candiani et al. 2001; Byrne et al. 2007; Stach 2014). Thus, based on these conserved features, chordate brains and ambulacrarian apical organs

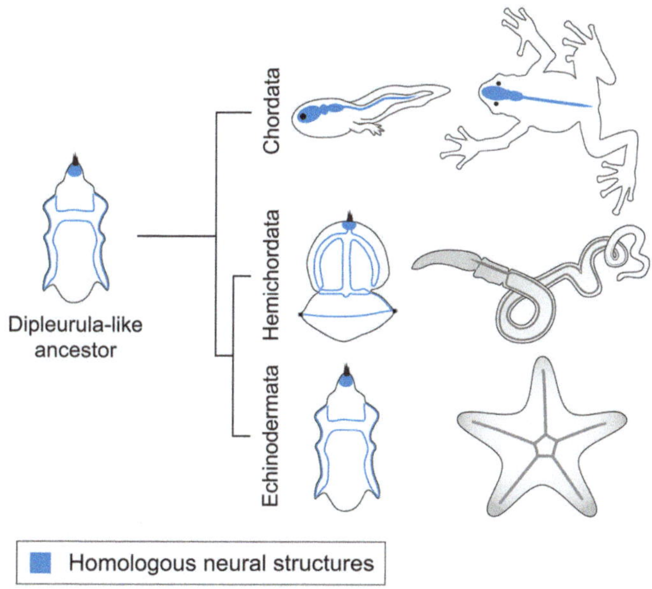

Fig. 3.3 Schematic representation of the auricularia hypothesis for the evolution of deuterostome nervous systems (adapted from Garstang 1894). Deuterostome phylogeny and main neural features are according to Fig. 3.1. Proposed homologous neural structures are highlighted in blue. Proposed derived neural structures are shown in gray

have frequently been described as being evolutionarily related (Wei et al. 2009; Tosches and Arendt 2013; Arendt et al. 2016).

One of the first theories about the evolutionary history of deuterostome nervous systems taking into account larval stages was that of Garstang (Garstang 1894). In this theory, called the auricularia hypothesis, the deuterostome ancestor was postulated to be a sexually mature dipleurula-like larva with an apical organ and a ciliary band (Fig. 3.3). From there, Garstang proposed that the chordate central nervous system evolved from the circumoral ciliary band of the ancestral dipleurula, with the two lateral sides of the ciliary band moving dorsally and fusing along the midline, thereby enclosing the apical organ and forming a dorsal neural tube. Although praised at first (Harvey 1961; Lacalli 1994) and supported by some lines of anatomical and gene expression data (Nielsen 2012), this theory has since then been pushed out of favor based on its incoherence with more recent phylogenetic, morphological, and molecular analyses (Nieuwenhuys 2002; Holland 2011, 2015a). For instance, studies in several ambulacrarian species have demonstrated that cells of the larval ciliary bands do not contribute to the formation of the adult nerve cords. Furthermore, there is no experimental evidence for a coordinated dorsal shift of the larval ciliary bands, which is generally very difficult to explain from an anatomical point of view given that the ciliary bands are firmly embedded in the general ectoderm of the larva. Despite this conflicting evidence from extant animals, Garstang's theory can

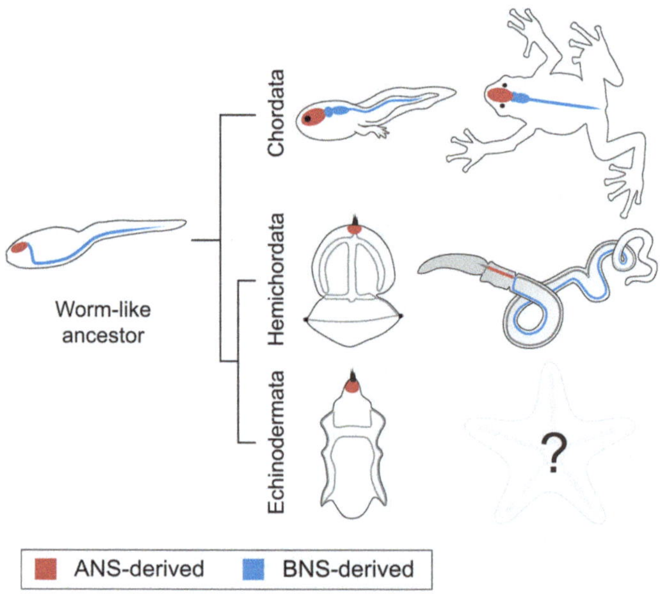

Fig. 3.4 Schematic representation of the chimeric brain hypothesis for the evolution of deuterostome nervous systems (adapted from Tosches and Arendt 2013). Deuterostome phylogeny and main neural features are according to Fig. 3.1. Proposed homologous neural structures are highlighted in identical colors. Proposed derived neural structures are shown in gray. The question mark indicates that the nervous system of adult echinoderms was not taken into consideration in the context of this hypothesis. Abbreviations: *ANS* apical nervous system, *BNS* blastoporal nervous system

however not be completely ruled out, since it is impossible to prove that the morphological modifications this theory assumes have indeed not occurred in the course of animal evolution.

A few years ago, an amended version of Garstang's theory was proposed by Tosches and Arendt (Tosches and Arendt 2013). Named the chimeric brain hypothesis, this theory postulates that bilaterian nervous systems evolved from an ancestral state characterized by two separate neuroectodermal territories: an apical neuroectodermal territory and a blastoporal neuroectodermal territory. These two territories then gave rise, respectively, to an apical nervous system (ANS) and a blastoporal nervous system (BNS), which in the course of bilaterian evolution would have merged to generate a single central nervous system with an anterior brain and a longitudinal nerve cord (Fig. 3.4). Interestingly, while this theory is currently well supported by various morphological and molecular analyses carried out in cnidarians, arthropods, annelids, and chordates (Denes et al. 2007; Steinmetz et al. 2010; Arendt et al. 2016), other comparative analyses have raised some unresolved issues that warrant further investigations. First, although strong scientific evidence supports the homology of the ANE, which in turn also supports the existence of an ancestral apical neuroectodermal territory, it cannot formally be ruled out that the molecular

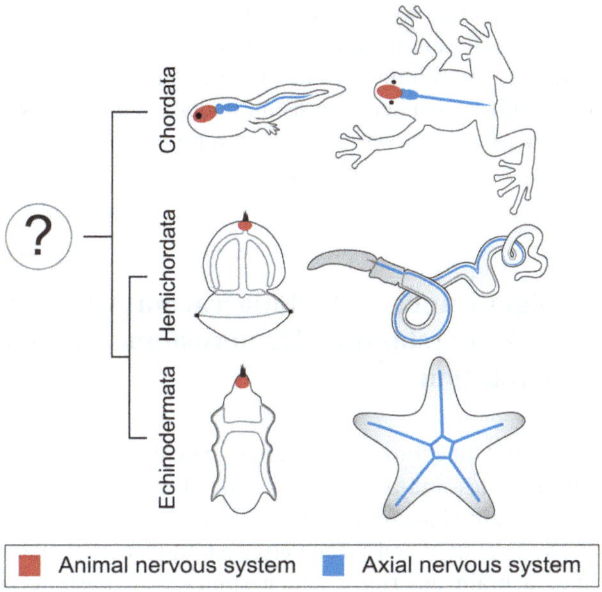

Fig. 3.5 Schematic representation of the "animal–axial" hypothesis for the evolution of deuterostome nervous systems (adapted from Burke 2011). Deuterostome phylogeny and main neural features are according to Fig. 3.1. Proposed homologous neural structures are highlighted in identical colors. Proposed derived neural structures are shown in gray. The question mark indicates that the ancestral deuterostome state was not defined in this hypothesis

mechanisms underlying ANE specification have simply been co-opted independently in several phyla (Raff 2008). For instance, similar mechanisms have also been shown to contribute to patterning other animal territories, such as the nonneural ectoderm (Holland et al. 2013). Second, the chimeric brain hypothesis has also been elaborated based on data from a very limited number of taxa. Therefore, it does not take into account the diversity of nervous systems observed amongst animals, such as, for instance, the nervous system of adult echinoderms, which has been excluded since it was judged too divergent (Benito-Gutiérrez and Arendt 2009). In addition, when considering additional data issued from a larger variety of animal taxa, the anatomy and molecular makeup of bilaterian nervous systems appears much more diverse than previously thought (Hejnol and Lowe 2015; Martín-Durán et al. 2018).

Focusing on the evolution of deuterostomes, Burke also recently proposed independently a theory on the evolution of their nervous systems, which, to a certain extent, is reminiscent of the chimeric brain hypothesis (Burke 2011). In this "animal–axial" theory, the ambulacrarian nervous system is considered as a bipartite system that has been separated both in time and space, with the larval apical organ of ambulacrarians being homologous to the chordate brain and the adult nerve cords of ambulacrarians being homologous to the chordate spinal cord (Fig. 3.5). Although this hypothesis remains highly contested, chiefly due to the same arguments that challenge the chimeric brain hypothesis (Holland 2015a), it nonetheless highlights the importance of considering the ambulacrarian adult nervous system. Indeed, this hypothesis points out the significance of investigating the anatomy and molecular organization of ambulacrarian nerve cords for understanding the evolutionary origin and history of deuterostome nervous systems, a point that has also been raised by other authors (Haag 2005; Mashanov et al. 2016).

3.4 On the Significance of Ambulacrarian Adult Nerve Cords for Understanding Deuterostome Nervous System Evolution

In chordates, the neural tube forms during embryonic development through a process called neurulation (Colas and Schoenwolf 2001; Lowery and Sive 2004). During neurulation, the ectodermal neuroepithelium, located dorsally in the embryo, folds in upon itself and closes over, thereby generating a hollow neural tube, which will later differentiate into a brain anteriorly and a spinal cord posteriorly. Interestingly, ambulacrarian adult nerve cord development has also been demonstrated to include, in some cases, a neurulation-like process. In enteropneust hemichordates, for instance, such a process takes place at the level of the collar cord (i.e., the portion of the dorsal nerve cord that is located in the collar). As in chordates, the neurulation-like process of the enteropneust collar cord involves the infolding of an ectodermal neuroepithelium and is mediated by transversely oriented ependymal cells possessing myofilaments (Bateson 1884; Morgan 1894; Bullock 1945; Dawydoff 1948; Knight-Jones 1952; Kaul and Stach 2010; Miyamoto and Wada 2013). In echinoderms, the five adult radial nerve cords of holothuroids, ophiuroids, and echinoids have also been demonstrated to be subepidermal, tubular neuroepithelia of ectodermal origin, and in echinoids and holothuroids, the radial nerve cords have been shown to form through a developmental process resembling chordate neurulation (MacBride 1903; von Ubisch 1913; Mashanov et al. 2007). Thus, based on these similarities, a homology has been proposed between the enteropneust collar cord, the echinoderm radial nerve cords, and the chordate neural tube (Fig. 3.6) (Haag 2005; Burke 2011; Luttrell et al. 2012).

Several additional anatomical and molecular lines of evidence further support the homology between the collar cord of enteropneust hemichordates, the radial nerve cords of echinoderms, and the neural tube of chordates. First, as in chordate brains, giant nerve cells with contralateral neurite projections and numerous synaptic connections have been identified in the collar cord of enteropneust hemichordates (Bullock 1945; Brown et al. 2008; Kaul and Stach 2010). In echinoderms, likewise, giant nerve fibers have been described in the radial nerve cords of at least ophiuroids (Stubbs and Cobb 1981). Second, comparable to the situation in chordates, Hedgehog signaling has also been suggested in enteropneust hemichordates to be involved in dorsal–ventral neural patterning (Miyamoto and Wada 2013). In chordates, the

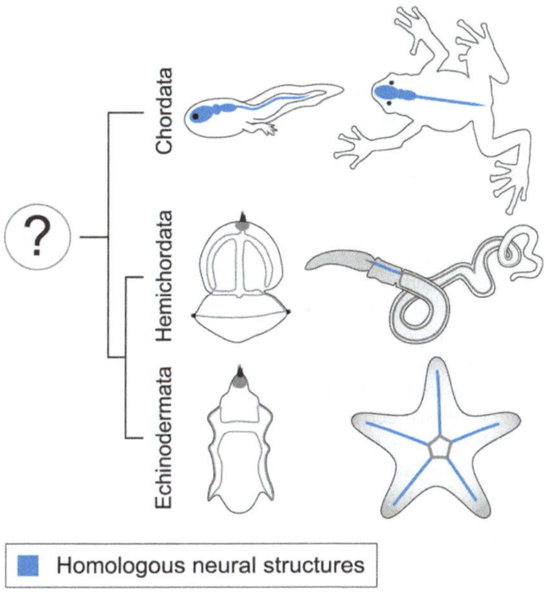

Fig. 3.6 Schematic representation illustrating the homology of the chordate neural tube, the enteropneust collar cord, and the echinoderm radial nerve cords (supported by Haag 2005; Burke 2011; Luttrell et al. 2012). Deuterostome phylogeny and main neural features are according to Fig. 3.1. Proposed homologous neural structures are highlighted in blue. Proposed derived neural structures are shown in gray. The question mark indicates that the ancestral deuterostome state was not considered in this theory

notochord expresses the signaling molecule Hedgehog and mediates dorsal–ventral patterning of the above neural tube, which expresses the gene encoding the Hedgehog receptor Patched. Similarly, in enteropneust hemichordates, the stomochord endoderm, which is located underneath the collar cord, expresses Hedgehog, and the collar cord expresses Patched. In echinoderms, based on anatomical evidence, the hydrocoel, which lies below the radial nerve cords, has also been proposed to be homologous to the chordate notochord (Heinzeller and Welsch 2001). Third, the molecular mechanisms regulating dorsal–ventral patterning within the collar cord of the enteropneust *Balanoglossus simodensis* and within the neural tube of chordates further share some similarities, as highlighted by the conserved, staggered expression of dorsal–ventral patterning markers, such as *Pax3/7* and *Pax2/5/8* (Miyamoto and Wada 2013).

Despite these robust similarities, the homology between the collar cord of enteropneust hemichordates, the radial nerve cords of echinoderms, and the neural tube of chordates has been challenged by a number of different observations. First, the morphogenetic movements driving the neurulation-like process in various ambulacrarians are significantly different. For instance, the formation of the collar cord in ptychoderid enteropneusts and of the radial nerve cords in echinoids involves tissue invagination (Morgan 1894; MacBride 1903; von Ubisch 1913; Dawydoff 1948), while the collar cord in harrimaniid enteropneusts develops by ingression (Bateson 1884), and the radial nerve cords in the holothuroid *Stichopus californicus* are formed by cavitation (Smiley 1986). However, it should be added that chordates also display different ways of forming a neural tube, including invagination, ingression, and cavitation. Some species even employ several of these modes for

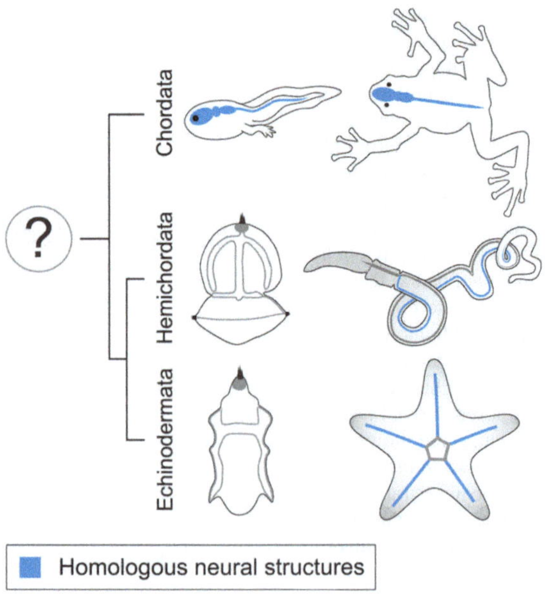

Fig. 3.7 Schematic representation illustrating the homology of the chordate neural tube, the enteropneust ventral nerve cord, and the echinoderm radial nerve cords (supported by Bullock 1940; Knight-Jones 1952; Cobb 1987). Deuterostome phylogeny and main neural features are according to Fig. 3.1. Proposed homologous neural structures are highlighted in blue. Proposed derived neural structures are shown in gray. The question mark indicates that the ancestral deuterostome state was not considered in this theory

developing distinct portions of their neural tube (Lowery and Sive 2004). Second, while the enteropneust collar cord has been shown to develop through a neurulation-like process, a similar mechanism does not seem to exist in pterobranchs, in which the cephalic ganglion and the main nerve tracts are simple basiepidermal neural condensations (Rehkämper et al. 1987; Stach et al. 2012). Similarly, in echinoderms, the nerve cords of asteroids and crinoids remain basiepidermal and are not internalized during their ontogeny (Smith 1965; Mashanov et al. 2016; Ezhova et al. 2017). It is thus possible that the neurulation processes identified in some ambulacrarian classes might simply reflect a secondary acquisition by convergent evolution rather than corresponding to an ancestral, inherited character. This notion is further supported by the fact that pterobranchs and crinoids constitute the most basal class of their respective phylum (Cannon et al. 2014). Taken together, this evidence has led some authors to reject the homology between the enteropneust collar cord, the echinoderm nerve cords, and the chordate neural tube (Ruppert 2005; Nielsen 2006; Kaul and Stach 2010).

The ventral nerve cord of enteropneust hemichordates has also been proposed to be homologous to the chordate neural tube (Fig. 3.7). Observations on injured animals have indicated that the responsiveness of the animal could only be negatively impacted when the ventral nerve cord was transected. Cutting of the dorsal nerve cord, including within the collar cord, did not lead to a loss of responsiveness (Bullock 1940; Knight-Jones 1952). These results have thus suggested that the ventral nerve cord, and not the collar cord, is likely to act as an integrative center, comparable to the chordate central nervous system. In accordance with this view, BMP2/4 ligands, which in bilaterian animals are generally expressed on the nonneural side of the dorsal–ventral axis (De Robertis and Sasai 1996; Mizutani and Bier 2008), are expressed dorsally in enteropneust hemichordates (Lowe et al.

2006). Thus, based on this regionalized expression of BMP2/4 and albeit the fact that BMP signaling appears not functionally required for dorsal–ventral neural patterning in enteropneust hemichordates (Lowe et al. 2006), some authors have considered either the ventral nerve cord or the ventral cord plus the collar cord to be homologous to the chordate neural tube (Nübler-Jung and Arendt 1996; Benito-Gutiérrez and Arendt 2009; Holland et al. 2013; Holland 2015b). In echinoderms, neurophysiological surveys have also determined that, following the reception of sensory input, any of the five radial nerve cords has the capacity to control the behavior of the whole animal (Cobb 1987), while the circumoral nerve ring has been suggested to simply relay neuronal activity between the radial nerve cords (Stubbs 1982; Moore and Cobb 1985). More recently, studies on ophiuroids have established that the circumoral nerve ring and the radial nerve cords actually cooperatively initiate and coordinate arm movements (Matsuzaka et al. 2017; Clark et al. 2019). Through local anesthesia and cuts of different body parts, these studies have demonstrated that the ophiuroid nervous system is organized into functional subunits (Matsuzaka et al. 2017) and that the nerve ring is required for the transmission of information between arms (Clark et al. 2019). Despite these observations, it is still a matter of debate, whether any neural structure of adult echinoderms acts as an integrative center, comparable to the chordate central nervous system (Cobb 1995; Haag 2006; Matsuzaka et al. 2017).

3.5 On the Significance of Ambulacrarian Basiepidermal Nerve Nets as Homologs of Chordate Neural Tubes

In chordates, patterning of the neural tube, along its anterior–posterior axis, is mediated by a specific core set of genes that are expressed in a precise sequential order along the axis (Holland 2009; Robertshaw and Kiecker 2012). Interestingly, in the directly developing enteropneust hemichordate *Saccoglossus kowalevskii*, homologs of these genes have been shown to display a similar organization of their expression along the anterior–posterior axis of the ectoderm (Lowe et al. 2003; Aronowicz and Lowe 2006). Thus, the transcription factors *Six3* and *Nkx2.1*, for instance, which in chordates are involved in the specification of the most anterior portion of the neural tube (Lagutin et al. 2003), are found expressed in the ectoderm of the proboscis of *S. kowalevskii* (i.e., the most anterior portion of the animal body). Likewise, the genes *Otx* and *Dbx*, known to be expressed in the vertebrate forebrain and midbrain (Nouri and Awatramani 2017), are confined to the ectoderm of the collar region of the enteropneust, and markers expressed in the vertebrate hindbrain and spinal cord, such as *Gbx* and the *Hox* genes (McGinnis and Krumlauf 1992; Rubenstein et al. 1998), are found in the trunk ectoderm of *S. kowalevskii*.

Similarly, comparable to the situation in the vertebrate hindbrain (Wassarman et al. 1997; Tümpel et al. 2009), *Hox* gene expression in *S. kowalevskii* displays spatial collinearity in the trunk ectoderm, with anterior *Hox* genes displaying a more

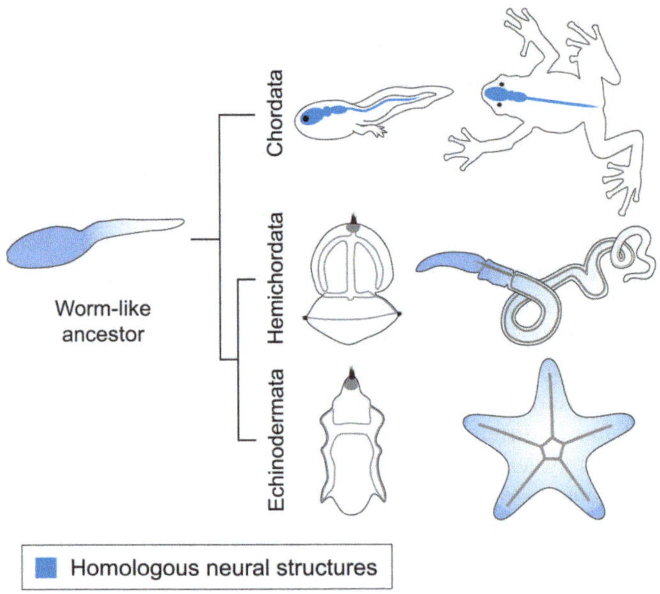

Fig. 3.8 Schematic representation of the ancestral nerve net hypothesis (adapted from Holland 2015a and supported by Lowe et al. 2003). Deuterostome phylogeny and main neural features are according to Fig. 3.1. Proposed homologous neural structures are highlighted in blue. Proposed derived neural structures are shown in gray

anterior limit of expression (close to the collar) than posterior *Hox* genes, presenting a more restricted posterior expression (close to tip of the trunk). This conservation of the deployment of this core set of transcription factors in the ectoderm of *S. kowalevskii* and the central nervous system of chordates is strongly suggestive of a homology and thus of a common ancestral origin of this genetic circuitry (Lowe et al. 2003; Holland 2003). This notion is further supported by similar findings obtained, through gene expression analyses, in two indirectly developing enteropneust hemichordates, the spengelid *Schizocardium californicum* (Gonzalez et al. 2017) and the ptychoderid *Balanoglossus misakiensis* (Kaul-Strehlow et al. 2017).

With this being said, in all surveys carried out so far on hemichordates, the expression of the anterior–posterior patterning genes was always found limited to circumferential domains in the ectoderm and never associated with the nerve cords (Lowe et al. 2003; Aronowicz and Lowe 2006; Gonzalez et al. 2017; Kaul-Strehlow et al. 2017). Thus, a homologous core set of genes seems to pattern two different tissues in hemichordates and chordates: the general ectoderm and the neurogenic ectoderm, respectively. Alternatively, given that the ectoderm of enteropneust hemichordates comprises a basiepidermal nerve net (Kaul-Strehlow and Röttinger 2015; Miyamoto and Wada 2018), it could be argued, as it has been by some authors, that this basiepidermal nerve net of hemichordates is actually homologous to the chordate neural tube (Fig. 3.8) (Holland 2003). Additional work carried out on

S. kowalevskii is in fact supportive of this latter notion (Pani et al. 2012; Yao et al. 2016). This work has established, for example, that genes required for controlling the specification of the three signaling centers of the developing vertebrate brain (the anterior neural ridge, the *zona limitans intrathalamica*, and the isthmic organizer) display similar co-expression domains in the enteropneust ectoderm (Pani et al. 2012). In addition, key components of the Wnt, FGF, and Hedgehog signaling pathways have been shown to be expressed in a vertebrate brain-like arrangement in the enteropneust ectoderm and to control homologous anterior–posterior patterning mechanisms (Pani et al. 2012). Finally, a *cis*-regulatory element mediating the localized expression of the signaling molecule Hedgehog in the *zona limitans intrathalamica* of mice has been demonstrated to be conserved in enteropneusts, where it functions to direct the expression of Hedgehog to the proboscis–collar boundary of the ectoderm (Yao et al. 2016).

Together, the discovery of a conserved molecular toolkit controlling anterior–posterior patterning of the chordate neural tube and of the enteropneust ectoderm has prompted a reassessment of the most plausible hypotheses for nervous system evolution in deuterostomes. To date, two scenarios are being discussed: (1) the deuterostome ancestor had a centralized, chordate-like nervous system or (2) the deuterostome ancestor had a diffuse, enteropneust-like nervous system. In the first scenario, the anterior–posterior patterning machinery of the ancestral centralized nervous system would have been co-opted to control ectodermal development in hemichordates. In the second scenario, the chordate central nervous system would have evolved independently and integrated the ancestral ectodermal anterior–posterior patterning system. In support of the first scenario, as mentioned above, BMP signaling does not seem to be required in enteropneust hemichordates for dorsal–ventral neural patterning and development of the ectodermal nerve net (Lowe et al. 2006). This contrasts markedly with the situation in other bilaterian animals and has thus been interpreted as evidence for the derived character of the hemichordate nervous system (Holland 2015a). In addition, anatomical and molecular data have suggested that in enteropneust hemichordates the radial nerve cords actually constitute a bona fide central nervous system, while the neurons scattered in the proboscis are more likely to be elements of a peripheral nervous system, rather than marking the presence of a cnidarian-like nerve net (Nomaksteinsky et al. 2009). In this context, it is important to highlight that the chordate ectoderm is also neurogenic, containing a sophisticated peripheral nervous system that is however never referred to as a nerve net (Holland et al. 2013). In support of the second scenario, by contrast, adult echinoderms also possess a basiepidermal nerve net (Smith 1965), and recent molecular analyses suggest that at least a subset of the anterior–posterior neural patterning toolkit is also expressed in the ectoderm of adult echinoderms (Tsuchimoto and Yamaguchi 2014; Koop et al. 2017; Adachi et al. 2018).

So far, adult echinoderms have largely remained at the margin of the debate on the evolution of deuterostome nervous systems, mainly due to their derived pentaradial body plan and peculiar anatomical characters, such as the water vascular system (Benito-Gutiérrez and Arendt 2009; Holland 2015a). Here, we argue however that they must be included to resolve the question of the evolutionary ancestry of deuterostome nervous systems. As of now, only very few studies have attempted to characterize the molecular signature of echinoderm nervous systems in developing rudiments or in adults (Tsuchimoto and Yamaguchi 2014; Koop et al. 2017; Adachi et al. 2018). Interestingly, though, when looking at the expression of a selection of anterior–posterior neural patterning genes in developing rudiments, this expression did not appear staggered along one of the body axes. Instead, it was radial around the body associated with either ambulacra or interambulacra. For instance, in the sand dollar *Peronella japonica*, the transcription factors *Six3* and *Otx* as well as some of the *Hox* genes were found expressed in all five ambulacra, while *Pax2/5/8* was detected in some and *Engrailed* in all interambulacra (Tsuchimoto and Yamaguchi 2014; Adachi et al. 2018). These data have been interpreted as evidence for a homology between the ambulacra of echinoderms and the proboscis and collar regions of enteropneust hemichordates (Adachi et al. 2018). However, given that the sand dollar *P. japonica* is characterized by a derived mechanism of rudiment formation, which is unique among echinoderms (Tsuchimoto and Yamaguchi 2014), this interpretation requires additional experimental scrutiny.

Despite this last point, expression of the neurogenic genes investigated to date in echinoderm rudiments has always been detected in either the ambulacral or interambulacral ectoderm and was never restricted to the radial nerve cord territory, which is similar to the situation in enteropneust hemichordates. Thus, regardless of the modified body plan of adult echinoderms, one could argue that a conserved core set of neurogenic genes is equivalently deployed in both the echinoderm and the hemichordate ectoderm to pattern the developing adult basiepidermal nerve net. Although this notion still requires additional experimental evidence, if corroborated, it would strongly support the idea that the deuterostome ancestor had a diffuse ectodermal nervous system patterned along the anterior–posterior axis by a conserved molecular toolkit. However, in echinoderms, the adult ambulacral ectoderm holds two neural structures: the basiepidermal nerve net and the radial nerve cords (Hyman 1955; Smith 1965). Thus, expression of the neurogenic genes in the rudiment might secondarily be modified in the course of development to become selectively associated with the adult radial nerve cords, a scenario that would then contradict the aforementioned hypothesis. As a matter of fact, a corresponding gene expression shift has already been reported, for instance, for the *Otx* gene in the sea urchin *Holopneustes purpurescens* (Morris et al. 2004). Thus, based on our current knowledge, the question of homology between ambulacrarian and chordate nervous systems remains unresolved, which in turn precludes any firm conclusion on the validity of either one of the two proposed scenarios for deuterostome nervous system evolution.

3.6 Conclusions and Future Directions

Understanding the ancestry and evolution of deuterostome nervous systems remains one of the key challenges of comparative biology. Several different theories have been proposed to reconcile the available experimental evidence with possible scenarios about the ancestral deuterostome state and how it evolved to generate the nervous systems of extant chordates and ambulacrarians. Here, we presented an overview of the most widely accepted theories and discussed anatomical, physiological, and molecular data from all deuterostome phyla in a comparative context. This work revealed that, albeit their derived pentaradial body plans, adult echinoderms enrich comparative analyses and should thus be considered for future studies. Furthermore, we found that strong arguments are still lacking to irrefutably establish homologies between the chordate central nervous system and any kind of ambulacrarian nervous system, larval or adult. Going forward, additional work on adult echinoderms and hemichordates will thus be particularly informative, and thorough investigations of the molecular patterning, anatomy, and neurophysiology of their adult nervous systems will be of tremendous help to obtain a better understanding of nervous system evolution in ambulacrarians.

To date, studies on the molecular signatures of adult ambulacrarian nervous systems remain limited to a couple of hemichordate and echinoderm species. For instance, most of the molecular work on hemichordates has so far been carried out only on the directly developing enteropneust *S. kowalevskii*, while studies on indirectly developing enteropneusts are rare and those on pterobranchs even rarer. In echinoderms, most of the molecular data on adults has also been obtained using chiefly either echinoid or holothuroid models. Yet, asteroids and crinoids are characterized by an adult nervous system that is anatomically very different from that of echinoids and holothuroids. A detailed molecular characterization of their basiepidermal nerve cords would thus be very useful for providing additional information on the possible architecture of the nervous system of the last ambulacrarian ancestor.

In terms of molecular analyses, it will be important to establish further, in representatives of all ambulacrarian taxa, whether the BMP signaling pathway is required for the specification of neural territories. While this is the case in most bilaterians, BMP signaling does not seem to control this process in the enteropneust hemichordate *S. kowalevskii*. In echinoids, BMP signaling participates in the positioning of the larval nervous system (Duboc et al. 2004). In addition, during adult rudiment formation, BMP2/4 ligands are expressed in the vestibule ectoderm and the hydrocoel-derived podia (Koop et al. 2017). However, whether the BMP signaling pathway is involved in echinoids in the establishment of the adult nervous system remains to be demonstrated. Similarly, the role of the Hedgehog signaling pathway in the development of the adult ambulacrarian nervous system needs to be characterized exhaustively. In chordates, the Hedgehog pathway is crucially involved in dorsal–ventral patterning of the neural tube (Patten and Placzek 2000; Dessaud et al. 2008), and it has been suggested that a comparable molecular mechanism coordinates collar cord development in enteropneust hemichordates (Miyamoto and Wada 2013). Furthermore, conserved *cis*-regulatory elements seem to control Hedgehog expression in the brain of chordates and in the ectoderm of enteropneust hemichordates (Pani et al. 2012). In echinoderms, mRNAs encoding Hedgehog are detected in the ambulacra of the adult rudiment of the sand dollar *P. japonica* (Adachi et al. 2018), but the biological relevance of this finding still remains elusive. Finally, it might also be important to assess the level of conservation, in deuterostomes, of the gene regulatory networks acting downstream of both the BMP and Hedgehog signaling pathways. As a matter of fact, gene expression profiles of transcription factors acting downstream of BMP and Hedgehog signaling in dorsal–ventral patterning have already proven to be critical for evaluating homologies of neural structures in protostomes and among bilaterians (Martín-Durán et al. 2018; Arendt 2018).

In sum, a more comprehensive understanding of the molecular programs controlling the development of ambulacrarian adult nervous systems is urgently required. With this information in hand, it should be possible to carry out exhaustive comparisons between ambulacrarians and chordates, which in turn will enable to reconstruct the evolutionary history of deuterostome nervous systems and reveal the origin of centralized nervous systems in bilaterians.

Acknowledgements The authors would like to thank Lucas Leclère for insightful discussions. Laurent Formery is indebted to the André Picard Network (ANR-11-IDEX-0004-02), the French Ministry of Higher Education and Research (2173/2015), and the Japan Society for the Promotion of Science (PE18758) for funding and support.

References

Adachi S, Niimi I, Sakai Y et al (2018) Anteroposterior molecular registries in ectoderm of the echinus rudiment. Dev Dyn 247:1297–1307. https://doi.org/10.1002/dvdy.24686

Arendt D (2018) Animal evolution: convergent nerve cords? Curr Biol 28:R225–R227. https://doi.org/10.1016/j.cub.2018.01.056

Arendt D, Tosches MA, Marlow H (2016) From nerve net to nerve ring, nerve cord and brain – evolution of the nervous system. Nat Rev Neurosci 17:61–72. https://doi.org/10.1038/nrn.2015.15

Aronowicz J, Lowe CJ (2006) Hox gene expression in the hemichordate *Saccoglossus kowalevskii* and the evolution of deuterostome nervous systems. Integr Comp Biol 46:890–901. https://doi.org/10.1093/icb/icl045

Baker CVH, Bronner-Fraser M (1997) The origins of the neural crest. Part II: An evolutionary perspective. Mech Dev 69:13–29. https://doi.org/10.1016/S0925-4773(97)00129-9

Bateson W (1884) The early stages in the development of balanoglossus (sp. incert.). J Cell Sci s2–s24:208–236

Benito-Gutiérrez È, Arendt D (2009) CNS evolution: new insight from the mud. Curr Biol 19: R640–R642. https://doi.org/10.1016/j.cub.2009.06.020

Blair JE, Hedges SB (2005) Molecular phylogeny and divergence times of deuterostome animals. Mol Biol Evol 22:2275–2284. https://doi.org/10.1093/molbev/msi225

Bourlat SJ, Juliusdottir T, Lowe CJ et al (2006) Deuterostome phylogeny reveals monophyletic chordates and the new phylum Xenoturbellida. Nature 444:85–88. https://doi.org/10.1038/nature05241

Brown FD, Prendergast A, Swalla BJ (2008) Man is but a worm: chordate origins. Genesis 46: 605–613. https://doi.org/10.1002/dvg.20471

Buchanan TW, Tranel D (2009) Central and peripheral nervous system interactions: from mind to brain to body. Int J Psychophysiol 72:1–4. https://doi.org/10.1016/j.ijpsycho.2008.09.002

Bullock TH (1940) The functional organization of the nervous system of Enteropneusta. Biol Bull 79:91–113. https://doi.org/10.2307/1537832

Bullock TH (1945) The anatomical organization of the nervous system of Enteropneusta. J Cell Sci s2–86:55–111

Burke RD (2011) Deuterostome neuroanatomy and the body plan paradox. Evol Dev 13:110–115. https://doi.org/10.1111/j.1525-142X.2010.00460.x

Burke RD, Angerer LM, Elphick MR et al (2006) A genomic view of the sea urchin nervous system. Dev Biol 300:434–460. https://doi.org/10.1016/j.ydbio.2006.08.007

Byrne M, Nakajima Y, Chee FC, Burke RD (2007) Apical organs in echinoderm larvae: insights into larval evolution in the Ambulacraria. Evol Dev 9:432–445. https://doi.org/10.1111/j.1525-142X.2007.00189.x

Candiani S, Augello A, Oliveri D et al (2001) Immunocytochemical localization of serotonin in embryos, larvae and adults of the lancelet, *Branchiostoma floridae*. Histochem J 33:413–420

Candiani S, Moronti L, Ramoino P et al (2012) A neurochemical map of the developing amphioxus nervous system. BMC Neurosci 13:59. https://doi.org/10.1186/1471-2202-13-59

Cannon JT, Kocot KM, Waits DS et al (2014) Phylogenomic resolution of the hemichordate and echinoderm clade. Curr Biol 24:2827–2832. https://doi.org/10.1016/j.cub.2014.10.016

Cannon JT, Vellutini BC, Smith J et al (2016) Xenacoelomorpha is the sister group to Nephrozoa. Nature 530:89–93. https://doi.org/10.1038/nature16520

Chia FS, Burke RD (1978) Echinoderm metamorphosis: fate of larval structures. In: Chia F-S, Rice ME (eds) Settlement and metamorphosis of marine invertebrates. Elsevier, New York, pp 219–234

Clark EG, Kanauchi D, Kano T et al (2019) The function of the ophiuroid nerve ring: how a decentralized nervous system controls coordinated locomotion. J Exp Biol 222:jeb192104. https://doi.org/10.1242/jeb.192104

Cobb JLS (1987) Neurobiology of the echinodermata. In: Ali MA (ed) Nervous systems in invertebrates. Springer, Boston, MA, pp 483–525

Cobb JLS (1995) The nervous systems of echinodermata: recent results and new approaches. In: Breidbach O, Kutsch W (eds) The nervous systems of invertebrates: an evolutionary and comparative approach: with a coda written by T.H. Bullock. Birkhäuser Basel, Basel, pp 407–424

Colas J-F, Schoenwolf GC (2001) Towards a cellular and molecular understanding of neurulation. Dev Dyn 221:117–145. https://doi.org/10.1002/dvdy.1144

Dawydoff C (1948) Embranchement des Stomochordes. In: Grasse PP (ed) Traite de Zoologie, Anatomie, Systematique, Biologie. Masson, Paris, pp 365–532

De Robertis EM, Sasai Y (1996) A common plan for dorsoventral patterning in bilateria. Nature 380:37–40. https://doi.org/10.1038/380037a0

Delsuc F, Brinkmann H, Chourrout D, Philippe H (2006) Tunicates and not cephalochordates are the closest living relatives of vertebrates. Nature 439:965–968. https://doi.org/10.1038/nature04336

Denes AS, Jékely G, Steinmetz PRH et al (2007) Molecular architecture of annelid nerve cord supports common origin of nervous system centralization in bilateria. Cell 129:277–288. https://doi.org/10.1016/j.cell.2007.02.040

Dessaud E, McMahon AP, Briscoe J (2008) Pattern formation in the vertebrate neural tube: a sonic hedgehog morphogen-regulated transcriptional network. Development 135:2489–2503. https://doi.org/10.1242/dev.009324

Dominguez P, Jacobson AG, Jefferies RPS (2002) Paired gill slits in a fossil with a calcite skeleton. Nature 417:841–844. https://doi.org/10.1038/nature00805

Duboc V, Röttinger E, Besnardeau L, Lepage T (2004) Nodal and BMP2/4 signaling organizes the oral-aboral axis of the sea urchin embryo. Dev Cell 6:397–410. https://doi.org/10.1016/S1534-5807(04)00056-5

Dunn CW, Hejnol A, Matus DQ et al (2008) Broad phylogenomic sampling improves resolution of the animal tree of life. Nature 452:745–749. https://doi.org/10.1038/nature06614

Ezhova OV, Lavrova EA, Malakhov VV (2013) Microscopic anatomy of the axial complex in the starfish *Asterias rubens* (Echinodermata, Asteroidea). Biol Bull Russ Acad Sci 40:643–653. https://doi.org/10.1134/S1062359013080049

Ezhova OV, Lavrova EA, Malakhov VV (2014) The morphology of the axial complex and associated structures in Asterozoa (Asteroidea, Echinoidea, Ophiuroidea). Russ J Mar Biol 40:153–164. https://doi.org/10.1134/S1063074014030043

Ezhova OV, Ershova NA, Malakhov VV (2017) Microscopic anatomy of the axial complex and associated structures in the sea cucumber *Chiridota laevis* Fabricius, 1780 (Echinodermata, Holothuroidea). Zoomorphology 136:205–217. https://doi.org/10.1007/s00435-016-0341-8

Garstang W (1894) Preliminary note on a new theory of the phylogeny of the Chordata. Zool Anz 17:122–125

Gillis JA, Fritzenwanker JH, Lowe CJ (2012) A stem-deuterostome origin of the vertebrate pharyngeal transcriptional network. Proc Biol Sci 279:237–246. https://doi.org/10.1098/rspb.2011.0599

Gonzalez P, Cameron CB (2009) The gill slits and pre-oral ciliary organ of Protoglossus (Hemichordata: Enteropneusta) are filter-feeding structures. Biol J Linn Soc 98:898–906. https://doi.org/10.1111/j.1095-8312.2009.01332.x

Gonzalez P, Uhlinger KR, Lowe CJ (2017) The adult body plan of indirect developing hemichordates develops by adding a Hox-patterned trunk to an anterior larval territory. Curr Biol 27:87–95. https://doi.org/10.1016/j.cub.2016.10.047

Gonzalez P, Jiang JZ, Lowe CJ (2018) The development and metamorphosis of the indirect developing acorn worm *Schizocardium californicum* (Enteropneusta: Spengelidae). Front Zool 15:26. https://doi.org/10.1186/s12983-018-0270-0

Grobben K (1908) Die systematische Einteilung des Tierreichs. Verhandlungen der Kaiserlich-Königlichen Zoologisch-Botanischen Gesellschaft in Wien 58:491–511

Haag ES (2005) Echinoderm rudiments, rudimentary bilaterians, and the origin of the chordate CNS. Evol Dev 7:280–281. https://doi.org/10.1111/j.1525-142X.2005.05032.x

Haag ES (2006) Reply to Nielsen. Evol Dev 8:3–5. https://doi.org/10.1111/j.1525-142X.2006.05069.x

Harvey LA (1961) New speculations on the origin of the chordates. Sci Prog 49:507–514

Heinzeller T, Welsch U (2001) The echinoderm nervous system and its phylogenetic interpretation. In: Roth G, Wullimann MF (eds) Brain evolution and cognition. Wiley, New York, pp 41–75

Hejnol A, Lowe CJ (2015) Embracing the comparative approach: how robust phylogenies and broader developmental sampling impacts the understanding of nervous system evolution. Philos Trans R Soc Lond Ser B Biol Sci 370. https://doi.org/10.1098/rstb.2015.0045

Hejnol A, Martindale MQ (2009) The mouth, the anus, and the blastopore—open questions about questionable openings. In: Telford MJ, Littlewood DTJ (eds) Animal evolution: genomes, fossils, and trees. Oxford Scholarship Press, Oxford, pp 33–40

Hejnol A, Obst M, Stamatakis A et al (2009) Assessing the root of bilaterian animals with scalable phylogenomic methods. Proc R Soc Lond B Biol Sci 276:4261–4270. https://doi.org/10.1098/rspb.2009.0896

Hinman VF, Burke RD (2018) Embryonic neurogenesis in echinoderms. Wiley Interdiscip Rev Dev Biol 7:e316. https://doi.org/10.1002/wdev.316

Holland ND (2003) Early central nervous system evolution: an era of skin brains? Nat Rev Neurosci 4:617–627. https://doi.org/10.1038/nrn1175

Holland LZ (2009) Chordate roots of the vertebrate nervous system: expanding the molecular toolkit. Nat Rev Neurosci 10:736–746. https://doi.org/10.1038/nrn2703

Holland ND (2011) Walter Garstang: a retrospective. Theory Biosci 130:247–258. https://doi.org/10.1007/s12064-011-0130-3

Holland LZ (2015a) Evolution of basal deuterostome nervous systems. J Exp Biol 218:637–645. https://doi.org/10.1242/jeb.109108

Holland LZ (2015b) The origin and evolution of chordate nervous systems. Philos Trans R Soc Lond Ser B Biol Sci 370. https://doi.org/10.1098/rstb.2015.0048

Holland LZ (2016) Tunicates. Curr Biol 26:R146–R152. https://doi.org/10.1016/j.cub.2015.12.024

Holland LZ, Carvalho JE, Escriva H et al (2013) Evolution of bilaterian central nervous systems: a single origin? EvoDevo 4:27. https://doi.org/10.1186/2041-9139-4-27

Houart C, Caneparo L, Heisenberg C-P et al (2002) Establishment of the telencephalon during gastrulation by local antagonism of Wnt signaling. Neuron 35:255–265. https://doi.org/10.1016/S0896-6273(02)00751-1

Hudson C (2016) The central nervous system of ascidian larvae. Wiley Interdiscip Rev Dev Biol 5:538–561. https://doi.org/10.1002/wdev.239

Hyman LH (1955) The invertebrates, vol 4. McGraw-Hill, New York

Kardong KV (2012) Vertebrates comparative anatomy, function, evolution, 6th edn. McGraw-Hill, New York

Kaul S, Stach T (2010) Ontogeny of the collar cord: neurulation in the hemichordate *Saccoglossus kowalevskii*. J Morphol 271:1240–1259. https://doi.org/10.1002/jmor.10868

Kaul-Strehlow S, Röttinger E (2015) Hemichordata. In: Wanninger A (ed) Evolutionary developmental biology of invertebrates 6: Deuterostomia. Springer, Vienna, pp 59–89

Kaul-Strehlow S, Urata M, Minokawa T et al (2015) Neurogenesis in directly and indirectly developing enteropneusts: of nets and cords. Org Divers Evol 15:405–422. https://doi.org/10.1007/s13127-015-0201-2

Kaul-Strehlow S, Urata M, Praher D, Wanninger A (2017) Neuronal patterning of the tubular collar cord is highly conserved among enteropneusts but dissimilar to the chordate neural tube. Sci Rep 7:7003. https://doi.org/10.1038/s41598-017-07052-8

Knight-Jones EW (1952) On the nervous system of *Saccoglossus cambrensis* (Enteropneusta). Phil Trans R Soc Lond B 236:315–354. https://doi.org/10.1098/rstb.1952.0004

Koop D, Cisternas P, Morris VB et al (2017) Nodal and BMP expression during the transition to pentamery in the sea urchin *Heliocidaris erythrogramma*: insights into patterning the enigmatic echinoderm body plan. BMC Dev Biol 17:4. https://doi.org/10.1186/s12861-017-0145-1

Lacalli TC (1994) Apical organs, epithelial domains, and the origin of the chordate central nervous system. Integr Comp Biol 34:533–541. https://doi.org/10.1093/icb/34.4.533

Lagutin OV, Zhu CC, Kobayashi D et al (2003) Six3 repression of Wnt signaling in the anterior neuroectoderm is essential for vertebrate forebrain development. Genes Dev 17:368–379. https://doi.org/10.1101/gad.1059403

Lowe CJ, Wu M, Salic A et al (2003) Anteroposterior patterning in hemichordates and the origins of the chordate nervous system. Cell 113:853–865. https://doi.org/10.1016/S0092-8674(03)00469-0

Lowe CJ, Terasaki M, Wu M et al (2006) Dorsoventral patterning in hemichordates: insights into early chordate evolution. PLoS Biol 4:e291. https://doi.org/10.1371/journal.pbio.0040291

Lowe CJ, Clarke DN, Medeiros DM et al (2015) The deuterostome context of chordate origins. Nature 520:456–465. https://doi.org/10.1038/nature14434

Lowery LA, Sive H (2004) Strategies of vertebrate neurulation and a re-evaluation of teleost neural tube formation. Mech Dev 121:1189–1197. https://doi.org/10.1016/j.mod.2004.04.022

Luttrell S, Konikoff C, Byrne A et al (2012) Ptychoderid hemichordate neurulation without a notochord. Integr Comp Biol 52:829–834. https://doi.org/10.1093/icb/ics117

MacBride EW (1903) VI. The development of *Echinus esculentus*, together some points the development of *E. miliaris* and *E. acutus*. Philos Trans R Soc Lond B 195:285–327. https://doi.org/10.1098/rstb.1903.0006

Marlow H, Tosches MA, Tomer R et al (2014) Larval body patterning and apical organs are conserved in animal evolution. BMC Biol 12:7. https://doi.org/10.1186/1741-7007-12-7

Martín-Durán JM, Janssen R, Wennberg S et al (2012) Deuterostomic development in the protostome *Priapulus caudatus*. Curr Biol 22:2161–2166. https://doi.org/10.1016/j.cub.2012.09.037

Martín-Durán JM, Pang K, Børve A et al (2018) Convergent evolution of bilaterian nerve cords. Nature 553:45–50. https://doi.org/10.1038/nature25030

Mashanov VS, Zueva OR, Heinzeller T et al (2007) Developmental origin of the adult nervous system in a holothurian: an attempt to unravel the enigma of neurogenesis in echinoderms. Evol Dev 9:244–256. https://doi.org/10.1111/j.1525-142X.2007.00157.x

Mashanov VS, Zueva OR, Rubilar T et al (2016) Echinodermata. In: Schmidt-Rhaesa A, Harzsch S, Purschke G (eds) Structure and evolution of invertebrate nervous systems. Oxford University Press, Oxford, pp 665–688

Matsuzaka Y, Sato E, Kano T et al (2017) Non-centralized and functionally localized nervous system of ophiuroids: evidence from topical anesthetic experiments. Biol Open 6:425–438. https://doi.org/10.1242/bio.019836

McGinnis W, Krumlauf R (1992) Homeobox genes and axial patterning. Cell 68:283–302. https://doi.org/10.1016/0092-8674(92)90471-N

Minsuk SB, Turner FR, Andrews ME, Raff RA (2009) Axial patterning of the pentaradial adult echinoderm body plan. Dev Genes Evol 219:89–101. https://doi.org/10.1007/s00427-009-0270-3

Miyamoto N, Wada H (2013) Hemichordate neurulation and the origin of the neural tube. Nat Commun 4:2713. https://doi.org/10.1038/ncomms3713

Miyamoto N, Wada H (2018) Hemichordate nervous system. Oxford Research Encyclopedia, Neuroscience 1–20. doi: https://doi.org/10.1093/acrefore/9780190264086.013.204

Miyamoto N, Nakajima Y, Wada H, Saito Y (2010) Development of the nervous system in the acorn worm *Balanoglossus simodensis*: insights into nervous system evolution. Evol Dev 12:416–424. https://doi.org/10.1111/j.1525-142X.2010.00428.x

Mizutani CM, Bier E (2008) EvoD/Vo: the origins of BMP signalling in the neuroectoderm. Nat Rev Genet 9:663–677. https://doi.org/10.1038/nrg2417

Mooi R, David B (2008) Radial symmetry, the anterior/posterior axis, and echinoderm Hox genes. Annu Rev Ecol Evol Syst 39:43–62. https://doi.org/10.1146/annurev.ecolsys.39.110707.173521

Moore A, Cobb JLS (1985) Neurophysiological studies on photic responses in *Ophiura ophiura*. Comp Biochem Physiol A Physiol 80:11–16. https://doi.org/10.1016/0300-9629(85)90669-3

Morgan TH (1894) The development of Balanoglossus. J Morphol 9:1–86. https://doi.org/10.1002/jmor.1050090102

Morris VB, Zhao J-T, Shearman DCA et al (2004) Expression of an Otx gene in the adult rudiment and the developing central nervous system in the vestibula larva of the sea urchin *Holopneustes purpurescens*. Int J Dev Biol 48:17–22. https://doi.org/10.1387/ijdb.15005570

Mwinyi A, Bailly X, Bourlat SJ et al (2010) The phylogenetic position of Acoela as revealed by the complete mitochondrial genome of *Symsagittifera roscoffensis*. BMC Evol Biol 10:309. https://doi.org/10.1186/1471-2148-10-309

Nakajima Y, Kaneko H, Murray G, Burke RD (2004) Divergent patterns of neural development in larval echinoids and asteroids. Evol Dev 6:95–104. https://doi.org/10.1111/j.1525-142X.2004.04011.x

Nakano H, Murabe N, Amemiya S, Nakajima Y (2006) Nervous system development of the sea cucumber *Stichopus japonicus*. Dev Biol 292:205–212. doi: https://doi.org/10.1016/j.ydbio.2005.12.038

Nielsen C (1987) Structure and function of metazoan ciliary bands and their phylogenetic significance. Acta Zool 68:205–262. https://doi.org/10.1111/j.1463-6395.1987.tb00892.x

Nielsen C (1998) Origin and evolution of animal life cycles. Biol Rev 73:125–155. https://doi.org/10.1111/j.1469-185X.1997.tb00027.x

Nielsen C (2006) Homology of echinoderm radial nerve cords and the chordate neural tube??? Evol Dev 8:1–2. https://doi.org/10.1111/j.1525-142X.2006.05068.x
Nielsen C (2012) Animal evolution: interrelationships of the living phyla. OUP, Oxford
Nielsen C (2015) Larval nervous systems: true larval and precocious adult. J Exp Biol 218: 629–636. https://doi.org/10.1242/jeb.109603
Nielsen C, Hay-Schmidt A (2007) Development of the enteropneust *Ptychodera flava*: ciliary bands and nervous system. J Morphol 268:551–570. https://doi.org/10.1002/jmor.10533
Nieuwenhuys R (2002) Deuterostome brains: synopsis and commentary. Brain Res Bull 57: 257–270. https://doi.org/10.1016/S0361-9230(01)00668-2
Nomaksteinsky M, Röttinger E, Dufour HD et al (2009) Centralization of the deuterostome nervous system predates chordates. Curr Biol 19:1264–1269. https://doi.org/10.1016/j.cub.2009.05.063
Northcutt RG (2012) Evolution of centralized nervous systems: two schools of evolutionary thought. PNAS 109:10626–10633. https://doi.org/10.1073/pnas.1201889109
Nouri N, Awatramani R (2017) A novel floor plate boundary defined by adjacent En1 and Dbx1 microdomains distinguishes midbrain dopamine and hypothalamic neurons. Development 144: 916–927. https://doi.org/10.1242/dev.144949
Nübler-Jung K, Arendt D (1996) Enteropneusts and chordate evolution. Curr Biol 6:352–353. https://doi.org/10.1016/S0960-9822(02)00491-8
Osborn KJ, Kuhnz LA, Priede IG et al (2012) Diversification of acorn worms (Hemichordata, Enteropneusta) revealed in the deep sea. Proc R Soc Lond B Biol Sci 279:1646–1654. https://doi.org/10.1098/rspb.2011.1916
Osugi T, Sasakura Y, Satake H (2017) The nervous system of the adult ascidian *Ciona intestinalis* Type A (*Ciona robusta*): insights from transgenic animal models. PLoS One 12:e0180227. https://doi.org/10.1371/journal.pone.0180227
Ou Q, Morris SC, Han J et al (2012) Evidence for gill slits and a pharynx in Cambrian vetulicolians: implications for the early evolution of deuterostomes. BMC Biol 10:81. https://doi.org/10.1186/1741-7007-10-81
Pani AM, Mullarkey EE, Aronowicz J et al (2012) Ancient deuterostome origins of vertebrate brain signalling centres. Nature 483:289–294. https://doi.org/10.1038/nature10838
Patten I, Placzek M (2000) The role of sonic hedgehog in neural tube patterning. CMLS, Cell Mol Life Sci 57:1695–1708. https://doi.org/10.1007/PL00000652
Peterson KJ, Arenas-Mena C, Davidson EH (2000) The A/P axis in echinoderm ontogeny and evolution: evidence from fossils and molecules. Evol Dev 2:93–101. https://doi.org/10.1046/j.1525-142x.2000.00042.x
Philippe H, Brinkmann H, Copley RR et al (2011) Acoelomorph flatworms are deuterostomes related to *Xenoturbella*. Nature 470:255–258. https://doi.org/10.1038/nature09676
Plouhinec J-L, Medina-Ruiz S, Borday C et al (2017) A molecular atlas of the developing ectoderm defines neural, neural crest, placode, and nonneural progenitor identity in vertebrates. PLoS Biol 15:e2004045. https://doi.org/10.1371/journal.pbio.2004045
Raff RA (1987) Constraint, flexibility, and phylogenetic history in the evolution of direct development in sea urchins. Dev Biol 119:6–19. https://doi.org/10.1016/0012-1606(87)90201-6
Raff RA (2008) Origins of the other metazoan body plans: the evolution of larval forms. Philos Trans R Soc B Biol Sci 363:1473–1479. https://doi.org/10.1098/rstb.2007.2237
Range R (2014) Specification and positioning of the anterior neuroectoderm in deuterostome embryos. Genesis 52:222–234. https://doi.org/10.1002/dvg.22759
Range RC, Wei Z (2016) An anterior signaling center patterns and sizes the anterior neuroectoderm of the sea urchin embryo. Development 143:1523–1533. https://doi.org/10.1242/dev.128165
Range RC, Angerer RC, Angerer LM (2013) Integration of canonical and noncanonical Wnt signaling pathways patterns the neuroectoderm along the anterior–posterior axis of sea urchin embryos. PLoS Biol 11:e1001467. https://doi.org/10.1371/journal.pbio.1001467
Rehkämper G, Welsch U, Dilly PN (1987) Fine structure of the ganglion of *Cephalodiscus gracilis* (pterobranchia, hemichordata). J Comp Neurol 259:308–315. https://doi.org/10.1002/cne.902590210

Reich A, Dunn C, Akasaka K, Wessel G (2015) Phylogenomic analyses of echinodermata support the sister groups of Asterozoa and Echinozoa. PLoS One 10:e0119627. https://doi.org/10.1371/journal.pone.0119627

Robertshaw E, Kiecker C (2012) Phylogenetic origins of brain organisers. Scientifica. https://doi.org/10.6064/2012/475017

Rouse GW, Wilson NG, Carvajal JI, Vrijenhoek RC (2016) New deep-sea species of *Xenoturbella* and the position of Xenacoelomorpha. Nature 530:94–97. https://doi.org/10.1038/nature16545

Rubenstein JLR, Shimamura K, Martinez S, Puelles L (1998) Regionalization of the prosencephalic neural plate. Annu Rev Neurosci 21:445–477. https://doi.org/10.1146/annurev.neuro.21.1.445

Ruppert EE (2005) Key characters uniting hemichordates and chordates: homologies or homoplasies? Can J Zool 83(1):8–23

Rychel AL, Swalla BJ (2007) Development and evolution of chordate cartilage. J Exp Zool B Mol Dev Evol 308B:325–335. https://doi.org/10.1002/jez.b.21157

Satoh N (2016) Chordate origins and evolution: the molecular evolutionary road to vertebrates. Academic Press, Amsterdam

Satoh N, Rokhsar D, Nishikawa T (2014) Chordate evolution and the three-phylum system. Proc R Soc B Biol Sci 281:20141729. https://doi.org/10.1098/rspb.2014.1729

Serafy DK, Fell FJ (1985) Marine flora and fauna of the Northeastern United States. Echinodermata: Echinoidea. In: National oceanic and atmospheric administration technical report NMFS 33, pp 1–27

Shepherd SV (2017) The Wiley handbook of evolutionary neuroscience. Wiley, Chichester

Shu D, Morris SC, Han J et al (2001) Primitive deuterostomes from the Chengjiang Lagerstätte (Lower Cambrian, China). Nature 414:419–424. https://doi.org/10.1038/35106514

Shu D, Morris SC, Zhang ZF et al (2003) A new species of Yunnanozoan with implications for deuterostome evolution. Science 299:1380–1384. https://doi.org/10.1126/science.1079846

Smiley S (1986) Metamorphosis of *Stichopus californicus* (Echinodermata: Holothuroidea) and its phylogenetic implications. Biol Bull 171:611–631. https://doi.org/10.2307/1541627

Smith JE (1965) Echinodermata. In: Bullock TH, Horridge GA (eds) Structure and function in the nervous systems of invertebrates, vol II. W. H. Freeman, London, pp 1519–1558

Smith AB (1997) Echinoderm larvae and phylogeny. Annu Rev Ecol Syst 28:219–241. https://doi.org/10.1146/annurev.ecolsys.28.1.219

Stach T (2014) Deuterostome phylogeny – a morphological perspective. In: Wägele JW, Bartolomaeus T (eds) Deep metazoan phylogeny: the backbone of the tree of life. De Gruyter, Berlin, pp 425–457

Stach T, Kaul S (2011) The postanal tail of the enteropneust *Saccoglossus kowalevskii* is a ciliary creeping organ without distinct similarities to the chordate tail. Acta Zool 92:150–160. https://doi.org/10.1111/j.1463-6395.2010.00462.x

Stach T, Gruhl A, Kaul-Strehlow S (2012) The central and peripheral nervous system of *Cephalodiscus gracilis* (Pterobranchia, Deuterostomia). Zoomorphology 131:11–24. https://doi.org/10.1007/s00435-011-0144-x

Steinmetz PR, Urbach R, Posnien N et al (2010) Six3 demarcates the anterior-most developing brain region in bilaterian animals. EvoDevo 1:14. https://doi.org/10.1186/2041-9139-1-14

Stubbs TR (1982) The neurophysiology of photosensitivity in ophiuroids (Tampa). In: Lawrence JM (ed) Proceedings of international echinoderms conference. A.A. Balkema, Rotterdam, pp 403–408

Stubbs TR, Cobb JLS (1981) The giant neurone system in ophiuroids. Cell Tissue Res 220: 373–385. https://doi.org/10.1007/BF00210515

Swalla BJ (2006) Building divergent body plans with similar genetic pathways. Heredity 97: 235–243. https://doi.org/10.1038/sj.hdy.6800872

Telford MJ, Lowe CJ, Cameron CB et al (2014) Phylogenomic analysis of echinoderm class relationships supports Asterozoa. Proc Biol Sci 281. https://doi.org/10.1098/rspb.2014.0479

Telford MJ, Budd GE, Philippe H (2015) Phylogenomic insights into animal evolution. Curr Biol 25:R876–R887. https://doi.org/10.1016/j.cub.2015.07.060

Tosches MA, Arendt D (2013) The bilaterian forebrain: an evolutionary chimaera. Curr Opin Neurobiol 23:1080–1089. https://doi.org/10.1016/j.conb.2013.09.005

Tsuchimoto J, Yamaguchi M (2014) Hox expression in the direct-type developing sand dollar *Peronella japonica*: Hox expression in an echinoid. Dev Dyn 243:1020–1029. https://doi.org/10.1002/dvdy.24135

Tümpel S, Wiedemann LM, Krumlauf R (2009) Chapter 8 Hox genes and segmentation of the vertebrate hindbrain. In: Current topics in developmental biology. Academic Press, San Diego, CA, pp 103–137

von Ubisch L (1913) Die Entwicklung con *Strongylocentrotus lividus*. Z Wiss Zool 106:409–448

Wassarman KM, Lewandoski M, Campbell K et al (1997) Specification of the anterior hindbrain and establishment of a normal mid/hindbrain organizer is dependent on Gbx2 gene function. Development 124:2923–2934

Webster D, Webster M (2013) Comparative vertebrate morphology. Academic Press, New York

Wei Z, Yaguchi J, Yaguchi S et al (2009) The sea urchin animal pole domain is a Six3-dependent neurogenic patterning center. Development 136:1179–1189. https://doi.org/10.1242/dev.032300

Willmer P (1990) Invertebrate relationships: patterns in animal evolution. Cambridge University Press, Cambridge

Wilson SW, Houart C (2004) Early steps in the development of the forebrain. Dev Cell 6:167–181. https://doi.org/10.1016/S1534-5807(04)00027-9

Yankura KA, Martik ML, Jennings CK, Hinman VF (2010) Uncoupling of complex regulatory patterning during evolution of larval development in echinoderms. BMC Biol 8:143. https://doi.org/10.1186/1741-7007-8-143

Yao Y, Minor PJ, Zhao Y-T et al (2016) *Cis*-regulatory architecture of a brain signaling center predates the origin of chordates. Nat Genet 48:575–580. https://doi.org/10.1038/ng.3542

Zhang G (2009) An evo-devo view on the origin of the backbone: evolutionary development of the vertebrae. Integr Comp Biol 49:178–186. https://doi.org/10.1093/icb/icp061

Zieger E, Candiani S, Garbarino G et al (2017) Roles of retinoic acid signaling in shaping the neuronal architecture of the developing amphioxus nervous system. Mol Neurobiol:1–20. https://doi.org/10.1007/s12035-017-0727-8

Part II
New and Emerging Model Systems in Evo-Devo Research

Chapter 4
Oikopleura dioica: An Emergent Chordate Model to Study the Impact of Gene Loss on the Evolution of the Mechanisms of Development

Alfonso Ferrández-Roldán, Josep Martí-Solans, Cristian Cañestro, and Ricard Albalat

Abstract The urochordate *Oikopleura dioica* is emerging as a nonclassical animal model in the field of evolutionary developmental biology (a.k.a. evo-devo) especially attractive for investigating the impact of gene loss on the evolution of mechanisms of development. This is because this organism fulfills the requirements of an animal model (i.e., has a simple and accessible morphology, a short generation time and life span, and affordable culture in the laboratory and amenable experimental manipulation), but also because *O. dioica* occupies a key phylogenetic position to understand the diversification and origin of our own phylum, the chordates. During its evolution, *O. dioica* genome has suffered a drastic process of compaction, becoming the smallest known chordate genome, a process that has been accompanied by exacerbating amount of gene losses. Interestingly, however, despite the extensive gene losses, including entire regulatory pathways essential for the embryonic development of other chordates, *O. dioica* retains the typical chordate body plan. This unexpected situation led to the formulation of the so-called inverse paradox of evo-devo, that is, when a genetic diversity is able to maintain a phenotypic unity. This chapter reviews the biological features of *O. dioica* as a model animal, along with the current data on the evolution of its genes and genome. We pay special attention to the numerous examples of gene losses that have taken place during the evolution of this unique animal model, which is helping us to understand to which the limits of evo-devo can be pushed off.

Alfonso Ferrández-Roldán and Josep Martí-Solans contributed equally to this work.

A. Ferrández-Roldán · J. Martí-Solans · C. Cañestro · R. Albalat (✉)
Facultat de Biologia, Departament de Genètica, Microbiologia i Estadística and Institut de Recerca de la Biodiversitat (IRBio), Universitat de Barcelona, Barcelona, Catalonia, Spain
e-mail: ralbalat@ub.edu

W. Tworzydlo, S. M. Bilinski (eds.), *Evo-Devo: Non-model Species in Cell and Developmental Biology*, Results and Problems in Cell Differentiation 68,
https://doi.org/10.1007/978-3-030-23459-1_4

4.1 Introduction

In 1872, the Swiss marine biologist Hermann Fol during his stay in Messina (Sicilia, Italy) described a new planktonic species of Urochordates (a.k.a. Tunicates), the appendicularian (a.k.a. Larvacean) *Oikopleura dioica* (Fol 1872). From the beginning, *O. dioica* exhibited interesting biological features because, using Fol's own words, "Je n'ai jamais trouvé sur un meme individu les organes male et femelle. Notre espece est réellement un tunicier a sexes distincts" (I have never found on the same individual the male and female organs. Our species is really a tunicate with distinct sexes) (Fol 1872). This was a remarkable finding because until then, it was assumed that Urochordates were hermaphrodites. At the beginning of the twentieth century, the Russian embryologist Vladimir Vladimirovich Salensky (commonly known as W. Salensky) described the anatomy of different Appendicularia species (Salensky 1903, 1904, 1905). Although in 1903 Richard Benedict Goldschmidt provided a first brief description of the larva of *O. dioica* (Goldschmidt 1903), the Dutch biologist Hendricus Christoffel Delsman can be considered the father of *O. dioica* embryogenesis. He, analyzing embryos directly collected from the field, was able to laboriously reconstruct its cleavage pattern with amazing exactitude (Delsman 1910, 1912), which a century later has been confirmed by confocal microscopy and live imaging (Fujii et al. 2008; Stach et al. 2008).

During the 1960s and 1970s, the *O. dioica* anatomy was described in deep detail, paying special attention to some conspicuous structures such as the notochord or the endostyle (Olsson 1963, 1965a, b; Welsch and Storch 1969) and to the organs required for the construction and use of the "house," a filtering complex structure that traps the food (Galt 1972). Almost at the same time, the first *O. dioica* cultures in the laboratory were successfully maintained through numerous generations (Paffenhöfer 1973), and over the next 30 years, numerous ecological and biogeographical studies were reported (e.g., Galt 1978; King et al. 1980; Gorsky et al. 1984; Bedo et al. 1993; Acuña et al. 1995; Hopcroft and Roff 1995; Uye and Ichino 1995; Nakamura et al. 1997; Hopcroft et al. 1998) along with multiple morphological, anatomical, and physiological descriptions (e.g., Last 1972; Bone and Mackie 1975; Olsson 1975; Fenaux 1976, 1986; Flood and Afzelius 1978; Fredriksson and Olsson 1981, 1991; Holmberg 1982, 1984; Fredriksson et al. 1985; Bollner et al. 1986; Georges et al. 1988; Holland et al. 1988; Olsson et al. 1990; Nishino and Morisawa 1998; Lopez-Urrutia and Acuña 1999; Acuña and Kiefer 2000).

All this accumulated knowledge led *O. dioica* to the twenty-first century, when shotgun sequencing techniques (Seo et al. 2001; Denoeud et al. 2010) together with the establishment of protocols for gene expression analysis (Bassham and Postlethwait 2000; Nishino et al. 2000) paved the groundwork for *O. dioica* as a promising nonclassical model species for different biological disciplines, including comparative genomics, molecular genetics, or evolutionary developmental biology (evo-devo). Here, we will review the main biological features of *O. dioica* as chordate model at the morphological, physiological, embryonic, ecological, methodological, evolutionary, genetic, and genomic levels, paying special attention to the fact that *O. dioica* has shrunk and compacted its genome retaining the archetypical chordate body plan. We will finish by analyzing the impact of gene loss, especially the loss of genes related to embryonic development, on the evolution of *O. dioica*.

4.2 *Oikopleura dioica* as a Nonclassical Model System for Evo-Devo

A model organism is a simplified and accessible system that represents a more complex entity, based on the notion summarized by Jacques Monod: "what is true for *E. coli* is true for the elephant." This notion reflects the evolutionary principle that all organisms share some degree of functional and genetic similarity due to common ancestry. There are several classical model species in cell and developmental biology, from non-vertebrates such as the roundworm *Caenorhabditis elegans* or the fruit fly *Drosophila melanogaster* to vertebrates such as zebrafish and mouse. The choice of a particular model organism depends, essentially, on the specific scientific question that needs to be investigated, and new scientific questions usually demand new model systems. Thus, when we became interested to understand the impact of the gene loss on the evo-devo of chordates, we had to look for a new "nonclassical" model system suitable for our studies. *O. dioica* was our first choice because it has many technical and biological advantages that made it attractive for our objectives: (1) *O. dioica* has a simple and accessible morphology, maintaining the typical chordate body plan throughout its life; (2) its generation time and life span is short with a fast, invariant, and determinative embryonic development; (3) *O. dioica* is easily accessible and affordably cultured; (4) *O. dioica* can be experimentally manipulated; (5) *O. dioica* occupies a privileged phylogenetic position within the chordates to better understand the evolution of our own phylum; and (6) its genome is small, compacted, and prone to lose genes. In the next sections, we will summarize these technical and biological advantages.

4.2.1 O. dioica *Is a Morphologically Simple Animal with a Typical Chordate Body Plan*

O. dioica has a simple but typical chordate body plan with organs and structures that are unequivocally homologous to those in vertebrates and other chordates, including a notochord anchoring muscle cells along a postanal tail, a hollow neural tube that becomes a central nervous system (CNS) organized in two big ganglia and a nerve cord, as well as a pair of gill slits and an endostyle in the pharyngeal region, which connects with the digestive tract (Fig. 4.1). The body of *O. dioica* is divided in two parts, the trunk (0.12–0.85 mm), which contains most of the organs, and the tail (1.2–3.2 mm), which contains the notochord, the tail nerve cord, and two rows of

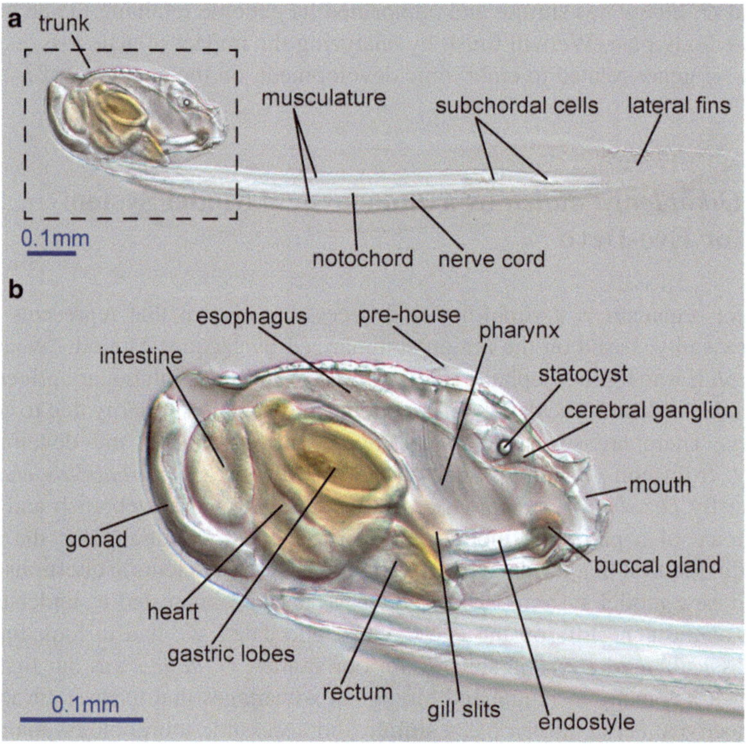

Fig. 4.1 Immature adult specimen of *O. dioica*. (**a**) Differential interference contrast (DIC) micrograph in right lateral view with anterior to the left and ventral down. (**b**) Magnification of the trunk showing the main organs and structures. Bar: 0.1 mm

paired muscle cells. The whole body is transparent so the inner structures can be visualized by simply changing the focal plane of the microscope.

4.2.1.1 Nervous System

The CNS of *O. dioica* is made by only approximately 100 cells. Despite this extremely simple structure, *O. dioica* CNS shows homology with the vertebrate forebrain, hindbrain, and spinal cord of vertebrates (but not the midbrain) (Cañestro et al. 2005). The CNS of *O. dioica* comprises a nerve cord running along the tail and two big ganglia (i.e., the cerebral ganglion in the most anterior part of the trunk and the caudal ganglion in the most proximal region of the tail) (Holmberg 1984). Both ganglia secrete gamma-aminobutyric acid (GABA) (Soviknes et al. 2005). The cerebral ganglion, proposed to be homologous to the vertebrate forebrain (Cañestro et al. 2005), is made of approximately 70 neurons and includes a statocyst, a balance sensory receptor in which a mineralized mass called statolith pushes a group of ciliary cells in response to gravity, providing therefore feedback to the animal on change in orientation (Holmberg 1984) (Fig. 4.1). The cerebral ganglion connects with sensory cells in the mouth, the pharynx, and the ventral organ (a simple structure of about 30 cells homologous to the olfactory placode and the pituitary of vertebrates) (Bollner et al. 1986; Cañestro et al. 2005) and integrates the information received from the sensory cells. The cerebral ganglion also connects to the caudal ganglion through the trunk nerve cord, which turns to the right before entering the tail. Caudal ganglion and trunk nerve cord have been proposed to be homologous to at least part of vertebrate hindbrain (Cañestro et al. 2005). The caudal ganglion and the nerve cord that runs along the tail consist of approximately 30 neurons. Along the nerve cord, neuronal nuclei are collected in groups of two to four forming small ganglia. Small fibers arising from these ganglia innervate the symmetric musculature of the tail with cholinergic motoneurons and coordinate its movements (Galt 1972; Soviknes and Glover 2007).

4.2.1.2 Epidermis and Secreted "House"

The epidermis is a monolayered epithelium without any underlying mesodermal tissue that covers the whole body of *O. dioica* (Nishida 2008). The epidermis of the trunk, called oikoblast or oikoplastic epithelium, contains a fixed number of cells (approximately 2000) organized in different domains defined by the shape of the cells and the morphology of their nuclei and showing a complex bilateral symmetric pattern (Thompson et al. 2001; Kishi et al. 2017; Mikhaleva et al. 2018) (Fig. 4.2a). The oikoplastic epithelium secretes the so-called house, a filter-feeding device made of glycopolysaccharides, mucopolysaccharides, and cellulose (Fig. 4.2b, c) (Kimura et al. 2001; Spada et al. 2001; Thompson et al. 2001). The distribution of the epidermal cells of the trunk has a direct correlation with the different architectonic structures of the house (Fig. 4.2a) (Spada et al. 2001; Thompson et al. 2001; Kishi et al. 2017). The adult specimens of *O. dioica* live inside this mucous house (Fig. 4.2b, c) that works as a food-trapping device by filtering picoplankton particles of different sizes (from 0.1 to 50 μm) from the water stream propelled by stylish and grooving tail movements (Fenaux 1986; Thompson et al. 2001). The house is abandoned when the filters are obstructed, and a new one is immediately inflated, which happens between four and ten times a day increasing with higher water temperatures (Flood and Deibel 1998; Sato et al. 2003). This energetically exhausting strategy has been proposed as an evolutionary adaptation to poor food environments (Acuña et al. 2002).

4.2.1.3 Pharyngeal Region and Digestive Tract

The food particles caught by the house enter through the mouth situated at the anterior region of the trunk to the pharynx, thanks to the water flux generated by the spiracles, a pair of gill slits with beating cilia that generate a water stream into the body to draw the food (Fenaux 1998a) (Fig. 4.3). Within the pharynx, food particles

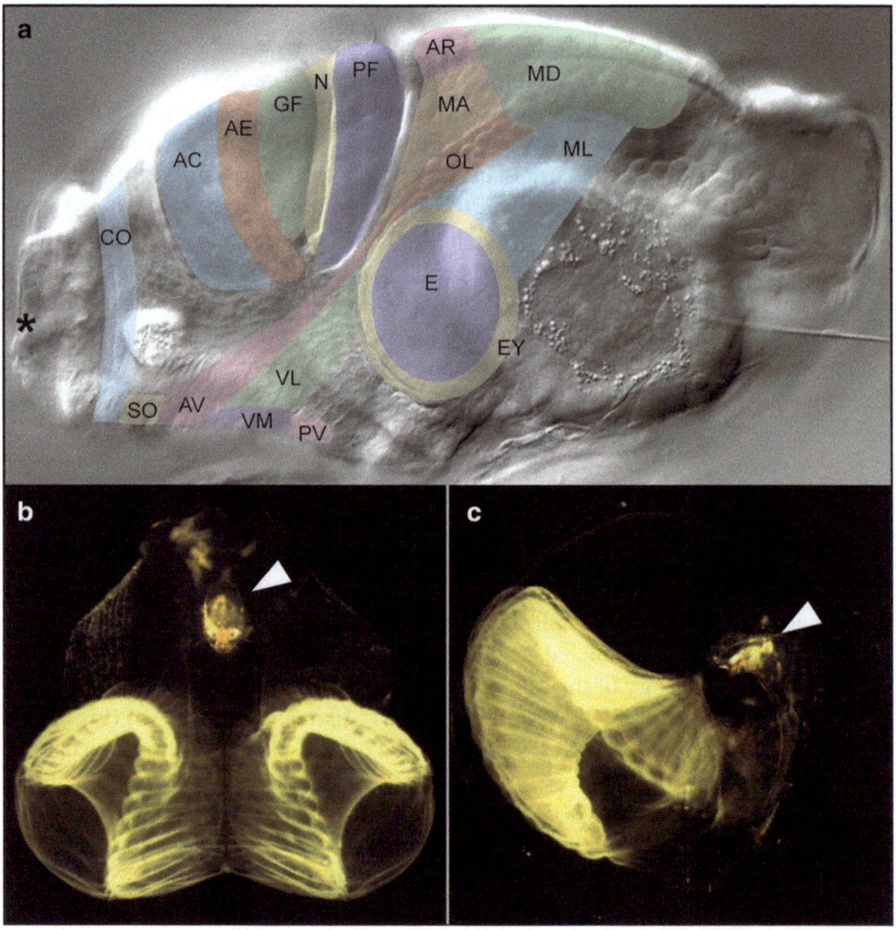

Fig. 4.2 The *O. dioica* epidermis and the secreted house. Lateral view of an *O. dioica* trunk, in which the main domains of the oikoplastic epithelium that correlate with the different architectonic structures of the house are depicted (**a**): *AC* anterior crescent, *AE* anterior elongated region, *AR* anterior rosette, *AV* anterior ventral domain, *CO* circumoral domain, *E* Eisen giant cells, *EY* Eisen's yard, *GF* giant Fol, *MA* martini, *ML* mid-lateral domain, *MD* middorsal domain, *N* nasse, *OL* oblique line, *PF* posterior fol, *PV* posterior ventral domain, *SO* sensory organ, *VL* ventrolateral domain, *VM* ventromedial domain. AC, AE, GF, N, and PF form the fol domain (Kishi et al. 2017). Anterior is to the left and ventral down. The asterisk indicates the mouth. Frontal (**b**) and lateral (**c**) views of the *O. dioica* house, which is visible thanks to the algae trapped in the structure. The trunk of the animal inside the house is also visible (arrowhead)

are trapped by the mucus that covers its walls. The mucus is secreted by the endostyle, an organ homologous to the thyroid of vertebrates, and localized on the ventral side of the pharynx (Olsson 1963; Cañestro et al. 2008). This mucus is conducted to the digestive tract by two symmetric rows of cilia at the lateral walls of the pharynx that finally meet at the superior region (Fenaux 1998a). The digestive

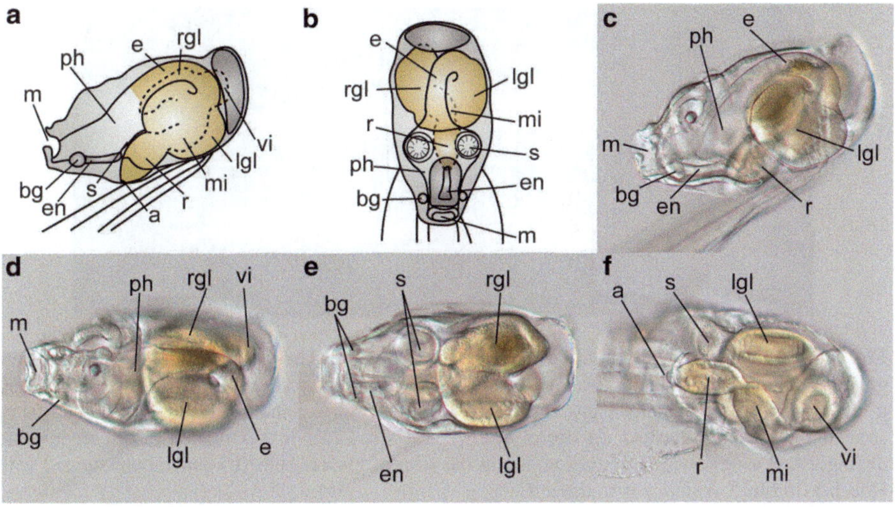

Fig. 4.3 Pharyngeal region and digestive system in the *O. dioica* trunk. Diagrams of the trunk and anatomy of the pharyngeal region and digestive tract of *O. dioica* from left (**a**) and dorsal (**b**) views. DIC micrographs of the trunk of an *O. dioica* adult specimen in left lateral view (**c**), dorsal views at two different focal planes (**d, e**), and ventral view (**f**). Anterior is to the left. Anus (a), buccal gland (bg), endostyle (en), esophagus (e), left gastric lobe (lgl), mid-intestine (mi), mouth (m), pharynx (ph), rectum (r), right gastric lobe (rgl), spiracles (s), and vertical intestine (vi)

tract is composed of an esophagus, two gastric lobes (right and left stomachs), and a bent intestine divided in vertical, mid, and distal intestine (or rectum) (Burighel and Brena 2001; Cima et al. 2002) (Fig. 4.3). Although the absorption of liquids, ions, and small molecules takes place all over the digestive system (Burighel and Brena 2001), different compartments have different functions. Digestive enzymatic activity, for instance, is led mainly by the left gastric lobule where different enzymes such as α-amylase, acid phosphatase, nonspecific esterase, 5′-nucleodidase, and amino-peptidase M are released (Cima et al. 2002). In contrast, accumulation of lipid droplets in ciliated cells of the right gastric lobule and the vertical and mid-intestines indicates a storage role for these compartments (Cima et al. 2002). Finally, fecal pellets are formed in the vertical intestine and transit along the intestine allowing protein absorption by the rectal granular cells (Burighel and Brena 2001; Cima et al. 2002), until reaching the anus where they are released. Functional compartmentalization is further supported by spatiotemporal differences in the onset of gene expression during the development of the digestive system (Martí-Solans et al. 2016), likely reflecting processes of functional differentiation and specialization of digestive cells.

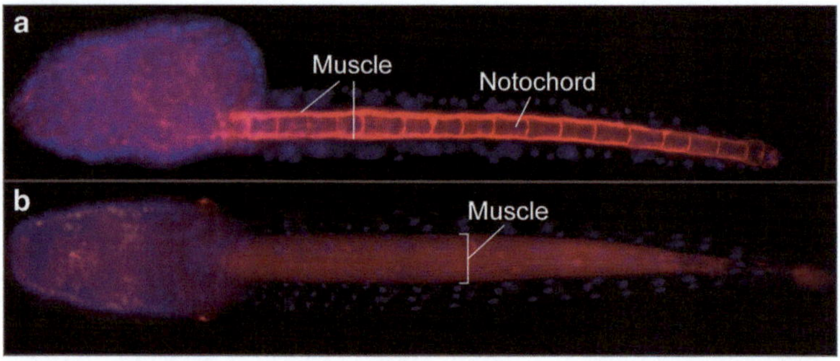

Fig. 4.4 Notochord and tail musculature in an *O. dioica* larva. These two tissues stand out over the general phalloidin staining of the actin filaments (in red), and they spread from the base of the trunk to the posterior tip. In lateral view, notochord is in the center of the tail flanked by two strips of striated muscle cells located on dorsal and ventral sides (**a**). In ventral view, sarcomeres are distinguished as repeated structures all along the strip of muscle cells (**b**). Nuclei are stained with Hoechst (in blue). Anterior is to the left

4.2.1.4 Notochord and Subchordal Cells

The notochord is a synapomorphic structure of the chordate phylum (Satoh et al. 2012) that has a dual function: providing structural support to the developing embryo and secreting patterning signals required for tissue specification and organogenesis (Cleaver and Krieg 2001). In *O. dioica*, the notochord is a row of cells or "stack of coins" that runs along the tail (Figs. 4.1 and 4.4). During larval development, notochord cells form vacuoles that eventually coalesce to form a hollow lumen (Soviknes and Glover 2008). In the adult, the notochord is a tube of cells with thin walls, whose cells secrete proteins rich in sulfur to the extracellular media to keep the stiffness of the notochord, but also flexible enough to facilitate swimming and other movements (Olsson 1965a). The genetic toolkit to build the notochord in *O. dioica* has been investigated, revealing not only similarities but also significant differences in the complement of genes employed by different chordates (Ferrier 2011; Kugler et al. 2011). The evolutionary origin of the notochord is a hot topic of discussion, and the analysis of the *O. dioica* notochord has contributed to corroborate that the urochordate noncontractile actin-expressing notochord may represent the ancestral condition in stem chordates (Almazán et al. 2019; Inoue and Satoh 2018).

At the right side of the notochord, there is a pair of conspicuous cells, the so-called subchordal cells (Fol 1872; Delsman 1910; Lohmann 1933) (Fig. 4.1). These cells migrate during *O. dioica* development from the trunk to the tip of the tail, after the endodermal strand cells enter the trunk following the same path but in opposite directions (Kishi et al. 2014). Subchordal cells together with endodermal strand cells and oral gland cells are three cell populations exhibiting long-distance migration during *O. dioica* development (Kishi et al. 2014), two of them—the subchordal and the endodermal strand cells—likely sharing the same origin (Fredriksson and Olsson 1991). The function of the subchordal cells remains elusive, although it has been proposed that they may be involved in the exchange of materials with the body fluid, from which they pick and transform substances in a hepatic-like process (Fredriksson and Olsson 1991). These substances can either be changed into low molecular weight products merely for detoxification of harmful substances or be metabolized into new, useful molecules that are released back to the hemolymph (Fredriksson and Olsson 1991; Fenaux 1998a).

4.2.1.5 Heart and Tail Musculature

In *O. dioica*, muscle cells are only found in the heart and in the tail. They are in both cases striated muscle cells showing the characteristic repeating functional units called sarcomeres (Onuma et al. 2017; Almazán et al. 2019) (Fig. 4.4b), while the existence of smooth muscle cells in *O. dioica* has not been established so far. *O. dioica* heart is the simplest chordate heart, and assisted by the tail movement, it contributes to the circulation of the hemolymph between the ectoderm and the internal organs in an open circulatory system (Fenaux 1998a). The heart consists of a posterior ventral bag wedged between the left stomach wall and the intestine (Fig. 4.1). The internal lumen of this bag is the only coelomic space in *O. dioica*, and it is completely lined by two types of flat mesodermal tissues, the myocardium composed by muscle fibers and the nonmuscular pericardium (Stach 2009; Onuma et al. 2017). Myocardial cells are connected by cell junctions, and their cytoplasm is full of actin filaments (Stach 2009; Onuma et al. 2017; Almazán et al. 2019). This muscle-structured tissue produces peristaltic contractions that periodically reverse, causing the hemolymph to course between the myocardium and the stomach wall reaching the rest of the organs in the trunk and circulating through the tail (Fenaux 1998a).

The muscular tissue of the tail spreads posteriorly from the base of the trunk in two strips of striated cells distributed in the dorsal and ventral sides of the notochord (Nishino et al. 2000; Nishino and Satoh 2001; Almazán et al. 2019) (Fig. 4.4). This paraxial musculature, combined with the notochord, provides the ability to drive movement to the tail, which is fundamental not only for locomotion but also for *O. dioica* feeding. The tail musculature of *O. dioica* consists of only ten multinucleated striated cells on the dorsal and ventral sides of the notochord (Nishino et al. 2000; Nishino and Satoh 2001; Soviknes et al. 2007; Almazán et al. 2019). In just hatched larvae, eight mononucleated muscular cells are easily recognizable along the anterior–posterior axes of the tail. Along the larval development, two additional small muscle cells appear at the tip of the tail although their origin remains unknown (Nishino and Satoh 2001).

Structurally, the actomyosin contractile complexes of the muscle cells in *O. dioica* have not been fully characterized, but actin and myosin genes can be identified in the *O. dioica* database (Almazán et al. 2019). *O. dioica* has four muscular-actin genes that appear to be expressed only in the muscle cell lineage (Almazán et al. 2019). The four muscular actins show differences in their expression domains during embryonic development, which suggests differences in their spatio-temporal regulation (Almazán et al. 2019). Muscular actin expression can be

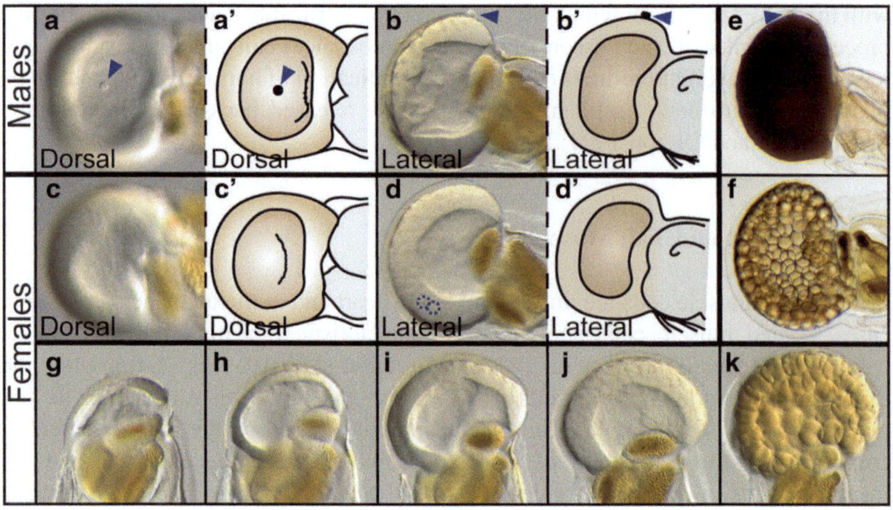

Fig. 4.5 Maturation of male and female gonads. DIC micrographs and schematic representations of dorsal (**a–a'**) and lateral (**b–b'**) views of a premature male gonad. The sperm duct appears as a small protuberance in the most dorsal part of the male gonad (arrowhead). DIC micrographs and schematic representations of dorsal (**c–c'**) and lateral (**d–d'**) views of a premature female ovary. Pro-oocytes can be distinguished inside the gonad (dashed circles). DIC micrographs of mature male (**e**) and female (**f**) gonads. Mature male gonads become brown and opaque as spermatogenesis progresses, while female gonads become yellowish and refringent until the round eggs are clearly visible (**g–k**)

detected as early as at 64-cell stage, although muscle fibers are formed later in the larvae. They are restricted to the inner part of the cell, while the epidermal side is full of mitochondria generating the energy for the tail movement (Nishino et al. 2000; Almazán et al. 2019). Comparative analyses among different chordate species, including *O. dioica*, have suggested that muscular actins originated from a cytoplasmic actin at the base of chordates, which was followed by the recruitment of the myosin motor-machinery that conferred contractile capability to muscle cells (Kusakabe et al. 1997; Almazán et al. 2019; Inoue and Satoh 2018).

4.2.1.6 Reproductive System

O. dioica is the sole dioecious species so far reported in the urochordate subphylum (Fol 1872; Nishida 2008). The gonad is localized in the posterior region of the trunk and is the only apparent character that allows us to differentiate males from females: male gonad is full of sperm and homogenously brown and dark, whereas the female one is yellowish and refringent as it is full of eggs (Fig. 4.5). The growth of the gonad is parallel to the growth of the animal, acquiring their final size and maturity in only some hours, when gametogenesis occurs (Nishino and Morisawa 1998; Ganot et al. 2007). Spermatogenesis takes place in the testicle, a syncytium with a substantial number of identical nuclei that become individual spermatozoa (Martinucci et al. 2005; Onuma et al. 2017). The ovary has also an initial phase of syncytial nuclear proliferation. Then, it becomes a coenocyst where half of the nuclei enter meiosis, whereas the other half became highly polyploid nurse nuclei (Ganot et al. 2007). Inside the coenocyst, a ramified structure composed of filamentous actin surrounds each of the pro-oocytes that are connected to the common cytoplasm via intercellular bridges termed also ring canals (Ganot et al. 2007). During oogenesis a subset of pro-oocytes grows by transferring the cytoplasm through the ring canals while the others degenerate (Ganot et al. 2007). Typically, a mature female spawns between 100 and 400 eggs, although as a clutch manipulator, *O. dioica* females might increase or decrease the oocyte number depending on favorable or unfavorable environmental conditions, respectively (Troedsson et al. 2002). This feature can be useful to manipulate the clutch size in laboratory conditions.

4.2.2 **O. dioica** *Has a Short Generation Time and a Determinative Embryonic Development*

A convenient characteristic shared by many model organisms is that their generation time (i.e., the time between two consecutive generations) is short, facilitating genetic analyses through generations. The generation times of *D. melanogaster* and *C. elegans*, for instance, are of only 9 and 3 days, respectively. The generation time of *O. dioica* is also short, from 5 days at 19 °C to 10 days at 13 °C (Nishida 2008; Bouquet et al. 2009; Martí-Solans et al. 2015). After the external fertilization, the zygote starts embryonic development, which lasts 3.5–6.0 h at 19 °C and 13 °C, respectively, and terminates when a larva hatches breaking the chorion (Fig. 4.6). Next, larval development starts; it takes 6–13 h (at 19 °C–13 °C, respectively) and terminates with a 120° twist of the tail at the tailshift stage. Ten minutes later, the juvenile makes its first house and starts feeding. During the next few days, animals filtrate water and feed arriving to the 1.5–4.0 mm of body size for an adult animal. At that point, the gonads complete their maturation, and males and females spawn the gametes and, as semelparous animals, die afterwards (Fig. 4.6).

In vitro fertilization protocols (Fenaux 1976; Holland et al. 1988; Nishino and Morisawa 1998) have enabled the understanding of processes that lead the egg to become a larva. Cleavage patterns, cell lineages, and morphogenetic movements, which are especially important for the understanding of the embryology of organisms that follow an invariant and determinative embryonic development such as those of *O. dioica*, have been extensively studied (reviewed in Nishida and Stach 2014). Thus, fate maps from one cell to the tail bud stage have been established (Fig. 4.7), similarly to what has been made in other organisms as the nematode *C. elegans* or the ascidian *Ciona robusta* (formerly *C. intestinalis*) (Sulston et al. 1983; Nishida 1987; Fujii et al. 2008; Stach et al. 2008; Nishida and Stach 2014; Stach and Anselmi 2015). After fertilization, the surface of the egg becomes rough,

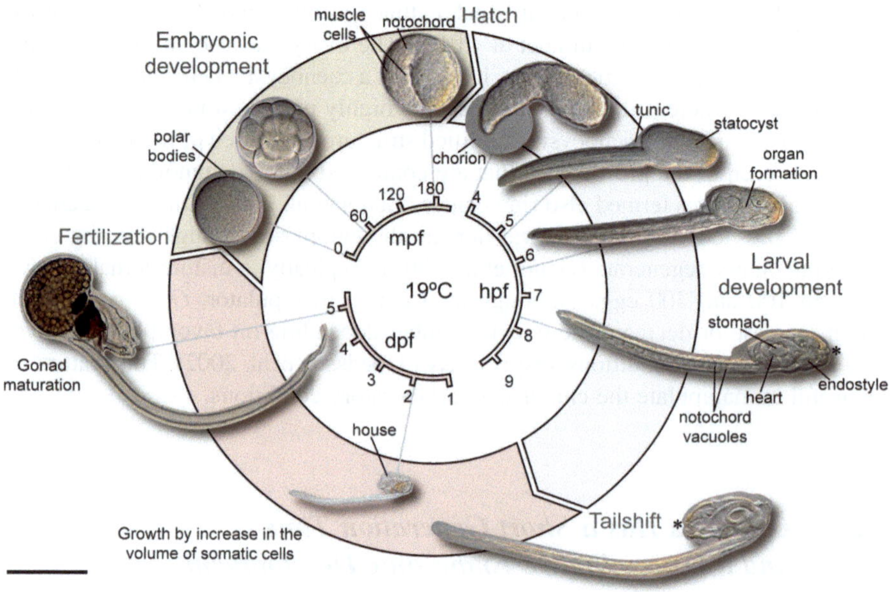

Fig. 4.6 Schematic representation of *O. dioica* life cycle. Embryonic development starts with the fertilization of the oocyte and ends when a larva hatches breaking the chorion (≈3.6 hpf at 19 °C). Larval development lasts for ≈6 h (at 19 °C) and ends when the tail of the larva changes 120° its orientation in a process called tailshift. Notice the change of the position of the mouth (asterisk) relative to the tail in juvenile animals. During the next ≈4.5 days (at 19 °C), juvenile animals feed, grow, and become mature males or females, which spawn the gametes closing the cycle. Scale bar represents 100 μm for all stages except for day 2 and mature adults (1 mm)

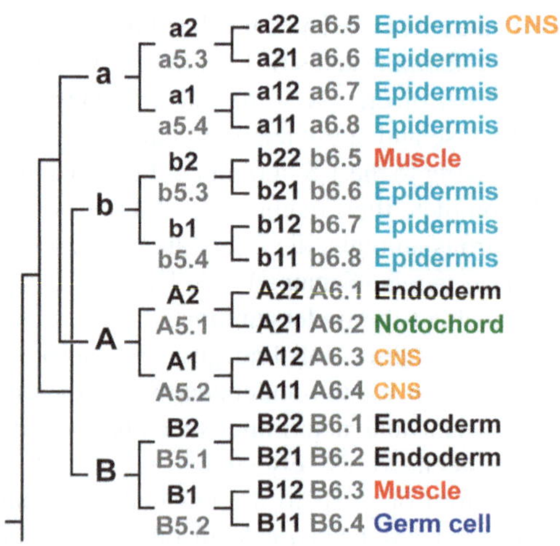

Fig. 4.7 Cell-lineage tree of *O. dioica* embryos. The tree is bilaterally symmetric, and therefore, only one-half is shown. The nomenclature agrees with that of Delsman (1910) and reviewed in Nishida (2008). The ascidian nomenclature system is shown in gray according to Stach et al. (2008) and Nishida and Stach (2014). Developmental fates of cells at 32-cell stage are indicated. CNS, central nervous system

and 9 min later, two polar bodies appear as an obvious signal of successful fertilization (Fig. 4.6) (Fenaux 1998a; Fujii et al. 2008). At 19 °C, the first division takes place after 23 min postfertilization (mpf), generating two morphologically equal cells that correspond to the right and left sides of the animal. At 32 mpf, another symmetric division takes place. This division is meridional and perpendicular to the previous one giving place to the anterior and posterior hemispheres. The third division (at 40 mpf) is asymmetric and leads to the formation of 8 cells, 4 big cells in the animal pole and 4 small cells in the vegetal pole. The morula stage appears 52 mpf, after the fourth division generates an embryo composed of 16 cells. Gastrulation starts 67 mpf, at 32-cell stage, and it consists in the ingression of the vegetal blastomeres that remain covered by the animal blastomeres (Nishida 2008). Neurulation starts at the 64-cell stage 80 mpf, when 8 cells from the anterior region are aligned in two rows of 4 cells that form a matrix that is internalized. Tail bud stage starts 135 mpf, when tail and trunk differentiate. During this stage, the tail elongates, and the embryo bents ventrally. Notochord cells align in a single row and become evident at 180 mpf. At 3.6 h postfertilization (hpf), a larva hatches, and embryogenesis terminates (Fig. 4.6) (Delsman 1910; Cañestro et al. 2005; Fujii et al. 2008; Nishida 2008; Stach et al. 2008).

Larval development is subdivided into six stages (I–VI) (Nishida 2008) that last 6 h at 19 °C, from the hatchling until the juvenile stage (Fig. 4.6). At stage I, the larva elongates and occasionally moves by tail beats. At stage II, the boundaries of the organs begin to appear, and at stage III, the organs are perceptible. At stage IV, organs are clearly recognizable, mouth opens, and the cilia of the digestive system and the ciliary rings of the gill slits start moving. Buccal glands locate at each side of the endostyle, and the heart starts beating. Notochord vacuoles fuse, and the tail flattens to form the lateral fins. At stage V, water current starts inside the larva, the lumen of all organs is continuous, and the trunk epidermis (oikoplastic epithelium) secretes house materials (pre-house). Swimming movements are vigorous. At stage VI, after a few seconds of intense movement, the tail changes its orientation 120° during a process called tailshift (Delsman 1910; Galt and Fenaux 1990; Fenaux 1998b; Nishida 2008). The tailshift event occurs when the larva reaches the juvenile stage, and it is thought comparable to metamorphosis in other urochordates (Fig. 4.6), although in *O. dioica* the definitive organs develop earlier than in ascidian species (Galt and Fenaux 1990).

After the tailshift, juvenile animals inflate the first house and start to feed. During the next days, animals filtrate water and feed at the time that they grow and their gonads mature. From day 1 to day 3, the trunk grows from 0.17 to 0.35 mm (Bouquet et al. 2009), and the gonad expands in the posterior and ventral part of the trunk to fill the whole posterior region (Fig. 4.5). On day 4, the gonad increases its size becoming wider than the trunk and becomes visible to the unaided eye. At day 5, the trunk of *O. dioica* measures almost 1 mm and the whole animal between 3 and 4 mm (Bouquet et al. 2009). At that time, the gonad is fully developed, and males and females are easily distinguishable as the gonad is full of sperm or eggs, respectively (Figs. 4.5 and 4.6).

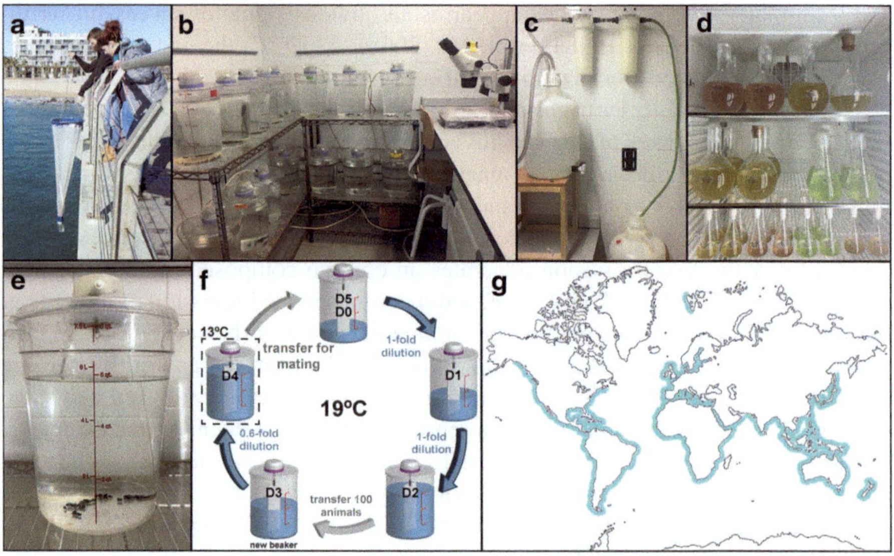

Fig. 4.8 *O. dioica* culture and geographic distribution. Animals might be collected in the coast using a plankton net or directly with a bucket (**a**) and maintained in the laboratory at 19 °C using a standard air conditioning device (**b**). Seawater is filtered through 10–0.5 μm polypropylene filters to remove major particles and organisms and sterilized using 0.2 μm filters (**c**). Three algal species and a cyanobacteria are individually cultured in a 13 °C incubator with a 12 h light/dark photoperiod and used to feed the animals (**d**). *O. dioica* animals are cultured at 19 °C in polycarbonate backers and maintained in suspension by the rotation (5–15 rpm) of a paddle driven by an electric motor mounted on the lid of the backer (**e**). A new animal culture starts by mating 20 females and 10 males close to spawn (D0) in a baker with 1.5 L of sterilized seawater (**f**, top). Next 2 days (D1 and D2), the culture is just diluted adding 1.5 and 3 L of sterilized seawater, respectively (**f**, right). At day 3 (D3), ≈100 animals are transferred to a new backer with 4 L of fresh sterilized seawater (**f**, left-down), and at day 4 (D4), cultures are diluted adding 2 L of sterilized seawater and moved to a 13 °C incubator (**f**, left-top). In the morning of day 5 (D5), most of the animals are mature and ready to set a new mating for new generation (**f**, top). For additional details, see Martí-Solans et al. (2015). Geographic distribution of the *O. dioica* species in the coast of most of the oceans (blue line) (reviewed in Fenaux et al. 1998, and supplemented with Bary 1960; Fenaux 1972; Costello and Stancyk 1983; Larson 1987; Gaughan and Potter 1994; Hwan Lee et al. 2001; Aravena and Palma 2002; Walkusz et al. 2003; Menéndez et al. 2011) (**g**)

4.2.3 **O. dioica** *Is Easily Accessible, with a Wide Geographic Distribution and Suitable for Lab Culturing*

4.2.3.1 Geographic Distribution and Abundance

O. dioica is a semi-cosmopolite free-swimming species present in the Atlantic, Pacific, and Indian Oceans as well as in the Mediterranean and Red Seas, but not in extreme latitude oceans as Arctic or Antarctic (Fenaux et al. 1998) (Fig. 4.8). Although *O. dioica* is a widespread species, its populations may vary in a seasonal way (Essenberg 1922; Raduan et al. 1985; Uye and Ichino 1995; Fenaux et al. 1998;

Tomita et al. 2003). In the west Mediterranean Sea, for instance, its maximum abundance is found during spring and autumn seasons (Raduan et al. 1985). *O. dioica* along with other appendicularian species are so abundant that they occupy an important trophic position in food webs. Appendicularians are the second most abundant species, after copepods, in marine mesozooplankton (Gorsky and Fenaux 1998; Captanio et al. 2008), and they graze about 10% of the ocean's primary production (Acuña et al. 2002). Acting as an important short-circuit that allows a rapid energy transfer from colloidal carbon and phytoplankton primary producers to zooplanktivorous predators such as fish larvae (Flood and Deibel 1998; Fernández et al. 2004), they contribute to at least 8% of the vertical carbon transport to the deep ocean (Davoll and Youngbluth 1990; Robison et al. 2005; Troedsson et al. 2013). Appendicularians are, indeed, a major contributor of "marine snow" (i.e., biological debris that originates from the top layers of the ocean and drifts to the seafloor) in euphotic and mesopelagic zones through the production of discarded houses and fecal pellets (Robison et al. 2005). Those particles, full of nutrients, contribute to 28–39% of total particulate organic carbon export to the deep oceans (Alldredge 2005). The ecological relevance of appendicularians is so high that human activities affecting their populations [e.g., global warming or an increase of toxins in oceans (Bouquet et al. 2018; Torres-Aguila et al. 2018)] might impact on marine food webs and in vertical carbon flux at a world scale. Therefore, the *O. dioica* populations might be valuable sentinels for monitoring marine ecosystems. In summary, *O. dioica* is an ecological relevant organism, abundant in the neritic zone in almost all marine costs, and easily accessible using ordinary plankton nets (Fig. 4.8a, g).

4.2.3.2 *O. dioica* Lab Culturing

In addition to its accessibility directly from nature, *O. dioica* cultures can be maintained in the laboratory (Fig. 4.8b, f). *O. dioica* was first cultured by Paffenhöfer in 1973, but the bases for a long-term maintenance system were developed by Fenaux and Gorsky (1985). Nowadays, to our knowledge, the procedures to culture *O. dioica* in the laboratory all year around have been published from three facilities in the world located in Norway, Japan, and Spain (our facility) (Nishida 2008; Bouquet et al. 2009; Martí-Solans et al. 2015). We culture the animals at 19 °C in 8 L polycarbonate backers containing 6 L of seawater and 10 grams of activated charcoal (see legend of Fig. 4.8 for details). Animals are maintained in suspension by the rotation of a polyvinyl carbonate paddle connected to an electric motor (5–15 rpm). We add 2 or 3 pearls of 1-hexadecanol to reduce surface tension and therefore avoiding that animals get trapped by the surface of the water. For feeding, a cocktail of three different species of algae (*Chaetoceros calcitrans*, *Isochrysis* sp., *Rhinomonas reticulata*) and a cyanobacterium (*Synechococcus* sp.) is added to the animal cultures every day. Because growing *O. dioica* specimens can trap only foods of a proper size, the amount of each alga and of the cyanobacterium in the cocktail is adjusted according to their cell sizes along the *O. dioica* life cycle. Animal cultures are a source of mature male and female specimens from which developmentally synchronic embryos can be obtained using several in vitro fertilization protocols (Clarke et al. 2007; Martí-Solans et al. 2015; Mikhaleva et al. 2015). Embryos and larvae at different developmental stages can be thereby easily collected, fixed, and stored for experiments on demand.

4.2.4 **O. dioica** *Can Be Experimentally Manipulated*

4.2.4.1 Pharmacological Treatments

Pharmacological treatments are a powerful tool to perform functional studies due to its experimental simplicity, and embryos are just transferred to a solution with the drug. The fact that we can easily obtain hundreds of synchronously developing *O. dioica* embryos by in vitro fertilization allows us to perform numerous treatments at different drug concentrations in a fast, simple, and reproducible way. The small size of the *O. dioica* embryos facilitates the diffusion of the drugs to the inner tissues, and their transparency allows the observation of the phenotypic effects of the treatments by DIC microscopy in live developing specimens without the need to perform histological sections. In addition, gene expression responses might be investigated by whole-mount in situ hybridization techniques or by high-density tiling arrays. Pharmacological treatments have been performed, for instance, with developmental morphogens such as all-trans-retinoic acid (RA), which yielded a range of morphological abnormalities, from mild to severe, depending on developmental stage, duration, and concentration of RA exposure (Cañestro and Postlethwait 2007). Treatments with xenobiotic compounds such as the carcinogenic polycyclic aromatic hydrocarbon BaP or with the lipid-lowering agent clofibrate have been also assayed to investigate the transcriptional regulation of the xenobiotic defense mechanisms and to identify the *O. dioica* chemical defensome (Yadetie et al. 2012). Finally, treatments with biotoxins such as *trans,trans*-2,4-decadienal, a model for polyunsaturated aldehydes produced during diatom blooms, have been used to analyze the impact on marine food webs of possible future intensification of algal blooms associated with climate change (Torres-Aguila et al. 2018).

4.2.4.2 Techniques for Altering Gene Function by Morpholino, dsRNA, or dsDNA Injections

Different knockdown approaches for altering gene function have been developed in *O. dioica* by injecting morpholinos (Sagane et al. 2010), double-stranded RNA (dsRNA; RNA interference-RNAi) (Omotezako et al. 2013; Mikhaleva et al. 2015), or double-stranded DNA (dsDNA; DNA interference-DNAi) (Omotezako et al. 2015, 2017), and new techniques for genome editing based on CRISPR-Cas9 are currently being developed (Deng et al. 2018). Most techniques rely on the injection of a given molecule into the gonad of premature females, when it is still

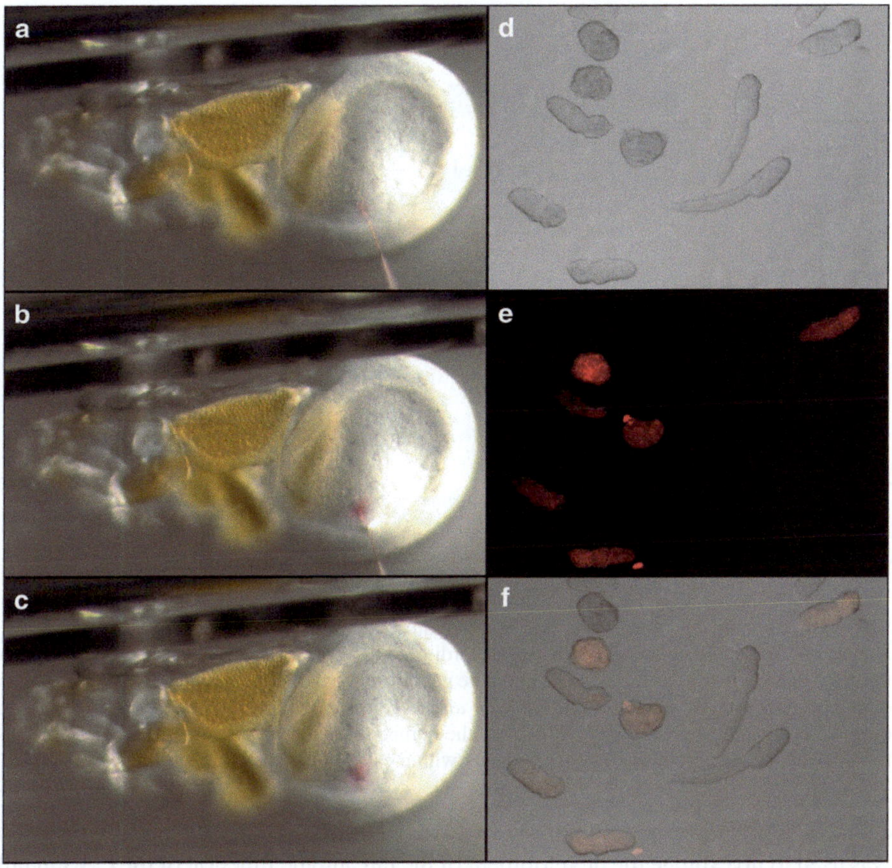

Fig. 4.9 Knocking down by dsDNA injection into the gonad of premature females. A dsDNA against the *brachyury* gene is injected into a premature female gonad (syncytium) (**a–c**), along with phenol red and *pSD64Flife-act-mCherry-mRNA*. Phenol red allows us to visualize the injected solution (red dot in panels **b** and **c**), while mCherry fluorescence highlights the embryos containing mRNA and, therefore, most likely also dsDNA after the cellularization process. In this regard, notice that malformed embryos showing the *brachyury*-dsDNA phenotype are fluorescent (compare panels **d–f**)

a syncytium of meiotic nuclei surrounded by common cytoplasm (Ganot et al. 2007). Then, the injected molecule might spread along the forming oocytes, generating tens of knockdown embryos with a single injection (Fig. 4.9). Alternatively, injections in spawned eggs are also possible (Mikhaleva et al. 2015, 2018; Deng et al. 2018), though they are more difficult and less productive.

Injection of morpholinos, synthetic molecules of approximately 25 nucleotides in length that bind to complementary sequences of RNA blocking translation or altering splicing, was the first knockdown approach in *O. dioica* (Sagane et al. 2010). Morpholinos were used to knock down a cellulose synthase gene necessary to build the house, obtaining different phenotypes related to the production of

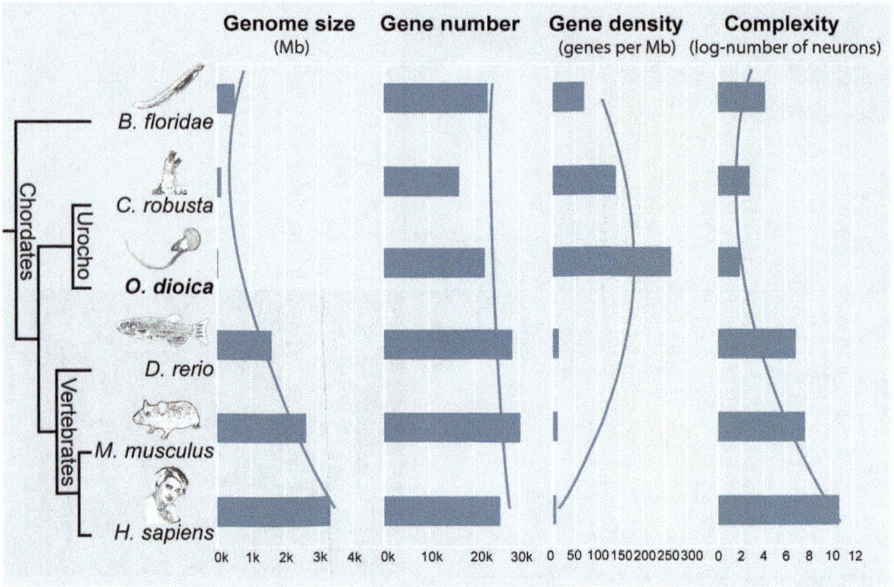

Fig. 4.10 Comparison of genome size, gene number, gene density, and biological complexity among selected chordate species arranged according to their phylogenetic relationships. Bar plots illustrate that *O. dioica* has the smallest chordate genome but has a number of genes rather similar to other chordate species, which implies that *O. dioica* gene density is the highest one. As a rough estimate of biological complexity, we plotted the number of neurons of each organism in logarithmic scale. Lines represent the overall tendency within each plot. Urocho, urochordates

cellulose fibrils (Sagane et al. 2010). In 2013, RNA interference by dsRNA injection was used to investigate the development of the tail and the function of the CNS in tail movement (Omotezako et al. 2013; Mikhaleva et al. 2015). Finally, in 2015, DNA interference by dsDNA injection, a gene silencing technique reported in plants, ciliates, and archaea, was successfully assayed in *O. dioica*, being the first case reported in any metazoan organism (Omotezako et al. 2015). Although the molecular bases for DNAi are unknown, *O. dioica* experiments suggest that dsDNA can induce sequence-specific transcription blocking and mRNA degradation.

4.2.5 **O. dioica** *Occupies a Privileged Phylogenetic Position Within the Chordate Phylum*

The urochordates (or tunicates), group to which the appendicularian *O. dioica* belongs, are the sister subphylum of the vertebrates, which together with basally branching cephalochordates form the chordate phylum (but see Satoh et al. (2014) for the chordate superphylum hypothesis) (Fig. 4.10). Urochordates are an extremely diverse taxonomic group that appears to have undergone a rapid evolution and speciation, spreading over many marine habitats, from shallow waters to deep sea (Holland 2016). Urochordates are classically divided into three classes, Appendicularia (larvaceans, ≈70 species including *O. dioica*), Ascidiacea (sea squirts, ≈3000 species), and Thaliacea (salps, doliolids, and pyrosomes, ≈100 species), although it has been recently proposed that Thaliacea species are nested within the Ascidiacea class (Delsuc et al. 2018; Kocot et al. 2018). In any case, Appendicularians—also known as Larvaceans because they maintain the larval morphology, including the chordate synapomorphies, throughout the entirely life cycle—appear to be the sister clade of all other urochordates, representing a branch of the subphylum that split from all other urochordates 450 million years ago (Delsuc et al. 2018). Appendicularians are divided into three families. The Fritillariidae family that includes 3 genera and 30 species, the Kowalevskiidae family, the smallest one with just 1 genus and 2 species, and the Oikopleuridae family, the most diverse group with 11 genera and 37 species (Bone 1998).

O. dioica occupies a privileged phylogenetic position within urochordates for evo-devo studies because it can be morphologically, functionally, genetically, and genomically compared with many other urochordate species (more than 10 ascidians have been totally or partially sequenced; http://www.aniseed.cnrs.fr/; http://octopus.obs-vlfr.fr/public/botryllus/blastbotryllus.php), including well-studied ascidian species of the *Ciona* or *Halocynthia* genera (Corbo et al. 2001; Dehal et al. 2002; Nishida 2002; Cañestro et al. 2003; Satoh 2003; Satoh et al. 2003; Brozovic et al. 2016, 2017). In addition, from a more general perspective, comparisons of *O. dioica* and ascidians with diverse vertebrate species (there are more than 100 vertebrate genome projects) are relevant for detecting subphylum-specific traits, which might be classified as evolutionary innovations or losses depending on their presence in the external cephalochordate subphylum (there are three *Branchiostoma* genome projects and one *Asymmetron* RNA-Seq project).

4.2.6 O. dioica *Has a Small and Compacted Genome*

With only 70 Mb, *O. dioica* genome is the smallest chordate genome and one of the smallest genomes in metazoans (Seo et al. 2001; Denoeud et al. 2010; Danks et al. 2013), only surpassed by the small genomes of some parasite animals (Burke et al. 2015; Chang et al. 2015; Mikhailov et al. 2016) (Table 4.1) (Fig. 4.10). The small size of the *O. dioica* genome is the result of a process of compaction mostly due to three factors: (1) a reduction of the length of intergenic regions, partly because of numerous operons; (2) a reduction of the abundance of transposable elements; and (3) the reduction of the size of introns (Denoeud et al. 2010; Berna and Alvarez-Valin 2014; reviewed in Chavali et al. 2011). As a result of the genome compaction, the gene density in *O. dioica* genome (1 gene per 3.9 kb) has become very high, 3 times higher than in ascidian *C. robusta*, 6 times higher than in cephalochordate amphioxus, and 35 times higher than in human genome (Table 4.1) (Fig. 4.10).

Table 4.1 Main features of *O. dioica* genes and genome (Volff et al. 2004; Denoeud et al. 2010; Chalopin et al. 2015)

Genome	
Size	70 Mb
Size of intergenic regions	53% of intergenic distances <1 kb
Population mutation rate	$\theta = 4N_e\mu = 0.0220$
Conserved synteny	Negligible
Diversity of transposable elements	4 class I retroelements + 3 class II DNA transposons
Abundance of transposable elements	10–20% genome (at most few hundreds of each TE)
Genes	
Number	18,020 predicted genes
Density	1 gene per 3.9 kb (257 genes per Mb)
Genes in operons	27% (1800 operons)
Number of introns per Gene	4.1
Intron size	<50 nt in 62% of the genes; >1 kb in 2.4% of the genes
Intron position	76% intron positions specie-specific
Noncanonical splicing sites	In 12% introns (9% GA-AG, 2.7% GC-AG, 0.7% GG-AG)

4.2.6.1 Intergenic Regions and Operons

O. dioica genes are densely packed, with 53% of their intergenic sequences smaller than 1 kb (Table 4.1). Contributing to the intragenic reduction, many *O. dioica* genes are organized in polycistronic transcription units (operons). The majority of operons are bicistronic (60%), but a substantial proportion contains three or more genes, and some have up to 11 genes. Operon organization reduces the intergenic space because it reduces the DNA segments containing transcription initiation and regulation signals. The fact that 27% of *O. dioica* genes are predicted to be organized in around 1800 operons (Denoeud et al. 2010), together with the observation that most genes have relatively small intergenic spaces, may favor the genome compaction of this species (Table 4.1).

4.2.6.2 Transposable Elements

Transposable elements (TEs or transposons) are main components of eukaryote genomes that can replicate and change to new locations. *O. dioica* genome is relatively poor in TEs, in terms of quantity and diversity, and most of the so-called pan-animal transposon families, many of them even present in *C. robusta* (Cañestro and Albalat 2012), are absent (Denoeud et al. 2010). *O. dioica* genome contains TEs of only six superfamilies—*Ty3/Gypsy*, *DIRS1*, and *Penelope*-like retrotransposons (Volff et al. 2004) and *Tc-Mariner*, *PyggyBac*, and *Maverik* DNA transposons (Chalopin et al. 2015)—plus a new family of non-LTR retrotransposon named *Odin* (Volff et al. 2004) (Table 4.1). Although the copy number of each TE is low (at most few hundreds), LTR retrotransposons appear to be the most abundant (>60%), while non-LTR retrotransposons and DNA transposons represent less than 20% each. The extent of TEs on the *O. dioica* genome size has been estimated about 10–20% (Chalopin et al. 2015). In summary, a massive purge of pan-animal TEs has occurred in *O. dioica* genome probably associated with an intense process of genome compaction. Some TEs show a low level of sequence corruption, suggesting a rather recent activity, but their low copy number and uneven genome distribution indicate that TE expansion in *O. dioica* is under tight genetic control (Denoeud et al. 2010).

4.2.6.3 Introns

Another factor contributing to *O. dioica* genome compaction is the small size of the introns of the majority of the genes (Table 4.1). Despite *O. dioica* having a typical number of introns per gene (4.1) as in other vertebrates, it has reduced the size of the introns to less than 50 nucleotides in 62% of the genes, and only 2.4% of the introns being larger than 1 kb (Seo et al. 2001; Denoeud et al. 2010). *O. dioica* shows a high intron turnover: 76% of the introns are in positions unique to *O. dioica* (newly acquired introns) and only 17% are in ancestral positions (7% remain unclassified) (Denoeud et al. 2010) (Table 4.1). Several mechanisms have been proposed for the gain and loss of introns in *O. dioica* (Denoeud et al. 2010), but since newly acquired introns tend to be shorter than old introns, intron turnover might have favored genome compaction. Surprisingly, canonical splicing signal (GT-AG intron boundaries) is not observed in about 12% of introns (1% in other species), which mostly (9%) have noncanonical GA-AG (GC-AG, 2.7%, or GG-AG, 0.7%) sequences (Edvardsen et al. 2004; Denoeud et al. 2010). Because *O. dioica* lacks the minor spliceosome and only has one type of each spliceosomal component (Denoeud et al. 2010), a single major spliceosome, permissive in terms of splicing signals, would take care of all splicing processes. It has been proposed that such permissive splicing could favor the intron turnover (Denoeud et al. 2010) and, hence, the compaction of the *O. dioica* genome.

4.3 Evolution of *O. dioica* Genes

The *O. dioica* genome project did not only reveal that this species has an extremely compacted genome with small intergenic regions, tiny introns, and few TEs, but that its genes have suffered many evolutionary changes affecting the conservation of the sequence, organization, and number of genes.

4.3.1 Gene Sequence Evolution: High Evolutionary Rates

The sequencing of the genome confirmed that *O. dioica* is a very fast evolver and its genes always show surprisingly long branches in phylogenetic tree reconstructions (Edvardsen et al. 2005; Denoeud et al. 2010). This high rate of sequence evolution appears to be the result of two independent evolutionary forces acting at different levels and with different strength.

First, affecting at a global genome scale with a moderate strength, we found that most genes show a high evolutionary rate (Table 4.1) as a result of a combination of a high population mutation rate and a reduced negative (purifying) selection. An estimate of *O. dioica* population mutation rate ($\theta = 4N_e$, $\mu = 0.0220$) shows it is high, which is consistent with a large effective population size (N_e) and/or a high mutation rate per generation (μ) (Denoeud et al. 2010). In addition, the short generation time of this species implies that the effective mutation rate per year is substantially increased. A reduced negative selection for *O. dioica* has been proposed based on the homogeneity of its amino acid rates that is a symptom of relaxation of selective—structural and/or functional—constraints (Berna et al. 2012; Berna and Alvarez-Valin 2014). Relaxation of selective constraints associated with rapid rates of sequence evolution has been actually proposed for other fast-evolving animal species such as *Ciona robusta* and *Caenorhabditis elegans* (Holland and Gibson-Brown 2003).

Second, some particular genes have even a higher evolutionary rate due to positive selection that contributes to the adaptation to new environments or to new challenges (Berna and Alvarez-Valin 2014). Many of these genes with this higher evolutionary rate have been described to be involved in regulatory and developmental functions (Berna et al. 2012; Berna and Alvarez-Valin 2014).

4.3.2 Evolution of Gene and Genome Organization

4.3.2.1 Negligible Synteny Conservation

Conserved synteny describes the colocalization of homologous genes on homologous chromosomic regions of different species. *O. dioica* genome appears to have suffered numerous chromosomal rearrangements during evolution, and thereby, it lacks any chromosomal synteny conservation with other animal genomes (Denoeud et al. 2010). Conservation of local synteny is also almost negligible since local gene order is indistinguishable from random for distances smaller than 30 genes, and a low level of conserved synteny is only detectable at a wider distance span (Denoeud et al. 2010; Irimia et al. 2012). These observations suggest that constraints that maintain gene order in metazoans may actually be relaxed in *O. dioica*.

4.3.2.2 Disintegration of *Hox* Cluster

Another significant feature of *O. dioica* genome is the disintegration of the cluster of *Hox* genes, which is generally well conserved in all bilaterians. *Hox* genes are a subset of homeobox genes involved in establishing morphological identities along the anterior–posterior axis, and although the biological significance of *Hox* clustering remains unclear, functional and structural explanations leading to a spatiotemporal coordinated transcriptional regulation of *Hox* genes have been proposed (Kmita and Duboule 2003). In many species, the position of the *Hox* genes in the cluster correlates with their temporal and spatial sequential expression along the anterior–posterior axis (reviewed in Duboule 2007). *O. dioica* does not have a *Hox* cluster since its complement of nine *Hox* genes—three anterior *Hox* genes (*Hox1*, *Hox2*, and *Hox4*) and six posterior genes (*Hox9A*, *Hox9B*, *Hox10*, *Hox11*, *Hox12*, and *Hox13*)—is totally dispersed in its genome (Seo et al. 2004). The disorganization of the *Hox* cluster in *O. dioica* has been related to the lineage-driven mode of development of *O. dioica*, in which *Hox* genes would contribute to tissue specification with separated domains of *Hox* expressions, rather than to axial patterning with overlapping *Hox* expressions (Seo et al. 2004).

4.3.2.3 Evolution of Operon Organization

A third important genomic feature in the organization of *O. dioica* genes is their grouping in operons (Table 4.1). The operons do not only contribute to the compaction of the genome (see above) but might also serve to group genes that have to be either efficiently coregulated—repressing or activating them as a group—or ubiquitously expressed with a low degree of transcriptional regulation (Blumenthal 2004). *O. dioica* has around 1800 operons transcribed as polycistronic pre-mRNAs, most of them processed to mature monocistronic mRNAs via spliced-leader RNA (SL RNA) *trans* splicing (Ganot et al. 2004; Denoeud et al. 2010). Functional annotation of the gene set in operons shows that they are significantly enriched for genes involved in housekeeping functions or general metabolic processes such as RNA, protein, DNA, lipid, and carbohydrate processing and transport. Genes involved in developmental processes such as morphogenesis and organogenesis are, in contrast, significantly underrepresented in the operon gene set (Denoeud et al. 2010).

4.3.3 Gene Number Evolution

Despite its reduced size, *O. dioica* genome contains 18,020 predicted genes, a similar number to other urochordates (e.g., *C. robusta* ≈15,300 genes) and only slightly below other chordates such as the cephalochordate *Branchiostoma floridae*

(≈22,000 genes) or the vertebrate *Fugu rubripes* (≈18,300 genes) (Table 4.1) (Fig. 4.10). The current number of genes is the result of the balance between gene gains and losses impacting a certain set of ancestral genes, and because *O. dioica* appears to be prone to lose genes (see next section), it has to be also prone to gain them.

Gene duplications have been proposed to be a major driving force for gene gains (Ohno 1970; Cañestro et al. 2013), and *O. dioica* seems to have retained many lineage-specific duplicates. For instance, among homeobox genes, *O. dioica* shows a high incidence of retention of lineage-specific duplicates of genes such as *Irx*, *Not*, and *Pax3/7* (Edvardsen et al. 2005). Additional examples of such duplicates are found in g-type lysozyme genes (Nilsen et al. 2003), caspase genes (Weill et al. 2005), metallothionein genes (Calatayud et al. 2018), notochord *Noto15* and *Noto9* genes (Kugler et al. 2011), *RdhE2* and *Cco* genes (Martí-Solans et al. 2016), and actin as well as other muscle structural genes (Almazán et al. 2019; Inoue and Satoh 2018), among many others predicted from the genome project analysis (Denoeud et al. 2010). Lineage-specific duplicates might contribute to both general evolutionary adaptations of the organisms or to innovations associated to the unique biology of the lineage. For example, homeobox genes expressed—and possibly patterning—in the oikoplastic epithelium, which represents a significant novelty of appendicularians for secreting the mucous house, mostly belong to the duplicated groups (Denoeud et al. 2010; Mikhaleva et al. 2018), while gene families with cell adhesion roles are overrepresented in *O. dioica*, most likely because of the extensive and assorted interactions required for building the house (Chavali et al. 2011).

4.4 Evolution by Gene Loss: *O. dioica* as a Model System for Evo-Devo Studies

Urochordate genomes (e.g., *C. robusta*) appear to have a "liberal" evolutionary pattern of gene loss (Dehal et al. 2002; Holland and Gibson-Brown 2003; Hughes and Friedman 2005; Somorjai et al. 2018), which contrasts with the "conservative" pattern of cephalochordates and vertebrates (Somorjai et al. 2018). *O. dioica* appears to have pushed this urochordate trend to its limits (Ferrier 2011) by having lost many genes or entire genetic pathways (Table 4.2). In this section, we review these losses grouped into three categories: Sect. 4.4.1, losses of genes involved in general characteristics of gene/genomic structure and expression, Sect. 4.4.2, losses of genes related to particular cellular and physiological functions, and Sect. 4.4.3, losses of genes essential for embryonic development in chordates.

Table 4.2 Absent genes[a] in *O. dioica* genome

Category	Genes	References
cNHEJ repair system	*Xrcc5 (Ku80), Xrcc6 (Ku70), Xrcc4, Lig4, NHEJ1 (Xlf), DNA-PKc, Dclre1c*	Denoeud et al. (2010)
Epigenetic machinery	*Dnmt1, Dnmt3, Mbd4/MeCP2, Gcn5/PcaF, Hat1, Mll1, Ash1, Rnf1, Suz12, Pcgf, Scmh1, Kdm2b*	Albalat et al. (2012), Navratilova et al. (2017)
Spliceosome machinery	snRNA U11, snRNA U12, snRNA U4atac, snRNA U6atac	Denoeud et al. (2010)
Caspase family	*Csp1/4/5, Csp6, Csp2/9, Csp8/10*	Weill et al. (2005)
Immune system	*NLRs, RLHs, MyD88-like, Sarm1-like, Tirap-like, Ticamp2-like*	Denoeud et al. (2010)
Defensome system	*AhR, AhRR, Nr1C (Ppar)*, CYP1 genes	Yadetie et al. (2012)
Peroxins	*Pex1, Pex2, Pex3, Pex5, Pex6, Pex7, Pex10, Pex11, Pex12, Pex13, Pex14, Pex16, Pex19, Pex26*	Zarsky and Tachezy (2015)
Retinoic acid signaling	*Rdh10, Rdh16, Bco1, Aldh1a, Cyp26, RAR, PPAR*	Cañestro et al. (2006), Cañestro and Postlethwait (2007), Martí-Solans et al. (2016)
Homeobox genes	*Hox3, Hox5, Hox6, Hox7, Hox8, Gbx, Nk3, Nk6, TGIF, POU VI, Lhx6/7, Vax, Cux, SATB, ZFH1, Sax, Xlox, Mox, Hlx, Bsh, Chox10, Otp, Prx, Goosecoid, Prox, Tlx*	Seo et al. (2004), Edvardsen et al. (2005)
Sox genes	*SoxC, SoxE, SoxF, SoxH*	Heenan et al. (2016)
miRNA	miR-9, miR-29, miR-33, miR-34, miR-96, miR-126, miR-133, miR-135, miR-153, miR-182, miR-183, miR-184, miR-196, miR-200, miR-216, miR-217, miR-218, miR-367	Wang et al. (2017)
Notochord genes	*Entactin, Fibrinogen-like, Multidom, Myomegalin, Noto1*, Noto2, *Noto3, Noto4, Noto5, Noto6, Noto7, Noto8, Noto11, Noto12, Noto13, Noto14, Noto16, Perlecan, Ptp, Slc, Swipi, Tropomyosin-like, Tune, Ube2*	Kugler et al. (2011)
General repair system	*Polb, Apex2, Lig3, Msh3, Atm, Chek2, Aptx, Nbn, Rad52*	Denoeud et al. (2010)
Apoptotic genes	*Bax, Bak,* $BCL\text{-}X_L$	Robinson et al. (2012)
Cyclins and CDK	*Cyclin J, Cyclin G, Cyclin F, Cdk14/15*	Campsteijn et al. (2012)
Rab GTPases	*Rab4, Rab7L1, Rab9, Rab15, Rab19/43, Rab20, Rab21, Rab22, Rab24, Rab26/37, Rab28, Rab30, Rab32LO, Rab40, Ift27, Rasef, EFcab44/Rab44, RabX1, RabX4, RabX6*	Coppola et al. (2019)

[a]Genes do not found in *O. dioica*, and, thereby, likely lost during the evolution of the *O. dioica* lineage

4.4.1 Loss of Genes Involved in Gene/Genomic Structure and Expression

4.4.1.1 Loss of Non-homologous End Joining Repair Genes

All organisms have the ability to repair the double-strand breaks (DSBs) that normally happen in DNA due to numerous external and internal factors. One fundamental mechanism present in all eukaryotes to repair these breaks is the canonical non-homologous end joining (cNHEJ) repair system. Unexpectedly, *O. dioica* (Denoeud et al. 2010) and other six species of appendicularians (Deng et al. 2018; Ferrier and Sogabe 2018) do not have the cNHEJ machinery (Table 4.2), meaning this repair system became dispensable during the evolution of this chordate lineage. In fact, *O. dioica* seems to use the alternative NHEJ (aNHEJ) system (a.k.a. alternative end joining, aEJ, Pannunzio et al. 2018) as a compensatory system that overcomes DSBs, although the existence of another so far undescribed pathway cannot be excluded (Deng et al. 2018). Because the aNHEJ system seems to be less faithful than cNHEJ (it often involves deletion of some intervening nucleotides and commonly leads to chromosome rearrangements, for example, translocations, (Deriano and Roth 2013)), the loss of the cNHEJ system might have had important consequences in the genome architecture. It may have contributed, for instance, to the compaction of the *O. dioica* genome by either favoring the deletions caused by the aNHEJ repair system or by restricting replication of autonomous retrotransposons that uses the cNHEJ system for their propagation (Deng et al. 2018). In addition, this loss may have also contributed to the reorganization of the genome through accumulated rearrangements, leading to negligible synteny conservation (Deng et al. 2018) that may be associated with modifications of the mechanisms of gene regulation, changing from long-range mechanisms acting on topologically associated genomic blocks that include several genes to short-range and gene-specific systems for gene regulation (Cañestro et al. 2007; Ferrier and Sogabe 2018).

4.4.1.2 Loss of Genes of the Epigenetic Machinery

Histone modifications and DNA methylation are epigenetic marks mainly associated with regulation of gene expression, replication, DNA repair, recombination, chromosome segregation, and other meiotic and mitotic processes. Histone proteins, responsible of packing DNA into nucleosomes, provide multiple sites for posttranslational modifications by evolutionarily conserved histone modifiers that establish the so-called histone code. Among the proteins that modify histones, *O. dioica* has lost several genes for histone acetyltransferases such as *Gcn5/Pcaf* and *Hat1* (Navratilova et al. 2017) (Table 4.2). In addition, homologs of core components of the canonical Polycomb complexes (*Rnf1*, *Suz12*, *Pcgf*, and *Scmh1*) and genes for several proteins of the trithorax group (*Mll1*, *Ash1*, and *Kdm2b*) have neither been found in *O. dioica*'s genome (Navratilova et al. 2017). Regarding DNA methylation, *O. dioica* has lost the two main DNA methyltransferases *Dnmt1* and *Dnmt3*, and one of the two *Mbd* genes, *Mbd4/MeCP2* (Cañestro et al. 2007; Albalat et al. 2012) (Table 4.2). Overall, it is through that the loss of components of the epigenetic machinery might have favored changes in the genome architecture and modifications of the mechanisms of gene regulation, leading to compacted regulatory spaces, reduced chromatin state domain widths, evolution of operons, and loss of synteny and dispersion of the Hox cluster (Cañestro et al. 2007; Navratilova et al. 2017).

4.4.1.3 Loss of Minor Spliceosome Genes

While the major spliceosome containing the snRNAs U1, U2, U4, U5, and U6 removes introns with canonical GT-AG boundaries, the minor spliceosome, which contains the snRNAs U11, U12, U4atac, U5, and U6atac, mainly acts on noncanonical AT-AC boundaries (Patel and Steitz 2003; Sheth et al. 2006). Both major and minor spliceosomes are present across most eukaryotic lineages tracing them back to the origin of eukaryote evolution. Although *O. dioica* has many noncanonical introns (see above), no AT-AC introns have been detected in cDNA resources (Denoeud et al. 2010), which may explain that none of the minor splicesomal snRNAs are found in *O. dioica* (Table 4.2). It seems therefore that in the absence of AT-AC introns, the components of the minor spliceosome became dispensable and lost during *O. dioica* evolution and at the same time that the major spliceosome became permissive and able to remove all—canonical and noncanonical—*O. dioica* introns, although with different efficiency (Denoeud et al. 2010). Whether these differences in splicing efficiency could reflect specific usage of noncanonical introns for gene expression regulation and what evolutionary impact these changes had needs further investigation.

4.4.2 Loss of Genes for Cellular and Physiological Functions

4.4.2.1 Loss of Genes of the Caspase Family

Caspases are a family of protease enzymes that play an important role not only in apoptosis but also in maturation of different immunity system proteins and in the control of the proliferation and differentiation of specific cell types (Weill et al. 2005). Caspases are present in all kingdoms (Uren et al. 2000), but their number has a taxon-dependent diversity ranging from 3 members in *C. elegans* to 10–13 in vertebrates and 17 in *C. robusta* (Weill et al. 2005). *O. dioica* has only three caspase genes deriving from a single founder distantly related to *Caspase 3/7* (Weill et al. 2005) (Table 4.2). *O. dioica* seems, therefore, to have lost all of the components of the caspase family except one. The reduced complexity of the caspase family might be associated with a low cell number at the adult stage, the absence of a major metamorphosis event, and a minimized immune system in this species (Weill et al. 2005).

4.4.2.2 Loss of Genes of the Immune System

As a marine organism, *O. dioica* should have an innate immune system prepared to fight against viruses and bacteria present in high amounts in seawater, but its short life span along with the absence of hemolymph cells—hemocytes or macrophages—points to a rapid immune system based on transcriptional upregulation response or on constitutively expressed effectors, rather than to a more slow system reliant on cell proliferation triggered by immune receptors (Denoeud et al. 2010). In agreement, *O. dioica* has lost almost all genes with domains corresponding to typical immune receptors or immune effectors and lacks homologous genes to interleukins or cytokines involved in the immunity response of more complex chordates (Denoeud et al. 2010) (Table 4.2). Interestingly, it has been proposed that a simplified immune system may be compensated by the antibacterial function of some oikosins with phospholipase A2 domains, which hydrolyze glycerophospholipids present in bacterial cell walls (Hosp et al. 2012). In summary, *O. dioica* seems to have a highly derived and simplified strategy of defense that together with its high fertility, short life cycle, and high polymorphism contributes to the survival of the *O. dioica* populations (Denoeud et al. 2010).

4.4.2.3 Loss of *Cyp* Genes for Xenobiotic Defensome Systems

The first step for elimination or inactivation of xenobiotic compounds often involves the oxidative modification of the toxic chemicals, mostly performed by enzymes of the cytochrome P450 (CYP) family, some of which are also required for embryonic development (Goldstone et al. 2006). Due to its variable functions, Cyp enzymes are quite abundant in animal genomes—up to 120 *Cyp* genes in sea urchin (Goldstone et al. 2006), 94 in zebrafish (Goldstone et al. 2010), or 236 in amphioxus (Nelson et al. 2013). *O. dioica*, however, has lost many *Cyp* genes, and with only 23 *Cyp* genes, this species has the smallest *Cyp* repertoire among sequenced metazoan genomes (Yadetie et al. 2012). *O. dioica* lacks, for instance, the CYP1 family genes, which play a central role in the metabolism of environmental toxicants (Table 4.2). *O. dioica* also lacks the aryl hydrocarbon receptor (*AhR*) gene (and its repressor *AhRR*) (Yadetie et al. 2012), which is the transcriptional regulator of CYP1 family genes and a major xenobiotic sensing receptor activated by pollutants. Overall, these data suggest the absence of an AhR-mediated xenobiotic biotransformation signaling pathway in *O. dioica* and suggest the evolution of alternative mechanisms of response to environmental xenobiotic compounds (Yadetie et al. 2012).

4.4.2.4 Loss of Peroxin Genes and Absence of Peroxisomes

Peroxisomes are single membrane-bound organelles with important functions in detoxification of reactive oxygen species, long-chain fatty acid beta-oxidation, plasmalogen synthesis, amino acid degradation, and purine metabolism. Peroxisome biosynthesis and protein import is mediated by a group of 13 highly conserved eukaryotic proteins called peroxins (Gabaldon et al. 2006; Schluter et al. 2006; Zarsky and Tachezy 2015). Peroxisomes are ubiquitous in eukaryotes, and only some groups of anaerobic protist and parasitic helminths lack peroxisomes (Schluter et al. 2006; Gabaldon and Capella-Gutierrez 2010). It is therefore extraordinary that *O. dioica* has lost all peroxin genes (Table 4.2), becoming the only known aerobic non-parasitic organism that does not have peroxisomes (Zarsky and Tachezy 2015). The evolutionary and physiological conditions in which *O. dioica* was able to lose the peroxisomes are still a matter of debate as this organism has high oxygen consumption and oxidizes fatty acids for ATP synthesis, unlike the anaerobic parasitic lineages missing these organelles (Zarsky and Tachezy 2015). It has been proposed that the loss of peroxisomes might be evolutionary adaptive, since it has been associated with reduced genomes and traits of r-reproductive strategies such as high fecundity, early maturation, and simplified ontogenesis, which *O. dioica* shares with parasite organisms as well as with some selective advantages by rendering organisms resistant to xenobiotics that become activated in the peroxisomal lumen by redox reactions (Zarsky and Tachezy 2015).

4.4.3 Loss of Genes Essential for Embryonic Development in Chordates

4.4.3.1 Loss of the RA Genetic Machinery

All-*trans*-RA is a vitamin A-derived compound that acts as a crucial signaling system involved in the differentiation and outgrowth of neurons in many metazoans (reviewed in Albalat 2009), which was adopted for Hox-controlled anterior–posterior patterning in the chordate phylum (Handberg-Thorsager et al. 2018). Despite the crucial role of RA in chordate axial patterning, it has been shown that *O. dioica* has lost most of the genes for RA production, degradation, and signaling (i.e., the RA genetic machinery, including *Rdh10*, *Aldh1a*, *Cyp26*, and *RAR* genes) (Cañestro et al. 2006, 2007; Cañestro and Postlethwait 2007; Martí-Solans et al. 2016) (Table 4.2) (Fig. 4.11). In the absence of mutational robustness (i.e., alternative pathways) capable of compensating them, these gene losses were accompanied by the loss of the RA signaling in this species (Martí-Solans et al. 2016).

The loss of the RA genetic machinery in *O. dioica* probably took place in the context of regressive evolution associated with the disintegration of the Hox cluster that, as mentioned in Sect. 4.3.2, has been related with the shift to a determinative mode of development. This shift would have released the restrictions to maintain the

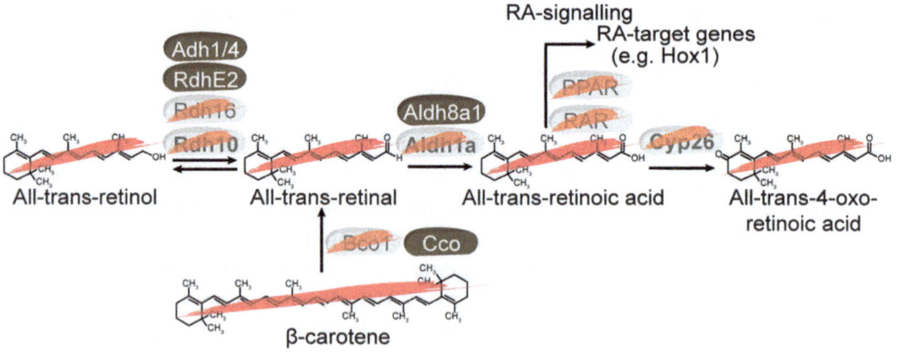

Fig. 4.11 The loss of the retinoic acid genetic machinery in *O. dioica* exemplifies how functionally linked genes are co-eliminated during evolution. In the absence of mutational robustness capable of compensating the loss, the co-elimination was accompanied by the loss of the RA signaling. *O. dioica* has lost genes for the Rdh10, Rdh16, Bco1, Aldh1a, and Cyp26 enzymes, as well as genes for the RAR and PPAR nuclear receptors (strikethrough in lighter boxes). The surviving genes (i.e., *Adh1/4*, *RdhE2*, *Aldh8a1*, and *Cco* in dark boxes) (Cañestro et al. 2010; Martí-Solans et al. 2016) do not constitute an alternative pathway for RA synthesis because neither RA nor RA precursors have been detected at concentrations that are likely to play any role in developmental or physiological processes of this species (strikethrough)

integrity of the Hox cluster (Seo et al. 2004) and led extracellular signals such as RA that establish embryonic coordinates and regulate the expression of *Hox* genes by gradually increasing the portion of the cluster exposed to transcription machinery over time to become dispensable and eventually lost (Cañestro and Postlethwait 2007; Cañestro et al. 2007).

4.4.3.2 Loss of Homeobox Gene Families

Homeobox genes encode transcription factors involved in many developmental processes in eukaryotes. Homeobox genes are classified into 11 classes and over 100 gene families/groups (Holland et al. 2007), some of them vertebrate innovations. *O. dioica* has lost 36 homeobox groups, 12 of which probably lost during the early urochordate evolution and 14 specifically lost in the *O. dioica* lineage (Edvardsen et al. 2005) (Table 4.2). Noteworthy, among the homeobox families, *O. dioica* has lost the anterior *Hox3* and all the central *Hox* genes of the cluster (Seo et al. 2004). In parallel with the losses, homeobox genes have also duplicated during the evolution of *O. dioica* lineage (Edvardsen et al. 2005). Interpreting these gains and losses in terms of the evolution of developmental mechanisms remains difficult until the function of *O. dioica* homeobox genes is revealed. It has been proposed that family losses may be the result of a relaxed selective pressure to conserve the full set of homeobox families in the *O. dioica* (or in the urochordate) lineage due to a simplification of its body plan in comparison with that of the chordate ancestor (Edvardsen et al. 2005). On the other hand, the preservation of duplicates in some families might be the result of a re-diversification of homeobox genes after major group losses (Edvardsen et al. 2005).

4.4.3.3 Loss of *Sox* Genes

Sox (Sry-type HMG box) genes are a family of transcription factors defined by a conserved sequence called the high-mobility group (HMG) box. *Sox* genes are involved in a number of essential functions during embryonic development, including sex determination and neuronal development (Wegner 2010). Ten different Sox groups have been described (A to J), some of them (B, C, D, E, F) conserved in almost all metazoans, and some groups being specific of certain lineages (A, G, and I of vertebrates, H of chordates, and J of nematodes) (Bowles et al. 2000; Heenan et al. 2016). *O. dioica*, however, has only four *Sox* genes, two of them belonging to the *SoxD* group and the other two to the *SoxB* group (Heenan et al. 2016; Torres-Aguila et al. 2018). These results imply that *O. dioica* has lost three ancient metazoan Sox groups (C, E, F) and the chordate Sox group (H) (Heenan et al. 2016) (Table 4.2). As in other examples reviewed above, the evolutionary impact of the gains and losses in the Sox family on the developmental mechanisms remains difficult to interpret since the expression of only two SoxB genes has been described (Torres-Aguila et al. 2018).

4.4.3.4 Loss of miRNAs

MicroRNAs (miRNAs) are noncoding RNAs of about 22 nucleotides involved in the regulation of different biological processes, including embryo development, cell differentiation, and growth. miRNA innovation has been correlated with increased developmental complexity during animal evolution (Hertel et al. 2006). Analysis of the miRNA repertoire of *O. dioica* has shown that this species has lost (or derived to the point they cannot be recognized anymore) at least eighteen highly conserved bilaterian miRNA families (Fu et al. 2008; Wang et al. 2017) (Table 4.2). On the other hand, at least 29 new miRNA families would have appeared in appendicularians (Fu et al. 2008), suggesting a profound reorganization of the miRNA repertoire due to recurrent events of gene losses and gains during *O. dioica* evolution. This scenario is consistent with the general notion that animal miRNAs are poorly conserved between distant taxa (Wang et al. 2017) and, therefore, that changes in miRNA repertoires have been important in shaping animal evolution (Fu et al. 2008). It is thought that the reorganization of *O. dioica* miRNA repertoire might have impacted on the temporal robustness of the rapid developmental program of this urochordate lineage (Fu et al. 2008), which would be in agreement with the idea that modifications in the miRNA repertoires have been important in adapting radically different life-history strategies from a common larval body plan (Fu et al. 2008).

4.4.3.5 Loss of Notochord Genes

The notochord, one of defining features of the chordate phylum, is a stiff rod of tissue located ventral to the neural tube that provides a rigid, but still bendy, structure for muscle attachment, as well as an important source of embryonic developmental signals (Stemple 2005). Actually, developing chordate embryos require a notochord as an organizer for secreting signals that pattern several organs such as the somites, heart, or pancreas. Comparison of a set of notochord genes known to be targets of the notochord-specific Brachyury transcription factor revealed that from 50 notochord genes of *C. robusta* used as a reference, 24 are absent in *O. dioica* (likely lost, e.g., *Noto2*), and 15 are not expressed (8 genes, e.g., *Asak*) or expressed at low levels (7 genes, e.g., *Noto10*) in the notochord (Kugler et al. 2011) (Table 4.2). Taken together, these results suggest a considerable divergence in the genetic toolkit for notochord development in *O. dioica*, which would have been impacted by events of gene loss or by the loss of the notochord function of some genes. Considering the morphological similarities between urochordate notochords, the divergence in the genetic toolkits used to develop them is surprising.

4.4.4 **O. dioica**, *Gene Loss, and the Inverse Paradox*

The original paradox in the field of evo-devo arose by the discovery that similar genetic toolkits were able to build a wide variety of morphologies in disparate animals, implying that the same genes (genetic unity) were used to build such different forms (phenotypic diversity) (Jacob 1977). The discovery of the pervasiveness of gene loss along evolution (Albalat and Cañestro 2016) affecting relevant developmental genes led to the formulation of the so-called inverse paradox of evo-devo. This hypothesis proposes that organisms might develop fundamentally similar morphologies (phenotypic unity) despite important differences in their genetic toolkits (genetic diversity) (Cañestro et al. 2007) (Fig. 4.12). Because differences between the genetic toolkits are often a consequence of the loss of some of their crucial genes, the study of *O. dioica* has become fundamental to build the new conceptual framework as this species has been able to maintain a phylotypic chordate-body plan after having lost many genes thought to be crucial for the archetypal chordate development (see Sect. 4.4.3). In this regard, the characterization of the *O. dioica* "lossosome" (the complete catalogue of gene losses in a phylogenetic context; Cañestro and Roncalli 2018) and the analysis of its functional consequences should provide the framework to investigate how gene loss might have been an important evolutionary force for generating differences in the developmental genetic toolkits of chordates and, thereby, to understand the evolution of the mechanisms of development of our own phylum.

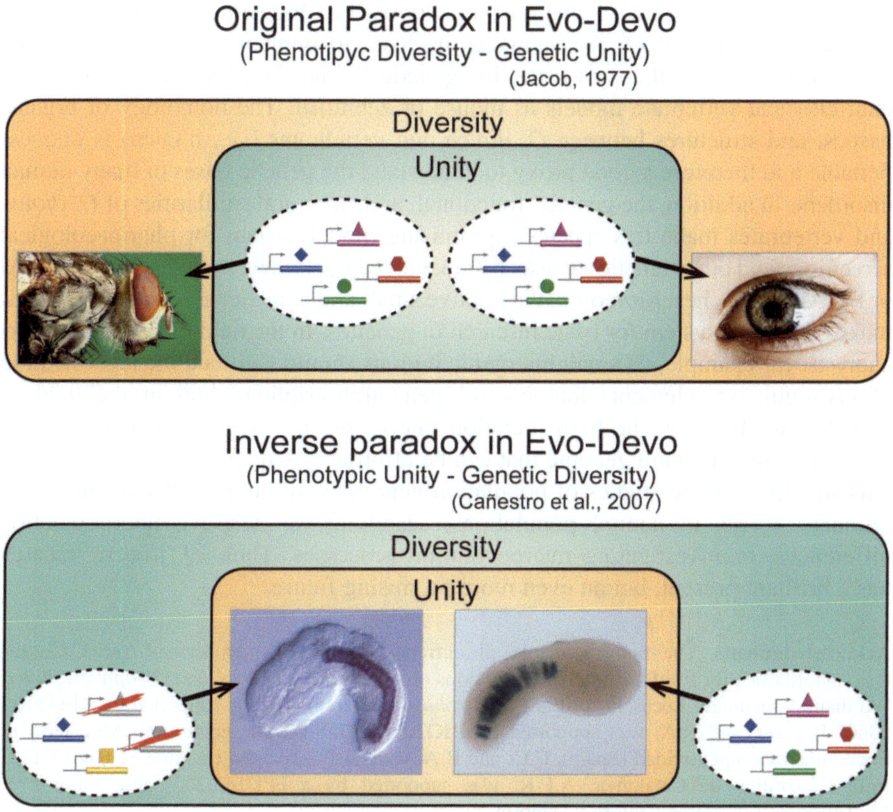

Fig. 4.12 In contrast to the original paradox, the inverse paradox in evo-devo proposes that organisms might develop fundamentally similar morphologies (phenotypic unity) despite important differences in their genetic toolkits (genetic diversity). Because differences in the genetic toolkits often are the consequence of the gene losses, the study of *O. dioica* has been fundamental to build this new conceptual framework

4.5 Future Directions

The tiny planktonic chordate *Oikopleura dioica* appears as an emerging nonclassical animal model, not only attractive to better understand the radiation of our own phylum in the field of evo-devo but also attractive in many other biological fields, such as ecology, environmental toxicology, biomedicine, or basic research. In ecology and toxicology, for instance, the relevance of *O. dioica* and other larvacean species in the marine food webs makes the study of its defense and detoxification mechanisms significant for understanding the impact of industrial pollutants or the effect of the climate change on marine environments, trophic webs, and ocean production. Monitoring *O. dioica* populations has been proposed to be used as valuable sentinels for following the health of marine ecosystems and the expression of its defensome genes as molecular biosensors that marine biologists could use to monitor stress situations of natural populations. For biomedical applications, the chordate condition of *O. dioica* makes it closer to humans than classical model animals such as fruit flies or worms, but genetically and functionally more tractable than classical vertebrate models as mouse or zebrafish. The homology of organs, tissues, and structures between *O. dioica* and vertebrates (i.e., human) is unquestionable and therefore a good proxy to understand the genetic bases of many human disorders. In addition, the genetic, functional, and structural similarities of *O. dioica* and vertebrates make this species a promising model system for pharmacological screenings and other functional tests by amenable knockdown systems such as RNAi or DNAi approaches. Moreover, the relative small and compact genome of *O. dioica* makes it a good system for basic research in genetics. In the field of gene regulation analysis, for example, its small intergenic regions should facilitate the identification of *cis*-regulatory elements that control gene transcription. And in the field of evolutionary biology, the high mutation rate of *O. dioica*, its propensity to lose genes, and its miniature genome that can be affordably sequenced in many individuals should facilitate studies of population genomics for better understanding evolutionary forces in natural populations, searching for adaptive interpopulation differences, or investigating microevolution processes. Thus, *O. dioica* research has a brilliant present, but an even more promising future.

Acknowledgments The authors thank all current and former members of the C.C. and R.A. laboratories for fruitful scientific discussions on evolution, gene loss, and *Oikopleura dioica*. We thank Annamaria Locascio for the *Ciona robusta* picture. C.C. was supported by BFU2016-80601-P grant, and R.A. was supported by BIO2015-67358-C2-1-P grant from Ministerio de Economía y Competitividad (Spain). C.C. and R.A. were also supported by grant SGR2017-1665 from Generalitat de Catalunya. A.F-R. was supported by a FPU14/02654 fellowship from Ministerio de Educación Cultura y Deporte.

References

Acuña JL, Kiefer M (2000) Functional response of the appendicularian *Oikopleura dioica*. Limnol Oceanogr 45:608–618

Acuña JL, Bedo AW, Harris RP, Anadón R (1995) The seasonal succession of appendicularians (Tunicata: Appendicularia) off Plymouth. J Mar Biol Assoc UK 75:755–758

Acuña JL, Deibel D, Saunders P, Booth B, Hatfield E, Klein B, Mei ZP, Rivkin R (2002) Phytoplankton ingestion by appendicularians in the North Water. Deep Sea Res Part II Top Stud Oceanogr 49:5101–5115

Albalat R (2009) The retinoic acid machinery in invertebrates: ancestral elements and vertebrate innovations. Mol Cell Endocrinol 313:23–35

Albalat R, Cañestro C (2016) Evolution by gene loss. Nat Rev Genet 17:379–391

Albalat R, Martí-Solans J, Cañestro C (2012) DNA methylation in amphioxus: from ancestral functions to new roles in vertebrates. Brief Funct Genomics 11:142–155

Alldredge AL (2005) The contribution of discarded appendicularian houses to the flux of particulate organic carbon from oceanic surface waters. In: Gorsky G, Youngbluth MJ, Deibel D (eds) Response of marine ecosystems to global change: ecological impact of appendicularians. Éditions Scientifiques, Paris

Almazán A, Ferrández-Roldán A, Albalat R, Cañestro C (2019) Developmental atlas of appendicularian *Oikopleura dioica* actins provides new insights into the evolution of the notochord and the cardio-paraxial muscle in chordates. Dev Biol 448:260–270

Aravena GP, Palma S (2002) Taxonomic identification of appendicularians collected in the epipelagic waters off northern Chile (Tunicata, Appendicularia). Rev Chilena Hist Nat 75:307–325

Bary BM (1960) Notes on ecology, distribution, and systematics of Pelagic Tunicata from New Zealand. Pac Sci 14(2):101–121

Bassham S, Postlethwait J (2000) Brachyury (T) expression in embryos of a larvacean urochordate, *Oikopleura dioica*, and the ancestral role of T. Dev Biol 220:322–332

Bedo AW, Acuña JL, Robins D, Harris RP (1993) Grazing in the micron and the sub-micron particle size range: the case of *Oikopleura dioica* (Appendicularia). Bull Mar Sci 52:2–14

Berna L, Alvarez-Valin F (2014) Evolutionary genomics of fast evolving tunicates. Genome Biol Evol 6:1724–1738

Berna L, D'onofrio G, Alvarez-Valin F (2012) Peculiar patterns of amino acid substitution and conservation in the fast evolving tunicate *Oikopleura dioica*. Mol Phylogenet Evol 62:708–717

Blumenthal T (2004) Operons in eukaryotes. Brief Funct Genomic Proteomic 3:199–211

Bollner T, Holmberg K, Olsson R (1986) A rostral sensory mechanism in *Oikopleura dioica* (Appendicularia). Acta Zool (Stockholm) 67:235–241

Bone Q (1998) The biology of pelagic tunicates. Oxford University Press, New York

Bone Q, Mackie GO (1975) Skin impulses and locomotion in *Oikopleura* (tunicata: larvacea). Biol Bull 149:267–286

Bouquet JM, Spriet E, Troedsson C, Ottera H, Chourrout D, Thompson EM (2009) Culture optimization for the emergent zooplanktonic model organism *Oikopleura dioica*. J Plankton Res 31:359–370

Bouquet JM, Troedsson C, Novac A, Reeve M, Lechtenborger AK, Massart W, Skaar KS, Aasjord A, Dupont S, Thompson EM (2018) Increased fitness of a key appendicularian zooplankton species under warmer, acidified seawater conditions. PLoS One 13:e0190625

Bowles J, Schepers G, Koopman P (2000) Phylogeny of the SOX family of developmental transcription factors based on sequence and structural indicators. Dev Biol 227:239–255

Brozovic M, Martin C, Dantec C, Dauga D, Mendez M, Simion P, Percher M, Laporte B, Scornavacca C, Di Gregorio A, Fujiwara S, Gineste M, Lowe EK, Piette J, Racioppi C, Ristoratore F, Sasakura Y, Takatori N, Brown TC, Delsuc F, Douzery E, Gissi C, Mcdougall A, Nishida H, Sawada H, Swalla BJ, Yasuo H, Lemaire P (2016) ANISEED 2015: a digital framework for the comparative developmental biology of ascidians. Nucleic Acids Res 44:D808–D818

Brozovic M, Dantec C, Dardaillon J, Dauga D, Faure E, Gineste M, Louis A, Naville M, Nitta KR, Piette J, Reeves W, Scornavacca C, Simion P, Vincentelli R, Bellec M, Aicha SB, Fagotto M, Gueroult-Bellone M, Haeussler M, Jacox E, Lowe EK, Mendez M, Roberge A, Stolfi A, Yokomori R, Brown CT, Cambillau C, Christiaen L, Delsuc F, Douzery E, Dumollard R, Kusakabe T, Nakai K, Nishida H, Satou Y, Swalla B, Veeman M, Volff JN, Lemaire P (2017) ANISEED 2017: extending the integrated ascidian database to the exploration and evolutionary comparison of genome-scale datasets. Nucleic Acids Res 46:D718–D725

Burighel P, Brena C (2001) Gut ultrastructure of the appendicularian *Oikopleura dioica* (Tunicata). Invertebr Biol 120:278–293

Burke M, Scholl EH, Bird DM, Schaff JE, Colman SD, Crowell R, Diener S, Gordon O, Graham S, Wang X, Windham E, Wright GM, Opperman CH (2015) The plant parasite *Pratylenchus coffeae* carries a minimal nematode genome. Nematology 17:621

Calatayud S, Garcia-Risco M, Rojas NS, Espinosa-Sanchez L, Artime S, Palacios O, Cañestro C, Albalat R (2018) Metallothioneins of the urochordate *Oikopleura dioica* have Cys-rich tandem repeats, large size and cadmium-binding preference. Metallomics 10:1585–1594

Campsteijn C, Ovrebo JI, Karlsen BO, Thompson EM (2012) Expansion of cyclin D and CDK1 paralogs in *Oikopleura dioica*, a chordate employing diverse cell cycle variants. Mol Biol Evol 29:487–502

Cañestro C, Albalat R (2012) Transposon diversity is higher in amphioxus than in vertebrates: functional and evolutionary inferences. Brief Funct Genomics 11:131–141

Cañestro C, Postlethwait JH (2007) Development of a chordate anterior-posterior axis without classical retinoic acid signaling. Dev Biol 305:522–538

Cañestro C, Roncalli V (2018) Gene losses did not stop the evolution of big brains. elife 7:e41912

Cañestro C, Bassham S, Postlethwait JH (2003) Seeing chordate evolution through the *Ciona* genome sequence. Genome Biol 4:208–211

Cañestro C, Bassham S, Postlethwait JH (2005) Development of the central nervous system in the larvacean *Oikopleura dioica* and the evolution of the chordate brain. Dev Biol 285:298–315

Cañestro C, Postlethwait JH, Gonzalez-Duarte R, Albalat R (2006) Is retinoic acid genetic machinery a chordate innovation? Evol Dev 8:394–406

Cañestro C, Yokoi H, Postlethwait JH (2007) Evolutionary developmental biology and genomics. Nat Rev Genet 8:932–942

Cañestro C, Bassham S, Postlethwait JH (2008) Evolution of the thyroid: Anterior-posterior regionalization of the *Oikopleura* endostyle revealed by Otx, Pax 2/5/8, and Hox1 expression. Dev Dyn 237:1490–1499

Cañestro C, Albalat R, Postlethwait JH (2010) *Oikopleura dioica* alcohol dehydrogenase class 3 provides new insights into the evolution of retinoic acid synthesis in chordates. Zool Sci 27:128–133

Cañestro C, Albalat R, Irimia M, Garcia-Fernandez J (2013) Impact of gene gains, losses and duplication modes on the origin and diversification of vertebrates. Semin Cell Dev Biol 24:83–94

Captanio FL, Curelovich J, Tresguerres M, Negri RM, Viñas MD, Esnal GB (2008) Seasonal cycle of appendicularians at a coastal station (38°28′S, 57°41′W) of the SW Atlantic Ocean. Bull Mar Sci 82:171–184

Chalopin D, Naville M, Plard F, Galiana D, Volff JN (2015) Comparative analysis of transposable elements highlights mobilome diversity and evolution in vertebrates. Genome Biol Evol 7:567–580

Chang ES, Neuhof M, Rubinstein ND, Diamant A, Philippe H, Huchon D, Cartwright P (2015) Genomic insights into the evolutionary origin of Myxozoa within Cnidaria. Proc Natl Acad Sci USA 112:14912–14917

Chavali S, Morais DA, Gough J, Babu MM (2011) Evolution of eukaryotic genome architecture: insights from the study of a rapidly evolving metazoan, *Oikopleura dioica*: non-adaptive forces such as elevated mutation rates may influence the evolution of genome architecture. BioEssays 33:592–601

Cima F, Brena C, Burighel P (2002) Multifarious activities of gut epithelium in an appendicularian (*Oikopleura dioica*: Tunicata). Mar Biol 141:479–490

Clarke T, Bouquet JM, Fu X, Kallesoe T, Schmid M, Thompson EM (2007) Rapidly evolving lamins in a chordate, *Oikopleura dioica*, with unusual nuclear architecture. Gene 396:159–169

Cleaver O, Krieg PA (2001) Notochord patterning of the endoderm. Dev Biol 234:1–12

Coppola U, Ristoratore F, Albalat R, D'aniello S (2019) The evolutionary landscape of the Rab family in chordates. Cell Mol Life Sci. https://doi.org/10.1007/s00018-019-03103-7

Corbo JC, Di Gregorio A, Levine M (2001) The ascidian as a model organism in developmental and evolutionary biology. Cell 106:535–538

Costello J, Stancyk SE (1983) Tidal influence upon appendicularian abundance in North Inlet estuary, South Carolina. J Plankton Res 5:263–277

Danks G, Campsteijn C, Parida M, Butcher S, Doddapaneni H, Fu B, Petrin R, Metpally R, Lenhard B, Wincker P, Chourrout D, Thompson EM, Manak JR (2013) OikoBase: a genomics and developmental transcriptomics resource for the urochordate *Oikopleura dioica*. Nucleic Acids Res 41:D845–D853

Davoll P, Youngbluth M (1990) Heterotrophic activity on appendicularian (Tunicata: Appendicularia) houses in mesopelagic regions and their potential contribution to particle flux. Deep-Sea Res 37:285–294

Dehal P, Satou Y, Campbell RK, Chapman J, Degnan B, De Tomoso A, Davidson B, Di Gregorio A, Gelpke M, Goodstein DM, Harafuji N, Hastings KEM, Ho I, Hotta K, Huang W, Kawashima T, Lemaire P, Martinez D, Meinertzhagen IA, Necula S, Nonaka M, Putnam N, Rash S, Saiga H, Satake M, Terry A, Yamada L, Wang H-G, Awazu S, Azumi K, Boore J, Branno M, Chin-Bow S, Desantis R, Doyle S, Francino P, Keys DN, Haga S, Hayashi H, Hino K, Imai KS, Inaba K, Kano S, Kobayashi K, Kobayashi M, Lee B-I, Makabe KW, Manohar C, Matassi G, Medina M, Mochizuki Y, Mount S, Morishita T, Miura S, Nakayama A, Nishizaka S, Nomoto H, Ohta F, Oishi K, Rigoutsos I, Sano M, Sasaki A, Sasakura Y, Shoguchi E, Shin-I T, Spagnuolo A, Stainier D, Suzuki MM, Tassy O, Takatori N, Tokuoka M, Yagi K, Yoshizaki F, Wada S, Zhang C, Hyatt PD, Larimer F, Detter C, Doggett N, Glavina T, Hawkins T, Richardson P, Lucas S, Kohara Y, Levine M, Satoh N, Rokhsar DS (2002) The draft genome of *Ciona intestinalis:* insights into chordate and vertebrate origins. Science 298:2157–2167

Delsman HC (1910) Beiträge zur Entwicklungsgeschichte von *Oikopleura dioica*. Verh Rijksinst Onderz Zee 3:1–24

Delsman HC (1912) Weitere beobachtungen über die entwicklung von *Oikopleura dioica*. Tijdschr ned dierk Ver 12:199–206

Delsuc F, Philippe H, Tsagkogeorga G, Simion P, Tilak MK, Turon X, Lopez-Legentil S, Piette J, Lemaire P, Douzery EJP (2018) A phylogenomic framework and timescale for comparative studies of tunicates. BMC Biol 16:39

Deng W, Henriet S, Chourrout D (2018) Prevalence of mutation-prone microhomology-mediated end joining in a chordate lacking the c-NHEJ DNA repair pathway. Curr Biol 28(3337–3341): e3334

Denoeud F, Henriet S, Mungpakdee S, Aury JM, Da Silva C, Brinkmann H, Mikhaleva J, Olsen LC, Jubin C, Cañestro C, Bouquet JM, Danks G, Poulain J, Campsteijn C, Adamski M, Cross I, Yadetie F, Muffato M, Louis A, Butcher S, Tsagkogeorga G, Konrad A, Singh S, Jensen MF, Cong EH, Eikeseth-Otteraa H, Noel B, Anthouard V, Porcel BM, Kachouri-Lafond R, Nishino A, Ugolini M, Chourrout P, Nishida H, Aasland R, Huzurbazar S, Westhof E, Delsuc F, Lehrach H, Reinhardt R, Weissenbach J, Roy SW, Artiguenave F, Postlethwait JH, Manak JR, Thompson EM, Jaillon O, Du Pasquier L, Boudinot P, Liberles DA, Volff JN, Philippe H, Lenhard B, Roest Crollius H, Wincker P, Chourrout D (2010) Plasticity of animal genome architecture unmasked by rapid evolution of a pelagic tunicate. Science 330:1381–1385

Deriano L, Roth DB (2013) Modernizing the nonhomologous end-joining repertoire: alternative and classical NHEJ share the stage. Annu Rev Genet 47:433–455

Duboule D (2007) The rise and fall of Hox gene clusters. Development 134:2549–2560

Edvardsen RB, Lerat E, Maeland AD, Flat M, Tewari R, Jensen MF, Lehrach H, Reinhardt R, Seo HC, Chourrout D (2004) Hypervariable and highly divergent intron-exon organizations in the chordate *Oikopleura dioica*. J Mol Evol 59:448–457

Edvardsen RB, Seo HC, Jensen MF, Mialon A, Mikhaleva J, Bjordal M, Cartry J, Reinhardt R, Weissenbach J, Wincker P, Chourrout D (2005) Remodelling of the homeobox gene complement in the tunicate *Oikopleura dioica*. Curr Biol 15:R12–R13

Essenberg CE (1922) The seasonal distribution of the Appendicularia in the region of San Diego, California. Ecology 3:55–64

Fenaux R (1972) A historical survey of the appendicularians from the area covered by the IIOE. Mar Biol 16:230–235

Fenaux R (1976) Cycle vital d'un appendiculaire: *Oikopleura dioica* Fol, 1872. Ann Inst Océanogr Paris 52:89–101

Fenaux R (1986) The house of *Oikopleura dioica* (Tunicata, Appendicularia): structure and functions. Zoomorphology 106:224–231

Fenaux R (1998a) Anatomy and functional morphology of the Appendicularia. In: Bone Q (ed) The biology of pelagic tunicates. Oxford University Press, New York

Fenaux R (1998b) Life history of the Appendicularia. In: Bone Q (ed) The biology of pelagic tunicates. Oxford University Press, Oxford

Fenaux R, Gorsky G (1985) Nouvelle technique d'élevage des appendiculaires. Rapp Comm Int Mer Médit 29:291–292

Fenaux R, Bone Q, Deibel D (1998) Appendicularian distribution and zoogeography. In: Bone Q (ed) The biology of pelagic tunicates. Oxford University Press, Oxford

Fernández D, López-Urrutia A, Fernández A, Acuña JL, Harris R (2004) Retention efficiency of 0.2 to 6 μm particles by the appendicularians *Oikopleura dioica* and *Fritillaria borealis*. Mar Ecol Prog Ser 266:89–101

Ferrier DE (2011) Tunicates push the limits of animal evo-devo. BMC Biol 9:3

Ferrier DEK, Sogabe S (2018) Genome biology: unconventional DNA repair in an extreme genome. Curr Biol 28:R1208–R1210

Flood PR, Afzelius BA (1978) The spermatozoon of *Oikopleura dioica* Fol (Larvacea, Tunicata). Cell Tissue Res 191:27–37

Flood PR, Deibel D (1998) The appendicularian house. In: Bone Q (ed) The biology of pelagic tunicates. Oxford University Press, Oxford

Fol H (1872) Etudes sur les Appendiculaires du Détroit de Messine. Mem Soc Physique Hist Nat Geneve 21:445

Fredriksson G, Olsson R (1981) The oral gland cells of *Oikopleura dioica* (Tunicata Appendicularia). Acta Zool (Stockholm) 62:195–200

Fredriksson G, Olsson R (1991) The subchordal cells of *Oikopleura dioica* and *Oikopleura albicans* (Appendicularia, Chrodata). Acta Zool (Stockholm) 72:251–256

Fredriksson G, Ofverholm T, Ericson LE (1985) Ultrastructural demonstration of iodine binding and peroxidase activity in the endostyle of *Oikopleura dioica* (Appendicularia). Gen Comp Endocrinol 58:319–327

Fu X, Adamski M, Thompson EM (2008) Altered miRNA repertoire in the simplified chordate, *Oikopleura dioica*. Mol Biol Evol 25:1067–1080

Fujii S, Nishio T, Nishida H (2008) Cleavage pattern, gastrulation, and neurulation in the appendicularian, *Oikopleura dioica*. Dev Genes Evol 218:69–79

Gabaldon T, Capella-Gutierrez S (2010) Lack of phylogenetic support for a supposed actinobacterial origin of peroxisomes. Gene 465:61–65

Gabaldon T, Snel B, Van Zimmeren F, Hemrika W, Tabak H, Huynen MA (2006) Origin and evolution of the peroxisomal proteome. Biol Direct 1:8

Galt CP (1972) Development of *Oikopleura dioica* (Urochordata: Larvacea): ontogeny of behavior and of organ systems related to construction and use of the house. PhD, University of Washington, Seattle

Galt C (1978) Bioluminescence: dual mechanism in a planktonic tunicate produces brilliant surface display. Science 200:70–72

Galt CP, Fenaux R (1990) Urochordata larvacea. In: Adiyodi KG, Adiyodi RG (eds) Reproductive biology of invertebrates. Oxford and IBH, New Delhi

Ganot P, Kallesoe T, Reinhardt R, Chourrout D, Thompson EM (2004) Spliced-leader RNA trans splicing in a chordate, *Oikopleura dioica*, with a compact genome. Mol Cell Biol 24:7795–7805

Ganot P, Bouquet JM, Kallesoe T, Thompson EM (2007) The Oikopleura coenocyst, a unique chordate germ cell permitting rapid, extensive modulation of oocyte production. Dev Biol 302:591–600

Gaughan D, Potter IC (1994) Relative abundance and seasonal changes in the macrozooplankton of the lower estuary in South-Western Australia. Rec West Aust Mus 16:461–474

Georges D, Holmberg K, Olsson R (1988) The ventral midbrain cells in *Oikopleura dioica* (Appendicularia). Acta Embryol Morphol Exp 9:39–47

Goldschmidt R (1903) Notiz über die Entwickekung der Appendicularien-Biologisches Centralblatt, Band

Goldstone JV, Hamdoun A, Cole BJ, Howard-Ashby M, Nebert DW, Scally M, Dean M, Epel D, Hahn ME, Stegeman JJ (2006) The chemical defensome: environmental sensing and response genes in the *Strongylocentrotus purpuratus* genome. Dev Biol 300:366–384

Goldstone JV, Mcarthur AG, Kubota A, Zanette J, Parente T, Jonsson ME, Nelson DR, Stegeman JJ (2010) Identification and developmental expression of the full complement of Cytochrome P450 genes in Zebrafish. BMC Genomics 11:643

Gorsky G, Fenaux R (1998) The role of Appendicularia in marine food webs. In: Bone Q (ed) The biology of pelagic tunicates. Oxford University Press, Oxford

Gorsky G, Fisher N, Fowler S (1984) Biogenic debris from the pelagic tunicate *Oikopleura dioica* and its role in the vertical transport of a transuranium element. Est Coast Shelf Sci 18:13–23

Handberg-Thorsager M, Gutierrez-Mazariegos J, Arold ST, Kumar Nadendla E, Bertucci PY, Germain P, Tomancak P, Pierzchalski K, Jones JW, Albalat R, Kane MA, Bourguet W, Laudet V, Arendt D, Schubert M (2018) The ancestral retinoic acid receptor was a low-affinity sensor triggering neuronal differentiation. Sci Adv 4:eaao1261

Heenan P, Zondag L, Wilson MJ (2016) Evolution of the Sox gene family within the chordate phylum. Gene 575:385–392

Hertel J, Lindemeyer M, Missal K, Fried C, Tanzer A, Flamm C, Hofacker IL, Stadler PF, Students of Bioinformatics Computer Labs A (2006) The expansion of the metazoan microRNA repertoire. BMC Genomics 7:25

Holland LZ (2016) Tunicates. Curr Biol 26:R146–R152

Holland LZ, Gibson-Brown J (2003) The *Ciona intestinalis* genome: when the constraints are off. BioEssays 25:529–532

Holland L, Gorsky G, Fenaux R (1988) Fertilization in *Oikopleura dioica* (Tunicata, Appendicularia): acrosome reaction, cortical reaction and sperm-egg fusion. Zoomorphology 108:229–243

Holland PW, Booth HA, Bruford EA (2007) Classification and nomenclature of all human homeobox genes. BMC Biol 5:47

Holmberg K (1982) The ciliated brain duct of *Oikopleura dioica* (Tunicata, Appendicularia). Acta Zool 63:101–109

Holmberg K (1984) A transmission electron microscopic investigation of the sensory vesicle in the brain of *Oikopleura dioica* (Appendicularia). Zoomorphology 104:298–303

Hopcroft RR, Roff JC (1995) Zooplankton growth rates: extraordinary production by the larvacean *Oikopleura dioica* in tropical waters. J Plankton Res 17:205–220

Hopcroft RR, Roff JC, Bouman HA (1998) Zooplankton growth rates: the larvaceans *Appendicularia*, *Fritillaria* and *Oikopleura* in tropical waters. J Plankton Res 20:539–555

Hosp J, Sagane Y, Danks G, Thompson EM (2012) The evolving proteome of a complex extracellular matrix, the *Oikopleura* house. PLoS One 7:e40172

Hughes AL, Friedman R (2005) Loss of ancestral genes in the genomic evolution of *Ciona intestinalis*. Evol Dev 7:196–200

Hwan Lee J, Chae J, Kim W-R, Won Jung S, Man Kim J (2001) Seasonal variation of phytoplankton and zooplankton communities in the Coastal Waters off Tongyeong in Korea. Ocean Polar Res 23:245–253

Inoue J, Satoh N (2018) Deuterostome genomics: lineage-specific protein expansions that enabled chordate muscle evolution. Mol Biol Evol 35:914–924

Irimia M, Tena JJ, Alexis MS, Fernandez-Minan A, Maeso I, Bogdanovic O, De La Calle-Mustienes E, Roy SW, Gomez-Skarmeta JL, Fraser HB (2012) Extensive conservation of ancient microsynteny across metazoans due to cis-regulatory constraints. Genome Res 22:2356–2367

Jacob F (1977) Evolution and tinkering. Science 196:1161–1166

Kimura S, Ohshima C, Hirose E, Nishikawa J, Itoh T (2001) Cellulose in the house of the appendicularian *Oikopleura rufescens*. Protoplasma 216:71–74

King KR, Hollibaugh JT, Azam F (1980) Predator-prey interactions between the larvacean *Oikopleura dioica* and bacterioplankton in enclosed water columns. Mar Biol 56:49–57

Kishi K, Onuma TA, Nishida H (2014) Long-distance cell migration during larval development in the appendicularian, *Oikopleura dioica*. Dev Biol 395:299–306

Kishi K, Hayashi M, Onuma TA, Nishida H (2017) Patterning and morphogenesis of the intricate but stereotyped oikoplastic epidermis of the appendicularian, *Oikopleura dioica*. Dev Biol 428:245–257

Kmita M, Duboule D (2003) Organizing axes in time and space; 25 years of colinear tinkering. Science 301:331–333

Kocot KM, Tassia MG, Halanych KM, Swalla BJ (2018) Phylogenomics offers resolution of major tunicate relationships. Mol Phylogenet Evol 121:166–173

Kugler JE, Kerner P, Bouquet JM, Jiang D, Di Gregorio A (2011) Evolutionary changes in the notochord genetic toolkit: a comparative analysis of notochord genes in the ascidian *Ciona* and the larvacean *Oikopleura*. BMC Evol Biol 11:21

Kusakabe T, Araki I, Satoh N, Jeffery WR (1997) Evolution of chordate actin genes: evidence from genomic organization and amino acid sequences. J Mol Evol 44:289–298

Larson RJ (1987) Daily ration and predation by medusae and ctenophores in Saanich Inlet, B.C., Canada. Neth J Sea Res 21:35–44

Last JM (1972) Egg development, fecundity and growth of *Oikopleura Dioica* Fol in the North Sea. ICES J Mar Sci 34:232–237

Lohmann H (1933) Erste Klasse der Tunicaten: Appendiculariae. In: Kükenthal W, Krumbach T (eds) Handbuch der Zoologie. Walter De Gruyter, Berlin

Lopez-Urrutia A, Acuña JL (1999) Gut throughput dynamics in the appendicularian *Oikopleura dioica*. Mar Ecol Prog Ser 191:195–205

Martinucci G, Brena C, Cima F, Burighel P (2005) Synchronous spermatogenesis in appendicularians. In: Gorsky G, Youngbluth M, Deibel D (eds) Response of marine ecosystems to global changes: ecological impact of appendicularians. Éditions des Archives Contempoaines, Paris

Martí-Solans J, Ferrández-Roldán A, Godoy-Marín H, Badia-Ramentol J, Torres-Aguila NP, Rodríguez-Marí A, Bouquet JM, Chourrout D, Thompson EM, Albalat R, Cañestro C (2015) *Oikopleura dioica* culturing made easy: a low-cost facility for an emerging animal model in EvoDevo. Genesis 53:183–193

Martí-Solans J, Belyaeva OV, Torres-Aguila NP, Kedishvili NY, Albalat R, Cañestro C (2016) Coelimination and survival in gene network evolution: dismantling the RA-signaling in a chordate. Mol Biol Evol 33:2401–2416

Menéndez M, Biancalana F, Berasategui A, Fernández Severini MD, Hoffmeyer MS, Esteves JL (2011) Mesozooplankton composition and spatial distribution, Nuevo Gulf, Patagonia, Argentina. Check List 7:101–107

Mikhailov KV, Slyusarev GS, Nikitin MA, Logacheva MD, Penin AA, Aleoshin VV, Panchin YV (2016) The genome of *Intoshia linei* affirms orthonectids as highly simplified spiralians. Curr Biol 26:1768–1774

Mikhaleva Y, Kreneisz O, Olsen LC, Glover JC, Chourrout D (2015) Modification of the larval swimming behavior in *Oikopleura dioica*, a chordate with a miniaturized central nervous system by dsRNA injection into fertilized eggs. J Exp Zool B Mol Dev Evol 324:114–127

Mikhaleva Y, Skinnes R, Sumic S, Thompson EM, Chourrout D (2018) Development of the house secreting epithelium, a major innovation of tunicate larvaceans, involves multiple homeodomain transcription factors. Dev Biol 443:117–126

Nakamura Y, Suzuki K, Suzuki S, Hiromi J (1997) Production of *Oikopleura dioica* (Appendicularia) following a picoplankton 'bloom' in a eutrophic coastal area. J Plankton Res 19:113–124

Navratilova P, Danks GB, Long A, Butcher S, Manak JR, Thompson EM (2017) Sex-specific chromatin landscapes in an ultra-compact chordate genome. Epigenetics Chromatin 10:3

Nelson DR, Goldstone JV, Stegeman JJ (2013) The cytochrome P450 genesis locus: the origin and evolution of animal cytochrome P450s. Philos Trans R Soc Lond Ser B Biol Sci 368:20120474

Nilsen IW, Myrnes B, Edvardsen RB, Chourrout D (2003) Urochordates carry multiple genes for goose-type lysozyme and no genes for chicken- or invertebrate-type lysozymes. Cell Mol Life Sci 60:2210–2218

Nishida H (1987) Cell lineage analysis in ascidian embryos by intracellular injection of a tracer enzyme. III Up to the tissue restricted stage. Dev Biol 121:526–541

Nishida H (2002) Patterning the marginal zone of early ascidian embryos: localized maternal mRNA and inductive interactions. BioEssays 24:613–624

Nishida H (2008) Development of the appendicularian *Oikopleura dioica*: culture, genome, and cell lineages. Develop Growth Differ 50(Suppl 1):S239–S256

Nishida H, Stach T (2014) Cell lineages and fate maps in tunicates: conservation and modification. Zool Sci 31:645–652

Nishino A, Morisawa M (1998) Rapid oocyte growth and artificial fertilization of the Larvaceans *Oikopleura dioica* and *Oikopleura longicauda*. Zool Sci 15:723–727

Nishino A, Satoh N (2001) The simple tail of chordates: phylogenetic significance of appendicularians. Genesis 29:36–45

Nishino A, Satou Y, Morisawa M, Satoh N (2000) Muscle actin genes and muscle cells in the appendicularian, *Oikopleura longicauda*: phylogenetic relationships among muscle tissues in the urochordates. J Exp Zool 288:135–150

Ohno S (1970) Evolution by gene duplication. Springer, New York

Olsson R (1963) Endostyles and endostylar secretions: a comparative histochemical study. Acta Zool (Stockholm) 44:299–329

Olsson R (1965a) Comparative morphology and physiology of the *Oikopleura* notochord. Isr J Zool 14:213–220

Olsson R (1965b) The cytology of the endostyle of *Oikopleura dioica*. Ann NY Acad Sci 118:1038–1051

Olsson R (1975) Primitive coronet cells in the brain of *Oikopleura* (Appendicularia, Tunicata). Acta Zool (Stockholm) 56:155–161

Olsson R, Holmberg K, Lilliemark Y (1990) Fine structure of the brain and brain nerves of *Oikopleura dioica* (Urochordata, Appendicularia). Zool Morphol 110:1–7

Omotezako T, Nishino A, Onuma TA, Nishida H (2013) RNA interference in the appendicularian *Oikopleura dioica* reveals the function of the Brachyury gene. Dev Genes Evol 223:261–267

Omotezako T, Onuma TA, Nishida H (2015) DNA interference: DNA-induced gene silencing in the appendicularian *Oikopleura dioica*. Proc Biol Sci 282:20150435

Omotezako T, Matsuo M, Onuma TA, Nishida H (2017) DNA interference-mediated screening of maternal factors in the chordate *Oikopleura dioica*. Sci Rep 7:44226

Onuma TA, Isobe M, Nishida H (2017) Internal and external morphology of adults of the appendicularian, *Oikopleura dioica*: an SEM study. Cell Tissue Res 367:213–227

Paffenhöfer G-A (1973) The cultivation of an appendicularian through numerous generations. Mar Biol 22:183–185

Pannunzio NR, Watanabe G, Lieber MR (2018) Nonhomologous DNA end-joining for repair of DNA double-strand breaks. J Biol Chem 293:10512–10523

Patel AA, Steitz JA (2003) Splicing double: insights from the second spliceosome. Nat Rev Mol Cell Biol 4:960–970

Raduan A, Blanco C, Soler E, Del Rio JG, Raga JA (1985) The seasonal distribution of the Appendicularia in the Bay of Cullera, Spain. Rapp Comm Int Mer Médit 29:293–294

Robinson AJ, Kunji ER, Gross A (2012) Mitochondrial carrier homolog 2 (MTCH2): the recruitment and evolution of a mitochondrial carrier protein to a critical player in apoptosis. Exp Cell Res 318:1316–1323

Robison BH, Reisenbichler KR, Sherlock RE (2005) Giant larvacean houses: rapid carbon transport to the deep sea floor. Science 308:1609–1611

Sagane Y, Zech K, Bouquet JM, Schmid M, Bal U, Thompson EM (2010) Functional specialization of cellulose synthase genes of prokaryotic origin in chordate larvaceans. Development 137:1483–1492

Salensky W (1903) Etudes anatomiques sur les Appendiculaires. I *Oikopleura vanhoeffeni* Lohmann. Mem Acad Sci St Petesbourg Ser 13:1–44
Salensky W (1904) Etudes anatomiques sur les Appendiculaires. II *Oikopleura refescens*. Fol Mem Acad Sci St Petesbourg Ser 15:1–54
Salensky W (1905) Zur Morphologie der Cardialorgane der Appendicularien. Sixième Congrès International de Zoologie, Berne. Kandiget fils. Genève, 381–383
Sato R, Tanaka Y, Ishimaru T (2003) Species-specific house productivity of appendicularians. Mar Ecol Prog Ser 259:163–172
Satoh N (2003) The ascidian tadpole larva: comparative molecular development and genomics. Nat Rev Genet 4:285–295
Satoh N, Satou Y, Davidson B, Levine M (2003) *Ciona intestinalis*: an emerging model for whole-genome analyses. Trends Genet 19:376–381
Satoh N, Tagawa K, Takahashi H (2012) How was the notochord born? Evol Dev 14:56–75
Satoh N, Rokhsar D, Nishikawa T (2014) Chordate evolution and the three-phylum system. Proc Biol Sci 281:20141729
Schluter A, Fourcade S, Ripp R, Mandel JL, Poch O, Pujol A (2006) The evolutionary origin of peroxisomes: an ER-peroxisome connection. Mol Biol Evol 23:838–845
Seo H-C, Kube M, Edvardsen RB, Jensen MF, Beck A, Spriet E, Gorsky G, Thompson EM, Lehrach H, Reinhardt R, Chourrout D (2001) Miniature genome in the marine chordate *Oikopleura dioica*. Science 294:2506
Seo HC, Edvardsen RB, Maeland AD, Bjordal M, Jensen MF, Hansen A, Flaat M, Weissenbach J, Lehrach H, Wincker P, Reinhardt R, Chourrout D (2004) Hox cluster disintegration with persistent anteroposterior order of expression in *Oikopleura dioica*. Nature 431:67–71
Sheth N, Roca X, Hastings ML, Roeder T, Krainer AR, Sachidanandam R (2006) Comprehensive splice-site analysis using comparative genomics. Nucleic Acids Res 34:3955–3967
Somorjai IML, Martí-Solans J, Diaz-Gracia M, Nishida H, Imai KS, Escriva H, Cañestro C, Albalat R (2018) Wnt evolution and function shuffling in liberal and conservative chordate genomes. Genome Biol 19:98
Soviknes AM, Glover JC (2007) Spatiotemporal patterns of neurogenesis in the appendicularian *Oikopleura dioica*. Dev Biol 311:264–275
Soviknes AM, Glover JC (2008) Continued growth and cell proliferation into adulthood in the notochord of the appendicularian *Oikopleura dioica*. Biol Bull 214:17–28
Soviknes AM, Chourrout D, Glover JC (2005) Development of putative GABAergic neurons in the appendicularian urochordate *Oikopleura dioica*. J Comp Neurol 490:12–28
Soviknes AM, Chourrout D, Glover JC (2007) Development of the caudal nerve cord, motoneurons, and muscle innervation in the appendicularian urochordate *Oikopleura dioica*. J Comp Neurol 503:224–243
Spada F, Steen H, Troedsson C, Kallesoe T, Spriet E, Mann M, Thompson EM (2001) Molecular patterning of the oikoplastic epithelium of the larvacean tunicate *Oikopleura dioica*. J Biol Chem 276:20624–20632
Stach TG (2009) Anatomy of the trunk mesoderm in tunicates: homology considerations and phylogenetic interpretation. Zoomorphology 128:97–109
Stach T, Anselmi C (2015) High-precision morphology: bifocal 4D-microscopy enables the comparison of detailed cell lineages of two chordate species separated for more than 525 million years. BMC Biol 13:113
Stach T, Winter J, Bouquet JM, Chourrout D, Schnabel R (2008) Embryology of a planktonic tunicate reveals traces of sessility. Proc Natl Acad Sci USA 105:7229–7234
Stemple DL (2005) Structure and function of the notochord: an essential organ for chordate development. Development 132:2503–2512
Sulston JE, Schierenberg E, White JG, Thomson JN (1983) The embryonic cell lineage of the nematode *Caenorhabditis elegans*. Dev Biol 100:64–119

Thompson EM, Kallesøe T, Spada F (2001) Diverse genes expressed in distinct regions of the trunk epithelium define a monolayer cellular template for construction of the oikopleurid house. Dev Biol 238:260–273

Tomita M, Shiga N, Ikeda T (2003) Seasonal occurrence and vertical distribution of appendicularians in Toyama Bay, southern Japan Sea. J Plankton Res 25:579–589

Torres-Aguila NP, Martí-Solans J, Ferrandez-Roldan A, Almazan A, Roncalli V, D'aniello S, Romano G, Palumbo A, Albalat R, Cañestro C (2018) Diatom bloom-derived biotoxins cause aberrant development and gene expression in the appendicularian chordate *Oikopleura dioica*. Commun Biol 1:121

Troedsson C, Bouquet J-M, Aksnes DL, Thompson EM (2002) Resource allocation between somatic growth and reproductive output in the pelagic chordate *Oikopleura dioica* allows opportunistic response to nutritional variation. Mar Ecol Prog Ser 243:83–91

Troedsson C, Bouquet J-M, Lobon C, Novac A, Nejstgaard J, Dupont S, Bosak S, Jakobsen H, Romanova N, Pankoke L, Isla A, Dutz JR, Sazhin A, Thompson E (2013) Effects of ocean acidification, temperature and nutrient regimes on the appendicularian *Oikopleura dioica*: a mesocosm study. Mar Biol 160:2175–2187

Uren AG, O'rourke K, Aravind LA, Pisabarro MT, Seshagiri S, Koonin EV, Dixit VM (2000) Identification of paracaspases and metacaspases: two ancient families of caspase-like proteins, one of which plays a key role in MALT lymphoma. Mol Cell 6:961–967

Uye S-I, Ichino S (1995) Seasonal variations in abundance, size composition, biomass and production rate of *Oikopleura dioica* (Fol) (Tunicata: Appendicularia) in a temperate eutrophic inlet. J Exp Mar Biol Ecol 189:1–11

Volff JN, Lehrach H, Reinhardt R, Chourrout D (2004) Retroelement dynamics and a novel type of chordate retrovirus-like elements in the miniature genome of the tunicate *Oikopleura dioica*. Mol Biol Evol 21:2022–2033

Walkusz W, Storemark K, Skau T, Gannefors C, Lundberg M (2003) Zooplankton community structure; a comparison of fjords, open water and ice stations in the Svalbard area. Pol Polar Res 24:149–165

Wang K, Dantec C, Lemaire P, Onuma TA, Nishida H (2017) Genome-wide survey of miRNAs and their evolutionary history in the ascidian, *Halocynthia roretzi*. BMC Genomics 18:314

Wegner M (2010) All purpose Sox: the many roles of Sox proteins in gene expression. Int J Biochem Cell Biol 42:381–390

Weill M, Philips A, Chourrout D, Fort P (2005) The caspase family in urochordates: distinct evolutionary fates in ascidians and larvaceans. Biol Cell 97:857–866

Welsch U, Storch V (1969) On the fine structure of the chorda dorsalis in lower chordata. (*Dendrodoa grossularia* (v. Beneden) and *Oikopleura dioica* Fol). Z Zellforsch Mikrosk Anat 93:547–549

Yadetie F, Butcher S, Forde HE, Campsteijn C, Bouquet JM, Karlsen OA, Denoeud F, Metpally R, Thompson EM, Manak JR, Goksoyr A, Chourrout D (2012) Conservation and divergence of chemical defense system in the tunicate *Oikopleura dioica* revealed by genome wide response to two xenobiotics. BMC Genomics 13:55

Zarsky V, Tachezy J (2015) Evolutionary loss of peroxisomes-not limited to parasites. Biol Direct 10:74

Chapter 5
Neuropeptides, Peptide Hormones, and Their Receptors of a Tunicate, *Ciona intestinalis*

Honoo Satake, Shin Matsubara, Akira Shiraishi, Tatsuya Yamamoto, Tomohiro Osugi, Tsubasa Sakai, and Tsuyoshi Kawada

Abstract The critical phylogenetic position of the ascidian, *Ciona intestinalis*, as the closest relative of vertebrates, suggested its potential applicability as a model organism in a wide variety of biological events including the nervous, neuroendocrine, and endocrine regulation. To date, approximately 40 neuropeptides and/or peptide hormones and several cognate receptors have been identified. These peptides are categorized into two types: (1) orthologs of vertebrate peptides, such as cholecystokinin, GnRH, tachykinin, vasopressin, and calcitonin, and (2) novel family peptides such as LF peptides and YFL/V peptides. *Ciona* GnRH receptors (Ci-GnRHR) were found to be multiplicated in the *Ciona*-specific lineages and to form unique heterodimers between Ci-GnRHR1 and R4 and between Ci-GnRHR2 and R4, leading to fine-tuning of the generation of second messengers. Furthermore, *Ciona* tachykinin was shown to regulate a novel protease-associated follicle growth pathway. These findings will pave the way for the exploration of both conserved and diversified endocrine, neuroendocrine, and nervous systems in the evolutionary lineage of invertebrate deuterostomes and/or chordates. In this chapter, we provide an overview of primary sequences, functions, and evolutionary aspects of neuropeptides, peptide hormones, and their receptors in *C. intestinalis*.

Abbreviations

AM	Adrenomedullin
AMY	Amylin
CCK	Cholecystokinin
CGRP	CT gene-related peptide
Ci	*Ciona intestinalis*

H. Satake (✉) · S. Matsubara · A. Shiraishi · T. Yamamoto · T. Osugi · T. Sakai · T. Kawada
Bioorganic Research Institute, Suntory Foundation for Life Sciences, Kyoto, Japan
e-mail: satake@sunbor.or.jp

W. Tworzydlo, S. M. Bilinski (eds.), *Evo-Devo: Non-model Species in Cell and Developmental Biology*, Results and Problems in Cell Differentiation 68,
https://doi.org/10.1007/978-3-030-23459-1_5

CRSP	CT receptor-stimulating peptide
CT	Calcitonin
GnRH	Gonadotropin-releasing hormone
OT	Oxytocin
TK	Tachykinin
TKRP	TK-related peptide
VP	Vasopressin

5.1 Introduction

Ascidians (sea squirts), filter-feeding marine invertebrate deuterostomes, belong to the phylum Chordata and subphylum Tunicata or Urochordata that constitutes the Olfactores with vertebrates (Satoh 2015). The invertebrate chordates occupy critical phylogenetic position as the sister group of vertebrates, and thus, they have been employed as useful and unique model organisms for studies in a wide range of biological fields including developmental biology, evolutionary biology, and genome science (Satoh 2015). In contrast to the advance in the investigation of developmental process and phylogenetic relationships, the neuroscience of ascidians largely remained to be explored. However, the neural network and the relevant connectomes have been verified in developing stages and adults of ascidians, particularly the cosmopolitan species, *Ciona intestinalis* (Osugi et al. 2017; Horie et al. 2018). Furthermore, over the past decade, various *Ciona* neuropeptides, peptide hormones, and their cognate receptors have been identified by purification, genome database searching, cDNA cloning, and peptidomic detection (Sherwood et al. 2006; Satake and Kawada 2006a; Kawada et al. 2010, 2011; Matsubara et al. 2016). These findings will lead to the exploration not only of neuropeptidergic and/or hormonal systems of *C. intestinalis* but also of the molecular mechanisms underlying the evolution and diversification of neuropeptides, peptide hormones, and their receptors. In this chapter, we provide basic and latest knowledge and the evolutionary aspects of sequences and several biological functions of neuropeptides, peptide hormones, and their receptors in ascidians.

5.2 Neuropeptides and Their Receptors in Ascidians

In *C. intestinalis*, 40 neuropeptides and peptide hormones (Table 5.1) of *C. intestinalis* have been characterized. Some of them were detected by the genome survey, and others were characterized by peptidomic analysis and cDNA cloning (Satake and Kawada 2006a; Sherwood et al. 2006; Kawada et al. 2010, 2011; Matsubara et al. 2016). To date, *Ciona* neuropeptides have been classified into two major groups. The one group includes homologs of vertebrate peptides:

Table 5.1 Major peptides detected by peptidomes of the *C. intestinalis* neural complex

Gene	Peptide	Sequence
Homologs or related peptides of vertebrate peptides		
ci-tk	Ci-TK-I	HVRHFYGLMa
	Ci-TK-II	SIGDQPSIFNERASFTGLMa
ci-gnrh-1	t-GnRH-3	pQHWSYEFMPGa
	t-GnRH-5	pQHWSYEYMPGa
	t-GnRH-6	pQHWSKGYSPGa
ci-gnrh-x	Ci-GnRH-X	pQHWSNWWIPGAPGYNGa
ci-vp	Ci-VP	CFFRDCSNMDWYR
ci-ntlp-A	Ci-NTLP-1	pQLHVPSIL
	Ci-NTLP-2	GMMGPSII
	Ci-NTLP-3	MMLGPGIL
	Ci-NTLP-4	FGMIPSII
ci-ntlp-B	Ci-NTLP-5	NKLLYPSVI
	Ci-NTLP-6	SRHPKLYFPGIV
ci-galp	Ci-GALP	PFRGQGGWTLNSVGYNAGLGALRKLFE
Novel peptides characterized from *Ciona*		
ci-lf	Ci-LF-1	FQSLF
	Ci-LF-2	YPGFQGLF
	Ci-LF-3	HNPHLPDLF
	Ci-LF-4	YNSMGLF
	Ci-LF-5	SPGMLGLF
	Ci-LF-6	SDARLQGLF
	Ci-LF-7	YPNFQGLF
	Ci-LF-8	GNLHSLF
ci-yfv/l	Ci-YFV-1	ELVVRDPYFV
	Ci-YFV-2	NNQESYFV
	Ci-YFV-3	DDEPRSYFV
	Ci-YFL-1	DAARPNYYFL

Pyro-glutamate and C-terminal amide are shown by “pQ” and “a”

cholecystokinin (CCK), gonadotropin-releasing hormones (GnRHs), tachykinins (TKs), vasopressin/oxytocin (VP/OT), insulin/relaxin, calcitonin (CT), and galanin. These peptides share the consensus motif with the respective vertebrate homologs. The other includes novel *Ciona*-specific peptides: Ci-YFV/Ls and Ci-LFs (Table 5.1). These have not ever been found in any other animal species and conserve a consensus sequence. These findings highlight both the molecular and functional evolution and diversity of neuropeptides of *Ciona* in chordates.

Table 5.2 Cionin and cholecystokinin/gastrin family peptides

Species	Peptide	Sequence
Deuterostome invertebrate		
Ciona intestinalis	**cionin**	**NYYGWMDFamide**
Vertebrate		
Mammals	CCK	DYMGWMDFamide
	gastrin	pQGPWLEEEEEAYGWMDFamide
Protostome		
fruit fly *Drosophila melanogaster*)	Dsk1	FDDYGHMRFamide
nematode (*Caenorhabditis elegans*)	NLP12	DGYRPLQFamide

Sulfated tyrosines are indicated by double line. The ascidian peptide (cionin) is indicated in boldface. The CCK/gastrin C-terminal consensus motif is shaded

5.2.1 Ciona *Cholecystokinin, Cionin*

Cholecystokinins (CCKs) are vertebrate brain/gut peptides and participate in the regulation of digestive functions (Johnsen 1998). Moreover, these peptides conserve the C-terminal consensus sequence Trp-Met-Asp-Phe-NH_2 (Table 5.2), which is critical for activities of CCKs. Furthermore, one sulfated Tyr is located at position 7 from the C-terminus of CCK. Two class A G protein-coupled receptors (GPCRs) for CCK/gastrin family peptides, CCK-A and -B, have been identified and shown to elicit intracellular calcium ion (Johnsen 1998).

Cionin is the first neuropeptide that was isolated from ascidians (Johnsen and Rehfeld 1990). Cionin completely conserves the CCK/gastrin C-terminal consensus motif and harbors two sulfated Tyr residues. These structural propensities indicate that cionin is an authentic member of the CCK/gastrin family (Kawada et al. 2010; Matsubara et al. 2016; Janssen et al. 2008). In insects and nematodes, an analogous sulfakinins were characterized. These peptides possess a sulfated tyrosine but different C-terminal consensus and do not bind to CCK/gastrin receptors (Kawada et al. 2010; Matsubara et al. 2016). Thus, sulfakinins are thought to have evolved in the different lineages from CCK and cionin.

Two cionin receptors, CioR1 and CioR2, were identified. Cionin was shown to activate intracellular calcium mobilization at both of the receptors to the same extent (Nilsson et al. 2003; Sekiguchi et al. 2012). Moreover, both of the two sulfated Tyr residues were shown to be essential for full activity via CioRs (Sekiguchi et al. 2012). Molecular phylogenetic tree analysis demonstrated that CioRs were generated from the common ancestor of vertebrate CCK gastrin receptors and that CioR1 and CioR2 were duplicated in the *Ciona*-specific lineage (Sekiguchi et al. 2012). CioRs are expressed in the neural complex, digestive organs, oral/atrial siphons, and ovary. The phylogenetic position of ascidians and the homologous sequence of cionin suggest that the CCK/gastrinergic system was established in ascidians.

5.2.2 GnRH

GnRHs are hypothalamic hormone peptides that stimulate release of gonadotropins from the pituitary. Gonadotropins trigger the steroidogenesis and stimulate gonadal maturation in vertebrates (Millar et al. 2004; Millar 2005; Millar and Newton 2013), and thus, GnRHs play crucial roles in the regulation of the reproductive systems in diverse neuroendocrine fashion. Furthermore, this hypothalamic hormone serves as a paracrine, autocrine, and neurotransmitter/neuromodulatory factors (Millar et al. 2004; Millar 2005; Millar and Newton 2013). GnRHs or their related peptides have also been identified in mollusks, echinoderms, and ascidians as well as nonmammalian vertebrates (Table 5.3). Gnathostomate GnRHs are all composed of ten amino acids and share the consensus sequence of pyro-Glu^1-His/Tyr^2-Trp^3-Ser^4, Gly^6, and Pro^9-Gly^{10}-amide (Millar et al. 2004; Millar 2005; Kah et al. 2007; Tello and Sherwood 2009; Kawada et al. 2009a, b; Millar and Newton 2013; Sakai et al. 2017). GnRH1 and -2 were characterized in all vertebrates, and GnRH3 was the teleost-specific paralog (Millar et al. 2004; Millar 2005; Millar and Newton 2013). In lamprey, one GnRH ortholog, l-GnRH-II, and two lamprey-specific paralogs, l-GnRH-I and -III, were identified (Kavanaugh et al. 2008).

In protostomes and echinoderms, GnRH-like peptides are composed of 11 to 12 amino acids and possess two additional amino acid residues after position 1 (Table 5.3). The C-terminal Pro-Gly-amide in vertebrate GnRHs is replaced with Pro/Ser/Ala-amide in non-cephalopod (octopus and squid) GnRHs, whereas this consensus motif is found in echinoderm GnRHs (Table 5.3). In contrast, these GnRHs conserve Gly^8 that corresponds to Gly^6 in vertebrate GnRHs (Table 5.3), which is essential for the adaption of active structure (Millar et al. 2004; Millar 2005). A single GnRH sequence is encoded in vertebrates, echinoderms, and protostomes (Millar et al. 2004; Millar 2005; Kah et al. 2007; Kawada et al. 2009a, b; Millar and Newton 2013; Roch et al. 2014; Semmens et al. 2016; Sakai et al. 2017). Recently, other protostome peptides, corazonin and adipokinetic hormones, are thought to be orthologous to the GnRH family due to the presence of several amino acids typical of GnRHs and sequence similarity among their receptors and GnRH receptors, but the evolutionary relationship remains controversial (Sakai et al. 2017).

In ascidians, 11 GnRHs have so far been characterized: t-GnRH1 and -2 in *Chelyosoma productum* (Powell et al. 1996), t-GnRH3 to -8 in *Ciona intestinalis*, t-GnRH-5 to -9 in *Ciona savignyi* (Adams et al. 2003), and t-GnRH10 and 11 in *Halocynthia roretzi* (Hasunuma and Terakado 2013). These ascidian GnRHs share the complete vertebrate consensus sequences, pyro-Glu-His-Trp-Ser and Pro-Gly-amide (Table 5.3). The *Ciona* possesses two GnRH genes, *Ci-gnrh-1* and *-2*. Interestingly, these two *Ciona* GnRH genes encode three different GnRH peptide sequences (Adams et al. 2003). t-GnRH-3, -5, and -6 are encoded in *Ci-gnrh-1*, and t-GnRH-4, -7, and -8 are encoded in *Ci-gnrh-2*. Such structural organization of *C. intestinalis* GnRH precursors was also found in *Cs-gnrh-1* and *-2* of *C. savignyi* (Adams et al. 2003). In *Halocynthia roretzi*, the *H. roretzi* GnRH gene encodes

Table 5.3 GnRH family peptides

Species	Peptide	Sequence
Invertebrate chordate		
Chelyosoma productum	**t-GnRH-1**	**pQHWSDYFKPGa**
	t-GnRH-2	**pQHWSLCHAPGa**
Ciona intestinalis	**t-GnRH-3**	**pQHWSYEFMPGa**
	t-GnRH-4	**pQHWSNQLTPGa**
	t-GnRH-5	**pQHWSYEYMPGa**
	t-GnRH-6	**pQHWSKGYSPGa**
	t-GnRH-7	**pQHWSYALSPGa**
	t-GnRH-8	**pQHWSLALSPGa**
Ciona savignyi	**t-GnRH-9**	**pQHWSNKLAPGa**
Ciona intestinalis	**Ci-GnRH-X**	**pQHWSNWWIPGAPGYNGa**
Halocynthia roretzi	**t-GnRH-10**	**pQHWSYGFSPGa**
	t-GnRH-11	**pQHWSYGFLPGa**
amphioxus (*Branchiostoma floridae*)	Amph.GnRHv	pQEHWQYGHWYa
	Amph.GnRH	pQILCARAFTYTHTWa
Echinoderms		
sea urchin (*Strongylocentrotus purpuratus*)	Sp-GnRHP	pQVHHRFSGWRPGa
starfish (*Asterias rubens*)	Ar-GnRH	pQIHYKNPGWGPGa
Vertebrate		
Homo sapiens (human)	GnRH1	pQHWSYGLRPGa
	GnRH2	pQHWSHGWYPGa
Cavia porcellus (guinea pig)	GnRH1	pQHWSYGVRPGa
Oncorhynchus mykiss (rainbow trout)	GnRH3	pQHWSYGWLPGa
Petromyzon marinus (sea lamprey)	l-GnRH-I	pQHYSLEWKPGa
	l-GnRH-II	pQHWSHGWFPGa
	l-GnRH-III	pQHWSHDWKPGa
Protostome		
octopus (*Octopus vulgaris*), cuttlefish (*Sepia officinalis*) and swordtip squid (*Loligo edulis*)	Oct-GnRH	pQNYHFSNGWHPGa
oyster (*Crassostrea gigas*)	Cg-GnRH	pQNYHFSNGWQPa
yesso scallop (*Patinopecten yessoensis*)	Py-GnRH	pQNFHYSNGWQPa
sea hare (*Aplysia californica*)	Ap-GnRH	pQNYHFSNGWYAa
marine worm (*Capitella teleta*)	Ca-GnRH	pQAYHFSHGWFPa
leech (*Helobdella robusta*)	Hr-GnRH	pQSIHFSRSWQPa

The N-terminal pyroglutamic acid and C-terminal amide are shown by "pQ" and "a," respectively. The ascidian peptides are indicated in boldface. Chordate GnRH consensus motifs are shaded

t-GnRH-10 and -11 (Hasunuma and Terakado 2013). These findings indicate unique molecular diversity of GnRHs in ascidians. Furthermore, another *Ciona*-specific GnRH-like peptide, Ci-GnRH-X (Kawada et al. 2009a, b), was also identified in the neural complex. This peptide shares the N-terminal pGlu-His-Trp-Ser and a C-terminal amidated Gly. In contrast, Ci-GnRH-X is composed of 16 amino acids and lacks the common Pro at position 2 from the C-terminus of other chordate GnRHs (Table 5.3). Ci-GnRH-X is encoded in the *t-gnrh-X* gene as a single copy (Kawada et al. 2009a, b), like non-chordate GnRHs. Collectively, ascidian GnRHs might have evolved in the unique lineage.

Vertebrate GnRH receptors (GnRHRs) induce elevation of intracellular calcium ion (Millar et al. 2004; Millar 2005; Kah et al. 2007; Kawada et al. 2009a, b, 2013; Millar and Newton 2013; Sakai et al. 2017). Type-I GnRHRs bind to both GnRH1 and -2, whereas type-II GnRHRs are interacted with only GnRH2. The octopus GnRHR induces mobilization of intracellular cellular calcium ions upon binding to the endogenous ligand, oct-GnRH, and was the only protostome GnRH receptor that was shown to be activated by the cognate ligand (Kanda et al. 2006; Kawada et al. 2013; Sakai et al. 2017). Ar-GnRH, identified in the starfish, *Asterias rubens*, was also found to induce intracellular calcium mobilization via the cognate receptor, ArGnRHR (Tian et al. 2016).

C. intestinalis was shown to possess four GnRH receptors, Ci-GnRHR-1, -2, -3, and -4 (Kusakabe et al. 2003; Tello et al. 2005; Tello and Sherwood 2009). All Ci-GnRHRs displays 30% identity to human Type I GnRHR, and Ci-GnRHR-1 is 70%, 38%, and 36% similar to Ci-GnRHR-2, -3, and -4, respectively, (Kusakabe et al. 2003; Tello et al. 2005; Tello and Sherwood 2009). Ci-GnRHRs were found to be multiplicated in the *Ciona*-specific lineages (Kusakabe et al. 2003; Tello et al. 2005). These receptors show unique ligand selectivity and second-messenger generation. Intracellular calcium mobilization is observed only upon binding of t-GnRH-6 to Ci-GnRHR-1 (Tello et al. 2005). t-GnRH6, -7, and -8 upregulate the cAMP production via Ci-GnRHR-2 (Tello et al. 2005). Ci-GnRHR-3 activates cAMP production in response to t-GnRH3 and -5 (Tello et al. 2005). Ci-GnRHR-4 is devoid of generating any second messengers (Tello et al. 2005) but is involved in unique fine-tuning of t-GnRH signaling via heterodimerization of other subtypes. Heterodimerization of Ci-GnRHR4 with Ci-GnRHR-1 leads to the t-GnRH6-specific elevation of intracellular calcium, calcium-dependent and calcium-independent protein kinase C, and ERK phosphorylation (Sakai et al. 2010). In contrast, heterodimerization of Ci-GnRHR-4 with Ci-GnRHR-2 reduces cAMP production by 50% in response to t-GnRH-7 and -8 via shifting from Gs-coupling to Gi-coupling (Sakai et al. 2012). These findings indicate that Ci-GnRHR-4 functions not as a ligand-binding receptor but a protomer of GPCR heterodimers (Satake et al. 2013; Matsubara et al. 2016), suggesting unique molecular and functional diversification of GnRHs and GnRHRs in the ascidian-specific lineage.

Ci-GnRH-X moderately (10–50%) inhibits the elevation of the intracellular calcium and cAMP production by t-GnRH-6 at Ci-GnRHR-1 and cAMP generation by t-GnRH-3 and -5 at Ci-GnRHR-3, whereas Ci-GnRHR-2 was not affected (Kawada et al. 2009a). Such inhibitory GnRH peptides have not ever been found in any other species (Kawada et al. 2009a). Taken together, these findings suggest that both GnRHs and GnRHRs might have coevolved through the complexed and unique evolutionary pathways.

t-GnRH-3 and -5 were found to enhance the spawning of sperm and oocytes (Adams et al. 2003) and to arrest the growth of adult organs by suppression of cell cycle and tail absorption in the *Ciona* larva (Kamiya et al. 2014). Combined with the fact that *C. intestinalis* has not acquired the pituitary and gonadotropins, these findings suggest that t-GnRHs are involved in developmental processes and sexual behavior and that ancestral GnRH might have been responsible for the regulation of non-gonadotropic functions.

5.2.3 Tachykinin

Tachykinins (TKs) are vertebrate brain/gut peptides involved in a wide variety of biological functions such as smooth muscle contraction, vasodilation, nociception, inflammation, neurodegeneration, and neuroprotection (Millar and Newton, 2013; Steinhoff et al. 2014). In mammals, substance P (SP), neurokinin A (NKA), neurokinin B (NKB), and homeokinin-1/endokinins (HK-1/EKs) are the major TK family peptides (Table 5.4). The respective TKs are encoded by three genes: *TAC1* (or *PPTA*), *TAC3* (or *PPTB*), or *TAC4* (*PPTC*) gene (Satake and Kawada 2006b; Page 2006; Steinhoff et al. 2014). *TAC1* generates four splicing variants that yield only SP or SP and NKA (Satake and Kawada 2006b; Page 2006; Steinhoff et al. 2014). The tetrapod *TAC3* gene encodes only NKB (Satake and Kawada 2006b; Page 2006; Steinhoff et al. 2014), whereas the fish *TAC3* gene encodes an additional TK peptide, NKF. HK-1/EKs are liberated from the *TAC4* gene (Page 2006; Steinhoff et al. 2014). As shown in Table 5.4, vertebrate TKs conserve the C-terminal common sequence, -Phe-X-Gly-Leu-Met-NH_2.

TKs are interacted with specific class A GPCRs, NK1, NK2, and NK3 in vertebrates (Satake and Kawada 2006b; Page 2006; Steinhoff et al. 2014). All TK receptors induce both elevation of intracellular calcium and cAMP production in response to TK ligands in a moderate ligand-selective fashion: SP > NKA > NKB for NK1, NKA > NKB > SP for NK2, and NKB > NKA > SP for NK3, respectively (Satake and Kawada 2006b; Page 2006; Steinhoff et al. 2014). HK/EK elicited the highest selectivity to NK1 (Satake and Kawada 2006b; Page 2006; Steinhoff et al. 2014). In teleost, NKF displays the highest selectivity to NK3. The molecular phylogenetic tree demonstrates that all NK subtypes were generated in the common ancestors of vertebrates and are conserved during the evolutionary process of vertebrates.

Authentic TKs have not ever been found in protostomes. Instead, various TK-related peptides (TKRPs) have been identified in the central nervous system of protostomes (Satake et al. 2003, 2013; Satake and Kawada 2006b; Jiang et al. 2013). TKRPs shares the Phe-Xaa1-(Gly/Ala/Pro)-Xaa2-Arg-NH_2 consensus (Table 5.4) that is analogous to that of vertebrate TKs. TKRP genes encode multiple (five to ten) TKRP sequences, which is distinct from those of vertebrate TK genes encoding one or two TK sequences (Satake et al. 2003, 2013; Satake and Kawada 2006b; Jiang et al. 2013). These features suggest that TKRP and TK genes might have evolved in the respective evolutionary lineages.

Table 5.4 Tachykinin family peptides

Species	Peptide	Sequence
Ascidian tachykinins		
Ciona intestinalis	**Ci-TK-I**	**HVRHFYGLMa**
	Ci-TK-II	**SIGDQPSIFNERASFTGLMa**
Vertebrate tachykinins		
Mammals	Substance P	RPKPQQFFGLMa
	Neurokinin A	HKTDSFVGLMa
	Neurokinin B	DMHDFVGLMa
Rodents (*Rattus norvegicus* and *Mus musculus*)	Hemokinin-1	SRTRQFYGLMa
Homo sapiens	Endokinin A/B	GKASQFFGLMa
	Endokinin C	KKAYQLEHTFQGLLa
	Endokinin D	VGAYQLEHTFQGLLa
zebrafish (*Danio rerio*)	Neurokinin F	YNDIDYDSFVGLMa
Protostome tachykinin-related peptides (TKRPs)		
echiuroid worm (*Urechis unicinctus*)	Uru-TK-I	LRQSQFVGARa
locust (*Locusta migratoria*)	Lom-TK-I	GPSGFYGVRa
fruit fly (*Drosophila melanogaster*)	DTK-1	APTSSFIGMRa
	Natalisin-1	EKLFDGYQFGEDMSKENDPFIGPRa

C-terminal amide is shown by "a." The ascidian peptides are indicated in boldface. The chordate TK consensus motif and its related sequences are underlined by plain and dotted lines, respectively. The protostome TKRP consensus motif and its related sequence are indicated by wavy lines. Tachykinins and TKRPs consensus motifs are shaded

TKRP receptors were characterized from several protostomes (Satake et al. 2003, 2013; Satake and Kawada 2006b; Jiang et al. 2013). TKRP receptors display 30–40% sequence identity to NKs, and the molecular phylogenetic tree demonstrated that TK receptors and TKRP receptors originated from the common ancestral gene (Satake et al. 2003, 2013; Satake and Kawada 2006b; Jiang et al. 2013). TKRP receptors, like NKs, induce intracellular calcium or generation of cAMP in response to cognate ligands (Satake et al. 2003, 2013; Satake and Kawada 2006b; Jiang et al. 2013). Moreover, TKRP receptors show redundant ligand selectivity to TKRPs (Satake et al. 2003, 2013; Satake and Kawada 2006b; Jiang et al. 2013).

In *C. intestinalis*, Ci-TK-I and Ci-TK-II have been isolated from the neural complex (Satake et al. 2004; Kawada et al. 2011). Ci-TKs possess the vertebrate TK consensus (Table 5.4), and Ci-TKs, unlike TKRPs, exert a TK-typical contraction of the mammalian smooth muscle (Satake et al. 2004). Furthermore, the Ci-TK precursor gene encodes Ci-TK-I and Ci-TK-II, which is similar to *Tac1* (Satake et al. 2003; Satake and Kawada 2006b). However, Ci-TK-I and -II are encoded in the same exon, indicating that no alternative production of TKs via generation of splicing variants occurs in *Ciona* (Satake et al. 2004. Consequently, alternative production of TKs is likely to have been established during the evolution of vertebrates.

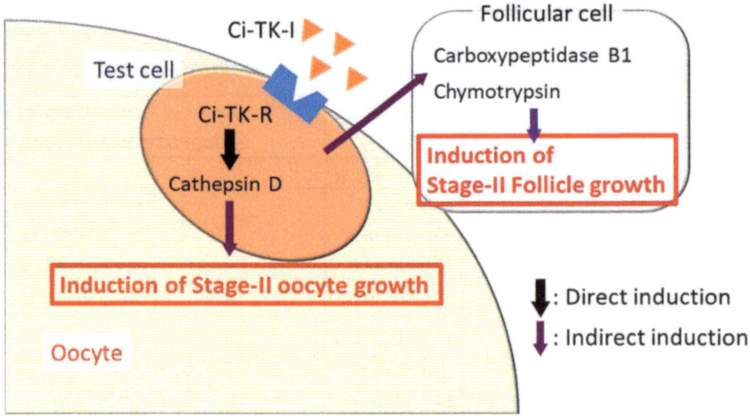

Fig. 5.1 Follicle growth induced by Ci-TK. Ci-TK-I acts on the cognate receptor, Ci-TK-R expressed on the plasma membrane of test cells, and upregulates the gene expression of cathepsin D in test cells and that of carboxypeptidase B1 and chymotrypsin in follicle cells. Activation of these proteases eventually leads to the growth of vitellogenic follicles

Ci-TK-R is the only endogenous Ci-TK receptor in *C. intestinalis*. Ci-TK-R displayed 30–43% sequence identity to vertebrate NKs (Satake et al. 2003, 2004; 2013; Satake and Kawada 2006b; Jiang et al. 2013; Steinhoff et al. 2014). The molecular phylogenetic tree indicated that Ci-TK-R is included in the vertebrate TK receptor clade, not in the protostome TKRPR clade (Satake et al. 2003, 2004; 2013; Satake and Kawada 2006b; Jiang et al. 2013; Steinhoff et al. 2014). Ci-TK-I induced intracellular calcium mobilization at Ci-TK-R (Satake et al. 2004).

The Ci-TK gene is expressed in the neural complex, unidentified small cells in the intestine, and secretory cells in the endostyle. The Ci-TK-R gene expression was observed abundantly in the brain, intestine, and test cells in the ovary (Satake et al., 2004). Notably, as shown in Fig. 5.1, Ci-TK-I specifically enhances the growth of vitellogenic follicles via upregulation of gene expression and the resultant enzymatic activities of cathepsin D, chymotrypsin, and carboxypeptidase B1 (Aoyama et al. 2008, 2012; Matsubara et al. 2016). Of interest is whether this TKergic follicle growth is conserved in vertebrates.

5.2.4 *Oxytocin/Vasopressin*

Oxytocin (OT) and vasopressin (VP) were characterized as neurohypophysial hormones and thereafter were found to serve as neuropeptides. The OT/VP family peptides share Cys^1, Asn^5, Cys^6, Pro^7, Gly^9, and C-terminal amidation. The disulfide bridge between the two cysteines is a prerequisite for activation of the cognate receptors (Table 5.5). The VP family peptides are featured by a basic amino acid (Lys or Arg) at position 8, whereas a neutral aliphatic one (Leu, Ile, Val, or Thr) is

Table 5.5 Oxytocin/vasopressin family peptides

Species	Peptide	Sequence
Deuterostome invertebrate		
Ciona intestinalis	**Ci-VP**	**CFFRDCSNMDWYR**
Styela plicata	**SOP**	**CYISDCPNSRFWSTa**
amphioxus (*Branchiostoma floridae*)	[Ile[4]]-vasotocin	CYIINCPRGa
sea urchin (*Strongylocentrotus purpuratus*)	echinotocin	CFISNCPKGa
Vertebrate		
Mammals	oxytocin	CYIQNCPLGa
	Arg-vasopressin	CYFQNCPRGa
	Lys-vasopressin	CYFQNCPKGa
Non-mammalian tetrapod, lungfish, etc.	mesotocin	CYIQNCPIGa
Teleost	isotocin	CYISNCPIGa
Dogfish (*Squalus acanthias*)	valitocin	CYISNCPVGa
	aspargtocin	CYINNCPLGa
Non-mammalian vertebrates	vasotocin	CYIQNCPRGa
Protostome		
leech (*Erpobdella octoculata*), sea hare (*Aplysia kurodai*)	Lys-conopressin	CFIRNCPKGa
earthworm (*Eisenia foetida*)	annetocin	CFVRNCPTGa
octopus (*Octopus vulgaris*)	cephalotocin	CYFRNCPIGa
	octopressin	CFWTSCPIGa
locust (*Locusta migratoria*), red flour beetle (*Tribolium castaneum*)	inotocin	CLITNCPRGa
nematode (*Caenorhabditis elegans*)	nematocin	CFLNSCPYRRYa

"a" denotes C-terminal amide. The ascidian peptides (Ci-VP and SOP) are indicated in boldface. Consensus cysteine residues are shaded

located at this position in vertebrates. They have so far been identified from a great variety of animal species from protostomes to mammals (Hoyle 1998; Satake et al. 1999; Kawada et al. 2008, 2009b; Stafflinger et al. 2008; Aikins et al. 2008; Elphick and Rowe 2009; Beets et al. 2012). VP plays a central role in osmoregulation, control of blood pressure, and anti-diuretic effect (Gimpl and Fahrenholz 2001; Frank and Landgraf 2008), and OT is responsible for reproductive functions including uterine contraction and milk ejection (Gimpl and Fahrenholz 2001; Kawada et al. 2009b). OT and VP are involved in the molecular mechanisms underlying learning, social behavior, anxiety, and autism (Gimpl and Fahrenholz 2001; Frank and Landgraf 2008). All gnathostomes possess one VP family peptide and one OT family peptide, whereas only a single OT/VP superfamily peptide has ever been found in agnathans and all invertebrates except octopus (Hoyle 1998; Kawada et al. 2009b). These findings suggested that OT and VP family might have occurred via gene duplication of the common ancestral gene during the evolution from agnathans to gnathostomes. As shown in Table 5.5, the OT/VP superfamily peptides have also been found in protostomes and non-ascidian deuterostome invertebrates.

Both of their precursors share structural organization comprising a signal peptide, an OT or VP sequence flanked by a glycine C-terminal amidation signal and dibasic endoproteolytic site, and a neurophysin containing 14 conserved cysteines (Hoyle 1998; Satake et al. 1999; Kawada et al. 2008, 2009b; Aikins et al. 2008; Stafflinger et al. 2008; Elphick and Rowe 2009; Beets et al. 2012). Seven intramolecular disulfide pairs by the 14 cysteines are responsible for adoption of the functional structure to bind to an OT/VP peptide (Hoyle 1998; Satake et al. 1999; Kawada et al. 2008, 2009b; Aikins et al. 2008; Stafflinger et al. 2008; Elphick and Rowe 2009; Beets et al. 2012). This structural organization is also conserved in invertebrates except for *C. intestinalis* counterparts.

The OT/VP peptide receptors display high sequence similarity with one another, indicating that they originated from a common ancestor (Hoyle 1998; Gimpl and Fahrenholz 2001; Frank and Landgraf 2008). Three VP receptor subtypes (V1aR, V1bR, and V2R) and one OT receptor (OTR) have been identified in mammals. V1aR, V1bR, and OTR induce the intracellular calcium ions, whereas V2R enhances the cAMP production (Hoyle 1998; Gimpl and Fahrenholz 2001; Frank and Landgraf 2008).

The OT/VP superfamily peptides were identified in two different ascidian species. *C. intestinalis* VP (Ci-VP) is the first OT/VP family peptide from ascidians (Kawada et al. 2008), and SOP was identified in a different ascidian, *Styela plicata* (Ukena et al. 2008). Unlike any other OT/VP superfamily peptides, Ci-VP and SOP are comprised of 13 and 14 amino acids, respectively (Table 5.5). Moreover, the C-terminus of Ci-VP lacks amidation (Table 5.5). Instead, the N-terminal circular region of Ci-VP and SOP is largely conserved (Table 5.5). The Ci-VP precursor also encodes an incomplete neurophysin that contains only ten cysteines (Kawada et al. 2008). In addition, the 13-residue and C-terminally non-amidated Ci-VP peptide sequence and the 10-cysteine neurophysin region were detected in the genome of *Ciona savignyi*, which is phylogenetically close to *C. intestinalis*. These findings indicate the ascidian-specific molecular evolution and divertification of the OT/VP superfamily.

Ci-VP-R is a single endogenous Ci-VP-receptor and displays 35–56% sequence similarity to other receptors for OT/VP superfamily peptide receptors. Ci-VP induces intracellular calcium mobilization at Ci-VP-R (Kawada et al. 2008). Although no biological roles of Ci-VP have been documented, Ci-VP-R is expressed in the neural complex, digestive tract, endostyle, and gonad (Kawada et al. 2008). These expression profiles for Ci-VP-R suggest various functions of Ci-VP in *Ciona*. Instead, the upregulation of SOP mRNA in hypotonic seawater and stimulation of the siphon contraction were observed in *Styela plicata* (Ukena et al. 2008).

Table 5.6 The CT/CGRP superfamily peptides

Species	Peptide	Sequence
Deuterostome invertebrate		
Ciona intestinalis	**Ci-CT**	**---------------CDGVSTCWLHELGNSVHATAGGKQNVGFGPa**
Branchiostoma floridae (amphioxus)	Bf-CTFP1	--------------DCSTLTCFNQKLAHELAMDNQRTDTANPYSPa
	Bf-CTFP2	-----------GKIACKTAWCMNNRLSHNLSSLDNPTDTGVGAPa
	Bf-CTFP3	--------------KCESGTCVQMHLADRLRLGLGHNMFTNTGPESPa
Vertebrate		
Mammals	CT	---------------CGNLSTCMLGTYTQDFNKFHTFPQTAIGVGAPa
	CGRP	--------------ACDTATCVTHRLAGLLSRSGGVVKNNFVPTNVGSKAFa
	Adrenomedullin	YRQSMNNFQGLRSFGCRFGTCTVQKLAHQIYQFTDKDKDNVAPRSKISPQGYa
	Amylin	--------------KCNTATCATQRLANFLVHSSNNFGAILSSTNVGSNTYa
pig	CRSP	--------------SCNTATCMTHRLVGLLSRSGSMVRSNLLPTKMGFKVFGa

"a" denotes C-terminal amide. The ascidian peptide (Ci-CT) is indicated in boldface. Consensus motif is underlined and consensus cysteine residues are shaded

5.2.5 *Calcitonin*

Calcitonin (CT) is released from various tissues including the C cells of the thyroid gland in mammals or the ultimobranchial gland in nonmammalian vertebrates expect cyclostomes (Conner et al. 2004). CTs play a central role in calcium metabolism via suppression of osteoclasts in the bones and teleost scales. Vertebrate CTs are composed of 32 amino acids and conserve a C-terminally amidated Pro and N-terminal circular structure formed by a disulfide bridge between Cys^1 and Cys^7. In vertebrates, the CT family (Table 5.6) includes CT, CT gene-related peptide (CGRP), amylin (AMY), adrenomedullin (AM), and CT receptor-stimulating peptide (CRSP) (Katafuchi et al. 2003, 2009; Conner et al. 2004). CGRPs are yielded by the CT gene via alternative splicing (α-CGRP) and another CGRP gene (β-CGRP) in the central and peripheral neuron and act both as a neuropeptide and as a vasodilator (Conner et al. 2004). CRSP is conserved in several mammals and is likely to have originated from the CGRP gene (Katafuchi et al. 2009). AMY and AM are encoded by the respective genes (Conner et al. 2004).

The CT family peptides activate two class B GPCRs, CT receptor (CTR) and CTR-like receptor (CRLR) (Conner et al. 2004). Furthermore, three receptor activity-modifying proteins 1 to 3 (RAMP1–3) heterodimerize with CTR or CRLR, which is crucial for the ligand specificity of these receptors (Conner et al. 2004).

Intriguingly, several CT family peptides have been identified in invertebrate deuterostomes (Sekiguchi et al. 2009; Rowe and Elphick 2012; Rowe et al. 2014; Sekiguchi et al. 2016; Semmens et al. 2016; Suwansa-Ard et al. 2018; Cai et al. 2018; Sekiguchi 2018). *C. intestinalis* CT (Ci-CT) was cloned from the adult neural complex (Sekiguchi et al. 2009). Ci-CT shares the CT-typical disulfide bridge between Cys^1 and Cys^7 and the C-terminal proline amide (Table 5.6). However, unlike the CT gene in vertebrates, the Ci-CT gene generates no splicing variant and encodes only a Ci-CT sequence (Sekiguchi et al. 2009; Sekiguchi 2018). Furthermore, no orthologs to CGRP AM, AMY, CRSP, and β-CGRP genes were found in the *Ciona* genome (Sekiguchi et al. 2009; Sekiguchi 2018), indicating that *C. intestinalis* possesses Ci-CT as the sole *Ciona* CT/CGRP family peptide.

The expression of the Ci-CT gene in the endostyle and the neural gland (a nonneuronal ovoid body of spongy texture lying immediately ventral to the cerebral ganglion) suggests the roles of Ci-CT as a hormonal factor released from the neural gland and endostyle rather than as a neuropeptide in *C. intestinalis* (Sekiguchi et al. 2009). Interaction of Ci-CT with the putative endogenous receptor, Ci-CT-R, has not been proved due to the failure of heterologous expression of Ci-CT-R (Sekiguchi et al. 2009). Recently, CGRP-related peptides, Bf-CTFP1 to -3 (Table 5.6), were characterized from an amphioxus *Branchiostoma floridae* (Sekiguchi et al. 2016, 2017; Sekiguchi 2018). Interestingly, all Bf-CTFP genes are expressed in nervous tissues (Sekiguchi et al. 2016), indicating that Bf-CTFPs serve as neuropeptides, unlike Ci-CT as a nonneural factor. Moreover, a Bf-CTFP receptor (Bf-CTFP-R) and RAMP-like proteins (Bf-RAMP-LPs 1-3) were found in *B. floridae* (Sekiguchi et al. 2016; Sekiguchi 2018). Bf-CTFP-R was shown to heterodimerize with Bf-RAMP-LPs for translocation to the plasma membrane and binding to Bf-CTFPs (Sekiguchi et al. 2016). Bf-RAMP-LPs modify the potency of Bf-CTFP but do not affect ligand selectivity of Bf-CTFP-R (Sekiguchi et al. 2016; Sekiguchi 2018). These findings indicate that the CTergic system evolved during different processes between ascidian and amphioxus. However, Ci-CT and Bf-CTFPs exhibited suppression of osteoclastic activity typical of vertebrate CTs, but not of other CT family peptides (Sekiguchi et al. 2009, 2017; Sekiguchi 2018). These results suggest that Ci-CT and Bf-CTFPs share essential active conformation with vertebrate CTs.

5.2.6 Ciona-*Specific Peptides*

Peptidomics of the *C. intestinalis* neural complex have identified not only the aforementioned homologs of vertebrate neuropeptides and peptide hormones but also two novel peptides (Kawada et al. 2011). The *Ciona*-specific peptides, Ci-LF1 to -8 (Table 5.1), conserve the Leu-Phe sequence at their C-termini. The Ci-LF precursor gene encodes all Ci-LF sequences flanked by a typical mono-Arg or dibasic endoproteolytic sites at their both termini (Fig. 5.2a). Other novel peptides are Ci-YFV/Ls (Table 5.1). Ci-YFV1 to -3 and Ci-YFL1 possess the Tyr-Phe-Val or Tyr-Phe-Leu sequence at their C-termini. Moreover, all of these peptides are encoded by a single precursor gene, in which their sequences are flanked by a typical mono-Arg or dibasic endoproteolytic sites at their both termini (Fig. 5.2b). No orthologs of Ci-LFs or Ci-YFV/Ls have been found in any other organisms. Moreover, the Ci-LF gene and Ci-YFV/L gene were shown to be expressed in neurons in the neural complex (Kawada et al. 2011). Taken together, these findings provide

a Ci-LF precursor sequence

MKLVKSFSILAAFVVCYFGCIADAVPVDTVEKQLLRREGTGNPENFLDW
TNQLNSTDVDETDEELYDQLLYNILIPMQQKMIAEENGQLDNAEDNPFL
AENRNSDEENNDYSPGQAESIQSDKR**FQSLF**KR**YPGFQGLF**KR**HNPHLP**
DLFKR**YNSMGLF**KR**SPGMLGLF**KRGLLGLFKR**SDARLQGLF**KRDSATQG
SFKRSSEAQALPKR**YPNFQGLF**KRLSEATEYPEDDSSNDDTKQR**GNLHS**
LFKRDTSAHYLEDRGESIPFLFRRS

b Ci-YFV/L precursor sequence

MTPTRIFSLLVCVVLIARSNSKPIETSQLEDNVSKQDDYSSAIDTRSH
LFDRHFPVHSSEILRGPEDDDNNNNGNLATPYLRAFTHGAHDDPEP
SLEYEDLLKAINPLQNENTADTDTTKR**ELVVRDPYFV**KK**NNQESYFV**K
K**DDEPRSYFV**KK**DAARPNYYFL**KKAFPSKLFNSFGSSRACPCMKKRDN
SDEQNPPFGSLIKSLSANNPEAGLPSYLPKKTAGAFWSIDTPAWIASQ
NRHMQKSPQAMKMLEFN

Fig. 5.2 Sequences of Ci-LF precursors (**a**) and Ci-YFV/L precursors (**b**). Signal peptide regions are indicated in boldface. Peptide sequences are indicated in red, and endoproteolytic mono- or dibasic sites are underlined

evidence that Ci-LFs and Ci-YFV/Ls are *Ciona*-specific and novel neuropeptides. Exploration of their cognate receptors and biological functions are currently in progress.

5.3 Conclusion Remarks

As stated above, homologs and prototypes of typical vertebrate neuropeptides and peptide hormones, including CCKs, GnRHs, TKs, CT, VP, and galanin, were characterized from *C. intestinalis*. Moreover, their cognate receptors, such as CioRs, Ci-GnRHRs, Ci-TK-R, and Ci-VP-R, and their signaling pathways have also been explored. Sequence identity and molecular phylogenetic trees demonstrated that these *Ciona* peptides and receptors share common ancestors with vertebrates. Furthermore, species-specific peptides, including Ci-LFs and Ci-YFV/Ls, have been found in *C. intestinalis*. These findings, combined with the crucial position of ascidians as the closest relatives of vertebrates, highlight the advantages of *C. intestinalis* for research not only on neuropeptidergic and hormonal molecular mechanisms underlying various biological events but also on the pathways of the molecular and functional evolution and diversification of the endocrine, neuroendocrine, and nervous system in chordates. Although Ci-TK has been shown to induce the growth of vitellogenic (stage II) follicles via activation of several proteases, biological roles of other *Ciona* peptides have yet to be elucidated. Furthermore, many cognate receptors for *C. intestinalis*-specific peptides remain to be identified. Identification of receptors for these peptides and exploration of their biological functions will eventually pave the way for investigating the total peptidergic regulatory systems in ascidians and the evolution and diversification of the endocrine, neuroendocrine, and nervous system throughout chordates.

Acknowledgments All ascidians were provided by Kyoto University through the National Bio-Resource Project of the MEXT, Japan. This study was funded by the Japan Society for the Promotion of Sciences, grant 16K07430 to H.S.

References

Adams BA, Tello JA, Erchegyi J, Warby C, Hong DJ, Akinsanya KO, Mackie GO, Vale W, Rivier JE, Sherwood NM (2003) Six novel gonadotropin-releasing hormones are encoded as triplets on each of two genes in the protochordate, *Ciona intestinalis*. Endocrinology 144:1907–1919

Aikins MJ, Schooley DA, Begum K, Detheux M, Beeman RW, Park Y (2008) Vasopressin-like peptide and its receptor function in an indirect diuretic signaling pathway in the red flour beetle. Insect Biochem Mol Biol 38:740–748

Aoyama M, Kawada T, Fujie M, Hotta K, Sakai T, Sekiguchi T, Oka K, Satoh N, Satake H (2008) A novel biological role of tachykinins as an up-regulator of oocyte growth: identification of an evolutionary origin of tachykininergic functions in the ovary of the ascidian, *Ciona intestinalis*. Endocrinology 149:4346–4356

Aoyama M, Kawada T, Satake H (2012) Localization and enzymatic activity profiles of the proteases responsible for tachykinin-directed oocyte growth in the protochordate, *Ciona intestinalis*. Peptides 34:186–192

Beets I, Janssen T, Meelkop E, Temmerman L, Suetens N, Rademakers S, Jansen G, Schoofs L (2012) Vasopressin/oxytocin-related signaling regulates gustatory associative learning in *C. elegans*. Science 338:543–545

Cai W, Kim CH, Go HJ, Egertová M, Zampronio CG, Jones AM, Park NG, Elphick MR (2018) Biochemical, anatomical, and pharmacological characterization of calcitonin-type neuropeptides in starfish: discovery of an ancient role as muscle relaxants. Front Neurosci 12:382

Conner AC, Simms J, Hay DL, Mahmoud K, Howitt SG, Wheatley M, Poyner DR (2004) Heterodimers and family-B GPCRs: RAMPs, CGRP and adrenomedullin. Biochem Soc Trans 32:843–846

Elphick MR, Rowe ML (2009) NGFFFamide and echinotocin: structurally unrelated myoactive neuropeptides derived from neurophysin-containing precursors in sea urchins. J Exp Biol 212:1067–1077

Frank E, Landgraf R (2008) The vasopressin system—from antidiuresis to psychopathology. Eur J Pharmacol 583:226–242

Gimpl G, Fahrenholz F (2001) The oxytocin receptor system: structure, function, and regulation. Physiol Rev 81:629–683

Hasunuma I, Terakado K (2013) Two novel gonadotropin-releasing hormones (GnRHs) from the urochordate ascidian, *Halocynthia roretzi*: implications for the origin of vertebrate GnRH isoforms. Zool Sci 30:311–318

Horie T, Horie R, Chen K, Cao C, Nakagawa M, Kusakabe TG, Satoh N, Sasakura Y, Levine M (2018) Regulatory cocktail for dopaminergic neurons in a protovertebrate identified by whole-embryo single-cell transcriptomics. Genes Dev 32:1297–1302

Hoyle CHV (1998) Neuropeptide families: evolutionary perspectives. Regul Pept 73:1–33

Janssen T, Meelkop E, Lindemans M, Verstraelen K, Husson SJ, Temmerman L, Nachman RJ, Schoofs L (2008) Discovery of a cholecystokinin-gastrin-like signaling system in nematodes. Endocrinology 149:2826–2839

Jiang H, Lkhagva A, Daubnerová I, Chae HS, Šimo L, Jung SH, Yoon YK, Lee NR, Seong JY, Žitňan D, Park Y, Kim YJ (2013) Natalisin, a tachykinin-like signaling system, regulates sexual activity and fecundity in insects. Proc Natl Acad Sci USA 110:E3526–E3534

Johnsen AH (1998) Phylogeny of the cholecystokinin/gastrin family. Front Neuroendocrinol 19:73–99

Johnsen AH, Rehfeld JF (1990) Cionin: a disulfotyrosyl hybrid of cholecystokinin and gastrin from the neural ganglion of the protochordate *Ciona intestinalis*. J Biol Chem 265:3054–3058

Kah O, Lethimonier C, Somoza G, Guilgur LG, Vaillant C, Lareyre JJ (2007) GnRH and GnRH receptors in metazoa: a historical, comparative, and evolutive perspective. Gen Comp Endocrinol 153:346–364

Kamiya C, Ohta N, Ogura Y, Yoshida K, Horie T, Kusakabe TG, Satake H, Sasakura Y (2014) Nonreproductive role of gonadotropin-releasing hormone in the control of ascidian metamorphosis. Dev Dyn 243:1524–1535

Kanda A, Takahashi T, Satake H, Minakata H (2006) Molecular and functional characterization of a novel gonadotropin-releasing-hormone receptor isolated from the common octopus (*Octopus vulgaris*). Biochem J 395:125–135

Katafuchi T, Kikumoto K, Hamano K, Kangawa K, Matsuo H, Minamino N (2003) Calcitonin receptor-stimulating peptide, a new member of the calcitonin gene-related peptide family. Its isolation from porcine brain, structure, tissue distribution, and biological activity. J Biol Chem 278:12046–12054

Katafuchi T, Yasue H, Osaki T, Minamino N (2009) Calcitonin receptor-stimulating peptide: its evolutionary and functional relationship with calcitonin/calcitonin gene-related peptide based on gene structure. Peptides 30:1753–1762

Kavanaugh SI, Nozaki M, Sower SA (2008) Origins of gonadotropin-releasing hormone (GnRH) in vertebrates: identification of a novel GnRH in a basal vertebrate, the sea lamprey. Endocrinology 149:3860–3869

Kawada T, Sekiguchi T, Itoh Y, Ogasawara M, Satake H (2008) Characterization of a novel vasopressin/oxytocin superfamily peptide and its receptor from an ascidian, *Ciona intestinalis*. Peptides 29:1672–1678

Kawada T, Aoyama M, Okada I, Sakai T, Sekiguchi T, Ogasawara M, Satake H (2009a) A novel inhibitory gonadotropin-releasing hormone-related neuropeptide in the ascidian, *Ciona intestinalis*. Peptides 30:2200–2205

Kawada T, Sekiguchi T, Sugase K, Satake H (2009b) Evolutionary aspects of molecular forms and biological functions of oxytocin family peptides. In: Jastrow H, Feuerbach D (eds) Handbook of oxytocin research: synthesis, storage and release, actions and drug forms. Nova Science, Hauppage, NY, pp 59–85

Kawada T, Sekiguchi T, Sakai T, Aoyama M, Satake H (2010) Neuropeptides, hormone peptides, and their receptors in *Ciona intestinalis*: an update. Zool Sci 27:134–153

Kawada T, Ogasawara M, Sekiguchi T, Aoyama M, Hotta K, Oka K, Satake H (2011) Peptidomic analysis of the central nervous system of the protochordate, *Ciona intestinalis*: homologs and prototypes of vertebrate peptides and novel peptides. Endocrinology 152:2416–2427

Kawada T, Aoyama M, Sakai T, Satake H (2013) Structure, function, and evolutionary aspects of invertebrate GnRHs and their receptors. In: Scott-Sills E (ed) Gonadotropin-releasing hormone (GnRH): production, structure and functions. Nova Science, New York, pp 1–16

Kusakabe T, Mishima S, Shimada I, Kitajima Y, Tsuda M (2003) Structure, expression, and cluster organization of genes encoding gonadotropin-releasing hormone receptors found in the neural complex of the ascidian *Ciona intestinalis*. Gene 322:77–84

Matsubara S, Kawada T, Sakai T, Aoyama M, Osugi T, Shiraishi A, Satake H (2016) The significance of *Ciona intestinalis* as a stem organism in integrative studies of functional evolution of the chordate endocrine, neuroendocrine, and nervous systems. Gen Comp Endocrinol 227:101–108

Millar RP (2005) GnRHs and GnRH receptors. Anim Reprod Sci 88:5–28

Millar RP, Newton CL (2013) Current and future applications of GnRH, kisspeptin and neurokinin B analogues. Nat Rev Endocrinol 9:451–466

Millar RP, Lu ZL, Pawson AJ, Flanagan CA, Morgan K, Maudsley SR (2004) Gonadotropin-releasing hormone receptors. Endocr Rev 25:235–275

Nilsson IB, Svensson SP, Monstein HJ (2003) Molecular cloning of a putative *Ciona intestinalis* cionin receptor, a new member of the CCK/gastrin receptor family. Gene 323:79–88

Osugi T, Sasakura Y, Satake H (2017) The nervous system of the adult ascidian *Ciona intestinalis* Type A (*Ciona robusta*): insights from transgenic animal models. PLoS One 12:e0180227

Page NM (2006) Characterization of the gene structures, precursor processing and pharmacology of the endokinin peptides. Vasc Pharmacol 45:200–208

Powell JF, Reska-Skinner SM, Prakash MO, Fischer WH, Park M, Rivier JE, Craig AG, Mackie GO, Sherwood NM (1996) Two new forms of gonadotropin-releasing hormone in a protochordate and the evolutionary implications. Proc Natl Acad Sci USA 93:10461–10464

Roch GJ, Tello JA, Sherwood NM (2014) At the transition from invertebrates to vertebrates, a novel GnRH-like peptide emerges in amphioxus. Mol Biol Evol 31:765–778

Rowe ML, Elphick MR (2012) The neuropeptide transcriptome of a model echinoderm, the sea urchin *Strongylocentrotus purpuratus*. Gen Comp Endocrinol 179:331–344

Rowe ML, Achhala S, Elphick MR (2014) Neuropeptides and polypeptide hormones in echinoderms: new insights from analysis of the transcriptome of the sea cucumber *Apostichopus japonicus*. Gen Comp Endocrinol 197:43–55

Sakai T, Aoyama M, Kusakabe T, Tsuda M, Satake H (2010) Functional diversity of signaling pathways through G protein-coupled receptor heterodimerization with a species-specific orphan receptor subtype. Mol Biol Evol 27:1097–1106

Sakai T, Aoyama M, Kawada T, Kusakabe T, Tsuda M, Satake H (2012) Evidence for differential regulation of GnRH signaling via heterodimerization among GnRH receptor paralogs in the protochordate, *Ciona intestinalis*. Endocrinology 153:1841–1849

Sakai T, Shiraishi A, Kawada T, Matsubara S, Aoyama M, Satake H (2017) Invertebrate gonadotropin-releasing hormone-related peptides and their receptors: an update. Front Endocrinol (Lausanne) 8:217

Satake H, Kawada T (2006a) Neuropeptides, hormones, and their receptors in ascidians: emerging model animals. In: Satake H (ed) Invertebrate neuropeptides and hormones: basic knowledge and recent advances. Transworld Research Network, Trivandrum, pp 253–276

Satake H, Kawada T (2006b) Overview of the primary structure, tissue-distribution, and functions of tachykinins and their receptors. Curr Drug Targets 7:963–974

Satake H, Takuwa K, Minakata H, Matsushima O (1999) Evidence for conservation of the vasopressin/oxytocin superfamily in Annelida. J Biol Chem 274:5605–5611

Satake H, Kawada T, Nomoto K, Minakata H (2003) Insight into tachykinin-related peptides, their receptors, and invertebrate tachykinins: a review. Zool Sci 20:533–549

Satake H, Ogasawara M, Kawada T, Masuda K, Aoyama M, Minakata H, Chiba T, Metoki H, Satou Y, Satoh N (2004) Tachykinin and tachykinin receptor of an ascidian, *Ciona intestinalis*: evolutionary origin of the vertebrate tachykinin family. J Biol Chem 279:53798–53805

Satake H, Matsubara S, Aoyama M, Kawada T, Sakai T (2013) GPCR heterodimerization in the reproductive system: functional regulation and implication for biodiversity. Front Endocrinol (Lausanne) 4:100

Satoh N (2015) Two decades of ascidian developmental biology: a personal research story. Curr Top Dev Biol 117:289–300

Sekiguchi T (2018) The calcitonin/calcitonin gene-related peptide family in invertebrate deuterostomes. Front Endocrinol (Lausanne) 9:695

Sekiguchi T, Suzuki N, Fujiwara N, Aoyama M, Kawada T, Sugase K, Murata Y, Sasayama Y, Ogasawara M, Satake H (2009) Calcitonin in a protochordate, *Ciona intestinalis*—the prototype of the vertebrate calcitonin/calcitonin gene-related peptide superfamily. FEBS J 276:4437–4447

Sekiguchi T, Ogasawara M, Satake H (2012) Molecular and functional characterization of cionin receptors in the ascidian, *Ciona intestinalis*: the evolutionary origin of the vertebrate cholecystokinin/gastrin family. J Endocrinol 213:99–106

Sekiguchi T, Kuwasako K, Ogasawara M, Takahashi H, Matsubara S, Osugi T, Muramatsu I, Sasayama Y, Suzuki N, Satake H (2016) Evidence for conservation of the calcitonin superfamily and activity-regulating mechanisms in the basal chordate *Branchiostoma floridae*: insights into the molecular and functional evolution in chordates. J Biol Chem 291:2345–2356

Sekiguchi T, Shiraishi A, Satake H, Kuwasako K, Takahashi H, Sato M, Urata M, Wada S, Endo M, Ikari T, Hattori A, Srivastav AK, Suzuki N (2017) Calcitonin-typical suppression of osteoclastic activity by amphioxus calcitonin superfamily peptides and insights into the evolutionary conservation and diversity of their structures. Gen Comp Endocrinol 246:294–300

Semmens DC, Mirabeau O, Moghul I, Pancholi MR, Wurm Y, Elphick MR (2016) Transcriptomic identification of starfish neuropeptide precursors yields new insights into neuropeptide evolution. Open Biol 6:150224

Sherwood NM, Tello JA, Roch GJ (2006) Neuroendocrinology of protochordates: insights from *Ciona* genomics. Comp Biochem Physiol A Mol Integr Physiol 144:254–271

Stafflinger E, Hansen KK, Hauser F, Schneider M, Cazzamali G, Williamson M, Grimmelikhuijzen CJ (2008) Cloning and identification of an oxytocin/vasopressin-like receptor and its ligand from insects. Proc Natl Acad Sci USA 105:3262–3267

Steinhoff MS, von Mentzer B, Geppetti P, Pothoulakis C, Bunnett NW (2014) Tachykinins and their receptors: contributions to physiological control and the mechanisms of disease. Physiol Rev 94:265–301

Suwansa-Ard S, Chaiyamoon A, Talarovicova A, Tinikul R, Tinikul Y, Poomtong T, Elphick MR, Cummins SF, Sobhon P (2018) Transcriptomic discovery and comparative analysis of neuropeptide precursors in sea cucumbers (Holothuroidea). Peptides 99:231–240

Tello JA, Sherwood NM (2009) Amphioxus: beginning of vertebrate and end of invertebrate type GnRH receptor lineage. Endocrinology 150:2847–2856

Tello JA, Rivier JE, Sherwood NM (2005) Tunicate gonadotropin-releasing hormone (GnRH) peptides selectively activate *Ciona intestinalis* GnRH receptors and the green monkey type II GnRH receptor. Endocrinology 146:4061–4073

Tian S, Zandawala M, Beets I, Baytemur E, Slade SE, Scrivens JH, Elphick MR (2016) Urbilaterian origin of paralogous GnRH and corazonin neuropeptide signaling pathways. Sci Rep 6:28788

Ukena K, Iwakoshi-Ukena E, Hikosaka A (2008) Unique form and osmoregulatory function of a neurohypophysial hormone in a urochordate. Endocrinology 149:5254–5261

Chapter 6
Emergence of Embryo Shape During Cleavage Divisions

Alex McDougall, Janet Chenevert, Benoit G. Godard, and Remi Dumollard

Abstract Cells are arranged into species-specific patterns during early embryogenesis. Such cell division patterns are important since they often reflect the distribution of localized cortical factors from eggs/fertilized eggs to specific cells as well as the emergence of organismal form. However, it has proven difficult to reveal the mechanisms that underlie the emergence of cell positioning patterns that underlie embryonic shape, likely because a systems-level approach is required that integrates cell biological, genetic, developmental, and mechanical parameters. The choice of organism to address such questions is also important. Because ascidians display the most extreme form of invariant cleavage pattern among the metazoans, we have been analyzing the cell biological mechanisms that underpin three aspects of cell division (unequal cell division (UCD), oriented cell division (OCD), and asynchronous cell cycles) which affect the overall shape of the blastula-stage ascidian embryo composed of 64 cells. In ascidians, UCD creates two small cells at the 16-cell stage that in turn undergo two further successive rounds of UCD. Starting at the 16-cell stage, the cell cycle becomes asynchronous, whereby the vegetal half divides before the animal half, thus creating 24-, 32-, 44-, and then 64-cell stages. Perturbing either UCD or the alternate cell division rhythm perturbs cell position. We propose that dynamic cell shape changes propagate throughout the embryo via cell-cell contacts to create the ascidian-specific invariant cleavage pattern.

A. McDougall · J. Chenevert · R. Dumollard (✉)
Laboratoire de Biologie du Développement de Villefranche-sur-mer (LBDV),
Sorbonne Université/CNRS, Villefranche-sur-Mer, France
e-mail: dumollard@obs-vlfr.fr

B. G. Godard
Institute of Science and Technology Austria, Klosterneuburg, Austria

W. Tworzydlo, S. M. Bilinski (eds.), *Evo-Devo: Non-model Species in Cell and Developmental Biology*, Results and Problems in Cell Differentiation 68,
https://doi.org/10.1007/978-3-030-23459-1_6

6.1 Introduction

Although we have acquired a great deal of knowledge about how genes control cell fate and axial patterning during embryonic development (Wieschaus and Nüsslein-Volhard 2016), less progress has been made in understanding how the shape of the early embryo is generated. Even though embryo shape is species-specific and thus encoded in the DNA, understanding the phenomenon of how shape emerges during embryonic development is complex and requires going beyond the role of one or many gene products, not least because cell and embryo shape are modulated at a higher level of organization and will require a systems-level or cross-disciplinary level approach. In embryos, cells can have different sizes; some cells can divide before others, while other groups of cells can divide in oriented ways during which time the cells remain adherent in the embryo. The interplay between all these factors influences the shape of the embryo. Moreover, at the cellular level, apicobasal polarity becomes apparent following cell division, while many maternal RNAs, protein products, granules, and organelles that are localized to specific regions of the oocyte and fertilized egg are often distributed to specific cells during early cleavage divisions, and some of these factors can in turn influence spindle position (Sardet et al. 1994; Weitzel et al. 2004).

When considering the regulation of cell and embryo shape, it should be borne in mind that DNA is subject to evolutionary change, while the physical forces acting upon cells that constitute embryos have remained unchanged. For example, cell adhesion, surface tension (for embryos, effective surface tension, manifested as an elastic property at the plasma membrane/cortex water interface leading to a minimization of surface area), and the viscoelastic properties (for embryos, the cytoplasm including the cortex: a material that exhibits both viscous and elastic characteristics when deformed) give shape to cells, embryos, and tissues and thus have to be taken into account (Heisenberg and Bellaïche 2013; Winklbauer 2015). Several proteins that regulate cell division orientation, cell shape, and cell adhesion have been used in manners that create different sizes and arrangements of cells in embryos during the evolution of the extant animal phyla, and these proteins can be considered molecular tool kits that are used in sometimes subtly different ways, creating the myriad array of embryonic and larval forms. These include many proteins (and indeed organelles) that organize the positioning of the mitotic spindle, proteins that influence cell cycle duration, as well as proteins involved in generating cell shape such as cell adhesion proteins and the actomyosin cortex and its regulators.

Another important consideration is what organism to choose to answer the biological question that is being posed. It is an old concept in biology that "the fortunate choice of animal often suffices to resolve general questions of the greatest importance": Claude Bernard (1865). This concept (plus the translation of Bernard's original phrase) is discussed in Cook et al. (2016), where the idea of model organisms is covered in great detail. In addition, by using the magnifying glass of evo-devo or comparative embryology, interesting conserved (and non-conserved) features are beginning to emerge by studying a wide array of invertebrates.

However, the progress of biology has greatly benefited from studying model organisms, those organisms that received intense focus creating genetic, genomic/transcriptomic, and molecular tools to dissect the principles and identify molecules that control some aspects of embryonic development (Gehring 1996; Wieschaus and Nüsslein-Volhard 2014). These organisms include representatives of the invertebrates, the fly *Drosophila melanogaster* and the worm *Caenorhabditis elegans* (members of the Ecdysozoa), and the vertebrates: *Xenopus laevis*, zebrafish, chicken, and mouse. Notwithstanding the progress made with canonical model organisms, embryonic development has also been studied in marine invertebrate organisms for more than a century, including members of the cnidarians (eumetazoans), the superphylum Lophotrochozoa (mollusks, annelids, and many more), echinoderms such as the sea urchin (a deuterostome), invertebrate chordates such as amphioxus (a cephalochordate), and the ascidian (a tunicate). Now, with the emergence of techniques for gene editing such as CRISPR/Cas9 coupled with the availability of transcriptomic and genomic data, techniques and genomic resources that were once restricted to the canonical model organisms are now accessible for many diverse marine invertebrate species (Cook et al. 2016). A brief example of some genomic and transcriptomic resources available is included below (not an exhaustive list):

Jellyfish, sea urchin, and amphioxus: http://marimba.obs-vlfr.fr/downloads/
Echinoderms: http://www.echinobase.org/Echinobase/
Amphioxus: https://genome.jgi.doe.gov/Brafl1/Brafl1.home.html
Ascidians: https://www.aniseed.cnrs.fr/ or http://ghost.zool.kyoto-u.ac.jp/cgi-bin/gb2/gbrowse/kh/

Finally, a number of computational models have been provided that explain cell positioning mechanisms in different embryos and common to these models are that cell adhesion and cortical sites influence spindle position. In nematodes, both the egg shell and attractive forces provided by cell adhesion can explain cellular position of different species at the four-cell stage where cells are aligned to form pyramid, diamond, T-shapes, or linear shapes (Yamamoto and Kimura 2017). A computational model of spiral cleavage depends on a set of rules including cell division polarized along the animal-vegetal axis, Sachs' rule (see definition later), cortical rotation, and cell adhesion (Brun-Usan et al. 2017). A mathematical model of the first four cleavage divisions for sea urchin embryos is based on cortical sites (Akiyama et al. 2010). In the ascidian, an apical cell shape computational model that integrates unequal cell division (UCD), asynchronous cell cycles, and apical cell shape can explain spindle positioning up to 64-cell stage (Dumollard et al. 2017).

6.1.1 The Importance of Cell Position During Early Embryogenesis

The position of cells during the earliest stages of embryogenesis is vitally important for subsequent events in many embryos. In some species of mollusk, the cleavage direction at the four- to eight-cell stage has been shown to influence the spiral direction of the adult shell, coiling to the right or left (Kuroda et al. 2009). Another important cue for setting up the ordered pattern of cell division is often provided by the primary axis of the egg (the animal vegetal axis), since this axis frequently prefigures the anterior–posterior axis of different larva (Peng et al. 2017). One reason for orchestrating cell division axes relative to the primary egg axis is that gastrulation-inducing factors, localized, for example, to the vegetal cortex of the fertilized egg, can be delivered to specific cells. These factors may be RNAs tethered to the cortex as in the ascidian (Prodon et al. 2007) or the protein Dishevelled that is localized to the egg vegetal cortex in the sea urchin (Weitzel et al. 2004; Peng and Wikramanayake 2013). These vegetal domains are segregated in a nonrandom manner to daughter cells, and many embryos position the first three cleavages such that at the eight-cell stage vegetal determinants are segregated to the vegetal tier of four cells (Nishida 1996; Nishida and Sawada 2001; Peng and Wikramanayake 2013). The three examples below highlight how the early cell division patterns can influence either the adult body shape or the developmental potential of the embryo:

1. If surface tension controlled cell placement in spiralian embryos at the eight-cell stage, we might expect the early embryo to look like an aggregate of soap bubbles, as proposed by Robert in 1903. But instead, the small bubbles do not align like micromeres do, so something is missing (Fig. 6.1a). Furthermore, recent investigations have revealed that a remarkable correlation exists between spindle orientation pattern, position of the first tier of four micromeres at the eight-cell stage, and the left- or right-hand spiral direction of the shell in the adult pond snail *Lymnaea stagnalis* (Shibazaki et al. 2004; Kuroda et al. 2009). Identification of the locus underlying this chiral behavior has led to the actin-binding protein diaphanous, one form of which is expressed maternally only in dextral animals, as likely playing a crucial role in creating the dominant or dextral form of the adult (Kuroda et al. 2016). Diaphanous is thought to create the kink in the cortex (Fig. 6.1b) that in turn alters spindle position creating a right-handed spiral rather than the left-handed one. Of course, cell shape, cell–cell contact, and UCD also play important roles.
2. Sea urchin embryos bisected at the eight-cell stage to create four blastomere animal or vegetal halves of the embryo behave quite differently from one another (Driesch 1892, 1908). For example, the animal blastomeres create an embryo that never gastrulates (a dauerblastula), while the vegetal blastomeres do gastrulate and create small but relatively normal larva (Fig. 6.1b). One reason for this difference is that the vegetal cortex is inherited by the vegetal blastomeres and

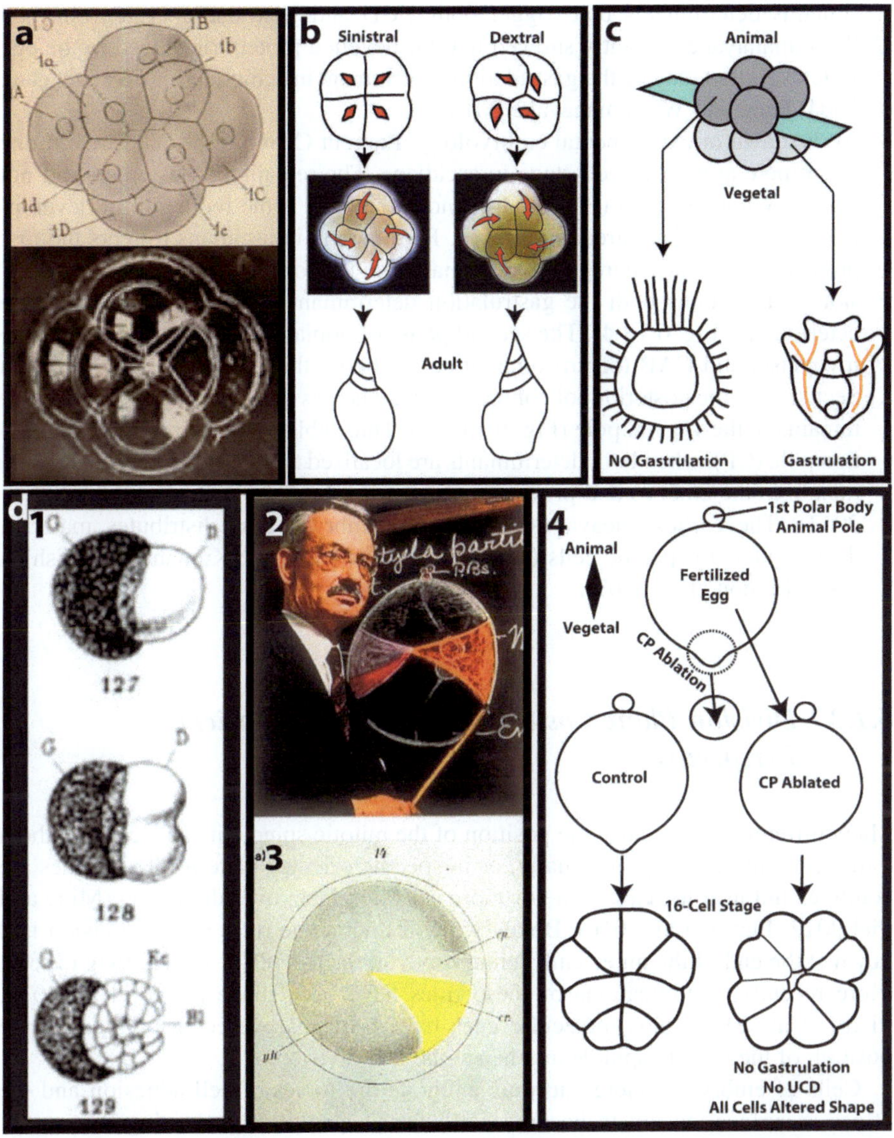

Fig. 6.1 Cell position, maternal determinants, and developmental potential. (**a**) Drawing and soap bubble pattern that resembles eight-cell stage spiralian (from Robert 1903). Note that upper small four bubbles do not align in grooves of lower four larger bubbles. (**b**) Pond snail. Spindle position (red) and cleavage orientation (red arrows) at the four- to eight-cell stage in pond snail blastomeres correlate with positioning of first four micromeres and ultimately with spiral direction of the adult shell. (**c**) Sea urchin. Bisecting sea urchin embryos at the eight-cell stage to create an animal tier or a vegetal tier of four cells. The animal tier does not gastrulate and forms a dauerblastula, while the vegetal tier does gastrulate and forms a small pluteus larva. (**d**) Ascidian. (1) Ascidian blastomere ablation at two-cell stage (Chabry 1887) gives half embryo rather than whole embryo. (2) Conklin (1939, Time magazine cover) indicating the location of the crescent (mesoplasm) which contains the muscle determinant macho-1 and the UCD determinant or CAB. (3) Yellow crescent in

contains determinants that trigger both UCD and gastrulation. Ettensohn and Wikramanayake elegantly showed that Dishevelled protein (localized to the egg vegetal cortex) may be the gastrulation determinant in echinoderms (Weitzel et al. 2004; Peng and Wikramanayake 2013).

3. At the dawn of experimental embryology, Laurent Chabry (1887) destroyed one blastomere at the two-cell stage in ascidians. The remaining blastomere did not make a whole embryo (Fig. 6.1d-1) indicating that some reduction in developmental potential had already occurred. Fertilization in ascidians triggers the first phase of ooplasmic segregation culminating in the formation of a vegetal protrusion that contains both the gastrulation determinant and the unequal cleavage determinant (Fig. 6.1d-4). The second phase of ooplasmic segregation moves the myoplasm and CAB (centrosome-attracting body: this causes UCD, covered in Sect. 6.5) to the posterior pole of the zygote, whereas the gastrulation determinant remains in the vegetal pole (Fig. 6.1d-2,3). Thus, ablation of the vegetal contraction pole (CP), when both determinants are localized together in the fertilized egg, radializes the embryo and prevents gastrulation (Nishida 1996; Dumollard et al. 2017). The earliest cleavages in ascidian embryos thus distributes maternal determinants to specific cells (Nishida and Sawada 2001; Kumano and Nishida 2007; Prodon et al. 2007).

6.1.2 Division Plane Positioning Rules: A Historical Perspective

Many different cues control the position of the mitotic spindle in cells causing them to divide symmetrically, unequally, or in specific oriented patterns. These cues are additive, and in some cases, one or more cues dominate over the others (Minc and Piel 2012; Pietro et al. 2016). Below we will review the rules of cell division laid down at the end of the nineteenth century by von Sachs (1887) and Hertwig (1884). More recently, modifications of these rules have been made (Minc et al. 2011; Pierre et al. 2016), plus molecular cues have been discovered that influence the position of the mitotic spindle and hence the cell division axis.

Cells of embryos adhere and thus integrate the forces of cell adhesion and the effective surface tension to become particular shapes thereby creating the overall shape of the early embryo (Turlier and Maître 2015), but embryos also alter their overall shapes through cell division and other dynamic cellular processes as they develop. See the insightful review by D'Arcy Thompson (1942), Chaps. VII and VIII) and also the beautiful soap bubble aggregates that almost completely

---

Fig. 6.1 (continued) fertilized ascidian egg (Conklin 1905). (4) Contraction pole (CP) ablation in fertilized egg abolishes UCD and radializes the embryo. CP ablation also prevents gastrulation

recapitulate the pattern of cells in early spiralian embryos [Fig. 6.1a, Robert, A. "Sur le Développement des Troques" (Robert 1903)].

E.B. Wilson considered the sphere to be the most important starting form since it represents the typical shape of the egg, even though other egg shapes are also known. He considered the pattern of cell division in the sea cucumber to conform to Sachs's rule (Sachs 1878, 1887). Sachs's rule states that "Each successive cleavage division is at right angle to the earlier." First formulated for plant cells, the embryos of the sea cucumber display a beautiful example of Sachs's rule. If first and second divisions are considered vertical and at right angles to each other and the third cleavage is horizontal, again at right angles to the first two divisions, this creates an eight-cell stage embryo which is composed of two tiers of four cells. However, Sach's rule was descriptive rather than mechanistic. Oscar Hertwig provided the missing piece of mechanistic information concerning the spindle (Hertwig 1884). Hertwig's rule stated that "The axis of a mitotic spindle occupies the longest axis of the protoplasmic mass in which it lies and division therefore tends to cut this axis transversely." Thus, with reference to sea cucumbers, the first three cell divisions were caused by the Hertwig rule which has come to be known as the "long axis rule" of cell division. Hertwig further tested this rule experimentally by compressing *Xenopus* eggs showing that the cleavage plane could be artificially redirected to bisect the cell mass according to the new long axis in these compressed cells [see English review of work in Wilson (1896, 1905, 1915)]. A recent reevaluation of the long axis rule demonstrated that the cleavage plane actually cuts through the geometric center of the cell rather than precisely at right angles to the longest axis (Wilson 1896). However, although the first three divisions often follow Sachs's rules, embryos then deviate from this rule and display cell divisions that can be unequal and/or oriented away from the Hertwig long axis, as well as being asynchronous, all of which can control the final position of cells in the embryo.

In the following sections, we will review three aspects of cell division: *unequal cell division (UCD)*, *oriented cell division (OCD)*, and *cell cycle asynchrony*. In the final section, we will consider all three aspects of cell division together, the redistribution of forces throughout the embryo via cell–cell contacts, and how they relate to formation of the ascidian blastula-stage embryo.

6.2 Unequal Cell Division

Cells display many types of UCD that are mediated by a variety of mechanisms and organelles including positioning by yolk, formation of polar lobes by cortical contraction (in some spiralians), cortical microtubule attracting sites (e.g., *Drosophila* neuroblasts and *C. elegans* zygotes), mitotic aster asymmetries (some spiralians), and cortical sites that favor microtubule depolymerization (ascidians; see discussion in Sect. 6.5).

6.2.1 *Effect of Yolk on UCD*

Many early embryos display UCD at various points in their embryogenesis which violate Hertwig's rule. These UCDs can be extreme or more subtle. Pfluger's rule (see Wilson, EB. (Wilson 1896) for a commentary) applies to the more subtle types of UCD, displayed, for example, by the formation of larger vegetal blastomeres in *Xenopus* eight-cell stage embryos. Pfluger's rule states that "During cleavage, formation of spindle fibres takes place in the region of lesser resistance or less yolk." *Xenopus* eggs are large (circa 1 mm in diameter) and have copious amounts of yolk that displays an increasing gradient from the animal to the vegetal pole. During early cleavage divisions, vegetal or yolk-rich blastomeres are created larger than animal blastomeres. Cephalopods, birds, reptiles, and fish embryos illustrate the extreme case for the effect of yolk, whereby cell division occurs in a meroblastic fashion atop a mass of nondividing yolk (Hasley et al. 2017). Hertwig tested the effect of yolk on cell division by centrifuging *Xenopus* eggs thus concentrating the yolk to one side of the egg. He was able to create meroblastic cleavages akin to fish embryos in these centrifuged *Xenopus* eggs (see Wilson (1896) for a discussion). A recent reevaluation of the cell shape rule combined with yolk distribution could recapitulate the ordered pattern of early cell divisions in sea urchin, ascidian, *Xenopus*, and fish embryos (Pierre et al. 2016). In fish embryos, an elegant model based on cell shape dictating cell division axis (that is robust yet evolvable since it adapts, e.g., to the size of the yolk sac) recapitulated the ordered sequence of cell divisions of all the blastomeres sitting atop the mass of nondividing yolk (Xiong et al. 2014). The authors suggested that altering molecular properties controlling cell shape during development could play a pivotal role in the evolution of epithelial thickness and buckling (Xiong et al. 2014). In addition to control of division axis, by altering cell shape or yolk distribution, cells have also evolved additional more extreme versions of UCD that also clearly violate Hertwig's long axis rule.

6.2.2 *UCD in* Drosophila *Neuroblast (Apicobasal Alignment and Cortical Contraction)*

Much of what we know of UCD comes from work carried out on *Drosophila* neuroblasts. Neuroblasts are the founder stem cell population from which most neurons are derived in *Drosophila*, and they form during several rounds of asymmetric cell division that produce one large neuroblast and one smaller ganglion mother cell (Siller and Doe 2009). The mitotic spindle orients along the apicobasal axis during neuroblast UCD (Savoian and Rieder 2002; Siller et al. 2006; Siller and Doe 2008). A protein called Pins (partner of inscuteable) is a key component of UCD in many cell types. Pins was first identified using a two-hybrid screen with inscuteable as bait in *Drosophila* (Yu et al. 2000). Pins is normally localized to basolateral cell surfaces but in neuroblasts becomes apical by interactions with inscuteable and Bazooka (PAR3) (Schaefer et al. 2000). NuMA is another key protein in spindle orientation. First identified as a nuclear matrix protein (Lydersen and Pettijohn 1980), NuMA was subsequently found to bind cytoplasmic dynein and dynactin (Merdes et al. 1996) and to the mammalian Pins homologue LGN (Du et al. 2001). In neuroblasts, the Pins/Mud (NuMA) complex is tethered to the apical plasma membrane via a small G protein, and the Pins/Mud (NuMA) complex was proposed to recruit dynein which pulls on microtubules thus attracting one pole of the mitotic spindle or anchoring one centrosome near the apical cortex causing nonplanar spindle alignment. After the first cell division in neuroblasts, where the whole mitotic spindle rotates with one pole being attracted to the apical cortex (Kaltschmidt et al. 2000), the mother centrosome that contains centrobin becomes anchored to the apical patch of Pins/NuMA, while the smaller daughter centrosome (smaller due to the absence of centrobin) rotates though 180° to cause mitotic spindle alignment with the apicobasal axis (Januschke et al. 2013). This centrosome anchoring causes alignment but not displacement of the spindle, and in neuroblasts, UCD is due to an additional mechanism. At the basal cortex of the neuroblast, a cap rich in myosin-2 is capable of causing UCD even when the mitotic spindle is experimentally rotated parallel to the apical cortex, resulting in a three daughter cell outcome: two daughter cells are created by a mitotic spindle-based mechanism and thus receive half of the chromatids each, while a third cytoplasmic mass that lacks chromatids buds off due to the cortical myosin-2 cap (Cabernard et al. 2010). One view of UCD in neuroblasts is thus that the spindle orients according to the apical-basal polarity of the cell due to the action of the apical Pins/NuMA complex and that UCD is created by the basal myosin-2 cap that directs the cleavage furrow away from the spindle mid-zone.

6.2.3 UCD in C. elegans *Zygotes (Cortical Pulling)*

C. elegans provided the evidence that cortical pulling forces move the spindle toward the cortex to create unequal-sized daughter cells. The first cell division in *C. elegans* zygotes creates a smaller posterior blastomere due to mitotic spindle movement toward the posterior cortex brought about by pulling forces generated at the posterior cortex (Grill and Hyman 2005). By cutting the mitotic spindle in half using a laser, pulling forces were clearly demonstrated that could move half of the severed mitotic spindle toward the posterior cortex (Grill et al. 2001, 2003; Fielmich et al. 2018). Genetic and RNAi-based experiments revealed that a posterior cortical Galpha/GPR-1/2 (Pins)/LIN-5 (NuMA) complex pulls the mitotic spindle toward the posterior of the dividing cell (Gönczy 2008; Kotak and Gönczy 2013). It has been postulated that the actomyosin cortex could provide a stable platform for the plasma membrane force-generating complexes (Pins/NuMA) to interact with the plus ends of astral microtubules (via dynein) thus causing the spindle to be pulled rather than the plasma membrane to invaginate into the asters. Indeed, by weakening the actomyosin cortex, plasma membrane invaginations that represent a proxy for force-generating cortical sites could be observed at the posterior end of the one-cell embryo rather than spindle movement (Redemann et al. 2010). These membrane invaginations can also be observed in *C. elegans* zygotes naturally, in the absence of actomyosin perturbation (Fielmich et al. 2018).

6.2.4 UCD in Zygotes of Spiralians (Cortical Contraction or Asymmetric Asters/Pushing Model)

Some spiralian embryos display UCD during their first cell cycle and also common in the literature are examples of UCD occurring at the four- to eight-cell stage to create a tier of four smaller animal micromeres (discussed above for pond snails).

The first cell division in mollusks can be unequal and caused by the formation of a large polar lobe during mitosis that transmits determinants to the larger CD cell (Conrad and Williams 1974). Among the annelids, a different type of UCD occurs in the zygote, where the segregation of teloplasm to the larger daughter cell can be generated though a mechanism independent of the cortex that depends on centrosome asymmetry. For example, UCD in *Tubifex hattai* during the first cell division is caused by a highly asymmetric mitotic apparatus that contains only one aster (Shimizu et al. 1998) of maternal centrosome origin (Ishii and Shimizu 1997). In another species of annelid such as in the leech *Helobdella robusta*, the mitotic aster is also asymmetric during the first cell division. However, in leech the centrosomes are of paternal origin, and the mitotic aster asymmetry occurs at metaphase when one centrosome of the mitotic spindle loses gamma-tubulin (Ren and Weisblat 2006). This leads to the transient disassembly of microtubules of one aster causing that aster to become smaller than the other (Ren and Weisblat 2006). Low doses of the microtubule-destabilizing drug nocodazole, which perturbs microtubule polymerization but not aster asymmetry, abolished spindle migration (indicating that displacement in the direction of the smaller aster is powered by microtubule polymerization) and thus led to equal cell division rather than UCD (Ren and Weisblat 2006). Moreover, no cortical site could be identified that influenced spindle alignment (Nelson and Weisblat 1992). One possibility is that the intact array of astral microtubules at the other pole pushes the spindle; however, one current view is that microtubule-based pushing cannot power spindle displacement in large cells because long microtubules buckle. In support of pushing, a more recent reevaluation of large asters has indicated that they also contain branched microtubules (Ishihara et al. 2014), thus suggesting that a pushing-based mechanisms could potentially operate. Moreover, it has been argued that even in large cells (circa 50 μm), microtubule pushing forces can center centrosomes and indeed that the pushing force per individual polymerizing microtubule (5–10pN) is equivalent to dynein-based (1–7pN) pulling forces (Howard and Garzon-Coral 2017). Thus, in leech zygotes, the cytoplasmic pulling-based mechanism for mitotic spindle centering may not operate, since the whole mitotic apparatus moves in the direction of the shortest microtubules, i.e., in the opposite direction to the prediction (Minc et al. 2011).

6.2.5 *UCD in Sea Urchin (Cortical Pulling?)*

Sea urchin embryos create small vegetal micromeres at the 8- to 16-cell stage. The pattern of cell division in sea urchins is predictable in that vegetal blastomeres inherit the vegetal cortex of the egg where Dishevelled protein is localized (Weitzel et al. 2004; Peng and Wikramanayake 2013). However, it is not clear how the orientation of the first two cleavage divisions are specified such that the vegetal cortex is partitioned to the lower tier of four cells at the eight-cell stage (the orientation of the third cell division is cued by the first two divisions as predicted by Sachs). It is also unknown how micromeres are formed at the vegetal pole of the embryo at the 8- to 16-cell stage. However, an intrinsic timing mechanism appears to exist since the vegetal cortex loses pigment granules during micromere formation at the 8- to 16-cell stage, even in cleavage arrested fertilized eggs (Schroeder 1980). Concerning the molecular mechanism controlling UCD in sea urchin, an orthologue of the LGN/Pins protein was found to be involved (Voronina and Wessel 2006). How the vegetal cortical enrichment of Pins/LGN is triggered at the eight-cell stage to cause UCD is not known, although it is tempting to speculate that it has something to do with the intrinsic timing mechanism that induces the loss of pigment granules from the vegetal cortex. It is also interesting to note that the cortical polarity protein aPKC, which is generally apical (outside cortex), is specifically excluded from the micromeres and the vegetal cortex to which the centrosome migrates (Prulière et al. 2011). Centrosomes also display a striking asymmetry during UCD in sea urchin embryos, whereby one centrosome is flattened upon the cortex that will become the micromere (Holy and Schatten 1991). It is currently unknown what causes this extreme form of centrosome asymmetry somewhat reminiscent of the centrosome asymmetry caused by the absence of Centrobin from the daughter centrosome in *Drosophila* neuroblasts (Januschke et al. 2011).

Overall, many of the extreme (non-yolk-based) forms of UCD employ either alignment of the spindle with apicobasal polarity cued by Pins/NuMA (neuroblast), cortical pulling of mitotic spindles (*C. elegans*), the generation of asymmetric asters via centrosome-based mechanisms (spiralians and neuroblasts), or cortical sites that depolymerize microtubules of one aster (ascidians; see Sect. 6.5).

6.3 Oriented Cell Division

Another feature of cell division in embryos is that spindles often align parallel (or planar) to the apical (outside) surface of the blastomeres even in defiance of the long axis rule. Here OCD is defined as a cell division that does not always obey Hertwig's long axis rule yet still produces daughter cells of roughly equal size (e.g., see OCDs for ascidian embryos in Sect. 6.5). Since planar cell division is also a conserved aspect of epithelia/endothelia, where cells divide within the sheet of the tissue, it is possible that some aspects of planar cell division in embryos are similar to planar cell division in epithelial sheets. However, since early embryos do not display mature epithelial subcellular architecture (e.g., junctions are maturing) and contain yolk, there are also likely to be key differences.

6.3.1 Planar OCD in Epithelia/Endothelia

Two key aspects of planar spindle orientation in epithelia/endothelia are that (1) LGN is excluded from the apical membrane and (2) the Dlg/LGN/NuMA complex becomes enriched on basolateral membranes during mitosis.

In mammalian tissue culture of MDCK cells, Gαi GDP (the plasma membrane tethered partner of LGN) is found both on the apical and basolateral cortical domains, yet LGN is localized only to basolateral domains. It has been shown that apical aPKC phosphorylates LGN (on ser 401, causing 14-3-3 proteins to associate with LGN), thus displacing LGN from the apical cortex (Hao et al. 2010). In Caco-2 cells, LGN is also removed from the apical cortex by the aPKC complex (Par3/Par6/aPKC). For example, aPKC or Par6B depletion both result in spindle misorientation in Caco-2 cells (Durgan et al. 2011). Similarly, in MDCK cells, silencing of Par3 also causes spindle misalignment and LGN to become recruited to the apical cortex (Hao et al. 2010). Finally, aPKC-mediated Pins (LGN) apical exclusion appears to be conserved in *Drosophila* wing disks, since temperature-sensitive mutants of aPKC revealed that aPKC is required for planar spindle orientation and apical exclusion of Pins (Guilgur et al. 2012). These various findings reveal how LGN is removed from the apical membrane, but they do not explain how LGN becomes enriched in the basolateral membrane in mitosis.

In *Drosophila* epithelia (wing and follicular), Aurora kinase (A and B) phosphorylation of basolateral Lgl (lethal giant larvae) leads to the basolateral recruitment of Pins/LGN in mitosis. During interphase, Lgl is removed from the apical cortex by aPKC and localizes exclusively to the basolateral cortex where it binds Dlg (disks large), an apicobasal polarity protein (Bell et al. 2015). During mitosis, additional phosphorylation of Lgl by Aurora kinases A and B causes dissociation of Lgl from Dlg thus providing a temporal switch for Pins to bind Dlg now free of Lgl (Bell et al. 2015; Carvalho et al. 2015).

This Dlg mechanism is conserved in chick and human cells. In chick neuroepithelia, LGN binds to basolateral Dlg1 (Saadaoui et al. 2014). Moreover, Dlg1 is also recruited to the cortex in HeLa cells during mitosis, and its recruitment is necessary for LGN cortical recruitment opposite each spindle pole (Saadaoui et al. 2014). Thus, basolateral mitotic cortical recruitment of the Dlg/LGN/NuMA complex together with apical exclusion of LGN appears to be well-conserved mechanisms for spindle orientation.

However, there are exceptions to this general rule involving apical aPKC and the lateral Dlg/Pins/Lgl pathways that align spindles along the planar axis in epithelia. A reinvestigation of *Drosophila* wing disk epithelium revealed that neither aPKC nor the Dlg/Pins/Lgl pathways are involved in planar spindle orientation, but Mud/NuMA is still required (Bergstralh et al. 2016). Another example comes from *Drosophila* pupal notum epithelia. Here the cells round up during mitosis and cells that were previously stretched in one direction during interphase due to tissue tension lose this long cell shape cue for cell division along that axis. How then do mitotic cells remember the direction of tissue tension/stretch during mitosis so that they divide along that axis? In the pupal notum, a mechanism transmits a memory of the anisotropic interphase cell shape to the rounded-up mitotic cell. This memory consists of tricellular junctions (TJs) which align with tissue tension during interphase. Mud/NuMA is localized to these interphase TJs independently of Pins/LGN (Bosveld et al. 2016). Importantly, the foci of cortical Mud/NuMA remain in place during mitotic cell rounding thus providing a mitotic memory of interphase cell shape to align the spindle during mitosis along the axis of tissue stretch (Bosveld et al. 2016). It is interesting to note that this function of maintaining an interphase memory of some spatial cue during mitosis is somewhat similar to the effect retraction fibers have on spindle alignment in adherent HeLA cells during mitosis (Théry et al. 2007).

6.3.2 Planar-Oriented Cell Division in Embryos

What processes cause spindles to align parallel to the outside or apical surface of the embryo? Is apicobasal polarity involved as it is in many epithelia?

In *Nematostella vectensis* (a cnidarian) embryos, apicobasal polarity is evident at or before the 16-cell stage (Ragkousi et al. 2017). These embryos display a hollow blastula in which spindles align parallel to the outside of the embryo. Lgl was demonstrated to localize to the basolateral cortex both in interphase and mitosis (Ragkousi et al. 2017); however, it is not known whether the LGN/NuMA complex is involved in planar spindle orientation in these embryos.

In syncytial *Drosophila* embryos, mitotic spindles also align parallel to the outside of the embryo. Even though individual cells are not formed, the cortex curves out from each nucleus, and Dlg is found to be localized at the intersection between these dimples of the cortex (Hirashima et al. 2018).

During gastrulation in *Xenopus,* the mechanism of planar spindle orientation during epiboly depends on apicobasal polarity. Here planar spindle orientation is precisely controlled by counteracting forces involving an apical directed force provided by microtubules and myosin-10 and a basal directed force provided by actin and myosin-2 (Woolner and Papalopulu 2012). Again, it is not clear what role the LGN/NuMA complex plays in these embryos.

Although planar spindle orientation is a well-conserved feature of many embryos, it should be noted that not all early embryos display planar-oriented spindles. In mouse embryos, the first symmetry-breaking event creates the inner cell mass (ICM) and the trophectoderm (TE). Unlike epithelia and most early embryos, spindles can align perpendicular rather than parallel to that apical cortex. Daughter cells that inherit the apical domain become TE while those that do not become ICM (Korotkevich et al. 2017). It was demonstrated that blastomeres that inherit the apical domain have less contractility and remain in the TE, while apolar cells are more contractile and have a tendency to become actively internalized (Maître et al. 2016). This creates a dynamic form of cell sorting allowing mouse embryos to regulate the total number of ICM versus TE. Therefore, the position of cells in mouse embryos does not depend only on cell division axis but also upon an interphase cell internalization mechanism unlike spiralian, urchin, or ascidian embryos.

6.4 Cell Cycle Asynchrony

Perhaps less obvious than UCD and OCD, asynchronous cell cycles can also affect spindle position and thus cell position in some embryos. Cell division involves drastic changes in cell shape caused by mitotic cell rounding. In epithelia, such intrinsic cell shape changes can generate tissue stress and promote, for example, epithelial invagination (Kondo and Hayashi 2013). Alternatively, mitotic cell rounding is not complete in some epithelia and can be biased/inhibited by tissue tension resulting in OCDs driven by anisotropic cell shape (Campinho et al. 2013; Kong et al. 2017). Cell cycle asynchrony is long known to be necessary for gastrulation movements, whether they consist of epithelial invagination (*Drosophila*, *Xenopus*, ascidians, sea urchin primary invagination), some jellyfish (*Nematostella*), or cell ingression/delamination (zebrafish, sea urchin, some cnidarian, e.g., *Clytia hemisphaerica*). In contrast, whether cell cycle asynchrony contributes to shaping the blastula prior to gastrulation is less well known.

Cell cycle asynchrony arises during the midblastula transition (MBT) in the form of meta-synchronous mitotic waves in zebrafish (Olivier et al. 2010). These mitotic waves are also observed in sea urchin blastula, but they do not seem to play a role in embryonic morphogenesis (Parisi et al. 1978). At the early gastrula transition (EGT), mitotic domains (i.e., part of the embryo where the cell cycle is regulated as an independent unit) appear in the freshly patterned germ layers. The first recognizable mitotic domain is a cluster of blastomeres undergoing gastrulation movements (bottle cells) which delay entry in mitosis by introduction of a long G2 phase (Bouldin and Kimelman 2014).

The cell cycle asynchrony observed at MBT in *Xenopus* or zebrafish is variable between individuals and reflects the period during which maternal control wanes and zygotic gene expression starts to support cell proliferation (Newport and Kirschner 1982; Kane and Kimmel 1993). During this time, the largest blastomeres cycle faster than the smallest ones as they contain more maternal factors for DNA replication such as Cut5, RecQ4, Treslin, and Drf1 in *Xenopus* (Newport and Kirschner 1982; Collart et al. 2013) or more maternal cyclin B3 transcripts as in the ascidian (Treen et al. 2018) to support DNA replication and mitotic entry. This correlation between cell size and cell cycle timing has also been observed in sea urchin, but the mechanisms are completely unknown (Duncan and Whiteley 2011).

6.5 The Ascidian Model

Ascidians belong to the Tunicata subphylum and are a sister group of vertebrates (Delsuc et al. 2006). Ascidian embryos have taken the concept of regulating cell and spindle position to the extreme among animals, whereby every cell follows an invariant cleavage pattern so that individual cells occupy the same specific territory of the embryo at the 64-cell stage (Fig. 6.2).

The three axes of the embryo (dorsoventral, anteroposterior, and left-right) can even be recognized in the fertilized ascidian egg due to the presence of colored cytoplasmic domains (Fig. 6.1d-2,3) (Conklin 1905). By following the distribution of the colored cytoplasm (in *Styela partita*), Edwin Conklin was able to produce a complete nomenclature and lineage map up to the gastrula stage (Conklin 1905). Given that the invariant cleavage pattern is the same among different species, this nomenclature and lineage map is used by the community working on different ascidians (Fig. 6.2, e.g., *Halocynthia roretzi*, a stolidobranch, or *Ciona robusta* and *Phallusia mammillata*, phlebobranchs). Although ascidian embryos were once considered completely mosaic (or cell autonomous) in their mode of development, we now know that neural induction at the 32-cell stage requires FGF signaling from vegetal blastomeres to their overlying animal blastomeres (Tassy et al. 2006), and this is the first of many inductions occurring during subsequent development (Lemaire 2009; Sobral et al. 2009). Thus, cells must be precisely positioned by the cleavage program to enable these subsequent inductive interactions.

We are interested in how the cleavage pattern is regulated during development, not least because the form of the early embryo is a product of cell size, cell number, and the interaction between cells. Despite being composed of only 64 cells (one cell cycle before gastrulation begins), the ascidian blastula displays the same fate map as other chordates such as *Xenopus*, with regions of both embryos displaying the same fate-restricted territories (Sobral et al. 2009), although cell number is less in ascidians by more than one order of magnitude. Ascidians accomplish precise cell positioning in the absence of cell displacement and death, indicating that cell

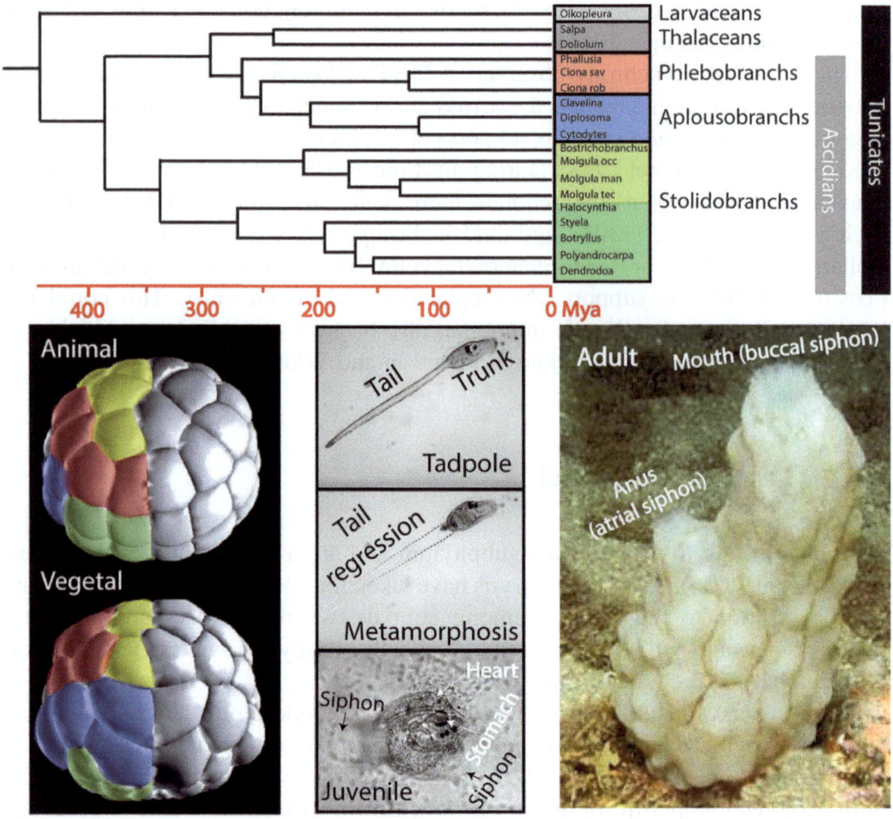

Fig. 6.2 Tunicate phylogeny, invariant cleavage pattern, and life cycle. Tunicate tree showing the three groups of ascidians shown (phlebobranchs, aplousobranchs, and stolidobranchs). Timeline is displayed below tree; Mya is million years ago (from Delsuc et al. 2018). Interestingly, the pelagic non-ascidian thalaceans group with the benthic ascidians (see Delsuc et al. 2018 for details). Ascidian embryo (*Phallusia mammillata*) at the 64-cell stage. Cells of the same color represent the four granddaughter cells from the 16-cell stage. Note that some cells form patterns of squares, other lines, and other a T-pattern. See Fig. 6.7 and Dumollard et al. (2017) for a thorough explanation of this phenomenon. About 12 h after fertilization, the embryo forms a swimming tadpole larva. After a variable swimming yet nonfeeding stage (1–2 days), the tadpole sticks down (here to glass) and undergoes metamorphosis accompanied by tail regression (note, this is the same larva as image above). A feeding juvenile then forms, and after about 5–6 months for *Phallusia mammillata*, a sexually mature hermaphrodite adult is created that is covered in a thick tunic composed partly of cellulose and feeds by sucking in seawater and organic matter via the buccal siphon

division orientation is key for blastomere positioning. When cell number is so low, UCD of just 2 cells at the 8- to 16-cell stage can affect the overall shape of the embryo (e.g., this would not be the case if it were 2 cells in an embryo with 2000 cells). As we have discussed, UCD has thus been exploited by many early embryos not only for segregation of maternal determinants to one of the two daughter cells but also for shaping the early embryo (Fig. 6.1). Incredibly, even though phlebobranch and stolidobranch ascidians last shared a common ancestor about 400 Mya (Delsuc et al. 2018), they display the same invariant cleavage pattern. Either because of conservation or convergence, this is quite remarkable since it has been estimated that these two groups of ascidians are as different from each other genetically as fish are from humans (Stolfi et al. 2014).

Implicit in the term invariant cleavage is that every cell in the gastrula will have the same relative position with exactly the same neighboring blastomeres. Curiously, in addition to cell positioning being invariant, cell cycle asynchrony and UCD are also invariant among distantly related ascidians. We therefore exploit these three invariant features of ascidian embryos as a robust experimental system to uncover the cell biological mechanisms that control *cell size inequality (UCD)*, *cell number (cell cycle asynchrony)*, and *cell position (OCD)*.

6.5.1 Unequal Cell Division (Cortical Pulling and Microtubule Depolymerization)

UCD in ascidian embryos begins at the eight-cell stage and is caused by the centrosome-attracting body or CAB (Nishikata et al. 1999). The CAB precursor forms during the two phases of ooplasmic segregation following fertilization. The first three cell divisions result in the CAB precursor being inherited by two vegetal posterior blastomeres (B4.1 pair) at the eight-cell stage (Fig. 6.3).

The CAB causes UCD and patterns the embryo. For example, surgical removal of the CAB precursor (CP, contraction pole) at the one-cell stage abolishes unequal cleavage and radializes embryos (Nishikata et al. 1999; Dumollard et al. 2017). The CAB first causes UCD starting at the eight-cell stage when two posterior vegetal blastomeres (B4.1 pair) divide unequally (in terms of size) and asymmetrically (in terms of fate) to form two smaller germ-cell precursors at the 16-cell stage. A further two successive rounds of UCD occur in the smaller blastomeres at the 16- and 32-cell stages creating two very small germ cell precursors at the 64-cell stage that each contain one CAB inherited from the B4.1 cells (Fig. 6.4).

The CAB generates UCD by causing one spindle pole to move toward the CAB cortex (Nishikata et al. 1999; Prodon et al. 2010). The precise molecular mechanism that causes UCD is not known, although posterior end mark/PEM1 (Negishi et al. 2007), polo-like kinase 1/Plk1 (Negishi et al. 2011), and the microtubule depolymerase Kif2 (Costache et al. 2017) are all thought to be involved. In addition to PEM1's role in UCD, PEM1 also silences transcription (Kumano et al. 2011) in a manner that shares similarity to germ-line silencing mediated by Pgc in *Drosophila* (Kumano et al. 2011). Finally, around 40 maternal RNAs termed PEM (posterior end mark) RNAs are tethered to the CAB, some of which like Vasa are involved in germ cell function (Prodon et al. 2007).

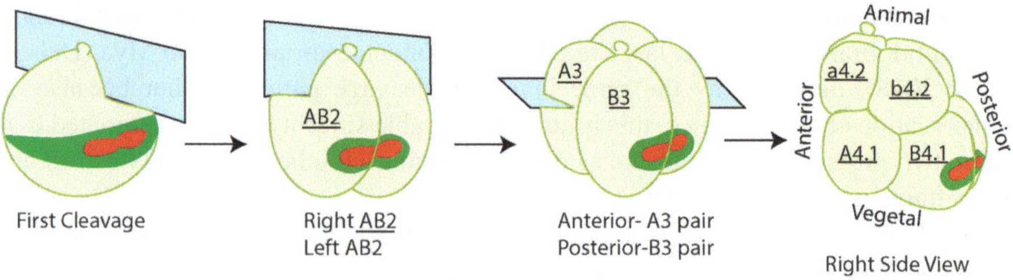

Fig. 6.3 First three cleavages in ascidian embryos. The CAB is colored red and the myoplasm green

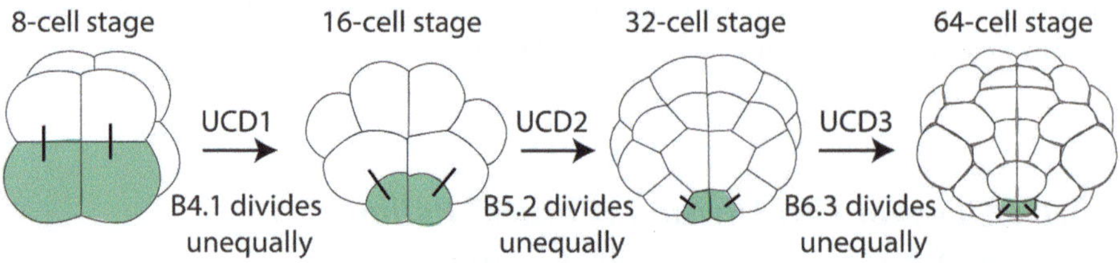

Fig. 6.4 Vegetal view of ascidian embryos showing three successive unequal cell divisions. The pairs of cells containing the 2 CABs are germ-line precursors and are colored green. Dark bars indicate sister cells that give rise to muscle. UCD is unequal cell division

Recently we discovered that a microtubule depolymerase, Kif2 a member of the kinesin-13 family of proteins, is localized to the cortical endoplasmic reticulum that is accumulated in the CAB (Fig. 6.5). This depolymerizing kinesin lacks motor function and is required for shrinking of the microtubule aster nearest the CAB, which we think is permissive for the whole mitotic spindle to approach the CAB since microtubule polymerization would act to counteract spindle movement toward the CAB (Costache et al. 2017). This may be a pushing-based mechanism powered, we think, by microtubule polymerization of the distant aster (which increases in size as the CAB aster shrinks, Fig. 6.5), thus pushing the spindle toward the CAB (Costache et al. 2017). However, we also have preliminary evidence for an additional pulling-based mechanism that operates during late mitosis to precisely position one pole of the mitotic spindle adjacent to the CAB near the midline in B5.2 cells (Fig. 6.5 and manuscript in preparation).

6.5.2 *Cell Cycle Asynchrony*

Ascidian embryos show an extreme example of how cell cycle asynchrony is tightly regulated already at the MBT and can be compared to *C. elegans* (which also have an invariant cleavage pattern). In ascidian embryos, the maternal to zygotic transition (MZT) occurs at the 16-cell stage (Treen et al. 2018) at exactly the time

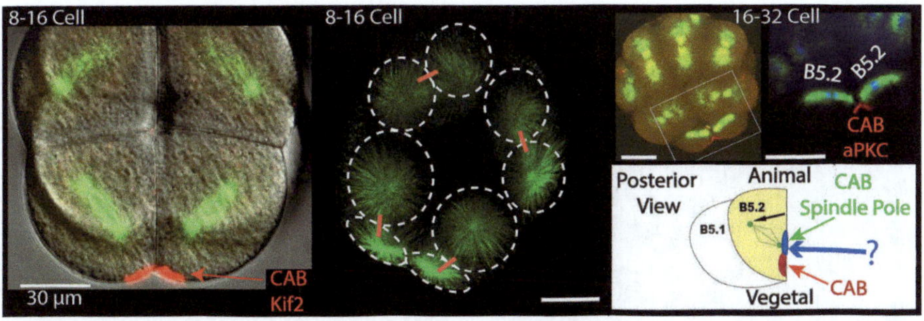

Fig. 6.5 Kif2 localizes to CAB but not to site where spindle poles align. Left image shows the typical localization pattern of Kif2 at the CAB in live embryos (Kif2::tdTomato) during early mitosis at eight-cell stage before spindle displacement (microtubule label EB3::Venus, light green) (see Costache et al. 2017 for full movie). Middle confocal image displays microtubule distribution during late mitosis at eight-cell stage in live embryo (Ens::3GFP). Note size of asters (dotted white circles) and size inequality in two pairs of lower cells. Red bars represent sister cells. Right pair of images (fixed embryos, anti-tubulin: green and anti-aPKC: red) show spindle position during mitosis at 16-cell stage. Note that the spindle does not point toward the CAB but rather above the CAB to an as yet unidentified midline cortical domain. The drawing indicates spindle position during mitosis at 16-cell stage and also our hypothesis as to the location of an additional midline cortical domain adjacent to the CAB-spindle pole (blue domain, blue arrow). All scale bars = 30 μm

endomesoderm and ectoderm are specified. At this stage, germ layer patterning by β-catenin also patterns the cell cycle in the endomesoderm where nuclear β-catenin maintains a rapid cell cycle, while the overlying animal ectoderm slows down (Dumollard et al. 2013). Such patterned cell cycle asynchrony gives rise to a brief yet crucial 24-cell stage that is found in both stolidobranch and phlebobranch ascidians even though these ascidians last shared a common ancestor approximately 400 million years ago (Delsuc et al. 2018). As will be discussed below, such cell cycle asynchrony is informative to implement not only the temporal aspect of the invariant cleavage pattern but also for spatial aspects (i.e., cell positioning).

Cell cycle asynchrony begins at the 16-cell stage in the ascidian embryo when endomesoderm and ectoderm are specified (Dumollard et al. 2013). At this stage, the 6 endomesoderm blastomeres in the vegetal half of the embryo enter mitosis 11 min before the 8 ectoderm cells of the animal half giving rise to a transient 24-cell stage. At the 32-cell stage, the 14 vegetal/endomesoderm blastomeres enter mitosis 15 min before the 16 ectoderm cells to give rise to a 44-cell stage preceding the 64 cell blastula (Fig. 6.6). Strikingly, the 24- or 44-cell stages have been observed in different orders of ascidians suggesting that such precise cell cycle asynchrony is of vital importance for subsequent ascidian development. The first difference between different ascidian orders is that stolidobranch embryos starts to gastrulate at the 110-cell stage, whereas gastrulation initiates at the 112-cell stage in phlebobranchs (Nishida 2005; Lemaire 2009).

At the 64-cell stage, some gene regulatory networks (GRNs) such as the GRN specifying the notochord differs between *Halocynthia* (Takatori et al. 2010) and *Ciona* (Hudson et al. 2013), illustrating how conservation of embryonic shape at this

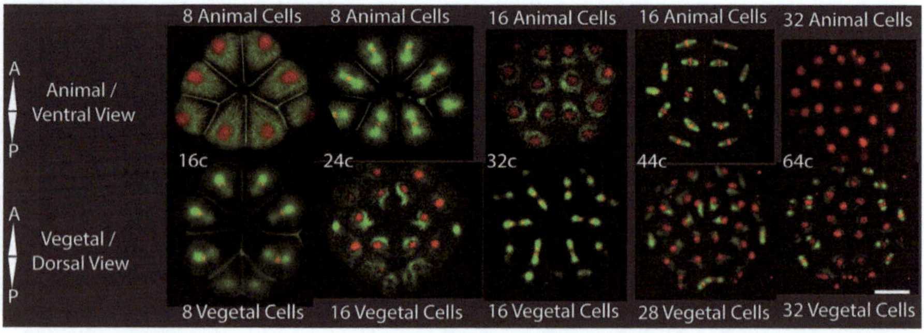

Fig. 6.6 Asynchronous cell cycle in ascidians. Animal and vegetal views of the same embryo (confocal images) from 16- to 64-cell stage. Nuclei and chromosomes are red (histone H2B::Rfp1) and microtubules green (MAP 7::GFP). Note that vegetal cells divide before the animal cells at the 16-cell stage creating a 24-cell stage. This asynchrony is maintained thus creating a 44-cell stage. Note also that at the 32-cell stage, the 14 endomesoderm vegetal cells divide before the 16 animal ectoderm cells. The 2 germ lineage cells are slightly delayed thus creating a brief 44-cell stage rather than a 48-cell stage. Scale bar = 30 μm

stage is stronger than GRN conservation. Thus, a short delay in cell cycle timing equivalent to the duration of mitosis (15 min in *Phallusia*) appears to have been conserved for around 400 million years (unless of course it has appeared via convergence). It is worth noting that while a mechanism relying on absolute time may vary at differing temperatures, a mechanism relying on cell cycle asynchrony is more robust since it operates over a range of temperatures (*Halocynthia* embryos develop at 12 °C, while *Phallusia* embryos can develop at 25 °C). We were thus intrigued to understand the importance of this short delay and found that a 15 min delay between vegetal and animal halves of the embryo is necessary to maintain an invariant cleavage pattern (Dumollard et al. 2017). Because these alternate cell divisions are shared by distantly related ascidian species, we surmise that alternate cell division has been exploited by ascidian embryos to maintain a precise spatial pattern of cell divisions, which are known to be required to implement cell-contact-based neural inductions starting at the 32-cell stage (Tassy et al. 2006; Lemaire 2009).

6.5.3 *Oriented Cell Division*

In contrast to many cells, ascidian blastomeres do not completely round up during mitosis (since they maintain some adhesion), which is not entirely without precedent among embryos. Spindle orientation in the ascidian embryo thus follows mitotic cell shape but crucially with an apical constraint.

The ascidian-specific pattern of cell divisions generating the blastula relies on OCDs. All spindles align parallel to the outside or apical surface of blastomeres in ascidians (Fig. 6.7a), so no cells are internal before gastrulation. However, this is

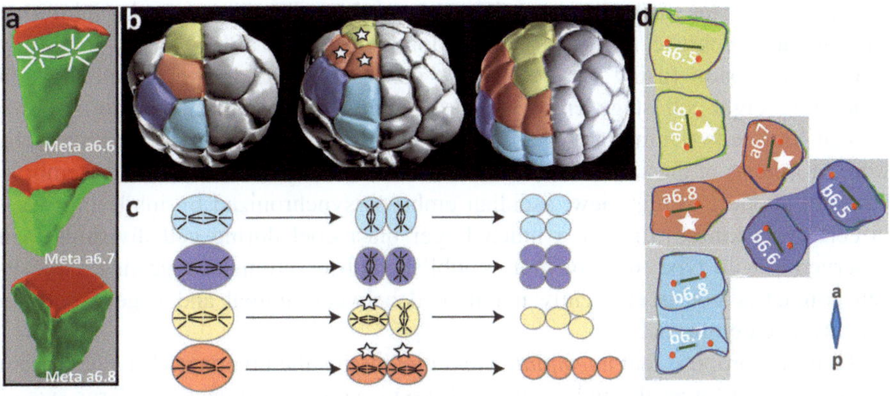

Fig. 6.7 Oriented cell division (OCD). (**a**) Three segmented animal cells in metaphase at 32-cell stage (a6.6, a6.8, and a6.8), apical is red and basolateral green. Spindle position parallel to apical surface is indicated (white spindle). Note that spindle does not align with the longest cell length in mitosis but rather with the longest apical length. (**b**) Animal view of 3D representations of fixed embryos at 24-, 32-, and 64-cell stages. Color coding represents granddaughter cell quartets at 64-cell stage. White stars indicate a6.6, a6.7, and a6.8. (**c**) Cartoon indicating OCD in all cells as spindles align with longest apical length: default spindle position where spindles are already aligned with longest apical length (blue and violet cells) give squares of 4 cells at 64-cell stage; or spindle rotation into longest apical length (in three cells, white stars) that give T-shape or line shape at 64-cell stage. Three white stars indicate cells that display spindle rotations (a6.6, 6.7, and 6.8) so that spindles align in same axis as the spindles of their mother cell. (**d**) Segmented apical projections from live 32-cell embryos in which actual spindle pole position is indicated by red circles and where green bars indicate predicted spindle positions based on longest apical length (from computational model Dumollard et al. 2017). White stars again indicate cells in which spindle rotated. Nomenclature of all cells are indicated (a, anterior; p, posterior)

oftentimes not the longest length of the cell (Fig. 6.7a). Thus, these cell divisions can be considered OCDs as they seem cued by apicobasal polarity rather than Hertwig's long axis rule. We found that apical cell shape dictates spindle orientation, and so all spindles align with the longest length of the apical cell shape. However, at the 32-cell stage, some spindles rotate away from their starting position (a6.6, a6.7, and a6.8), while other spindles remain in their starting orientation (e.g., all remaining animal cells, Fig. 6.7b–d). This is because some spindles are already aligned in the long length of the apical cell shape at prophase while others are not, and these latter spindles thus rotate into the longest mitotic apical cell shape. Thus, here we define *all* cell divisions at this stage as oriented by apical cell shape as examples of OCD (McDougall et al. 2015; Dumollard et al. 2017). Strikingly, all these OCDs are generated by the shape of the apical surface of blastomeres, and preventing cell cycle asynchrony changes the invariant blueprint of apical cell shapes (Dumollard et al. 2017).

With regard to the dynamic cell and embryo shape changes occurring during cleavage stages, it is still puzzling how such highly stereotyped cell shapes at metaphase can be so robustly obtained during ascidian blastula formation. Since cell–cell contact observed in ascidian embryos are partly maintained during mitosis, this might explain such robustness. First, removing cell-cell contact by inhibiting basolateral membrane formation (using dominant active aPKC) or by inhibiting cadherin adhesion (using Ca^{2+}/Mg^{2+}-free seawater) alters the third cell division generating eight-cell stage embryos of different shapes rather than normal two tiers of four cells (Dumollard et al. 2017). We also hypothesize that if all the cells were in mitosis at the same time, cell-cell contact would be reduced at the embryo level. In support of this view, ascidian embryos synchronized by inhibiting wee1 (a cell cycle kinase) display a much larger blastocoel during cell divisions than observed in wild-type embryos (our unpublished observations), suggesting that cell-cell contact is not as efficiently maintained between animal and vegetal cells in synchronized embryos.

Thus, we propose that a *template model* maintains the invariant cleavage pattern by propagating forces throughout the embryo via cell–cell contact. The large size of the cell–cell interfaces in the ascidian blastula could allow efficient force transmission throughout the whole embryo. So, alternate vegetal/animal cell divisions and UCD of just two cells in one part of the embryo impacts the global cleavage pattern (Figs. 6.1, 6.5, 6.6 and 6.7). Cell–cell contacts in the ascidian embryo may then be necessary to transduce forces into changes in apical cell shape at metaphase to drive OCDs. Such cell–cell contacts likely also counteract forces generated through mitotic cell rounding [mitotic cell rounding occurs in dissociated blastomeres (Prodon et al. 2010)] in order to maintain an anisotropic apical cell shape during metaphase.

6.5.4 *Future Directions*

Ascidians thus offer a unique systems-level model where the forces that shape early embryos and their constituent cells can be analyzed using biophysical and cell biological techniques to understand how cleavage patterns lead to embryo shape. Our working model so far is that the effects of UCD (Costache et al. 2017) and cell cycle asynchrony (Dumollard et al. 2013) are propagated throughout the embryo from the 24- to 64-cell stages through cell adhesion and cell shape (Dumollard et al. 2017). Further research is required to determine whether asynchronous cell cycles bring about more internal forces. For example, differential cell contractility between interphasic and mitotic cells would generate a specific pattern of forces throughout the embryo. Moreover, to understand how forces are generated, it would be of great importance to take into account the contractility at the apical surface and both contractility and adhesion at the basolateral surface. However, it may prove difficult to measure contractility of basolateral cortices that might be very different from contractility of apical cortices. We surmise that because cells adhere, all the constituent cells of the embryo are affected by what at first appears to be small localized changes such as UCD in two posterior blastomeres or the emergence of the 24-cell stage. We have come to the hypothesis that the shape of all the cells in the early embryo and in particular the shapes of their apical surfaces are affected by both UCD and asynchronous cell cycles leading to spindle positioning in the longest apical cell length (Dumollard et al. 2017). The only cells that do not obey the apical cell shape rule are those that divide unequally (Dumollard et al. 2017) due to the localization of molecular cues that position the spindle (Costache et al. 2017). Thus, we note that this is not a full explanation since we do not know how cell adhesion and the mechanical properties of the cells lead to cell and ultimately embryo shape. Another important unresolved question is how all the spindles align parallel with the apical cell surface: ascidian embryos do not have any internal cells before gastrulation. Our thoughts are that this could involve Pins/LGN and NuMA as in epithelia, but we must not forget about subcellular distribution of cytoplasmic domains including the yolk, which occupies the deeper cytoplasm of the blastomeres at this stage of embryogenesis.

Acknowledgments We would like to thank Zak Swartz for thoughtful comments that improved the manuscript. We would also like to thank the funding agencies that support our work: the French government funding agency Agence National de la Recherche (ANR "MorCell": ANR-17-CE 13-0028) and Sorbonne University for supporting the Réseau André Picard. We also thank the CRBM of Institut de la Mer de Villefranche (IMEV) that is supported by EMBRC-France, whose French state funds are managed by the ANR within the Investments of the Future program under reference ANR-10-INBS-02.

References

Akiyama M, Tero A, Kobayashi R (2010) A mathematical model of cleavage. J Theor Biol 264: 84–94. https://doi.org/10.1016/j.jtbi.2009.12.016

Bell GP, Fletcher GC, Brain R, Thompson BJ (2015) Aurora kinases phosphorylate Lgl to induce mitotic spindle orientation in *Drosophila epithelia*. Curr Biol 25:61–68. https://doi.org/10.1016/j.cub.2014.10.052

Bergstralh DT, Lovegrove HE, Kujawiak I et al (2016) Pins is not required for spindle orientation in the Drosophila wing disc. Development 143:2573–2581. https://doi.org/10.1242/dev.135475

Bosveld F, Markova O, Guirao B et al (2016) Epithelial tricellular junctions act as interphase cell shape sensors to orient mitosis. Nature 530:495–498. https://doi.org/10.1038/nature16970

Bouldin CM, Kimelman D (2014) Cdc25 and the importance of G2 control: insights from developmental biology. Cell Cycle 13:2165–2171. https://doi.org/10.4161/cc.29537

Brun-Usan M, Marín-Riera M, Grande C et al (2017) A set of simple cell processes is sufficient to model spiral cleavage. Development 144:54–62. https://doi.org/10.1242/dev.140285

Cabernard C, Prehoda KE, Doe CQ (2010) A spindle-independent cleavage furrow positioning pathway. Nature 467:91–94. https://doi.org/10.1038/nature09334

Campinho P, Behrndt M, Ranft J et al (2013) Tension-oriented cell divisions limit anisotropic tissue tension in epithelial spreading during zebrafish epiboly. Nat Cell Biol 15:1405–1414. https://doi.org/10.1038/ncb2869

Carvalho CA, Moreira S, Ventura G et al (2015) Aurora A triggers Lgl cortical release during symmetric division to control planar spindle orientation. Curr Biol 25:53–60. https://doi.org/10.1016/j.cub.2014.10.053

Chabry L (1887) Contribution a l'embryologie normale et teratologique des Ascidies simples. J Anat Physiol 23:167–319

Collart C, Allen GE, Bradshaw CR et al (2013) Titration of four replication factors is essential for the Xenopus laevis midblastula transition. Science 341:893–896. https://doi.org/10.1126/science.1241530

Conklin EG (1905) The organization and cell-lineage of the ascidian egg. J Acad Natl Sci Phila 13:1–119

Conrad GW, Williams DC (1974) Polar lobe formation and cytokinesis in fertilized eggs of *Ilyanassa obsoleta*. I Ultrastructure and effects of cytochalasin B and colchicine. Dev Biol 36: 363–378

Cook CE, Chenevert J, Larsson TA et al (2016) Old knowledge and new technologies allow rapid development of model organisms. Mol Biol Cell 27:882–887. https://doi.org/10.1091/mbc.E15-10-0682

Costache V, Hebras C, Pruliere G et al (2017) Kif2 localizes to a subdomain of cortical endoplasmic reticulum that drives asymmetric spindle position. Nat Commun 8:s41467–s41017. https://doi.org/10.1038/s41467-017-01048-8

Delsuc F, Brinkmann H, Chourrout D, Philippe H (2006) Tunicates and not cephalochordates are the closest living relatives of vertebrates. Nature 439:965–968. https://doi.org/10.1038/nature04336

Delsuc F, Philippe H, Tsagkogeorga G et al (2018) A phylogenomic framework and timescale for comparative studies of tunicates. BMC Biol 16:39. https://doi.org/10.1186/s12915-018-0499-2

di Pietro F, Echard A, Morin X (2016) Regulation of mitotic spindle orientation: an integrated view. EMBO Rep 17:1106–1130. https://doi.org/10.15252/embr.201642292

Driesch H (1892) Entwicklungsmechanische Studien 1. Der Wert der erster Furchungszellen in der Echinodermentwicklung Experimentelle Erzeugung von Teil-und Doppelbildungen Z Zool 53

Driesch H (1908) The 1907 Gifford Lectures. The Science and Philosophy of the Organism. Adam & Charles Black, London

Du Q, Stukenberg PT, Macara IG (2001) A mammalian partner of inscuteable binds NuMA and regulates mitotic spindle organization. Nat Cell Biol 3:1069–1075. https://doi.org/10.1038/ncb1201-1069

Dumollard R, Hebras C, Besnardeau L, McDougall A (2013) Beta-catenin patterns the cell cycle during maternal-to-zygotic transition in urochordate embryos. Dev Biol 384:331–342. https://doi.org/10.1016/j.ydbio.2013.10.007

Dumollard R, Minc N, Salez G et al (2017) The invariant cleavage pattern displayed by ascidian embryos depends on spindle positioning along the cell's longest axis in the apical plane and relies on asynchronous cell divisions. elife 6. https://doi.org/10.7554/eLife.19290

Duncan REL, Whiteley AH (2011) The echinoid mitotic gradient: effect of cell size on the micromere cleavage cycle. Mol Reprod Dev 78:868–878. https://doi.org/10.1002/mrd.21373

Durgan J, Kaji N, Jin D, Hall A (2011) Par6B and atypical PKC regulate mitotic spindle orientation during epithelial morphogenesis. J Biol Chem 286:12461–12474. https://doi.org/10.1074/jbc.M110.174235

Fielmich L-E, Schmidt R, Dickinson DJ et al (2018) Optogenetic dissection of mitotic spindle positioning in vivo. elife 7. https://doi.org/10.7554/eLife.38198

Gehring WJ (1996) The master control gene for morphogenesis and evolution of the eye. Genes Cells 1:11–15

Gönczy P (2008) Mechanisms of asymmetric cell division: flies and worms pave the way. Nat Rev Mol Cell Biol 9:355–366. https://doi.org/10.1038/nrm2388

Grill SW, Hyman AA (2005) Spindle positioning by cortical pulling forces. Dev Cell 8:461–465. https://doi.org/10.1016/j.devcel.2005.03.014

Grill SW, Gönczy P, Stelzer EH, Hyman AA (2001) Polarity controls forces governing asymmetric spindle positioning in the *Caenorhabditis elegans* embryo. Nature 409:630–633. https://doi.org/10.1038/35054572

Grill SW, Howard J, Schäffer E et al (2003) The distribution of active force generators controls mitotic spindle position. Science 301:518–521. https://doi.org/10.1126/science.1086560

Guilgur LG, Prudêncio P, Ferreira T et al (2012) Drosophila aPKC is required for mitotic spindle orientation during symmetric division of epithelial cells. Development 139:503–513. https://doi.org/10.1242/dev.071027

Hao Y, Du Q, Chen X et al (2010) Par3 controls epithelial spindle orientation by aPKC-mediated phosphorylation of apical Pins. Curr Biol 20:1809–1818. https://doi.org/10.1016/j.cub.2010.09.032

Hasley A, Chavez S, Danilchik M et al (2017) Vertebrate embryonic cleavage pattern determination. Adv Exp Med Biol 953:117–171. https://doi.org/10.1007/978-3-319-46095-6_4

Heisenberg C-P, Bellaïche Y (2013) Forces in tissue morphogenesis and patterning. Cell 153:948–962. https://doi.org/10.1016/j.cell.2013.05.008

Hertwig O (1884) Untersuchungen zur Morphologie und Physiologie der Zelle: Das Problem der Befruchtung und der Isotropie des Eies, eine Theorie der Vererbung. Fischer

Hirashima T, Tanaka R, Yamaguchi M, Yoshida H (2018) The ABD on the nascent polypeptide and PH domain are required for the precise Anillin localization in Drosophila syncytial blastoderm. Sci Rep 8:12910. https://doi.org/10.1038/s41598-018-31106-0

Holy J, Schatten G (1991) Differential behavior of centrosomes in unequally dividing blastomeres during fourth cleavage of sea urchin embryos. J Cell Sci 98(Pt 3):423–431

Howard J, Garzon-Coral C (2017) Physical limits on the precision of mitotic spindle positioning by microtubule pushing forces. BioEssays 39:1700122. https://doi.org/10.1002/bies.201700122

Hudson C, Kawai N, Negishi T, Yasuo H (2013) β-catenin-driven binary fate specification segregates germ layers in ascidian embryos. Curr Biol. https://doi.org/10.1016/j.cub.2013.02.005

Ishihara K, Nguyen PA, Groen AC et al (2014) Microtubule nucleation remote from centrosomes may explain how asters span large cells. Proc Natl Acad Sci USA 111:17715–17722. https://doi.org/10.1073/pnas.1418796111

Ishii R, Shimizu T (1997) Equalization of unequal first cleavage in the Tubifex egg by introduction of an additional centrosome: implications for the absence of cortical mechanisms for mitotic spindle asymmetry. Dev Biol 189:49–56. https://doi.org/10.1006/dbio.1997.8653

Januschke J, Llamazares S, Reina J, Gonzalez C (2011) Drosophila neuroblasts retain the daughter centrosome. Nat Commun 2:243. https://doi.org/10.1038/ncomms1245

Januschke J, Reina J, Llamazares S et al (2013) Centrobin controls mother-daughter centriole asymmetry in Drosophila neuroblasts. Nat Cell Biol 15:241–248. https://doi.org/10.1038/ncb2671

Kaltschmidt JA, Davidson CM, Brown NH, Brand AH (2000) Rotation and asymmetry of the mitotic spindle direct asymmetric cell division in the developing central nervous system. Nat Cell Biol 2:7–12. https://doi.org/10.1038/71323

Kane DA, Kimmel CB (1993) The zebrafish midblastula transition. Development 119:447–456

Kondo T, Hayashi S (2013) Mitotic cell rounding accelerates epithelial invagination. Nature 494:125–129. https://doi.org/10.1038/nature11792

Kong D, Wolf F, Großhans J (2017) Forces directing germ-band extension in Drosophila embryos. Mech Dev 144:11–22. https://doi.org/10.1016/j.mod.2016.12.001

Korotkevich E, Niwayama R, Courtois A et al (2017) The apical domain is required and sufficient for the first lineage segregation in the mouse embryo. Dev Cell 40:235–247.e7. https://doi.org/10.1016/j.devcel.2017.01.006

Kotak S, Gönczy P (2013) Mechanisms of spindle positioning: cortical force generators in the limelight. Curr Opin Cell Biol 25:741–748. https://doi.org/10.1016/j.ceb.2013.07.008

Kumano G, Nishida H (2007) Ascidian embryonic development: an emerging model system for the study of cell fate specification in chordates. Dev Dyn 236:1732–1747. https://doi.org/10.1002/dvdy.21108

Kumano G, Takatori N, Negishi T et al (2011) A maternal factor unique to ascidians silences the germline via binding to P-TEFb and RNAP II regulation. Curr Biol 21:1308–1313. https://doi.org/10.1016/j.cub.2011.06.050

Kuroda R, Endo B, Abe M, Shimizu M (2009) Chiral blastomere arrangement dictates zygotic left-right asymmetry pathway in snails. Nature 462:790–794. https://doi.org/10.1038/nature08597

Kuroda R, Fujikura K, Abe M et al (2016) Diaphanous gene mutation affects spiral cleavage and chirality in snails. Sci Rep 6:34809. https://doi.org/10.1038/srep34809

Lemaire P (2009) Unfolding a chordate developmental program, one cell at a time: invariant cell lineages, short-range inductions and evolutionary plasticity in ascidians. Dev Biol 332:48–60. https://doi.org/10.1016/j.ydbio.2009.05.540

Lydersen BK, Pettijohn DE (1980) Human-specific nuclear protein that associates with the polar region of the mitotic apparatus: distribution in a human/hamster hybrid cell. Cell 22: 489–499

Maître J-L, Turlier H, Illukkumbura R et al (2016) Asymmetric division of contractile domains couples cell positioning and fate specification. Nature 536:344–348. https://doi.org/10.1038/nature18958

McDougall A, Chenevert J, Pruliere G et al (2015) Centrosomes and spindles in ascidian embryos and eggs. Methods Cell Biol 129:317–339. https://doi.org/10.1016/bs.mcb.2015.03.006

Merdes A, Ramyar K, Vechio JD, Cleveland DW (1996) A complex of NuMA and cytoplasmic dynein is essential for mitotic spindle assembly. Cell 87:447–458. https://doi.org/10.1016/S0092-8674(00)81365-3

Minc N, Piel M (2012) Predicting division plane position and orientation. Trends Cell Biol 22:193–200. https://doi.org/10.1016/j.tcb.2012.01.003

Minc N, Burgess D, Chang F (2011) Influence of cell geometry on division-plane positioning. Cell 144:414–426. https://doi.org/10.1016/j.cell.2011.01.016

Negishi T, Takada T, Kawai N, Nishida H (2007) Localized PEM mRNA and protein are involved in cleavage-plane orientation and unequal cell divisions in ascidians. Curr Biol 17:1014–1025. https://doi.org/10.1016/j.cub.2007.05.047

Negishi T, Kumano G, Nishida H (2011) Polo-like kinase 1 is required for localization of posterior end mark protein to the centrosome-attracting body and unequal cleavages in ascidian embryos. Dev Growth Differ 53:76–87. https://doi.org/10.1111/j.1440-169X.2010.01231.x

Nelson BH, Weisblat DA (1992) Cytoplasmic and cortical determinants interact to specify ectoderm and mesoderm in the leech embryo. Development 115:103–115

Newport J, Kirschner M (1982) A major developmental transition in early Xenopus embryos: I. characterization and timing of cellular changes at the midblastula stage. Cell 30:675–686

Nishida H (1996) Vegetal egg cytoplasm promotes gastrulation and is responsible for specification of vegetal blastomeres in embryos of the ascidian *Halocynthia roretzi*. Development 122:1271–1279

Nishida H (2005) Specification of embryonic axis and mosaic development in ascidians. Dev Dyn 233:1177–1193. https://doi.org/10.1002/dvdy.20469

Nishida H, Sawada K (2001) macho-1 encodes a localized mRNA in ascidian eggs that specifies muscle fate during embryogenesis. Nature 409:724–729. https://doi.org/10.1038/35055568

Nishikata T, Hibino T, Nishida H (1999) The centrosome-attracting body, microtubule system, and posterior egg cytoplasm are involved in positioning of cleavage planes in the ascidian embryo. Dev Biol 209:72–85. https://doi.org/10.1006/dbio.1999.9244

Olivier N, Luengo-Oroz MA, Duloquin L et al (2010) Cell lineage reconstruction of early zebrafish embryos using label-free nonlinear microscopy. Science 329:967–971. https://doi.org/10.1126/science.1189428

Parisi E, Filosa S, De Petrocellis B, Monroy A (1978) The pattern of cell division in the early development of the sea urchin, *Paracentrotus lividus*. Dev Biol 65:38–49

Peng CJ, Wikramanayake AH (2013) Differential regulation of disheveled in a novel vegetal cortical domain in sea urchin eggs and embryos: implications for the localized activation of canonical Wnt signaling. PLoS One 8:e80693. https://doi.org/10.1371/journal.pone.0080693

Peng JC-F, Wang L, Wikramanayake AH (2017) Origins of anterior-posterior polarity by localized activation of Disheveled. Mol Reprod Dev 84:443. https://doi.org/10.1002/mrd.22839

Pierre A, Sallé J, Wühr M, Minc N (2016) Generic theoretical models to predict division patterns of cleaving embryos. Dev Cell 39:667–682. https://doi.org/10.1016/j.devcel.2016.11.018

Prodon F, Yamada L, Shirae-Kurabayashi M et al (2007) Postplasmic/PEM RNAs: a class of localized maternal mRNAs with multiple roles in cell polarity and development in ascidian embryos. Dev Dyn 236:1698–1715. https://doi.org/10.1002/dvdy.21109

Prodon F, Chenevert J, Hébras C et al (2010) Dual mechanism controls asymmetric spindle position in ascidian germ cell precursors. Development 137:2011–2021. https://doi.org/10.1242/dev.047845

Prulière G, Cosson J, Chevalier S et al (2011) Atypical protein kinase C controls sea urchin ciliogenesis. Mol Biol Cell 22:2042–2053. https://doi.org/10.1091/mbc.E10-10-0844

Ragkousi K, Marr K, McKinney S et al (2017) Cell-cycle-coupled oscillations in apical polarity and intercellular contact maintain order in embryonic epithelia. Curr Biol 27:1381–1386. https://doi.org/10.1016/j.cub.2017.03.064

Redemann S, Pecreaux J, Goehring NW et al (2010) Membrane invaginations reveal cortical sites that pull on mitotic spindles in one-cell *C. elegans* embryos. PLoS One 5:e12301. https://doi.org/10.1371/journal.pone.0012301

Ren X, Weisblat DA (2006) Asymmetrization of first cleavage by transient disassembly of one spindle pole aster in the leech *Helobdella robusta*. Dev Biol 292:103–115. https://doi.org/10.1016/j.ydbio.2005.12.049

Robert A (1903) Sur le développement des Troches. Thesis, Faculté des Sciences de L'Université de Paris

Saadaoui M, Machicoane M, di Pietro F et al (2014) Dlg1 controls planar spindle orientation in the neuroepithelium through direct interaction with LGN. J Cell Biol 206:707–717. https://doi.org/10.1083/jcb.201405060

Sachs J (1878) Uber die Anordnung der Zellen in jungsten Pflanzentheilen. Arb Bot Inst Wurzburg 2:46–104

Sardet C, McDougall A, Houliston E (1994) Cytoplasmic domains in eggs. Trends Cell Biol 4:166–172

Savoian MS, Rieder CL (2002) Mitosis in primary cultures of *Drosophila melanogaster* larval neuroblasts. J Cell Sci 115:3061–3072

Schaefer M, Shevchenko A, Shevchenko A, Knoblich JA (2000) A protein complex containing inscuteable and the Gα-binding protein Pins orients asymmetric cell divisions in Drosophila. Curr Biol 10:353–362

Schroeder TE (1980) Expressions of the prefertilization polar axis in sea urchin eggs. Dev Biol 79: 428–443

Shibazaki Y, Shimizu M, Kuroda R (2004) Body handedness is directed by genetically determined cytoskeletal dynamics in the early embryo. Curr Biol 14:1462–1467. https://doi.org/10.1016/j.cub.2004.08.018

Shimizu T, Ishii R, Takahashi H (1998) Unequal cleavage in the early Tubifex embryo. Dev Growth Differ 40:257–266

Siller KH, Doe CQ (2008) Lis1/dynactin regulates metaphase spindle orientation in Drosophila neuroblasts. Dev Biol 319:1–9. https://doi.org/10.1016/j.ydbio.2008.03.018

Siller KH, Doe CQ (2009) Spindle orientation during asymmetric cell division. Nat Cell Biol 11:365–374. https://doi.org/10.1038/ncb0409-365

Siller KH, Cabernard C, Doe CQ (2006) The NuMA-related mud protein binds Pins and regulates spindle orientation in Drosophila neuroblasts. Nat Cell Biol 8:594–600. https://doi.org/10.1038/ncb1412

Sobral D, Tassy O, Lemaire P (2009) Highly divergent gene expression programs can Lead to similar chordate larval body plans. Curr Biol 19:2014–2019. https://doi.org/10.1016/j.cub.2009.10.036

Stolfi A, Lowe EK, Racioppi C et al (2014) Divergent mechanisms regulate conserved cardiopharyngeal development and gene expression in distantly related ascidians. elife 3. https://doi.org/10.7554/eLife.03728

Takatori N, Kumano G, Saiga H, Nishida H (2010) Segregation of germ layer fates by nuclear migration-dependent localization of Not mRNA. Dev Cell 19:589–598. https://doi.org/10.1016/j.devcel.2010.09.003

Tassy O, Daian F, Hudson C et al (2006) A quantitative approach to the study of cell shapes and interactions during early chordate embryogenesis. Curr Biol 16:345–358. https://doi.org/10.1016/j.cub.2005.12.044

Théry M, Jiménez-Dalmaroni A, Racine V et al (2007) Experimental and theoretical study of mitotic spindle orientation. Nature 447:493–496. https://doi.org/10.1038/nature05786

Thompson DW (1942) On growth and form, new ed. Cambridge University Press/Macmillan, Cambridge/New York

Treen N, Heist T, Wang W, Levine M (2018) Depletion of maternal cyclin B3 contributes to zygotic genome activation in the ciona embryo. Curr Biol 28:1330–1331. https://doi.org/10.1016/j.cub.2018.03.058

Turlier H, Maître J-L (2015) Mechanics of tissue compaction. Semin Cell Dev Biol 47–48: 110–117. https://doi.org/10.1016/j.semcdb.2015.08.001

von Sachs J (1887) Lecture XXVII. Relations between growth and cell-division in the embryonic tissues. Lectures in plant physiology. Clarendon, Oxford, 431–459

Voronina E, Wessel GM (2006) Activator of G-protein signaling in asymmetric cell divisions of the sea urchin embryo. Dev Growth Differ 48:549–557. https://doi.org/10.1111/j.1440-169X.2006.00895.x

Weitzel HE, Illies MR, Byrum CA et al (2004) Differential stability of beta-catenin along the animal-vegetal axis of the sea urchin embryo mediated by dishevelled. Development 131: 2947–2956. https://doi.org/10.1242/dev.01152

Wieschaus E, Nüsslein-Volhard C (2014) Walter Gehring (1939–2014). Curr Biol 24:R632–R634. https://doi.org/10.1016/j.cub.2014.06.039

Wieschaus E, Nüsslein-Volhard C (2016) The Heidelberg screen for pattern mutants of drosophila: a personal account. Annu Rev Cell Dev Biol 32:1–46. https://doi.org/10.1146/annurev-cellbio-113015-023138

Wilson EB (1896, 1905, 1915) The cell in development and inheritance. Macmillan, London

Winklbauer R (2015) Cell adhesion strength from cortical tension – an integration of concepts. J Cell Sci 128:3687–3693. https://doi.org/10.1242/jcs.174623

Woolner S, Papalopulu N (2012) Spindle position in symmetric cell divisions during epiboly is controlled by opposing and dynamic apicobasal forces. Dev Cell 22:775–787. https://doi.org/10.1016/j.devcel.2012.01.002

Xiong F, Ma W, Hiscock TW et al (2014) Interplay of cell shape and division orientation promotes robust morphogenesis of developing epithelia. Cell 159:415–427. https://doi.org/10.1016/j.cell.2014.09.007

Yamamoto K, Kimura A (2017) An asymmetric attraction model for the diversity and robustness of cell arrangement in nematodes. Development 144:4437–4449. https://doi.org/10.1242/dev.154609

Yu F, Morin X, Cai Y et al (2000) Analysis of partner of inscuteable, a novel player of Drosophila asymmetric divisions, reveals two distinct steps in inscuteable apical localization. Cell 100:399–409

Chapter 7
Sex Determination, Sexual Development, and Sex Change in Slipper Snails

Maryna P. Lesoway and Jonathan Q. Henry

Sex is the queen of problems in evolutionary biology.
Graham Bell 1982
However, what is more interesting is the mollusc's sex life.
Monty Python's Flying Circus, Episode 32, "The War Against Pornography," 1972

Abstract Sex determination and sexual development are highly diverse and controlled by mechanisms that are extremely labile. While dioecy (separate male and female functions) is the norm for most animals, hermaphroditism (both male and female functions within a single body) is phylogenetically widespread. Much of our current understanding of sexual development comes from a small number of model systems, limiting our ability to make broader conclusions about the evolution of sexual diversity. We present the calyptraeid gastropods as a model for the study of the evolution of sex determination in a sequentially hermaphroditic system. Calyptraeid gastropods, a group of sedentary, filter-feeding marine snails, are sequential hermaphrodites that change sex from male to female during their life span (protandry). This transition includes resorption of the penis and the elaboration of female genitalia, in addition to shifting from production of spermatocytes to oocytes. This transition is typically under environmental control and frequently mediated by social interactions. Males in contact with females delay sex change to transition at larger sizes, while isolated males transition more rapidly and at smaller sizes. This phenomenon has been known for over a century; however, the mechanisms that control the switch from male to female are poorly understood. We review here our current understanding of sexual development and sex determination in the calyptraeid gastropods and other molluscs, highlighting our current understanding of factors implicated in the timing of sex change and the potential mechanisms. We also

M. P. Lesoway · J. Q. Henry (✉)
Department of Cell and Developmental Biology, University of Illinois, Urbana-Champaign, Urbana, IL, USA
e-mail: j-henry4@illinois.edu

W. Tworzydlo, S. M. Bilinski (eds.), *Evo-Devo: Non-model Species in Cell and Developmental Biology*, Results and Problems in Cell Differentiation 68,
https://doi.org/10.1007/978-3-030-23459-1_7

consider the embryonic origins and earliest expression of the germ line and the effects of environmental contaminants on sexual development.

7.1 Introduction

The most intriguing questions in biology often focus on the fundamentals of life: food, relationships, and sex. Sex in biology has led to foundational questions of why sex exists and why most species have separate sexes and has proven to be a fruitful area of research across evolution, development, and ecology. As intriguing as these questions are, the focus of most research has been on separate sexes, the dominant expression of sexuality in metazoans (Jarne and Auld 2006). Hermaphroditism, where individuals function as both male and female either at the same time (simultaneous hermaphroditism) or in series (sequential hermaphroditism), is less common as a reproductive strategy, occurring in approximately 5% of animals (Jarne and Auld 2006). Despite the relative rarity of hermaphroditism as a sexual strategy in animals, it is phylogenetically widespread, found in 70% of phyla (Policansky 1982; Jarne and Auld 2006). For example, sequential hermaphroditism is known from systems as diverse as reef fishes (e.g., wrasse, Robertson 1972; clownfish, Buston 2003), shrimp (Bauer 2006), and molluscs (Heller 1993), including the slipper snails, described below. These exceptions to the norm have stimulated exploration of how individuals determine optimal sexual roles and what factors influence the timing and direction of sexual transitions. As hermaphroditism has emerged multiple times, this mode of reproduction is of interest for understanding the evolution of sex and repeated transitions between dioecy and hermaphroditism. Sequential hermaphroditism, in which an individual functions as different sexes through time, is particularly valuable as a system in which to explore sexual development in an evolutionary context.

While most groups have some level of sexual diversity, molluscs (e.g., snails, squid and octopus, bivalves, etc.) exhibit nearly as much diversity in their sexual types as they do in their body plan. As the second most speciose phylum after the arthropods, molluscs are widely distributed and often commercially important. As members of the spiralian lophotrochozoans, which include over one third of recognized phyla (e.g., annelids, nemerteans, platyhelminths, etc.), most molluscs develop via a stereotyped pattern of early cleavage. This shared developmental pattern allows cellular homologies in early development to be traced across large phylogenetic distances and is a powerful tool for understanding the origins of body plan diversity. These features make molluscs an attractive system for exploring the evolution of sexual development. As in other animals, most molluscs are dioecious with separate sexes (around 75%, Auld and Jarne 2016). Many groups are hermaphroditic. This includes simultaneous hermaphroditism as in most pulmonates and opisthobranchs, where both sexes are expressed at the same time, and sequential hermaphroditism, as in many bivalves and certain gastropod groups, where an individual transitions from

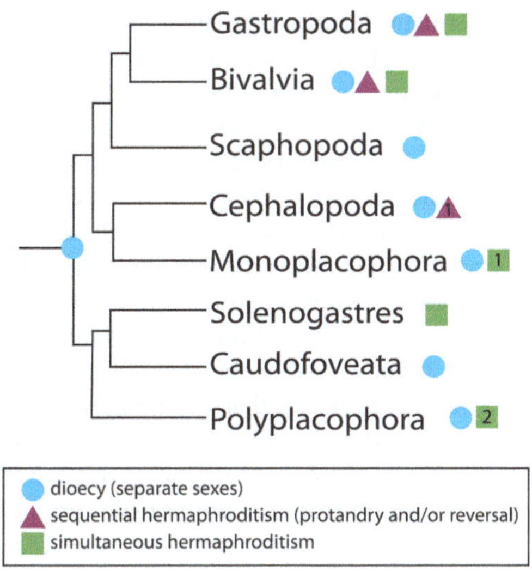

Fig. 7.1 Phylogram of molluscan relationships. Symbols indicate sexual type (see legend), with numbers indicating number of species where this type of sexuality is not widely distributed within the group. After Collin (2013). Based on Kocot et al. (2011) and Smith et al. (2011)

one sex to another (Fig. 7.1) (Heller 1993). Sequential hermaphroditism can be expressed as male to female sexual transition (protandry), as found in the calyptraeid gastropods, or as female to male transition (protogyny), which is common in other sex-changing animals such as fish (Robertson 1972; Buston 2003) but is not known to occur in molluscs (Heller 1993; Hoagland 1978). Some molluscs, namely, bivalves, have alternating sexuality, where multiple sex changes and reversals in the same individual are possible (Coe 1943). It should be noted that the terminology surrounding the sexual diversity in animals has historically been muddled (see Collin 2013). Here, we restrict the term "sex change" to a unidirectional, nonreversible change in sex and "sexual reversal" to repeated changes in sexual characters.

While sexual types in the molluscs are diverse, most of this diversity is limited to the gastropods and bivalves (Fig. 7.1) (see Collin 2013). Outside of these two classes, molluscs are almost exclusively described as dioecious with separate sexes, with few exceptions. For example, all aplacophoran solenogasters described to date are simultaneous hermaphrodites. Other exceptions include two members of one polyplacophoran genus with simultaneous hermaphroditism (*Lepidochitona*, Eernisse 1988), one case of potential sequential hermaphroditism in a cephalopod (*Ancistrocheirus lesueurii*, Hoving et al. 2006) and one case of simultaneous hermaphroditism in a monoplacophoran (*Micropilina arntzi*, Haszprunar and Schaefer 1996). The infrequency of these exceptions does not rule out the possibility that there are more cases yet to be described. Molluscan phylogenetic relationships and outgroup comparisons suggest that dioecy is the ancestral state, with multiple independent origins of hermaphroditism (Fig. 7.1) (Ponder and Lindberg 1997; Kocot et al. 2011; Smith et al. 2011). Basal groups of gastropods and bivalves are dioecious, lending support to this hypothesis (Collin 2013). Thus, the evolution of molluscs includes repeated parallel emergence of different types of hermaphroditism, mainly restricted to a few groups within the gastropods and bivalves.

Hermaphroditism is thought to emerge under certain selective conditions. Theoretical models, most prominently sex allocation theory, have been used to explore how selection has led to the repeated evolution of hermaphroditism. Sex allocation asks how individuals and populations divide resources between male and female functions (Charnov 1982). In dioecious species with separate sexes, sex allocation determines the sex ratio (ratio of males to females) in a population. It can also be used to understand how individuals and populations balance costs of male and female function with reproductive output. In hermaphroditic species, sex allocation relates to the timing and direction of sex change in sequential hermaphrodites and the allocation of male and female reproductive resources in simultaneous hermaphrodites (Charnov 1982). The size-advantage model of sex allocation (Ghiselin 1969; Warner 1988) posits that changing sex should be advantageous if one sex has a higher reproductive output at larger sizes than the other sex. For example, larger males that produce more offspring relative to females of the same size should select for a transition from female to male (protogyny). Conversely, when larger females produce more offspring relative to males of comparable sizes, this should select for a transition from male to female (protandry) (Munday et al. 2006; Warner 1975; Warner et al. 1975). When conspecifics are scarce and/or widely scattered, it appears to be advantageous for an individual to perform both male and female functions simultaneously, thereby increasing reproductive output (Ghiselin 1969). Sedentary lifestyles, combined with patchy distribution and environmental heterogeneity, are thought to select for sequential hermaphroditism (Wright 1988). Sequentially hermaphroditic molluscs such as oysters and some gastropod groups, like limpets and slipper snails, are often sedentary, reducing the chances of encountering a mate (Hoagland 1978; Wright 1988). Reproductive behavior does not appear to influence this, as sequential hermaphrodites reproduce via either free spawning or copulation (Hoagland 1978); however, sexual reversals in molluscs appear to be associated with free spawning (Collin 2013). While sequential hermaphrodites generally function as only one sex at a time, nearly all heterobranchs (opisthobranchs and the pulmonates) are simultaneous hermaphrodites with mutual insemination (Heller 1993; Jarne and Auld 2006). While marine opisthobranchs do not typically self-fertilize, it is common for terrestrial pulmonates to do so, and some species do so preferentially (Jarne and Auld 2006; Auld and Jarne 2016; Koene 2017). Simultaneous hermaphroditism, including the role of sex allocation and sexual selection, has been reviewed elsewhere (Auld and Jarne 2016; Nakadera and Koene 2013).

Within the molluscs, slipper snails (calyptraeid gastropods, Fig. 7.2) provide an excellent system for exploration of sequential hermaphroditism, including the molecular mechanisms and evolution of sex determination, sexual differentiation, and sex change. For more than a century, scientists have studied the sex lives of these animals. The influence of environment in sex determination makes the calyptraeid gastropods of particular interest. While all calyptraeids change sex from male to female (Fig. 7.2), different species show greater or lesser influence from the presence of conspecifics on the timing of sex change. New molecular techniques are opening non-model systems to mechanistic exploration, and the calyptraeids, with an abundance of information on sexual development and the ability to make comparisons

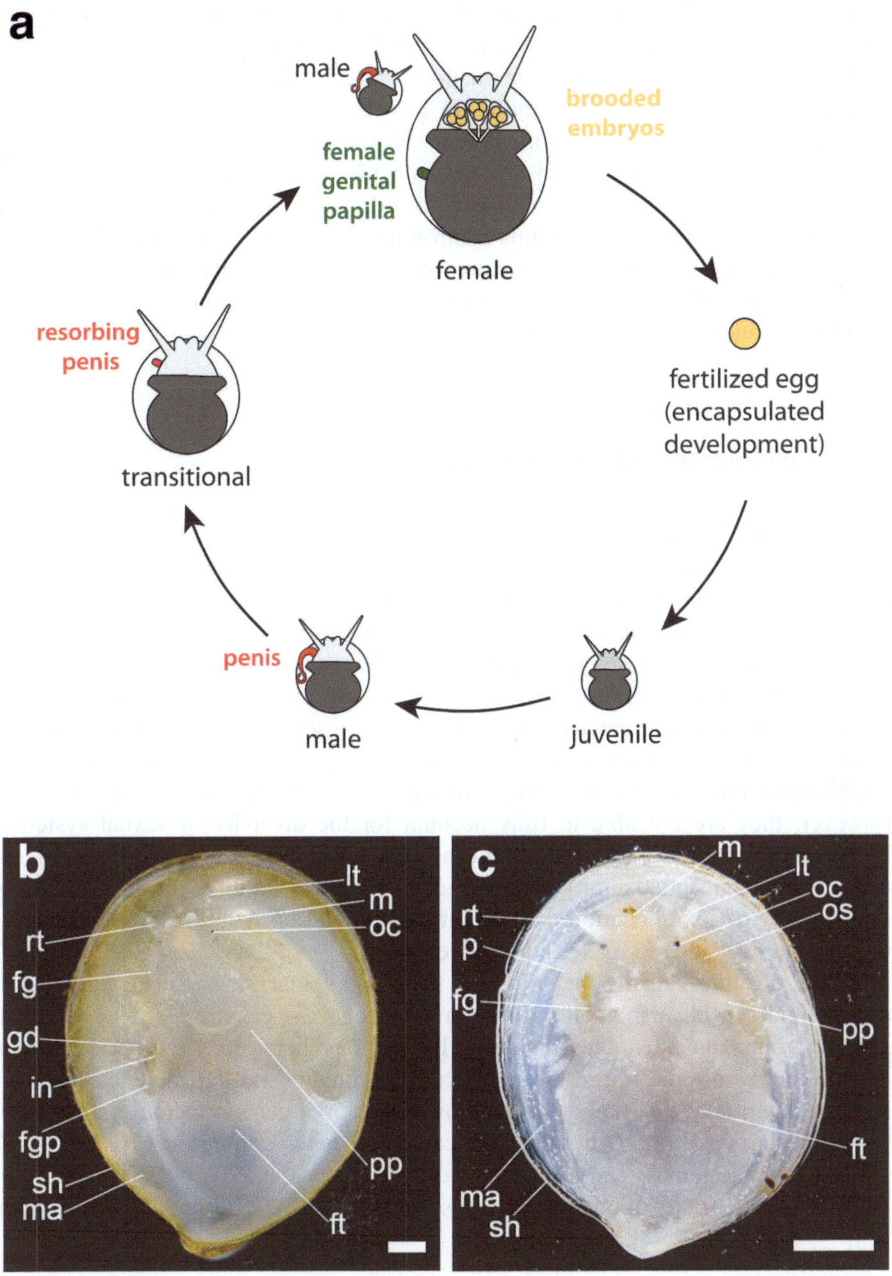

Fig. 7.2 Generalized life cycle and external anatomy of calyptraeid gastropods. (**a**) Life cycle of sequentially hermaphroditic slipper snails. Females brood embryos, which in the case of direct developers, complete the development to juvenile stages within the capsule. Embryos emerge either as juvenile snails (direct developers) or veliger larvae (indirect developers, not shown), which complete their development in the water column prior to settlement and metamorphosis. Juvenile snails first mature into males, producing sperm and grow an external penis. Males eventually resorb the penis, convert to oocyte production, and produce female external genitalia. The timing of sexual

between closely related species, make them an excellent model to study the development of sexual systems (Henry et al. 2017; Henry and Lyons 2016). While the underlying mechanisms of these interactions remain poorly known, new tools and techniques are increasing our ability to answer questions about all aspects of development in this group of animals (e.g., Henry et al. 2017). In this chapter, we will discuss sexual systems and sex determination in molluscs, with a focus on environmental sex determination in calyptraeids. This will include discussion of some known proximate mechanisms controlling sex determination in molluscs. We will also briefly discuss the earliest expression of sexual development, the germ line. We conclude with anthropogenic impacts on sexual development in molluscan populations and suggest directions for future research.

7.2 Mechanisms of Sex Determination

Sex determination, the developmental processes that produce male or female individuals or cells (eggs and sperm), is fundamental to questions of sex allocation (Beukeboom and Perrin 2014). Exploration of mechanisms of sex determination has led to the realization that diversity is the rule (Beukeboom and Perrin 2014; Bull 1983). The diversity of sex determination mechanisms has prompted a search for commonalities at multiple levels of biological organization and has shown that there are at least some shared mechanisms. The lability of sex determination mechanisms and the action of selection in producing this variety are exciting questions. Well-established model systems have taught us a great deal about how sex is determined; however, they are not able to fully account for the diversity of sexual systems. Increasing accessibility of genetic and molecular techniques in non-model organisms and the conceptual approach of evo-devo are powerful tools for understanding the diversity of sex allocation, sex determination, and how the underlying mechanisms have been altered by selective forces.

While the mechanisms that produce male and female forms are staggeringly diverse, most mechanisms fall into two main categories, genetic sex determination and environmental sex determination. Interactions between environmental and genetic factors are the norm. Genetic sex determination (GSD) includes chromosomally determined sex and determination by multiple genetic factors. In chromosomally determined sex, either the male can be the heterogametic sex (i.e., XX/XY systems), or the female can be the heterogametic sex (i.e., ZW/WW systems), or males may lack a copy of one sex chromosome (i.e., XX/XO systems). Systems with

Fig. 7.2 (continued) transitions typically depends on environment. Relevant factors include age, nutrition, size, and the presence/absence and sex of conspecifics. (**b**) Ventral view of *Crepidula atrasolea* female. (**c**) Ventral view of smaller *C. atrasolea* male. *fg* food groove, *fgp* female genital papilla, *ft* foot, *gd* gonoduct, *in* intestine, *lt* left tentacle, *m* mouth, *ma* mantle, *oc* ocellus, *os* ospharadium, *p* penis, *pp* propodium, *rt* right tentacle, *sh* shell. Scale bars represent 1 mm

sex determination under the control of multiple genetic factors may be oligogenic, with control by a few genes, or polygenic, with control by many genes of small effect. Environmental sex determination (ESD) includes sexual systems that are determined by external factors such as nutritional state, temperature, social structure, or some combination of environmental triggers. This diversity has prompted a great deal of speculation about how and why diversity of sex determination has arisen (Bachtrog et al. 2014; Beukeboom and Perrin 2014; Bull 1983; Haag and Doty 2005) and prompted the search for underlying commonalities. Molluscs include examples of all types of genetic sex determination. However, despite the diversity of sex determination in molluscs, the mechanistic bases have only been explored in detail for relatively few species.

Much of our current understanding of genetic sex determination comes from studies of model organisms (e.g., *Drosophila*, Oliver and Clough 2012; *C. elegans*, Riddle et al. 1997; and mouse, Koopman et al. 1991). Evidence for the genetic control of sex determination in molluscs has come from chromosomal analyses, analysis of sex ratios in natural and experimental populations, and more recently transcriptomic and molecular expression studies. These studies are increasing our understanding of the diversity of molluscan sex determination and started to hint at key players in the underlying gene regulatory networks. Many studies of sex determination in molluscs rely on the use of sex ratios to infer underlying mechanisms (see Yusa 2007). Sex ratios have the advantage of being relatively easy to measure, at least in species with obvious, external sexual characters. While the sex ratio approach has been useful, it has important caveats. Simple tallies of sexes in a given population can miss important influences on sex ratios such as differential mortality at various life stages, or broader population differences, and can miss cases of sequential hermaphroditism (Yusa 2007). While the best method of determining sexual patterns is to follow animals throughout the life cycle, this remains impractical for long-lived or rare species.

7.2.1 Genetic Sex Determination in Molluscs

Chromosomal studies have found heterogametic chromosomes in a number of gastropods, presumed to be sex chromosomes (reviewed by Thiriot-Quiévreux 2003). This includes the XX–XY system, with males as the heterogametic sex (i.e., *Viviparus* spp., *Littorina saxatalis*, two neogastropod species), the XO–XX system where males lack a sex chromosome (e.g., *Neotricula aperta*), and the ZW–ZZ system, with females as the heterogametic sex (e.g., *Viviparus* spp.). No sex chromosomes have been reported in either Pulmonata or Opisthobranchia (Thiriot-Quiévreux 2003). Comparable studies of bivalves have also found no evidence of heterogamety (see Breton et al. 2018 for review). Many members of these groups are true hermaphrodites, expressing both male and female gametes at the same time, making chromosomally based sex determination mechanisms unlikely. Karyotypes continue to be explored (e.g., García-Souto et al. 2018, *Littorina* spp.); however, these studies generally have been limited to chromosome number and description. The actual mechanisms of sex determination in relation to chromosomally expressed genes remain unexplored in molluscs.

Genetic control of sex allocation and sex determination more generally has been best studied in bivalves, owing to their commercial importance and value in aquaculture (see Breton et al. 2018 for review). Controlled crosses in the Pacific oyster, *Crassostrea gigas*, suggest that sex determination is controlled by a single locus (Guo et al. 1998). Analysis of 86 families suggested the presence of a dominant male allele (M) and recessive protandrous hermaphroditic allele (F) (Guo et al. 1998). However, this two-locus model does not fully account for the ratios observed, and a three-locus model proposes that FF produces true females, MM true males, and FM protandric females (Hedgecock and Hedrick 2010). Similar lineage experiments in the dioecious pulmonate, *Pomacea canaliculata*, have suggested that a small number of genes are required to produce the sex ratios observed in full- and half-sibling crosses (Yusa 2007; Yusa and Kumagai 2018).

Chromosomal and genomic information is generally lacking in the calyptraeid gastropods. Conklin (1902) reported 30 chromosomes ($n = 30$) in maturing oocytes of *Crepidula plana*. The karyotype of *Crepidula unguiformis* is reported as $n = 17$ (Libertini et al. 2009). Interestingly, the authors of the latter report a heteromorphic chromosome in approximately one third of the cases examined but doubt that this is a sex chromosome, stating unequivocally "the possibility that it may be linked to a sex determining mechanism is easily ruled out since hermaphrodite organisms cannot have sex-linked chromosomes" (Libertini et al. 2009, p. 112). Be that as it may, several aspects of calyptraeid sexual development and behavior likely have a genetic component. For example, juveniles and small males are typically highly motile and become more sedentary as they increase in size and age. This behavioral shift includes changes in pedal gland morphology (Chaparro et al. 1998, 2001). These shifts in motility are linked to the initiation of the sexual transition (Coe 1935). However, interindividual differences in behavior and motility of males in relation to the timing of sexual transition could be also under genetic control (Coe 1935). Indeed, some more sedentary males delay their transition to females sometimes for very lengthy periods (Coe 1935, 1938a, b). Lack of mechanistic understanding limits the conclusions that we can make about the genetic control of sex determination. Chromosomal studies and multi-generation crosses, as described previously, could be used to demonstrate the genetic component of sex allocation in calyptraeids.

7.2.2 Environmental Sex Determination in Molluscs

While no system can realistically be said to be under either exclusive genetic or environmental control, environmental factors can be the controlling factor in sex determination. For example, temperature is known to determine sex in turtles and other reptiles, with higher temperatures producing females and lower temperatures resulting in development of males (Bull and Vogt 1979; Miyagawa et al. 2018). Molluscan examples of ESD are found primarily in sequential hermaphrodites, which allocate resources to male or female function in response to environmental factors. Potential environmental influences include (but are not limited to) temperature, nutritional status, and social environment. For example, some patellogastropods (true limpets) change sex from male to female depending on population density (*Lottia gigantea*, Lindberg and Wright 1985). Detailed observations have also shown that seasonal sex reversals are possible in limpets (*Patella ferruginea*, Guallart et al. 2013). The oyster, *Ostrea edulis*, and other bivalves also change and reverse sex in response to nutritional and temperature cues (Coe 1943; Breton et al. 2018).

All calyptraeids are protandric hermaphrodites, changing sex once from male to female (Fig. 7.2). Fertilization is internal via copulation, so the sexual transition includes resorption of the penis, elaboration of a female genital papilla and associated reproductive structures, as well as a shift from production of spermatocytes to oocytes in the gonad (Fig. 7.2). While Conklin (1897) speculated that *C. fornicata* might be hermaphroditic, Orton's (1909) examination of the gonadal tissues was the first to confirm sequential hermaphroditism in *Crepidula*. Subsequent studies have confirmed this pattern in all members of the family (Coe 1936, 1938a, b; Collin 1995, 2000, 2006; Collin et al. 2005; Gould 1917a, b, 1919, 1952; Hoagland 1978; Warner et al. 1996). While all individuals pass through a male phase before transitioning to female maturity, the precise timing of sexual development is influenced by the presence of conspecifics. Males associated with females delay their sexual transition, changing sex at a larger size relative to isolated males or males associated with other males (Coe 1938b; Collin 1995). This has been confirmed in multiple species (Coe 1935, 1936, 1938a, b; Ishiki 1936, 1938; Warner et al. 1996; Collin 2000, 2006; Collin et al. 2005). Empirical evidence of size at sex change in the calyptraeids has generally shown that individuals make decisions about transitions based on local environmental factors. These include population density, local mating population size and composition, as well as the age, size, and nutritional status of an individual.

One of the best studied examples in the calyptraeids is the Atlantic slipper snail, *Crepidula fornicata*, which famously forms stacks or chains of up to tens of animals (Fig. 7.3a). Small males are motile and are found at the top or along the sides of stacks, while large, mature females are sedentary and found at the bottom of a stack attached to an empty shell or other substrate (Coe 1936; Collin 1995). Reduced motility in larger males is thought to reflect reduced opportunity for copulation; however, paternity studies have shown that the male closest to the female in a stack has a reproductive advantage (Dupont et al. 2006; Proestou et al. 2008; Broquet et al. 2015). Furthermore, paternity is generally restricted to members of a stack (Dupont et al. 2006; Le Cam et al. 2009; Proestou et al. 2008). Stack or aggregate size, motility, and mode of development are all associated with variation in sex ratio and size at sex change (Collin 2006). Adult behavior in other calyptraeid species varies. Adults of some species may be found forming stacks as in *C. fornicata*; others are more commonly found in isolation or in mating pairs (Fig. 7.3b) or in

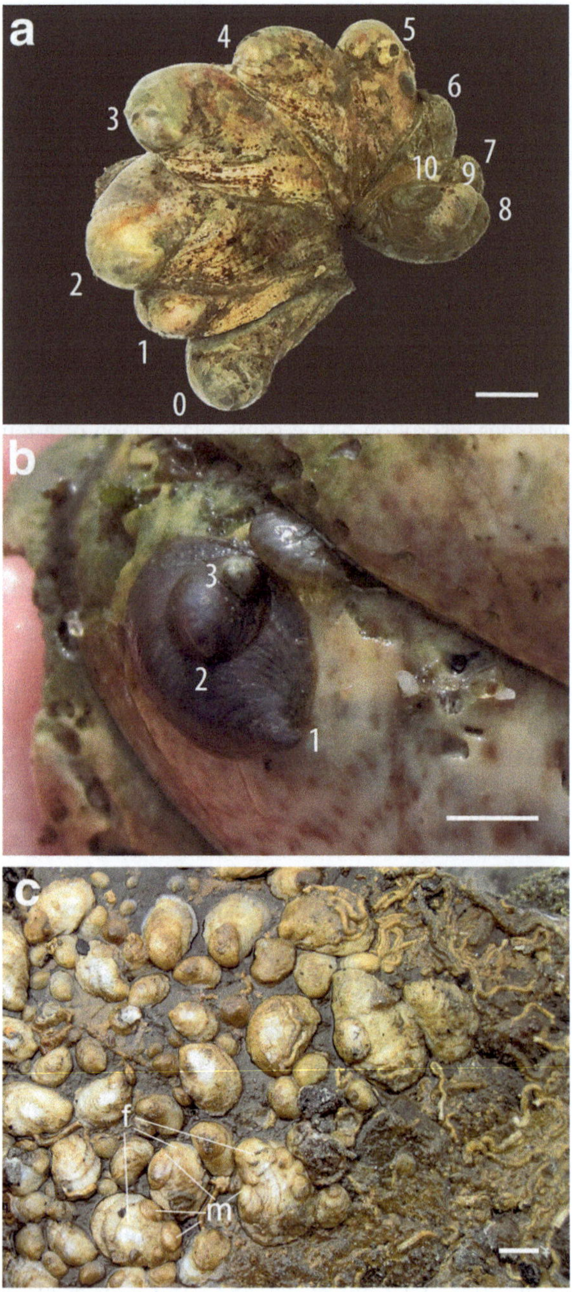

Fig. 7.3 Social behavior and sexual dimorphism in calyptraeid gastropods. (**a**) Gregarious stacking as exemplified by *C. fornicata*. This stack contains at least ten animals, anchored by an empty shell (0). The largest animals at the bottom of the stack are female (1, etc.), and smaller animals at the top of the stack are males (9, 10). (**b**) Substrate-restricted animals such as *C. convexa* are typically found singly or as breeding pairs. In this case, three animals form a mating group, with a large female at the bottom (1) and two smaller males above (2, 3). (**c**) Where substrate is not limiting, calyptraeids can cover all available hard substrate, as seen in this group of mainly *C. lessoni*. Note the presence of smaller animals (likely males, m) on top of much larger females (f). Scale bars in (**a**, **c**) represent 1 cm, 0.5 cm in (**b**)

masses covering available substrate (Fig. 7.3c). These behaviors influence local mating populations and therefore sexual transitions, as the size of a mating group influences size at sex change and the sex ratio of a population (Collin 2006).

Experimental manipulation of *Crepidula norrisarium* also found evidence of social control of sex change in a large, direct-developing species restricted to limited habitats on the shells of other gastropods (Warner et al. 1996). When multiple hosts are available, males of *C. norrisarium* tend to move to hosts where females are present, thereby increasing their reproductive potential (Hobday and Riser 1998). Male reproductive success has been shown to directly influence the timing of sexual transition in *Crepidula coquimbensis*, where males with more access to females and longer periods in copulatory positions showed increased levels of paternity and reduced growth rates and rates of sex change (Brante et al. 2016).

Not all calyptraeids appear to be influenced by the presence of conspecifics. For example, *C. convexa* are small and brood their direct-developing offspring until they hatch as juveniles. Field and laboratory manipulations show little influence of conspecifics on the timing of sex change (Coe 1938b; Hoagland 1978). Similarly, *Calyptraea chinensis* does not show any changes in the rate of male growth due to female presence or absence (Bacci 1951). Likewise, social environment is not the only factor that can influence the timing of sex change in calyptraeids. Decreased food availability appears to lengthen the transitional period at the expense of male function in *C.* cf. *marginalis* and *C. incurva* (Mérot and Collin 2012a, b), while desiccation stress and increased early mortality produce smaller females (Mérot and Collin 2012b).

These studies highlight the role of plasticity in response to social and environmental cues in controlling the timing of sex change. Considering sexual development and sex change as plastic responses to environmental inputs provides us with a conceptual framework to ask and address these questions (Capel 2017). Few systems to date present examples of known mechanisms of environmentally mediated polyphenic shifts. One example is the nematode worm, *Pristionchus pacificus*, which produces either a wide-mouthed, predatory morph or a narrow-mouthed, grazing morph in response to nutritional and social cues (Ragsdale et al. 2013). This developmental switch appears to be controlled at the site of the polyphenism by the activity of a sulfotransferase (SEUD-1) and NHR-40, with a threshold set by the opposing activity of a sulfatase (EUD-1) (Bui et al. 2018). The integration of multiple inputs (e.g., nutrition, social environment, etc.) in calyptraeid sex change suggests the integration of hormonal/neural signals with molecular switches based on certain thresholds. The availability of multispecies comparisons within the calyptraeids and differences in the degree to which conspecifics influence sex change will allow us to explore the commonality (or lack) of these environmentally mediated mechanisms.

7.3 Mechanisms of Sexual Differentiation

While the role of conspecifics on the timing of sex change in many calyptraeids is well recognized, the proximate mechanisms controlling these changes have only recently begun to be described. Early workers suggested either that females secreted a waterborne pheromonal cue that could act at a distance (Gould 1917b, 1919, 1952) or that males were delaying their sexual transition in response to cues that required direct contact with females (Coe 1938a, 1942). However, recent experiments have revealed that social control of sex change is contact-mediated. In *C. fornicata* (Cahill et al. 2015) and *C.* c.f. *marginalis* (Carrillo-Baltodano and Collin 2015), animals were reared in conditions that allowed for the free flow of water and potential waterborne cues and either allowed or blocked from direct physical contact. Males of *C. fornicata* were also exposed to clean, empty shells (Cahill et al. 2015; see also Coe 1938a), and males of *C.* c.f. *marginalis* were exposed to mucus and substrate that was previously in contact with conspecifics (Carrillo-Baltodano and Collin 2015). Males of both species were inhibited from transitioning to female only by direct physical contact with females. While these experiments confirm the importance of physical contact in influencing the timing of sex change, the precise nature of the signal remains unknown, though Carrillo-Baltodano and Collin (2015) speculate that signals during copulation may play a role in mediating this response (see also Coe 1938a).

Male development may be controlled by secretions of the pedal ganglion at the base of the right tentacle, adjacent to the penis outgrowth region (Le Gall and Streiff 1975; Le Gall 1981; Le Gall and Feral 1982; Féral et al. 1987). In a series of experiments, Le Gall (1981) found that removal of the right tentacle will quickly result in tentacle regrowth. Similarly, if the penis is removed, it can be regrown. When both the tentacle and penis are removed, the tentacle will regenerate first, followed by the penis. However, a second removal of the right tentacle prior to penis regrowth blocks its regeneration (Le Gall 1981). Removal of the pedal ganglion blocks regeneration of the penis, but not the tentacles (Le Gall 1981). Interestingly, the insertion of another pedal ganglion from another male individual of the same species (*C. fornicata*) or even the pedal ganglion from males of another species (*Calyptraea*, *Littorina* spp., *Buccinum* spp., Le Gall 1981) is able to induce penis growth. Similar results were reported in *Calyptraea chinesis* (Streiff 1966, 1967). This is reminiscent of the nervous control of optic gland secretions in octopus. Severing innervation to the optic gland initiates the growth of the reproductive organs in males and females (Wells and Wells 1959), and similar neurohormonal controls may be shared among molluscs. The broad action of the pedal ganglion in regeneration and penis growth suggests a hormonal secretion. While a potential signaling compound has been partially purified (Le Gall et al. 1987), its identity remains unknown.

Despite these tantalizing glimpses of mechanisms controlling sex determination and sex change in the calyptraeid gastropods, the precise molecular and hormonal mechanisms remain elusive. However, a recent spate of transcriptomic studies in molluscs, primarily bivalves, has implicated several candidate genes, including conserved sex-determining genes. For example, in the oyster *Crassostrea gigas*, Zhang et al. (2014) showed expression of several conserved sex-determining genes, including *Cgdsx* (*C. gigas doublesex*) and *CgSoxH* (*C. gigas Sry-related HMG box*), which are expressed specifically in gonadal tissues (Zhang et al. 2014). Similar transcriptomic studies in other bivalves have suggested a number of conserved genes involved in sexual differentiation. This includes *dmrt* in *Pinctada margaritifera* (Teaniniuraitemoana et al. 2014) and *dmrt*, *SoxH*, and *FoxL2* in *Patinopecten yessoensis* (Li et al. 2016). More fine-grained approaches are beginning to identify novel candidates and suggest important roles for neurotransmitters and long non-coding RNAs (*C. gigas*, Yue et al. 2018). Although their functions in sexual differentiation and sex change have yet to be confirmed, these studies provide a useful starting point. Further, the molecular components identified in these studies may also play important roles in sexual development and differentiation in the calyptraeid gastropods.

The role of broadly conserved components, such as DM-domain containing *dmrt-like* genes in gonadal tissues of molluscs, is of particular interest. The transcription factor, *dmrt* (*doublesex/male-abnormal-3 related*), was first described in studies of the fruit fly, where alternately spliced forms of the *Drosophila melanogaster* transcript *dsx* (*doublesex*) lead to differentiation between male and female somatic forms (Burtis and Baker 1989; Hildreth 1965). *Dsx* contains a DM domain, which binds DNA. The discovery of DM-like domains in *Caenorhabditis elegans mab-3 (male abnormal*) and their role in sex determination demonstrated that despite abundant variation in sex differentiation mechanisms, some commonalities are present (Raymond et al. 1998). DM domains have subsequently been found to play a role in sex determination in vertebrates (e.g., *dmrt1* in mouse, Raymond et al. 2000) and throughout the Metazoa (Picard et al. 2015), including lophotrochozoans (Chong et al. 2013). *Dmrt* is so far the only gene family identified with a conserved role in sex determination (Kopp 2012). While DM domain containing *dmrt* homologues are found throughout the animal kingdom, its position within the hierarchy of sex-determining gene regulatory networks is highly variable (Kopp 2012). Exploring the role of this conserved sex-determining gene in a group with environmental sex determination will be a useful comparison to existing models of genetic sex determination.

7.4 Developmental Origins of the Germ Line

The earliest expression of sexual development in the individual is the primordial germ cell. The germ line is either determined autonomously early in development as in the fly, or induced by surrounding cells later in development, as in the mouse (Extavour and Akam 2003). It is thought that induction is the ancestral mode of germ cell determination in the Metazoa, and this has also been suggested for spiralian animals (Extavour and Akam 2003; Rebscher 2014). However, evidence increasingly suggests that many spiralians have autonomous determination of the germ line early in development (Rebscher 2014).

Spiralian animals (lophotrochozoans) share a highly conserved, stereotyped early cleavage pattern, referred to as spiral cleavage (Henry 2014). This pattern of early development allows for comparisons of cell lineages across the group and is a powerful tool for understanding how changes in early development have led to diversity in adult body plans. One of the best characterized cells in all spiralians is the 4d cell, which plays an important role in early embryonic organization, gives rise to much of the mesoderm, and is often the first cell to break the spiral cleavage pattern (Henry 2014). This cell arises from one of the four cell quadrants, the D quadrant (Fig. 7.4a) after the three earlier tiers of micromeres have been born. The 4d cell also appears to give rise to the primordial germ cells (Fig. 7.4 and Table 7.1) (Rebscher 2014). Several lines of evidence have been used to identify the putative primitive germ cells (PGCs) in spiralians. Classical studies used the presence of distinct cytoplasmic markers and arrest of cell divisions to identify primordial germ cells. For example, dense aggregations of mitochondria are localized to the primordial germ cells of the bivalve *Sphaerium striatinum*, which are incorporated into the developing gonad (Woods 1931, 1932). Similarly, the primordial germ cells of the polychaete *Salmacina dysteri* are identified by their arrest of cell divisions in addition to their distinctive cytoplasmic contents and size (Malaquin 1925). In both these cases, the putative germ cells appear to arise from the 4d cell (Malaquin 1925; Woods 1931, 1932). Modern cell lineage studies, using long-lived vital dyes to trace cell lineage fates, have similarly highlighted cell populations that arrest division shortly after their birth and localize to positions where they are likely to contribute to the development of germ cells. For example, In *C. fornicata* and *C. convexa*, the origins of the PGCs have been traced to the 4d lineage. Cell behavior suggests that two daughter cells, $2mL^{2.2}$ and $2mR^{2.2}$, are the source of the germ line in these calyptraeids (Fig. 7.4b, c) (Lyons et al. 2012). These cells do not divide, and from their initial location on bilateral sides of the embryo, they migrate to a position adjacent to the intestinal rudiment, where they are ultimately relocated to the region where the gonads form, deep to the heart (Lyons et al. 2012).

In addition to cell lineage studies, widely conserved germ line markers (e.g., *vasa*, *piwi*, and *nanos*) have been used to define PGCs (Ewen-Campen et al. 2010). For example, *vasa* is initially expressed throughout embryos of *Tritia* (*Ilyanassa*) *obsoleta*, becoming progressively more concentrated in the 4d lineage, until it is finally restricted to the 3ML and 3MR teloblasts (Fig. 7.4) (Swartz et al. 2008). Likewise, *nanos* expression is initially widespread, becoming localized to the 4d lineage and persisting in the $2mL^2/R^2$ cells (Rabinowitz et al. 2008). Expression of *vasa* and *nanos* in the abalone *Haliotis asinina* suggests that the primordial germ cells arise from the 4d lineage; however, expression is not restricted to the gonadal region during larval development in these equal-cleaving animals (Kranz et al. 2010). The authors suggest that the PGCs are not formed strictly by maternal preformation but that induction is also involved. *Vasa* expression in the bivalve *C. gigas* is restricted to two clusters of cells that appear to arise from the 4d lineage and localize to the gonad after metamorphosis (Fabioux et al. 2004). *Cg-nanos-like* is restricted to bilateral expression in what appears to be the ML and MR cells at gastrulation, following ubiquitous expression during early development, supporting the 4d origin of the PGCs in *C. gigas* (Xu et al. 2018). A combination of lineage tracing, expression of conserved germ line markers, cell deletion, and cell behavior have been used to suggest the identity of putative PGCs in other spiralians, highlighting the diversity of PGC origins possible within the conserved spiral cleavage

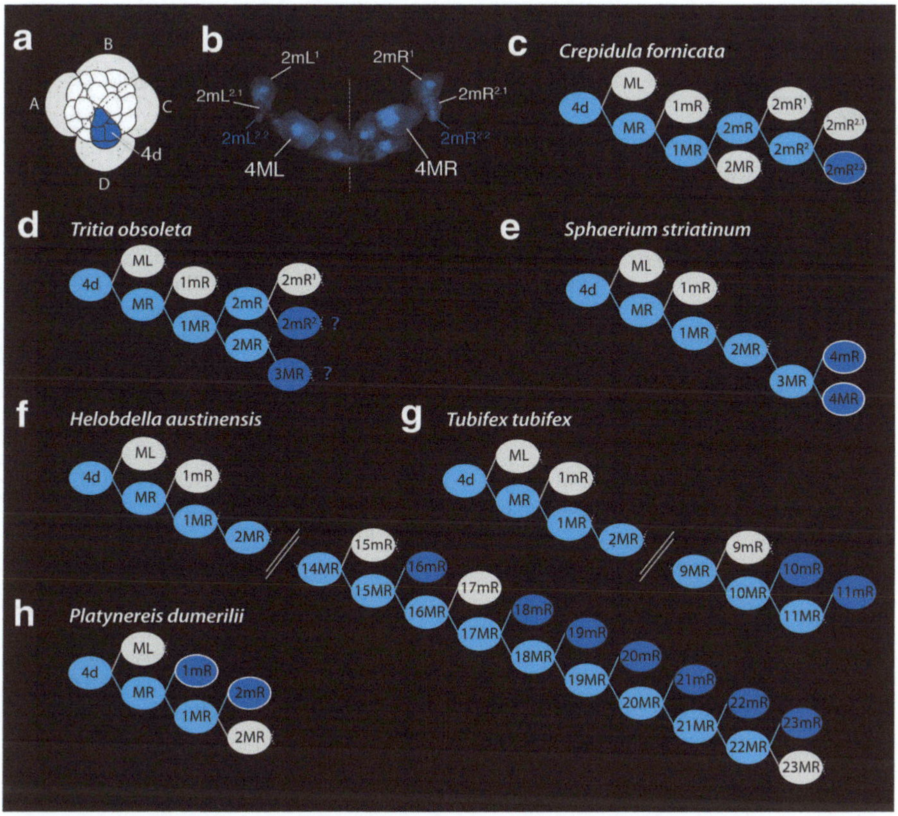

Fig. 7.4 Origins of primordial germ cells from the 4d cell lineage of selected molluscs and annelids, where sublineages of the teleoblasts have been reported. (**a**) Cartoon of the 25-cell stage in a generalized embryo of *Crepidula*. The 4d cell is indicated in blue. The relative positions of the four embryonic quadrants (A–D) are also shown. (**b**) 4d sublineage of *C. fornicata* at the birth of the putative primordial germ cells, $2mL^{2.2}$ and $2mR^{2.2}$. Some sister cells are also labeled including the teleoblasts 4ML and 4MR. Central dashed line indicates the plane of bilateral symmetry. The 4d cell has been injected with a fluorescent lineage tracer at the 25-cell stage (**a**) to permit visualization of the lineage. (**c**–**h**) Origins of the primordial germ cells as reported in various spiralian embryos. Diagrams in (**c**–**h**) show only half of the bilaterally symmetric 4d lineages with primordial germ cell lineages represented in light blue, putative primordial germ cells in dark blue, and other lineages in gray. Shown here are the gastropod molluscs (**c**) *Crepidula fornicata* and (**d**) *Tritia obsoleta*; the bivalve mollusc (**e**) *Sphaerium striatinium*; the clitellate annelids (**f**) *Helobdella austinensis* and (**g**) *Tubifex tubifex*; and the polychaete annelid (**h**) *Platynereis dumerilii*. Dashed lines indicate continuation of lineages not detailed here. Cells with white outlines indicate cleavage arrest where reported. 4d cleavage nomenclature follows that used in Lyons et al. (2012). See Table 7.1 and references therein for further details

pattern. For example, the presumptive PGCs of the polychaete *Platynereis dumerilii* are the first lineage to segregate from 4d (Fischer and Arendt 2013; Rebscher et al. 2007, 2012; Özpolat et al. 2017) (Fig. 7.4 and Table 7.1). In contrast, clitellates such as the sludge worm *Tubifex tubifex* and the leech *Helobdella* show much later

Table 7.1 Origins, specification, and behavior of the primordial germ cells within spiralians

Phylum/ class	Species	Embryonic origins (#)	Specification/ behavior	References
Annelida				
Clitellata	*Helobdella robusta* *Helobdella austinensis*	4d (16mL/R, 18–23mL/R)	Autonomous	Kang et al. (2002) Cho et al. (2013)
	Tubifex tubifex	4d (10mL/R, 11mL/R)	Autonomous	Oyama et al. (2007) Gline et al. (2011) Kato et al. (2013)
Polychaeta	*Capitella teleta*	4d (?)	Autonomous Can be regenerated	Dill and Seaver (2008) Meyer et al. (2010) Giani et al. (2011) Dannenberg and Seaver (2018)
	Platynereis dumerilii	4d (1mL/R, 2mL/R) (4)	Autonomous Cleavage arrested	Rebscher et al. (2007, 2012) Özpolat et al. (2017)
	Salmacina dysteri	4d (1mL/R)	–	Malaquin (1925)
Mollusca				
Bivalvia	*Crassostrea gigas*	4d (ML/R)	Autonomous	Fabioux et al. (2004) Xu et al. (2018)
	Sphaerium striatinum	4d (4ML/R, 4mL/R) (4)	Autonomous Cleavage arrested	Woods (1931, 1932)
Gastropoda	*Crepidula fornicata*	4d ($2mL/R^{2.2}$) (2)	Autonomous Cleavage arrested	Lyons et al. (2012)
	Haliotis asinina	4d (ML/R)	Conditional	Kranz et al. (2010)
	Tritia obsoleta	4d ($2mL/R^2$ or 3ML/R)	Autonomous	Swartz et al. (2008) Rabinowitz et al. (2008)

Table modified from Rebscher (2014). In all cases the germ line arises from the 4d lineage. The 4d sublineages are provided where this is known. Lineage nomenclature is based on that of Lyons et al. (2012). For some cases, the initial number (#) of PGCs is provided. In a few cases, it has been shown that the PGCs are initially cleavage arrested after they are born. These cells eventually undergo divisions much later during development or in the adult. Cases exhibiting early unequal spiral cleavages with early segregation of the germ line in the 4d lineage are assumed to have autonomous specification. *Haliotis asinina* exhibits early equal spiral cleavages, and thus the germ line is argued to be conditionally specified. In some cases, it is known that the germ line can be regenerated from other cell lineages, such as in the annelid *Capitella teleta* (Dannenberg and Seaver 2018)

segregation of germ cells from 4d (Fig. 7.4 and Table 7.1) (Kang et al. 2002; Oyama and Shimizu 2007; Gline et al. 2011; Cho et al. 2013; Kato et al. 2013). This variation, particularly evident in the annelids, may be related to regenerative capacity. For example, in the polychaete, *Capitella teleta*, cell lineage data and expression of *vasa*, *piwi*, and *nanos* suggest that PGCs derive from 4d (Meyer et al. 2010; Dill and Seaver 2008; Giani et al. 2011). Cell deletion experiments show that while expression of the PGC markers *vasa*, *piwi1*, and *nanos* is lost following cell ablation, juveniles regain *vasa* expression and are able to regenerate the germ line and produce offspring (Dannenberg and Seaver 2018). Despite the diversity in primordial germ cell origins, these and other examples all point to 4d as the source of the germ line in spiralians (Fig. 7.4 and Table 7.1) (Rebscher 2014). What remains lacking is an understanding of how germ cell origins differ (or remain the same) within more closely related groups, such as the calyptraeids. Other outstanding questions include the timing of segregation of male and female functions in hermaphroditic animals. For example, in *Helobdella robusta*, the putative PGCs segregate relatively late in embryonic development (Fig. 7.4 and Table 7.1). In these simultaneous hermaphrodites, *nanos* is differentially expressed at the site of the future testes, while *vasa* and *piwi* are expressed at the site of the future ovaries (Cho et al. 2013). Similar segregation of male and female germ cells may be possible in the sequential hermaphrodite *Crepidula*, where two PGCs are initially formed (Lyons et al. 2012), but it is not known if these cells eventually contribute to the production of either eggs or sperm, or both.

7.5 Environmental Contaminants and Sexual Development in Molluscs: Imposex

Adult reproductive structures remain understudied in molluscs, despite their widespread use as bioindicators in marine environments. In the 1970s, researchers began to recognize widespread abnormalities in marine mollusc populations. Shell growth abnormalities and fisheries collapses in commercially important bivalves in the European Atlantic (Evans et al. 1995) and penis growth in normally dioecious females, particularly in neogastropods (Blaber 1970; Jenner 1979; Smith 1971), were linked to exposure to the organotin compound, tributyltin (TBT) (Smith 1981). TBT, called the most toxic compound ever knowingly introduced to the aquatic environment (Goldberg 1986), was widely used as a highly effective biocide in antifouling paints on ships and docks. The abnormal growth of male sexual characteristics in females without becoming functional males is termed imposex ("superimposed sex") (Smith 1971), exemplified by the sentinel species, *Nucella lapillus* (Gibbs et al. 1987). Imposex can be measured on a scale determined by the extent of growth of the penis (Fig. 7.5) (Gibbs et al. 1987; Stroben et al. 1992) and by the extent of vas deferens growth (Gibbs et al. 1987). Imposex eventually blocks the female genital tract, resulting in sterility and in severe cases, death. Local efforts to ban TBT were initiated relatively quickly, and a worldwide ban on TBT has been in effect since September 2008 (see Sousa et al. 2014 for review). In some areas, this has proven to be an effective strategy for reducing TBT concentration in the environment and concomitant imposex rates (Schøyen et al. 2019). However, TBT

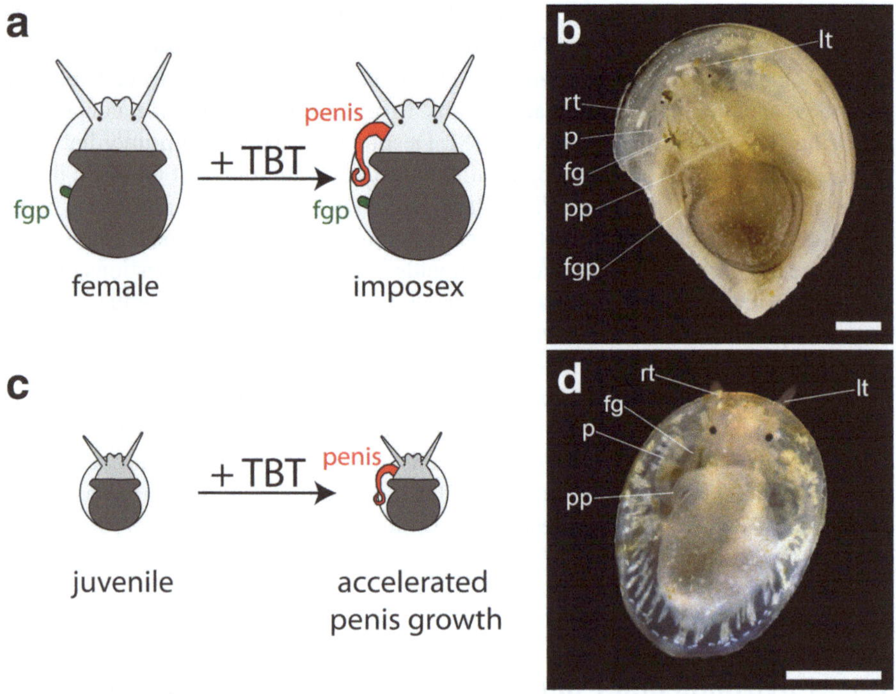

Fig. 7.5 The effects of tribultyltin (TBT) exposure on *C. atrasolea*. (**a**) Adult females exposed to TBT show imposex, superimposed growth of male characters, most obviously the penis. (**b**) Ventral view showing imposex female of *C. atrasolea* after 8 days of exposure to TBT. Note the presence of both the female genital papilla and a penis on the right side of the animal. (**c**) Juvenile snails exposed to TBT initiate penis growth within a few days. TBT can induce accelerated penis development in juveniles immediately after hatching. Typical males grown in the laboratory show penis development after approximately 1 month. (**d**) Ventral view of juvenile snail exposed to tributyltin for 8 days post-hatching. Note the growth of the external penis on the right side of the neck. Comparable results are seen in other calyptraeids (not shown). *fg* food grove, *fgp* female genital papilla, *lt* left tentacle, *p* penis, *pp* propodium, *rt* right tentacle. Scale bars represent 1 mm

continues to effect marine populations in some areas, particularly in non-signatory countries or where local regulations are lacking (e.g., Batista et al. 2016).

Since exposure to TBT and related compounds were first linked to imposex in molluscs, there has been interest in elucidating the underlying mechanisms. Proposed hypotheses fall into three major categories (Pascoal et al. 2013; Horiguchi 2017). The steroid hypothesis suggests that TBT increases the activity of androgenic steroidal hormones, increasing concentrations of androgens via interference with steroid metabolism (Bettin et al. 1996; Gooding et al. 2003; Ronis and Mason 1996). However, endogenous production of vertebrate steroid hormones in molluscs remains questionable, as suggested by the ready uptake of environmental hormones and the lack of conserved biosynthesis pathways in many invertebrates (see Horiguchi 2017; Scott 2018, for review). The neuroendocrine hypothesis suggests that TBT interferes with excretion of neuropeptides that control reproductive development (Féral and Le Gall 1983), such as APGWamide, which induces penis growth in *T. obsoleta* (Oberdörster and McClellan-Green 2000, 2002). However, the effect of APGWamide is weak compared to TBT and does not induce imposex in other species, including *N. lapillus* (Castro et al. 2007). The retinoid hypothesis is the most promising mechanistic explanation to date, suggesting that TBT actively binds to the nuclear receptor, retinoid X receptor, RXR. TBT has been shown to bind specifically to RXR in molluscs (Nishikawa et al. 2004), and treatment with RXR agonists induces imposex in *N. lapillus* (Castro et al. 2007; Lima et al. 2011) and *Thais clavigera* (Nishikawa et al. 2004). Transcriptomic expression in *N. lapillus* exposed to TBT confirmed the expression of RXR in imposex animals and further suggested a role for peroxisome proliferator-activated receptors (PPAR) (Pascoal et al. 2013). Application of rosiglitazone, a PPAR agonist, induced imposex at levels comparable to TBT (Pascoal et al. 2013). Both RXR and PPAR belong to NR subfamily 1 and form homo- and heterodimers with each other and other members of the retinoic acid pathway, including the retinoic acid receptor, RAR. TBT is capable of binding to both RXR (Nishikawa et al. 2004) and PPAR (le Maire et al. 2009). While nuclear receptors such as RAR (Babonis and Martindale 2017) are broadly conserved across the Metazoa, care must be taken to understand their specific roles in molluscs and non-vertebrate animals. For example, molluscan RAR shows no binding activity to known agonists, although RAR–RXR heterodimers show changes in transcriptional activity in in vitro transactivation assays (André et al. 2019). In contrast, in the polychaete *P. dumerilii*, RAR binds retinoic acid and functions in neurogenesis, although this requires much higher concentrations than in chordates (Handberg-Thorsager et al. 2018).

The majority of studies on mechanisms of imposex induction have typically been carried out in temperate gastropod species with separate sexes. For example, *N. lapillus* is used as an environmental sentinel and experimental model for many TBT-imposex studies (as mentioned above), and *T. obsoleta* was among the first species where imposex was described (Jenner 1979; Smith 1981). Experimental work with these temperate species often requires long periods of study, and the effects of environmentally relevant doses of TBT can require months to be visible. In contrast, in sequentially hermaphroditic calyptraeid gastropods, the effects of TBT exposure can be seen within a few days (Fig. 7.5, MPL, pers. obs.), making them a useful model not only for testing the effects of potential reproductive disrupting toxins but also for understanding the development of the reproductive organs in gastropods and molluscs more broadly. The growth, resorption, and remodeling of the reproductive organs during the normal course of individual development (see Fig. 7.2) makes *Crepidula* ideal for laboratory manipulation of sexual differentiation. This includes functional testing of the signaling pathways implicated in reproductive development using environmental toxins such as TBT and related compounds. For example, the connection between TBT and retinoic acid signaling suggests a key role for nuclear receptors in regulating sexual differentiation in the calyptraeids and other molluscs.

7.6 Conclusion

Sex determination is an interesting and important area of study for evolutionary developmental biology and the calyptraeid gastropods are an excellent system to explore the evolution of sex determination and sexual development. As lophotrochozoans, they share with more than a third of known phyla a common pattern of early development, namely, spiral cleavage, allowing direct comparisons of early development across large evolutionary distances. As molluscs, they are one example of sequential hermaphroditism and provide comparative data with other examples of this type of development, such as bivalves. The calyptraeid gastropods provide a powerful system of related species with differences in the timing of sex change and varying environmental influence on sex determination. Understanding the role of plasticity and other developmental mechanisms in the evolution of sequential hermaphroditism and ESD will benefit from multiple comparisons of closely related species as offered by the calyptraeids. In addition, the calyptraeids are amenable to experimental manipulation, including initiation of sexual development in the embryo, the effects of conspecifics and other environmental factors on sexual transitions, and the effects of industrial pollutants on sex determination and sexual development. The calyptraeids, using *C. atrasolea* as a central hub from which to explore development of these sequential hermaphrodites (Henry et al. 2017), as well as through making broader comparisons across the diversity of molluscan sexual systems (Collin 2013), show great promise in providing a better understanding the diversity of sexual systems.

Acknowledgements The authors acknowledge the invaluable support of the National Science Foundation (NSF) and were supported by NSF grant IOS-1558061 and NSF EDGE IOS-1827533 to JQH (JJH). MPL was supported by a postdoctoral fellowship from the Fonds de recherche du Québec-Nature et technologies (FRQ-NT), and thanks the Society for Developmental Biology for an Emerging Models Grant.

References

André A, Ruivo R, Fonseca E, Froufe E, Castro LFC, Santos MM (2019) The retinoic acid receptor (RAR) in molluscs: function, evolution and endocrine disruption insights. Aquat Toxicol 208:80–89. https://doi.org/10.1016/j.aquatox.2019.01.002

Auld JR, Jarne P (2016) Sex and recombination in snails. In: Kliman RM (ed) Encyclopedia of evolutionary biology. Academic, Oxford, pp 49–60. https://doi.org/10.1016/B978-0-12-800049-6.00149-9

Babonis LS, Martindale MQ (2017) Phylogenetic evidence for the modular evolution of metazoan signalling pathways. Philos Trans R Soc B Biol Sci 372(1713):20150477. https://doi.org/10.1098/rstb.2015.0477

Bacci G (1951) L'ermafrodismo di *Calyptraea chinensis* L. e di altri Calyptraeidae. Pubbl Stn Zool Napoli 23(1):66–90

Bachtrog D, Mank JE, Peichel CL, Kirkpatrick M, Otto SP, Ashman T-L, Hahn MW, Kitano J, Mayrose I, Ming R, Perrin N, Ross L, Valenzuela N, Vamosi JC, The tree of sex consortium

(2014) Sex determination: why so many ways of doing it? PLoS Biol 12(7):e1001899. https://doi.org/10.1371/journal.pbio.1001899
Batista RM, Castro IB, Fillmann G (2016) Imposex and butyltin contamination still evident in Chile after TBT global ban. Sci Total Environ 566-567:446–453. https://doi.org/10.1016/j.scitotenv.2016.05.039
Bauer RT (2006) Same sexual system but variable sociobiology: evolution of protandric simultaneous hermaphroditism in *Lysmata* shrimps. Integr Comp Biol 46(4):430–438. https://doi.org/10.1093/icb/icj036
Bell G (1982) The masterpiece of nature: the evolution and genetics of sexuality. University of California Press, Berkeley
Bettin C, Oehlmann J, Stroben E (1996) TBT-induced imposex in marine neogastropods is mediated by an increasing androgen level. Helgoländer Meeresun 50(3):299–317. https://doi.org/10.1007/bf02367105
Beukeboom LW, Perrin N (2014) The evolution of sex determination. Oxford University Press, New York
Blaber SJ (1970) The occurrence of a penis-like outgrowth behind the right tentacle in spent females of *Nucella lapillus* (L.). J Molluscan Stud 39(2–3):231–233
Brante A, Quiñones A, Silva F (2016) The relationship between sex change and reproductive success in a protandric marine gastropod. Sci Rep 6(29439):1–10. https://doi.org/10.1038/srep29439
Breton S, Capt C, Guerra D, Stewart D (2018) Sex-determining mechanisms in bivalves. In: Leonard J (ed) Transitions between sexual systems. Springer, Cham. https://doi.org/10.1007/978-3-319-94139-4_6
Broquet T, Barranger A, Billard E, Bestin A, Berger R, Honnaert G, Viard F (2015) The size advantage model of sex allocation in the protandrous sex-changer *Crepidula fornicata*: role of the mating system, sperm storage, and male mobility. Am Nat 186(3):404–420. https://doi.org/10.1086/682361
Bui LT, Ivers NA, Ragsdale EJ (2018) A sulfotransferase dosage-dependently regulates mouthpart polyphenism in the nematode *Pristionchus pacificus*. Nat Commun 9(1):4119. https://doi.org/10.1038/s41467-018-05612-8
Bull JJ (1983) Evolution of sex determining mechanisms. The Benjamin/Cummings, Menlo Park, CA
Bull J, Vogt R (1979) Temperature-dependent sex determination in turtles. Science 206 (4423):1186–1188. https://doi.org/10.1126/science.505003
Burtis KC, Baker BS (1989) *Drosophila* doublesex gene controls somatic sexual differentiation by producing alternatively spliced mRNAs encoding related sex-specific polypeptides. Cell 56 (6):997–1010. https://doi.org/10.1016/0092-8674(89)90633-8
Buston P (2003) Social hierarchies: size and growth modification in clownfish. Nature 424 (6945):145–146
Cahill AE, Juman AR, Pellman-Isaacs A, Bruno WT (2015) Physical and chemical interactions with conspecifics mediate sex change in a protandrous gastropod *Crepidula fornicata*. Biol Bull 229 (3):276–281. https://doi.org/10.1086/BBLv229n3p276
Capel B (2017) Vertebrate sex determination: evolutionary plasticity of a fundamental switch. Nat Rev Genet 18:675. https://doi.org/10.1038/nrg.2017.60
Carrillo-Baltodano A, Collin R (2015) *Crepidula* slipper limpets alter sex change in response to physical contact with conspecifics. Biol Bull 229(3):232–242. https://doi.org/10.1086/BBLv229n3p232
Castro LFC, Lima D, Machado A, Melo C, Hiromori Y, Nishikawa J, Nakanishi T, Reis-Henriques MA, Santos MM (2007) Imposex induction is mediated through the Retinoid X Receptor signalling pathway in the neogastropod *Nucella lapillus*. Aquat Toxicol 85(1):57–66. https://doi.org/10.1016/j.aquatox.2007.07.016

Chaparro OR, Bahamondes-Rojas I, Vergara AM, Rivera AA (1998) Histological characteristics of the foot and locomotory activity of *Crepidula dilatata* Lamarck (Gastropoda: Calyptraeidae) in relation to sex changes. J Exp Mar Biol Ecol 223(1):77–91

Chaparro OR, Pereda SV, Bahamondes-Rojas I (2001) Effects of protandric sex change on radula, pedal morphology, and mobility in *Crepidula fecunda* (Gastropoda: Calyptraeidae). NZ J Mar Fresh 35(5):881–890

Charnov EL (1982) The theory of sex allocation. Monographs in population biology, vol 18. Princeton University Press, Princeton, NJ

Cho S-J, Vallès Y, Weisblat DA (2013) Differential expression of conserved germ line markers and delayed segregation of male and female primordial germ cells in a hermaphrodite, the leech *Helobdella*. Mol Biol Evol 31(2):341–354. https://doi.org/10.1093/molbev/mst201

Chong T, Collins JJ, Brubacher JL, Zarkower D, Newmark PA (2013) A sex-specific transcription factor controls male identity in a simultaneous hermaphrodite. Nat Commun 4:1814. https://doi.org/10.1038/ncomms2811

Coe WR (1935) Sexual phases in prosobranch mollusks of the genus *Crepidula*. Science 81 (2110):570–572

Coe WR (1936) Sexual phases in *Crepidula*. J Exp Zool 72(3):455–477. https://doi.org/10.1002/jez.1400720306

Coe WR (1938a) Conditions influencing change of sex in mollusks of the genus *Crepidula*. J Exp Zool 77(3):401–424

Coe WR (1938b) Influence of association on the sexual phases of gastropods having protandric consecutive sexuality. Biol Bull 75(2):274–285

Coe WR (1942) Influence of natural and experimental conditions in determining shape of shell and rate of growth in gastropods of the genus *Crepidula*. J Morphol 71(1):35–51

Coe WR (1943) Sexual differentiation in mollusks. I. Pelecypods. Q Rev Biol 18(2):154–164. https://doi.org/10.1086/394673

Collin R (1995) Sex, size, and position: a test of models predicting size at sex change in the protandrous gastropod *Crepidula fornicata*. Am Nat 146(6):815–831

Collin R (2000) Sex change, reproduction, and development of *Crepidula adunca* and *Crepidula lingulata* (Gastropoda: Calyptraeidae). Veliger 43(1):24–33

Collin R (2006) Sex ratio, life-history invariants, and patterns of sex change in a family of protandrous gastropods. Evolution 60(4):735–745

Collin R (2013) Phylogenetic patterns and phenotypic plasticity of molluscan sexual systems. Integr Comp Biol 53(4):723–735. https://doi.org/10.1093/icb/ict076

Collin R, McLellan M, Gruber K, Bailey-Jourdain C (2005) Effects of conspecific associations on size at sex change in three species of calyptraeid gastropods. Mar Ecol Prog Ser 293:89–97

Conklin EG (1897) The embryology of *Crepidula*: a contribution to the cell lineage and early development of some marine gasteropods. J Morphol 13(1):1–226

Conklin EG (1902) Karyokinesis and cytokinesis in the maturation, fertilization and cleavage of *Crepidula* and other gasteropoda. J Acad Nat Sci 12:1–121

Dannenberg LC, Seaver EC (2018) Regeneration of the germline in the annelid *Capitella teleta*. Dev Biol 440(2):74–87. https://doi.org/10.1016/j.ydbio.2018.05.004

Dill KK, Seaver EC (2008) Vasa and nanos are coexpressed in somatic and germ line tissue from early embryonic cleavage stages through adulthood in the polychaete *Capitella* sp. I. Dev Genes Evol 218(9):453–463

Dupont L, Richard J, Paulet YM, Thouzeau G, Viard F (2006) Gregariousness and protandry promote reproductive insurance in the invasive gastropod *Crepidula fornicata*: evidence from assignment of larval paternity. Mol Ecol 15(10):3009–3021. https://doi.org/10.1111/j.1365-294X.2006.02988.x

Eernisse DJ (1988) Reproductive patterns in six species of *Lepidochitona* (Mollusca: Polyplacophora) from the Pacific Coast of North America. Biol Bull 174(3):287–302. https://doi.org/10.2307/1541955

Evans SM, Leksono T, McKinnell PD (1995) Tributyltin pollution: a diminishing problem following legislation limiting the use of TBT-based anti-fouling paints. Mar Pollut Bull 30(1):14–21. https://doi.org/10.1016/0025-326X(94)00181-8

Ewen-Campen B, Schwager EE, Extavour CGM (2010) The molecular machinery of germ line specification. Mol Reprod Dev 77(1):3–18. https://doi.org/10.1002/mrd.21091

Extavour CG, Akam M (2003) Mechanisms of germ cell specification across the metazoans: epigenesis and preformation. Development 130(24):5869–5884. https://doi.org/10.1242/dev.00804

Fabioux C, Huvet A, Lelong C, Robert R, Pouvreau S, Daniel JY, Minguant C, Le Pennec M (2004) Oyster vasa-like gene as a marker of the germline cell development in *Crassostrea gigas*. Biochem Biophys Res Commun 320(2):592–598. https://doi.org/10.1016/j.bbrc.2004.06.009

Féral C, Le Gall S (1983) The influence of a pollutant factor (tributyltin) on the neuroendocrine mechanisms responsible for the occurrence of a penis in the females of *Ocenebra erinacea*. In: International minisymposium on molluscan endocrinology. North Holland, Amsterdam

Féral C, Le Gall S, Martin MC, Lengronne C (1987) The neuroendocrine mechanism responsible for sexual inversion of the gonad in the protandrous hermaphroditic mollusk, *Crepidula fornicata* L. Gen Comp Endocr 65(3):432–438

Fischer AHL, Arendt D (2013) Mesoteloblast-like mesodermal stem cells in the polychaete annelid *Platynereis dumerilii* (Nereididae). J Exp Zool B Mol Dev Evol 320(2):94–104. https://doi.org/10.1002/jez.b.22486

García-Souto D, Alonso-Rubido S, Costa D, Eirín-López JM, Rolán-Álvarez E, Faria R, Galindo J, Pasantes JJ (2018) Karyotype characterization of nine periwinkle species (Gastropoda, Littorinidae). Genes 9(11):517

Ghiselin MT (1969) The evolution of hermaphroditism among animals. Q Rev Biol 44(2):189–208. https://doi.org/10.1086/406066

Giani VC, Yamaguchi E, Boyle MJ, Seaver EC (2011) Somatic and germline expression of piwi during development and regeneration in the marine polychaete annelid *Capitella teleta*. EvoDevo 2(1):10. https://doi.org/10.1186/2041-9139-2-10

Gibbs PE, Bryan GW, Pascoe PL, Burt GR (1987) The use of the dog-whelk, *Nucella lapillus*, as an indicator of tributyltin (TBT) contamination. J Mar Biol Assoc UK 67(3):507–523. https://doi.org/10.1017/S0025315400027260

Gline SE, Nakamoto A, Cho S-J, Chi C, Weisblat DA (2011) Lineage analysis of micromere 4d, a super-phylotypic cell for Lophotrochozoa, in the leech *Helobdella* and the sludgeworm *Tubifex*. Dev Biol 353(1):120–133. https://doi.org/10.1016/j.ydbio.2011.01.031

Goldberg ED (1986) TBT: an environmental dilemma. Environ Sci Policy Sustain Dev 28(8):17–44. https://doi.org/10.1080/00139157.1986.9928814

Gooding MP, Wilson VS, Folmar LC, Marcovich DT, LeBlanc GA (2003) The biocide tributyltin reduces the accumulation of testosterone as fatty acid esters in the mud snail (*Ilyanassa obsoleta*). Environ Health Persp 111(4):426–430. https://doi.org/10.1289/ehp.5779

Gould HN (1917a) Studies on sex in the hermaphrodite mollusc *Crepidula plana* I. History of the sexual cycle. J Exp Zool 23(1):1–69

Gould HN (1917b) Studies on sex in the hermaphrodite mollusk *Crepidula plana*. II. Influence of the environment on sex. J Exp Zool 23(1):225–250

Gould HN (1919) Studies on sex in the hermaphrodite mollusc *Crepidula plana*. III. Transference of the male-producing stimulus through sea-water. J Exp Zool 29:113–120

Gould H (1952) Studies on sex in the hermaphrodite mollusk *Crepidula plana*: IV. Internal and external factors influencing growth and sex development. J Exp Zool 119:93–163. https://doi.org/10.1002/jez.1401190107

Guallart J, Calvo M, Acevedo I, Templado J (2013) Two-way sex change in the endangered limpet *Patella ferruginea* (Mollusca, Gastropoda). Invertebr Reprod Dev 57(3):247–253

Guo X, Hedgecock D, Hershberger WK, Cooper K, Allen SKJ (1998) Genetic determinants of protandric sex in the Pacific oyster, *Crassostrea gigas* Thunberg. Evolution 52(2):394–402. https://doi.org/10.1111/j.1558-5646.1998.tb01640.x

Haag ES, Doty AV (2005) Sex determination across evolution: connecting the dots. PLoS Biol 3(1): e21. https://doi.org/10.1371/journal.pbio.0030021

Handberg-Thorsager M, Gutierrez-Mazariegos J, Arold ST, Kumar Nadendla E, Bertucci PY, Germain P, Tomançak P, Pierzchalski K, Jones JW, Albalat R, Kane MA, Bourguet W, Laudet V, Arendt D, Schubert M (2018) The ancestral retinoic acid receptor was a low-affinity sensor triggering neuronal differentiation. Sci Adv 4(2):eaao1261. https://doi.org/10.1126/sciadv.aao1261

Haszprunar G, Schaefer K (1996) Anatomy and phylogenetic significance of *Micropilina arntzi* (Mollusca, Monoplacophora, Micropilinidae Fam. Nov.). Acta Zool 77(4):315–334. https://doi.org/10.1111/j.1463-6395.1996.tb01276.x

Hedgecock D, Hedrick PW (2010) Sex determination: genetic models for oysters. J Hered 101 (5):602–611. https://doi.org/10.1093/jhered/esq065

Heller J (1993) Hermaphroditism in molluscs. Biol J Linn Soc 48(1):19–42. https://doi.org/10.1111/j.1095-8312.1993.tb00874.x

Henry JQ (2014) Spiralian model systems. Int J Dev Biol 58:389–401

Henry JQ, Lyons DC (2016) Molluscan models: *Crepidula fornicata*. Curr Opin Gen Dev 39:138–148. https://doi.org/10.1016/j.gde.2016.05.021

Henry JQ, Lesoway MP, Perry KJ, Osborne CC, Shankland M, Lyons DC (2017) Beyond the sea: *Crepidula atrasolea* as a spiralian model system. Int J Dev Biol 61(8–9):479–493. https://doi.org/10.1387/ijdb.170110jh

Hildreth PE (1965) Doublesex, a recessive gene that transforms both males and females of *Drosophila* into intersexes. Genetics 51(4):659

Hoagland K (1978) Protandry and the evolution of environmentally-mediated sex change: a study of the Mollusca. Malacologia 17(2):365–391

Hobday AJ, Riser KL (1998) The role of movement by *Crepidula norrisiarum* Williamson in altering reproductive potential. J Exp Mar Biol Ecol 225(1):139–154. https://doi.org/10.1016/S0022-0981(97)00218-9

Horiguchi T (2017) Mode of action of organotins to induce the development of imposex in gastropods, focusing on steroid and the retinoid X receptor activation hypotheses. In: Horiguchi T (ed) Biological effects by organotins. Springer, Tokyo, pp 199–219. https://doi.org/10.1007/978-4-431-56451-5_9

Hoving HJT, Roeleveld MAC, Lipinski MR, Videler JJ (2006) Nidamental glands in males of the oceanic squid *Ancistrocheirus lesueurii* (Cephalopoda: Ancistrocheiridae) – sex change or intersexuality? J Zool 269(3):341–348. https://doi.org/10.1111/j.1469-7998.2006.00093.x

Ishiki H (1936) Sex-changes in Japanese slipper limpets, *Crepidula aculeata* and *Crepidula walshi*. J Sci Hiroshima Univ Ser B 4:91–99

Ishiki H (1938) Histological studies on the sexual organs during sex changes of *Crepidula aculeata* and *Crepidula walshi*. J Sci Hiroshima Univ Ser B 6:103–113

Jarne P, Auld JR (2006) Animals mix it up too: the distribution of self-fertilization among hermaphroditic animals. Evolution 60(9):1816–1824. https://doi.org/10.1111/j.0014-3820.2006.tb00525.x

Jenner MG (1979) Pseudohermaphroditism in *Ilyanassa obsoleta* (Mollusca: Neogastropoda). Science 205(4413):1407. https://doi.org/10.1126/science.472758

Kang D, Pilon M, Weisblat DA (2002) Maternal and zygotic expression of a nanos-class gene in the leech *Helobdella robusta*: primordial germ cells arise from segmental mesoderm. Dev Biol 245 (1):28–41. https://doi.org/10.1006/dbio.2002.0615

Kato Y, Nakamoto A, Shiomi I, Nakao H, Shimizu T (2013) Primordial germ cells in an oligochaete annelid are specified according to the birth rank order in the mesodermal teloblast lineage. Dev Biol 379(2):246–257. https://doi.org/10.1016/j.ydbio.2013.04.028

Kocot KM, Cannon JT, Todt C, Citarella MR, Kohn AB, Meyer A, Santos SR, Schander C, Moroz LL, Lieb B, Halanych KM (2011) Phylogenomics reveals deep molluscan relationships. Nature 477(7365):452–456. https://doi.org/10.1038/nature10382

Koene JM (2017) Sex determination and gender expression: reproductive investment in snails. Mol Reprod Dev 84(2):132–143. https://doi.org/10.1002/mrd.22662

Koopman P, Gubbay J, Vivian N, Goodfellow P, Lovell-Badge R (1991) Male development of chromosomally female mice transgenic for Sry. Nature 351(6322):117–121. https://doi.org/10.1038/351117a0

Kopp A (2012) Dmrt genes in the development and evolution of sexual dimorphism. Trends Genet 28(4):175–184. https://doi.org/10.1016/j.tig.2012.02.002

Kranz AM, Tollenaere A, Norris BJ, Degnan BM, Degnan SM (2010) Identifying the germline in an equally cleaving mollusc: vasa and Nanos expression during embryonic and larval development of the vetigastropod *Haliotis asinina*. J Exp Zool B Mol Dev Evol 314B(4):267–279

Le Cam S, Pechenik JA, Cagnon M, Viard F (2009) Fast versus slow larval growth in an invasive marine mollusc: does paternity matter? J Hered 100(4):455–464. https://doi.org/10.1093/jhered/esp007

Le Gall S (1981) Etude expérimentale du facteur morphogénétique contrôlant la différenciation du tractus génital mâle externe chez *Crepidula fornicata* L. (Mollusque hermaphrodite protandre). Gen Comp Endocr 43(1):51–62. https://doi.org/10.1016/0016-6480(81)90031-9

Le Gall S, Feral C (1982) Experimental study of in vitro development of the male genital tract in *Crepidula fornicata* L (Protandric Hermaphrodite Mesogastropod). Int J Invertebr Reprod 5 (1):31–42

Le Gall S, Streiff W (1975) Protandric hermaphroditism in prosobranch gastropods. In: Reinboth R (ed) Intersexuality in the animal kingdom. Springer, Berlin, pp 170–178. https://doi.org/10.1007/978-3-642-66069-6_17

Le Gall S, Feral C, Lengronne C, Porchet M (1987) Partial purification of the neuroendocrine mitogenic factor in the mollusk *Crepidula fornicata* L. Comp Biochem Phys B 86(2):393–396

le Maire A, Grimaldi M, Roecklin D, Dagnino S, Vivat-Hannah V, Balaguer P, Bourguet W (2009) Activation of RXR–PPAR heterodimers by organotin environmental endocrine disruptors. EMBO Rep 10(4):367–373. https://doi.org/10.1038/embor.2009.8

Li Y, Zhang L, Sun Y, Ma X, Wang J, Li R, Zhang M, Wang S, Hu X, Bao Z (2016) Transcriptome sequencing and comparative analysis of ovary and testis identifies potential key sex-related genes and pathways in scallop *Patinopecten yessoensis*. Mar Biotechnol 18(4):453–465. https://doi.org/10.1007/s10126-016-9706-8

Libertini A, Vitturi R, Gregorini A, Colomba M (2009) Karyotypes, banding patterns and nuclear DNA content in *Crepidula unguiformis* Lamarck, 1822, and *Naticarius stercusmuscarum* Gmelin, 1791 (Mollusca, Caenogastropoda). Malacologia 51(1):111–118

Lima D, Reis-Henriques MA, Silva R, Santos AI, Filipe C. Castro L, Santos MM (2011) Tributyltin-induced imposex in marine gastropods involves tissue-specific modulation of the Retinoid X Receptor. Aquat Toxicol 101(1):221–227. https://doi.org/10.1016/j.aquatox.2010.09.022

Lindberg DR, Wright W (1985) Patterns of sex change of the protandric patellacean limpet *Lottia gigantea* (Mollusca: Gastropoda). Veliger 27(3):261–265

Lyons DC, Perry KJ, Lesoway MP, Henry JQ (2012) Cleavage pattern and fate map of the mesentoblast, 4d, in the gastropod *Crepidula*: a hallmark of spiralian development. EvoDevo 3(1):1–26

Malaquin A (1925) La ségrégation, au cours de l'ontogenèse, de deux cellules sexuelles primordiales, souches de la lignée germinale, chez *Salmacina dysteri* (Huxley). CR Acad Sci D Nat 180:324–327

Mérot C, Collin R (2012a) Effects of food availability on sex change in two species of *Crepidula* (Gastropoda: Calyptraeidae). Mar Ecol Prog Ser 449:173–181

Mérot C, Collin R (2012b) Effects of stress on sex change in *Crepidula cf. marginalis* (Gastropoda: Calyptraeidae). J Exp Mar Biol Ecol 416–417:68–71. https://doi.org/10.1016/j.jembe.2012.02.012

Meyer NP, Boyle MJ, Martindale MQ, Seaver EC (2010) A comprehensive fate map by intracellular injection of identified blastomeres in the marine polychaete *Capitella teleta*. EvoDevo 1 (1):8. https://doi.org/10.1186/2041-9139-1-8

Miyagawa S, Yatsu R, Iguchi T (2018) Environmental control of sex determination and differentiation in reptiles. In: Kobayashi K, Kitano T, Iwao Y, Kondo M (eds) Reproductive and developmental strategies: the continuity of life. Springer, Tokyo, pp 367–390. https://doi.org/10.1007/978-4-431-56609-0_18

Munday PL, Buston PM, Warner RR (2006) Diversity and flexibility of sex-change strategies in animals. Trends Ecol Evol 21(2):89–95. https://doi.org/10.1016/j.tree.2005.10.020

Nakadera Y, Koene JM (2013) Reproductive strategies in hermaphroditic gastropods: conceptual and empirical approaches. Can J Zool 91(6):367–381. https://doi.org/10.1139/cjz-2012-0272

Nishikawa J, Mamiya S, Kanayama T, Nishikawa T, Shiraishi F, Horiguchi T (2004) Involvement of the Retinoid X Receptor in the development of imposex caused by organotins in gastropods. Environ Sci Technol 38(23):6271–6276. https://doi.org/10.1021/es049593u

Oberdörster E, McClellan-Green P (2000) The neuropeptide APGWamide induces imposex in the mud snail, *Ilyanassa obsoleta*. Peptides 21(9):1323–1330. https://doi.org/10.1016/S0196-9781(00)00274-6

Oberdörster E, McClellan-Green P (2002) Mechanisms of imposex induction in the mud snail, *Ilyanassa obsoleta*: TBT as a neurotoxin and aromatase inhibitor. Mar Environ Res 54 (3–5):715–718. https://doi.org/10.1016/S0141-1136(02)00118-6

Oliver B, Clough E (2012) Genomics of sex determination in *Drosophila*. Brief Funct Genomics 11 (5):387–394. https://doi.org/10.1093/bfgp/els019

Orton JH (1909) On the occurrence of protandric hermaphroditism in the Mollusc *Crepidula fornicata*. Proc R Soc Lond B Biol 81(550):468–484

Oyama A, Shimizu T (2007) Transient occurrence of vasa-expressing cells in nongenital segments during embryonic development in the oligochaete annelid *Tubifex tubifex*. Dev Genes Evol 217 (10):675–690. https://doi.org/10.1007/s00427-007-0180-1

Özpolat BD, Handberg-Thorsager M, Vervoort M, Balavoine G (2017) Cell lineage and cell cycling analyses of the 4d micromere using live imaging in the marine annelid *Platynereis dumerilii*. elife 6. https://doi.org/10.7554/eLife.30463

Pascoal S, Carvalho G, Vasieva O, Hughes R, Cossins A, Fang Y, Ashelford K, Olohan L, Barroso C, Mendo S, Creer S (2013) Transcriptomics and in vivo tests reveal novel mechanisms underlying endocrine disruption in an ecological sentinel, *Nucella lapillus*. Mol Ecol 22 (6):1589–1608. https://doi.org/10.1111/mec.12137

Picard MA-L, Cosseau C, Mouahid G, Duval D, Grunau C, Toulza È, Allienne J-F, Boissier J (2015) The roles of Dmrt (double sex/male-abnormal-3 related transcription factor) genes in sex determination and differentiation mechanisms: ubiquity and diversity across the animal kingdom. C R Biol 338(7):451–462. https://doi.org/10.1016/j.crvi.2015.04.010

Policansky D (1982) Sex change in plants and animals. Annu Rev Ecol Syst 13(1):471–495

Ponder WF, Lindberg DR (1997) Towards a phylogeny of gastropod molluscs: an analysis using morphological characters. Zool J Linn Soc-Lond 119:83–265

Proestou DA, Goldsmith MR, Twombly S (2008) Patterns of male reproductive success in *Crepidula fornicata* provide new insight for sex allocation and optimal sex change. Biol Bull 214(2):192–200

Rabinowitz JS, Chan XY, Kingsley EP, Duan Y, Lambert JD (2008) Nanos is required in somatic blast cell lineages in the posterior of a mollusk embryo. Curr Biol 18(5):331–336. https://doi.org/10.1016/j.cub.2008.01.055

Ragsdale EJ, Muller MR, Rodelsperger C, Sommer RJ (2013) A developmental switch coupled to the evolution of plasticity acts through a sulfatase. Cell 155(4):922–933. https://doi.org/10.1016/j.cell.2013.09.054

Raymond CS, Shamu CE, Shen MM, Seifert KJ, Hirsch B, Hodgkin J, Zarkower D (1998) Evidence for evolutionary conservation of sex-determining genes. Nature 391:691. https://doi.org/10.1038/35618

Raymond CS, Murphy MW, O'Sullivan MG, Bardwell VJ, Zarkower D (2000) Dmrt1, a gene related to worm and fly sexual regulators, is required for mammalian testis differentiation. Genes Dev 14(20):2587–2595

Rebscher N (2014) Establishing the germline in spiralian embryos. Int J Dev Biol 58:403–411

Rebscher N, Zelada-González F, Banisch TU, Raible F, Arendt D (2007) Vasa unveils a common origin of germ cells and of somatic stem cells from the posterior growth zone in the polychaete *Platynereis dumerilii*. Dev Biol 306(2):599–611. https://doi.org/10.1016/j.ydbio.2007.03.521

Rebscher N, Lidke AK, Ackermann CF (2012) Hidden in the crowd: primordial germ cells and somatic stem cells in the mesodermal posterior growth zone of the polychaete *Platynereis dumerilii* are two distinct cell populations. EvoDevo 3(1):9. https://doi.org/10.1186/2041-9139-3-9

Riddle D, Blumenthal T, Meyer B, Priess J (1997) Introduction to *C. elegans*. In: Riddle D, Blumenthal T, Meyer B, Priess J (eds) *C. elegans* II, 2nd edn. Cold Spring Harbor Laboratory, Cold Spring Harbor

Robertson DR (1972) Social control of sex reversal in a coral-reef fish. Science 177 (4053):1007–1009. https://doi.org/10.1126/science.177.4053.1007

Ronis MJJ, Mason AZ (1996) The metabolism of testosterone by the periwinkle (*Littorina littorea*) in vitro and in vivo: effects of tributyltin. Mar Environ Res 42(1):161–166. https://doi.org/10.1016/0141-1136(95)00069-0

Schøyen M, Green NW, Hjermann DØ, Tveiten L, Beylich B, Øxnevad S, Beyer J (2019) Levels and trends of tributyltin (TBT) and imposex in dogwhelk (*Nucella lapillus*) along the Norwegian coastline from 1991 to 2017. Mar Environ Res 144:1–8. https://doi.org/10.1016/j.marenvres.2018.11.011

Scott AP (2018) Is there any value in measuring vertebrate steroids in invertebrates? Gen Comp Endocr 265:77–82. https://doi.org/10.1016/j.ygcen.2018.04.005

Smith BS (1971) Sexuality in the American mud snail, *Nassarius obsoletus* Say. J Molluscan Stud 39(5):377–378

Smith BS (1981) Tributyltin compounds induce male characteristics on female mud snails *Nassarius obsoletus* = *Ilyanassa obsoleta*. J Appl Toxicol 1(3):141–144. https://doi.org/10.1002/jat.2550010302

Smith SA, Wilson NG, Goetz FE, Feehery C, Andrade SC, Rouse GW, Giribet G, Dunn CW (2011) Resolving the evolutionary relationships of molluscs with phylogenomic tools. Nature 480 (7377):364–367. https://doi.org/10.1038/nature10526

Sousa ACA, Pastorinho MR, Takahashi S, Tanabe S (2014) History on organotin compounds, from snails to humans. Environ Chem Lett 12(1):117–137. https://doi.org/10.1007/s10311-013-0449-8

Streiff W (1966) Endocrinological study of determinism of sexual cycle in a protandric hermaphrodite mollusc *Calyptraea sinensis* (L.). Ann Endocrinol 3:385

Streiff W (1967) Endocrinologic study of determinism of the sexual cycle in a protandrous hermaphroditic mollusk, *Calyptraea sinensis* (L). II. Demonstration by in vitro cultivation of hormonal factors conditioning the development of the female tract. Ann Endocrinol 4:461

Stroben E, Oehlmann J, Fioroni P (1992) The morphological expression of imposex in *Hinia reticulata* (Gastropoda: Buccinidae): a potential indicator of tributyltin pollution. Mar Biol 113 (4):625–636. https://doi.org/10.1007/bf00349706

Swartz SZ, Chan XY, Lambert JD (2008) Localization of Vasa mRNA during early cleavage of the snail Ilyanassa. Dev Genes Evol 218(2):107–113. https://doi.org/10.1007/s00427-008-0203-6

Teaniniuraitemoana V, Huvet A, Levy P, Klopp C, Lhuillier E, Gaertner-Mazouni N, Gueguen Y, Le Moullac G (2014) Gonad transcriptome analysis of pearl oyster *Pinctada margaritifera*: identification of potential sex differentiation and sex determining genes. BMC Genomics 15 (1):491. https://doi.org/10.1186/1471-2164-15-491

Thiriot-Quiévreux C (2003) Advances in chromosomal studies of gastropod molluscs. J Molluscan Stud 69(3):187–202. https://doi.org/10.1093/mollus/69.3.187

Warner RR (1975) The adaptive significance of sequential hermaphroditism in animals. Am Nat 109(965):61–82. https://doi.org/10.1086/282974

Warner RR (1988) Sex change and the size-advantage model. Trends Ecol Evol 3(6):133–136. https://doi.org/10.1016/0169-5347(88)90176-0

Warner RR, Robertson DR, Leigh EG (1975) Sex change and sexual selection. Science 190 (4215):633–638

Warner RR, Fitch DL, Standish JD (1996) Social control of sex change in the shelf limpet, *Crepidula norrisiarum*: size-specific responses to local group composition. J Exp Mar Biol Ecol 204(1):155–167

Wells MJ, Wells J (1959) Hormonal control of sexual maturity in *Octopus*. J Exp Biol 36(1):1–33

Woods FH (1931) History of the germ cells in *Sphaerium striatinum* (Lam.). J Morphol 51 (2):545–595. https://doi.org/10.1002/jmor.1050510208

Woods FH (1932) Keimbahn determinants and continuity of the germ cells in *Sphaerium striatinum* (Lam.). J Morphol 53(2):345–365. https://doi.org/10.1002/jmor.1050530207

Wright WG (1988) Sex change in the Mollusca. Trends Ecol Evol 3(6):137–140. https://doi.org/10.1016/0169-5347(88)90177-2

Xu R, Li Q, Yu H, Kong L (2018) Oocyte maturation and origin of the germline as revealed by the expression of *Nanos*-like in the Pacific oyster *Crassostrea gigas*. Gene 663:41–50. https://doi.org/10.1016/j.gene.2018.04.021

Yue C, Li Q, Yu H (2018) Gonad transcriptome analysis of the Pacific oyster *Crassostrea gigas* identifies potential genes regulating the sex determination and differentiation process. Mar Biotechnol 20(2):206–219. https://doi.org/10.1007/s10126-018-9798-4

Yusa Y (2007) Causes of variation in sex ratio and modes of sex determination in the Mollusca – an overview. Am Malacol Bull 23(1–2):89–98

Yusa Y, Kumagai N (2018) Evidence of oligogenic sex determination in the apple snail *Pomacea canaliculata*. Genetica 146(3):265–275. https://doi.org/10.1007/s10709-018-0017-z

Zhang N, Xu F, Guo X (2014) Genomic analysis of the Pacific Oyster (*Crassostrea gigas*) reveals possible conservation of vertebrate sex determination in a Mollusc. G3 Genes Genom Genet 4 (11):2207–2217. https://doi.org/10.1534/g3.114.013904

Chapter 8
The Cricket *Gryllus bimaculatus*: Techniques for Quantitative and Functional Genetic Analyses of Cricket Biology

Arpita Kulkarni and Cassandra G. Extavour

Abstract All extant species are an outcome of nature's "experiments" during evolution, and hence multiple species need to be studied and compared to gain a thorough understanding of evolutionary processes. The field of evolutionary developmental biology (evo-devo) aspires to expand the number of species studied, because most functional genetic studies in animals have been limited to a small number of "traditional" model organisms, many of which belong to the same phylum (Chordata). The phylum Arthropoda, and particularly its component class Insecta, possesses many important characteristics that are considered favorable and attractive for evo-devo research, including an astonishing diversity of extant species and a wide disparity in body plans. The development of the most thoroughly investigated insect genetic model system to date, the fruit fly *Drosophila melanogaster* (a holometabolous insect), appears highly derived with respect to other insects and indeed with respect to most arthropods. In comparison, crickets (a basally branching hemimetabolous insect lineage compared to the Holometabola) are thought to embody many developmental features that make them more representative of insects. Here we focus on crickets as emerging models to study problems in a wide range of biological areas and summarize the currently available molecular, genomic, forward and reverse genetic, imaging and computational tool kit that has been established or adapted for cricket research. With an emphasis on the cricket species *Gryllus bimaculatus*, we highlight recent efforts made by the scientific community in establishing this species as a laboratory model for cellular biology and developmental genetics. This broad toolkit has the potential to accelerate many traditional areas of cricket research, including studies of adaptation, evolution,

A. Kulkarni
Department of Organismic and Evolutionary Biology, Harvard University, Cambridge, MA, USA

C. G. Extavour (✉)
Department of Organismic and Evolutionary Biology, Harvard University, Cambridge, MA, USA

Department of Molecular and Cellular Biology, Harvard University, Cambridge, MA, USA
e-mail: extavour@oeb.harvard.edu

W. Tworzydlo, S. M. Bilinski (eds.), *Evo-Devo: Non-model Species in Cell and Developmental Biology*, Results and Problems in Cell Differentiation 68,
https://doi.org/10.1007/978-3-030-23459-1_8

neuroethology, physiology, endocrinology, regeneration, and reproductive behavior. It may also help to establish newer areas, for example, the use of crickets as animal infection model systems and human food sources.

8.1 Introduction

All cellular life forms share a last common ancestor. Every extant species is a current outcome of an evolutionary process that has taken place over hundreds of millions of years. Thus, each species that exists today can be used as a data point toward increasing our understanding of the living world. Comparative study of the developmental biology of multicellular organisms motivates the field of evolutionary developmental biology, or "evo-devo." However, to date only a relatively small number of species has been used to study the functions of genes that regulate animal developmental processes, which limits our understanding of how developmental genetic changes underpin evolution. If the idealized goal of evo-devo research is to do a species comparison that includes representatives of all major evolutionary transitions, then this calls for the establishment of many more animals as laboratory model organisms than is currently the case. As a step in this direction, developmental biologists are increasingly choosing new animal models that are suitable to address thus far neglected areas of research, while simultaneously selecting these candidates for their ability to fulfill other criteria relevant for evo-devo work. Some criteria that could maximize the scientific gains achieved by establishing new models include choosing organisms that belong to clades that are representative of a wide range of ecological niches, and those belonging to phyla that are species-rich, display high diversity in form and function, have interdisciplinary scientific appeal, are economical to maintain in the laboratory, and could inform issues that directly affect humans (e.g., disease or agriculture). Strategically choosing and studying examples satisfying some or all of these criteria may therefore be impactful, evolutionarily informative, and a good use of limited resources.

The phylum Arthropoda contains multiple species that satisfy many of the above criteria and has thus played a prominent role in modern evo-devo research. Importantly, good fossil records exist for this phylum, providing researchers with snapshots into the evolutionary past and aiding in comparative work. For example, the EDNA fossil insect database lists over 23,000 species (Giribet and Edgecombe 2013; Mitchell 2013), with the earliest records of arthropod fossils dating back to nearly 555 million years ago (Mya) during the Cambrian era (Harvey et al. 2012; Zhang et al. 2010; Vaccari et al. 2004). Undoubtedly, access to such fossil records is essential to understanding the key phenotypic innovations that have made Arthropoda species-rich and evolutionarily successful (Mayhew 2007).

Extensive studies on insects, which account for the majority of all species described on earth (Wheeler 1990; Grimaldi and Engel 2005) and for ~85% of all arthropod diversity (Giribet and Edgecombe 2013), have pioneered and shaped the evo-devo field. This work has informed us of the genetic and evolutionary basis of pivotal developmental mechanisms. For example, the description of the first homeotic mutant (Bridges and Morgan 1923), and the realization of the significance of conserved developmental genes in body patterning and in the evolution of different body plans across animals, come from studies in the insect *Drosophila melanogaster* (Lewis 1978; Nüsslein-Volhard and Wieschaus 1980; Duboule and Dolle 1989; Graham et al. 1989; Panganiban et al. 1997; Heffer et al. 2013). Other insect-based research that has broadened our understanding of evolutionary processes includes work on key evolutionary innovations such as the insect body plan (Grimaldi and Engel 2005), metamorphosis (Truman and Riddiford 1999), development of wings (Nicholson et al. 2014; Alexander 2018; Bruce and Patel 2018; Linz and Tomoyasu 2018), morphological novelties (Kijimoto et al. 2013), and insect eusociality (Toth and Rehan 2017). The widespread scope of such research is a result of the practical advantages that come with working on insects: insect phylogeny is well established (Giribet and Edgecombe 2013), many species are easy to maintain and culture in the laboratory, are often amenable to functional genetic analysis, in many instances produce easily accessible large broods suitable for external manipulation, and often have life cycles sufficiently short to allow laboratory rearing and multigenerational analysis.

To date, popular insect models to study the genetic basis of development have included the fruit fly *Drosophila melanogaster* (Diptera) (Demerec 1950), the flour beetle *Tribolium castaneum* (Coleoptera) (Sokoloff 1966, 1972, 1974, 1977; Denell 2008), the honeybee *Apis mellifera* (Hymenoptera) (Gould and Grould 1995; Oldroyd and Thompson 2006), the wasp *Nasonia vitripennis* (Hymenoptera) (Werren and Loehlin 2009), and the silk moth *Bombyx mori* (Lepidoptera) (Xia et al. 2004; Goldsmith et al. 2005; Meng et al. 2017). All of these insects, however, share a commonality: they belong to the same insect superorder of Holometabola, or insects that undergo complete metamorphosis during development. Complete metamorphosis is characterized by a pupal stage in the transition from larvae to adults, with neither the larval nor the pupal stages resembling the final adult form (Truman and Riddiford 1999). Such insects display a number of evolutionarily derived developmental characters that are not generally representative of all insects, let alone all arthropods (Mito and Noji 2008). To correct the overrepresentation of holometabolous insects in modern comparative developmental literature, insects branching basally to the Holometabola should be studied, as they, based on parsimony, appear to display characters that are likely ancestral to insects and in some cases also to arthropods. These insects are the Hemimetabola, insects displaying incomplete metamorphosis and lacking pupal stages during development. The embryo in such insects develops into a miniature adult (referred to as a nymph or a juvenile) which then undergoes several successive molts before reaching adulthood and sexual maturity. Orthopterans (crickets, grasshoppers, and locusts), one of the most abundant and dominant terrestrial insect groups, are in this category (Grimaldi and Engel 2005). Orthopterans display extraordinary diversity in developmental processes and are also economically important herbivores, which has resulted in them becoming popular for functional genetic research in recent years.

For the rest of this chapter, we focus on crickets as promising evo-devo models, deserving of serious regard. We discuss *Gryllus bimaculatus*, a field cricket species, introduce this model system, and discuss recent advances in establishing this animal for cell, developmental, and genetic research.

8.2 A Hemimetabolous Insect Model: The Cricket *Gryllus bimaculatus* De Geer

G. bimaculatus is a cosmopolitan orthopteran belonging to the family Gryllidae and is, to our knowledge, the most widespread of all *Gryllus* species for laboratory studies (Otte and Cade 1984). Although its use for developmental genetics is relatively recent, this species is by no means new to biological research: *G. bimaculatus* has been extensively used to inform areas such as neurobiology, insect physiology, reproduction, and behavior since the 1960s (Huber et al. 1989; Engel and Hoy 1999; Paydar et al. 1999; Wenzel and Hedwig 1999; Hedwig and Poulet 2004; Nakamura et al. 2008a, b; Horch et al. 2017b). The discovery of RNA interference (RNAi) as a mechanism for abrogating gene function (reviewed by Sen and Blau 2006) has greatly accelerated *G. bimaculatus* research (Mito et al. 2011), yielding important information about the developmental biology of this organism and unveiling its potential as an upcoming functional genetics laboratory model.

This species was first described by Baron Charles de Geer in 1773 (Geer 1773) and named *Gryllus* (meaning "cricket" in Latin) *bimaculatus* (from the Latin "macula" for "spot"). Indeed, this species is commonly referred to as the "two spotted field cricket," for the white spot that this species displays on the dorsal surface of the forewings next to the pronotal margin (Fig. 8.1a) (Otte and Cade 1984). Some of the morphological keys used to distinguish adult *G. bimaculatus* from other similar looking field cricket species, both within and outside the genus *Gryllus* (Otte and Cade 1984), include the white spots, a black colored adult body size of ~30 mm lacking any bands, forewings nearly covering and large hindwings extending well beyond the abdomen, and an ovipositor (in females) slightly longer than the hind femora (Fig. 8.1a). As an additional tool in the field, entomologists have documented that *G. bimaculatus* does not undergo an obligatory or facultative winter diapause (i.e., developmental arrest in response to adverse environmental conditions such as temperature and/or photoperiod), at any stage of its life cycle (Bigelow 1962). Such a lifestyle is called a homodynamic life cycle, and is in contrast to the heterodynamic lifestyle observed in many other *Gryllus* species entering diapause. Examples of overwintering species in this genus include *G. pennsylvanicus*, *G. campestris*, *G. fultoni*, *G. veletis*, *G. vernalis*, and *G. firmus* (Bigelow 1962). This means that it is possible to collect *G. bimaculatus* in the field across seasons and makes it easy to culture and breed this species in the laboratory all year-round. Additionally, in the field, *G. bimaculatus* is often found close to human settlements, on the ground surface or in soil cracks (Otte and Cade 1984).

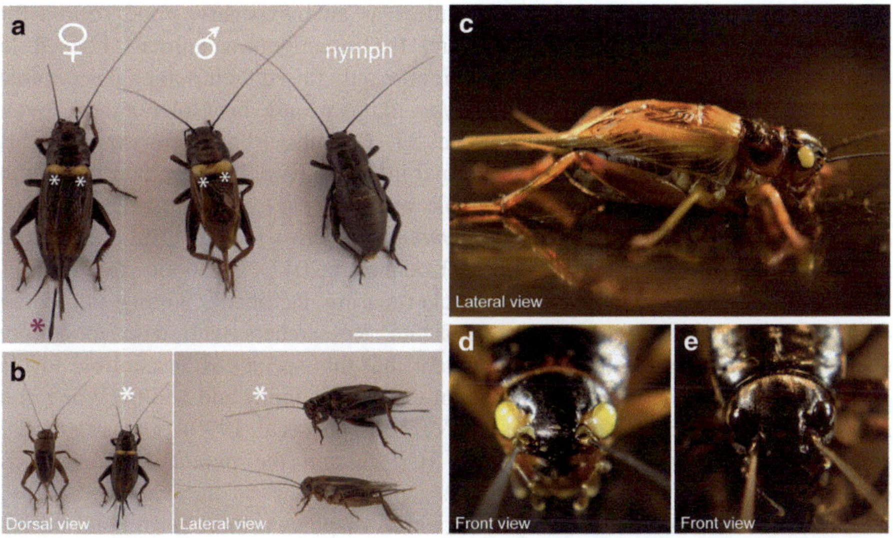

Fig. 8.1 *G. bimaculatus* morphology. (**a**) *G. bimaculatus* adult female (left), adult male (center), and male nymph (right) displaying some morphological keys used to identify this species. The characteristic white spots (marked with white asterisks on the adult female and male) on the forewings next to the pronotal margin are shown. Animals are black in color and are therefore vernacularly also known as the "black cricket" in some parts of the world. The body (in both sexes) lacks any bands of contrasting pigmentation; forewings nearly cover the abdomen, and hindwings extend well beyond the abdomen; females possess an ovipositor (marked with pink asterisk), used to deposit eggs. The presence or absence of an ovipositor can be used to sex animals in nymphal stages (developmental stages prior to becoming an adult). Note: The adult female and male photographed in this picture have broken left antennae; normally both antennae are of similar lengths. Scale bar is 1.5 cm. (**b**) Comparison (dorsal and lateral view) between an adult *G. assimilis* (unmarked) and *G. bimaculatus* female (marked with white asterisk). Note the brown body color, leaner and less bulky adult body, and the absence of the white spots on the forewings in *G. assimilis*. (**c**) A lateral view of an adult *G. bimaculatus* male white-eye mutant and (**d**) higher magnification images of the cricket head (front view) showing the white eye color in these mutants (left) compared to wild-type pigmented eyes shown in (**e**) (right)

While one could collect this species from the wild, the easiest way to establish a laboratory culture of *G. bimaculatus* is from animal pet suppliers. Multiple online pet suppliers based in various countries (e.g., Pets at Home, UK; Bugs International, Germany) culture and distribute this species as live crickets for captivity feeding. It is important to note, however, that at the time of writing, the United States is an exception: the United States Department of Agriculture (USDA) does not permit commercial distribution of *G. bimaculatus* in the United States, and the cricket species commonly commercially available for purchase are *Gryllodes sigillatus* and *Acheta domesticus* (examples of online retailers selling cricket species in the United States at the time of writing include Ghann's Crickets, Top Hat Cricket Farm, Fluker's Cricket Farm, and Premium Crickets). Being bulkier and meatier than other cricket species (Fig. 8.1b dorsal and lateral view), *G. bimaculatus* is preferentially exploited in multiple countries as an inexpensive food source for humans (described below, Sect. 8.8) and for insectivorous animals housed in captivity (Mito and Noji 2008). This has significantly raised their potential economic importance in recent years. Because *G. bimaculatus* is quite heat tolerant and one of the very few insects that can be reared at 37 °C (human body temperature), this species is also promising as a simple animal infection model system and has found application in studying human pathogenic bacteria (e.g., *Staphylococcus aureus*, *Pseudomonas aeruginosa*, and *Listeria monocytogenes*) (Kochi et al. 2016) and fungi including various *Candida* species (Kochi et al. 2017). Epizootic viral diseases are devastating in crickets (and for cricket-rearing facilities), wiping out entire colonies and becoming difficult to eradicate. Researchers who wish to culture *G. bimaculatus*, therefore, should be aware that this species is susceptible to the *G. bimaculatus* nudivirus (GbNV), known to infect nymphs and adults (Wang and Jehle 2009) and the cricket iridovirus (CrIV) (Kleespies et al. 1999), but is reportedly resistant to the cricket paralysis virus (CrPV) and the potent *A. domesticus* densovirus (AdDNV) (Szelei et al. 2011).

As a hemimetabolous insect, *G. bimaculatus* displays a short germ band during embryonic development and thus differs substantially from the well-studied long germ band characteristic of *Drosophila*. Short germ band development refers to a form of insect body patterning that is thought to be ancestral to arthropods (reviewed by Davis and Patel 2002) and present in many extant insects including crickets. In this form of development, only the anterior body segments (head only or the head and thorax) are specified in the early embryonic rudiment before gastrulation, whereas posterior segments (the thorax or the thorax and abdomen) are formed sequentially later in development during a secondary growth phase (reviewed by Krause 1939; Davis and Patel 2002; Liu and Kaufman 2005). In contrast, insects such as *Drosophila* follow the presumed derived long germ band type of development, whereby all segments are specified near simultaneously during the early blastoderm stage (Krause 1939; Campos-Ortega and Hartenstein 1985; Lohs-Schardin et al. 1979; Liu and Kaufman 2005). Another way in which crickets may display putative ancestral insect characteristics is in the structure of its ovaries. The *G. bimaculatus* ovary is panoistic, meaning that there are no germ-line-derived nurse cells that provide cytoplasmic content to growing oocytes (Büning 1994). Instead, every germ-line cell (i.e., every cystoblast) in the adult female ovary is thought to give rise to an oocyte (Büning 1994). In contrast, in the meriostic type of ovaries, as seen in *Drosophila* and nearly all other holometabolous insects (for details see Bilinski et al. 2017), the oocytes are connected to groups of germ-line cells called nurse cells. Panoistic ovary type and short germ development are, based on parsimony, thought to be features ancestral to insects and possibly to Pancrustacea. Consequently, it has been proposed that this species has the potential to serve as a representative study model for basally branching, hemimetabolous insect and arthropod lineages (Sander 1997; Mito and Noji 2008).

Sex determination in *G. bimaculatus* is thought to follow the XX/X0 system, with females being the homogametic sex and having a chromosome complement of $2n = 28 + XX$ (Yoshimura 2005) and having a predicted genome size of a few gigabases (Mito and Noji 2008). *G. bimaculatus* is polyandrous—females are known to mate with several males and exert postcopulatory mate choice (Tregenza and Wedell 1998). This polyandry is associated with increased egg-hatching rates and is hypothesized to prevent effects of inbreeding in wild populations (Simmons 1986, 1987; Tregenza and Wedell 1998, 2002; Bretman and Tregenza 2005). Indeed, *G. bimaculatus* females can lay many hundreds or thousands of eggs over their lifetime in the lab and, in our hands, have been maintained successfully as an inbred line (originally founded from a few dozen individuals) for over a decade, without any noticeable decline in health that might be attributed to inbreeding depression (Extavour lab, unpublished observations).

8.3 Cricket Sources, Animal Husbandry, Life Cycle, and Available Strains

At the time of writing and to the best of our knowledge (as described above), most current working laboratory cultures of *G. bimaculatus* were either established from adults purchased from commercial vendors (e.g., Tsukiyono Farm, Gunma, Japan; Scope Reptile Pet Store, Okayama, Japan; Livefood UK Ltd., UK; Kreca Ento-Feed BV, the Netherlands) or caught in the wild. However, this genus contains many species that are morphologically very similar, many species are known to overlap with *G. bimaculatus* in their local distribution, and adequate species-level, prominent morphological keys are lacking within this genus. Although some keys have been described (e.g., Nickle and Walker 1974), including the morphological characters described above, to an inexperienced eye, many of these features are often distinguishable only in comparison with another species present, or are easier to observe in preserved specimens than in live animals. We thus recommend performing molecular barcoding (e.g., using 16s ribosomal DNA or the *cytochrome b* mitochondrial DNA sequence) of the founding adults of a new colony, whether purchased commercially or captured in the wild, to ensure that all founding adults are indeed *G. bimaculatus* (Ferreira and Ferguson 2010).

Rearing *G. bimaculatus* is straightforward, and detailed cricket husbandry protocols are well described for this species (Mito and Noji 2008; Kainz et al. 2011; Kochi et al. 2016). Crickets (nymphs and adults) can be kept as inbred lines at 26–30 °C in well-ventilated plastic cages with egg cartons (Fig. 8.2a) or crumpled paper for shelter, and can be maintained on either a 12 h light/12 h dark (Kainz et al. 2011) or a 10 h day/14 h dark photoperiod (Mito and Noji 2008). They can be fed on general insect food or artificial insect diets (e.g., Oriental Yeast Co., Ltd., Tokyo, Japan), artificial fish food, finely ground dry cat food (e.g., Purina Kitten Chow), a mixture of oils and whole grain cereals (Kainz et al. 2011), or a combination of these food sources (Fig. 8.2b). Cricket Quencher water gel (Fluker Farms) can be used as a water source. Alternatively, a 50 mL falcon tube filled with water and stopped with cotton, or wet tissue or cotton in petri dishes, can also serve as water sources

Fig. 8.2 *G. bimaculatus* husbandry. (**a**) (side view) and (**b**) (top view) showing a well-ventilated plastic container used for housing a cricket colony. Note the use of egg cartons for providing shelter, ground cat food, and a 50 ml Falcon filled with water and stopped with cotton as a food and water source, respectively. A wet cotton plate (seen in **a**, **b**, and **b** inset) is placed in the adult cages for females to oviposit fertilized eggs. Oviposited eggs (higher magnification shown in (**d**)) need to be kept moist and clean until the eggs hatch. (**c**) A close-up of a cotton plate showing newly emerged cricket hatchlings, which can then be transferred into new plastic cages with food, water, and shelter until they reach adulthood. (**e**) Two *G. bimaculatus* eggs (6 days after egg laying) imaged under bright field white light (top) and green florescent light (bottom). Both eggs are progeny obtained from a cross between the histone2B-GFP (H2B-GFP) transgenic and wild-type *G. bimaculatus* line. Embryos carrying the H2B-GFP transgene (bottom egg marked with white asterisk) can be distinguished from non-transgenic embryos (top egg) based on the presence of bright florescent nuclei, which is detectable from day 5 until day 10 after egg laying

(Fig. 8.2b). Crickets will oviposit fertilized eggs into damp sand (e.g., Sandtastik Sparkling White Play Sand, Product Code PLA0050), wet paper towels, Whatman paper, or wet cotton placed in petri plates in cricket cages (Fig. 8.2b inset, d). These eggs will develop successfully and hatch in 12–14 days (Fig. 8.2c) under the following conditions: incubation at ~28–29 °C with 70% humidity, dead or moldy

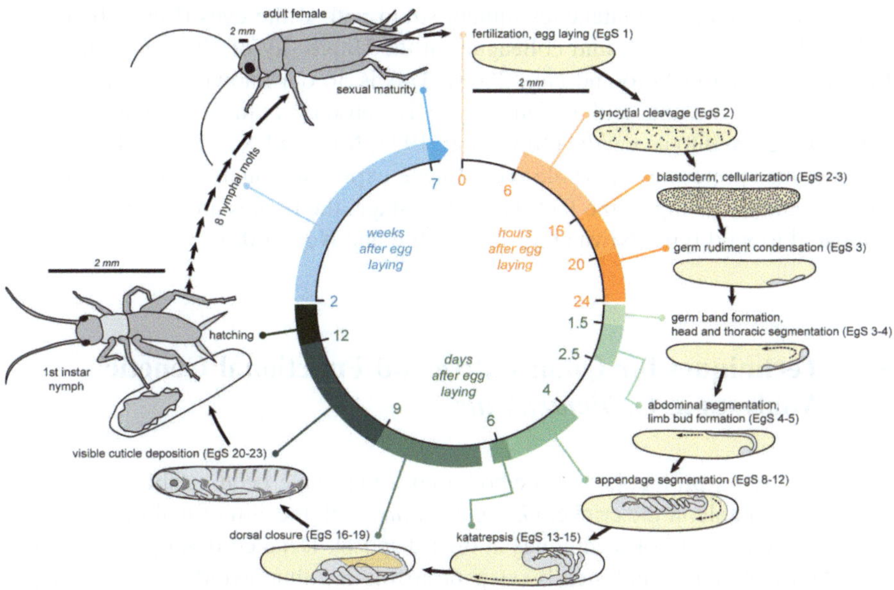

Fig. 8.3 A schematic showing an overview of the *G. bimaculatus* life cycle. Selected embryonic and nymphal developmental stages are shown, alongside the duration of each developmental stage depicted in hours (orange arc and lines), days (green arc and lines), or weeks (blue arc and lines) after egg laying. The colored arcs indicate the entire duration of time occupied by the indicated developmental stage, whereas the lines show a cartoon schematic of the selected stages within this time window. Each displayed embryonic stage during embryogenesis shows the position of the embryo (in gray) relative to the yolk (yellow) within the egg and has a small description of features that are characteristic of that developmental stage (depicted as egg stage "EgS"). Dotted lines with arrowheads indicate the different movements that the embryo makes during the course of development in this species. Upon hatching, nymphs undergo eight nymphal molts (indicated by solid black arrows) to reach adulthood. Newly emerged adult animals are sexually mature at molting and begin mating soon after. This figure is modified from Donoughe and Extavour (2016)

embryo removal on a regular basis, and maintenance of a moist and clean substrate (Mito and Noji 2008; Donoughe and Extavour 2016). Embryonic development for this species has been divided into 16 stages based on morphological features of eggs, developing embryos and their appendages (Niwa et al. 1997; Donoughe and Extavour 2016). After hatching, nymphs undergo eight nymphal molts to finally become adults over the next 5 weeks. The generation time (total time to adulthood and sexual maturity) of *G. bimaculatus* is thus approximately 7 weeks at 29 °C (Fig. 8.3). Adults are thought to reach maximum fecundity 1 week after the final molt (Mito and Noji 2008). Sexing and identification of virgin males and females are straightforward: late-stage male and female nymphs can be separated based on the presence or absence of an ovipositor and then isolated until they undergo the final molt to sexual maturity (Fig. 8.1a). This also helps in setting up single mating crosses, for example, to establish genetically modified lines. Similarly, precisely timed egg collections are possible by placing egg collection petri dishes in the cages and removing them at desired intervals (described in Donoughe and Extavour 2016).

A *G. bimaculatus* spontaneous mutant strain with white eyes (Fig. 8.1c–e) was isolated by Isao Nakatani and colleagues at the University of Yamagata, Japan, in 1989 (referenced in Mito and Noji 2008). This is, to our knowledge, currently the only available mutant, and its phenotype is caused by an autosomal recessive mutation, referred to as *gwhite* (Niwa et al. 1997; Mito and Noji 2008). This mutant is sometimes preferred for whole-mount gene expression analysis at late stages of embryogenesis, owing to the fact that at this stage, the tissues are more transparent compared to wild type (Niwa et al. 1997; Mito and Noji 2008).

8.4 Techniques for Quantitative and Functional Genetic Analyses in *G. bimaculatus*

Here we discuss protocols and methodologies that have been established and are currently available in the cricket *G. bimaculatus*, with the aim of making new users aware of the plethora of techniques at their disposal. Detailed descriptions of these published techniques and step-by-step protocols are thus avoided in this chapter (we refer the reader to Horch et al. 2017a, b for detailed protocols). While these protocols are now well established in crickets, these tools are not limited to them and could in principle be modified or adapted for use in other hemimetabolous insects to further species-specific research.

8.4.1 Precise Embryonic Staging System

To make meaningful observations of deviations from normal embryonic development, one first needs a wild-type reference for any given species. Donoughe and Extavour (2016) have reported a detailed embryonic staging system for *G. bimaculatus* (Figs. 8.4 and 8.5). This system is based on externally observable characters of the developing cricket embryo that are visible through the eggshell, thereby circumventing the need for embryonic dissections to ascertain embryonic developmental stage. This is especially informative for studying early embryos of insects such as crickets, which are embedded within a large amount of opaque yolk, making direct observations through the eggshell difficult if not impossible. *G. bimaculatus* development, based on this staging, is presented as 24 "egg stages" and encompasses the entire development of the animal from fertilization to hatching, based solely on external observable egg characters (called stage identifiers). Each of the 24 "egg stages" described here corresponds to one or more of 16 "embryonic stages." For each stage, the authors provide a list of embryonic developmental features defining that stage, including features of body segmentation, mesoderm, and appendage formation. Determining the stage of embryogenesis through the eggshell is a useful complement to earlier described staging schemes that require dissection of the embryo (Mito and Noji 2008; Kainz 2009).

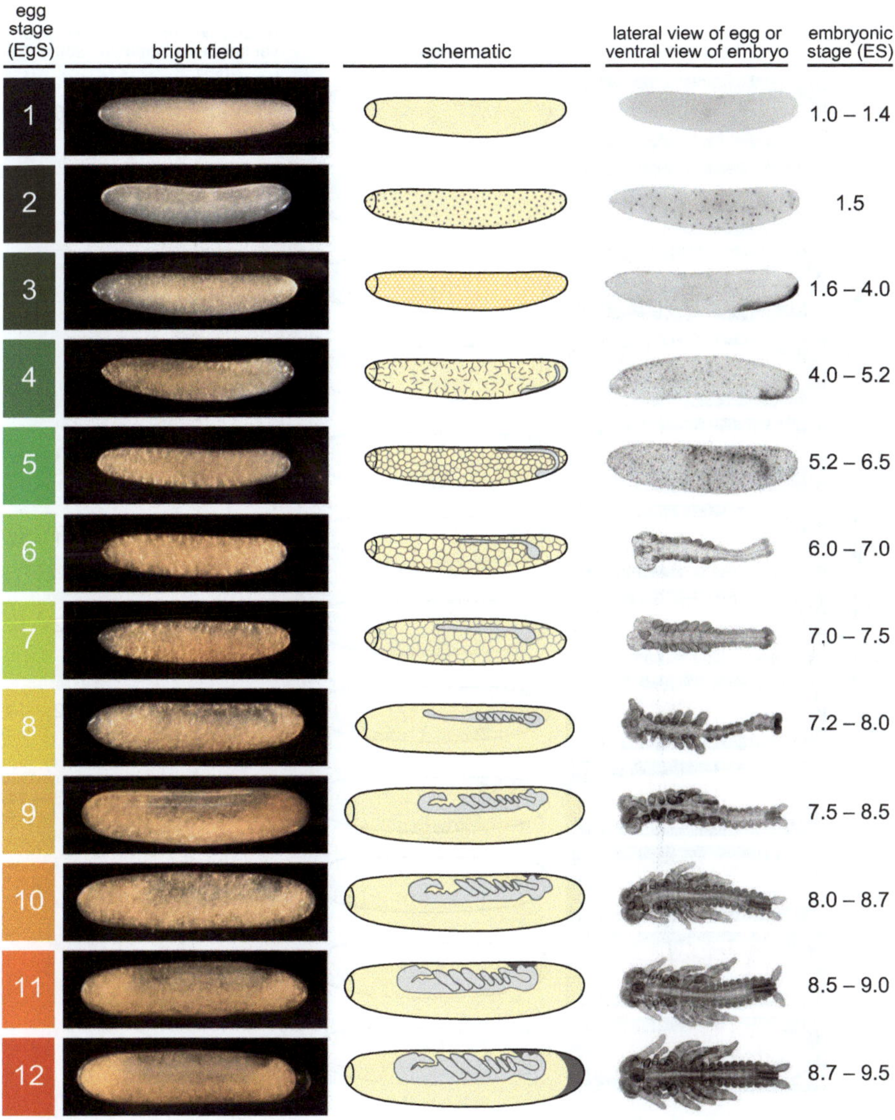

Fig. 8.4 A detailed description of the egg stages (EgS 1–12) in *G. bimaculatus.* Micrographs in second column from left help display the morphological features of the egg that can be used to assign embryos to an egg stage (EgS) and also describe the corresponding morphological features and embryonic stage (ES) of the embryo within the egg. In the schematic, the embryo is depicted in gray and yolk in yellow. Micrographs in the right second column from the right are not to scale, and are taken using lateral views of either a H2B-GFP transgenic live embryo (EgS 1–5) or ventral views of dissected and fixed, Hoechst 33342-stained embryos (EgS 6-182). Micrographs are not to scale. This figure is modified from Donoughe and Extavour (2016)

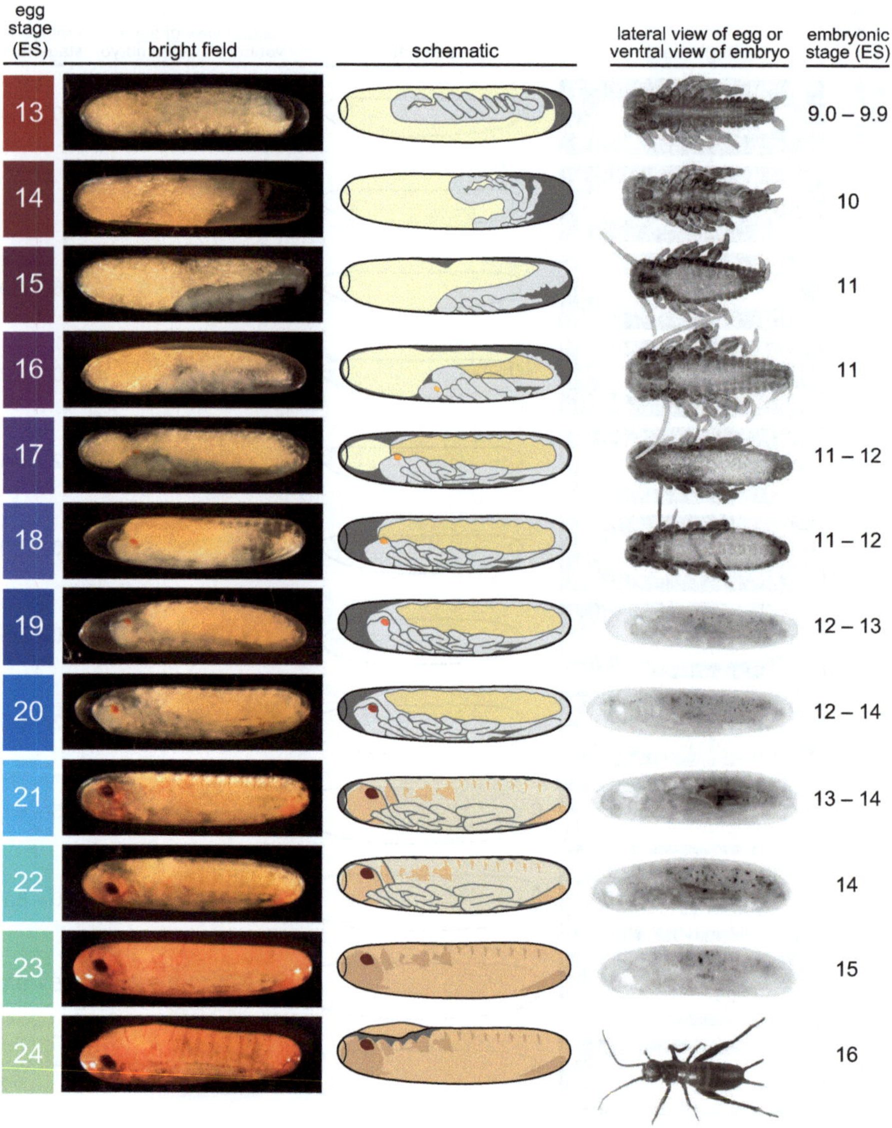

Fig. 8.5 A detailed description of the egg stages (EgS 12–24) in *G. bimaculatus.* Staging system for egg stages (EgS) 12–24 continued from Fig. 8.4. This staging system ends at hatching and does not include postembryonic development of nymphs to adulthood. This figure is modified from Donoughe and Extavour (2016)

8.4.2 Injection Methods for Eggs, Nymphs, and Adults

A basic requirement for many experimental procedures in modern developmental biology, including live imaging, RNA interference, and gene editing, is the delivery of synthetic or biological materials into the body of an animal, without disrupting its health or sacrificing its life. Direct manual injection is one such method and, in the case of crickets, has been well established and found effective in egg, nymphal, and adult stages. Two methodological variations are commonly in use for *G. bimaculatus* egg injections, differing essentially in the number and arrangement of embryos for injection, and are described in great detail in Horch et al. (2017a) and Barry et al. (2019). Both methods are thus described below in brief.

The first variant of the egg injection method, developed by the Noji lab (University of Tokushima), involves the construction of a mold to house eggs for injections (Horch et al. 2017a). Watson chambers (which resemble a rectangular mold) are glued onto microscopic glass slides using double-sided tape. Embryos are then lined up end to end along the length of the chamber, using a small stainless-steel spatula. The wall of the chamber and the adhesive of the double-sided tape (on the slide) help secure and hold the embryos in place during the injections. The second variant, optimized in the Extavour lab, uses rectangular troughs made using plastic molds set in low-melting agarose that hold eggs in place (Kainz et al. 2011; Barry et al. 2019). Using this setup, over 35 embryos per slide can be prepared for injection simultaneously, making it efficient in terms of preparation time and the number of embryos injected in one sitting.

Regardless of the egg injection method used, eggs are injected under a dissecting or compound microscope, using a needle held by a micromanipulator (Fig. 8.6a, b). The needle must be loaded with the desired injection material and connected to a pressure source, which may be manual (e.g., a syringe) or electronically controlled compressed gas (e.g., using a commercial micro-injector). The injected material may be mixed with a dye that is visible under white light (e.g., phenol red or fast green) or fluorescent light (e.g., fluorescein- or rhodamine-conjugated dextrans) depending on the user's preference. The choice of dye will determine the best microscopy and light regime to be used for injections. Following injections, embryos are allowed to develop normally in humid incubators at 28 °C on wet paper towels or are submerged in 1× phosphate buffered saline in closed petri dishes and monitored daily until embryos hatch. Eggs of developmental stages that are turgid and under high pressure, including very early stages in the first few hours following fertilization and middle stages following elongation of the germ band, are more difficult to inject than earlier stages.

For nymphal and adult injections, a manually held Hamilton syringe or any automated micromanipulator/microinjector system specifically designed for delicate microinjections and capable of injecting nanoliter volumes can be used effectively. Nymphs and adults are prepared for injections by first cooling them on ice to temporarily immobilize them, and then injected either between the abdominal segments (A2 and A3) or in the soft tissue between the T3 coxa and the thorax. For site- or tissue-specific injections (e.g., the leg or brain), the injection site should

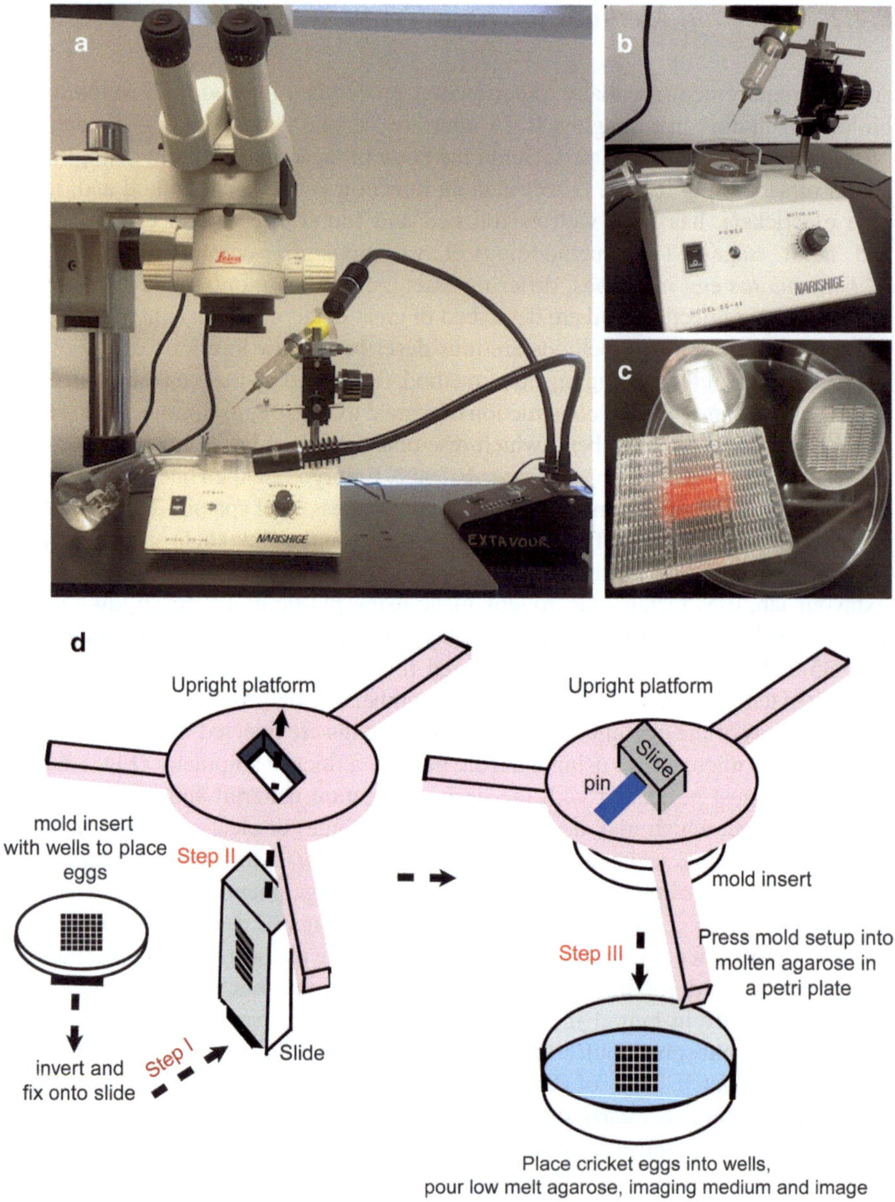

Fig. 8.6 Injection and OMMAwell setup. (**a**) The apparatus used for beveling glass needles for use in injecting crickets, consisting of a light source, a dissecting microscope, a micromanipulator, and a beveling stone (Narishige model EG-45 is shown here). (**b**) Higher magnification of the beveling setup shown in (**a**). (**c**) Different types of agarose mold inserts used for mounting cricket embryos for application in OMMAwell or, alternatively, for cricket egg injections. (**d**) OMMAwell (Donoughe et al. 2018) schematic for top-loaded microwells, used for injecting and imaging cricket embryos using a configuration for upright objectives. Different assembly components are shown: the mold insert (white) consists of wells that will house the cricket embryos and is inverted and attached to the base of the slide (gray). This is then placed into the upright platform (pink) and

be modified accordingly. However, locales of abundant fat tissue should be avoided as injection sites in crickets, to help facilitate dispersal of liquid into the body upon injection, prevent backflow of injected material or hemolymph into the needle, and prevent blockage of fine needle tips with insect tissue. Irrespective of the injection site, care should be taken while injecting the needle into the nymphal or adult body, so as to prevent injuring the internal organs, which could potentially kill the animal or disrupt recovery. Such injuries can be easily avoided by inserting the needle only deep enough into the animal body to prevent oozing of material at injection.

Other recommendations for successful injections of adult or juvenile *G. bimaculatus* include inserting the needle parallel to the body of the insect, rather than at a perpendicular angle, injecting larger volumes (relative to the insect size) as multiple pulses of smaller doses rather than all at once, injecting slowly to prevent leakage, minimizing handling stress for the animal, making sure the needle is not blocked prior to insertion, and maintaining basic cleanliness and sterility during the procedure. Following injection, animals should be allowed to recover in isolated cages with food and water at room temperature before proceeding with the desired study.

8.4.3 High-Throughput Live Imaging of Embryos Using OMMAwell

Open Modular Mold for Agarose Microwells (OMMAwell) is a simple, reusable, all-in-one device that allows users to easily mount and simultaneously image dozens of live *G. bimaculatus* embryos consistently and economically for 2D and/or 3D time-lapse analyses of early development (Donoughe et al. 2018). OMMAwell has the added advantage of being adaptable and customizable: it has been made to accommodate the imaging needs of researchers with different experimental designs, can be used on diverse species (OMMAwell has been successfully designed for and tested on nine animal species, including many traditional model organisms), and can be used for both inverted and upright objective microscopes (Fig. 8.6d). With this device, embryos can be efficiently and quickly lined up in arrays of agarose microwells, whose dimensions and spacing can also be customized as per individual user needs (Fig. 8.6c) (see Donoughe et al. 2018). In addition, OMMAwell has reservoirs to hold live imaging media and help maintain specimen-specific humidity, osmolarity, and oxygen levels during time-lapse live imaging, thereby enhancing embryonic survival and data quality. This device also allows positional tracking of

Fig. 8.6 (continued) secured at the desired height with the help of a pin (blue). The assembled components are then lowered into a petri plate containing molten low-melt agarose and allowed to set. Following the removal of the mold insert from the cooled and set agarose, specimens are added into the wells in the petri plate either individually or in bulk and covered with low-melt agarose (in microliter volumes of up to 100 μl) to hold them in place. Embryos can be oriented carefully using forceps prior to this step. Once the agarose sets, live-imaging media is poured into the dish

individual embryos and permits users to control sample orientation for imaging. The OMMAwell microwell array arrangement is also convenient to hold embryos in place during injections. OMMAwell has been used to image the development of as many as 102 *G. bimaculatus* live embryos simultaneously for 12 consecutive days (Donoughe et al. 2018). Hatching rates of these embryos were not significantly different from the hatching rates of controls, suggesting no lingering effects of phototoxicity, developmental delays, or defects on these embryos from the use of OMMAwell.

8.4.4 Gene Expression Analyses Using Embryonic or Whole Mount In Situ Hybridization and Immunohistochemistry

The ability to detect mRNAs [using in situ hybridization (ISH)] and proteins [immunohistochemistry using labeled antibodies (IAb)] of interest is central to whole mount gene expression analyses in any organism. Standard protocols to study gene expression in other vertebrate and invertebrate embryos have been applied successfully in *G. bimaculatus* (Niwa et al. 2000; Mito and Noji 2008). These include whole mount in situ hybridization using digoxigenin (DIG)-labeled antisense RNA probes (as per Wilkinson 1992), protein detection (Patel 1994), and double in situ hybridization using probes labeled with different haptens (e.g., Dietrich et al. 1997). Optimized ISH and IAb protocols have also been developed in this species for specific tissues including the brain, nymphal legs, and wings. Automated medium- or high-throughput gene expression assays on *G. bimaculatus* tissues using specialized robots (e.g., Intavis InsituPro VSi) are also possible (Extavour lab, unpublished).

8.4.5 RNA Interference

The Noji lab pioneered the establishment of RNA interference (RNAi) technology in the cricket *G. bimaculatus* (Miyawaki et al. 2004), and many researchers have since used this technique successfully to deplete mRNAs of multiple target genes in this species. Four main types of RNAi techniques have been developed for use in crickets: embryonic, nymphal, parental, and regenerative RNAi (Miyawaki et al. 2004; Mito et al. 2005; Nakamura et al. 2008a; Mito and Noji 2008; Ronco et al. 2008).

To perform RNAi, double-stranded RNA (dsRNA), preferably 300–500 nucleotides in length, complementary to a region in the *G. bimaculatus* gene of interest, is injected into the eggs (embryonic RNAi) or into the body cavity of nymphs (nymphal RNAi) or adults (parental RNAi). Successful concentrations of dsRNA have been reported to range from 2 to 6 μg/μl (e.g., Kainz et al. 2011). It is recommended that the dsRNAs designed should match a region close to or including the 3′ UTR of the target *G. bimaculatus* gene, which may minimize off-target effects. Typical specificity controls may include testing at least one other dsRNA designed against a nonoverlapping fragment of this same gene. Injecting dsRNA against exogenous genes not encoded by the cricket genome (e.g., DsRed) and injecting the buffer alone can also serve as meaningful controls and are thus strongly recommended for every RNAi experiment. Together, these measures can help researchers distinguish between specific and nonspecific effects of RNAi, allowing meaningful interpretation of their results.

RNAi is systemic in *G. bimaculatus*, such that RNAi-induced phenotypes may be detected throughout the body of the embryo, nymph, or adult, regardless of the site of injection. Moreover, the injection of dsRNA into sexually mature adult females allows for observation of RNAi effects not only in the adult animal itself but also in its progeny (i.e., eggs) that the animal will lay over the weeks following a postinjection mating as long as the gene knockdown does not interfere with oogenesis, fertilization, or egg laying. Alternatively, nymphal RNAi can be conveniently used to determine gene functions in postembryonic stages. Regenerative RNAi was optimized in the Noji lab and has been performed as a specific application of nymphal RNAi in the cricket (Nakamura et al. 2008a). For this procedure, a leg of a third instar nymph is amputated following dsRNA injection, and the effects of RNAi are then assessed during the regeneration of the lost leg (which normally occurs over subsequent molts). Based on these observations, the RNAi response in crickets can be robust, stable, and even transmissible through subsequent molts (Nakamura et al. 2008a; Hamada et al. 2015). However, it is recommended that the robustness of RNAi response, its stability, and duration be determined on a case-to-case basis; in our hands, there have been instances where the RNAi response for some genes has lasted only a few days (see Kainz et al. 2011).

8.4.6 Calcium Imaging to Study Neurobiology and Neuroethology

The cricket has been an important model for neurobiology and neuroethological studies, and many physiological techniques are easily applicable to the cricket (Ogawa and Miller 2017). One such technique is that of calcium imaging, which uses florescent dyes and optical methods to monitor the changes in intracellular levels of calcium ions in live cells and tissues (Neubauer and MacLean 2010), including cricket neurons. Information on selection of calcium indicators, dye loading protocols, experimental designs, and calcium imaging techniques in the cricket are well described (Ogawa and Miller 2017). In 2013, Matsumoto and colleagues successfully expressed Yellow Chameleon (YC) 3.60, a genetically encoded calcium indicator (GECI) in the cricket brain via electroporation (Matsumoto et al. 2013), enabling prolonged deep imaging of the cricket brain for the first time. Together with high-resolution microscopy and gene editing techniques, calcium imaging is expected to facilitate major advances in our understanding of cricket neurobiology (Ogawa and Miller 2017). Calcium imaging is not limited to neurobiology, so its successful establishment in the study of cricket nervous tissue suggests that this technique can now also be used to study physiology in other tissues and cell types in this animal.

8.4.7 High-Sensitivity Trackball Recording Systems for Studying Phonotaxis and Auditory Neuronal Plasticity

Acoustic communication is paramount in insects, both within and between species. Making precise recordings of insect locomotory behavior in response to auditory stimuli (phonotaxis), such as male calling songs, is often challenging under natural settings. Various laboratory assays for measuring cricket phonotactic behavior have been developed, making such studies possible. Examples of such assays include analyzing the number of crickets that reach an acoustic stimuli or sound target within a defined time period (Tschuch 1976; Stout et al. 1983), studying cricket behavior in mazes (Popov and Shuvalov 1977; Rheinlaender and Blätgen 1982), or steering responses of tethered flying female crickets (Pollack and Hoy 1979). The development of two different trackball recording systems in crickets has been paramount in enhancing our understanding of *G. bimaculatus* auditory steering behavior (Hedwig 2017) and provided detailed insights into insect locomotory behavior in general. All trackball recording systems measure the movements of the trackball, based on which insect velocity and direction of insect movement (walking) are inferred, without allowing the insect to reach the auditory target. In closed-loop trackball systems, the cricket is allowed to walk and turn freely on the trackball during recordings, with the trackball compensating for cricket movement by having the ability to counter-rotate. By contrast, in open-loop systems, the tethered cricket has the ability to walk but not change its orientation in an acoustic field. Due to their sophisticated design, these trackball recording systems can now easily be integrated into experimental setups using other forms of recording, including neuro- or electrophysiological and high-speed video recording experiments. Thus, combined with the GECI YC3.60 discussed above, and alongside other sophisticated imaging and video recording techniques (see below), trackball recording systems are expected to provide new insights not only into cricket biology but also into the study of insect phonotaxis in general.

8.4.8 Automated and Customizable Video Tracking Systems, Artificial Crickets, and Cricket Robots for Synthetic Neuroethology and Social Behavior

Crickets have been used over the past several decades as systems to study behaviors including mating, flight, aggression, wandering, obstacle avoidance, and importantly, to study the neurophysiology underlying these processes. When investigating, quantifying, and qualifying animal behavior, dependable and accurate measurement systems are needed to record animal responses to external stimuli, at both behavioral and physiological levels. The advent of engineering approaches in crickets, especially robotics, is expected to greatly facilitate such research and is the genesis of the field of cricket synthetic neuroethology. Aonuma and colleagues have described a novel approach developed for crickets, where provoked animal behavior in response to computer-generated simulation and robots is captured to effectively bridge the gap between insect behavior and physiology (Aonuma 2017). Different commercially available automated video tracking systems designed to follow cricket movement have also been previously described (Noldus et al. 2001). Recently, another customizable tracking system based on a simple open-source solution called SwisTrack (Lochmatter et al. 2008) has been introduced for use in crickets. Using this system, multiple crickets can be video recorded and tracked simultaneously. Because the entire process is semiautomatic, data collection and its interpretation are more efficient than previous methods that were based exclusively on manual tracking. Using artificial crickets (e.g., Funato et al. 2011; Kawabata et al. 2012; Mizuno et al. 2012) or cricket robots (Funato et al. 2008, 2011) alongside computer modeling is another way of analyzing cricket behavior that has recently been reported. Further detailed information on artificial crickets, cricket robots, biomimetic robots (Ritzmann et al. 2000), and behavioral modeling in this species can be found in Aonuma (2017).

8.4.9 Standardized Protocols for Assessing Learning and Memory

G. bimaculatus has been reported to have a robust memory and thus has been exploited for studying the neural mechanisms underlying olfactory, auditory, and visual learning. Mizunami and colleagues have published detailed protocols for classical conditioning, operant testing, associative learning, memory retention, and subsequent data analyses in *G. bimaculatus* (Mizunami and Matsumoto 2017a). A "classical conditioning and operant testing" procedure has also been developed in crickets by these researchers. The establishment of such protocols has resulted in the elucidation of detailed cellular mechanisms and signaling cascades that are important for memory formation in crickets. These studies have additionally revealed that crickets display unexpected diversity in the mechanisms underlying these processes in comparison to other insects including *Drosophila* (Mizunami and Matsumoto 2017b). The use of such classical conditioning paradigms and their variants in crickets may provide novel breakthroughs in our understanding of learning, cognition, and memory across animals.

8.4.10 Transgenic Lines

Stable transgenic lines are an invaluable tool for developmental genetics and contribute to the successful establishment of a model animal system. Transgenesis using P elements, which are the transposon of choice for *Drosophila* transgenesis (Rubin and Spradling 1982), have been found ineffective in crickets, such that other transposable elements need to be used to achieve transformation in this species. Zhang and colleagues (2002) showed that *Minos* transposons (Pavlopoulos et al. 2007) are active in *G. bimaculatus* embryos and highlighted the possibility of using these as gene vectors for germ line transformation in this species. However, to our knowledge, this transposon has not yet been used to establish stable transgenic cricket lines. Shinmyo et al. (2004) succeeded in somatic transformation of *G. bimaculatus* embryos, using the *piggyBac* transposon (Handler et al. 1998) to achieve somatic insertion of a construct containing an enhanced green florescent protein (eGFP) coding region driven by a *G. bimaculatus actin 3/4* promoter. Construction of plasmids and injection protocols for this line are as described in Shinmyo et al. (2004) and Zhang et al. (2002). Subsequently, this technique has been optimized to achieve germ line transmission of transgenes (Nakamura et al. 2010).

At the time of writing, a histone2B-GFP (H2B-GFP) transgenic line is stably maintained in multiple laboratories (Nakamura et al. 2010). In this line, the promoter of the *G. bimaculatus* actin orthologue (*Gb-Actin*) drives the expression of the Gb-histone2B protein tagged with eGFP. This transgene is ubiquitously and constitutively expressed and is maternally contributed to eggs. Based on viability ratios of the embryos laid by this line, it is likely that the transgene is sublethal in homozygosis (Extavour lab, unpublished observations). As there are no balancer chromosomes for *G. bimaculatus*, heterozygotes must be manually selected at every generation to maintain the transgene (Extavour lab, unpublished observations). Zygotic expression of this transgene begins at approximately the fourth day after egg laying (AEL) at 28 °C, and eggs expressing the transgene can then be easily identified and selected between 5 and 10 days AEL, based on the presence of brightly fluorescent nuclei under a fluorescent stereomicroscope (Fig. 8.2e top and bottom).

8.4.11 Genome Editing Using CRISPR/Cas9, TALEN, and Zinc-Finger Nucleases

Sophisticated functional genetics techniques commonly used to modify genomes in vivo at a specific site include *c*lustered *r*egularly *i*nter*s*paced *p*alindromic *r*epeats (CRISPR)/CRISPR-*as*sociated nuclease 9 (Cas9), collectively known as the CRISPR/Cas9 system (Cong et al. 2013), *t*ranscription *a*ctivator-*l*ike (TAL) *e*ffector *n*ucleases (TALENs), and *z*inc-*f*inger *n*ucleases (ZFNs) (Porteus and Carroll 2005; Moscou and Bogdanove 2009; Remy et al. 2010; Miller et al. 2011; Jinek et al. 2012). All of these approaches work by generating double-stranded breaks in target DNA sequences, which in turn trigger the cell's DNA damage response (Remy et al. 2010), and this cellular response can then generate mutations (insertions or deletions) in the targeted gene. All of these techniques are now available and functional in crickets. The Mito lab established and reported the use of ZFNs and TALENs in crickets in 2012, by successfully creating homozygous genetic knockouts (Watanabe et al. 2012). CRISPR/Cas9 has also now been effectively used for the generation of both knock-ins (Horch et al. 2017b) and knockouts (Awata et al. 2015) of cricket genes. Detailed protocols for knocking-in or knocking-out cricket genes using the CRISPR/Cas9 method are available in Horch et al. (2017b).

8.4.12 Genomics and Transcriptomics

While no genome sequence is currently publicly available for *G. bimaculatus*, a number of de novo transcriptomes have been published for this species, providing gene expression datasets for a number of different specific tissue types and developmental stages. To date, these include transcriptomes of the ovaries, embryos, the prothoracic ganglion, and regenerating legs (Zeng and Extavour 2012; Bando et al. 2013; Zeng et al. 2013; Fisher et al. 2018). Moreover, several transcriptomes reflecting gene expression at different life stages (Berdan et al. 2016), in the male accessory gland (Andres et al. 2013), fat body and flight muscles of different ecological morphs (Vellichirammal et al. 2014), and under cold-acclimation conditions (Des Marteaux et al. 2017; Toxopeus et al. 2019), are available for other species of the genus.

8.5 Novel Insights into Biological Processes Using Forward Genetics in a Hemimetabolous Insect

G. bimaculatus has been effectively used to study various disciplines of biological sciences over the past decades. These include early embryonic development and body patterning, tissue and organ system specification, regeneration, body size regulation, memory and learning, reproductive biology, ecology, physiology, and endocrinology (reviewed in Horch et al. 2017b). Here, we will therefore refrain from reiterating the contributions that these results have made to our understanding of biology. Instead, we will briefly discuss one example of a novel insight that the biological community has gained through the use of functional genetics in the cricket.

8.6 The Evolution of the *oskar* Gene and Its Implications for Germ-Line Research

Germ cells are the cells that give rise to eggs and sperm. They are therefore an important cell type in sexually reproducing organisms and are sometimes referred to as the ultimate totipotent stem cell (Cinalli et al. 2008), because they alone maintain a genetic link between generations. In a developing embryo, the first cells to give rise to the germ cells by clonal mitotic divisions are known as the primordial germ cells (PGCs). Across metazoans, PGCs are specified using one of two mechanisms (Extavour and Akam 2003). In some animals, including *G. bimaculatus* and *Mus musculus* inductive cell–cell signaling among neighboring somatic cells instructs certain cells to adopt PGC fate; this method of PGC specification is known as "induction." In other animals, including the fruit fly *D. melanogaster*, the nematode worm *Caenorhabditis elegans*, the zebrafish *Danio rerio*, and the clawed frog *Xenopus laevis*, PGCs are instead specified by "inheritance." In this mechanism, PGCs are specified very early in development through the cytoplasmic inheritance of a maternally derived special cytoplasm called "germ plasm," which often contains determinants that confer germ-line fate.

oskar is an insect-specific gene critical for the establishment of germ plasm in *D. melanogaster* and is the only gene reported in the animal kingdom to be both necessary and sufficient for germ cell formation (Lehmann and Nüsslein-Volhard 1986; Ephrussi et al. 1991; Kim-Ha et al. 1991; Ephrussi and Lehmann 1992; Smith et al. 1992). Following its discovery in *D. melanogaster*, *oskar* orthologues were reported in the genomes of other holometabolous insects known to specify their germ cells using germ plasm (Goltsev et al. 2004; Juhn and James 2006; Juhn et al. 2008; Lynch et al. 2011). Interestingly, *oskar* appears to be absent from many insect genomes that are known to lack germ plasm, including the bee *A. mellifera*, the beetle *T. castaneum*, and the silk moth *B. mori* (summarized by Quan and Lynch 2016). Based on this observation and the fact that hemimetabolous insects reportedly lack germ plasm (see Ewen-Campen et al. 2013 and references therein), it was hypothesized that *oskar* was a novel gene that arose at the base of the Holometabola, concurrent with the advent of insect germ plasm (Lynch et al. 2011). However, Ewen-Campen and colleagues discovered an *oskar* orthologue in the cricket *G. bimaculatus* genome and demonstrated that in this species, *oskar* is neither expressed at high levels in PGCs nor required for PGC formation (Ewen-Campen et al. 2012). Instead, cricket *oskar* is expressed in the neuroblasts (stem cells that arise from the neural ectoderm and give rise to the nervous system in Pancrustacea) of the brain and central nervous system (CNS) of the developing embryo, and is required for proper embryonic CNS patterning (Ewen-Campen et al. 2012). This observation, taken together with the reports that *D. melanogaster oskar* also plays a neural role (Xu et al. 2013), suggests two novel hypotheses: (1) *oskar* arose at least 50 million years earlier in insect evolution than previously hypothesized, before the divergence of Hemimetabola and Holometabola, and (2) *oskar*'s ancestral role in insects may have been in the nervous system and not in the germ line. This implies that *oskar* may have been co-opted for its essential role in holometabolous germ plasm assembly rather than having originated concurrently with germ plasm as had been previously suggested. This significantly changes our understanding of the evolutionary origins and functional evolution not only of this gene but perhaps also of insect germ plasm. Moreover, it constitutes an important example of how novel genes may arise and become co-opted, across evolutionary time scales, to perform different biological roles in animals.

8.7 Genomic Resources in Other Orthopterans

For an organism to become widely used as a research model for comparative or evolutionary studies, an important contributing factor is whether resources and tools are also available to study its close relatives. With this in mind, we will discuss available resources in other orthopterans that may be of use in aiding comparative work with *G. bimaculatus*. To our knowledge, at the time of writing, large-scale genomic resources are available for only two other orthopterans, a locust and a Hawaiian cricket species.

8.7.1 Resources in Field Crickets

Several transcriptomes are available for tissues and stages of many species of the genus *Gryllus*, including (1) gene expression at different life stages in *G. rubens* (Berdan et al. 2016), (2) in the male accessory gland of *G. firmus* and *G. pennsylvanicus* (Braswell et al. 2006; Andres et al. 2013), (3) fat body and flight muscles of different ecological morphs of *G. firmus* (Vellichirammal et al. 2014), (4) under cold-acclimation in *G. veletis* (Des Marteaux et al. 2017; Toxopeus et al. 2019), or (5) adult femur-derived transcriptomes from *G. assimilis* (Palacios-Gimenez et al. 2018). Genomic resources are also available for an inbred line of *G. assimilis* (Palacios-Gimenez et al. 2018). Outside of the genus *Gryllus*, large-scale genomic resources are also available for the Hawaiian cricket *Laupala kohalensis* in the form of an EST resource from a nerve cord cDNA library (Danley et al. 2007) and a de novo draft genome (Blankers et al. 2018). In fact, the *L. kohalensis* genome is, to our knowledge, the only published cricket genome to date. Transcriptomic data are also available for *Allonemobius fasciatus* embryos (Reynolds and Hand 2009), male accessory glands of *Gryllodes sigillatus* (Pauchet et al. 2015), and the testis, accessory glands, and adult body of *Teleogryllus oceanicus* (Bailey et al. 2013).

8.7.2 Resources in Grasshoppers and Locusts

A de novo transcriptome spanning several stages is available for the grasshoppers *Chorthippus biguttulus* and *Oxya chinensis sinuosa* (Kim et al. 2016) and for nymphs, adult females and males of *Xenocatantops brachycerus* (Zhao et al. 2018). An organ-specific transcriptome is available for the gut of *Oedaleus asiaticus* (Huang et al. 2017), an EST database exists for transcripts from the central nervous system of *Schistocerca gregaria* (Badisco et al. 2011), and a de novo transcriptome for *Tetrix japonica* (Qiu et al. 2017) is also available. In addition, other tools including RNAi have been tested and reported to be successful in *S. americana*

(Dong and Friedrich 2005). *Locusta migratoria* is a well-studied locust species that has large-scale genomic resources available in the form of a de novo genome and transcriptome (Wang et al. 2014) and an EST database from whole body and dissected organs (Kang et al. 2004; Ma et al. 2006). In addition, RNAi has been established and reported to be successful in adults, nymphs, and embryos for this species (He et al. 2006).

8.8 Commercial Importance of Crickets as Edible Insects and "Food of the Future"

In this chapter we have primarily focused on crickets as emerging evo-devo models. In this final section, we would like to briefly highlight other reasons that crickets are gaining popularity as study systems. Given their cosmopolitan distribution (all areas of the world, except the arctic and subarctic regions) and over 2400 documented species, crickets represent the most diverse lineage of "jumping or leaping" insects (Horch et al. 2017b). While the chirping sounds made by males have historically given them acoustic appeal as pets and in research, many species are now becoming economically important as an alternative food source for humans (Huis et al. 2013; Horch et al. 2017b), as feedstock for poultry (Ravindran and Blair 1993), or as fish bait (Huis et al. 2013), all of which are multibillion dollar industries. With the human population predicted to reach nine billion by the year 2050 (Huis et al. 2013), meat production and consumption is soon expected to reach unsustainable levels (Boland et al. 2013). Insects, especially crickets, have therefore been proposed and marketed as a novel, alternative, environmentally efficient food source with high nutritional value (Oonincx and de Boer 2012; Huis et al. 2013; Deroy et al. 2015).

Crickets are reportedly common street snacks in some parts of the world and have been part of the traditional diet in Thailand, the Lao People's Democratic Republic, Vietnam, the Democratic Republic of Congo, and Nigeria for hundreds of years (Kuhnlein et al. 2009; Huis et al. 2013). As an example, over 20,000 farmers are reported to rear crickets in Thailand, resulting in an estimated production of over 7500 tons per year in this country alone (Hanboonsong et al. 2013). To increase their appeal in the West, crickets are now being advertised and marketed to be eaten whole, in granular or in paste form, and as ingredients in commercially available flours and protein bars (e.g., Aspire Food Group USA, Inc.). While many species of crickets are edible (e.g., *G. bimaculatus*, *Gryllodes sigillatus*, *A. domesticus*, *A. testacea*, *T. occipitalis*, *T. mitratus*, and *Brachytrupes portentosus*), to our knowledge currently only *G. bimaculatus*, *G. sigillatus*, and *A. domesticus* are farmed economically for human consumption (Huis et al. 2013).

The use of crickets has the potential to change the future of the food industry, because of how effective they are in maximizing nutrition for minimal resources. Cricket rearing is comparatively inexpensive, requires a fraction of input resources, and has fewer negative environmental impacts than rearing traditional vertebrate protein sources (Halloran et al. 2017). Crickets produce only 1% of greenhouse gases compared to cattle and pigs, in addition to showing an approximate tenfold reduction in ammonia emission (Oonincx et al. 2010), all relevant factors when considering sustainable production in the age of climate change. They act as a complete protein source and consist of over 50% protein by volume (Wang et al. 2004). Other advantages of eating crickets include their high edible weight: Nakagaki and DeFoliart (1991) have estimated that over 80% of a cricket is edible and digestible compared to 55% for chicken and pigs and 40% for cattle. This translates into making crickets twice as efficient as chicken, at least 4 times as efficient as pigs, and 12 times more efficient as cattle in converting feed into meat (Huis et al. 2013). As a specific example, the food conversion efficiency of the house cricket *Acheta domesticus* has been reported to be five times higher than beef, and when their fecundity is considered, this has been shown to increase as much as 15- to 20-fold (Horch et al. 2017b; Nakagaki and Defoliart 1991). Farming crickets is projected to become a multimillion dollar industry, with the US market for edible insects alone expected to exceed $50 million by as early as 2023 (Ahuja and Deb 2018).

The development of a novel food source like crickets must include assessing the potential risks involved with consumption of such sources. Therefore, there is increased interest in understanding the biology of these insects. While studies addressing entomophagy-induced food allergies (especially ones arising from eating crickets alone) are few, there is some preliminary evidence to confirm that crustacean-allergic individuals (or people with seafood allergies) may also show cross-reactivity to edible insects in general (Srinroch et al. 2015; Pener 2016; Ribeiro et al. 2018). Another study has reported that some individuals can develop asthmatic symptoms upon ingesting insects belonging to Orthoptera (Auerswald and Lopata 2005). Overall, however, eating and/or exposure to insects is not expected to pose significant risks of allergenic reactions for most people, especially if the individual has no prior history of arthropod or insect allergen sensitivity (Huis et al. 2013). In summary, the disadvantages associated with eating insects like crickets currently seem few and the advantages many. Consuming reared insects is potentially more environmentally friendly, nutritious, cheap, and affordable for people in all parts of the world. Cricket rearing is one way to use land efficiently, reduce or lower pesticide use and greenhouse gas emissions, may boost human and/or animal immunity (Goodman 1989; Muzzarelli 2010; Taufek et al. 2016), and finally improve the livelihood of women and children in rural areas by supporting local economies (Huis et al. 2013).

8.9 Conclusion

The successful establishment of the many functional and genetic manipulation tools in crickets has contributed to a new era of non-drosophilid insect research, not limited to evo-devo research. We hope that scientists from various disciplines feel encouraged to use the cricket as a system to address intriguing questions in their respective fields.

Acknowledgments We thank Extavour lab members Aracely Newton and Maitreyi Upadhyay for helpful comments and Leo Blondel for technical support on the manuscript. This work was supported by Harvard University.

References

Ahuja K, Deb S (2018) Edible insects market size by product (beetles, caterpillars, grasshoppers, bees, wasps, ants, scale insects & true bugs), by application (flour, protein bars, snacks), industry analysis report, Regional Outlook (U.S., Belgium, Netherlands, UK, France, China, Thailand, Vietnam, Brazil, Mexico), application potential, price trends, competitive market share & forecast, 2018–2024

Alexander DE (2018) A century and a half of research on the evolution of insect flight. Arthropod Struct Dev 47:322–327

Andres JA, Larson EL, Bogdanowicz SM, Harrison RG (2013) Patterns of transcriptome divergence in the male accessory gland of two closely related species of field crickets. Genetics 193:501–513

Aonuma H (2017) Chapter 20: Synthetic approaches for observing and measuring cricket behaviors. In: Horch HW, Mito T, Popadic A, Ohuchi H, Noji S (eds) The cricket as a model organism: development, regeneration and behaviour. Springer, New York

Auerswald L, Lopata AL (2005) Insects – diversity and allergy. Curr Allergy Clin Immunol 18:58–60

Awata H, Watanabe T, Hamanaka Y, Mito T, Noji S, Mizunami M (2015) Knockout crickets for the study of learning and memory: dopamine receptor Dop1 mediates aversive but not appetitive reinforcement in crickets. Sci Rep 5:15885

Badisco L, Huybrechts J, Simonet G, Verlinden H, Marchal E, Huybrechts R, Schoofs L, De Loof A, Vanden Broeck J (2011) Transcriptome analysis of the desert locust central nervous system: production and annotation of a *Schistocerca gregaria* EST database. PLoS One 6: e17274

Bailey NW, Veltsos P, Tan YF, Millar AH, Ritchie MG, Simmons LW (2013) Tissue-specific transcriptomics in the field cricket *Teleogryllus oceanicus*. G3 (Bethesda) 3:225–230

Bando T, Ishimaru Y, Kida T, Hamada Y, Matsuoka Y, Nakamura T, Ohuchi H, Noji S, Mito T (2013) Analysis of RNA-Seq data reveals involvement of JAK/STAT signalling during leg regeneration in the cricket *Gryllus bimaculatus*. Development 140:959–964

Barry S, Nakamura T, Matsuoka Y, Straub C, Horch HW, Extavour CG (2019) Injecting *Gryllus bimaculatus* eggs. J Vis Exp (in press)

Berdan EL, Blankers T, Waurick I, Mazzoni CJ, Mayer F (2016) A genes eye view of ontogeny: de novo assembly and profiling of the *Gryllus rubens* transcriptome. Mol Ecol Resour 16:1478–1490

Bigelow RS (1962) Factors affecting developmental rates and diapause in field crickets. Evolution 16:396–406

Bilinski SM, Jaglarz MK, Tworzydło W (2017) The pole (germ) plasm in insect oocytes. In: Kloc M (ed) Results and problems in cell differentiation, oocytes, vol 63. Springer, New York, pp 103–126

Blankers T, Oh KP, Bombarely A, Shaw KL (2018) The genomic architecture of a rapid island radiation: recombination rate variation, chromosome structure, and genome assembly of the Hawaiian Cricket *Laupala*. Genetics 209:1329–1344

Boland MJ, Rae AN, Vereijken JM, Meuwissen MPM, Fischer ARH, Boekel MAJSv, Rutherfurd SM, Gruppen H, Moughan PJ, Hendriks WH (2013) The future supply of animal-derived protein for human consumption. Trends Food Sci Technol 29:62–73

Braswell WE, Andres JA, Maroja LS, Harrison RG, Howard DJ, Swanson WJ (2006) Identification and comparative analysis of accessory gland proteins in Orthoptera. Genome 49:1069–1080

Bretman A, Tregenza T (2005) Measuring polyandry in wild populations: a case study using promiscuous crickets. Mol Ecol 14:2169–2179

Bridges CB, Morgan TH (1923) The third-chromosome group of mutant characters of *Drosophila melanogaster*. Carnegie Inst Wash Publ 327:1–251

Bruce HS, Patel NH (2018) Insect wings and body wall evolved from ancient leg segments. bioRxiv, 244541

Büning J (1994) The insect ovary: ultrastructure, previtellogenic growth and evolution. Chapman & Hall, London

Campos-Ortega JA, Hartenstein V (1985) The embryonic development of *Drosophila melanogaster*. Springer, Heidelberg

Cinalli RM, Rangan P, Lehmann R (2008) Germ cells are forever. Cell 132:559–562

Cong L, Ran FA, Cox D, Lin S, Barretto R, Habib N, Hsu PD, Wu X, Jiang W, Marraffini LA, Zhang F (2013) Multiplex genome engineering using CRISPR/Cas systems. Science 339:819–823

Danley PD, Mullen SP, Liu F, Nene V, Quackenbush J, Shaw KL (2007) A cricket gene index: a genomic resource for studying neurobiology, speciation, and molecular evolution. BMC Genomics 8:109

Davis GK, Patel NH (2002) Short, long, and beyond: molecular and embryological approaches to insect segmentation. Annu Rev Entomol 47:669–699

Demerec M (1950) Biology of *Drosophila*, Facsimile edn. Cold Spring Harbor Laboratory, Cold Spring Harbor, NY

Denell R (2008) Establishment of *Tribolium* as a genetic model system and its early contributions to evo-devo. Genetics 180:1779–1786

Deroy O, Reade B, Spence C (2015) The insectivore's dilemma, and how to take the West out of it. Food Qual Prefer 44:44–55

Des Marteaux LE, McKinnon AH, Udaka H, Toxopeus J, Sinclair BJ (2017) Effects of cold-acclimation on gene expression in fall field cricket (*Gryllus pennsylvanicus*) ionoregulatory tissues. BMC Genomics 18:357

Dietrich S, Schubert FR, Lumsden A (1997) Control of dorsoventral pattern in the chick paraxial mesoderm. Development 124:3895–3908

Dong Y, Friedrich M (2005) Nymphal RNAi: systemic RNAi mediated gene knockdown in juvenile grasshopper. BMC Biotechnol 5:25

Donoughe S, Extavour CG (2016) Embryonic development of the cricket *Gryllus bimaculatus*. Dev Biol 411:140–156

Donoughe S, Kim C, Extavour CG (2018) High-throughput live-imaging of embryos in microwell arrays using a modular specimen mounting system. Biol Open 7:bio031260

Duboule D, Dolle P (1989) The structural and functional organization of the murine HOX gene family resembles that of *Drosophila* homeotic genes. EMBO J 8:1497–1505

Engel JE, Hoy RR (1999) Experience-dependent modification of ultrasound auditory processing in a cricket escape response. J Exp Biol 202:2797–2806

Ephrussi A, Lehmann R (1992) Induction of germ cell formation by *oskar*. Nature 358:387–392

Ephrussi A, Dickinson LK, Lehmann R (1991) Oskar organizes the germ plasm and directs localization of the posterior determinant *nanos*. Cell 66:37–50

Ewen-Campen B, Srouji JR, Schwager EE, Extavour CG (2012) Oskar predates the evolution of germ plasm in insects. Curr Biol 22:2278–2283

Ewen-Campen B, Donoughe S, Clarke DN, Extavour CG (2013) Germ cell specification requires zygotic mechanisms rather than germ plasm in a basally branching insect. Curr Biol 23:835–842

Extavour CG, Akam ME (2003) Mechanisms of germ cell specification across the metazoans: epigenesis and preformation. Development 130:5869–5884

Ferreira M, Ferguson JW (2010) Do Mediterranean crickets *Gryllus bimaculatus* De Geer (Orthoptera: Gryllidae) come from the Mediterranean? Large-scale phylogeography and regional gene flow. Bull Entomol Res 100:49–58

Fisher HP, Pascual MG, Jimenez SI, Michaelson DA, Joncas CT, Quenzer ED, Christie AE, Horch HW (2018) De novo assembly of a transcriptome for the cricket *Gryllus bimaculatus* prothoracic ganglion: an invertebrate model for investigating adult central nervous system compensatory plasticity. PLoS One 13:e0199070

Funato T, Kurabayashi D, Nara M, Aonuma H (2008) Switching mechanism of sensor-motor coordination through an oscillator network model. IEEE Trans Syst Man Cybern B Cybern 38:764–770

Funato T, Nara M, Kurabayashi D, Ashikaga M, Aonuma H (2011) A model for group-size-dependent behaviour decisions in insects using an oscillator network. J Exp Biol 214:2426–2434

Geer CD (1773) Mémoires pour servir a l'histoire des insectes. Pierre Hesselberg, Stockholm

Giribet G, Edgecombe GD (2013) The Arthropoda: a phylogenetic framework. In: Minelli A, Boxshall G, Fusco G (eds) Arthropod biology and evolution – molecules, development, morphology. Springer, New York, pp 17–40

Goldsmith MR, Shimada T, Abe H (2005) The genetics and genomics of the silkworm, *Bombyx mori*. Annu Rev Entomol 50:71–100

Goltsev Y, Hsiong W, Lanzaro G, Levine M (2004) Different combinations of gap repressors for common stripes in *Anopheles* and *Drosophila* embryos. Dev Biol 275:435–446

Goodman WG (1989) Chitin: a magic bullet? The Food Insects Newsletter 2

Gould JL, Grould CG (1995) The honey bee. Scientific American Library, New York

Graham A, Papalopulu N, Krumlauf R (1989) The murine and *Drosophila* homeobox gene complexes have common features of organization and expression. Cell 57:367–378

Grimaldi D, Engel MS (2005) Evolution of the insects. Cambridge University Press, Cambridge

Halloran A, Hanboonsong Y, Roos N, Bruun S (2017) Life cycle assessment of cricket farming in north-eastern Thailand. J Clean Prod 156:83–94

Hamada Y, Bando T, Nakamura T, Ishimaru Y, Mito T, Noji S, Tomioka K, Ohuchi H (2015) Leg regeneration is epigenetically regulated by histone H3K27 methylation in the cricket *Gryllus bimaculatus*. Development 142:2916–2927

Hanboonsong Y, Jamjanya T, Durst PB (2013) Six-legged livestock: edible insect farming, collection and marketing in Thailand. Food and Agriculture Organization of the United Nations Regional Office for Asia and the Pacific, Bangkok

Handler AM, McCombs SD, Fraser MJ, Saul SH (1998) The lepidopteran transposon vector, piggyBac, mediates germ-line transformation in the Mediterranean fruit fly. Proc Natl Acad Sci USA 95:7520–7525

Harvey TH, Velez MI, Butterfield NJ (2012) Exceptionally preserved crustaceans from western Canada reveal a cryptic Cambrian radiation. Proc Natl Acad Sci USA 109:1589–1594

He ZB, Cao YQ, Yin YP, Wang ZK, Chen B, Peng GX, Xia YX (2006) Role of *hunchback* in segment patterning of *Locusta migratoria manilensis* revealed by parental RNAi. Dev Growth Differ 48:439–445

Hedwig B (2017) Chapter 19: Trackball systems for analysing cricket phonotaxis. In: Horch HW, Mito T, Popadic A, Ohuchi H, Noji S (eds) The cricket as a model organism: development, regeneration and behaviour. Springer, New York

Hedwig B, Poulet JF (2004) Complex auditory behaviour emerges from simple reactive steering. Nature 430:781–785

Heffer A, Xiang J, Pick L (2013) Variation and constraint in Hox gene evolution. Proc Natl Acad Sci USA 110:2211–2216

Horch HW, Liu J, Mito T, Popadic A, Watanabe T (2017a) Chapter 21: Protocols in the cricket. In: Horch HW, Mito T, Popadic A, Ohuchi H, Noji S (eds) The cricket as a model organism: development, regeneration and behaviour. Springer, New York, pp 327–370

Horch HW, Mito T, Popadic A, Ohuchi H, Noji S (2017b) The cricket as a model organism: development, regeneration and behaviour. Springer, New York, p 376

Huang X, Ma J, Qin X, Tu X, Cao G, Wang G, Nong X, Zhang Z (2017) Biology, physiology and gene expression of grasshopper *Oedaleus asiaticus* exposed to diet stress from plant secondary compounds. Sci Rep 7:8655

Huber F, Moore TE, Loher W (1989) Cricket behavior and neurobiology. Cornell University Press, Ithaca, NY

Huis Av, Itterbeeck JV, Klunder H, Mertens E, Halloran A, Muir G, Vantomme P (2013) Edible insects: future prospects for food and feed security. Food and Agriculture Organization of the United Nations, Rome, p 201

Jinek M, Chylinski K, Fonfara I, Hauer M, Doudna JA, Charpentier E (2012) A programmable dual-RNA-guided DNA endonuclease in adaptive bacterial immunity. Science 337:816–821

Juhn J, James AA (2006) *oskar* gene expression in the vector mosquitoes, *Anopheles gambiae* and *Aedes aegypti*. Insect Mol Biol 15:363–372

Juhn J, Marinotti O, Calvo E, James AA (2008) Gene structure and expression of *nanos* (*nos*) and *oskar* (*osk*) orthologues of the vector mosquito, *Culex quinquefasciatus*. Insect Mol Biol 17:545–552

Kainz F (2009) Cell communication during patterning: Notch and FGF signalling in *Gryllus bimaculatus* and their role in segmentation. Department of Zoology, University of Cambridge, Cambridge, p 161

Kainz F, Ewen-Campen B, Akam M, Extavour CG (2011) Delta/Notch signalling is not required for segment generation in the basally branching insect *Gryllus bimaculatus*. Development 138:5015–5026

Kang L, Chen X, Zhou Y, Liu B, Zheng W, Li R, Wang J, Yu J (2004) The analysis of large-scale gene expression correlated to the phase changes of the migratory locust. Proc Natl Acad Sci USA 101:17611–17615

Kawabata K, Fujii T, Aonuma H, Suzuki T, Ashikaga M, Ota J, Asama H (2012) A neuromodulation model of behavior selection in the fighting behavior of male crickets. Robot Auton Syst 60:707–713

Kijimoto T, Pespeni M, Beckers O, Moczek AP (2013) Beetle horns and horned beetles: emerging models in developmental evolution and ecology. Wiley Interdiscip Rev Dev Biol 2:405–418

Kim IW, Markkandan K, Lee JH, Subramaniyam S, Yoo S, Park J, Hwang JS (2016) Transcriptome profiling and in silico analysis of the antimicrobial peptides of the grasshopper oxya chinensis sinuosa. J Microbiol Biotechnol 26:1863–1870

Kim-Ha J, Smith JL, Macdonald PM (1991) *oskar* mRNA is localized to the posterior pole of the *Drosophila* oocyte. Cell 66:23–35

Kleespies RG, Tidona CA, Darai G (1999) Characterization of a new iridovirus isolated from crickets and investigations on the host range. J Invertebr Pathol 73:84–90

Kochi Y, Miyashita A, Tsuchiya K, Mitsuyama M, Sekimizu K, Kaito C (2016) A human pathogenic bacterial infection model using the two-spotted cricket, *Gryllus bimaculatus*. FEMS Microbiol Lett 363:fnw163

Kochi Y, Matsumoto Y, Sekimizu K, Kaito C (2017) Two-spotted cricket as an animal infection model of human pathogenic fungi. Drug Discov Ther 11:259–266

Krause G (1939) Die Eitypen der Insekten. Biol Zentralbl 50:495–536

Kuhnlein HV, Erasmus B, Spigelski D (2009) Indigenous Peoples' food systems: the many dimensions of culture, diversity and environment for nutrition and health. Food and Agriculture Organization of the United Nations Center for Indigenous Peoples' Nutrition and Environment, Rome

Lehmann R, Nüsslein-Volhard C (1986) Abdominal segmentation, pole cell formation, and embryonic polarity require the localized activity of *oskar*, a maternal gene in *Drosophila*. Cell 47:144–152

Lewis EB (1978) A gene complex controlling segmentation in *Drosophila*. Nature 276:565–570

Linz DM, Tomoyasu Y (2018) Dual evolutionary origin of insect wings supported by an investigation of the abdominal wing serial homologs in *Tribolium*. Proc Natl Acad Sci U S A 115: E658–E667

Liu PZ, Kaufman TC (2005) Short and long germ segmentation: unanswered questions in the evolution of a developmental mode. Evol Dev 7:629–646

Lochmatter T, Roduit P, Cianci CM, Correll N (2008) SwisTrack – a flexible open source tracking software for multi-agent systems. In: IEEE/RSJ international conference on intelligent robots and systems, Acropolis Convention Center, Nice

Lohs-Schardin M, Cremer C, Nusslein-Volhard C (1979) A fate map for the larval epidermis of *Drosophila melanogaster*: localized cuticle defects following irradiation of the blastoderm with an ultraviolet laser microbeam. Dev Biol 73:239–255

Lynch JA, Özüak O, Khila A, Abouheif E, Desplan C, Roth S (2011) The phylogenetic origin of *oskar* coincided with the origin of maternally provisioned germ plasm and pole cells at the base of the Holometabola. PLoS Genet 7:e1002029

Ma Z, Yu J, Kang L (2006) LocustDB: a relational database for the transcriptome and biology of the migratory locust (*Locusta migratoria*). BMC Genomics 7:11

Matsumoto CS, Shidara H, Matsuda K, Nakamura T, Mito T, Matsumoto Y, Oka K, Ogawa H (2013) Targeted gene delivery in the cricket brain, using in vivo electroporation. J Insect Physiol 59:1235–1241

Mayhew PJ (2007) Why are there so many insect species? Perspectives from fossils and phylogenies. Biol Rev Camb Philos Soc 82:425–454

Meng X, Zhu F, Chen K (2017) Silkworm: a promising model organism in life science. J Insect Sci 17:97, 1–6

Miller JC, Tan S, Qiao G, Barlow KA, Wang J, Xia DF, Meng X, Paschon DE, Leung E, Hinkley SJ, Dulay GP, Hua KL, Ankoudinova I, Cost GJ, Urnov FD, Zhang HS, Holmes MC, Zhang L, Gregory PD, Rebar EJ (2011) A TALE nuclease architecture for efficient genome editing. Nat Biotechnol 29:143–148

Mitchell AA (2013) EDNA. The Fossil Insect Database. https://fossilinsectdatabase.co.uk/

Mito T, Noji S (2008) The two-spotted cricket *Gryllus bimaculatus*: an emerging model for developmental and regeneration studies. Cold Spring Harbor Protoc. https://doi.org/10.1101/pdb.emo110

Mito T, Sarashina I, Zhang H, Iwahashi A, Okamoto H, Miyawaki K, Shinmyo Y, Ohuchi H, Noji S (2005) Non-canonical functions of *hunchback* in segment patterning of the intermediate germ cricket *Gryllus bimaculatus*. Development 132:2069–2079

Mito T, Nakamura T, Bando T, Ohuchi H, Noji S (2011) The advent of RNA interference in entomology. Entomol Sci 14:1–8

Miyawaki K, Mito T, Sarashina I, Zhang H, Shinmyo Y, Ohuchi H, Noji S (2004) Involvement of Wingless/Armadillo signaling in the posterior sequential segmentation in the cricket, *Gryllus bimaculatus* (Orthoptera), as revealed by RNAi analysis. Mech Dev 121:119–130

Mizunami M, Matsumoto Y (2017a) Chapter 9: Leaning and memory. In: Horch HW, Mito T, Popadic A, Ohuchi H, Noji S (eds) The cricket as a model organism: development, regeneration and behaviour. Springer, New York, pp 129–140

Mizunami M, Matsumoto Y (2017b) Roles of octopamine and dopamine neurons for mediating appetitive and aversive signals in Pavlovian conditioning in crickets. Front Physiol 8:1027

Mizuno T, Sakura M, Ashikaga M, Aonuma H, Chiba R, Ota J (2012) Model of a sensory–behavioral relation mechanism for aggressive behavior in crickets. Robot Auton Syst 60:700–706

Moscou MJ, Bogdanove AJ (2009) A simple cipher governs DNA recognition by TAL effectors. Science 326:1501

Muzzarelli RAA (2010) Chitins and chitosans as immunoadjuvants and non-allergenic drug carriers. Mar Drugs 8:292–312

Nakagaki BJ, Defoliart GR (1991) Comparison of diets for mass-rearing *Acheta domesticus* (Orthoptera: Gryllidae) as a novelty food, and comparison of food conversion efficiency with values reported for livestock. J Econ Entomol 84:891–896

Nakamura T, Mito T, Bando T, Ohuchi H, Noji S (2008a) Dissecting insect leg regeneration through RNA interference. Cell Mol Life Sci 65:64–72

Nakamura T, Mito T, Miyawaki K, Ohuchi H, Noji S (2008b) EGFR signaling is required for re-establishing the proximodistal axis during distal leg regeneration in the cricket *Gryllus bimaculatus* nymph. Dev Biol 319:46–55

Nakamura T, Yoshizaki M, Ogawa S, Okamoto H, Shinmyo Y, Bando T, Ohuchi H, Noji S, Mito T (2010) Imaging of transgenic cricket embryos reveals cell movements consistent with a syncytial patterning mechanism. Curr Biol 20:1641–1647

Neubauer FB, MacLean JN (2010) Calcium imaging in neuroscience. In: Encycopedia of Life Sciences (ELS). John Wiley & Sons, Ltd, Chichester

Nicholson DB, Ross AJ, Mayhew PJ (2014) Fossil evidence for key innovations in the evolution of insect diversity. Proc Biol Sci 281:20141823

Nickle DA, Walker TJ (1974) A morphological key to field crickets of southeastern United States (Orthoptera: Gryllidae: Gryllus). Fla Entomol 57:8–12

Niwa N, Saitoh M, Ohuchi H, Yoshioka H, Noji S (1997) Correlation between distal-less expression patterns and structures of appendages in development of the two-spotted cricket, *Gryllus bimaculatus*. Zool Sci 14:115–125

Niwa N, Inoue Y, Nozawa A, Saito M, Misumi Y, Ohuchi H, Yoshioka H, Noji S (2000) Correlation of diversity of leg morphology in *Gryllus bimaculatus* (cricket) with divergence in *dpp* expression pattern during leg development. Development 127:4373–4381

Noldus LP, Spink AJ, Tegelenbosch RA (2001) EthoVision: a versatile video tracking system for automation of behavioral experiments. Behav Res Methods Instrum Comput 33:398–414

Nüsslein-Volhard C, Wieschaus EF (1980) Mutations affecting segment number and polarity in *Drosophila*. Nature 287:795–801

Ogawa H, Miller JP (2017) Chapter 18: Optical recording methods: how to measure neural activities with calcium imaging. In: Horch HW, Mito T, Popadic A, Ohuchi H, Noji S (eds) The cricket as a model organism: development, regeneration and behaviour. Springer, New York

Oldroyd BP, Thompson GJ (2006) Behavioural genetics of the honey bee *Apis mellifera*. Adv Insect Physiol 33:1–49

Oonincx DG, de Boer IJ (2012) Environmental impact of the production of mealworms as a protein source for humans – a life cycle assessment. PLoS One 7:e51145

Oonincx DG, van Itterbeeck J, Heetkamp MJ, van den Brand H, van Loon JJ, van Huis A (2010) An exploration on greenhouse gas and ammonia production by insect species suitable for animal or human consumption. PLoS One 5:e14445

Otte D, Cade W (1984) African crickets (Gryllidae). 5. East and South African species of *Modicogryllus* and several related genera (Gryllinae, Modicogryllini). Proc Acad Natl Sci Phila 136:67–97

Palacios-Gimenez OM, Bardella VB, Lemos B, Cabral-de-Mello DC (2018) Satellite DNAs are conserved and differentially transcribed among *Gryllus* cricket species. DNA Res 25:137–147

Panganiban G, Irvine SM, Lowe C, Roehl H, Corley LS, Sherbon B, Grenier JK, Fallon JF, Kimble J, Walker M, Wray GA, Swalla BJ, Martindale MQ, Carroll SB (1997) The origin and evolution of animal appendages. PNAS 94:5162–5166

Patel NH (1994) Imaging neuronal subsets and other cell types in whole-mount *Drosophila* embryos and larvae using antibody probes. Methods Cell Biol 44:445–487

Pauchet Y, Wielsch N, Wilkinson PA, Sakaluk SK, Svatos A, ffrench-Constant RH, Hunt J, Heckel DG (2015) What's in the gift? Towards a molecular dissection of nuptial feeding in a cricket. PLoS One 10:e0140191

Pavlopoulos A, Oehler S, Kapetanaki MG, Savakis C (2007) The DNA transposon Minos as a tool for transgenesis and functional genomic analysis in vertebrates and invertebrates. Genome Biol 8(Suppl 1):S2
Paydar S, Doan CA, Jacobs GA (1999) Neural mapping of direction and frequency in the cricket cercal sensory system. J Neurosci 19:1771–1781
Pener MP (2016) Allergy to crickets: a review. J Orthop Res 25:91–95
Pollack GS, Hoy RR (1979) Temporal pattern as a cue for species-specific calling song recognition in crickets. Science 204:429–432
Popov AV, Shuvalov VF (1977) Phonotactic behavior of crickets. J Comp Physiol 119:111–126
Porteus MH, Carroll D (2005) Gene targeting using zinc finger nucleases. Nat Biotechnol 23:967–973
Qiu Z, Liu F, Lu H, Huang Y (2017) Characterization and analysis of a de novo transcriptome from the pygmy grasshopper *Tetrix japonica*. Mol Ecol Resour 17:381–392
Quan H, Lynch JA (2016) The evolution of insect germline specification strategies. Curr Opin Insect Sci 13:99–105
Ravindran V, Blair R (1993) Feed resources for poultry production in Asia and the Pacific. III. Animal protein sources. Worlds Poult Sci J 49:219–235
Remy S, Tesson L, Menoret S, Usal C, Scharenberg AM, Anegon I (2010) Zinc-finger nucleases: a powerful tool for genetic engineering of animals. Transgenic Res 19:363–371
Reynolds JA, Hand SC (2009) Embryonic diapause highlighted by differential expression of mRNAs for ecdysteroidogenesis, transcription and lipid sparing in the cricket *Allonemobius socius*. J Exp Biol 212:2075–2084
Rheinlaender J, Blätgen G (1982) The precision of auditory lateralization in the cricket, *Gryllus bimaculatus*. Physiol Entomol 7:209–218
Ribeiro JC, Cunha LM, Sousa-Pinto B, Fonseca J (2018) Allergic risks of consuming edible insects: a systematic review. Mol Nutr Food Res 62:1700030
Ritzmann RE, Quinn RD, Watson JT, Zill SN (2000) Insect walking and biorobotics: a relationship with mutual benefits. Bioscience 50:23–33
Ronco M, Uda T, Mito T, Minelli A, Noji S, Klingler M (2008) Antenna and all gnathal appendages are similarly transformed by *homothorax* knock-down in the cricket *Gryllus bimaculatus*. Dev Biol 313:80–92
Rubin GM, Spradling AC (1982) Genetic transformation of *Drosophila* with transposable element vectors. Science 218:348–353
Sander K (1997) Pattern formation in insect embryogenesis: the evolution of concepts and mechanisms. Int J Insect Morphol Embryol 25:349–367
Sen GL, Blau HM (2006) A brief history of RNAi: the silence of the genes. FASEB J 20:1293–1299
Shinmyo Y, Mito T, Matsushita T, Sarashina I, Miyawaki K, Ohuchi H, Noji S (2004) piggyBac-mediated somatic transformation of the two-spotted cricket, *Gryllus bimaculatus*. Dev Growth Differ 46:343–349
Simmons LW (1986) Female choice in the field cricket *Gryllus bimaculatus* (De Geer). Anim Behav 34:1463–1470
Simmons LW (1987) Female choice contributes to offspring fitness in the field cricket, *Gryllus bimaculatus* (De Geer). Behav Ecol Sociobiol 21:313–321
Smith JL, Wilson JE, Macdonald PM (1992) Overexpression of *oskar* directs ectopic activation of *nanos* and presumptive pole cell formation in *Drosophila* embryos. Cell 70:849–859
Sokoloff A (1966) The genetics of *Tribolium* and related species. Academic Press, New York
Sokoloff A (1972) The biology of *Tribolium* with special emphasis on genetic aspects, vol 1. Clarendon, Oxford
Sokoloff A (1974) The biology of *Tribolium* with special emphasis on genetic aspects, vol 2. Clarendon, Oxford
Sokoloff A (1977) The biology of *Tribolium* with special emphasis on genetic aspects, vol 3. Clarendon, Oxford

Srinroch C, Srisomsap C, Chokchaichamnankit D, Punyarit P, Phiriyangkul P (2015) Identification of novel allergen in edible insect, *Gryllus bimaculatus* and its cross-reactivity with Macrobrachium spp. allergens. Food Chem 184:160–166
Stout JF, DeHaan CG, McGhee RW (1983) Attractiveness of the male *Acheta domesticus* calling song to females. J Comp Physiol 153:509–521
Szelei J, Woodring J, Goettel MS, Duke G, Jousset FX, Liu KY, Zadori Z, Li Y, Styer E, Boucias DG, Kleespies RG, Bergoin M, Tijssen P (2011) Susceptibility of North-American and European crickets to *Acheta domesticus* densovirus (AdDNV) and associated epizootics. J Invertebr Pathol 106:394–399
Taufek NM, Aspani F, Muin H, Raji AA, Razak SA, Alias Z (2016) The effect of dietary cricket meal (*Gryllus bimaculatus*) on growth performance, antioxidant enzyme activities, and haematological response of African catfish (*Clarias gariepinus*). Fish Physiol Biochem 42:1143–1155
Toth AL, Rehan SM (2017) Molecular evolution of insect sociality: an eco-evo-devo perspective. Annu Rev Entomol 62:419–442
Toxopeus J, Des Marteaux LE, Sinclair BJ (2019) How crickets become freeze tolerant: the transcriptomic underpinnings of acclimation in *Gryllus veletis*. Comp Biochem Physiol Part D Genomics Proteomics 29:55–66
Tregenza T, Wedell N (1998) Benefits of multiple mates in the cricket *Gryllus bimaculatus*. Evolution 52:1726–1730
Tregenza T, Wedell N (2002) Polyandrous females avoid costs of inbreeding. Nature 415:71–73
Truman JW, Riddiford LM (1999) The origins of insect metamorphosis. Nature 401:447–452
Tschuch G (1976) Der Einfluss synthetischer Gesänge auf die Weibchen von *Gryllus bimaculatus* De Geer. Zool Jahrb Abt allgemeine Zool Physiol Tiere 80:383–388
Vaccari NE, Edgecombe GD, Escudero C (2004) Cambrian origins and affinities of an enigmatic fossil group of arthropods. Nature 430:554–557
Vellichirammal NN, Zera AJ, Schilder RJ, Wehrkamp C, Riethoven J-JM, Brisson JA (2014) De novo transcriptome assembly from fat body and flight muscles transcripts to identify morph-specific gene expression profiles in *Gryllus firmus*. PLoS One 9:e82129
Wang Y, Jehle JA (2009) Nudiviruses and other large, double-stranded circular DNA viruses of invertebrates: new insights on an old topic. J Invertebr Pathol 101:187–193
Wang D, Bai Y-Y, Li J-H, Zhang C-X (2004) Nutritional value of the field cricket (*Gryllus testaceus* Walker). Entomol Sin 11:275–283
Wang X, Fang X, Yang P, Jiang X, Jiang F, Zhao D, Li B, Cui F, Wei J, Ma C, Wang Y, He J, Luo Y, Wang Z, Guo X, Guo W, Wang X, Zhang Y, Yang M, Hao S, Chen B, Ma Z, Yu D, Xiong Z, Zhu Y, Fan D, Han L, Wang B, Chen Y, Wang J, Yang L, Zhao W, Feng Y, Chen G, Lian J, Li Q, Huang Z, Yao X, Lv N, Zhang G, Li Y, Wang J, Wang J, Zhu B, Kang L (2014) The locust genome provides insight into swarm formation and long-distance flight. Nat Commun 5:2957
Watanabe T, Ochiai H, Sakuma T, Horch HW, Hamaguchi N, Nakamura T, Bando T, Ohuchi H, Yamamoto T, Noji S, Mito T (2012) Non-transgenic genome modifications in a hemimetabolous insect using zinc-finger and TAL effector nucleases. Nat Commun 3:1017
Wenzel B, Hedwig B (1999) Neurochemical control of cricket stridulation revealed by pharmacological microinjections into the brain. J Exp Biol 202:2203–2216
Werren JH, Loehlin DW (2009) The parasitoid wasp *Nasonia*: an emerging model system with haploid male genetics. Cold Spring Harb Protoc. https://doi.org/10.1101/pdb.emo134
Wheeler QD (1990) Insect diversity and cladistic constraints. Ann Entomol Soc Am 83:1031–1047
Wilkinson DG (1992) In situ hybridization: a practical approach. Oxford University Press, Oxford
Xia Q, Zhou Z, Lu C, Cheng D, Dai F, Li B, Zhao P, Zha X, Cheng T, Chai C, Pan G, Xu J, Liu C, Lin Y, Qian J, Hou Y, Wu Z, Li G, Pan M, Li C, Shen Y, Lan X, Yuan L, Li T, Xu H, Yang G, Wan Y, Zhu Y, Yu M, Shen W, Wu D, Xiang Z, Yu J, Wang J, Li R, Shi J, Li H, Li G, Su J, Wang X, Li G, Zhang Z, Wu Q, Li J, Zhang Q, Wei N, Xu J, Sun H, Dong L, Liu D, Zhao S, Zhao X, Meng Q, Lan F, Huang X, Li Y, Fang L, Li C, Li D, Sun Y, Zhang Z, Yang Z, Huang Y,

Xi Y, Qi Q, He D, Huang H, Zhang X, Wang Z, Li W, Cao Y, Yu Y, Yu H, Li J, Ye J, Chen H, Zhou Y, Liu B, Wang J, Ye J, Ji H, Li S, Ni P, Zhang J, Zhang Y, Zheng H, Mao B, Wang W, Ye C, Li S, Wang J, Wong GK, Yang H (2004) A draft sequence for the genome of the domesticated silkworm (*Bombyx mori*). Science 306:1937–1940

Xu X, Brechbiel JL, Gavis ER (2013) Dynein-dependent transport of nanos RNA in Drosophila sensory neurons requires rumpelstiltskin and the germ plasm organizer oskar. J Neurosci 33:14791–14800

Yoshimura A (2005) Karyotypes of two American field crickets: *Gryllus rubens* and *Gryllus* sp. (Orthoptera: Gryllidae). Entomol Sci 8:219–222

Zeng V, Extavour CG (2012) ASGARD: an open-access database of annotated transcriptomes for emerging model arthropod species. Database. https://doi.org/10.1093/database/bas048

Zeng V, Ewen-Campen B, Horch HW, Roth S, Mito T, Extavour C (2013) Developmental gene discovery in a hemimetabolous insect: *de novo* assembly and annotation of a transcriptome for the cricket *Gryllus bimaculatus*. PLoS One 8:e61479

Zhang H, Shinmyo Y, Hirose A, Mito T, Inoue Y, Ohuchi H, Loukeris TG, Eggleston P, Noji S (2002) Extrachromosomal transposition of the transposable element *Minos* in embryos of the cricket *Gryllus bimaculatus*. Dev Growth Differ 44:409–417

Zhang XG, Maas A, Haug JT, Siveter DJ, Waloszek D (2010) A eucrustacean metanauplius from the lower Cambrian. Curr Biol 20:1075–1079

Zhao L, Zhang X, Qiu Z, Huang Y (2018) De novo assembly and characterization of the *Xenocatantops brachycerus* transcriptome. Int J Mol Sci 19:520

Chapter 9
The Rove Beetle *Creophilus maxillosus* as a Model System to Study Asymmetric Division, Oocyte Specification, and the Germ-Somatic Cell Signaling

Malgorzata Kloc

Abstract *Creophilus maxillosus* (Staphylinidae, Coleoptera, Polyphaga) has a meroistic-telotrophic ovary composed of tropharium, which contains trophocytes (nurse cells) and vitellarium, which contains growing oocytes. The trophocytes are connected to the oocytes by cytoplasmic nutritive cords, which deliver nutrients to the oocytes. The formation/differentiation of the oocytes and trophocytes takes place in the pupal ovary within linear chains of sibling cells. Each chain is composed of a single oocyte connected to a linear chain of sister trophocytes. The nuclei of the oocytes contain an extrachromosomal DNA body (extra DNA body) consisting of amplified ribosomal DNA (rDNA). During oogenesis, the prospective oocyte, located at the base (posterior) of each chain, is the only cell within the chain that amplifies rDNA and retains permanent contact with the somatic pre-follicular cells. The oogonial divisions leading to the formation of the oocyte/trophocytes chain are asymmetric, and during consecutive divisions, the rDNA body always segregates basally (posteriorly) to the prospective oocyte abutted on the somatic cells. However, the segregation of rDNA is imperfect, and within each oocyte/trophocytes chain, there is a gradient of rDNA: the prospective oocyte has the highest amount of rDNA and the trophocyte that is most distant (most anterior) from the oocyte has no or the lowest amount of rDNA. In addition, the divisions within each chain are parasynchronous, with the pro-oocyte being the most mitotically advanced cell in the chain. These observations indicate the presence of a signaling gradient emanating from the somatic cells and/or oocyte; this gradient diminishes in strength with the

M. Kloc (✉)
The Houston Methodist Research Institute, Houston, TX, USA

Weill Cornell Medical College, New York, NY, USA

Department of Genetics, The University of Texas, MD Anderson Cancer Center, Houston, TX, USA

Department of Surgery, The Houston Methodist Hospital, Houston, TX, USA
e-mail: MKloc@houstonmethodist.org

W. Tworzydlo, S. M. Bilinski (eds.), *Evo-Devo: Non-model Species in Cell and Developmental Biology*, Results and Problems in Cell Differentiation 68,
https://doi.org/10.1007/978-3-030-23459-1_9

increasing distance from its source, i.e., the oocyte/somatic cells. Because of this phenomenon, *C. maxillosus* is the perfect model in which to study the germ-somatic cell interactions and signaling. This chapter describes the methods for the collection and laboratory culture of *C. maxillosus* and the analysis of divisions and signaling in the *C. maxillosus* ovary.

9.1 Collection and Laboratory Culture of *Creophilus maxillosus*

Creophilus maxillosus occurs on nearly all continents, mainly in the Northern Hemisphere (Fig. 9.1). *Creophilus* imago (10–20 mm in length) and larvae (20–25 mm in length) are scavengers and feed on other insects' larvae. In the wild, *C. maxillosus* is very abundant between spring and early autumn on/and in the vicinity of decaying carrions, where it lays eggs and feeds on the carrion's maggots. The presence of *C. maxillosus* on carrions also makes this species a valuable tool for a forensic postmortem analysis during crime scene investigations (Frątczak and Matuszewski 2014).

Creophilus eggs hatch in 2–5 days. The larval stage (Fig. 9.1) lasts approximately 14–19 days and consists of three instars, and the pupal stage lasts approximately 10–16 days (Dajoz and Caussanel 1968; Kloc and Matuszewski 1977).

The adult *Creophilus* and its larvae can be collected from the decaying animal remains found in the wild or by using traps containing decaying meat setup in the field or forest. Imago and larvae are easily raised in the laboratory in Petri dishes lined with moist blotting paper or in moist sand-filled containers. Larvae and imago are fed either with commercially available live insect larvae or with small pieces of raw animal fat (lard). Newly laid eggs should be transferred to a separate dish to prevent egg cannibalism. The last instar larvae at the end of the larval stage should be transferred to a container filled with moist sand where the larvae will bury and pupate.

9.2 *Creophilus maxillosus* Oocyte and Trophocyte Formation

C. maxillosus has a meroistic-telotrophic ovary [for the classification of insect ovaries/ovarioles, see Buning (1993, 1994)]. The ovary of the adult contains many ovarioles composed of an anteriorly located tropharium, which contains the syncytium of trophocytes (nurse cells), and a posteriorly located vitellarium containing oocytes in consecutive stages of growth and vitellogenesis (Fig. 9.2) (Buning 1979a, b, 1993, 1994). The cytoplasmic trophic cords connect the tropharium to the oocytes and deliver the nutrients necessary for oocyte growth (Fig. 9.2). The formation and differentiation of trophocytes and oocytes begin in the late larval/early pupal stage in the region of undifferentiated ovariole called the germarium. In the 1-day-old

Fig. 9.1 *Creophilus maxillosus* imago and larva. (**a**) *Creophilus maxillosus* from Poland. Photograph courtesy of Anna Mądra-Bielewicz, M.Sc. and Szymon Matuszewski, Ph.D., Laboratory of Criminalistics, Adam Mickiewicz University, Św. Marcin 90, 61–809, Poznań, Poland. (**b**) *Creophilus maxillosus* from Bolton, Worcester County, Massachusetts, USA, August 8, 2005. Size: 10–12 mm. Photograph courtesy of Tom Murray. (**c**) *Creophilus maxillosus* from Hwy 9, Dona Ana County, New Mexico, USA. Coordinates: 31.798390, −106.935599. May 3, 2016. Size: 20 mm. Photograph courtesy of Salvador Vitanza-Hedman, PhD, USDA-APHIS-PPQ 200 N Mariposa Rd., B-500, Nogales, AZ 85621. (**d**) *Creophilus maxillosus*, stage L1 larva, from Poland. Photograph courtesy of Anna Mądra-Bielewicz, MSc and Szymon Matuszewski, PhD. Laboratory of Criminalistics, Adam Mickiewicz University, Poznań, Poland

Creophilus pupa, the ovariole's germarium contains one layer of large germline cells (prospective oocytes/pro-oocytes/cystocytes) located on the layer of somatic pre-follicular cells (Fig. 9.2). The pro-oocytes are slightly elongated, with the

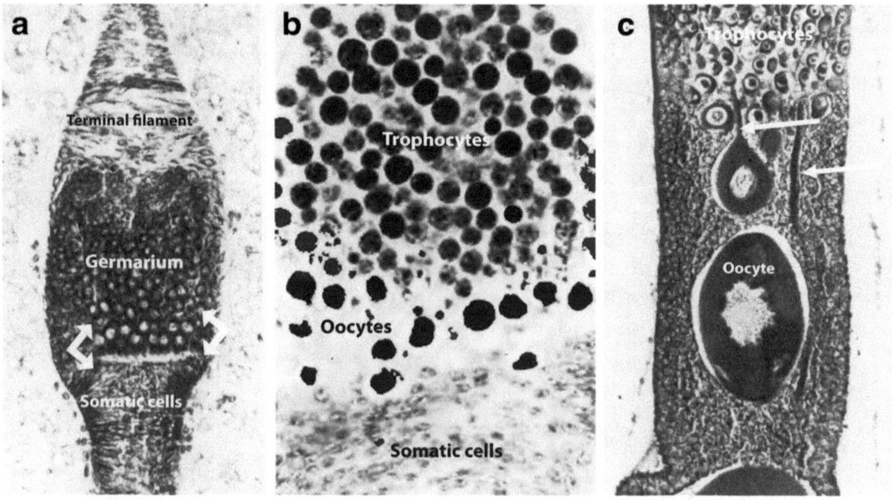

Fig. 9.2 *Creophilus maxillosus* ovary. (**a**) Longitudinal section of an ovariole from a 1-day-old pupa. The basal part of the germarium shows a layer of columnar pro-oocytes (brackets) placed on the somatic pre-follicular cells. Toluidine blue. Magnification ×140. (**b**) Longitudinal section through the distal region of the germarium from (**a**) 9-day-old pupa showing a basally located layer of oocytes containing compact extra rDNA bodies and anteriorly located nurse cells. Heidenhein's hematoxylin. Magnification ×650. From Kloc M, Eur J Cell Biol (1980) 21:328–334. (**c**) Longitudinal section through a fragment of the telotrophic ovariole of an adult showing the nurse cells (at the anterior) connected by trophic cords (arrows) to the oocytes (located at the posterior). Methyl green/pyronin. Magnification ×140. Panels (**a**, **c**) from Kloc M and Matuszewski B, Wilhelm Roux's Archives (1977) 183:351–368; reproduced with permission from Springer Nature

nucleus positioned in the anterior part and the spectrosome (in early *Creophilus* papers erroneously identified as a germ plasm) located in the posterior cytoplasm (Figs. 9.2, 9.3, 9.4 and 9.5; Kloc and Matuszewski 1977; Matuszewski et al. 1985, 1999). The spectrosome is a cytoplasmic organelle present in the germline stem cells and cytoblasts/cystocytes of invertebrates and vertebrates (Deng and Lin 1997, 2001; Kloc et al. 1998, 2004; Lin and Spradling 1995; Megraw and Kaufman 2000; Miyauchi et al. 2013). The spectrosome contains cytoskeletal proteins such as α- and β-spectrin, actin, the adducin-like Hts protein, ankyrin, and Orbit, a *Drosophila* ortholog of the microtubule plus-end-enriched protein (CLASP). It is believed that the spectrosome is a precursor of the fusome, which anchors the mitotic spindle and determines the spindle orientation during cystoblast/cystocyte divisions (Deng and Lin 2001; Fichelson and Huynh 2007; Huynh 2018; Kloc et al. 1998; Miyauchi et al. 2013; Snapp et al. 2004). The fusome contains the endoplasmic reticulum-like cisternae and a transitional endoplasmic reticulum ATPase (TER94), which suggests that the fusome grows by vesicle/cisternae fusion, and represents a germ cell modification of endoplasmic reticulum (León and McKearin 1999). Recent studies indicate that in *Drosophila* spermatocytes the integrity of the fusome, and its role in centriole/spindle anchoring, require inhibition of cyclin A/Cdk1 by Myt1

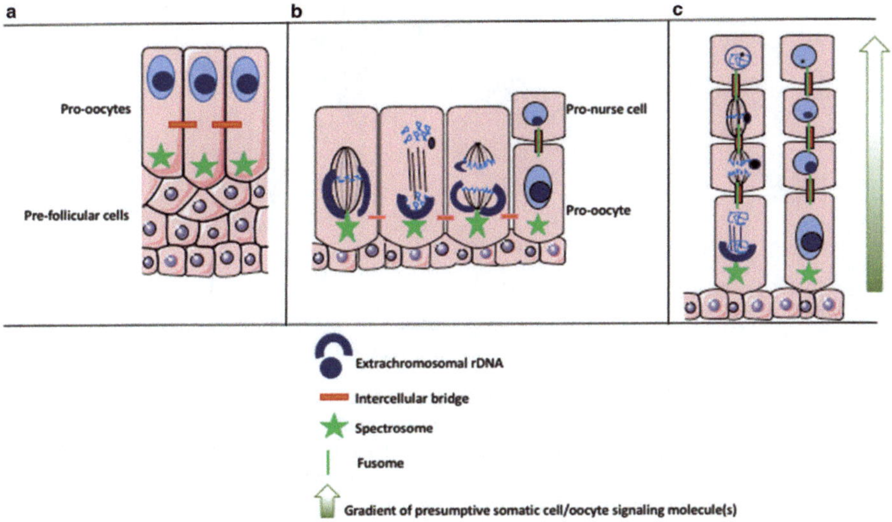

Fig. 9.3 Diagram of the pro-oocyte/pro-nurse cell chain formation. (**a**) In the early pupal ovariole, the pro-oocytes (cystocytes) form a layer of elongated cells located on the somatic pre-follicular cells. The pro-oocytes, connected by the transverse intercellular bridges, are polar and have an anteriorly located nucleus containing amplified extrachromosomal ribosomal DNA (rDNA) and a posteriorly located spectrosome. (**b**) Despite being connected by intercellular bridges the pro-oocytes divide asynchronously. During consecutive mitotic divisions with incomplete cytokinesis, the extra rDNA always segregates to the posterior pole. Consecutive divisions of the pro-oocyte create a chain (cyst) of sister cells (cystocytes) connected by intercellular bridges, spanned by the spectrosome-derived fusome. (**c**) The cells within each chain divide parasynchronously, with the pro-oocyte being the most advanced in mitosis. Because of the imperfect segregation, some of the lagging extra rDNA ends up in the pro-nurse cells. Extra rDNA does not disperse during mitosis but forms blocks or rings of heterochromatin on the surface of the mitotic spindle. The existence of posterior–anterior mitotic and rDNA gradients within the cystocyte chain suggests the presence of a signal that originates in the somatic cells and gradually dissipates along the posterior–anterior axis of the ovariole

kinase (Varadarajan et al. 2016). It has been shown in *Drosophila* that the knockout of the spectrosome components disrupts the proper orientation of mitotic spindles during cystocyte divisions (Bang and Cheng 2015). Interestingly, the divisions of the cystocytes in the polychaete worm *Ophryotrocha labronica*, which lacks a spectrosome or fusome, are asynchronous and indeterminate (Brubacher and Huebner 2011). *Creophilus* spectrosome contains a membranous reticulum (Fig. 9.5; Matuszewski et al. 1999), and both spectrosome and fusome contain spectrin (Fig. 9.5; Matuszewski et al. 1999).

The nuclei of *Creophilus* pro-oocytes contain many nucleoli and extrachromosomal rDNA in the form of large extrachromosomal rDNA body (Figs. 9.2, 9.3, 9.4 and 9.6; Matuszewski and Kloc 1976; Kloc and Matuszewski 1977; Matuszewski et al. 1985, 1999; Kloc 1980). The pro-oocytes are interconnected by transverse intercellular bridges, which indicate that they are the progeny of a single progenitor cell (the cystoblast/chordoblast). Interestingly, despite being connected, the

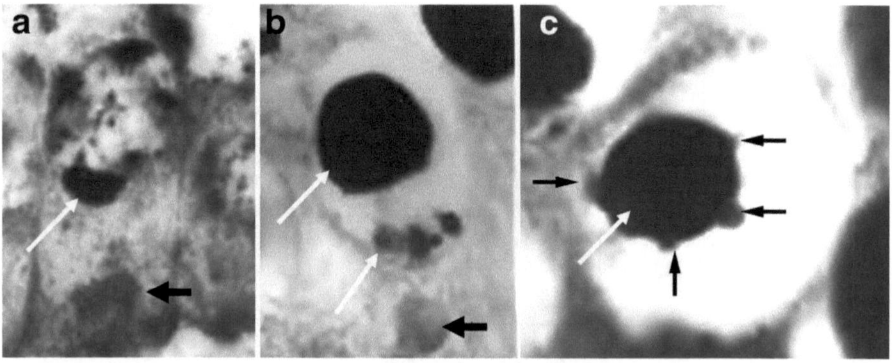

Fig. 9.4 Ribosomal DNA amplification. (**a**) The pro-oocyte from a 2-day-old pupa showing blocks of extra ribosomal DNA (rDNA) (white arrow) in the nucleus and spectrosome ("germ plasm," black arrow) in the posterior cytoplasm. Feulgen and light green. Magnification ×2000. From Kloc M, Eur J Cell Biol (1980) 21:328–334. (**b**) The pro-oocyte from a 6-day-old pupa showing large rDNA and several smaller extra rDNA aggregates (white arrows) in the nucleus and the spectrosome ("germ plasm") in the posterior region (black arrow). Heidenhain's hematoxylin. Magnification ×2000. From Kloc M, Eur J Cell Biol (1980) 21:328–334. (**c**) Pro-oocyte nucleus from a 7-day-old pupa. The extra rDNA body (white arrow) has nucleoli on its surface (black arrows). Heidenhain's hematoxylin. Magnification ×2000. From Kloc M, Eur J Cell Biol (1980) 21:328–334

pro-oocytes divide asynchronously of each other (Matuszewski et al. 1985). The first division of each pro-oocyte is perpendicular to the longitudinal axis of the ovariole and creates two daughter cells (cystocytes/chordocytes): the anterior pro-trophocyte and the posterior pro-oocyte that retains contact with the somatic pre-follicular cells (Fig. 9.3). The cytokinesis of the division is incomplete, and the pro-trophocyte and the pro-oocyte remain connected by the intercellular bridge (Fig. 9.3). During the second and each consecutive division, always perpendicular to the longitudinal axis of the ovariole, and with incomplete cytokinesis, the pro-trophocyte forms two daughter pro-trophocytes, and the pro-oocyte forms the anterior pro-trophocyte and the posterior pro-oocyte. This process creates a chain of interconnected cells (cystocytes). The cytoplasmic bridges connecting cystocytes contain spectrin-rich fusome (Matuszewski et al. 1985, 1999). Each chain has one pro-oocyte, located at the base (posterior region) of the chain and abutted on the somatic cells, and several pro-trophocytes located anteriorly (Matuszewski et al. 1985, 1999). Although the exact number of cystocyte divisions has not been determined in *Creophilus*, the observation that each cystocyte chain contains more than 12 cells indicates that there are probably at least four consecutive divisions that produce a cluster (cyst) of 16 sibling cells, similar to other invertebrate and vertebrate species (Kloc et al. 2004). However, because in *Creophilus* each ovariole contains at least several interconnected pro-oocytes and each of the pro-oocyte is associated with a chain of at least 12 pro-trophocytes, the number of consecutive divisions of a single progenitor cell must be much higher than four. The analysis of Fig. 9.2 indicates that the number of cells in each huge "ovariolar" cluster must exceed a hundred of

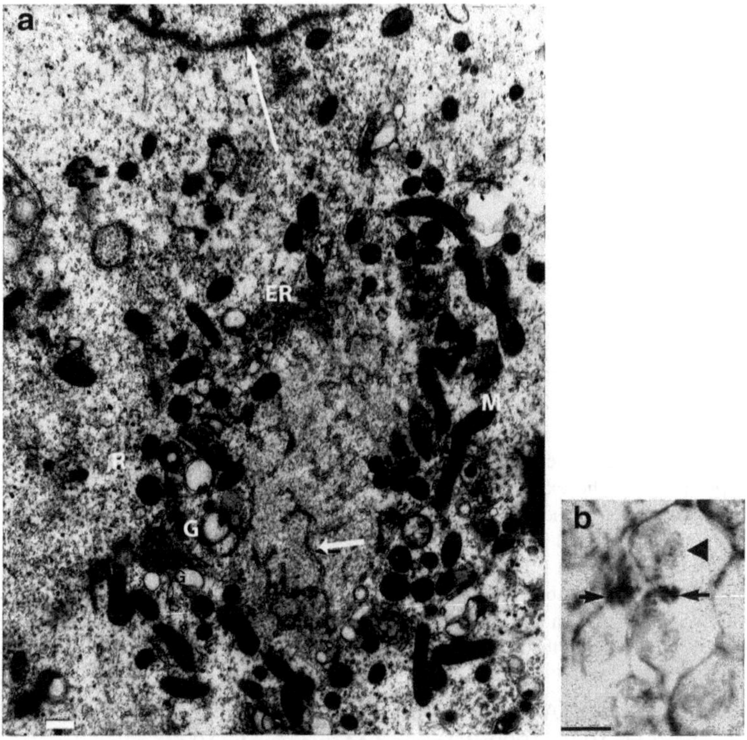

Fig. 9.5 Spectrosome and fusome. (**a**) Ultrastructure of *Creophilus* spectrosome present in the posterior pole of the pro-oocyte in a 2-day-old pupa. A central part of the spectrosome contains membranous reticulum (short arrow). Mitochondria (M), cisternae of smooth ER (ER), ribosomes (R), and Golgi (G) are located at the spectrosome periphery. Long arrow denotes a fragment of the pro-oocyte nucleus. Bar is equal to 500 nm. (**b**) Immunostaining with anti-spectrin antibody showing the presence of spectrin in the fusomes in the cytoplasmic bridges connecting pro-nurse cells in the developing ovary of a 5-day-old pupa. Arrows point to the spectrin-positive fusomes. Arrowhead points to the cell nucleus. Magnification ×2000. From Matuszewski B Ciechomski K and Kloc M (1999) Folia Histochem Cytobiol 37:179–190; reproduced with permission from Via Medica, publishers of FHC

cells. Therefore, it seems that the number of consecutive divisions must be higher than six. Such a number of divisions is not as rare as one may expect among insects (Jaglarz 1992, 1998; Szklarzewicz et al. 2000). In *Creophilus*, the first and second divisions of cystocytes are synchronous, but later divisions of cystocytes within each chain are parasynchronous, with the pro-oocyte always the most mitotically advanced cell of the chain (Kloc and Matuszewski 1977; Matuszewski et al. 1985). This indicates the presence of the signaling gradient decreasing in strength with the growing distance from its pro-oocyte/somatic cell source (Fig. 9.3). After the last division, the pro-oocyte, which contains a large extrachromosomal rDNA body, rounds up and becomes the oocyte, and the pro-trophocytes start to

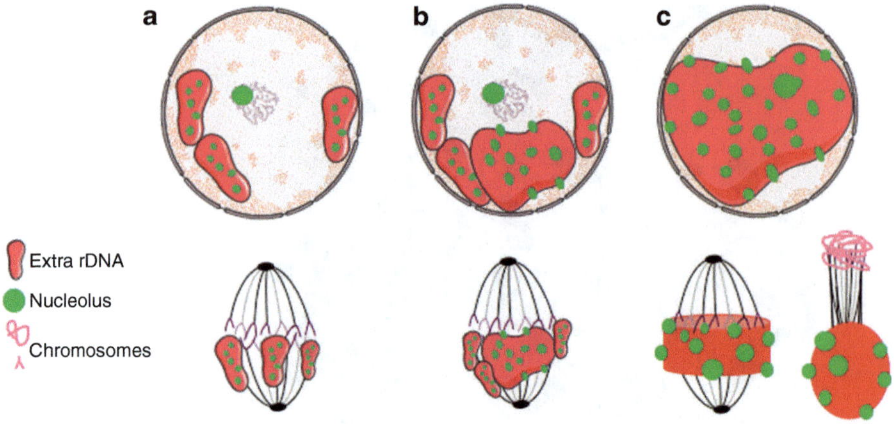

Fig. 9.6 Diagram of amplified rDNA during interphase and in the dividing pro-oocytes. (**a**, **b**) In the early pro-oocytes, the extrachromosomal ribosomal DNA (rDNA) forms several blocks of heterochromatin located at the nuclear membrane. Extra rDNA contains multiple nucleoli. During mitosis, the extra rDNA and nucleoli do not disperse. (**c**) In the late pro-oocyte, the extra rDNA forms a single aggregate that fills the entire nucleus. The multiple nucleoli occupy the interior and the surface of the extra DNA body and do not disperse during division. During metaphase of the last pro-oocyte division, the extra rDNA forms a heterochromatin ring surrounding the spindle and the chromosomes. In telophase, the extra rDNA body segregates posteriorly to the prospective oocyte

endoreplicate their DNA and become the nurse cells (Figs. 9.2, 9.3 and 9.6; Matuszewski et al. 1985, 1999; Kloc and Matuszewski 1977).

9.3 Amplification and Asymmetric Segregation of rDNA

Insects have two major types of the ovaries: the panoistic ovary, which contains only oocytes without the nurse cells, and the meroistic (polytrophic or telotrophic) ovary, where oocytes are accompanied by the nurse cells/trophocytes (Buning 1993, 1994). In the panoistic ovary, because of the lack of nurse cells, the oocyte must be nutritionally and synthetically self-sufficient, i.e., it must be very active transcriptionally and translationally to produce enough nutrients to sustain embryo development. In the meroistic ovary, the main burden of nutrient production is switched from the oocytes to the nurse cells/trophocytes, and in theory, the oocyte can be transcriptionally and translationally latent. However, in reality, probably to speed up the accumulation of nutrients and subsequent development of the embryo, both panoistic and meroistic species amplify ribosomal DNA (rDNA) in the oocyte nucleus (Cave 1972; Trendelenburg et al. 1977; Mazurkiewicz and Kubrakiewicz 2002; Kloc 1980). One such species is *Creophilus maxillosus* (Kloc 1980).

9.3.1 Amplification of rDNA

In *Creophilus* the amplification of rDNA begins in the pro-oocytes before they start dividing. The pro-oocytes of the same generation in the 1- to 2-day-old pupae contain varying amounts of rDNA, which indicates that they amplify rDNA asynchronously and independently of each other (Matuszewski et al. 1999; Kloc 1980). At the beginning of the amplification, the small blocks of heterochromatic extrachromosomal rDNA underlie the nuclear envelope (Figs. 9.4, 9.6 and 9.7). As the amplification progresses during subsequent interphases of oogonial divisions, the blocks of extra rDNA become larger and fuse. In the last generation of the pro-oocytes (starting in a 5-day-old pupa), these fused rDNA blocks form one large, spherical, highly condensed extrachromosomal rDNA body, which occupies nearly the entire oocyte nucleus and contains the equivalent of 24C DNA (Figs. 9.2, 9.4 and 9.6; Kloc and Matuszewski 1977; Matuszewski et al. 1999; Kloc et al. 1995; Kloc 1980). The amplification process is paralleled by an increase in the number of nucleoli, which occupy the interior and the surface of the extra DNA body (Figs. 9.4 and 9.6). During the subsequent oocyte growth and vitellogenesis, the extra rDNA body slowly decondenses and disperses, producing more and more nucleoli (Fig. 9.2c). In fully grown vitellogenic oocytes, the nucleus (extremely large and lobulated) is filled with thousands of nucleoli and lacks any remnants of the condensed rDNA body (Matuszewski and Kloc 1976; Kloc et al. 1995; Matuszewski et al. 1999).

9.3.2 Asymmetric Divisions and Segregation of Amplified rDNA

During the consecutive divisions of the pro-oocyte, the extrachromosomal rDNA always segregates to the posterior pole, i.e., to the pro-oocyte of the next generation (Figs. 9.3 and 9.7). During the divisions, the extra rDNA does not disperse in the cytoplasm but remains compact (Figs. 9.3 and 9.7). During the early divisions of the pro-oocytes, the independent blocks of rDNA are located on the surface of the mitotic spindle, and in the late divisions, the extra rDNA forms a continuous cylinder surrounding the spindle and the chromosomes (Figs. 9.3 and 9.7; Kloc 1980). Similarly, the multiple nucleoli present in the extra rDNA body do not disperse but remain attached to its surface during the divisions (Fig. 9.7; Kloc 1980). The asymmetric segregation of the extra rDNA is especially well visible during telophase, when the posterior pole of the dividing pro-oocyte contains chromosomes surrounded by the extra rDNA and the anterior pole contains only the chromosomes (Figs. 9.6 and 9.7; Kloc 1980).

Although the extrachromosomal rDNA always segregates to the posterior pole of the pro-oocyte, sometimes, especially during early divisions when the extra rDNA forms several independent blocks, the segregation of extra rDNA to the posterior is

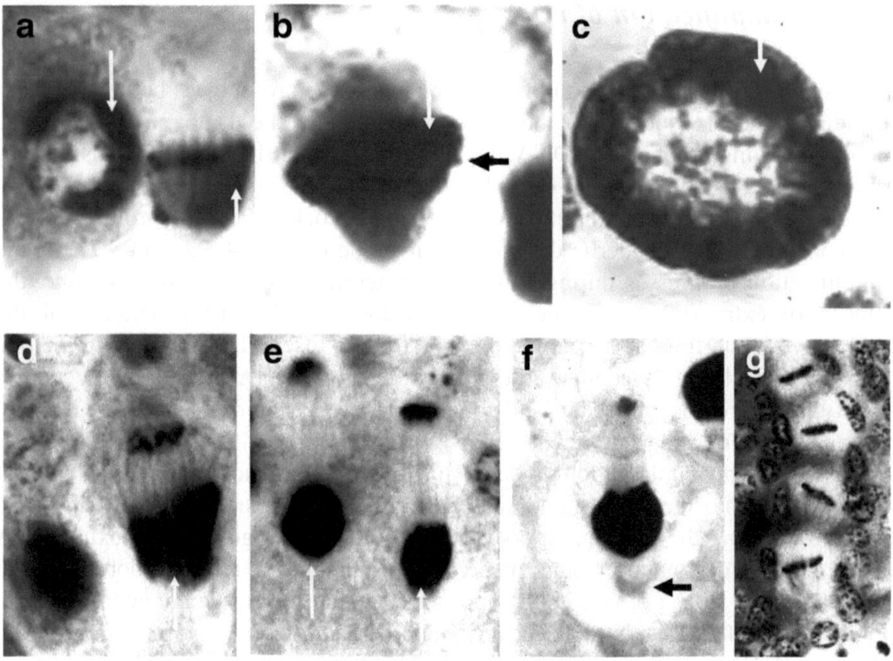

Fig. 9.7 Extrachromosomal ribosomal DNA in the dividing pro-oocytes. (**a**) Pro-oocytes from a 4-day-old pupa. The nucleus of the interphase cell (on the left) shows blocks of extrachromosomal ribosomal DNA (rDNA) (arrow). The cell shown on the right is in metaphase. The extra rDNA (arrows) is located in the posterior half of the spindle. Heidenhain's hematoxylin. Magnification ×2000. From Kloc M, Eur J Cell Biol (1980) 21:328–334. (**b**) Pro-oocyte in metaphase from a 6-day-old pupa. The extrachromosomal rDNA (white arrow) surrounds metaphase plate chromosomes. The black arrow denotes the nucleolus on the surface of the extra rDNA body. From Kloc M, Eur J Cell Biol (1980) 21:328–334. Toluidine blue. Magnification ×2000. (**c**) Polar view of a pro-oocyte in metaphase from a 6-day-old pupa. A ring of extra rDNA (arrow) surrounds the chromosomes in the metaphase plate. Aceto-orcein squash preparation. Magnification ×2000. From Kloc M, Eur J Cell Biol (1980) 21:328–334. (**d**) A pro-oocyte in anaphase from a 5-day-old pupa. The anterior pole contains only chromosomes, and the posterior pole contains chromosomes and extra rDNA (arrow). Heidenhain's hematoxylin. Magnification ×2000. From Kloc M and Matuszewski B, Wilhelm Roux's Archives (1977) 183:351–368; reproduced with permission from Springer Nature. (**e**) Two pro-oocytes in early telophase from a 5-day-old pupa showing preferential segregation of the extra rDNA (arrows) to the posterior pole. Magnification ×2000. From Kloc M and Matuszewski B, Wilhelm Roux's Archives (1977) 183:351–368; reproduced with permission from Springer Nature. (**f**) Pro-oocyte in late telophase from a 6-day-old pupa. Anterior pole contains chromosomes only, and the posterior pole contains chromosomes, extra rDNA (white arrow), and the spectrosome (black arrow). Heidenhain's hematoxylin. Magnification ×2000. From Kloc M and Matuszewski B, Wilhelm Roux's Archives (1977) 183:351–368; reproduced with permission from Springer Nature. (**g**) The 4-cell chain of sibling cells in the metaphase of the second division of the pro-oocyte. The second division is still synchronous. In the subsequent divisions while the chain elongates and anterior cells in the chain become more and more distant from the pro-oocyte/somatic cells signal source, the divisions of sibling cells in the chain become less and less synchronous. Feulgen and light green. Magnification ×960. From Kloc M and Matuszewski B, Wilhelm Roux's Archives (1977) 183:351–368; reproduced with permission from Springer Nature

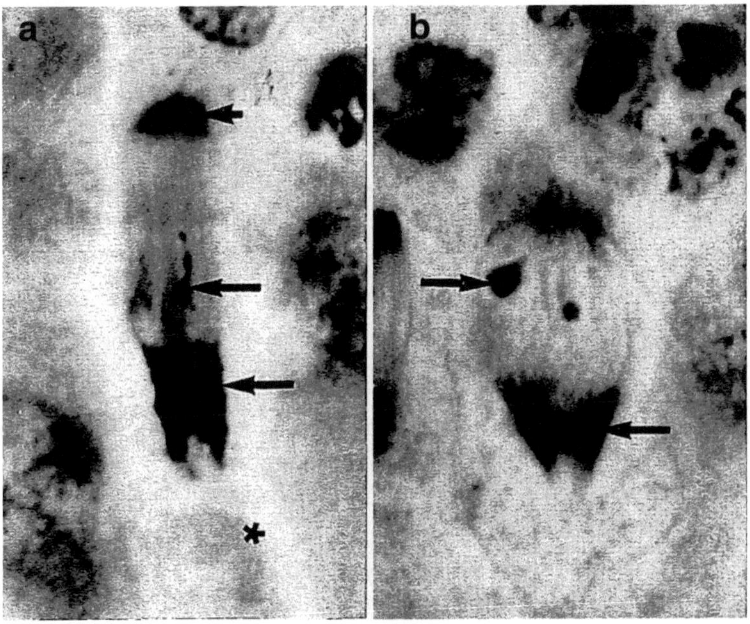

Fig. 9.8 Imperfect segregation of extra rDNA. During the pro-oocyte divisions, the extra rDNA segregates preferentially to the posterior pole of the spindle. However, the preferential segregation is imperfect, and very often some of the extra rDNA lags behind in the anaphase movement and becomes inherited by the pro-nurse cell. (**a**, **b**) Dividing pro-oocytes from a 1-day-old pupa showing posteriorly segregating and lagging behind extra rDNA (long arrows). Short arrow points to the chromosomes of the pro-nurse cell, and asterisk indicates the spectrosome. Long arrows indicate extra rDNA. Feulgen and light green. Magnification ×2000. From Matuszewski B, Ciechomski K and Kloc M (1999) Folia Histochem Cytobiol 37:179–190; reproduced with permission from Via Medica

imperfect (Figs. 9.3 and 9.8). As a result, different-sized fragments of extra rDNA end up in the middle and the anterior half of the spindle and, eventually, in the pro-trophocyte (Fig. 9.8). When this pro-trophocyte divides, this extra rDNA will also segregate preferentially to the posterior pro-trophocyte, but sometimes the small fragments of rDNA will end up in the anterior daughter pro-trophocyte (Fig. 9.3). Because this imperfect segregation might be repeated at each consecutive division, after the last division, the resulting chain of cells will contain a posterior–anterior gradient of extrachromosomal rDNA content, with the pro-oocyte containing the bulk of extra rDNA and the most anterior pro-trophocyte containing no or minimal quantity of the extra rDNA (Fig. 9.3; Kloc 1980; Kloc and Matuszewski 1977; Matuszewski et al. 1985, 1999).

9.4 Signaling from the Somatic Cells

There is no direct, experimental proof that the somatic pre-follicular cells emanate molecular signal(s) that cause the asymmetry of cystocyte divisions and rDNA amplification in *Creophilus*. However, the conclusions drawn from cytological observations of dividing cystocytes strongly support this hypothesis.

The observations pointing to the existence of the somatic cell-derived signal(s), which gradually dissipates with the increasing distance from the somatic cell source, are as follows:

1. The pro-oocytes are the only germline cells in a direct contact with the somatic pre-follicular tissue.
2. Only the pro-oocytes amplify rDNA.
3. There is a posterior–anterior gradient of mitotic phases along the chain of cells, with the pro-oocyte being the most advanced in mitosis.
4. The extra rDNA always segregates toward the posterior (both in the pro-oocyte and in the pro-nurse cells).

Although we do not know the identity of the somatic cell-derived signal(s) in *Creophilus,* studies conducted in *Drosophila* suggest several candidate molecules and pathways. In *Drosophila,* the interaction of the oocyte with the posterior somatic cells is mediated by DE-cadherin (Godt and Tepass 1998; Gonzalez-Reyes et al. 1997; Gonzalez-Reyes and St. Johnston 1998), while the polarity of the oocyte and its microtubules depends on somatic follicle cells (van Eeden and St. Johnston 1999) and Hedgehog signaling (Zhang and Kalderon 2000). Recent studies also indicate that the partitioning-defective (PAR) proteins and somatic signaling involving piwi, fs(1)Yb, and the Dpp pathways regulate asymmetric cell division in gametogenesis (Fichelson and Huynh 2007; Deng and Lin 2001). Oocyte determination appears to be regulated by egalitarian, Bicaudal D, stonewall, and encore gene pathways (Deng and Lin 2001).

I believe that two other phenomena observed in the *Creophilus* ovary, i.e., that only the pro-oocytes contain the spectrosome and that the longitudinal axes of the mitotic spindles in all cells of the chain are always parallel to the longitudinal axis of the ovariole—are not directly imposed by the somatic-derived signal(s) but rather are an intrinsic property of the cystoblast/pro-oocyte. It is known that the spectrosome is a precursor of the fusome, which spans the intercellular bridges that connect the cystocytes and anchors the poles of the spindles, thus imposing spindle directionality (Deng and Lin 2001). Thus, a directionality of cystocyte divisions depends on the spectrosome/fusome, which is a germ cell-specific structure. Of course, we cannot exclude the possibility that the formation of the spectrosome and/or the arrangement and architecture of the fusome between the cystocytes also partially depend on somatic signaling, but this notion requires further study.

In summary, because *Creophilus maxillosus* is distributed globally, easily cultured in the laboratory, and contains oocytes (and prospective oocytes) characterized by prominent extrachromosomal rDNA, this insect is especially well suited as a novel model system to study various aspects of the asymmetry and polarity of germline cells, differential mitotic divisions, and the germ-somatic cell interaction during oocyte specification and oogenesis.

Acknowledgments I am very grateful to Tom Murray, Dr. Salvador Vitanza-Hedman, Dr. Szymon Matuszewski and Anna Mądra-Bielewicz, MSc for allowing me to use their photographs of Creophilus.

References

Bang C, Cheng J (2015) Dynamic interplay of spectrosome and centrosome organelles in asymmetric stem cell divisions. PLoS One 10(4):e0123294. https://doi.org/10.1371/journal.pone.0123294

Brubacher JL, Huebner E (2011) Evolution and development of polarized germ cell cysts: new insights from a polychaete worm, *Ophryotrocha labronica*. Dev Biol 357:96–107

Buning J (1979a) The trophic tissue of telotrophic ovarioles in polyphage Coleoptera. Zoomorphology 93:33–50

Buning J (1979b) The telotrophic nature of ovarioles of polyphage Coleoptera. Zoomorphology 93:51–57

Buning J (1993) Germ cell cluster formation in insect oocytes. Int J Insect Morphol Embryol 22:237–253

Buning J (1994) The insect ovary: ultrastructure, previtellogenic growth and evolution. Chapman & Hall, London

Cave MD (1972) Localization of ribosomal DNA within oocytes of the house cricket, *Acheta domesticus* (Orthoptera: Gryllidae). J Cell Biol 55:310–321

Dajoz R, Caussanel C (1968) Morphologic et biologie d'un Coleoptere predateur: *Creophilus maxillosus* (L.) (Staphylinidae). Cah Nat (N ser) 24:65–102

Deng W, Lin H (1997) Spectrosomes and fusomes anchor mitotic spindles during asymmetric germ cell divisions and facilitate the formation of a polarized microtubule array for oocyte specification in *Drosophila*. Dev Biol 189:79–94

Deng W, Lin H (2001) Asymmetric germ cell division and oocyte determination during *Drosophila* oogenesis. Int Rev Cytol 203:93–138

Fichelson P, Huynh JR (2007) Asymmetric divisions of germline cells. Prog Mol Subcell Biol 45:97–120

Frątczak K, Matuszewski S (2014) Instar determination in forensically useful beetles *Necrodes littoralis* (Silphidae) and *Creophilus maxillosus* (Staphylinidae). Forensic Sci Int 241:20–26

Godt D, Tepass U (1998) *Drosophila* oocyte localization is mediated by differential cadherin-based adhesion. Nature 395:387–391

Gonzalez-Reyes A, St. Johnston D (1998) The *Drosophila* AP axis is polarized by the cadherin-mediated positioning of the oocyte. Development 125:3635–3644

Gonzalez-Reyes A, Elliott H, St. Johnston D (1997) Oocyte determination and the origin of polarity in *Drosophila*: the role of the spindle genes. Development 124:4927–4937

Huynh J-R (2018) Fusome as a cell-cell communication channel of *Drosophila* ovarian cyst. Madame Curie Bioscience Database [Internet]

Jaglarz M (1992) Peculiarities of the organization of egg chambers in carabid ground beetles and their phylogenetic implications. Tissue Cell 24:397–409

Jaglarz M (1998) The number that counts. Phylogenetic implications of the number of nurse cells in ovarian follicles of Coleoptera-Adephaga. Folia Histochem Cytobiol 36:167–178

Kloc M (1980) Extrachromosomal DNA and its activity in RNA synthesis in oogonia and oocytes in the pupal ovary of *Creophilus maxillosus* (Staphylinidae, Coleoptera-Polyphaga). Eur J Cell Biol 21:328–334

Kloc M, Matuszewski B (1977) Extrachromosomal DNA and the origin of oocytes in the telotrophic-meroistic ovary of *Creophilus maxillosus* (L) (Staphylinidae Coleoptera-Polyphaga). Wilhelm Roux's Arch 183:351–368

Kloc M, Matuszewski B, Nurkowska J (1995) Ribosomal gene amplification in the oocytes of *Creophilus maxillosus* (Staphylinidae, Coleoptera-polyphaga) – an insect with telotrophic ovaries. Folia Histochem Cytobiol 33:267–276

Kloc M, Larabell C, Chan AP, Etkin LD (1998) Contribution of METRO pathway localized molecules to the organization of the germ cell lineage. Mech Dev 75:81–93

Kloc M, Bilinski S, Dougherty MT, Brey EM, Etkin LD (2004) Formation, architecture and polarity of female germline cyst in *Xenopus*. Dev Biol 266:43–61

León A, McKearin D (1999) Identification of TER94, an AAA ATPase protein, as a bam-dependent component of the *Drosophila* fusome. Mol Biol Cell 10:3825–3834

Lin H, Spradling AC (1995) Fusome asymmetry and oocyte determination in drosophila. Dev Genet 16:6–12

Matuszewski B, Kloc M (1976) Gene amplification in oocytes of the rove beetle *Creophilus maxillosus* (Staphylinidae, Coleoptera-Polyphaga). Experientia 32:34–36

Matuszewski B, Ciechomski K, Nurkowska J, Kloc M (1985) The linear clusters of oogonial cells in the development of telotrophic ovarioles in polyphage Coleoptera. Wilhelm Roux's Arch Dev Biol 194:462–469

Matuszewski B, Ciechomski K, Kloc M (1999) Extrachromosomal rDNA and polarity of pro-oocytes during ovary development in *Creophilus maxillosus* (Coleoptera, Staphylinidae). Folia Histochem Cytobiol 37:179–190

Mazurkiewicz M, Kubrakiewicz J (2002) Nucleolar activity of germ cells in polytrophic ovaries of crane flies (Diptera: Tipulidae). Ultrastructural and cytochemical studies. Folia Histochem Cytobiol 40:47–50

Megraw TL, Kaufman TC (2000) The centrosome in *Drosophila* oocyte development. Curr Top Dev Biol 49:385–407

Miyauchi C, Kitazawa D, Ando I, Hayashi D, Inoue YH (2013) Orbit/CLASP is required for germline cyst formation through its developmental control of fusomes and ring canals in *Drosophila* males. PLoS One 8(3):e58220. https://doi.org/10.1371/journal.pone.0058220

Snapp EL, Iida T, Frescas D, Lippincott-Schwartz J, Lilly MA (2004) The fusome mediates intercellular endoplasmic reticulum connectivity in *Drosophila* ovarian cysts. Mol Biol Cell 15:4512–4521

Szklarzewicz T, Wnek A, Bilinski SM (2000) Structure of ovarioles in *Adelges laricis*, a representative of primitive family Adelgidae. Acta Zool (Stockholm) 81:307–313

Trendelenburg MF, Franke WW, Scheer U (1977) Frequencies of circular units of nucleolar DNA in oocytes of two insects, *Acheta domesticus* and *Dytiscus marginalis*, and changes of nucleolar morphology during oogenesis. Differentiation 7:133–158

van Eeden F, St. Johnston D (1999) The polarization of the anterior-posterior and dorsal-ventral axes during *Drosophila* oogenesis. Curr Opin Genet Dev 9:396–404

Varadarajan R, Ayeni J, Jin Z, Homola E, Campbell SD (2016) Myt1 inhibition of cyclin a/Cdk1 is essential for fusome integrity and premeiotic centriole engagement in *Drosophila* spermatocytes. Mol Biol Cell 27:2051–2063

Zhang Y, Kalderon D (2000) Regulation of cell proliferation and patterning in *Drosophila* oogenesis by hedgehog signaling. Development 127:2165–2176

Chapter 10
Cell Biology of the Tardigrades: Current Knowledge and Perspectives

K. Ingemar Jönsson, Ingvar Holm, and Helena Tassidis

Abstract The invertebrate phylum Tardigrada has received much attention for containing species adapted to the most challenging environmental conditions where an ability to survive complete desiccation or freezing in a cryptobiotic state is necessary for persistence. Although research on tardigrades has a long history, the last decade has seen a dramatic increase in molecular biological ("omics") studies, most of them with the aim to reveal the biochemical mechanisms behind desiccation tolerance of tardigrades. Several other aspects of tardigrade cell biology have been studied, and we review some of them, including karyology, embryology, the role of storage cells, and the question of whether tardigrades are eutelic animals. We also review some of the theories about how anhydrobiotic organisms are able to maintain cell integrity under dry conditions, and our current knowledge on the role of vitrification and DNA protection and repair. Many aspects of tardigrade stress tolerance have relevance for human medicine, and the first transfers of tardigrade stress genes to human cells have now appeared. We expect this field to develop rapidly in the coming years, as more genomic information becomes available. However, many basic cell biological aspects remain to be investigated, such as immunology, cell cycle kinetics, cell metabolism, and culturing of tardigrade cells. Such development will be necessary to allow tardigrades to move from a nonmodel organism position to a true model organism with interesting associations with the current models *C. elegans* and *D. melanogaster*.

K. I. Jönsson (✉) · I. Holm · H. Tassidis
Department of Environmental Science and Bioscience, Kristianstad University, Kristianstad, Sweden
e-mail: ingemar.jonsson@hkr.se

W. Tworzydlo, S. M. Bilinski (eds.), *Evo-Devo: Non-model Species in Cell and Developmental Biology*, Results and Problems in Cell Differentiation 68,
https://doi.org/10.1007/978-3-030-23459-1_10

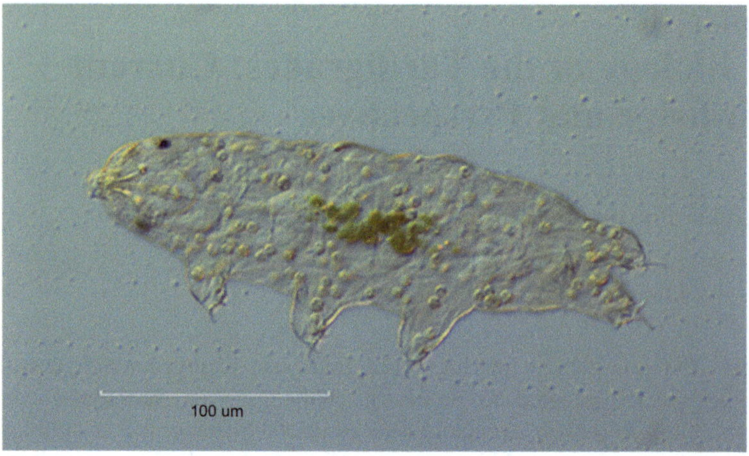

Fig. 10.1 Light microscope image of the tardigrade *Hypsibius exemplaris*. Note the two black eyes in the head region, green food (algae) in the central midgut part, and circular storage cells distributed throughout the body

10.1 Introduction

10.1.1 The General Biology of Tardigrades

Tardigrades are small aquatic invertebrates representing a separate phylum (Tardigrada) within the animal kingdom. Currently, more than 1300 species of tardigrades have been described (Degma et al. 2018) from a variety of ecosystems and microhabitats in terrestrial, freshwater, and marine environments (Nelson et al. 2015). Tardigrades have generally been considered closely related to the phylum Arthropoda, with support from both morphological and molecular data (Campbell et al. 2011), but a recent molecular analysis found a closer relationship with the phylum Nematoda (Yoshida et al. 2017). Two main taxonomic classes of tardigrades have been described, Eutardigrada and Heterotardigrada, which are well distinguished morphologically. Tardigrades belong to the smallest animals with a body length that rarely exceeds 1 mm, and with eight pairs of legs with claws (Fig. 10.1). Their general internal anatomy is simple, with a complete digestive tract (foregut, midgut, hindgut), a nervous system with a ventral chain of four bilobed nerve ganglia and a dorsal lobed brain, and a muscular system with somatic, pharyngeal, stylet, and visceral muscles (Dewel et al. 1993). Tardigrades have no special respiration organ, and oxygen is taken up directly through the cuticle from the surrounding water; thus, continuous metabolism requires that the animal is surrounded by water. The circulation system is represented by a large number of free-floating single coelomocytes ("storage cells") in the body cavity fluid (see more information on these cells below). Several functions of excretion have been reported, including specific excretory glands (Malpighian organs), excretion through the midgut, and excretion connected with shedding of the cuticle connected with molting (Nelson et al. 2015).

The general cytology and cytogenetics of tardigrades have recently been reviewed by Bertolani and Rebecchi (2018), and a detailed review of the ultrastructure of different tissues and cells is provided by Dewel et al. (1993). The genomes of two tardigrade species have so far been reported, *Hypsibius exemplaris*[1] (Koutsovoulos et al. 2016; Arakawa 2016) and *Ramazzottius varieornatus* (Hashimoto et al. 2016), and several more genomes are expected to be described in the near future as the sequencing technology has improved dramatically (Arakawa et al. 2016). The genome size of tardigrades shows considerable variation (Bertolani and Rebecchi 2018), and in the two sequenced species, the *H. exemplaris* genome has been estimated at about 100 Mb while the *R. varieornatus* genome is estimated at 55 Mb (Yoshida et al. 2017).

There are few estimates of the total number of cells in tardigrades. Møbjerg et al. (2011) reported an estimate of around 1060 cells in the marine tardigrade *Halobiotus crispae* based on four specimens, but due to variability in the number of storage cells (see next section), the total number of cells can vary considerably. We have found no other compiled information on the number of somatic cells in tardigrades excluding storage cells, but information on some types of tissues can be found in Bertolani (1970b).

Given the variety of environmental conditions that tardigrades inhabit, it is not surprising that there is also a large variation among species in general ecological and evolutionary aspects such as reproductive system, life history strategies, and food preference. Thus, there are parthenogenetic, bisexual, and hermaphroditic tardigrades (Bertolani 2001), a range in life span among species from a few months to a few years (Nelson et al. 2015), and tardigrades feeding on plant cells, algae, protozoa, bacteria, and other micrometazoa such as nematodes and rotifers (Schill et al. 2011).

10.1.2 Tolerance to Environmental Stress in Tardigrades

Tardigrades have a reputation of being among the most stress-tolerant animals on Earth, related to the ability of some taxa within the phylum to survive some of the most challenging environmental stress agents for living organisms, such as desiccation, freezing, and radiation. Tardigrades share these tolerances with many species within the invertebrate phyla Rotifera and Nematoda, and with a much more limited number of species within the Arthropoda classes Insecta and Crustacea. Tolerance to

[1]The species referred to *Hypsibius dujardini* in many studies reviewed in this chapter originates from the same strain/population collected from a pond in England, and has recently been described by Gąsiorek et al. (2018) as a new species, *Hypsibius exemplaris*, distinguished from the originally described species *Hypsibius dujardini*. In this chapter, we have chosen to consistently use the new name, *H. exemplaris,* although the publications referred to use *H. dujardini*.

desiccation includes complete dehydration under extremely dry conditions (space vacuum, Jönsson et al. 2008), and prolonged dehydration under ambient laboratory conditions (20 years, Jørgensen et al. 2007). Under subzero conditions, tardigrades have been revived from Antarctic moss kept at −20 °C for 30 years (Tsujimoto et al. 2016). Survival under desiccated or deeply frozen conditions requires that the animal can enter a reversible state of arrested metabolism called cryptobiosis (Keilin 1959; Wright 2001). Many studies have also shown that adult tardigrades survive ionizing radiation at very high doses, in the order of several kilogray, both in the desiccated cryptobiotic state and in the hydrated active state (see review in Jönsson et al. 2018). Tardigrades are also so far the only animals that have survived the space conditions in low Earth orbit, with combined exposure to space vacuum, cosmic radiation, and high doses of radiation in the UV_A and UV_B spectra (Jönsson et al. 2008). The cellular and molecular mechanisms behind these tolerances are not yet understood, although many omics studies of tardigrades in the last decade have considerably improved our knowledge of the molecular responses connected with exposures to desiccation in particular (e.g., Schokraie et al. 2012; Wang et al. 2014; Yoshida et al. 2017). In contrast, few studies have taken a more classical cell biology approach, and in general, the cell biology and physiology of tardigrades is a neglected field. This chapter reviews some aspects of tardigrade cell biology with a special focus on areas of interest for understanding the tolerance of tardigrades to extreme environmental conditions.

10.2 Chromosomes and Ploidy

Many tardigrade populations are morphologically similar to each other but differ in sex ratio, ploidy, and mode of reproduction. The amphimictic populations are almost always diploid and the parthenogenetic ones are often polyploid. Ameiotic parthenogenesis is much more common than meiotic parthenogenesis. Some species have both diploid (amphimictic) and polyploid (parthenogenetic) karyotypes (Bertolani and Rebecchi 2018). Sympatric occurrence of diploid and polyploid populations is not uncommon in tardigrades (Rebecchi et al. 2003) and may even in rare cases be syntopic (Bertolani 1994). This suggests that tardigrades have both intra- and interpopulation ploidy polymorphism, which has also been reported in, for example, rotifers (Walsh and Zhang 1992). Intra- and interpopulation variability at chromosome levels has frequently been shown in animal populations in variable environment, and polyploidy seems to be favored in harsher environments compared to their diploid relatives (Collares-Pereira et al. 2013). Parthenogenetic tardigrade populations are also more common in low-stability environments. This could be due to higher enzymatic activity in polyploid animals (Levin 1983), and higher levels of gene expression (Schoenfelder and Fox 2015) compared to diploid relatives.

Most of the karyological studies of tardigrades have been conducted on oocytes. However, variation of ploidy levels is also common within individuals where subsets

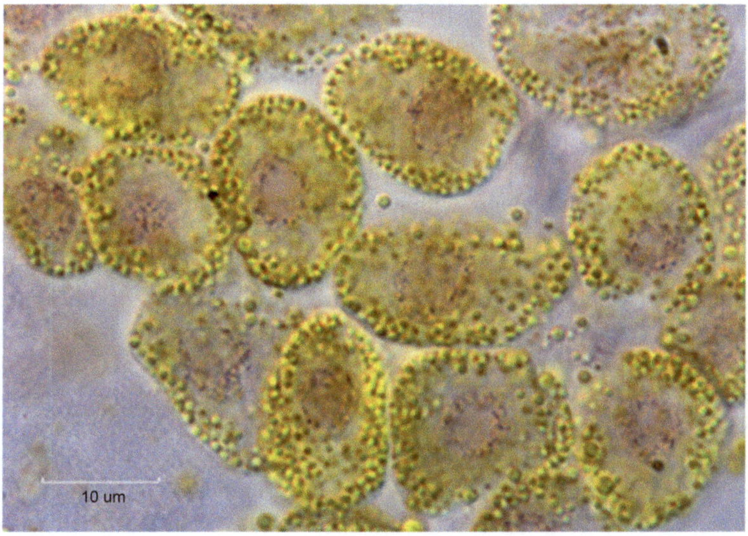

Fig. 10.2 Storage cells of the tardigrade *Richtersius* cf. *coronifer*, with the nucleus in reddish stain by acetic lactic orcein. Light microscopy with differential interference contrast

of cells can have higher ploidy level than represented of the organism as a whole (Parfrey et al. 2008). This phenomenon is known as endopolyploidy, which broadly can be defined as some somatic cells being polyploid. Endopolyploid cells undergo endoreduplication (also called endoreplication), which is achieved by amplifying their genome during the S-phase of the cell cycle but not undergoing mitosis (Neiman et al. 2017). *Richtersius* cf. *coronifer* has been shown to be diploid irrespective of reproductive mode, and the karyotyping has been done in both oocyte (Rebecchi et al. 2003) and in male germ cells (Altiero and Rebecchi 2003). However, preliminary data from flow cytometry analysis of whole animal conducted in our laboratory indicates that *R.* cf. *coronifer* could have the ability to endoreduplicate its genome, since the results indicate different ploidy in somatic cells. This data would be interesting to confirm by karyotyping the somatic cells, but this requires actively dividing cells which are sparse in tardigrades, and no methods of culturing tardigrade cells have so far been reported.

10.3 The Storage Cells of Tardigrades

As mentioned above, tardigrades do not have any specialized circulatory system, and distribution of nutrients from the gut relies on free-floating coelomocytes in the body cavity. In tardigrades, these are often referred to as “storage cells” and they attach to the gut and take up nutrients, followed by detachment and supposedly random distribution of the cells and nutrients within the body cavity fluid as the animal moves about (Fig. 10.2). The storage cells accumulate polysaccharides, lipids, and proteins (Rosati 1968; Węglarska 1975; Szymańska 1994), but the relative amounts of these nutritional components differ among tardigrade species (Hyra et al. 2016).

The number of storage cells in a single animal varies both among tardigrade species, among individuals, and within individuals (over time), but the number is in the range of a few hundred up to more than 1000 cells (Reuner et al. 2010; Møbjerg et al. 2011; Czernekova and Jönsson 2016). Also the size of the storage cells varies between species and individuals within species. In particular, the size of the cells changes along with the oocyte developmental cycle (Szymańska 1994; Jönsson and Rebecchi 2002), from small size at the beginning of the cycle to a maximum in the intermediate part, and again a reduction in size toward the end of the cycle. These changes in cell size clearly represent the accumulation and use of nutrients in the storage cells related to the demand of the developing eggs. Starvation of three tardigrade species (*Milnesium tardigradum, Paramacrobiotus tonollii, Macrobiotus sapiens*) for 7 days also led to reductions in the cell size, as shown by Reuner et al. (2010). In addition, Jönsson and Rebecchi (2002) documented a decline in storage cell size in the eutardigrade *R.* cf. *coronifer* after induction of a single period of anhydrobiosis. However, Czernekova and Jönsson (2016) did not observe a change in the cell size over a sequence of repeated anhydrobiotic periods in the same species; instead, cell number tended to decline. Reuner et al. (2010) also reported differences among the three tardigrade species above in the response in cell size to induced anhydrobiosis, with decline in cell size in one species and no change in the other two. The energetic cost of the physiological processes connected with anhydrobiosis and the role of the energetic status of the storage cells therefore remain unclear and deserve more studies.

Although the main function of storage cells is to store and distribute nutrients, they might also have an immunological role. Volkmann and Greven (1993) reported the presence of tyrosinase activity in storage cells of *Macrobiotus hufelandi*, perhaps related to similar immunological defense functions (melanization) as in insects (Nakhleh et al. 2017). No studies have followed up this observation, and whether tardigrade storage cells indeed serve immunological functions remains to be shown.

10.4 Developmental Biology of Tardigrades

Tardigrade developmental biology was studied as early as a hundred years ago (von Erlanger 1895; von Wenck 1914; Marcus 1929). In these early studies, cleavage was described as more or less equal and the formation of epithelia to occur as early as after five or six divisions.

Later studies showed that cleavage was total and equal but asynchronous as early as from the second division and derived from a spiral pattern, and that gastrulation took place very early in development (Eibye-Jacobsen 1997). Hejnol and Schnabel (2005) performed cell lineage studies on the species *Thulinius stephaniae*, but could not see a stereotyped cleavage pattern, thus prohibiting the reproducible identification of cells early in development.

Gabriel et al. (2007) identified *H. exemplaris* as a suitable species for the study of the evolution of development. This species was chosen based on its optically clear embryos, short early embryonic cell cycle time, and its ability to be kept in culture for prolonged periods and stored using cryopreservation. Using differential interference microscopy time-lapse photography, Gabriel et al. (2007) constructed a cell lineage through the first seven divisions. Features of the early development included a stereotyped nuclear migration that could be used to predict the orientation of the embryonic axis. They further showed that the first division formed equally sized anterior and posterior blastomeres, much as in *Caenorhabditis elegans*. The second division produces two dorsal and two ventral blastomeres, also of equal size. They further showed that the future ventral side could be predicted already at this stage by a specific pattern of nuclear migration. Six of the cells at the eight-cell stage gave rise to cell lineages in more or less synchronous divisions through the 128-cell stage. An embryonic staging series was described which may be useful for further studies of the embryology of this species.

Relatively little is known about the specific gene expression during tardigrade development. However, Gabriel et al. (2007) speculate that Notch-dependent signaling between cells may be of importance for pattern formation in tardigrades. Gabriel and Goldstein (2007) further describe the segmented expression of Pax3/7 and engrailed homologs in *H. exemplaris* development. It appears that Pax3/7 does not show pair-rule pattern (as it does in, e.g., *D. melanogaster*) in *H. exemplaris*, but rather appears in areas where neurons will form. They speculate that both these genes may have had a role in neurogenesis in early bilaterians.

10.5 Are Tardigrades Eutelic Animals?

Eutely is the characteristic of a constant number of somatic cells within a species, either of the total soma or of particular organs or tissues (partial eutely) (Van Cleave 1932). Eutely is usually connected with a lack of cell regeneration after the animal has reach maturity, implying that growth in body size after maturity has to rely on increased cell size rather than increased cell number. Well-known eutelic animals are the nematode *C. elegans* (Sulston et al. 1983) and rotifers (Gilbert 1983). The notion that nematodes are eutelic animals has, however, been challenged by, for example, Cunha et al. (1999). Early studies on tardigrades reported them as eutelic animals (e.g., Baumann 1920; Martini 1923) but as shown by Bertolani (1970a, b), tardigrades are certainly not eutelic in the complete soma, and show cell number variability in several types of organs such as the claw glands, ventral ganglia (Marcus 1929), and the ovary (Poprawa et al. 2015). As first shown by Bertolani (1970a), and later by Czernekova and Jönsson (2016), tardigrade storage cells are mitotically active, that is, the cells divide (Fig. 10.3). The latter study found mitotic storage cells in about 18% of examined adult specimens of *R.* cf. *coronifer*, but a much higher frequency (38%) in juveniles; however, the frequency of mitotic cells (among all storage cells) in specimens where mitosis was found was higher in adults

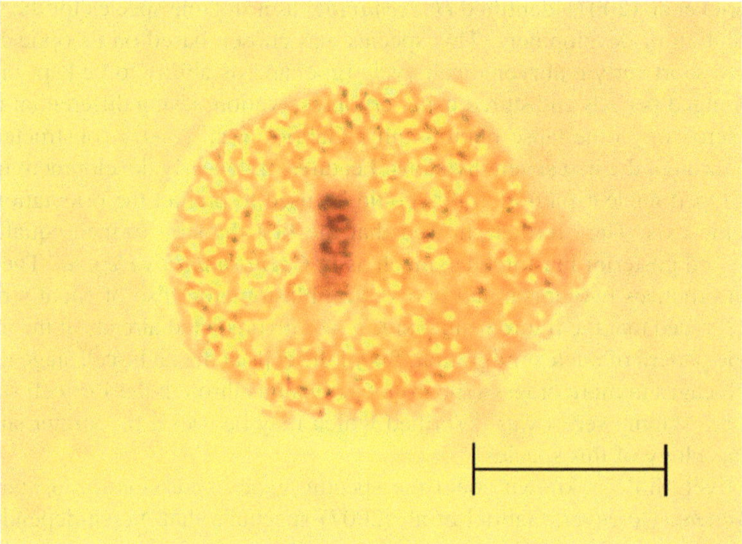

Fig. 10.3 Mitotic storage cell of *Richtersius* cf. *coronifer* with condensed chromosomes in metaphase. Scale bar = 10 um. Reprinted from Czernekova and Jönsson (2016, Fig. 1) with permission from Oxford University Press

(1.47% compared to 0.76% in juveniles). The occurrence of mitotic cells was also much more frequent in adult animals that were in the process of molting (88%), compared to nonmolting animals (12%). Since the molting and egg developmental cycles are coordinated in tardigrades, this pattern also corresponded with a higher level of mitotic cells in adult animals carrying eggs in late developmental stage. The reason for the higher occurrence of mitotic cells in molting/late egg development stage individuals is currently unclear, but one possibility discussed by Czernekova and Jönsson (2016) is that it may represent a preparation for the increased body size (and a corresponding increased demand of storage cells) that normally is the result of molting. However, these studies estimated mitotic cells at one time, and showed that the total number of dividing cells per specimens was low. Gross et al. (2018) used another method (labeling with the thymidine analog EdU) to analyze cell division in *H. exemplaris* over a period of several days to estimate the rate of mitosis. They focused on the digestive system and found that the number of cells increased consistently in the midgut over 4 days, at a rate of 6–10 cell divisions per day. In contrast to Bertolani (1970a), Poprawa et al. (2015), and Czernekova and Jönsson (2016), Gross et al. (2018) did not find evidence of mitosis in any other tissues of the animals than the gut. Possible reasons for this discrepancy in results could be differences in labeling methods or in the state of the analyzed animals due to environmental conditions during the study or in the season of animal collection (Bertolani 1970a), potentially affecting the frequency of cell division. It could also be due to natural differences in patterns of mitosis among tardigrade species, since all mentioned studies used different species.

Since tardigrades are not characterized by cell constancy and mitosis has been documented in most organs, they should not be considered eutelic animals. Even if cell numbers would be found to be constant in some organs or tissues, it may well represent a secondary character resulting from renewal/replacement of cells, rather than a fixed number of cells developed during the preadult stage.

There is very little information about stem cells in tardigrades. We have found no reports of true embryonic stem cells, but primordial germ cells (PGCs) of tardigrades were described by Hejnol and Schnabel (2005) in the freshwater species *T. stephaniae*. We have found no information about adult stem cells in tardigrades, but Czernekova et al. (2018) reported two types of storage cells in female *R.* cf. *coronifer*, one of which was interpreted as young undifferentiated cells that could represent stem cells.

10.6 Maintaining Cell Integrity Under Extreme Dehydration

The environmental challenges that tardigrades and other cryptobiotic organisms are exposed to and are able to handle and survive are remarkable, and represent boundary conditions for living organisms. The ability to allow the cells to dry out completely without permanent cellular damage is one such condition. Removal of bulk water in cells and to some extent also the structural water bound to membranes leads to membrane disruption and cell death in "normal" cells. Still, studies of the ultrastructure of desiccated storage cells in tardigrades have not revealed any apparent disruption of membranes or organelles, even after desiccation for 6 months followed by heating at 50 °C for 24 h (Czernekova et al. 2017, 2018). Obviously, tardigrades and other desiccation-tolerant organisms have found evolutionary pathways to overcome the challenges of desiccation.

Since membrane damage is one of the main effects of dehydration, studies on membrane stability and factors that may prevent membrane damage have been one of the major research fields related to anhydrobiotic organisms (Crowe 2015). Early work on anhydrobiotic nematodes revealed that the disaccharide trehalose was induced to high levels (15–20% dry weight) during the dehydration process, suggesting that this sugar could be involved in protection against dehydration damage (Madin and Crowe 1975). This research quickly turned to model membrane systems with the aim to evaluate the "Water Replacement Hypothesis," suggesting that trehalose could replace the water bound to membranes and provide structural membrane integrity under dry conditions (Crowe et al. 1998b). The ability of trehalose to preserve membrane integrity in in vitro cell systems is now well documented, and proposed to relate to lowering of the gel to liquid crystalline transition temperature (T_m), allowing membranes to remain in a liquid-crystalline phase in the dry state (Crowe 2008). Several possible mechanisms behind this depression of T_m are reviewed in Crowe (2015). Trehalose also has the property to form a glassy (vitrified) state under low water levels and at natural temperatures, and this property has been proposed as another adaptive protection mechanism in desiccation-tolerant organisms (Crowe et al. 1998a). By forming a glass-like matrix in the cell, vitrification would prevent protein denaturation and aggregation, as well as fusion of membranes. Using differential scanning calorimetry (DSC), Hengherr et al. (2008) provided evidence that some species of tardigrades were in a glassy state when desiccated, but underwent glass transition when heated above 75–90 °C. This level of glass transition temperatures also correlated with the upper heat tolerance of the same species, indicating that the glassy state may have allowed survival up to these high temperatures. However, some other tardigrade species did not show any evidence of glass transition, while temperature tolerance was as high or higher compared to species entering a glassy state. Interestingly, the species that showed evidence of glass transition all belong to the tardigrade family Macrobiotidae, which is also the only taxa where induction of trehalose by desiccation has been shown.

Although tardigrades were included in the initial work on trehalose in anhydrobiotic animals (e.g., Crowe and Madin 1974; Crowe 1975), it took almost two decades until quantitative estimates of trehalose levels were reported in tardigrades (Westh and Ramløv 1991). The amount of desiccation-induced trehalose reported by Westh and Ramlöv (1991) and in later studies (Hengherr et al. 2008; Jönsson and Persson 2010) is much lower (not exceeding 3%, and in most species investigated <1%) than in nematodes, and in some tardigrade taxa, trehalose levels do respond to desiccation. Thus, although trehalose may have some protective role against dehydration damage in some tardigrade species, other compounds must clearly be involved in other species. A study providing evidence of this was reported by Boothby et al. (2017), who showed that tardigrade-specific intrinsically disordered proteins (TDPs) were induced by desiccation in the tardigrade *H. exemplaris*. They also reported evidence that the animals were in a vitrified state when dried and that the temperature for glass transition (Tg) corresponded with the upper level of heat tolerance of the tardigrade. Since the TDPs were also confirmed to vitrify in vitro, Boothby et al. (2017) proposed TDP as the glass-forming protectant molecules in this species, functionally comparable to the suggested role of trehalose in some other cryptobiotic species. Accumulating evidence indicates that combinations of synergistically interacting biomolecules, including sugars, heat-shock proteins (HSP), intrinsically disordered proteins (IDPs), including late embryogenesis abundant (LEA) proteins), and antioxidant proteins represent the adaptive molecular protection system in most, if not all, desiccation-tolerant invertebrates (Schokraie et al. 2012; Crowe 2015).

10.7 Genetic Stability Under Extreme Environmental Stress

All organisms must be able to handle damage to DNA, by damage-detection mechanisms and mechanisms of DNA repair. Organisms that are exposed to extraordinary environmental conditions with potential to damage DNA, such as desiccation, freezing, and radiation, must have corresponding cell mechanisms to prevent and/or repair damage. Understanding how tardigrades and other cryptobiotic micrometazoans handle potential DNA damage is therefore of considerable interest, not only for understanding their evolutionary adaptations, but also for possible applications to medical fields where DNA integrity is of central importance, such as cancer and aging. There are so far few studies evaluating how much damage is induced to tardigrade DNA by agents such as desiccation, freezing, or ionizing radiation. However, Neumann et al. (2009) and Rebecchi et al. (2009a) reported that desiccation of tardigrades per se did not induce significant damage to DNA. Instead, DNA fragmentation of storage cells tends to increase with increased time spent in a desiccated state (Neumann et al. 2009; Fig. 10.4). Similar results were reported by Rebecchi et al. (2009a) for desiccated tardigrades under combined high temperature and humidity conditions. This suggests that damage to DNA accumulates over time since no repair mechanisms are activated in the dry state, only at rehydration.

DNA repair mechanisms have not been studied in details in tardigrades, but a number of molecular studies have provided some information. Jönsson and Schill (2007) showed that the heat-shock protein Hsp70 was induced by heat (37 °C for 90 min) and ionizing radiation (gamma ray, 500 Gy) in *R.* cf. *coronifer*, and elevated levels were also observed in rehydrated animals after desiccation. Hsp70 is a highly conserved protein with many known functions related to genomic stability under stress (Pandita et al. 2004) and chaperone functions related to folding and repair of damaged proteins (Mayer and Bukau 2005). Schokraie et al. (2012) also showed that heat shock proteins from the Hsp70 and Hsp90 families as well as two members of the DnaJ family are present in the desiccated ("tun") state of *M. tardigradum.* Beltrán-Pardo et al. (2013) showed that the conserved Rad51 protein, known to be important in homologous recombination of damaged DNA, was highly induced by gamma radiation in *M.* cf. *tardigradum.*

Using expressed sequence tags, Förster et al. (2012) showed that both major DNA repair pathways (e.g., Rad50) and heat-shock response pathways and protective pathways (including superoxide dismutase and DNA helicase) are involved in the stress response of tardigrades. They suggest that a unique combination of resistance and repair pathways may play a role in making tardigrades so stress tolerant.

Oxidative stress is one of the major causes of cellular damage, caused by reactive oxygen species (ROS) that arise from metabolic processes or exogenous factors such as radiation (Reisz et al. 2014) and dehydration (França et al. 2007; Rebecchi 2013). Several macromolecules of cells, including proteins, lipids, nucleic acids, and sugars, may be damaged by ROS in a reaction called carbonylation, and

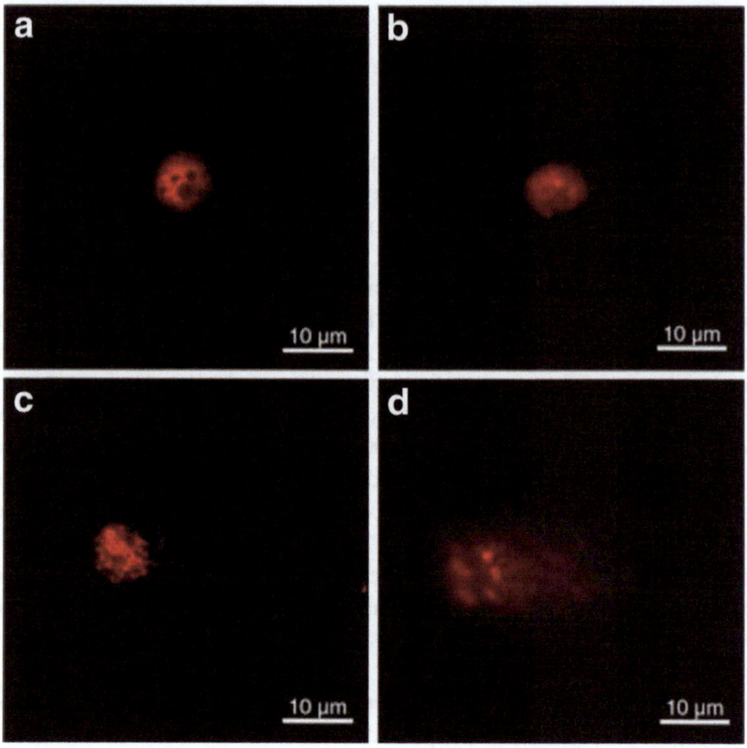

Fig. 10.4 Quantification of nuclear DNA fragmentation of storage cells from *Milnesium tardigradum* using comet assay after anhydrobiosis for different periods of time. (**a**) Negative control (untreated animals), (**b**) 2 days of anhydrobiosis, (**c**) 6 weeks of anhydrobiosis, (**d**) 10 months of anhydrobiosis. A clearly higher fragmentation is seen in the cells exposed to the longest time in a dry state. Reprinted from Neumann et al. (2009, Fig. 2) with permission from Elsevier

quantification of carbonylation is used as a biomarker of oxidative stress. In particular, carbonylation of proteins has received much attention due to its documented connection with aging and various human diseases (Dalle-Donne et al. 2003; Fedorova et al. 2014). There are few studies on carbonylation in tardigrades, but Kuzmic et al. (2018) reported increased levels of carbonylated proteins with increased dose of UVC in both active hydrated and desiccated *H. exemplaris*, and carbonylation also increased with time spent in the desiccated state (up to 73 days). The challenge of oxidative damage and carbonylation is counteracted by the antioxidant response system, and Rizzo et al. (2010) reported increased levels of the antioxidants superoxide dismutase, glutathione peroxidase, and glutathione in desiccated *Paramacrobiotus richtersi* compared to hydrated animals. Daly (2012) suggested that proteins, rather than DNA, may be the main target of ROS, and that antioxidant protection of proteins involved in repair and replication of DNA determines toxicity of radiation. Along the same line, Krisko et al. (2012) reported that the

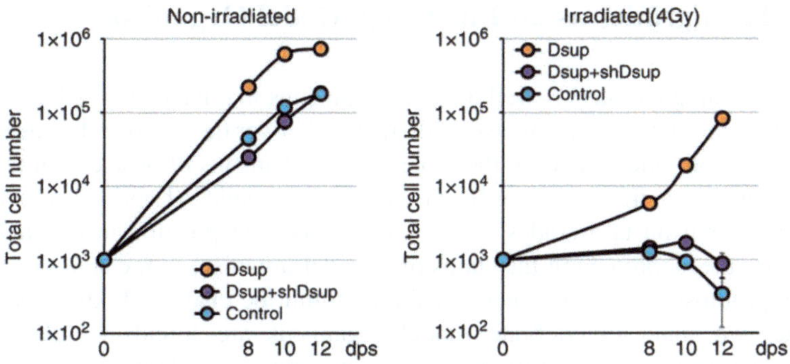

Fig. 10.5 Growth curves of human embryonic kidney cells (HEK293) in untransfected cells (Control), transfected cells expressing the tardigrade-specific Dsup gene (Dsup), and transfected cells without expression of Dsup (Dsup+shDsup). The right panel shows cell growth after irradiation with 4 Gy of X-rays, and the left panel shows nonirradiated cells. Reprinted from Hashimoto et al. (2016, Fig. 6b) under a Creative Commons Attribution-NonCommercial-ShareAlike 4.0 International License

desiccation- and radiation-tolerant rotifer *Adineta vaga* expressed lower levels of protein carbonylation in response to ionizing radiation compared to the more sensitive nematode *C. elegans*.

10.8 Application of Tardigrade Cellular Adaptations to Human Medicine

The potential of research on tardigrades and other cryptobiotic invertebrates to contribute to development of applied human fields has been recognized for a long time (Keilin 1959). This potential was also a central aspect of the research on the properties of trehalose as a cell protectant. It is, however, only recently that applications of tardigrade omics on human cells have been reported. Tanaka et al. (2015) reported that transfection of the mitochondrial heat-soluble proteins RvLEAM and MAHS of the tardigrade *R. varieornatus* improved the tolerance of human cells (HEp-2) to hyperosmotic stress. In the same species, Hashimoto et al. (2016) reported another tardigrade-unique protein, Dsup, associating with nuclear DNA. They showed that human embryonic kidney cells (HEK293) transfected with Dsup had up to 40% reduced damage to DNA from X-ray and improved viability compared to cells without Dsup (Fig. 10.5). Thus, the Dsup protein seemed to protect DNA of human cells from strand breaks, and Hashimoto and Kunieda (2017) suggested that Dsup may physically shield the DNA from damage. Also the transcriptomic study by Förster et al. (2012) identified potentials for transferring tardigrade stress responses to human cells, noting that several identified pathways in tardigrade stress response were known also from humans.

10.9 Tardigrades as an Emerging Model Organism

Research on invertebrates as well as other organisms that have evolved adaptations for cryptobiotic survival has increased dramatically over the last two decades, with a clear bias toward understanding the molecular mechanisms of desiccation tolerance (anhydrobiosis), a field that has been called *desiccomics* (Leprince and Buitink 2010). Research on tardigrades has taken an important position in this field over the last decade, approaching the status of a model organism in the study of invertebrate cryptobiosis. Its phylogenetic position within the clade of Ecdysozoa, where also the two main invertebrate model organisms are found, *C. elegans* and *D. melanogaster*, makes tardigrades an interesting group both for phylogenetic studies of the early development of metazoa and for evolutionary comparative studies of cryptobiotic taxa within Ecdysozoa. An example of the former is the evo-devo study of Hox genes in *H. exemplaris* by Smith et al. (2016). While *D. melanogaster* is a noncryptobiotic animal, *C. elegans* has been shown to be able to enter anhydrobiosis under restricted dehydration conditions in its dauer larvae form (Erkut et al. 2011). The tardigrade phylum contains a variety of species, from full featured cryptobionts surviving dehydration under relatively low humidity conditions (e.g., *M. tardigradum* and *R. varieornatus*) to more sensitive species that require preconditioning dehydration at high humidity (e.g., *H. exemplaris*, Kondo et al. 2015), and species that are not able to survive complete dehydration (e.g., *Borealibius zetlandicus*, Rebecchi et al. 2009b). This genetic and phenotypic variety provides rich opportunities to explore both the proximate (mechanisms) and the ultimate (phylogenetic) origin of cryptobiotic adaptations. Tardigrades have also been proposed as a metazoan model organism in astrobiological research (Jönsson 2007; Horikawa 2008), a position supported by being the first animal taxa surviving under space conditions (Jönsson et al. 2008).

Although recent studies on tardigrades have considered several functional aspects of tardigrade cell biology, such as various omics, ultrastructure, dynamics of storage cells, and embryological development, many basic aspects of tardigrade cell biology remain to be elucidated. So far, there are no reports of cell cultures of tardigrade cells. Establishing cell cultures would facilitate studies on many aspects of tardigrade cell biology, including the induction of cell division, the study and manipulation of gene and protein expression (silencing, over-expression and the effect of external stimuli), cell cycle kinetics as well as individual cell metabolism. The availability of tardigrade cell cultures may also facilitate studies on membrane structure and transport, as well as immunology.

References

Altiero T, Rebecchi L (2003) First evidence of achiasmatic male meiosis in the water bears *Richtersius coronifer* and *Macrobiotus richtersi* (Eutardigrada, Macrobiotidae). Hereditas 139:116–120

Arakawa K (2016) No evidence for extensive horizontal gene transfer from the draft genome of a tardigrade. Proc Natl Acad Sci USA 113(22):E3057

Arakawa K, Yoshida Y, Tomita M (2016) Genome sequencing of a single tardigrade *Hypsibius dujardini* individual. Sci Data 3:160063

Baumann H (1920) Mitteilungen zum feineren bau der tardigraden. Zool Anz 52:56–66

Beltrán-Pardo EA, Jönsson KI, Wojcik A, Haghdoost S, Bermúdez Cruz RM, Bernal Villegas JE (2013) Sequence analysis of the DNA-repair gene *rad51* in the tardigrades *Milnesium* cf. *tardigradum*, *Hypsibius dujardini* and *Macrobiotus* cf. *harmsworthi*. J Limnol 72(s1):80–91

Bertolani R (1970a) Mitosi somatiche e costanza cellular numerica nei Tardigradi. Atti Accad Naz Lincei Rc Ser 8a 48:739–742

Bertolani R (1970b) Variabilità numerica cellulare in alcuni tessuti di Tardigradi. Atti Accad Naz Lincei Rc Ser 8a 49:442–446

Bertolani R (1994) Tardigrada. In: Adiyodi KG, Adiyodi RG (eds) Reproductive biology of invertebrates. Asexual propagation and reproductive strategies, vol VI. Oxford and IBH, New Delhi, pp 25–37

Bertolani R (2001) Evolution of the reproductive mechanisms in tardigrades – a review. Zool Anz 240:247–252

Bertolani R, Rebecchi L (2018) Cytologi and cytogenetics. In: Schill RO (ed) Water bears: the biology of tardigrades. Springer, New York, pp 145–161

Boothby TC, Tapia H, Brozena AH, Piszkiewicz S, Smith AE, Giovannini I, Rebecchi L, Pielak GJ, Koshland D, Goldstein B (2017) Tardigrades use intrinsically disordered proteins to survive desiccation. Mol Cell 65(6):975–984.e5

Campbell LI, Rota-Stabelli O, Edgecombe GD, Marchioro T, Longhorn SJ, Telford MJ et al (2011) MicroRNAs and phylogenomics resolve the relationships of Tardigrada and suggest that velvet worms are the sister group of Arthropoda. Proc Natl Acad Sci USA 108(38):15920–15924

Collares-Pereira MJ, Matos I, Morgado-Santos M, Coelho MM (2013) Natural pathways towards polyploidy in animals: the *Squalius alburnoides* fish complex as a model system to study genome size and genome reorganization in polyploids. Cytogenet Genome Res 140:97–116

Crowe JH (1975) The physiology of cryptobiosis in tardigrades. Mem Ist Idrobiol 32(Suppl):37–59

Crowe JH (2008) Trehalose as a "chemical chaperone": fact and fantasy. Adv Exp Med Biol 594:143–158

Crowe JH (2015) Anhydrobiosis: an unsolved problem with applications in human welfare. In: Disalvo EA (ed) Membrane hydration: the role of water in the structure and function of biological membranes. Subcellular biochemistry 71. Springer, Basel, pp 263–280

Crowe JH, Madin KA (1974) Anhydrobiosis in tardigrades and nematodes. Trans Am Microsc Soc 93:513–524

Crowe JH, Carpenter IE, Crowe LM (1998a) The role of vitrification in anhydrobiosis. Annu Rev Physiol 60:73–103

Crowe JH, Clegg JS, Crowe LM (1998b) Anhydrobiosis: the water replacement hypothesis. In: Reid DS (ed) The properties of water in foods. Chapman & Hall, New York, pp 440–455

Cunha A, Azevedo RBR, Emmons SW, Leroi AM (1999) Variable cell number in nematodes. Nature 402(6759):253

Czernekova M, Jönsson KI (2016) Experimentally induced repeated anhydrobiosis in the eutardigrade *Richtersius coronifer*. PLoS One 11(11):e0164062

Czernekova M, Jönsson KI, Chajec L, Student S, Poprawa I (2017) The structure of the desiccated *Richtersius coronifer* (Richters, 1903). Protoplasma 254(3):1367–1377

Czernekova M, Janelt K, Student S, Jönsson KI, Poprawa I (2018) A comparative ultrastructure study of storage cells in the eutardigrade *Richtersius coronifer* in the hydrated state and after desiccation and heating stress. PLoS One 13(8):e0201430

Dalle-Donne I, Giustarini D, Colombo R, Rossi R, Milzani A (2003) Protein carbonylation in human diseases. Trends Mol Med 9:169–176

Daly MJ (2012) Death by protein damage in irradiated cells. DNA Repair 11(1):12–21

Degma P, Bertolani R, Guidetti R (2018) Actual checklist of Tardigrada species. Ver. 25: 10 05-2014, p 48. http://www.tardigrada.modena.unimo.it/miscellanea/Actual%20checklist%20of%20Tardigrada.pdf. Accessed 7 Nov 2018

Dewel RA, Nelson DR, Dewel WC (1993) Tardigrada. In: Harrison FW, Rice EM (eds) Microscopic anatomy of invertebrates, Onychophora, Chilopoda and Lesser Protostomata, vol 12. Wiley-Liss, New York, pp 143–183

Eibye-Jacobsen J (1997) New observations on the embryology of the Tardigrada. Zool Anz 235:201–216

Erkut C, Penkov S, Khesbak H, Vorkel D, Verbavatz JM, Fahmy K, Kurzchalia TV (2011) Trehalose renders the dauer larva of *Caenorhabditis elegans* resistant to extreme desiccation. Curr Biol 21:1331–1336

Fedorova M, Bollineni RC, Hoffmann R (2014) Protein carbonylation as a major hallmark of oxidative damage: update of analytical strategies. Mass Spectrom Rev 33:79–97

Förster F, Beisser D, Grohme MA, Liang C, Mali B, Siegl AM, Engelmann JC, Shkumatov AV, Schokraie E, Müller T, Schnölzer M, Schill RO, Frohme M, Dandekar T (2012) Transcriptome analysis in tardigrade species reveals specific molecular pathways for stress adaptations. Bioinform Biol Insights 6:69–96

França MB, Panek AD, Eleutherio ECA (2007) Oxidative stress and its effects during dehydration. Comp Biochem Physiol A 146:621–631

Gabriel WN, Goldstein B (2007) Segmental expression of Pax3/7 and engrailed homologs in tardigrade development. Dev Genes Evol 217:421–433

Gabriel WN, McNuff R, Patel SK, Gregory TR, Jeck WR, Jones CD, Goldstein B (2007) The tardigrade *Hypsibius dujardini*, a new model for studying the evolution of development. Dev Biol 312:545–559

Gąsiorek P, Stec D, Morek W, Michalczyk Ł (2018) An integrative redescription of *Hypsibius dujardini* (Doyère, 1840), the nominal taxon for Hypsibioidea (Tardigrada: Eutardigrada). Zootaxa 4415(1):45

Gilbert JJ (1983) Rotifera. In: Adiyodi KG, Adiyodi RG (eds) Reproductive biology of invertebrates, Oogenesis, oviposition, and oosorption, vol I. Wiley, Chichester, pp 181–209

Gross V, Bährle R, Mayer G (2018) Detection of cell proliferation in adults of the water bear *Hypsibius dujardini* (Tardigrada) via incorporation of a thymidine analog. Tissue Cell 51:77–83

Hashimoto T, Kunieda T (2017) DNA protection protein, a novel mechanism of radiation tolerance: lessons from tardigrades. Life 7(2):26

Hashimoto T, Horikawa DD, Saito Y, Kuwahara H, Kozuka-Hata H, Shin-I T, Minakuchi Y, Ohishi K, Motoyama A, Aizu T, Enomoto A, Kondo K, Tanaka S, Hara Y, Koshikawa S, Sagara H, Miura T, Yokobori S-I, Miyagawa K, Suzuki Y, Kubo T, Oyama M, Kohara Y, Fujiyama A, Arakawa K, Katayama T, Toyoda A, Kunieda T (2016) Extremotolerant tardigrade genome and improved radiotolerance of human cultured cells by tardigrade-unique protein. Nat Commun 7:12808

Hejnol A, Schnabel R (2005) The eutardigrade *Thulinia stephaniae* has an indeterminate development and the potential to regulate early blastomere ablations. Development 132:1349–1361

Hengherr S, Heyer AG, Köhler HR, Schill RO (2008) Trehalose and anhydrobiosis in tardigrades – evidence for divergence in response to dehydration. FEBS J 275:281–288

Horikawa DD (2008) The Tardigrade *Ramazzottius varieornatus* as a model animal for astrobiological studies. Biol Sci Space 22(3):93–98

Hyra M, Rost-Roszkowska MM, Student S, Włodarczyk A, Deperas M, Janelt K, Poprawa I (2016) Body cavity cells of Parachela during their active life. Zool J Linnean Soc 178(4):878–887

Jönsson KI (2007) Tardigrades as a potential model organism in space research. Astrobiology 7:757–766

Jönsson KI, Persson O (2010) Trehalose in three species of desiccation tolerant tardigrades. Open Zool J 3:1–5

Jönsson KI, Rebecchi L (2002) Experimentally induced anhydrobiosis in the tardigrade *Richtersius coronifer*: phenotypic factors affecting survival. J Exp Zool 293:578–584

Jönsson KI, Schill RO (2007) Induction of Hsp70 by desiccation, ionising radiation and heat-shock in the eutardigrade *Richtersius coronifer*. Comp Biochem Physiol B Biochem Mol Biol 146:456–460

Jönsson KI, Rabbow E, Schill RO, Harms-Ringdahl M, Rettberg P (2008) Tardigrades survive exposure to space in low earth orbit. Curr Biol 18:R729–R731

Jönsson KI, Levine EB, Wojcik A, Haghdoost S, Harms-Ringdahl M (2018) Environmental adaptations – radiation tolerance. In: Schill RO (ed) Water bears: the biology of tardigrades. Springer, New York, pp 311–330

Jørgensen A, Møbjerg N, Kristensen RM (2007) A molecular study of the tardigrade *Echiniscus testudo* (Echiniscidae) reveals low DNA sequence diversity over a large geographical area. J Limnol 66(Suppl 1):77–83

Keilin D (1959) The problem of anabiosis or latent life: history and current concept. Proc R Soc Lond B 150:149–191

Kondo K, Kubo T, Kunieda T (2015) Suggested involvement of PP1/PP2A activity and *de novo* gene expression in anhydrobiotic survival in a tardigrade, *Hypsibius dujardini*, by chemical genetic approach. PLoS One 10(12):e0144803

Koutsovoulos G, Kumar S, Laetsch DR, Stevens L, Daub J, Conlon C, Maroon H, Thomas F, Aboobaker AA, Blaxter M (2016) No evidence for extensive horizontal gene transfer in the genome of the tardigrade *Hypsibius dujardini*. Proc Natl Acad Sci USA 113(18):5053–5058

Krisko A, Leroya M, Radman M, Meselson M (2012) Extreme anti-oxidant protection against ionizing radiation in bdelloid rotifers. Proc Natl Acad Sci USA 109(7):2354–2357

Kuzmic M, Richaud M, Cuq P, Frelon S, Galas S (2018) Carbonylation accumulation of the *Hypsibius exemplaris* anhydrobiote reveals age-associated marks. PLoS One 13(12):e0208617

Leprince O, Buitink J (2010) Desiccation tolerance: from genomics to the field. Plant Sci 179:554–564

Levin DA (1983) Polyploidy and novelty in flowering plants. Am Nat 122(1):1–25

Madin KAC, Crowe J (1975) Anhydrobiosis in nematodes: carbohydrate and lipid metabolism during drying. J Exp Zool 193:335–342

Marcus E (1929) Tardigrada. In: Bronns HG (ed) Klassen und Ordungen des Tierreichs, vol 5, Section 4, Part 3. Akademische Verlagsgesellschaft, Leipzig, pp 1–608

Martini E (1923) Die zellkonstanz und ihre beziehungen zu anderen zoologischen vorwürfen. Z Anat Entwicklungsgesch 70(1–3):179–259

Mayer MP, Bukau B (2005) Hsp70 chaperones: cellular functions and molecular mechanism. Cell Mol Life Sci 62(6):670–684

Møbjerg N, Halberg KA, Jørgensen A, Persson D, Bjørn M, Ramløv H, Kristensen RM (2011) Survival in extreme environments – on the current knowledge of adaptations in tardigrades. Acta Physiol 202:409–420

Nakhleh J, El Moussawi L, Osta MA (2017) The melanization response in insect immunity. Adv Insect Physiol 52:83–109

Neiman M, Beaton MJ, Hessen DO, Jeyasingh PD, Weider LJ (2017) Endopolyploidy as a potential driver of animal ecology and evolution. Biol Rev 92(1):234–247

Nelson DR, Guidetti R, Rebecchi L (2015) Phylum Tardigrada. In: Thorp J, Rogers DC (eds) Ecology and general biology: Thorp and Covich's freshwater invertebrates. Academic, London, pp 347–380

Neumann S, Reuner A, Brümmer F, Schill RO (2009) DNA damage in storage cells of anhydrobiotic tardigrades. Comp Biochem Physiol A Mol Integr Physiol 153:425–429

Pandita TK, Higashikubo R, Hunt CR (2004) HSP70 and genomic stability. Cell Cycle 3 (5):591–592

Parfrey LW, Lahr DJG, Katz LA (2008) The dynamic nature of eukaryotic genomes. Mol Biol Evol 25(4):787–794

Poprawa I, Hyra M, Rost-Roszkowska MM (2015) Germ cell cluster organization and oogenesis in the tardigrade *Dactylobiotus parthenogeneticus* Bertolani, 1982 (Eutardigrada, Murrayidae). Protoplasma 252:1019–1029

Rebecchi L (2013) Dry up and survive: the role of antioxidant defences in anhydrobiotic organisms. J Limnol 72(s1):62–72

Rebecchi L, Rossi V, Altiero T, Frigieri A, Bertolani R, Menozzi P (2003) Reproductive modes and genetic polymorphism in the tardigrade *Richtersius coronifer* (Eutardigrada, Macrobiotidae). Invertebr Biol 122(1):19–27

Rebecchi L, Cesari M, Altiero T, Frigieri A, Guidetti R (2009a) Survival and DNA degradation in anhydrobiotic tardigrades. J Exp Biol 212:4033–4039

Rebecchi L, Boschini D, Cesari M, Lencioni V, Bertolani R, Guidetti R (2009b) Stress response of a boreo-alpine species of tardigrade, *Borealibius zetlandicus* (Eutardigrada, Hypsibiidae). J Limnol 68(1):64–70

Reisz JA, Bansal N, Qian J, Zhao W, Furdui CM (2014) Effects of ionizing radiation on biological molecules – mechanisms of damage and emerging methods of detection. Antioxid Redox Signal 21(2):260–291

Reuner A, Hengherr S, Brümmer F, Schill RO (2010) Comparative studies on storage cells in tardigrades during starvation and anhydrobiosis. Curr Zool 56(2):259–263

Rizzo AM, Negroni M, Altiero T, Montorfano G, Corsetto P, Berselli P, Berra B, Guidetti R, Rebecchi L (2010) Antioxidant defences in hydrated and desiccated states of the tardigrade *Paramacrobiotus richtersi*. Comp Biochem Physiol B Biochem Mol Biol 156:115–121

Rosati F (1968) Ricerche di microscopia elettronica sui Tardigradi, 2. I globuli cavitari. Atti Accad Fisiocritici, Siena 17:1439–1452

Schill RO, Jönsson KI, Brümmer F, Pfannkuchen M (2011) Food of tardigrades: a case study to understand food choice, intake and digestion. J Zool Syst Evol Res 49(Suppl 1):66–70

Schoenfelder KP, Fox DT (2015) The expanding implications of polyploidy. J Cell Biol 209:485–491

Schokraie E, Warnken U, Hotz-Wagenblatt A, Grohme MA, Hengherr S, Förster F, Schill RO, Frohme M, Dandekar T, Schnölzer M (2012) Comparative proteome analysis of *Milnesium tardigradum* in early embryonic state *versus* adults in active and anhydrobiotic state. PLoS One 7:e45682

Smith FW, Boothby TC, Giovannini I, Rebecchi L, Jockusch EL, Goldstein B (2016) The compact body plan of tardigrades evolved by the loss of a large body region. Curr Biol 26:224–229

Sulston JE, Schierenberg E, White JG, Thomson JN (1983) The embryonic cell lineage of the nematode *Caenorhabditis elegans*. Dev Biol 100(1):64–119

Szymańska B (1994) Interdependence between storage bodies and egg developmental stages in *Macrobiotus richtersi* Murray, 1911 (Tardigrada). Acta Biol Cracov 36:41–48

Tanaka S, Tanaka J, Miwa Y, Horikawa DD, Katayama T, Arakawa K, Toyoda A, Kubo T, Kunieda T (2015) Novel mitochondria-targeted heat-soluble proteins identified in the anhydrobiotic tardigrade improve osmotic tolerance of human cells. PLoS One 10(2):e0118272

Tsujimoto M, Imura S, Kanda H (2016) Recovery and reproduction of an Antarctic tardigrade retrieved from a moss sample frozen for over 30 years. Cryobiology 72:78–81

Van Cleave HJ (1932) Eutely or cell constancy in its relation to body size. Q Rev Biol 7(1):59–67

Volkmann A, Greven H (1993) Ultrastructural localization of tyrosinase in the tardigrade cuticle. Tissue Cell 25:435–438

von Erlanger R (1895) Beiträge zur Morphologie der Tardigraden: I. Zur Embryologie eines Tardigraden: *Macrobiotus macronyx* Dujardin. Morph Jb 22:491–513

von Wenck W (1914) Entwicklungsgeschichtliche Untersuchungen an Tardigraden (*Macrobiotus lacustris* Duj.). Zool Jb Anat 37:465–514

Walsh EJ, Zhang L (1992) Polyploidy and body size variation in a natural population of the rotifer *Euchlanis dilatata*. J Evol Biol 5:345–353

Wang C, Grohme MA, Mali B, Schill RO, Frohme M (2014) Towards decrypting cryptobiosis – analyzing anhydrobiosis in the tardigrade *Milnesium tardigradum* using transcriptome sequencing. PLoS One 9(3):e92663

Węglarska B (1975) Studies on the morphology of *Macrobiotus richtersi* Murray, 1911. Mem Ist Ital Idrobiol 32:445–464

Westh P, Ramløv H (1991) Trehalose accumulation in the tardigrade *Adorybiotus coronifer* during anhydrobiosis. J Exp Zool 258:303–311

Wright JC (2001) Cryptobiosis 300 years from van Leuwenhoek: what have we learned about tardigrades? Zool Anz 240:563–582

Yoshida Y, Koutsovoulos G, Laetsch DR, Stevens L, Kumar S, Horikawa DD, Ishino K, Komine S, Kunieda T, Tomita M, Blaxter M, Arakawa K (2017) Comparative genomics of the tardigrades *Hypsibius dujardini* and *Ramazzottius varieornatus*. PLoS Biol 15(7):e2002266

[illegible]

Chapter 11
Development of Xenoturbellida

Hiroaki Nakano

Abstract Xenoturbellida is a group of benthic marine invertebrates with a very simple bilaterian body plan. Together with Acoelomorpha, it constitutes the clade Xenacoelomorpha, recently interpreted as a sister group to all other extant bilaterians (=*Nephrozoa*). Therefore, it occupies an important phylogenetic position for studies on origins and evolution of *Bilateria*. However, due to a number of reasons, developmental and reproductive studies on *Xenoturbella* have been scarce. In this chapter, I will summarize what is known concerning the reproduction and development of Xenoturbellida, including information from five species that were recently discovered. Gonads are absent in xenoturbellids, and gametes are found in various parts of the body. The gametes are released through ruptures of the body wall, and therefore, the fertilization is presumably external. After holoblastic radial cleavage, embryos hatch as free-swimming hatchlings with uniform ciliation over the surface. Apical tuft forms later in development, and 5 days after hatching, the larvae begin to settle on the substrate. Comparison with the other marine invertebrate larvae suggests that the morphologically simple swimming larvae described in *Xenoturbella* represent an ancestral larval type of all metazoans and bilaterians.

11.1 Introduction

Xenoturbella is a benthic marine worm with a simple body plan (Nakano et al. 2017; Rouse et al. 2016; Westblad 1949). Two furrows are morphologically recognizable on the body surface: the ring furrow runs across the mid-region of the body (Fig. 11.1), and the side furrow runs down along lateral sides from the anterior end. It has been suggested that these furrows have a sensory function (Martín-Durán et al. 2018; Westblad 1949). The mouth opens at the midventral position, anteriorly to the ring furrow. A frontal pore is present below the side furrow at the anterior end. This pore leads into the ventral glandular network (Nakano et al. 2017; Rouse et al.

H. Nakano (✉)
Shimoda Marine Research Center, University of Tsukuba, Shimoda, Japan
e-mail: h.nakano@shimoda.tsukuba.ac.jp

W. Tworzydlo, S. M. Bilinski (eds.), *Evo-Devo: Non-model Species in Cell and Developmental Biology*, Results and Problems in Cell Differentiation 68,
https://doi.org/10.1007/978-3-030-23459-1_11

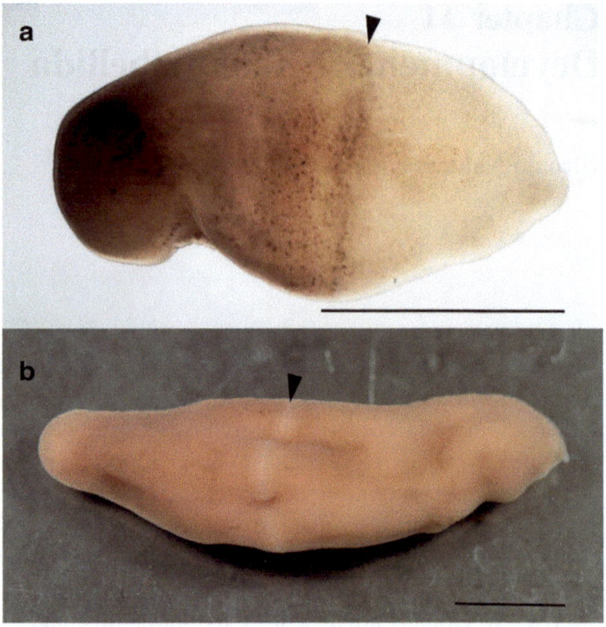

Fig. 11.1 Xenoturbellids. (**a**) *Xenoturbella bocki*. (**b**) *Xenoturbella japonica*. Both animals with anterior to the left. Arrowhead: ring furrow. Bars: 1 cm

2016). The network is a branching structure that reaches the posterior tip in some species. It cannot be observed externally in *X. bocki*, but its presence has been confirmed with micro-CT imaging (Nakano et al. 2017).

Internal observations reveal that the animal has a double bag-like structure (Nakano et al. 2017; Rouse et al. 2016; Westblad 1949). The outer epidermal layer is composed of epidermal cells, intraepidermal nerve network, basal lamina, and muscle cells. Along the ring and side furrows and near the statocyst, concentrations of the nerve cells are present within the nerve network, but centralized nervous systems, such as a brain, are lacking (Martín-Durán et al. 2018; Raikova et al. 2000; Stach et al. 2005; Westblad 1949). A statocyst, regarded as a georeceptor, is present near the anterior tip. The inner layer constitutes the intestine surrounding the digestive cavity. It has only one opening, the mouth, since the anus is not present in xenoturbellids.

There are six known species of xenoturbellids. The first reported species (therefore, the type species of the genus *Xenoturbella*) is *Xenoturbella bocki*, with a body length no larger than 4 cm (Westblad 1949). It can be collected regularly on the west coast of Sweden at a depth of 50–200 m, but fluctuations in its populations have been known (Nakano 2015). It has also been reported from other parts of Northeast Atlantic, such as off the coasts of Norway and the UK. Reported in 1949, it remained the only known xenoturbellid species for over 65 years. In 2016, four species were reported from the Eastern Pacific near the coasts of the USA and Mexico at a depth of 630–3700 m (Rouse et al. 2016). The discovery of *X. profunda*, *X. monstrosa*, *X. churro*, and *X. hollandorum* revealed more diversity within the genus. Some species are much larger than *X. bocki*; e.g., *X. monstrosa* exceeds 20 cm and

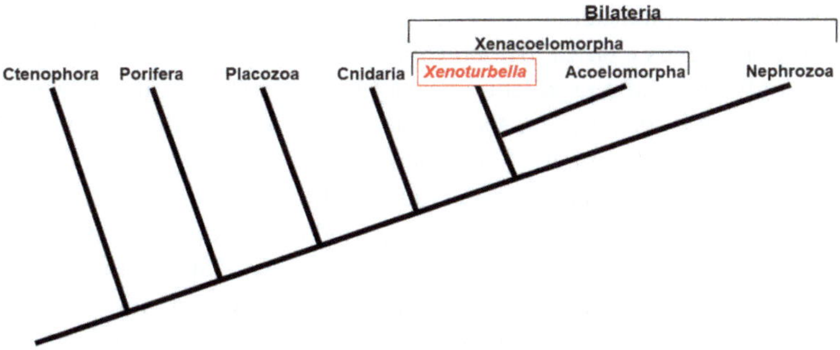

Fig. 11.2 Phylogenetic position of *Xenacoelomorpha*

X. profunda reaches 15 cm. The ventral glandular network was first reported from studies on these four species. The sixth species, *X. japonica*, was reported from the coast of Japan, western Pacific, at a depth of 380–560 m (Nakano et al. 2017, 2018). The frontal pore was first reported from this species and also found in *X. bocki*. Molecular phylogenetic analyses suggest that the six species can be divided into two groups (Nakano et al. 2017; Rouse et al. 2016). Species in the first group, *X. bocki*, *X. hollandorum*, and *X. japonica*, have body length no larger than 6 cm and can be collected at depth shallower than 640 m. The second group consists of *X. profunda*, *X. monstrosa*, and *X. churro*, with body length exceeding 10 cm when fully grown and inhabiting seafloors deeper than 1700 m.

Recent phylogenomic studies support a sister group relationship between Xenoturbellida and Acoelomorpha, and this clade, termed *Xenacoelomorpha*, has been suggested to be either a deuterostome taxon (Philippe et al. 2011; Robertson et al. 2017) or a sister group to all other extant bilaterians, the *Nephrozoa* (Hejnol et al. 2009; Cannon et al. 2016), with the latter having stronger support (Fig. 11.2) (reviewed in Ruiz-Trillo and Paps 2016). Acoelomorphans, comprising acoels and nemertodermatids, are marine benthic worms with a simple body plan similar to that of *Xenoturbella* (Achatz et al. 2013; Hauszprunar 2016; Jondelius et al. 2011). Some important differences between the aforementioned sister taxa are as follows:

1. The digestive organ of some acoelomorphans is syncytial, lacking epithelial organization as found in Xenoturbellida.
2. Centralized nervous systems have been reported in some acoelomorph species.
3. Symbiotic algae are sometimes present in acoelomorph bodies.

11.2 Asexual Reproduction

Remarkable regenerative ability and asexual reproduction by fission or budding have been reported in acoelomorphs (reviewed in Gehrke and Srivastava 2016; Achatz et al. 2013). Moreover, a wide diversity of asexual reproduction modes, such as transverse fission, longitudinal fission, and reversed polarity budding, has been described in the genus *Convolutriloba* (Sikes and Bely 2008).

Surprisingly, asexual reproduction has not been experimentally proven in any of the six xenoturbellid species. This fact might be related to the difficulties in obtaining and keeping sufficient number of adult animals for a relatively long period under laboratory conditions needed for observations of asexual reproduction. Long-term survival in laboratory tanks has only been reported for *X. bocki*, and rupture into two pieces has occasionally been observed in this species (Nakano 2015). However, the pieces failed to regenerate into two complete animals with ring and side furrows and a statocyst. Although future improvements in culture methods for *X. bocki* and studies on other species may reveal fission or budding, it is well possible that xenoturbellids are not capable of asexual reproduction.

11.3 Brief Overview of Sexual Reproduction and Development of Acoelomorpha

In comparison to the six described species of xenoturbellids, about 400 species of acoelomorphs are known, with many of them living in shallow coastal areas. Therefore, much information on sexual reproduction and development is available for these animals (reviewed in Achatz et al. 2013; Hauszprunar 2016; Hejnol 2015; Hejnol and Pang 2016). I would first like to describe the reproduction and development of Acoelomorpha before going into those of *Xenoturbella*.

Some species of acoels occur in dense swarms at beaches and can easily be collected during the breeding season. Moreover, the development of zygotes can be artificially induced in laboratory kept cultures of other species, making mature acoelomorph specimens much more accessible than those of xenoturbellids. Acoels are hermaphrodites, and the fertilization occurs internally through copulation. Acoel sperm cells have a peculiar structure with two flagella incorporated into the cell body. Eggs are laid through the female gonopore, the mouth, or rupture in the body wall. Development proceeds through a unique mode of cleavage called a spiral duet cleavage, somewhat similar to the quartet spiral cleavage of protostomes. Gastrulation, or the segregation of cells that mainly form internal and external structures, occurs by delamination or migration of cells inward. Subsequently, the embryos are structured with the epidermal cell layer surrounding a cavity filled with undifferentiated cells that later form the reproductive organs, digestive syncytium, muscles, and nerve cells. After hatching, hatchlings of many species glide on substrate, whereas those of others swim in the water column before settling. Various organs are gradually formed prior to and after hatching, with the timing depending on the species.

11.4 Sexual Reproduction and Development of Xenoturbellida

The breeding season of xenoturbellids is only known for the *X. bocki* population on the west coast of Sweden. It takes place during winter, as mature gametes are only observed during this season (Nakano 2015; Westblad 1949). A species of Orthonectida parasitizes individuals of this population (Nakano and Miyazawa 2019), and some species of the parasite are known to have a negative effect on the reproduction of its host (Deheyn et al. 1998). If this is also the case for *X. bocki* and its parasitic orthonectid, it could be an obstacle for developmental studies of this *X. bocki* population.

X. bocki has been reported to be a hermaphrodite (Westblad 1949), while *X. profunda* was described as gonochoric (Rouse et al. 2016). Gametes can be found distributed in various parts of the body: inside the digestive cavity, within the cells of the intestine, on the outer surface of the intestine, between the intestine and the epidermal layer, and within the epidermal layer without any differences between the anterior and posterior or the ventral and the dorsal sides of the body (Nakano et al. 2017; Rouse et al. 2016; Westblad 1949). In some *X. bocki* specimens, the gametes are more numerous on the outer surface of the intestine, suggesting that germ cells are originally situated at this site (Nakano 2015).

Oocytes have been reported from five of the six known *Xenoturbella* species (Nakano et al. 2017; Rouse et al. 2016; Westblad 1949), but mature egg has been described only in *X. bocki* (Nakano et al. 2013; Nakano 2015; Westblad 1949). Eggs are opaque, pale orange, and surrounded by a transparent layer and measure about 205 μm in diameter (Nakano et al. 2013). Oocytes of *X. profunda* reach 450 μm (Rouse et al. 2016), suggesting that their mature eggs can be even larger. Mature sperm cells have been described in *X. bocki* and *X. profunda* (Obst et al. 2011; Rouse et al. 2016; Westblad 1949). They have spherical heads (about 1.5 μm in diameter) with mitochondria basally incorporated. The tail flagellum is about 42 μm long. The axoneme has a typical 9 + 2 microtubule doublet arrangement (Nakano et al. 2013; Nakano 2015; Rouse et al. 2016; Westblad 1949). As a rule, the gametes are released through the rupture of the body wall (Nakano et al. 2013); however, it cannot be excluded that gametes present inside the digestive cavity are released through the mouth.

Cleavage of *X. bocki* has been interpreted as holoblastic and radial (Nakano 2015). Gastrulation has not been described. However, based on the morphological similarities between embryos of acoelomorphs and xenoturbellids around the time of hatching, it is likely that gastrulation in *Xenoturbella* shares similarities with that of acoelomorphs.

Before hatching, uniform ciliation arises on the surface of the *X. bocki* embryos (Nakano et al. 2013). In consequence, they begin to rotate inside the fertilization envelope using the cilia. At this stage, the embryonic cells differentiate into two large groups: the outside epidermal cell layer and a cluster of round undifferentiated cells present inside the embryo.

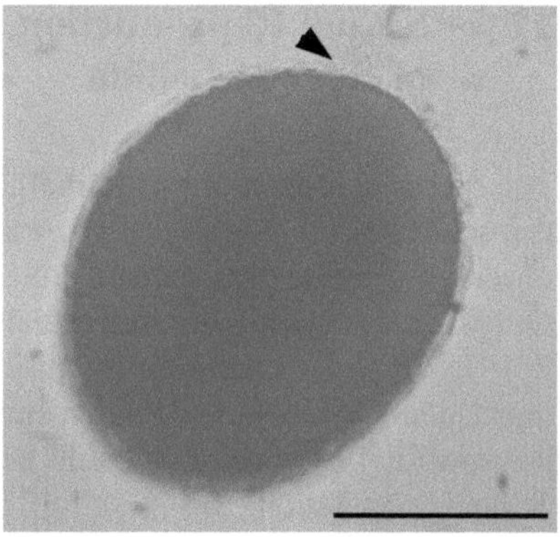

Fig. 11.3 *X. bocki* free-swimming stage embryo. Three days after hatching. Arrowhead: apical tuft. Bar: 100 μm

After hatching, the hatchlings swim inside the water column using their cilia and rotate clockwise when seen from the posterior. Three days after hatching, an apical tuft, a group of cilia longer than those found on other parts of the embryo, arises at the anterior tip of the hatchling (Fig. 11.3). Interestingly, hatchlings are able to change the direction of swimming when the apical tuft comes in contact with an obstacle (e.g., walls of dishes). This observation suggests that the nervous system and coordination of swimming behavior are already developed. At this stage, the cells of the epidermal layer start to differentiate into epidermal cells, the nerve cells, and the developing muscle cells. In contrast, the cells present inside the embryo do not differentiate, and consequently the digestive system is not yet developed. No mouth, vestibule, anus, ciliary bands, larval arms/ridges, or a coelom has been observed during the swimming stage, i.e., prior to settlement. Five days after hatching, xenoturbellids begin to glide on the bottom using their cilia. They become capable of constricting and elongating using their internal musculature.

Although our present knowledge on xenoturbellid reproduction and development is still incomplete, the accumulated data indicate that the embryogenesis of xenoturbellids and acoelomorphs differs. The differences are as follows:

1. The sperm morphology: the sperm cells of acoelomorphs show derived features.
2. Xenoturbellids lack male or female copulatory organs.
3. Most acoelomorphs are hermaphrodites, whereas *X. profunda* has been reported to be gonochoric.
4. Mode of fertilization (internal versus external).
5. Cleavage pattern (spiral duet cleavage versus holoblastic radial).

Further studies on xenoturbellids, especially on species other than *X. bocki*, are necessary for clarifying the condition (ancestral or derived) of above "developmental" characters for *Xenacoelomorpha*.

Interestingly, Acoelomorpha and *X. bocki* share one important character, i.e., the absence of a free-swimming feeding larval stage. Even in species possessing a free-swimming stage, the larvae have very simple morphology and consist of differentiated epidermal cells and a cavity filled with spherical undifferentiated cells. Species devoid of a free-swimming stage also pass through a similar stage prior to or just after hatching. Moreover, no larval mouth, vestibule, anus, or intestine has been observed during development of studied xenacoelomorphs. Based on the similarities between the larva of *Xenacoelomorpha* and the larvae of cnidarians (planula) as well as sponges (parenchymella and amphiblastula), it might be suggested that this type of morphologically simple larva (external epidermal layer filled with undifferentiated cells) represents the ancestral larval type of all metazoans and also bilaterians.

11.5 Conclusions

Xenoturbella occupies a crucial phylogenetic position as a sister group to all other extant bilaterians, and studies on its development shall yield new insights into the origins and evolution of metazoan development. The latest species to be discovered, *X. japonica*, can be collected rather easily at 380–560 m depth using marine biological dredges or benthic sleds (Nakano et al. 2017). Further research on this species, especially discoveries of populations in which regular collections are possible, promises to be invaluable for developmental studies of xenoturbellids.

References

Achatz JG, Chiodin M, Salvenmoser W, Tyler S, Martinez P (2013) The Acoela: on their kind and kinships, especially with nemertodermatids and xenoturbellids (Bilateria incertae sedis). Org Divers Evol 13:267–286

Cannon JT, Vellutini BC, Smith J III, Ronquist F, Jondelius U, Hejnol A (2016) Xenacoelomorpha is the sister group to Nephrozoa. Nature 530:89–93

Deheyn D, Watson NA, Jangoux M (1998) Symbioses in *Amphipholis squamata* (Echinodermata, Ophiuroidea, Amphiuridae): geographical variation of infestation and effect of symbionts on the host's light production. Int J Parasitol 28:1413–1424

Gehrke AR, Srivastava M (2016) Neoblasts and the evolution of whole-body regeneration. Curr Opin Genet Dev 40:131–137

Hauszprunar G (2016) Review of data for a morphological look on Xenacoelomorpha (Bilateria incertae sedis). Org Divers Evol 16:363–389

Hejnol A (2015) Acoelomorpha and Xenoturbellida. In: Wanninger A (ed) Evolutionary developmental biology of invertebrates, vol 1. Springer, Wien, pp 203–214

Hejnol A, Pang K (2016) Xenacoelomorpha's significance for understanding bilaterian evolution. Curr Opin Genet Dev 39:48–54

Hejnol A, Obst M, Stamatakis A, Ott M, Rouse GW, Edgecombe GD, Martinez P, Baguna J, Bailly X, Jondelius U, Wiens M, Muller WEG, Seaver E, Wheeler WC, Martindale MQ, Giribet G, Dunn CW (2009) Assessing the root of bilaterian animals with scalable phylogenomic methods. Proc R Soc Biol Sci B 276:4261–4270

Jondelius U, Wallberg A, Hooge M, Raikova OI (2011) How the worm got its pharynx: phylogeny, classification and Bayesian assessment of character evolution in Acoela. Syst Biol 60:845–871

Martín-Durán JM, Pang K, Børve A, Lê HS, Furu A, Cannon JT, Jondelius U, Hejnol A (2018) Convergent evolution of bilaterian nerve cords. Nature 553:45–50

Nakano H (2015) What is *Xenoturbella*? Zool Lett 1:22
Nakano H, Miyazawa H (2019) A new species of Orthonectida that parasitizes *Xenoturbella bocki*: implications for studies on *Xenoturbella*. Biol Bull 236:66–73
Nakano H, Lundin K, Bourlat SJ, Telford MJ, Funch P, Nyengaard JR, Obst M, Thorndyke MC (2013) *Xenoturbella bocki* exhibits direct development with similarities to Acoelomorpha. Nat Commun 4:1537
Nakano H, Miyazawa H, Maeno A, Shiroishi T, Kakui K, Koyanagi R, Kanda M, Satoh N, Omori A, Kohtsuka H (2017) A new species of *Xenoturbella* from the western Pacific Ocean and the evolution of *Xenoturbella*. BMC Evol Biol 17:245
Nakano H, Miyazawa H, Maeno A, Shiroishi T, Kakui K, Koyanagi R, Kanda M, Satoh N, Omori A, Kohtsuka H (2018) Correction to: a new species of *Xenoturbella* from the western Pacific Ocean and the evolution of *Xenoturbella*. BMC Evol Biol 18:83
Obst M, Nakano H, Bourlat SJ, Thorndyke MC, Telford MJ, Nyengaard JR, Funch P (2011) Spermatozoon ultrastructure of *Xenoturbella bocki* (Westblad 1949). Acta Zool 92:109–115
Philippe H, Brinkmann H, Copley RR, Moroz LL, Nakano H, Poustka AJ, Wallberg A, Peterson KJ, Telford MJ (2011) Acoelomorph flatworms are deuterostomes related to *Xenoturbella*. Nature 470:255–258
Raikova O, Reuter M, Jondelius U, Gustafsson M (2000) An immunocytochemical and ultrastructural study of the nervous and muscular systems of *Xenoturbella westbladi* (Bilateria inc. sed.). Zoomorphology 120:107–118
Robertson HE, Lapraz F, Egger B, Telford MJ, Schiffer PH (2017) The mitochondrial genomes of the acoelomorph worms *Paratomella rubra*, *Isodiametra pulchra* and *Archaphanostoma ylvae*. Sci Rep 7:1847
Rouse GW, Wilson NG, Carvajal JI, Vrijenhoek RC (2016) New deep-sea species of *Xenoturbella* and the position of Xenacoelomorpha. Nature 530:94–97
Ruiz-Trillo I, Paps J (2016) Acoelomorpha; earliest branching bilaterians or deuterostomes? Org Divers Evol 16:391–399
Sikes JM, Bely AE (2008) Radical modification of the A-P axis and the evolution of asexual reproduction in *Convolutriloba* acoels. Evol Dev 10:619–631
Stach T, Dupont S, Israelsson O, Fauville G, Nakano H, Kånneby T, Thorndyke M (2005) Nerve cells of *Xenoturbella bocki* (phylum uncertain) and *Harrimania kupfferi* (Enteropneusta) are positively immunoreactive to antibodies raised against echinoderm neuropeptides. J Mar Biol Assoc UK 85:1519–1524
Westblad E (1949) *Xenoturbella bocki* n.g, n.sp. a peculiar, primitive turbellarian type. Ark Zool 1:3–29

Chapter 12
Cellular and Molecular Mechanisms of *Hydra* Regeneration

Puli Chandramouli Reddy, Akhila Gungi, and Manu Unni

Abstract Regeneration of lost body parts is essential to regain the fitness of the organism for successful living. In the animal kingdom, organisms from different clades exhibit varied regeneration abilities. *Hydra* is one of the few organisms that possess tremendous regeneration potential, capable of regenerating complete organism from small tissue fragments or even from dissociated cells. This peculiar property has made this genus one of the most invaluable model organisms for understanding the process of regeneration. Multiple studies in *Hydra* led to the current understanding of gross morphological changes, basic cellular dynamics, and the role of molecular signalling such as the Wnt signalling pathway. However, cell-to-cell communication by cell adhesion, role of extracellular components such as extracellular matrix (ECM), and nature of cell types that contribute to the regeneration process need to be explored in depth. Additionally, roles of developmental signalling pathways need to be elucidated to enable more comprehensive understanding of regeneration in *Hydra*. Further research on cross communication among extracellular, cellular, and molecular signalling in *Hydra* will advance the field of regeneration biology. Here, we present a review of the existing literature on *Hydra* regeneration biology and outline the future perspectives.

12.1 Introduction

Replacement of lost body parts or tissues is an extraordinary phenomenon. The demonstration of the ability to regenerate missing parts after amputation in an animal by Trembley is the earliest record of this phenomenon. He called this animal *Hydra*, after a Greek mythological serpent which can regenerate multiple heads upon decapitation. This initiated a great interest in search for animals with regeneration abilities by biologists. Regeneration of damaged or diseased tissue or body part is

P. C. Reddy (✉) · A. Gungi · M. Unni
Department of Biology, Indian Institute of Science Education and Research, Pune, Maharashtra, India

W. Tworzydlo, S. M. Bilinski (eds.), *Evo-Devo: Non-model Species in Cell and Developmental Biology*, Results and Problems in Cell Differentiation 68,
https://doi.org/10.1007/978-3-030-23459-1_12

important for successful survival of the organism. Therefore, it has fascinated the field of biomedical science for its potential applications and has led to the emergence of regenerative medicine.

Occurrence of regeneration process varies in different animal phyla (Table 12.1). In certain animals, ability of regenerating the whole organism is observed, and in many others it is restricted to specific organs or tissue types. Whole body regeneration ability is evident from the earliest divergent multicellular organisms such as sponges (Ayling 1983; Borisenko et al. 2015). Members of the phylum Cnidaria, which are early divergent animals with defined body column and cell types similar to bilaterians, also exhibit regeneration ability (Schmid and Tardent 1971; Bossert et al. 2013). Especially the freshwater polyp *Hydra* exhibits tremendous regeneration capacity, and it is the earliest record of animal regeneration (Trembley 1744). In *Bilateria*, members of basal phylum Platyhelminthes which belong to free living flatworm groups possess the whole organismal regeneration ability (Egger et al. 2007; Ritter and Congdon 1900; Stevens and Boring 1905; Monti 1900; Graff 1882; Dalyell 1814; Ruhl 1927; Spallanzani 1769). Until now there is no evidence of whole body regeneration in Ecdysozoa (Arthropoda and Nematoda) (Bely and Nyberg 2010). However, regeneration of limb has been well studied in Arthropoda, and this ability varies across the taxa with good regeneration abilities observed in crustaceans (Bohn 1970; Minelli et al. 2013). In annelids, regeneration of anterior and posterior parts has been shown with varying capacity while the regeneration potential is completely absent in some animals like leeches (Bely 2006; Hyman 1940). In Mollusca, siphonophores of bivalves (Meyer and Byers 2005) and arms of cephalopods (Tressler et al. 2014) can regenerate, and this ability is not conserved to the same extent in all the members of the phylum. Echinoderms are interesting invertebrates which can regenerate most complex organs/body parts, and this ability is well conserved in the phylum (Cuénot 1948; Goss 1969; Hyman 1955; Carnevali 2006).

Chordates show different degrees of regeneration potential across the phylum. Among these, urochordates, especially ascidians such as *Botrylloides violaceus* (complete body) (Berrill 1951; Brown et al. 2009) and *Ciona intestinalis* (partial body) (Dahlberg et al. 2009; Auger et al. 2010), display considerable regeneration capacities. Within the Pisces, selected members were reported to exhibit regeneration (Unguez 2013; Goss 1969; Broussonet 1786), and recent studies in zebrafish have demonstrated the ability to regenerate different organs and fins (Johnson and Weston 1995). In Amphibia, salamanders such as newt and axolotl can regenerate lost limb (a relatively a complex structure) and this property is very unique and not reported in any other vertebrates (Spallanzani 1769; Goss 1969). Noticeable regeneration power has been demonstrated in lizard tail (Reptilia), and they are the closest animals to mammals that exhibit significant replacement potential of complex tissue (structure) (Lozito and Tuan 2017; Simpson 1964; Bellairs and Bryant 1985; Bryant and Bellairs 1967; Etheridge 1967). Aves (Stone and Rubel 2000; Cotanche 1987) and mammals (Iismaa et al. 2018) possess very poor regeneration abilities and can regenerate relatively simple tissue types in their adult life. They cannot regenerate more complex structures such as appendages. An interesting exception is found in many species of deer which can regenerate their antlers (Goss 1983).

Table 12.1 Animals with diverse ability to regenerate complex tissues across the animal phyla

	Sub-clade	Animal	Whole organism	Body part/structure
Non-chordates	Homoscleromorpha	*Oscarella lobularis*	Yes	
Porifera	Demospongiae	*Halisarca dujardini*	Yes	
	Calcaria	*Sycon ciliatum*	Yes	
		Leucosolenia complicata	Yes	
Ctenophora	Tentaculates	*Mnemiopsis leidyi*	Yes	
		Vallicula multiformis	Yes	
Cnidaria	Hydrozoa	*Hydra*	Yes	
		Clytia hemisphaerica	No	Partial medusa
	Anthozoa	*Nematostella*	Yes	
	Scyphozoa	*Aurelia aurita*	Only juvenile medusa	
	Cubozoa	*Carybdea marsupialis*	Polyp stage	
Platyhelminthes	Trematoda	*Macrostomum lignano*	Yes	
		Monocelis	Yes	
		Schmidtea mediterranea	Yes	
		Dugesia japonica	Yes	
		Girardia tigrina	Yes	
Arthropoda				
Crustacea	Decapoda	*Menippe mercenaria*	No	Chela
Chelicerata	Arachnida	*Latrodectus variolus*	No	Limb
Hexapoda	Insecta	*Gryllus bimaculatus*	No	Limb
Hexapoda	Insecta	*Tribolium castaneum*	No	Limb
Hexapoda	Insecta	*Drosophila*	No	Wing imaginal discs of larva
Annelida	Oligochaeta	*Pristina leidyi*	Yes	
		Lumbriculus sp.	Yes	
	Polychaeta	*Chaetopterus variopedatus*	Yes	
		Myxicola aesthetica	Yes	
Mollusca	Cephalopoda	*Sepia officinalis*	No	Limb
		Loligo pealei	No	Limb
		Argonauta	No	Limb
		Octopus sp.	No	Limb
	Bivalvia	*Mesodesma mactroides*	No	Inhalant siphon

(continued)

Table 12.1 (continued)

	Sub-clade	Animal	Whole organism	Body part/structure
Echinodermata	Asteroidea	*Leptasterias hexactis*	Yes	
		Asterias rubens	Yes	
		Coscinasterias muricata	Yes	
		Asterina gibbosa	No	Arm
	Ophiuroidea	*Amphiura filiformis*	No	Arm, pyloric caeca and cardiac stomach
	Crinoidea	*Antedon mediterranea*	No	Arm
	Holothuroidea	*Holothuria glaberrima*	No	Appendages and visceral organs
	Echinoidea	*Paracentrotus lividus*	No	External appendages
		Lytechinus variegates	Larva	
		Strongylocentrotus purpuratus	Larva	
Hemichordata				
	Enteropneusta	*Ptychodera flava*	No	Proboscis
Chordata				
Tunicata	Ascidiacea	*Botrylloides violaceus*	Yes	
	Ascidiacea	*Ciona intestinalis*	No	Oral siphon, branchial sac, and pharynx
Cephalochordata	Leptocardii	*Branchiostoma lanceolatum*	posterior and anterior body parts	
Pisces	Actinopterygii	*Danio rerio*	No	Heart, lens, fins
		Sternopygus macrurus	No	Tail
Amphibia	Urodela	*Notophthalmus viridescens*		Limb, tail, spinal cord, lens, jaw, and apex of heart
		Cynops pyrrhogaster		Limb, tail, spinal cord, lens, jaw, and apex of heart
		Ambystoma mexicanum		Limb, tail, spinal cord, lens, jaw, and apex of heart
	Anura	*Xenopus laevis*		Juvenile limb
Reptilia	Squamata	*Anolis carolinensis*	No	Tail
Mammals		*Dama dama*	No	Antlers
		Cervus sp.	No	Antlers
		Rangifer tarandus	No	Antlers
		Mus musculus	No	Digit tips in foetal and juvenile stages
		Homo sapiens	No	Finger tips of children

The regeneration ability was lost in different animals which are closely related or belonging to sister clades of organisms which exhibit this property. Many hypotheses such as adaptive, epiphenomenal, proximate causes, etc. have been proposed to explain the loss of regeneration ability. However, most of these cannot be applied universally, thus providing no satisfactory explanation (Goss 1963, 1992; Wagner and Misof 1992).

Among the mentioned animals with varied regeneration capacities, only a few representative members have gained popularity. Invertebrate model systems such as *Hydra* and planaria are widely used for understanding the whole organism regeneration, whereas zebrafish and axolotl are used as vertebrate model systems to understand the regeneration potential limited to few structures or organs. Apart from the regeneration abilities (*Hydra* and Planaria) and phylogenetic closeness to mammals (zebrafish and axolotl), these also have other major advantages as mentioned below.

1. Established protocols and ease of maintaining in controlled laboratory conditions
2. Amenable to genetic manipulations using transgenic technology and siRNA-mediated gene expression knockdown
3. Availability of genome, transcriptome, and other molecular information
4. Concerted efforts and sharing of reagents by active research groups

In this chapter, we aim to review the current understanding of the regeneration process in *Hydra*, the oldest regeneration model system. We believe it is important to consolidate the recent findings and compare them in light of present understanding in other regenerative model systems which will also help us identify future directions for the studies to unravel the complex process of regeneration. Especially, in vivo model systems are ideal due to limited success of research focused majorly on stem cell-based regenerative therapies. Further, since tissue regeneration is a complex phenomenon that depends on different cell types and extracellular factors/niche, it is important to investigate regeneration at an organism level. Previous reports have provided significant insights into the organismal regeneration. These studies suggest that the formation of signalling centre is a critical step in the regeneration process. Such signalling centres called 'organizers' were reported in axis induction of *Hydra* and newt embryo. The organizer comprises of a group of cells that emit signals and regulate the patterning of neighbouring tissue. However, the cells that contribute to formation of signalling centre and the process of evoking tissue patterning required for regeneration remain elusive. Due to these reasons, organisms such as *Hydra* and Planaria which can exhibit complete organismal regeneration serve as important model systems.

12.2 Hydra

Hydra is a freshwater living organism and belongs to the class Hydrozoa of phylum Cnidaria (Marques and Collins 2004; Campbell 1987). Cnidaria is a sister group to Bilateria and an early divergent phylum comprising multicellular animals with tissue

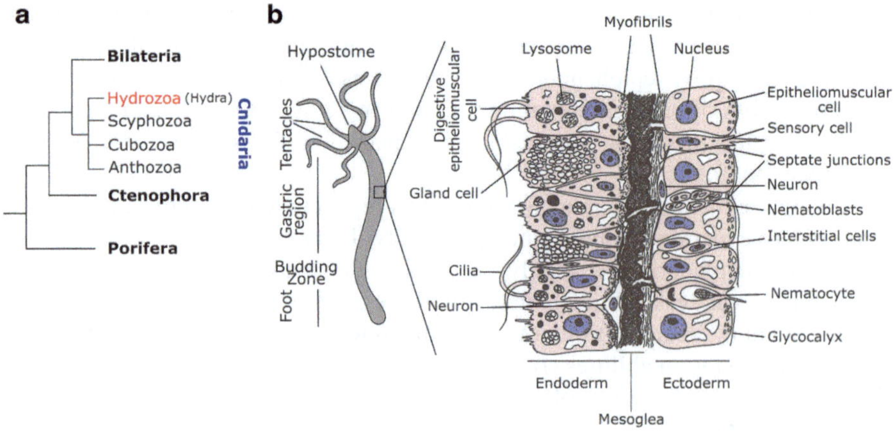

Fig. 12.1 Phylogenetic position of *Hydra* and general morphology. (**a**) Phylogenetic relation of basal metazoans and different classes of phylum Cnidaria. (**b**) General body plan and histology cartoon depiction with germ layers and cell types

level organization (Fig. 12.1a). Members of the phylum are radially symmetrical and evolved with two germ layers. Animals from this phylum share bilaterian cell types such as neurons and muscular epithelial cells (Fig. 12.1b). These properties have positioned the members of this phylum as favourite models to understand the molecular evolution determining organized body plan and evolution of cell types conserved in *Bilateria*.

Hydra is a diploblast possessing two layers of the cells which form a cylindrical body containing a gastric lumen. These polyps are among the earliest animals to have evolved a defined body plan and has an oral–aboral axis with the oral end consisting of a dome-shaped hypostome (mouth) being surrounded by a ring of 5–6 tentacles which help them in feeding. The body column of a polyp can be divided into a gastric region, a budding region, and a peduncle region. The aboral end of the polyp is called basal disc which can secrete mucus that helps in the attachment of *Hydra* to various substrates (Hoffmeister and Schaller 1985). As seen in Fig. 12.1b, the cell layers are mainly constituted of epithelial cells with either ectodermal or endodermal stem cell origin. The epithelial cells have myofibrils at the base, due to which they are called as epitheliomuscular cells (Hess et al. 1957). These myofibrils are oriented perpendicular to each other in ectoderm and endoderm. This organization of myofibrils helps in contraction or elongation of the polyp (Passano and McCullough 1964). Endodermal cells are ciliated and have phagocytic capabilities while the ectodermal cells secrete a glycocalyx layer on the outermost side of the cells to protect the polyp from environment (Hess et al. 1957). Apart from epithelial cells, *Hydra* also possess multipotent interstitial stem cells located in the interstitial spaces of epithelial cells. These stem cells differentiate into cells with special functions such as gland cells, mucus cells, neuronal cells, nematocytes, and gametes (oocytes or sperms) (Campbell and David 1974). The ectoderm and endoderm are divided by an acellular mesoglea which is essentially the extracellular matrix (ECM) (Sarras et al. 1993).

12.3 Gross Morphological Changes During *Hydra vulgaris* Regeneration

Hydra can regenerate missing body parts upon transverse or longitudinal amputation (Fig. 12.2a). The *Hydra* polyps are also capable of regenerating from reaggregated cells. These polyps when dissociated into single cells can reorganize and regenerate into a whole polyp when these cells are pelleted (Fig. 12.2b). The cells reorganize into a lumen within first 12 h with ectodermal cells outside and endodermal cells facing inside the lumen (Technau and Holstein 1992; Gierer et al. 1972). This reorganization (possibly) arises from differential cell adhesion between ectodermal and endodermal cells (Technau and Holstein 1992). This process is followed by establishment of head organizer de novo within 18–30 h (Sato et al. 1990). First appearance of the head and tentacle structures can be seen by 48–72 h, and finally the whole adult *Hydra* body forms within 4–7 days (Technau and Holstein 1992).

In case of regeneration in response to amputation, the site of injury responds by reorganization of the epithelial cells to close the wound within an hour. This is then followed by lack of any gross morphological changes until 30 h post-amputation. After 30–36 h, small tentacle buds begin to emerge from the regenerating tip beginning the process of morphological differentiation of cells. The emergence of the tentacles happens over the next 24 h with the completion of the whole process taking place by 72 h, where the tentacles are mature enough to catch the prey. On the other hand, basal disc regeneration is completed within 30–36 h (Hoffmeister and Schaller 1985). During regeneration wound healing is the initial step towards the morphological changes leading to major cellular reorganization. This requires remodelling of ECM to facilitate tissue morphogenesis.

12.4 ECM Dynamics During Axis Patterning and Regeneration

ECM in animals has varied functions like cell adhesion, mechanical support, cell signalling, scaffolding for other extracellular molecules, etc. The transition of unicellular organism to multicellular organisms requires ECM as a conduit for integration of various information between cells to function together. Recent studies have shown that the basic 'toolkit' required to form the basal membrane coincided with emergence of multicellularity. ECM components have been found to be encoded in eukaryotes as early as the unicellular choanoflagellate *Monosiga brevicollis*. More recently, it was reported that a unicellular organism named

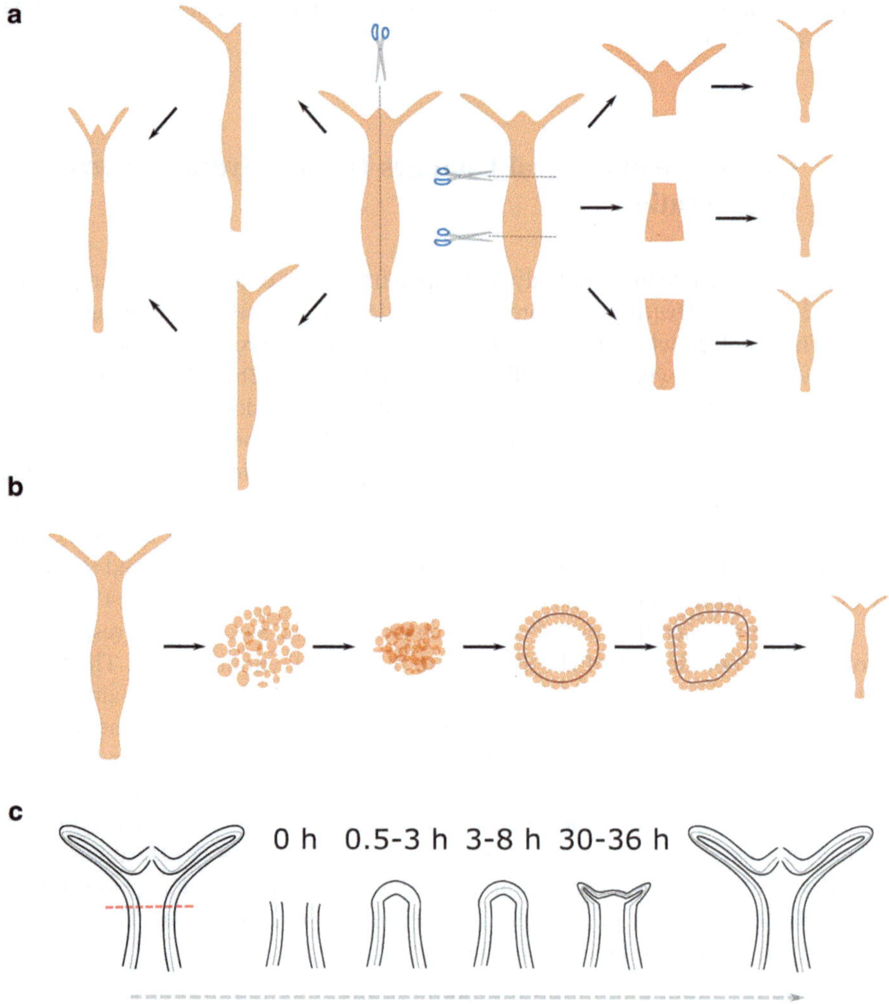

Fig. 12.2 Regeneration capacities of *Hydra vulgaris*. The ability of *Hydra* to regenerate when dissected in various manners and the morphological changes observed are shown. (**a**) *Hydra* can regenerate missing parts upon both transverse and longitudinal dissection. (**b**) Regeneration from dissociated and reaggregated cells. (**c**) Kinetics of gross morphological changes during *Hydra* head regeneration

Ministeria vibrans belonging to a filozoan clade of Filasterea which diverged earlier to Choanozoa contains a gene similar to collagen IV (Grau-Bove et al. 2017). This premetazoan origin of collagen type IV indicates that it is possibly the most primitive form of collagens and possibly other components of ECM evolved from this. Earliest phylum reported to have a complete functional set of ECM components and related proteins is Cnidaria. The fact that ECM components expanded their repertoire from phylum Cnidaria onwards indicates their eumetazoan specific functions. Hence, studying cnidarian ECM can shed light on the role of ECM components in the organization of the body plans of early multicellular organisms.

ECM has been reported to play a crucial role in regulating cell fate and tissue morphology in multiple organisms. The chemical and physical properties of ECM can influence cell behaviour and fate. It was reported that solely by varying the stiffness properties of ECM, the differentiation fate of mesenchymal stem cells could be altered (Engler et al. 2006). The regulation of cell fate by ECM occurs as a function of its interaction with ECM-specific cell receptors such as integrins which in turn transmit the information either through chemical signal transduction or mechanically through force distribution via actomyosin networks (Mao and Baum 2015). ECMs can also regulate tissue morphogenesis by controlling and sequestering morphogenic ligands including BMP (Nistala et al. 2010). These capabilities of ECM can therefore make them an important player in 'sculpting' an organism during embryogenesis. A recent study showed that during *Drosophila* embryogenesis, ECM components in the basement membrane are differentially deposited to form a morphogen-like gradient during tissue elongation (Crest et al. 2017). This creates an anisotropic resistance to isotropic tissue expansion and hence helps in shaping organs by creating a mechanical imbalance. Another study in sea urchins showed how spatio-temporal incorporation of the hygroscopic chondroitin sulphate proteoglycan causes swelling of ECM which creates hydrostatic pressure for folding of epithelial sheet during embryonic invagination (Lane et al. 1993). The branching morphogenesis in mice salivary gland has been shown to be regulated by fibronectin deposition which signals the tightly packed epithelial cells to lose the E-cadherin contacts and help in conversion of cell–cell adhesions to cell–matrix adhesions (Sakai et al. 2003).

In mammalian systems, ECM has been reported to be crucial for regeneration and sustaining cell proliferation and differentiation post injury. Mammals possess very limited potency to regenerate their lost or injured tissues. Skeletal muscles are among the few tissues that mammals can regenerate, and it was found that upon muscle degeneration, it leaves behind a hull of basement membrane of ECM which acts as a scaffold to facilitate myofiber fusion (Vracko and Benditt 1972). In mammals, early stages of skeletal muscle regeneration are associated with downregulation of laminin and collagen type I to facilitate dedifferentiation of satellite cells to form myoblasts (Gulati et al. 1983). Fibronectin, hyaluronic acid, and tenascin-C are shown to be upregulated during the same period. Another study showed that by altering the ECM stiffness of a 3-day-old mouse, it was possible to restore the ability to regenerate the heart which is lost after day 1 (Notari et al. 2018). Certain species of mammals are reported to have an astonishing capability to regenerate, for example, the African spiny mouse (*Acomys* sp.) which can lose and regenerate up to 60% of its dorsal surface including hair, follicle, skin, sweat glands, etc. Here, it was shown that laminin and collagen type I molecules are downregulated, while the cells dedifferentiate and later are upregulated during cellular differentiation (Seifert et al. 2012). Another study which looked at the regeneration of skeletal muscle of tibialis anterior in *Acomys* found that it has higher collagen type VI in the regenerating tissue as compared to *Mus* (Maden et al. 2018). This difference renders a softer biomechanical property to the *Acomys* tissue which is believed to be important to achieve a faster, scar-free regeneration with higher macrophage infiltration, low expression of inflammatory, and fibrotic genes at the site of injury. A recent study has demonstrated that fibronectin in the niche was necessary for maintaining a functional muscular stem cell population capable of regenerating lost muscular tissue. The reconstitution of fibronectin levels in aged stem cell niche is enough to remobilize and restore functional muscular stem cells (Lukjanenko et al. 2016).

In axolotls, skeletal muscle regeneration follows similar regulation of collagen type I, laminins, fibronectin, hyaluronic acid, and tenascin-C as in mammals in the early phase of regeneration to allow dedifferentiation of satellite cells (Mailman and Dresden 1976; Tassava et al. 1996; Contreras et al. 2009). In *Xenopus* tadpoles, it has been reported that hyaluronic acid plays an important role in regeneration of tail bud after amputation (Contreras et al. 2009). Adequate level of hyaluronic acid in the regenerating tail bud is required for sustaining mesenchymal cell proliferation.

Studies in zebrafish also indicate a similar role of ECM in regeneration. Fin regeneration studies indicate an early upregulation of fibronectin, hyaluronic acid, and tenascin-C which seems to be a conserved mechanism across different organisms in the vertebrates (Contreras et al. 2009). Additionally, Hapln1a, hyaluronan, and proteoglycan aggrecan are also upregulated during the process. On the other hand, laminin levels do not show any appreciable changes during fin regeneration in zebrafish. Fibronectin has been shown to be necessary for the heart regeneration and is shown to be expressed by epicardial cells. The cardiomyocytes start to express itgb3 receptors which can respond to presence of fibronectin in the site of injury for successful heart regeneration (Wang et al. 2013). A study on the caudal fin regeneration indicates a crucial role of collagen type IV and specific matrix remodellers from the MMP and TIMP family of proteins (Bai et al. 2005).

Hydra mesoglea is 0.5–2 μm thick depending upon its position in the body. The mesoglea is a trilaminar structure dividing the two epithelia (Davis and Haynes 1968). The epithelial cells are in direct contact with the basal lamina which sandwiches a thick central layer of interstitial matrix. The mesoglea also has trans-mesogleal pores which enable inter-epithelial cellular interactions via cell processes running through them (Shimizu et al. 2008). ECM secretory dynamics have been studied using the *Hydra* regeneration model. It is known that post-amputation there is a retraction of Mesoglea, and its restoration is required for initiation of regeneration. Post-amputation, loss of ECM is followed by sealing of wound which causes direct contact of ectoderm and endoderm cells, leading to change of morphology of these cells from being cuboidal or columnar to flattened (Fig. 12.3) (Shimizu et al. 2002, 2008). During regeneration, within 3 h, all the components of ECM are actively transcribed by their respective cells and then by 7 h components of the basal lamina—endoderm secretes laminin and HyCol IV such that basal lamina is formed hence leading to regaining the normal cuboidal morphology of the cells. Contact of the basal lamina with receptors of the ectodermal cells results in secretion of components of the interstitial matrix hence leading to complete formation of mesoglea within 20 h post-amputation (Sarras 2012). During the regeneration of reaggregated cells, it is reported that mesoglea formation takes place immediately

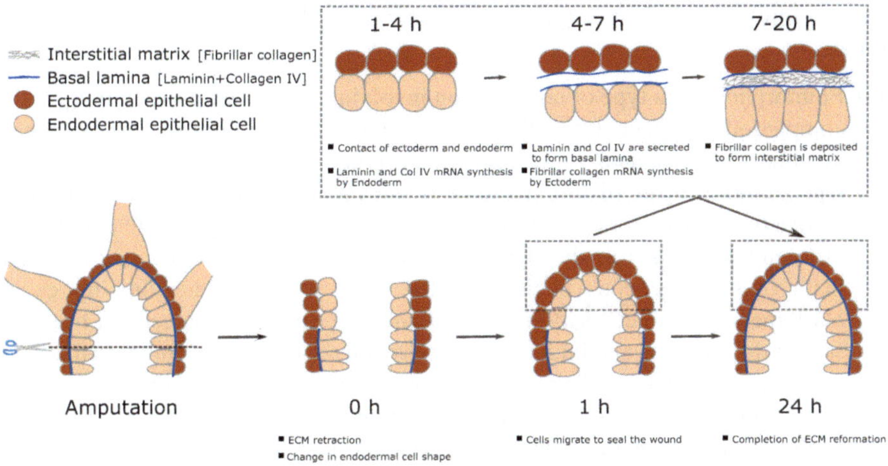

Fig. 12.3 Cellular and structural changes during head regeneration in *Hydra*. ECM is retracted immediately upon decapitation followed by change in endodermal cell shape. This is succeeded by wound closure and secretion of ECM components completing the early stages of head regeneration

after reorganization of ectoderm and endoderm cells into a lumen (Technau and Holstein 1992). The mesogleal components are first secreted by 12–17 h of aggregation, and the maturation of the mesogleal structure is reported to be completed by 48–72 h (Sarras et al. 1993). The morphogenetic processes required for regeneration of reaggregated cells succeeds the mesoglea formation and the formation of a mature mesoglea is an absolute necessity for success of regeneration (Sarras et al. 1993). It has also been reported that if there is a disruption of any of this process either by knockdown of the ECM components, inhibiting their biosynthesis chemically or by preventing ECM-receptor interaction, the process of regeneration is arrested (Sarras et al. 1993). Therefore, it seems very plausible that the important role of ECM in regulating regeneration and other tissue function evolved very early during the evolution. As can be noted above, the precise mechanistic role of ECM during regeneration in more complex organisms is unclear despite strong evidence showing its importance. A highly regenerative model system such as *Hydra* with its less complicated tissue system and signalling networks could be crucial to shedding light to ECM-mediated regulation. Studies focussing on biomechanical aspects of ECM in *Hydra* could unravel various aspects of their role in physiological regulation and how it evolved to perform more complex function in a system such as human.

12.5 Cellular Basis of Regeneration

Regeneration in animals can be either for homeostatic cell replacement like the replacement of blood cells or for reparative replacement wherein a damaged tissue or organ needs to be regenerated. In the context of this chapter, discussion will be limited to the latter type. Regeneration can be broadly divided into two types:

(a) Epimorphosis is a type of regeneration involving active cell proliferation for the completion of regeneration. This process can further be divided into two types, i.e. blastema-mediated or compensatory regeneration. In blastema-mediated regeneration, repair of lost tissue or organ proceeds by recruitment of either preexisting progenitor cells in the vicinity of the injury or by dedifferentiation of differentiated cells from the vicinity. These cells then start proliferating to form a mass of heterogeneous undifferentiated cells at the regenerating tip called blastema. Once a necessary number of cells are amassed, the cells in blastema start to differentiate and initiate the process of morphogenesis similar to that seen during embryogenesis. This process then culminates with restoration of the lost tissue either completely or partially depending upon the organism, e.g. *Planaria*, axolotls, etc. (Alvarado 2006). In the compensatory regeneration, the regeneration proceeds without any blastema formation or requirement of stem cells. In this process, differentiated cells from the vicinity are recruited to site of injury and proliferate to replace the lost tissue. Liver regeneration is a prime example of compensatory regeneration (Michalopoulos and DeFrances 1997).
(b) Morphallaxis is a type of regeneration hallmarked by absence of cellular proliferation. The existing tissue is re-patterned to replace the lost tissue. This kind of regeneration is frequently observed in lower invertebrate organisms such as *Hydra* (Lenhoff et al. 1986; Trembley 1744).

Epimorphic type of regeneration is the most prevalent type of regeneration across the animal kingdom. It would be wrong to assume dividing cells as the sole contributors towards the new tissue during regeneration. There exist possibilities of contribution from transdifferentiated and dedifferentiated cells which might contribute to a new population of dividing and differentiating cells (Jopling et al. 2011). Hence to understand how regeneration process is regulated, it becomes important to dissect the origins of cells involved as a source of new cells required for regeneration upon specific injury signals. In *Planaria*, multiple studies have shown that a continuously dividing population of cells called neoblasts contribute to formation of new tissue (Newmark and Sanchez Alvarado 2000). These cells are clonal and pluripotent in nature and have the capability to rescue regeneration deficit planarians to regenerate with as few as 3–5 cells (Wagner et al. 2011). These clonogenic neoblasts are recruited to site of injury where they proliferate into a mass of undifferentiated cells to form blastema. The blastema cells then undergo differentiation process for patterning the lost tissue (Newmark and Sanchez Alvarado 2000). In vertebrate systems, blastema formation doesn't consist in entirely of one type of clonogenic pluripotent cells like neoblasts. There seems to have a predetermined heterogeneous group of cells having unipotent or multipotent capabilities. For example, in *Xenopus* the muscle regeneration is solely dependent on dormant population of satellite cells found adjacent to muscle fibres (Le Grand and Rudnicki 2007). These are the muscle stem cells which, upon activation, differentiate into muscle fibres. In axolotl, it was shown using lineage tracing experiments that muscular satellite cells and Schwann cells are found to be unipotent and were responsible for muscular and Schwann cells, respectively, while dermis cells of mesodermal origin contribute to a diverse repertoire of cartilage and connective tissues (Kragl et al. 2009). In the zebrafish caudal fin regeneration model, it was shown that most of the cell types required for regenerating the fin were derived from specific unipotent progenitors, and the same is the case during the fin development during embryogenesis (Tu and Johnson 2011).

In *Hydra*, regeneration is manifested in two different modes depending on the region of amputation. Any amputation away from mid-gastric region will exhibit morphallactic mode of regeneration. In this mode, the wounded tissue undergoes wound healing by bringing together the ectodermal and endodermal cells which is initiated by the endodermal cells from severed edge in a fashion similar to blastopore closure (Takaku et al. 2005). This is then followed by cellular rearrangements and differentiation required for regeneration without any proliferation. It has also been reported that there is no involvement of interstitial stem cells or their derivatives towards the regeneration process and hence only requiring epithelial stem cells for the whole process (Sugiyama and Fujisawa 1978). On the other hand, it was reported that head regeneration after a mid-gastric cut proceeds through a process similar to the epimorphosis (Chera et al. 2009). Upon amputation, there are detectable levels of activation of apoptotic pathway in interstitial cells such as neurons, nematocytes, etc. present near the area of injury. These cells start secreting Wnt3a in a yet unknown fashion. These molecules of Wnt ligand then activate the Wnt signalling in nearby interstitial stem cells and induce them to do *compensatory cell proliferation*. This proliferation however does not form a very large mass of cells like in blastema. Simultaneously, the epithelial stem cells which are located at the apical end starts secreting Wnt3a and assume the role of head organizer after which follows the process of regeneration as seen in morphallactic regeneration (Galliot and Chera 2010).

Several studies have shown in *Hydra* that certain strains of *Hydra* do not exhibit any signs of aging or senescence while other strains can be induced to senesce upon induction of gametes (Martínez and Bridge 2012; Kaliszewicz 2018). The immortality in *Hydra* is attributed to the ability of *Hydra* to maintain a steady state of shedding of cells and cell proliferation. The stem cells in the body column divide continuously to either replenish themselves or differentiate into specialized cells. The epithelial cells divide roughly every 3–4 days and interstitial cells divide every 1.5 days (Martínez and Bridge 2012). These cells produce differentiated cells which are pushed to either end of the body column and are sloughed off. The differentiated cells proceed to move towards their destination depending on the origin of the cells. Cells originated from the peduncle region move towards the basal disc. The tentacle cells originate from the region just below the tentacle ring while, cells are pushed to form new bud if they originate from region above the budding region and below the tentacle ring. The cells moving towards their respective destinations have variable rate of movement depending on their position. Cells at the base of the tentacle takes on an average of 4 days to reach the tip of the tentacles and being pushed off. The cells at the base of the hypostome and the cells at the peduncle reach their respective

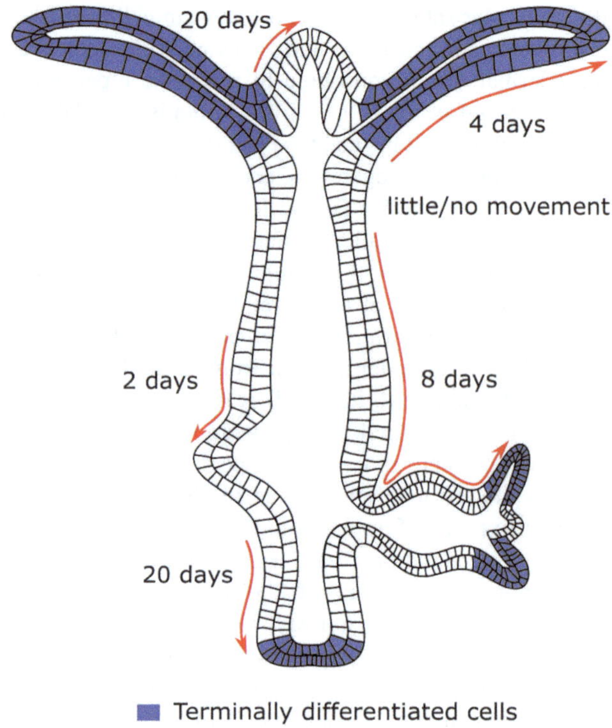

Fig. 12.4 Cellular dynamics in *Hydra*. The different cell movements that take place during maintenance of the *Hydra* body plan are depicted here. Cells undergo continuous proliferation and movement in both directions along the oral-aboral axis. The empty cells denote proliferative cells and the blue-coloured cells are terminally differentiated cells. The number of days taken for cells to reach their respective destinations such as buds or the hypostome and peduncle termini is shown. The arrows depict the direction of cell movement

destination on 20 days, and the cells in the body column take about 2–8 days to reach the budding zone and to bud depending on their position with respect to the bud (Fig. 12.4) (Campbell 1967). In conclusion, it is evident that specific cell types play important role in regeneration and tissue homeostasis.

12.6 Molecular Signalling

For the development of a zygote into an animal, there are seven major signalling pathways that act in a concerted manner and organize the proliferating cells into the final morphology. These different signalling pathways have been classified based on either ligands or signal transducers involved. These are Wnt/Wingless (Wg), transforming growth factor-β (TGF-β)/BMPs, Notch, RTK, Hedgehog (Hh), JAK-STAT [signal transducers and activators of transcription] pathway), and nucleic acid receptor/hormone receptor-mediated signalling (Barolo and Posakony 2002; Gerhart 1999). All the signalling pathways work towards a unified goal which is to create axes and break symmetry in the developing embryo. From early metazoans like sponges that have a single oral–aboral axis to mammals that have multiple axes like dorsoventral (D-V), anterior–posterior (A-P), and left–right (L-R) planes, these pathways are critical in the patterning of all organisms. During the process of development, these signalling pathways act in a spatio-temporally regulated manner to regulate various cellular functions like cell division, migration, differentiation, organization into tissues, recycling cellular components, enabling cells to respond to other signalling pathways, and apoptosis (Sanz-Ezquerro et al. 2017). These developmental signalling pathways are also reactivated when an organism must undergo the process of regeneration. The reformation of the body axes and proper patterning of lost tissues is a result of the cumulative action of these pathways.

12.6.1 Injury Response

When there is loss of tissues, the first event to occur is the activation of injury-mediated signalling response. In multiple organisms such as *Planaria* (Petersen and Reddien 2009), axolotl (Bryant et al. 2017), newt, and salamander (Yokoyama 2008), it has been reported that the earliest responses upon wounding were essential for initiation of the cascade of developmental signalling needed to regenerate lost structures. In *Hydra*, following a mid-gastric amputation, the initial ROS signalling responses upregulated by H_2O_2 were shown to activate multiple pathways. Majority of these are immune response genes, and it has also been shown that multiple signalling pathways such as STK, Pi3K, ERK 1–2, and MAPK pathways that are a part of injury-induced responses are necessary for the formation of the head organizer (Arvizu et al. 2006; Tischer et al. 2013) (Fig. 12.6).

12.6.2 Wnt Signalling

The Wnt signalling pathway is highly conserved in evolution and is reported to play roles in proliferation, cell polarity, cell fate determination, migration, and apoptosis during the development of both invertebrates and vertebrates (Miller 2002; Loh et al. 2016). This pathway is essentially a short-range communication system that uses the Wnt lipid-modified glycoproteins and has roles in both axis patterning (canonical) and cell-polarity (non-canonical) via two distinct signalling cascades (Alexandre et al. 2014). The canonical pathway is activated when the Frizzled receptors are bound by the Wnt ligands and the coreceptor Lrp5/6 is ligated to it. This leads to inhibition of the GSK3-β kinase which prevents phosphorylation of β-catenin resulting in its stabilization and nuclear localization. Nuclear β-catenin binds to the Tcf/Lef transcription factors and activates the Wnt target genes resulting in determination of cell fate. The non-canonical component of Wnt signalling is often β-catenin independent and, via multiple intracellular effector proteins, plays a role in planar cell polarity (PCP), convergent extension (CE), and directed cell migration (Loh et al. 2016).

Wnt signalling is thought to predate the 'Hox patterning system' and is possibly therefore the original symmetry breaking mechanism and primary axis organizer (Larroux et al. 2007; Martindale 2005). Traditionally, Wnt has been considered as a posterior organizer signal, although there exists a debate due to the expression of few Wnt targets that play a role in sensory epithelium differentiation in the oral region of pre-bilaterian organisms (Nichols et al. 2006; Ryan et al. 2013).

The different components of the Wnt signalling pathway are differentially present in various animals across the animal phyla. Even in basal metazoans like sponges, it has been observed that the multiple Wnt ligands express in a polarized manner across the larvae from the anterior to posterior extremes (Windsor Reid et al. 2018; Adamska et al. 2007, 2010). In the ctenophore *Mnemiopsis leidyi*, it was observed that the Wnt ligands were expressed in the aboral pole, tentacles, and the apical organ indicating their role in polarized pattern formation (Pang et al. 2010). In *Nematostella*, the presence of multiple Wnt ligands and their conservation at sequence level has been shown (Rigo-Watermeier et al. 2012). In *Nematostella*, the Wnt/β-catenin signalling is important for patterning the oral–aboral axis during the development of the organism using a crosstalk with the Hox patterning system (Leclere et al. 2016; DuBuc et al. 2018).

In bilaterians, deuterostomes (echinoderms to vertebrates), Wnt signalling has been shown to play a role during early embryogenesis in anterior-posterior patterning (Schubert and Holland 2013; Darras et al. 2018). Similar roles have been demonstrated in protostome clades such as Arthropoda (*Drosophila, Tribolium, Gryllus*, and *Achaearanea*) (Nusslein-Volhard and Wieschaus 1980; Bolognesi et al. 2008; Miyawaki et al. 2004; McGregor et al. 2008). Wnt is also known to regulate the *caudal* gene and in turn the Hox patterning system in mouse and zebrafish (Chawengsaksophak et al. 2004; Shimizu et al. 2005). The Hox patterning system is another set of genes conserved from cnidarians important in axis patterning (Ryan et al. 2006; Shenk et al. 1993; Gauchat et al. 2000; Chourrout et al. 2006; Naito et al. 1993; Schummer et al. 1992; Reddy et al. 2015).

Besides playing a role in embryonic development, the Wnt/β-catenin signalling also plays a posteriorizing role in the process of homeostasis and regeneration. In sponges, Wnt signalling was found to be important for regeneration (Windsor Reid et al. 2018). In *Planaria*, it has been observed that the Wnt/β-catenin signalling elicited by wounding is necessary for posterior tissue patterning (Petersen and Reddien 2008, 2009). While the canonical Wnt signalling is necessary for proper posterior regeneration, even anterior regeneration has been shown to require nuclear translocation of β-catenin and signalling through unknown mechanisms (Sureda-Gomez et al. 2016). Among the vertebrates, different Wnt ligands have been shown to act in the regulation of the regeneration process. In zebrafish tailfin regeneration, Wnt/β-catenin signalling acts upstream of FGF signalling and is necessary for the cell proliferation involved in blastema formation (Kawakami et al. 2006). During heart regeneration in adult zebrafish and juvenile mice, Wnt signalling was observed to be induced upon cardiomyocyte injury and necessary to switch between fibrosis and relatively scar-free regeneration (Ozhan and Weidinger 2015). Wnt signalling is active and multiple Wnt ligands express throughout adulthood in urodele tails. Such an expression pattern suggests multiple roles in urodele tail regeneration as well (Caubit et al. 1997). In axolotl, which can regenerate whole appendages, Wnt signalling has been shown to be necessary (Kawakami et al. 2006). Multiple targets and effectors of the Wnt signalling pathway were found to be differentially regulated and were implicated in the lizard tail regeneration (Hutchins et al. 2014).

In *Hydra*, the Wnt signalling pathway was identified to be a key player in the organization of head, bud formation, and head regeneration in *Hydra* (Hobmayer et al. 2000; Bode 2003; Broun et al. 2005). A total of 11 Wnt ligands have been identified and six of them have been shown to be a part of the organizer centre of the *Hydra* hypostome (Lengfeld et al. 2009) (Fig. 12.5). There were also autoregulatory mechanisms discovered to localize the Wnt expression to the hypostome (Nakamura et al. 2011). Systemic activation of the Wnt signalling pathway results in ectopic tentacle formation all over the body column of *Hydra*, turning the whole organism into a head-like structure confirming its role in head patterning (Broun et al. 2005). The regulation of Hox genes by Wnt signalling was observed in *Hydra* and these were also upregulated during the process of head regeneration (Reddy et al. 2015; Schummer et al. 1992) (Fig. 12.5). Multiple components of Wnt signalling have been shown to exhibit dynamic expression pattern during *Hydra* head regeneration (Petersen et al. 2015) (Fig. 12.6).

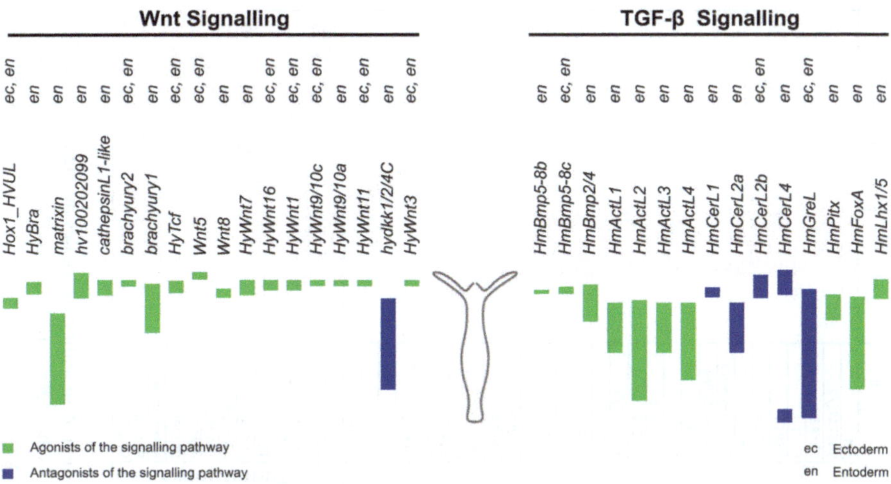

Fig. 12.5 Expression profile of Wnt and TGF-beta superfamily signalling components along the Hydra body axis. Molecular components that have been shown to be regulated by Wnt (left) and TGF-β (right) signalling pathways have been depicted here. The coloured bars represent the localization of gene expression along the body column of the *Hydra* polyp. Green bars represent molecules that play a positive regulatory role and the blue bars represent negative regulators of the pathways. The expression of the genes in the ectoderm and endoderm of *Hydra* has been denoted as ec and en, respectively

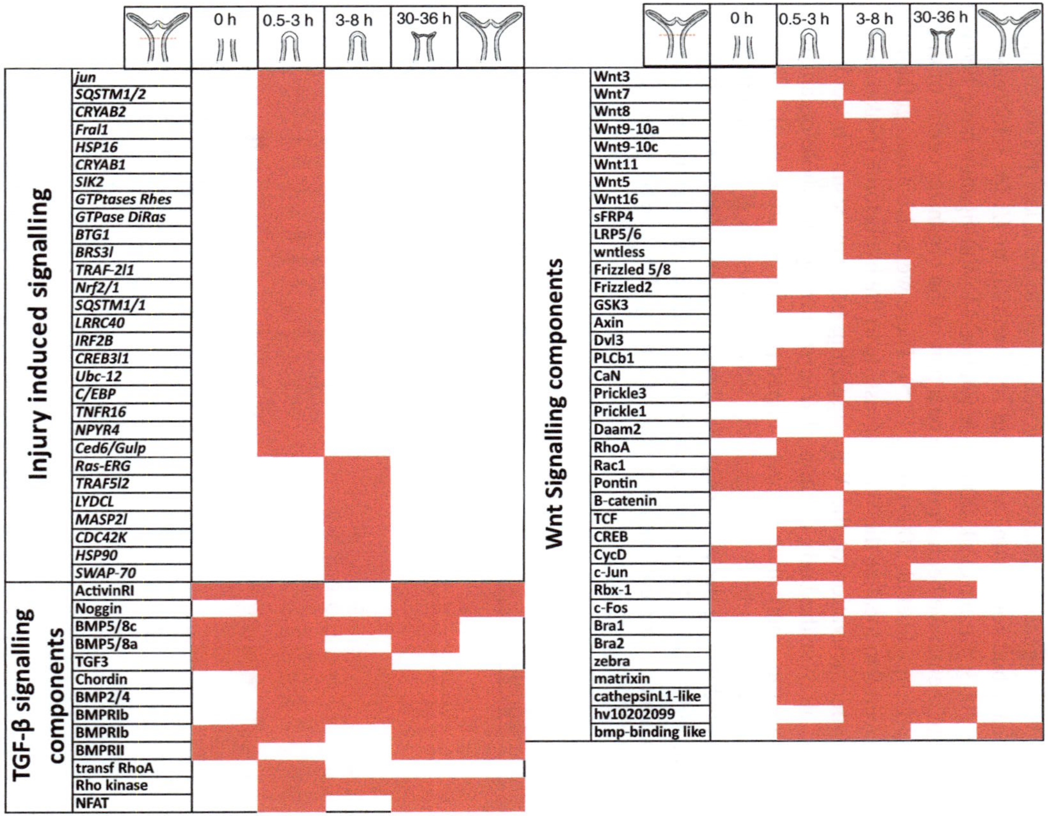

Fig. 12.6 Dynamic expression of signalling molecules during *Hydra* head regeneration. Components of multiple signalling pathways that are reported to be differentially regulated across the regeneration time course have been depicted. The signalling pathways shown are injury responses, TGF-β signalling pathway members, and Wnt signalling pathway members. The regeneration timeline is depicted above the table. The shaded region for each gene denotes the time when mRNA expression of the respective molecule has been recorded

12.6.3 TGF-β Superfamily Signalling

The TGF-β superfamily is a very large group of molecules that play a vital role in development of metazoans. The canonical form of signal transduction is through Ser-Thr kinases to intracellular mediator class of Smad proteins. The superfamily of ligands consists of more than 30 members including TGF-βs, BMPs, GDFs, Activin, and Nodal (Feng and Derynck 2005; Kitisin et al. 2007). The effector Smad proteins are classified into three classes, the R-Smads, Co-Smads, and the Anti-Smads, based on their structure and function. The R-Smads including Smad-1, 2, 3, 5, and 8 are the receptor-regulated Smads and are directly phosphorylated by binding of the Activins, BMPs, and TGF-β ligands. The co-Smads or the common Smads include one member, the Smad 4, and are required for all distinct pathways as they form heteromeric complexes with R-Smads and translocate to the nucleus for target gene activation. The anti-Smads include the Smad 6 and 7 classes and inhibit signalling by the R-Smads and co-Smads by stably associating with the receptors (Kawabata and Miyazono 1999).

The components of the TGF-β signalling pathway are conserved across metazoans starting from placozoans (Huminiecki et al. 2009). The members of the TGF-β superfamily are morphogens generating gradients and playing an important role in both development of body axes and tissues patterning. Nodal and BMPs initiate regulatory loops in the zygote to form the first embryonic axis. While the Wnt signalling pathway plays an important role in the formation of the anteroposterior axis, the TGF-β members play a critical role in the development of the dorsoventral (D-V) axis of the organism (Wu and Hill 2009). Components of the TGF-β pathway have been identified in the ctenophore, *Mnemiopsis leidyi*, and play a role in morphogenesis but not in early axis patterning (Pang et al. 2011). TGF-β members specify the dorsal regions of a developing embryo in both invertebrate (arthropods) and vertebrate (*Xenopus* and zebrafish) systems (De Robertis and Kuroda 2004; O'Connor et al. 2006).

In different classes of echinoderms, there are studies reporting TGF-β members playing a role in regeneration of the arms (Bannister et al. 2005; Patruno et al. 2002, 2003). In the zebrafish heart regeneration, TGF-β signalling pathway plays a role in switching between the scar-based repair and cardiomyocyte-based regeneration (Chablais and Jaźwińska 2012; Choi et al. 2013; Dogra et al. 2017). The TGF-β/ALk4/Smad3 pathway has been shown to play important role in the spinal cord development during regeneration (Casari et al. 2014). TGF-β has been shown to play a role in cell proliferation, activation of the BMP and ERK signalling cascades, and reversible prevention of wound epithelium resulting in the formation of regeneration structures (Ho and Whitman 2008; Beck et al. 2003). This signalling pathway is also critical for initiation and control of limb regeneration in axolotl (Levesque et al. 2007).

Multiple studies have elucidated the role of BMP signalling in body patterning and regeneration in *Hydra*. A BMP5–8 homolog was identified in *Hydra* and was shown to play a role both in tentacle formation and foot patterning during regeneration (Reinhardt et al. 2004) (Fig. 12.5). A chordin-like protein has been reported to play a conserved role in the formation of the head organizer during budding and regeneration (Rentzsch et al. 2007) (Fig. 12.5). The Nodal signalling pathway via *Pitx* in *Hydra* has been shown to play a role in breaking the radial symmetry during the process of budding and allow for the organizer pathways to form a bud on one side of the body column (Watanabe et al. 2014) (Fig. 12.5). This suggests a possible role in organizer formation during head regeneration. This is further supported by a report of multiple TGF ligands being downregulated during regeneration at transcript level and the upregulation of Smads (Petersen et al. 2015) (Fig. 12.6).

12.6.4 Notch Signalling

The Notch signalling pathway is a mediator of cell-cell communication in a juxtacrine manner and plays a crucial role in embryonic development. Like other signalling pathways, it controls cell proliferation, differentiation, and cell fate determination during embryonic development in spatio-temporal manner (Liao et al. 2016). A very important function of the Notch signalling pathway is the ability to form boundaries within developing tissues. The Notch signalling pathway is one of the simplest with very few components in its core network. The canonical Notch signalling pathway consists of a Notch transmembrane receptor that contains a Notch intracellular domain (NICD) and is activated by transmembrane ligands like Delta and Serrate/Jagged on neighbouring cells. This leads to the cleavage of the NICD which moves to the nucleus and regulates transcription of target genes (Kopan 2012). The lack of an amplification mechanism in this signalling pathway makes the stoichiometry and the signalling strength, which can be modulated by post-translational modifications, important for manifestation of the effect of the signalling pathway. Such dosage-dependent action and consequences have been studied in *Drosophila*, mice, and human diseases (Andersson et al. 2011).

Across the animal kingdom, this pathway has been shown to play a role in the specification of germ layers and differentiation of cells into specific subtypes (Shi and Stanley 2006). Notch signalling has a well-characterized role in the development of nervous system, and this is conserved from sponges (Richards et al. 2008; Richards and Degnan 2012). In *Nematostella*, disruption of the Notch signalling pathway affects the development of neurons and cnidocytes and indicates a conserved ancient origin for the multifunctional sensory cells/neurons (Marlow et al. 2012). In amphioxus embryos, Notch signalling was shown to be co-opted into the anterior-posterior boundary patterning of the developing somites (Onai et al. 2015). The role of the Notch signalling pathway during early mouse embryonic development in axial mesoderm formation, ectodermal anterior–posterior patterning, and neural tube formation has also been established (Souilhol et al. 2015).

Notch signalling components (like *delta-1*) were among the genes that have been shown to be upregulated in a translation-independent manner, early in the regeneration of planarians (Wenemoser et al. 2012). This is necessary in maintaining the midline identity of the *Planaria* (Sasidharan et al. 2017). Notch signalling is necessary for zebrafish tailfin regeneration in the formation of a blastema (Münch et al. 2013; Grotek et al. 2013). In addition to the tailfin, Notch signalling has been shown to be critical for the proliferation of cardiomyocytes and atrial to ventricular transdifferentiation during heart regeneration (Zhao et al. 2014; Zhang et al. 2013; Raya et al. 2003). In the *Xenopus* tadpole, it was shown that the Notch signalling pathway, acting downstream of BMP signalling, is required and positively correlates with regenerative capacity (Beck et al. 2003).

The Notch signalling pathway is also conserved in *Hydra* and was shown to govern differentiation of cells originating from the interstitial stem cell lineage of the polyps (Kasbauer et al. 2007). It also plays an important role in determining the boundaries in *Hydra* polyps. These boundaries include the ones between tentacles and the head region and between the newly formed bud and the parent polyp (Munder et al. 2013). The process of boundary formation between tissues/cells is an important part of development and determines the morphogenetic processes. The Notch target is expressed just before the foot cells begin to differentiate at the boundary between the bud and the parent polyp. This signalling was necessary for the completion of the bud development and detachment from the parent polyp (Munder et al. 2010). The Notch signalling pathway in *Hydra* is necessary for the formation of the *Wnt3a* expressing head organizer during the regeneration of *Hydra*. Further, inhibition of Notch signalling led to the loss of boundary between head and tentacle and repression of the head formation. This finally ended in non-regenerating tips because of perturbed spatio-temporal expression of head- and tentacle-specific genes and networks (Munder et al. 2013).

12.6.5 RTK Signalling

The receptor tyrosine kinases (RTKs) are autocatalytic receptors that function by phosphorylating various tyrosine residues. All RTKs consist of an extracellular ligand binding domain, a transmembrane domain, and a cytoplasmic tyrosine kinase domain which harbours serine, threonine, and tyrosine residues that can be phosphorylated. When the ligand binds, the receptors dimerize or undergo allosteric transitions leading to activation. The phosphorylation of the receptors recruits other cytoplasmic adaptor proteins via multiple interaction domains. The signals are then transmitted via a network of cytoplasmic proteins to the nucleus resulting in activation of various target genes (Volinsky and Kholodenko 2013).

In planarians, the FGF signalling and Wnt signalling circuits are juxtaposed and regulate the anterior-posterior axis patterning (Scimone et al. 2016). Among other growth factors, the FGFs have been shown to play a critical role in regeneration in few organisms. EGFR-3 was shown to be important for the asymmetric cell divisions that take place in planarian neoblasts upon near-lethal irradiation and was important for regeneration of these animals by regulating the size of the blastema produced

(Fraguas et al. 2011; Lei et al. 2016). This signalling was also necessary for proper gut regeneration and homeostasis in planarians (Barberán et al. 2016). During the process of homeostasis of fins in adult zebrafish, the blastema markers including *shh*, *msxb*, and *mkp3* are expressed at low levels indicating that regenerative processes are important for the maintenance of the fins in these animals. Removal of Fgfs resulted in drastic atrophy of all fin types and showed a critical role for these growth factors in preservation of dermal bone, joint structures, and all supporting tissues (Wills et al. 2008). Fgfs are also necessary for the vascularization of the regenerating heart in zebrafish (Scimone et al. 2016). The FGFs were also shown to play a role in lizard tail regeneration (Fraguas et al. 2011; Lei et al. 2016).

Sponges have nine members belonging to the RTK family (Suga et al. 2001). In *Hydra*, 15 RTKs belonging to eight families have been identified, including three ephrin family members (Reddy et al. 2011). In *Hydra*, an FGFR called *kringelchen* has been shown to express at the boundary between parent polyp and bud. Additionally, multiple components of the FGFR signalling pathway have been shown to be essential for bud detachment by modulating the actin cytoskeleton at that location (Sudhop et al. 2004). Homologs of VEGF and FGF have been identified in *Hydra*, and VEGF has been shown to have roles in bud formation and regeneration indicating a conserved role in tube formation and branching morphogenesis (Krishnapati and Ghaskadbi 2013). *HyEph1* has been shown to be upregulated during the later stages of regeneration during which its role in boundary formation could be necessary (Tischer et al. 2013). The members of RTK signalling pathway in *Hydra* and their expression during regeneration have been characterized; however, their functional significance has not been established yet.

12.7 Conclusion and Future Perspectives

Studies on the early morphological changes during *Hydra* head regeneration have revealed changes in cell shape and cell-cell contacts established between ecto- and endodermal cells. However, the significance of these cell-cell contacts remains elusive. Therefore, a focused approach is required to analyse the molecular mechanism underlying this cell-cell contact during early stages of regeneration.

Another important component that appears to play crucial role in regeneration is the ECM. It retracts immediately after amputation facilitating the contact of outer ectoderm and inner endoderm. Additionally, experiments perturbing the ECM components indicate its role in the regeneration process. The questions that still remain unanswered are: What are the changes occurring in ECM in response to damage and how these changes are read or communicated to neighbouring cells? Such studies will also be important in understanding the molecular mechanism of wound healing and tissue repair in higher organisms.

Formation of a signalling centre (organizer) is important for successful regeneration. The nature of cell types that are competent and form the organizer is not understood. This needs to be addressed to gain insight into the cellular contribution of regeneration process.

The role of signalling pathways such as the Wnt and the TGF-β superfamily are relatively well characterized. At the same time, molecular regulation of other developmental signalling pathways such as Hedgehog, Hippo, etc. is not studied. Understanding the role of these signalling mechanisms and their crosstalk will prove invaluable towards elucidation of the regeneration process.

The signalling pathways lead to transcriptional changes in the cell required for both developmental and regenerative processes. To bring about these changes, there are multiple epigenetic regulators that are recruited by or recruit transcription factors to target regions. Very few studies have investigated the role of histone modifiers, modifications, and chromatin remodellers in the process of regeneration. The conservation of histones in *Hydra* (Reddy et al. 2017) and *Hydractinia* (Török et al. 2016) has been established. Since the histones are conserved, their modification machinery, at least in part, may also be conserved. Therefore, *Hydra* can be used as a model organism to study the epigenetic mechanism of regeneration and identify differences with other organisms that cannot regenerate. This will facilitate in identifying the presence or absence of epigenetic blockade in animals with no regeneration ability.

In *Hydra*, head regeneration process is well studied, whereas regeneration of foot is not addressed extensively. Comparing the head and foot regeneration process might provide clues towards a common molecular mechanism necessary for the regeneration process.

Acknowledgements The authors would like to thank Prof. Sanjeev Galande for the discussions and suggestions and Dr. Apurva Barve for reviewing this chapter. PCR is supported by Early Career Fellowship of the Wellcome Trust-DBT India Alliance (IA/E/16/1/503057). AG is supported by a fellowship from the Council of Scientific and Industrial Research, India. MU is supported by fellowship from the University Grants Commission (UGC), India.

References

Adamska M, Degnan SM, Green KM, Adamski M, Craigie A, Larroux C, Degnan BM (2007) Wnt and TGF-beta expression in the sponge *Amphimedon queenslandica* and the origin of metazoan embryonic patterning. PLoS One 2(10):e1031. https://doi.org/10.1371/journal.pone.0001031

Adamska M, Larroux C, Adamski M, Green K, Lovas E, Koop D, Richards GS, Zwafink C, Degnan BM (2010) Structure and expression of conserved Wnt pathway components in the demosponge *Amphimedon queenslandica*. Evol Dev 12(5):494–518. https://doi.org/10.1111/j.1525-142X.2010.00435.x

Alexandre C, Baena-Lopez A, Vincent JP (2014) Patterning and growth control by membrane-tethered Wingless. Nature 505(7482):180–185. https://doi.org/10.1038/nature12879

Alvarado AS (2006) Planarian regeneration: its end is its beginning. Cell 124(2):241–245

Andersson ER, Sandberg R, Lendahl U (2011) Notch signaling: simplicity in design, versatility in function. Development 138(17):3593. https://doi.org/10.1242/dev.063610

Arvizu F, Aguilera A, Salgado LM (2006) Activities of the protein kinases STK, PI3K, MEK, and ERK are required for the development of the head organizer in *Hydra magnipapillata*. Differentiation 74(6):305–312

Auger H, Sasakura Y, Joly JS, Jeffery WR (2010) Regeneration of oral siphon pigment organs in the ascidian *Ciona intestinalis*. Dev Biol 339(2):374–389. https://doi.org/10.1016/j.ydbio.2009.12.040

Ayling AL (1983) Growth and regeneration rates in thinly encrusting Demospongiae from temperate waters. Biol Bull 165(2):343–352. https://doi.org/10.2307/1541200

Bai S, Thummel R, Godwin AR, Nagase H, Itoh Y, Li L, Evans R, McDermott J, Seiki M, Sarras MP Jr (2005) Matrix metalloproteinase expression and function during fin regeneration in zebrafish: analysis of MT1-MMP, MMP2 and TIMP2. Matrix Biol 24(4):247–260

Bannister R, McGonnell IM, Graham A, Thorndyke MC, Beesley PW (2005) Afuni, a novel transforming growth factor-beta gene is involved in arm regeneration by the brittle star *Amphiura filiformis*. Dev Genes Evol 215(8):393–401. https://doi.org/10.1007/s00427-005-0487-8

Barberán S, Fraguas S, Cebrià F (2016) The EGFR signaling pathway controls gut progenitor differentiation during planarian regeneration and homeostasis. Development 143(12):2089. https://doi.org/10.1242/dev.131995

Barolo S, Posakony JW (2002) Three habits of highly effective signaling pathways: principles of transcriptional control by developmental cell signaling. Genes Dev 16(10):1167–1181. https://doi.org/10.1101/gad.976502

Beck CW, Christen B, Slack JM (2003) Molecular pathways needed for regeneration of spinal cord and muscle in a vertebrate. Dev Cell 5(3):429–439

Bellairs AA, Bryant S (1985) Autotomy and regeneration in reptiles. Biol Reptil 15(5):301–410

Bely AE (2006) Distribution of segment regeneration ability in the Annelida. Integr Comp Biol 46(4):508–518. https://doi.org/10.1093/icb/icj051

Bely AE, Nyberg KG (2010) Evolution of animal regeneration: re-emergence of a field. Trends Ecol Evol 25(3):161–170. https://doi.org/10.1016/j.tree.2009.08.005

Berrill NJ (1951) Regeneration and budding in tunicates. Biol Rev 26(4):456–475. https://doi.org/10.1111/j.1469-185X.1951.tb01207.x

Bode HR (2003) Head regeneration in Hydra. Dev Dyn 226(2):225–236. https://doi.org/10.1002/dvdy.10225

Bohn H (1970) Interkalare Regeneration und segmentale Gradienten bei den Extremitäten von Leucophaea-Larven (Blattaria): II. Coxa und Tarsus. Dev Biol 23(3):355–379

Bolognesi R, Farzana L, Fischer TD, Brown SJ (2008) Multiple Wnt genes are required for segmentation in the short-germ embryo of *Tribolium castaneum*. Curr Biol 18(20):1624–1629. https://doi.org/10.1016/j.cub.2008.09.057

Borisenko IE, Adamska M, Tokina DB, Ereskovsky AV (2015) Transdifferentiation is a driving force of regeneration in *Halisarca dujardini* (Demospongiae, Porifera). PeerJ 3:e1211. https://doi.org/10.7717/peerj.1211

Bossert PE, Dunn MP, Thomsen GH (2013) A staging system for the regeneration of a polyp from the aboral physa of the anthozoan Cnidarian *Nematostella vectensis*. Dev Dyn 242(11):1320–1331. https://doi.org/10.1002/dvdy.24021

Broun M, Gee L, Reinhardt B, Bode HR (2005) Formation of the head organizer in Hydra involves the canonical Wnt pathway. Development 132(12):2907–2916. https://doi.org/10.1242/dev.01848

Broussonet M (1786) Observations sur la régénérations de quelques parties du corps des poissons. Hist d l'Acad Roy des Sciences

Brown FD, Keeling EL, Le AD, Swalla BJ (2009) Whole body regeneration in a colonial ascidian, *Botrylloides violaceus*. J Exp Zool B Mol Dev Evol 312(8):885–900. https://doi.org/10.1002/jez.b.21303

Bryant SV, Bellairs AA (1967) Tail regeneration in the lizards *Anguis fragilis* and *Lacerta dugesii*. J Linn Soc London Zool 46(310):297–305

Bryant DM, Sousounis K, Payzin-Dogru D, Bryant S, Sandoval AGW, Martinez Fernandez J, Mariano R, Oshiro R, Wong AY, Leigh ND, Johnson K, Whited JL (2017) Identification of regenerative roadblocks via repeat deployment of limb regeneration in axolotls. NPJ Regen Med 2(1):30. https://doi.org/10.1038/s41536-017-0034-z

Campbell RD (1967) Tissue dynamics of steady state growth in *Hydra littoralis*. II Patterns of tissue movement. J Morphol 121(1):19–28

Campbell RD (1987) A new species of Hydra (Cnidaria: Hydrozoa) from North America with comments on species clusters within the genus. Zool J Linnean Soc 91(3):253–263

Campbell R, David CN (1974) Cell cycle kinetics and development of *Hydra attenuata*: II. Interstitial cells. J Cell Sci 16(2):349–358

Carnevali MC (2006) Regeneration in Echinoderms: repair, regrowth, cloning. Invertebr Surviv J 3 (1):64–76

Casari A, Schiavone M, Facchinello N, Vettori A, Meyer D, Tiso N, Moro E, Argenton F (2014) A Smad3 transgenic reporter reveals TGF-beta control of zebrafish spinal cord development. Dev Biol 396(1):81–93. https://doi.org/10.1016/j.ydbio.2014.09.025

Caubit X, Nicolas S, Le Parco Y (1997) Possible roles for Wnt genes in growth and axial patterning during regeneration of the tail in urodele amphibians. Dev Dyn 210(1):1–10. https://doi.org/10.1002/(SICI)1097-0177(199709)210:1<1::AID-AJA1>3.0.CO;2-L

Chablais F, Jaźwińska A (2012) The regenerative capacity of the zebrafish heart is dependent on TGFβ signaling. Development 139(11):1921. https://doi.org/10.1242/dev.078543

Chawengsaksophak K, de Graaff W, Rossant J, Deschamps J, Beck F (2004) Cdx2 is essential for axial elongation in mouse development. Proc Natl Acad Sci USA 101(20):7641–7645. https://doi.org/10.1073/pnas.0401654101

Chera S, Ghila L, Dobretz K, Wenger Y, Bauer C, Buzgariu W, Martinou J-C, Galliot B (2009) Apoptotic cells provide an unexpected source of Wnt3 signaling to drive Hydra head regeneration. Dev Cell 17(2):279–289

Choi W-Y, Gemberling M, Wang J, Holdway JE, Shen M-C, Karlstrom RO, Poss KD (2013) In vivo monitoring of cardiomyocyte proliferation to identify chemical modifiers of heart regeneration. Development 140(3):660. https://doi.org/10.1242/dev.088526

Chourrout D, Delsuc F, Chourrout P, Edvardsen RB, Rentzsch F, Renfer E, Jensen MF, Zhu B, De Jong P, Steele RE (2006) Minimal ProtoHox cluster inferred from bilaterian and cnidarian Hox complements. Nature 442(7103):684

Contreras EG, Gaete M, Sánchez N, Carrasco H, Larraín J (2009) Early requirement of hyaluronan for tail regeneration in Xenopus tadpoles. Development 136(17):2987–2996

Cotanche DA (1987) Regeneration of the tectorial membrane in the chick cochlea following severe acoustic trauma. Hear Res 30(2–3):197–206

Crest J, Diz-Muñoz A, Chen D-Y, Fletcher DA, Bilder D (2017) Organ sculpting by patterned extracellular matrix stiffness. Elife 6:e24958

Cuénot L (1948) Anatomie, éthologie et systématique des échinodermes. Traité de zoologie 11:1–363

Dahlberg C, Auger H, Dupont S, Sasakura Y, Thorndyke M, Joly JS (2009) Refining the *Ciona intestinalis* model of central nervous system regeneration. PLoS One 4(2):e4458. https://doi.org/10.1371/journal.pone.0004458

Dalyell J (1814) Observations on some interesting phenomena in animal physiology. Exhibited by several species of Planariae illustrated by coloured figures of living animals. Edinburgh, Archibald Constable

Darras S, Fritzenwanker JH, Uhlinger KR, Farrelly E, Pani AM, Hurley IA, Norris RP, Osovitz M, Terasaki M, Wu M, Aronowicz J, Kirschner M, Gerhart JC, Lowe CJ (2018) Anteroposterior axis patterning by early canonical Wnt signaling during hemichordate development. PLoS Biol 16(1):e2003698. https://doi.org/10.1371/journal.pbio.2003698

Davis LE, Haynes JF (1968) An ultrastructural examination of the mesoglea of Hydra. Z Zellforsch Mikrosk Anat 92(2):149–158

De Robertis EM, Kuroda H (2004) Dorsal-ventral patterning and neural induction in Xenopus embryos. Annu Rev Cell Dev Biol 20:285–308. https://doi.org/10.1146/annurev.cellbio.20.011403.154124

Dogra D, Ahuja S, Kim H-T, Rasouli SJ, Stainier DYR, Reischauer S (2017) Opposite effects of Activin type 2 receptor ligands on cardiomyocyte proliferation during development and repair. Nat Commun 8(1):1902. https://doi.org/10.1038/s41467-017-01950-1

DuBuc TQ, Stephenson TB, Rock AQ, Martindale MQ (2018) Hox and Wnt pattern the primary body axis of an anthozoan cnidarian before gastrulation. Nat Commun 9(1):2007. https://doi.org/10.1038/s41467-018-04184-x

Egger B, Gschwentner R, Rieger R (2007) Free-living flatworms under the knife: past and present. Dev Genes Evol 217(2):89

Engler AJ, Sen S, Sweeney HL, Discher DE (2006) Matrix elasticity directs stem cell lineage specification. Cell 126(4):677–689. https://doi.org/10.1016/j.cell.2006.06.044

Etheridge R (1967) Lizard caudal vertebrae. Copeia:699–721

Feng XH, Derynck R (2005) Specificity and versatility in tgf-beta signaling through Smads. Annu Rev Cell Dev Biol 21:659–693. https://doi.org/10.1146/annurev.cellbio.21.022404.142018

Fraguas S, Barberan S, Cebria F (2011) EGFR signaling regulates cell proliferation, differentiation and morphogenesis during planarian regeneration and homeostasis. Dev Biol 354(1):87–101. https://doi.org/10.1016/j.ydbio.2011.03.023

Galliot B, Chera S (2010) The Hydra model: disclosing an apoptosis-driven generator of Wnt-based regeneration. Trends Cell Biol 20(9):514–523

Gauchat D, Mazet F, Berney C, Schummer M, Kreger S, Pawlowski J, Galliot B (2000) Evolution of Antp-class genes and differential expression of Hydra Hox/paraHox genes in anterior patterning. Proc Natl Acad Sci USA 97(9):4493–4498

Gerhart J (1999) 1998 Warkany lecture: signaling pathways in development. Teratology 60 (4):226–239

Gierer A, Berking S, Bode H, David CN, Flick K, Hansmann G, Schaller H, Trenkner E (1972) Regeneration of Hydra from reaggregated cells. Nat New Biol 239(91):98–101

Goss RJ (1963) Adaptive growth. Lagos, London

Goss RJ (1969) Principles of regeneration. Elsevier, Amsterdam

Goss RJ (1983) Deer antlers: regeneration, function and evolution. Academic Press, New York

Goss RJ (1992) The evolution of regeneration: adaptive or inherent? J Theor Biol 159(2):241–260

Graff L (1882) VON. 1882. Monographie der Turbellarien. I Rhabdocoelida Ak der Wiss zu Berlin 442

Grau-Bove X, Torruella G, Donachie S, Suga H, Leonard G, Richards TA, Ruiz-Trillo I (2017) Dynamics of genomic innovation in the unicellular ancestry of animals. Elife 6:e26036. https://doi.org/10.7554/eLife.26036

Grotek B, Wehner D, Weidinger G (2013) Notch signaling coordinates cellular proliferation with differentiation during zebrafish fin regeneration. Development 140(7):1412–1423. https://doi.org/10.1242/dev.087452

Gulati AK, Reddi A, Zalewski A (1983) Changes in the basement membrane zone components during skeletal muscle fiber degeneration and regeneration. J Cell Biol 97(4):957–962

Hess A, Cohen A, Robson EA (1957) Observations on the structure of Hydra as seen with the electron and light microscopes. J Cell Sci 3(43):315–326

Ho DM, Whitman M (2008) TGF-beta signaling is required for multiple processes during Xenopus tail regeneration. Dev Biol 315(1):203–216. https://doi.org/10.1016/j.ydbio.2007.12.031

Hobmayer B, Rentzsch F, Kuhn K, Happel CM, von Laue CC, Snyder P, Rothbacher U, Holstein TW (2000) WNT signalling molecules act in axis formation in the diploblastic metazoan Hydra. Nature 407(6801):186–189. https://doi.org/10.1038/35025063

Hoffmeister S, Schaller HC (1985) A new biochemical marker for foot-specific cell differentiation in Hydra. Wilhelm Roux Arch Dev Biol 194(8):453–461

Huminiecki L, Goldovsky L, Freilich S, Moustakas A, Ouzounis C, Heldin CH (2009) Emergence, development and diversification of the TGF-beta signalling pathway within the animal kingdom. BMC Evol Biol 9:28. https://doi.org/10.1186/1471-2148-9-28

Hutchins ED, Markov GJ, Eckalbar WL, George RM, King JM, Tokuyama MA, Geiger LA, Emmert N, Ammar MJ, Allen AN, Siniard AL, Corneveaux JJ, Fisher RE, Wade J, DeNardo DF, Rawls JA, Huentelman MJ, Wilson-Rawls J, Kusumi K (2014) Transcriptomic analysis of tail regeneration in the lizard *Anolis carolinensis* reveals activation of conserved vertebrate developmental and repair mechanisms. PLoS One 9(8):e105004. https://doi.org/10.1371/journal.pone.0105004

Hyman LH (1940) Aspects of regeneration in annelids. Am Nat 74(755):513–527

Hyman LH (1955) The invertebrates: Echinodermata, the coelomate bilateria, vol 4. McGraw-Hill, New York

Iismaa SE, Kaidonis X, Nicks AM, Bogush N, Kikuchi K, Naqvi N, Harvey RP, Husain A, Graham RM (2018) Comparative regenerative mechanisms across different mammalian tissues. NPJ Regen Med 3:6. https://doi.org/10.1038/s41536-018-0044-5

Johnson SL, Weston JA (1995) Temperature-sensitive mutations that cause stage-specific defects in Zebrafish fin regeneration. Genetics 141(4):1583–1595

Jopling C, Boue S, Belmonte JCI (2011) Dedifferentiation, transdifferentiation and reprogramming: three routes to regeneration. Nat Rev Mol Cell Biol 12(2):79

Kaliszewicz A (2018) Sex ratio patterns and trade-off between sexual and asexual reproduction in the brown Hydra. Freshw Sci 37(3):551–561

Kasbauer T, Towb P, Alexandrova O, David CN, Dall'armi E, Staudigl A, Stiening B, Bottger A (2007) The Notch signaling pathway in the cnidarian Hydra. Dev Biol 303(1):376–390. https://doi.org/10.1016/j.ydbio.2006.11.022

Kawabata M, Miyazono K (1999) Signal transduction of the TGF-beta superfamily by Smad proteins. J Biochem 125(1):9–16

Kawakami Y, Rodriguez Esteban C, Raya M, Kawakami H, Marti M, Dubova I, Izpisua Belmonte JC (2006) Wnt/beta-catenin signaling regulates vertebrate limb regeneration. Genes Dev 20 (23):3232–3237. https://doi.org/10.1101/gad.1475106

Kitisin K, Saha T, Blake T, Golestaneh N, Deng M, Kim C, Tang Y, Shetty K, Mishra B, Mishra L (2007) Tgf-Beta signaling in development. Sci STKE 2007(399):cm1. https://doi.org/10.1126/stke.3992007cm1

Kopan R (2012) Notch signaling. Cold Spring Harb Perspect Biol 4(10):a011213. https://doi.org/10.1101/cshperspect.a011213

Kragl M, Knapp D, Nacu E, Khattak S, Maden M, Epperlein HH, Tanaka EM (2009) Cells keep a memory of their tissue origin during axolotl limb regeneration. Nature 460(7251):60

Krishnapati LS, Ghaskadbi S (2013) Identification and characterization of VEGF and FGF from Hydra. Int J Dev Biol 57(11–12):897–906. https://doi.org/10.1387/ijdb.130077sg

Lane MC, Koehl M, Wilt F, Keller R (1993) A role for regulated secretion of apical extracellular matrix during epithelial invagination in the sea urchin. Development 117(3):1049–1060

Larroux C, Fahey B, Degnan SM, Adamski M, Rokhsar DS, Degnan BM (2007) The NK homeobox gene cluster predates the origin of Hox genes. Curr Biol 17(8):706–710. https://doi.org/10.1016/j.cub.2007.03.008

Le Grand F, Rudnicki MA (2007) Skeletal muscle satellite cells and adult myogenesis. Curr Opin Cell Biol 19(6):628–633

Leclere L, Bause M, Sinigaglia C, Steger J, Rentzsch F (2016) Development of the aboral domain in Nematostella requires beta-catenin and the opposing activities of Six3/6 and Frizzled5/8. Development 143(10):1766–1777. https://doi.org/10.1242/dev.120931

Lei K, Thi-Kim Vu H, Mohan RD, McKinney SA, Seidel CW, Alexander R, Gotting K, Workman JL, Sanchez Alvarado A (2016) Egf signaling directs neoblast repopulation by regulating asymmetric cell division in planarians. Dev Cell 38(4):413–429. https://doi.org/10.1016/j.devcel.2016.07.012

Lengfeld T, Watanabe H, Simakov O, Lindgens D, Gee L, Law L, Schmidt HA, Ozbek S, Bode H, Holstein TW (2009) Multiple Wnts are involved in Hydra organizer formation and regeneration. Dev Biol 330(1):186–199. https://doi.org/10.1016/j.ydbio.2009.02.004

Lenhoff SG, Lenhoff HM, Trembley A (1986) Hydra and the birth of experimental biology, 1744: Abraham Trembley's Mémoires concerning the polyps. Boxwood Press, Grove

Levesque M, Gatien S, Finnson K, Desmeules S, Villiard E, Pilote M, Philip A, Roy S (2007) Transforming growth factor: beta signaling is essential for limb regeneration in axolotls. PLoS One 2(11):e1227. https://doi.org/10.1371/journal.pone.0001227

Liao BK, Jorg DJ, Oates AC (2016) Faster embryonic segmentation through elevated Delta-Notch signalling. Nat Commun 7:11861. https://doi.org/10.1038/ncomms11861

Loh KM, van Amerongen R, Nusse R (2016) Generating cellular diversity and spatial form: Wnt signaling and the evolution of multicellular animals. Dev Cell 38(6):643–655. https://doi.org/10.1016/j.devcel.2016.08.011

Lozito TP, Tuan RS (2017) Lizard tail regeneration as an instructive model of enhanced healing capabilities in an adult amniote. Connect Tissue Res 58(2):145–154. https://doi.org/10.1080/03008207.2016.1215444

Lukjanenko L, Jung MJ, Hegde N, Perruisseau-Carrier C, Migliavacca E, Rozo M, Karaz S, Jacot G, Schmidt M, Li L (2016) Loss of fibronectin from the aged stem cell niche affects the regenerative capacity of skeletal muscle in mice. Nat Med 22(8):897

Maden M, Brant JO, Rubiano A, Sandoval AGW, Simmons C, Mitchell R, Collin-Hooper H, Jacobson J, Omairi S, Patel K (2018) Perfect chronic skeletal muscle regeneration in adult spiny mice, *Acomys cahirinus*. Sci Rep 8(1):8920

Mailman ML, Dresden MH (1976) Collagen metabolism in the regenerating forelimb of *Notophthalmus viridescens*: synthesis, accumulation, and maturation. Dev Biol 50(2):378–394

Mao Y, Baum B (2015) Tug of war—the influence of opposing physical forces on epithelial cell morphology. Dev Biol 401(1):92–102

Marlow H, Roettinger E, Boekhout M, Martindale MQ (2012) Functional roles of Notch signaling in the cnidarian *Nematostella vectensis*. Dev Biol 362(2):295–308. https://doi.org/10.1016/j.ydbio.2011.11.012

Marques AC, Collins AG (2004) Cladistic analysis of Medusozoa and cnidarian evolution. Invertebr Biol 123(1):23–42

Martindale MQ (2005) The evolution of metazoan axial properties. Nat Rev Genet 6(12):917–927. https://doi.org/10.1038/nrg1725

Martínez DE, Bridge D (2012) Hydra, the everlasting embryo, confronts aging. Int J Dev Biol 56 (6–7–8):479–487

McGregor AP, Pechmann M, Schwager EE, Feitosa NM, Kruck S, Aranda M, Damen WG (2008) Wnt8 is required for growth-zone establishment and development of opisthosomal segments in a spider. Curr Biol 18(20):1619–1623

Meyer JJ, Byers JE (2005) As good as dead? Sublethal predation facilitates lethal predation on an intertidal clam. Ecol Lett 8(2):160–166

Michalopoulos GK, DeFrances MC (1997) Liver regeneration. Science 276(5309):60–66

Miller JR (2002) The Wnts. Genome Biol 3(1):REVIEWS3001

Minelli A, Boxshall G, Fusco G (2013) Arthropod biology and evolution: molecules, development, morphology. Springer Science and Business Media, Heidelberg

Miyawaki K, Mito T, Sarashina I, Zhang H, Shinmyo Y, Ohuchi H, Noji S (2004) Involvement of Wingless/Armadillo signaling in the posterior sequential segmentation in the cricket, *Gryllus bimaculatus* (Orthoptera), as revealed by RNAi analysis. Mech Dev 121(2):119–130. https://doi.org/10.1016/j.mod.2004.01.002

Monti R (1900) La rigenerazione nelle planarie marine. Milan R. Istituto Lombardo, Milan

Münch J, González-Rajal A, de la Pompa JL (2013) Notch regulates blastema proliferation and prevents differentiation during adult zebrafish fin regeneration. Development 140(7):1402. https://doi.org/10.1242/dev.087346

Munder S, Kasbauer T, Prexl A, Aufschnaiter R, Zhang X, Towb P, Bottger A (2010) Notch signalling defines critical boundary during budding in Hydra. Dev Biol 344(1):331–345. https://doi.org/10.1016/j.ydbio.2010.05.517

Munder S, Tischer S, Grundhuber M, Buchels N, Bruckmeier N, Eckert S, Seefeldt CA, Prexl A, Kasbauer T, Bottger A (2013) Notch-signalling is required for head regeneration and tentacle patterning in Hydra. Dev Biol 383(1):146–157. https://doi.org/10.1016/j.ydbio.2013.08.022

Naito M, Ishiguro H, Fujisawa T, Kurosawa Y (1993) Presence of eight distinct homeobox-containing genes in cnidarians. FEBS Lett 333(3):271–274

Nakamura Y, Tsiairis CD, Ozbek S, Holstein TW (2011) Autoregulatory and repressive inputs localize Hydra Wnt3 to the head organizer. Proc Natl Acad Sci USA 108(22):9137–9142. https://doi.org/10.1073/pnas.1018109108

Newmark PA, Sanchez Alvarado A (2000) Bromodeoxyuridine specifically labels the regenerative stem cells of planarians. Dev Biol 220(2):142–153. https://doi.org/10.1006/dbio.2000.9645

Nichols SA, Dirks W, Pearse JS, King N (2006) Early evolution of animal cell signaling and adhesion genes. Proc Natl Acad Sci USA 103(33):12451–12456. https://doi.org/10.1073/pnas.0604065103

Nistala H, Lee-Arteaga S, Smaldone S, Siciliano G, Carta L, Ono RN, Sengle G, Arteaga-Solis E, Levasseur R, Ducy P (2010) Fibrillin-1 and -2 differentially modulate endogenous TGF-β and BMP bioavailability during bone formation. J Cell Biol 190(6):1107–1121

Notari M, Ventura-Rubio A, Bedford-Guaus SJ, Jorba I, Mulero L, Navajas D, Martí M, Raya Á (2018) The local microenvironment limits the regenerative potential of the mouse neonatal heart. Sci Adv 4(5):eaao5553

Nusslein-Volhard C, Wieschaus E (1980) Mutations affecting segment number and polarity in Drosophila. Nature 287(5785):795–801

O'Connor MB, Umulis D, Othmer HG, Blair SS (2006) Shaping BMP morphogen gradients in the Drosophila embryo and pupal wing. Development 133(2):183–193. https://doi.org/10.1242/dev.02214

Onai T, Aramaki T, Inomata H, Hirai T, Kuratani S (2015) On the origin of vertebrate somites. Zoological Lett 1:33. https://doi.org/10.1186/s40851-015-0033-0

Ozhan G, Weidinger G (2015) Wnt/beta-catenin signaling in heart regeneration. Cell Regen (Lond) 4(1):3. https://doi.org/10.1186/s13619-015-0017-8

Pang K, Ryan JF, Mullikin JC, Baxevanis AD, Martindale MQ, Program NCS (2010) Genomic insights into Wnt signaling in an early diverging metazoan, the ctenophore *Mnemiopsis leidyi*. EvoDevo 1(1):10. https://doi.org/10.1186/2041-9139-1-10

Pang K, Ryan JF, Baxevanis AD, Martindale MQ (2011) Evolution of the TGF-beta signaling pathway and its potential role in the ctenophore, *Mnemiopsis leidyi*. PLoS One 6(9):e24152. https://doi.org/10.1371/journal.pone.0024152

Passano L, McCullough C (1964) Co-ordinating systems and behaviour in Hydra: I. Pacemaker system of the periodic contractions. J Exp Biol 41(3):643–664

Patruno M, Smertenko A, Carnevali MC, Bonasoro F, Beesley P, Thorndyke M (2002) Expression of transforming growth factor β-like molecules in normal and regenerating arms of the crinoid *Antedon mediterranea*: immunocytochemical and biochemical evidence. Proc R Soc Lond B Biol Sci 269(1502):1741–1747

Patruno M, McGonnell I, Graham A, Beesley P, Carnevali MC, Thorndyke M (2003) Anbmp2/4 is a new member of the transforming growth factor-β superfamily isolated from a crinoid and involved in regeneration. Proc R Soc Lond B Biol Sci 270(1522):1341–1347

Petersen CP, Reddien PW (2008) Smed-betacatenin-1 is required for anteroposterior blastema polarity in planarian regeneration. Science 319(5861):327–330. https://doi.org/10.1126/science.1149943

Petersen CP, Reddien PW (2009) A wound-induced Wnt expression program controls planarian regeneration polarity. Proc Natl Acad Sci USA 106(40):17061–17066. https://doi.org/10.1073/pnas.0906823106

Petersen HO, Hoger SK, Looso M, Lengfeld T, Kuhn A, Warnken U, Nishimiya-Fujisawa C, Schnolzer M, Kruger M, Ozbek S, Simakov O, Holstein TW (2015) A comprehensive transcriptomic and proteomic analysis of Hydra head regeneration. Mol Biol Evol 32(8):1928–1947. https://doi.org/10.1093/molbev/msv079

Raya Á, Koth CM, Büscher D, Kawakami Y, Itoh T, Raya RM, Sternik G, Tsai H-J, Rodríguez-Esteban C, Izpisúa-Belmonte JC (2003) Activation of Notch signaling pathway precedes heart regeneration in zebrafish. Proc Natl Acad Sci USA 100(suppl 1):11889. https://doi.org/10.1073/pnas.1834204100

Reddy PC, Bidaye SS, Ghaskadbi S (2011) Genome-wide screening reveals the emergence and divergence of RTK homologues in basal Metazoan *Hydra magnipapillata*. J Biosci 36 (2):289–296

Reddy PC, Unni MK, Gungi A, Agarwal P, Galande S (2015) Evolution of Hox-like genes in Cnidaria: study of Hydra Hox repertoire reveals tailor-made Hox-code for cnidarians. Mech Dev 138(Pt 2):87–96. https://doi.org/10.1016/j.mod.2015.08.005

Reddy PC, Ubhe S, Sirwani N, Lohokare R, Galande S (2017) Rapid divergence of histones in Hydrozoa (Cnidaria) and evolution of a novel histone involved in DNA damage response in Hydra. Zoology (Jena) 123:53–63. https://doi.org/10.1016/j.zool.2017.06.005

Reinhardt B, Broun M, Blitz IL, Bode HR (2004) HyBMP5-8b, a BMP5-8 orthologue, acts during axial patterning and tentacle formation in Hydra. Dev Biol 267(1):43–59. https://doi.org/10.1016/j.ydbio.2003.10.031

Rentzsch F, Guder C, Vocke D, Hobmayer B, Holstein TW (2007) An ancient chordin-like gene in organizer formation of Hydra. Proc Natl Acad Sci USA 104(9):3249–3254. https://doi.org/10.1073/pnas.0604501104

Richards GS, Degnan BM (2012) The expression of Delta ligands in the sponge *Amphimedon queenslandica* suggests an ancient role for Notch signaling in metazoan development. EvoDevo 3(1):15. https://doi.org/10.1186/2041-9139-3-15

Richards GS, Simionato E, Perron M, Adamska M, Vervoort M, Degnan BM (2008) Sponge genes provide new insight into the evolutionary origin of the neurogenic circuit. Curr Biol 18 (15):1156–1161. https://doi.org/10.1016/j.cub.2008.06.074

Rigo-Watermeier T, Kraft B, Ritthaler M, Wallkamm V, Holstein T, Wedlich D (2012) Functional conservation of Nematostella Wnts in canonical and noncanonical Wnt-signaling. Biol Open 1 (1):43–51. https://doi.org/10.1242/bio.2011021

Ritter WE, Congdon EM (1900) On the inhibition by artificial section of the normal fission plane in Stenostoma, vol 2, 6. Academy, San Francisco

Ruhl L (1927) Regenerationserscheinungen an Rhabdocoelen. Zool Anz 72:160–175

Ryan JF, Burton PM, Mazza ME, Kwong GK, Mullikin JC, Finnerty JR (2006) The cnidarian-bilaterian ancestor possessed at least 56 homeoboxes: evidence from the starlet sea anemone, *Nematostella vectensis*. Genome Biol 7(7):R64. https://doi.org/10.1186/gb-2006-7-7-R64

Ryan JF, Pang K, Schnitzler CE, Nguyen AD, Moreland RT, Simmons DK, Koch BJ, Francis WR, Havlak P, Program NCS, Smith SA, Putnam NH, Haddock SH, Dunn CW, Wolfsberg TG, Mullikin JC, Martindale MQ, Baxevanis AD (2013) The genome of the ctenophore *Mnemiopsis leidyi* and its implications for cell type evolution. Science 342(6164):1242592. https://doi.org/10.1126/science.1242592

Sakai T, Larsen M, Yamada KM (2003) Fibronectin requirement in branching morphogenesis. Nature 423(6942):876

Sanz-Ezquerro JJ, Munsterberg AE, Stricker S (2017) Editorial: Signaling pathways in embryonic development. Front Cell Dev Biol 5:76. https://doi.org/10.3389/fcell.2017.00076

Sarras MP Jr (2012) Components, structure, biogenesis and function of the Hydra extracellular matrix in regeneration, pattern formation and cell differentiation. Int J Dev Biol 56(6–7–8):567–576

Sarras MP Jr, Zhang X, Huff JK, Accavitti MA, John PS, Abrahamson DR (1993) Extracellular matrix (mesoglea) of *Hydra vulgaris*: III. Formation and function during morphogenesis of Hydra cell aggregates. Dev Biol 157(2):383–398

Sasidharan V, Marepally S, Elliott SA, Baid S, Lakshmanan V, Nayyar N, Bansal D, Sánchez Alvarado A, Vemula PK, Palakodeti D (2017) The miR-124 family of microRNAs is crucial for regeneration of the brain and visual system in the planarian *Schmidtea mediterranea*. Development 144(18):3211. https://doi.org/10.1242/dev.144758

Sato M, Bode HR, Sawada Y (1990) Patterning processes in aggregates of Hydra cells visualized with the monoclonal antibody, Ts19. Dev Biol 141(2):412–420

Schmid V, Tardent P (1971) The reconstitutional performances of the Leptomedusa *Campanularia jonstoni*. Mar Biol 8(2):99–104

Schubert M, Holland LZ (2013) The Wnt gene family and the evolutionary conservation of Wnt expression. In: Madame Curie bioscience database [Internet]. Landes Bioscience, Austin, TX

Schummer M, Scheurlen I, Schaller C, Galliot B (1992) HOM/HOX homeobox genes are present in Hydra (*Chlorohydra viridissima*) and are differentially expressed during regeneration. EMBO J 11(5):1815–1823

Scimone ML, Cote LE, Rogers T, Reddien PW (2016) Two FGFRL-Wnt circuits organize the planarian anteroposterior axis. Elife 5:e12845. https://doi.org/10.7554/eLife.12845

Seifert AW, Kiama SG, Seifert MG, Goheen JR, Palmer TM, Maden M (2012) Skin shedding and tissue regeneration in African spiny mice (Acomys). Nature 489(7417):561

Shenk MA, Bode HR, Steele RE (1993) Expression of Cnox-2, a HOM/HOX homeobox gene in Hydra, is correlated with axial pattern formation. Development 117(2):657–667

Shi S, Stanley P (2006) Evolutionary origins of Notch signaling in early development. Cell Cycle 5 (3):274–278. https://doi.org/10.4161/cc.5.3.2396

Shimizu H, Zhang X, Zhang J, Leontovich A, Fei K, Yan L, Sarras MP (2002) Epithelial morphogenesis in Hydra requires de novo expression of extracellular matrix components and matrix metalloproteinases. Development (Cambridge, England) 129:1521–1532

Shimizu T, Bae YK, Muraoka O, Hibi M (2005) Interaction of Wnt and caudal-related genes in zebrafish posterior body formation. Dev Biol 279(1):125–141. https://doi.org/10.1016/j.ydbio.2004.12.007

Shimizu H, Aufschnaiter R, Li L, Sarras MP Jr, Borza D-B, Abrahamson DR, Sado Y, Zhang X (2008) The extracellular matrix of Hydra is a porous sheet and contains type IV collagen. Zoology 111(5):410–418

Simpson SB Jr (1964) Analysis of tail regeneration in the lizard *Lygosoma laterale*. I. Initiation of regeneration and cartilage differentiation: the role of ependyma. J Morphol 114:425–435. https://doi.org/10.1002/jmor.1051140305

Souilhol C, Perea-Gomez A, Camus A, Beck-Cormier S, Vandormael-Pournin S, Escande M, Collignon J, Cohen-Tannoudji M (2015) NOTCH activation interferes with cell fate specification in the gastrulating mouse embryo. Development 142(21):3649–3660. https://doi.org/10.1242/dev.121145

Spallanzani L (1769) An essay on animal reproductions. T. Becket and PA de Hondt, London

Stevens N, Boring A (1905) Regeneration in polychœrus caudatus. Part I. Observations on living material. J Exp Zool 2(3):335–346

Stone JS, Rubel EW (2000) Cellular studies of auditory hair cell regeneration in birds. Proc Natl Acad Sci USA 97(22):11714–11721

Sudhop S, Coulier F, Bieller A, Vogt A, Hotz T, Hassel M (2004) Signalling by the FGFR-like tyrosine kinase, Kringelchen, is essential for bud detachment in Hydra vulgaris. Development 131(16):4001–4011. https://doi.org/10.1242/dev.01267

Suga H, Katoh K, Miyata T (2001) Sponge homologs of vertebrate protein tyrosine kinases and frequent domain shufflings in the early evolution of animals before the parazoan-eumetazoan split. Gene 280(1–2):195–201

Sugiyama T, Fujisawa T (1978) Genetic analysis of developmental mechanisms in Hydra. II. Isolation and characterization of an interstitial cell-deficient strain. J Cell Sci 29(1):35–52

Sureda-Gomez M, Martin-Duran JM, Adell T (2016) Localization of planarian beta-CATENIN-1 reveals multiple roles during anterior-posterior regeneration and organogenesis. Development 143(22):4149–4160. https://doi.org/10.1242/dev.135152

Takaku Y, Hariyama T, Fujisawa T (2005) Motility of endodermal epithelial cells plays a major role in reorganizing the two epithelial layers in Hydra. Mech Dev 122(1):109–122

Tassava RA, Nace JD, Wei Y (1996) Extracellular matrix protein turnover during salamander limb regeneration. Wound Repair Regen 4(1):75–81

Technau U, Holstein TW (1992) Cell sorting during the regeneration of Hydra from reaggregated cells. Dev Biol 151(1):117–127

Tischer S, Reineck M, Soding J, Munder S, Bottger A (2013) Eph receptors and ephrin class B ligands are expressed at tissue boundaries in *Hydra vulgaris*. Int J Dev Biol 57(9–10):759–765. https://doi.org/10.1387/ijdb.130158ab

Török A, Schiffer PH, Schnitzler CE, Ford K, Mullikin JC, Baxevanis AD, Bacic A, Frank U, Gornik SG (2016) The cnidarian *Hydractinia echinata* employs canonical and highly adapted histones to pack its DNA. Epigenetics Chromatin 9(1):36

Trembley A (1744) Mémoires pour servir à l'histoire d'un genre de polypes d'eau douce, à bras en forme de cornes. Par A. Trembley. Chez Jean & Herman Verbeek, Leide

Tressler J, Maddox F, Goodwin E, Zhang Z, Tublitz NJ (2014) Arm regeneration in two species of cuttlefish *Sepia officinalis* and *Sepia pharaonis*. Invertebr Neurosci 14(1):37–49. https://doi.org/10.1007/s10158-013-0159-8

Tu S, Johnson SL (2011) Fate restriction in the growing and regenerating zebrafish fin. Dev Cell 20 (5):725–732

Unguez GA (2013) Electric fish: new insights into conserved processes of adult tissue regeneration. J Exp Biol 216(13):2478. https://doi.org/10.1242/jeb.082396

Volinsky N, Kholodenko BN (2013) Complexity of receptor tyrosine kinase signal processing. Cold Spring Harb Perspect Biol 5(8):a009043. https://doi.org/10.1101/cshperspect.a009043

Vracko R, Benditt EP (1972) Basal lamina: the scaffold for orderly cell replacement: observations on regeneration of injured skeletal muscle fibers and capillaries. J Cell Biol 55(2):406–419

Wagner GP, Misof BY (1992) Evolutionary modification of regenerative capability in vertebrates: a comparative study on teleost pectoral fin regeneration. J Exp Zool 261(1):62–78. https://doi.org/10.1002/jez.1402610108

Wagner DE, Wang IE, Reddien PW (2011) Clonogenic neoblasts are pluripotent adult stem cells that underlie planarian regeneration. Science 332(6031):811–816

Wang J, Karra R, Dickson AL, Poss KD (2013) Fibronectin is deposited by injury-activated epicardial cells and is necessary for zebrafish heart regeneration. Dev Biol 382(2):427–435

Watanabe H, Schmidt HA, Kuhn A, Hoger SK, Kocagoz Y, Laumann-Lipp N, Ozbek S, Holstein TW (2014) Nodal signalling determines biradial asymmetry in Hydra. Nature 515 (7525):112–115. https://doi.org/10.1038/nature13666

Wenemoser D, Lapan SW, Wilkinson AW, Bell GW, Reddien PW (2012) A molecular wound response program associated with regeneration initiation in planarians. Genes Dev 26 (9):988–1002. https://doi.org/10.1101/gad.187377.112

Wills AA, Kidd AR, Lepilina A, Poss KD (2008) Fgfs control homeostatic regeneration in adult zebrafish fins. Development 135(18):3063. https://doi.org/10.1242/dev.024588

Windsor Reid PJ, Matveev E, McClymont A, Posfai D, Hill AL, Leys SP (2018) Wnt signaling and polarity in freshwater sponges. BMC Evol Biol 18(1):12. https://doi.org/10.1186/s12862-018-1118-0

Wu MY, Hill CS (2009) Tgf-beta superfamily signaling in embryonic development and homeostasis. Dev Cell 16(3):329–343. https://doi.org/10.1016/j.devcel.2009.02.012

Yokoyama H (2008) Initiation of limb regeneration: the critical steps for regenerative capacity. Develop Growth Differ 50(1):13–22. https://doi.org/10.1111/j.1440-169X.2007.00973.x

Zhang R, Han P, Yang H, Ouyang K, Lee D, Lin Y-F, Ocorr K, Kang G, Chen J, Stainier DYR, Yelon D, Chi NC (2013) In vivo cardiac reprogramming contributes to zebrafish heart regeneration. Nature 498:497. https://doi.org/10.1038/nature12322. https://www.nature.com/articles/nature12322#supplementary-information

Zhao L, Borikova AL, Ben-Yair R, Guner-Ataman B, MacRae CA, Lee RT, Burns CG, Burns CE (2014) Notch signaling regulates cardiomyocyte proliferation during zebrafish heart regeneration. Proc Natl Acad Sci USA 111(4):1403. https://doi.org/10.1073/pnas.1311705111

Chapter 13
Paramecium Biology

Judith Van Houten

Abstract Imagine that in 1678 you are Christiaan Huygens or Antonie van Leeuwenhoek seeing paramecia swim gracefully across the field of view of your new microscope. These unicellular, free-living, and swimming cells might have remained a curiosity if not for the ability of H.S. Jennings (Behavior of the lower organisms. Indiana University Press, Bloomington, 1906) and T.M. Sonneborn (Proc Natl Acad Sci USA 23:378–385, 1937) to recognize them for their behavior and genetics, both Mendelian and non-Mendelian. Following many years of painstaking work by Sonneborn and other researchers, *Paramecium* now serves as a modern model organism that has made specific contributions to cell and molecular biology and development. We will review the continuing usefulness and contributions of *Paramecium* species in this chapter.

Even without a microscope, *Paramecium species* is visible to the naked eye because of their size (50–300 μ long). Paramecia are holotrichous ciliates, that is, unicellular organisms in the phylum Ciliophora that are covered with cilia. It was the beating of these cilia that propelled them across the slides of the first microscopes and continue to fascinate us today. Over time, *Paramecium* became a favorite model organism for a large variety of studies. Denis Lyn has called *Paramecium* the "white rat" of the Ciliophora for their manipulability and amenity to research. We will touch upon the use of *Paramecium* species to examine swimming behavior, ciliary structure and function, ion channel function, basal body duplication and patterning, non-Mendelian cortical inheritance, programmed DNA rearrangements, regulated secretion and exocytosis, and cell trafficking. In particular, we will focus on the use of *P. tetraurelia* and *P. caudatum*.

J. Van Houten (✉)
Department of Biology, University of Vermont, Burlington, VT, USA
e-mail: Judith.vanhouten@uvm.edu

W. Tworzydlo, S. M. Bilinski (eds.), *Evo-Devo: Non-model Species in Cell and Developmental Biology*, Results and Problems in Cell Differentiation 68,
https://doi.org/10.1007/978-3-030-23459-1_13

13.1 General Considerations and Genetics

John Preer, an enormously important mind in the world of *Paramecium* biology, reflected on the advantages and disadvantages of *Paramecium* as a model in the foreword of H.D. Görtz's book on *Paramecium* (Görtz 1988). Among advantages are that these ciliates are fully organisms even though they are not multicellular. They divide to produce large numbers convenient for many kinds of studies; they have sex and can be analyzed for inheritance of traits; they carry out the processes that most multicellular organisms do such as find food, eat, and digest; and they swim by beating cilia and respond to multiple stimuli in their pond water or stream environment.

Figure 13.1 (right panel) shows an asymmetrical cell with star-shaped contractile vacuoles for osmotic control (cv), food vacuoles (fv) forming at the bottom of the mouth or gullet that is lined with beating cilia, small structures called trichocysts beneath the surface that can be ejected (tr), and two kinds of nuclei: macronucleus and micronucleus. The cell is covered with cilia seen better in (b) and (d) caught as they beat in metachronal waves that are mechanically entrained. Bacteria swept by cilia into the gullet to the food vacuole at the end provide the cells with food to eat. Not shown is the cytoproct that allows the debris from the digestion of the bacteria to be excreted.

Preer pointed out that disadvantages of using *Paramecium* in research have been mostly overcome: Most labs feed them inexpensively on bacteria, which then could carry over into and contaminate biochemical preparations. However, the bacteria can be efficiently eliminated with antibiotics and do not confound sensitive studies such as protein identification and mass spectrometry. Axenic cultures can be grown at much greater cost and probably are not necessary for most purposes. Maintenance of stocks was and remains "troublesome" because they cannot generally be frozen for long-term storage.

A hallmark of *Paramecium* that can be disconcerting for those unused to ciliate biology is the presence of two kinds of nuclei, one highly polyploid for transcription (800n in *P. tetraurelia*) and another diploid with the expected 2n number of chromosomes for germline transmission of the genome. In *P. tetraurelia*, there are two micronuclei (Mic) and one macronucleus (Mac); in *P. caudatum* there is one Mic nestled next to the large polypoid Mac.

This brings us back to the work of Sonneborn and others to tease apart the Mendelian and non-Mendelian inheritance in *Paramecium* and making *Paramecium* a model for non-Mendelian patterns. It is first necessary to understand the cellular and nuclear divisions focusing on *P. tetraurelia* (Fig. 13.2). In the middle of this figure, there is the vegetative cell with two micronuclei and one large Mac primarily for gene expression. On the left, the cell division cycle shows that the vegetative cell duplicates its micronuclei and elongates its Mac. The micronuclei undergo mitosis and move into the poles of the nascent new cells that will form by cytokinesis. The Mac also divides but not by the orderly process of the micronuclei. After cytokinesis there are two cells with their copies of the micronuclei and macronuclei. As we will

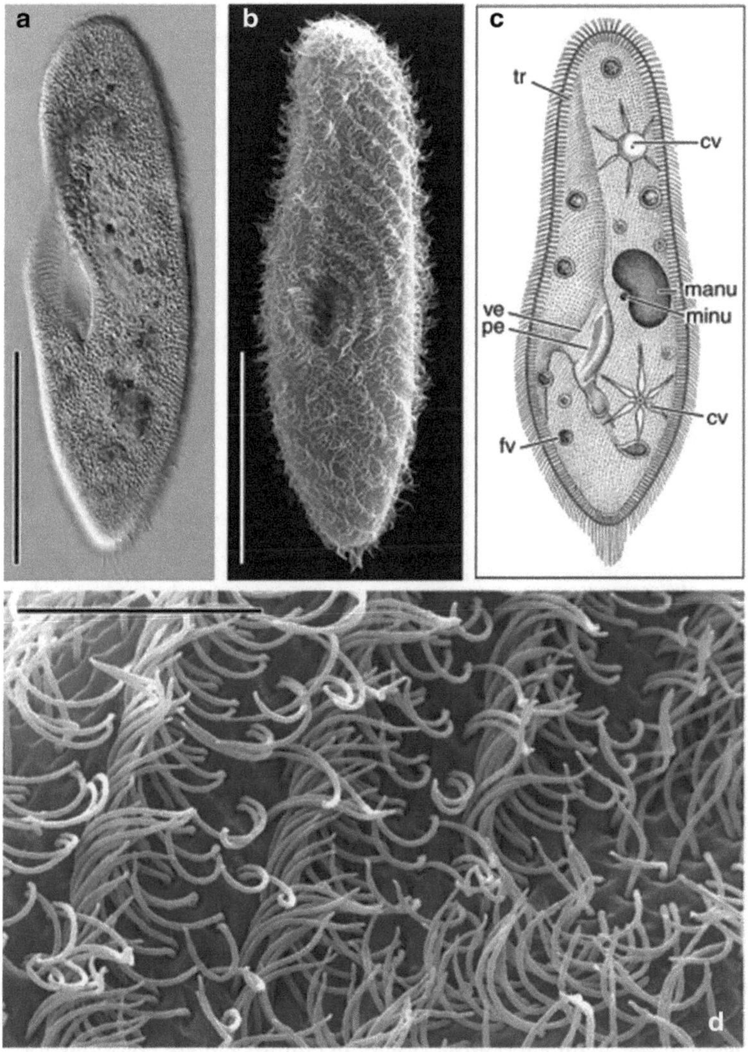

Fig. 13.1 (**a**, **b**) Light and scanning electron microscopic appearance of *Paramecium caudatum*. (**c**) Drawing of *Paramecium* illustrating might microscopic features: *cv* contractile vacuoles, *fv* food vacuole, *manu* macronucleus, *mino* micronucleus, *pe* peristome, *tr* trichocysts, *ve* vestibulum. (**d**) Higher magnification of the metachronal waves of the cilia. Scale bar in (**a**) and (**b**) 100 μ; (**d**) 10 μ. From Fig. 1 in Hausmann and Allen (2010); with permission

see in genetic studies, the micronuclei are identical, but the macronuclei in the product cells are not necessarily identical as they were not formed by mitosis.

Starvation of the cells initiates a nuclear reorganization by mating with the complementary mating type cell, or in the absence of such cells, by going through the process of autogamy (Figs. 13.2 and 13.3). The species of *Paramecium* that can

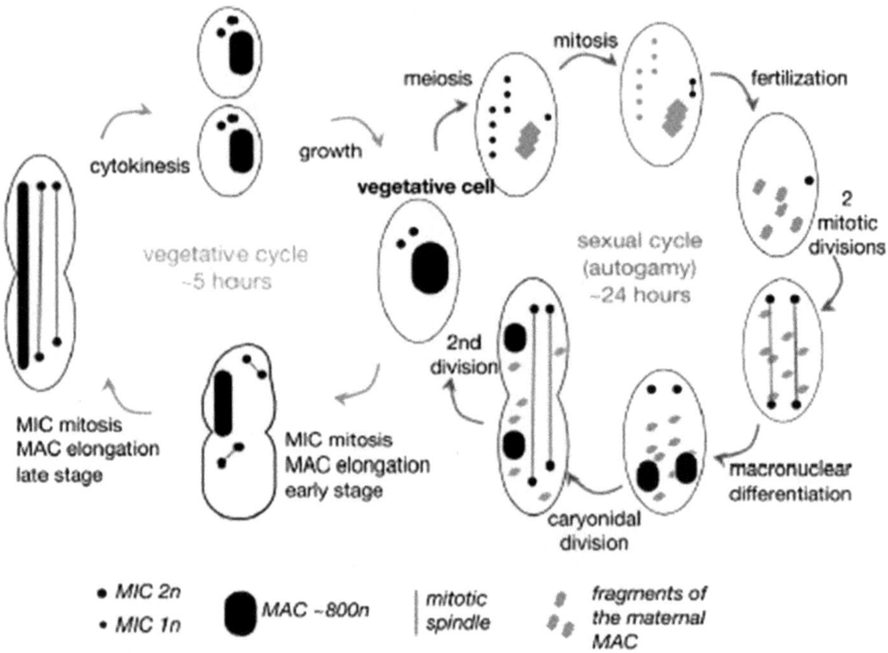

Fig. 13.2 Vegetative and sexual phases of the *Paramecium* life cycle. Vegetative cycle consists of mitotic division of the two Mic and non-mitotic division of the Mac. The right side of the figure shows nuclear reorganization during autogamy. After meiosis of the two Mic, seven of the eight products degenerate. An additional division of the surviving 1n nucleus, fusion of these two haploid nuclei (karyogamy), and two more mitotic divisions restores the cells to two 2n Mic each and two new Macs. Mics divide once more into two new cells, and result in two new identical cells homozygous at all loci. Figure 2 from Beisson et al. (2010); with permission

go through autogamy provide the advantage of making the cells homozygous at all loci in one step and make a second set of crosses to produce an F2 unnecessary. By the same process, autogamy fixes into place any mutations that creep in and are selected for if they provide an advantage to the cells in culture. The mutants *Pawn*, which we will see again later, grow slowly in culture and are notorious for accumulating revertant mutations (Chang et al. 1976).

Autogamy starts with the same vegetative cell (Fig. 13.2, the center). Seen in the right-hand part of the figure, the Mac breaks down and fragments; the micronuclei divide by meiosis to produce haploid nuclei. All but one of these disintegrates; the remaining one divides and fuses to form a diploid micronucleus, which divides again to produce four nuclei. The two at the posterior of the cell differentiate into macronuclei; the ones at the anterior pole remain micronuclei. These nuclei are distributed to the two daughter cells that are forming by cytokinesis, with each receiving two identical micronuclei and one macronucleus. The fragments of the old macronucleus continue to disintegrate. The important result is that these new cells are now homozygous at all loci.

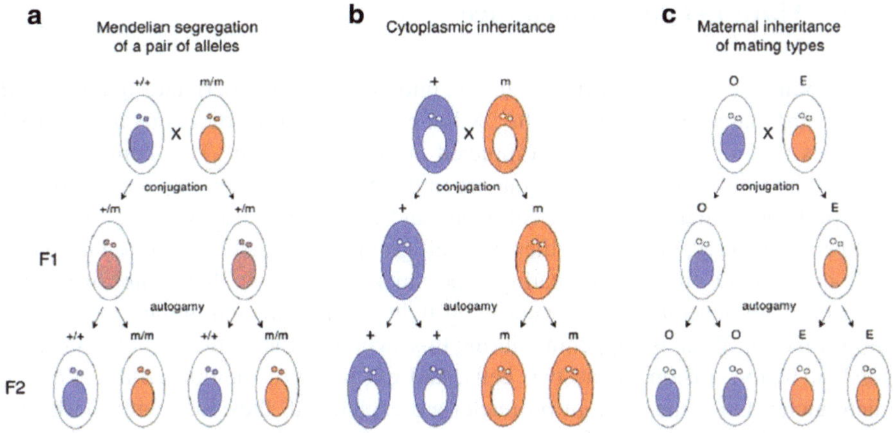

Fig. 13.3 Genetic analyses of Mendelian and non-Mendelian inheritance. (**a**) Conjugation between two cells homozygous for different alleles at one locus +/+ and m/m. F1 is heterozygous, and F2 are completely homozygous each with a 50% chance of becoming homozygous for +/+ or m/m. (**b**) Cytoplasmic inheritance of characters such as those determined by the mitochondrial genome. Little cytoplasm is exchanged to affect this. (**c**) Maternal inheritance follows the cytoplasmic line and not the nuclear genes for mating type inheritance. Figure 3 from Beisson et al. (2010); with permission

For mating and inheritance patterns in Fig. 13.3, we again focus on *P. tetraurelia*. The first example shows the mating of two homozygous cells for traits +/+ or m/m for wild type and mutant. With starvation, cells become mating reactive and conjugate with a cell of complementary mating type. The micronuclei divide by meiosis with only one product surviving; this one again divides to prepare a gamete that is transferred to the complementary cell during conjugation when the cells physically mate. After transfer of the gametic nuclei, the retained haploid micronuclei and the gametic nuclei fuse and create the diploid micronuclei, which again divide by mitosis to give each new cell two diploid (+/m) micronuclei. The old macronuclei disintegrate, and new macronuclei are formed from new Mic mitotic products that have the + and m alleles (standing in for the entire genomes).

At this point the F1 cells could undergo autogamy and on average distribute their genes + or m to create homozygous +/+ or m/m cells in about a 1:1 ratio. If you are working with a *Paramecium* species without autogamy, such as *P. caudatum*, a second set of crosses of the F1s would be necessary to create the F2.

An important feature of the diagrams in Fig. 13.3 is that examination of the F2 clearly delineates whether the pattern of inheritance is Mendelian, cytoplasmic (e.g., mitochondrial genes), or maternal (e.g., mating type genes). These patterns and deep understanding by Sonneborn made *Paramecium* attractive as a genetic model starting in 1937.

13.2 Mating-Type Inheritance

At this point we will take a brief excursion into mating-type inheritance diagrammed in the third pattern in Fig. 13.3. The "X" in each of the three diagrams in Fig. 13.3 indicates that a cross has been completed between two cells. This conjugation is initiated by mild starvation; cells that are in the presence of their "opposite" mating type find each other by random collisions and not by pheromone attraction. They will become sticky, clump together in small groups, and finally separate into pairs, with one cell O and one E mating type stuck together at their anterior-ventral surfaces. They resemble little swept wing fighter planes swimming around in their pair formation. What is happening during this swim together is the loss of cilia at the point of fusion, fusion of the two cells, and exchange of meiotic products of the micronuclei. Once the haploid pronuclei are exchanged, they fuse with the host's haploid nucleus left behind from meiosis and form a diploid nucleus that will then direct the formation of a new macronucleus once the cells separate. As we will see below in Figs. 13.5 and 13.6, the process of developing the new Mac is complex. Here we will address how the cells differentiate into only one of two possible mating types available to them.

Note that in Fig. 13.3, the cells are labeled O or E, odd or even, for the two mating types in *P. tetraurelia*. The inheritance of the trait O or E mating type appears to be determined by the cytoplasm that the new Mac finds itself in, hence the name maternal inheritance. There are 15 sibling species of the *P. aurelia* complex, and they are sexually incompatible with F1s sterile from any interspecies conjugation. Only the members of the *P. aurelia* complex have the capacity for autogamy. All 15 species have a binary mating-type system: O or E. How mating type is determined can differ.

For one of the *P. aurelia* species (*P. tredecaurelia*) and *P. caudatum*, mating type is determined in a Mendelian manner, with a dominant E and recessive O trait. For another seven of the *P. aurelia* complex species, the process is stochastic with the two karyonide products of the original mating being O or E at random with equal frequency. This leaves the four F2 products after autogamy (Fig. 13.3c) with a random assortment of O and E cells.

What is illustrated in Fig. 13.3c is the process for the remaining seven of the sibling species: After autogamy the cells become determined for the same mating type as the parent cell. Two will be O coming from the parent originally O and two E from the parent originally E. The influence of the mating type is passed down from the Mac and cytoplasm of the original maternal cell.

Figure 13.4 depicts a model for three of the *P. aurelia* species. Notable is that for *P. tetraurelia* and *P. octaurelia*, two proteins, mtC and mtB, are needed for transcription of mtA. The mtA protein is critical to produce an E cell. In the O cell, the mtA promoter has been excised with the small RNA system as with the internal eliminated sequences (IESs) discussed at length below. In the absence of mtA protein, the cell is phenotype O. The same process holds for *P. septaurelia* E cell phenotype, but the *mtB* gene has suffered an excision to make it impossible to transcribe the *mtA* gene in O cells. Again, in the absence of the mtA protein, the default phenotype is O mating type. Interestingly, the small RNAs (scn RNAs

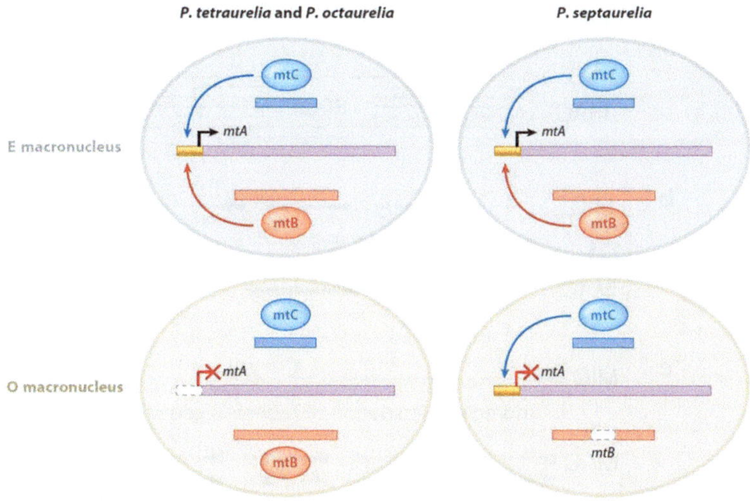

Fig. 13.4 General model for mating-type determination in *Paramecium aurelia* species. In these three species, mating type E depends on expression of the mtA protein while cells are mating reactive. The mtA transcription in turn requires the mtB and mtC gene products, although not all of these steps have been completely verified except in *P. tetraurelia*. Mating type O is determined in *P. tetraurelia* and *P. octaurelia*, when during Mac development, the promoter for the mtA gene is excised as an IES. In *P. septaurelia* mating type O is determined by excision of part of the mtB coding sequence as an IES. In both cases mating type O is determined when the mtB protein cannot be made. Figure 4 in Orias et al. (2017); with permission

below) have been co-opted (exapted from the evolutionary biology terminology) for the process of mating-type determination (Singh et al. 2014).

Ciliate mating-type determination research has a very long history. For much more detail, see the review in Orias et al. (2017).

13.3 DNA Amplification and Elimination

The current deep understanding of *Paramecium* genomes and how the Mac and Mic interact to safeguard the germline provides insights into the transgenerational inheritance of acquired traits and silencing of transposons for protection of the genome and epigenetic markers.

In Fig. 13.2, the development of the new macronucleus during sexual cycle is clearly delineated on a macroscopic level. However, the realities of how the old Mic is broken up and the new 800n Mac is created by endoduplication (Fig. 13.5a) is not addressed except by the arrow that points from the micronucleus to the new macronucleus. When one digs deeper (Fig. 13.5b), the second arrow from the Mic to the new Mac shows that there is a precise deletion of sequences (internal eliminated sequences or IESs) from the macronucleus, which is being streamlined to support gene expression (Allen and Nowacki 2017). There are about 45,000 of

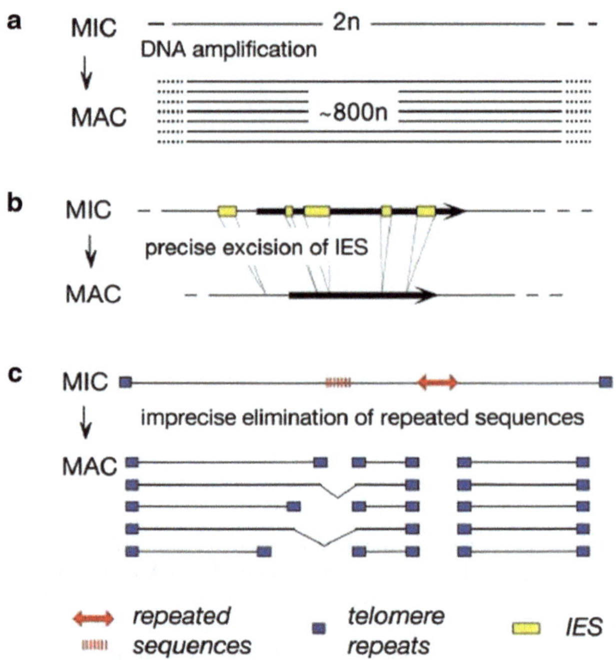

Fig. 13.5 Genome rearrangements during Mac development. (**a**) DNA amplification (2n to 800n), (**b**) precise excision of IES, and (**c**) imprecise elimination of repeated sequences. Figure 4 in Beisson et al. (2010); with permission

these short noncoding sequences in about half of the 30,000 genes in the *Paramecium* micronuclear genome (Arnaiz et al. 2012). They have been likened to introns but in DNA. They must be removed with precision while the Mac develops in order to retain a functional set of genes. The third arrow between the two types of nuclei in Fig. 13.5c shows further streamlining due to fragmentation and removal of repeated sequences such as transposable elements in the Mac and addition of telomeres.

Precise IES removal: Together six transposases from the piggyback family accomplish the precise removal of IESs (Bischerour et al. 2018). They are called domesticated transposases because they have been co-opted for a eukaryotic process. The IESs are flanked on their two ends by the sequence TA, but otherwise there are no outstanding signals for cleavage. The PiggyBac transposases start with double-strand breaks of the sequence flanked by the TAs followed by double-strand break repair. There is one catalytic subunit and five additional partners for this IES elimination process.

How the transposases target and recognize the correct sequences with little signal resident in the DNA sequence has led to a focus on scan and other small RNAs (Allen and Nowacki 2017). Genome rearrangements in macronuclear development are guided by RNAs for the global comparison of the Mic and Mac sequences, resulting in a new Mac devoid of germline-specific sequences that should only be in a Mic. These RNAs are small scan or scn RNAs and non-coding RNAs from the maternal Mac and from the new developing Mac.

Figure 13.6 summarizes a great deal of work from several laboratories [see Allen and Nowacki (2017) for a review]. The process of removing IES sequences from the Mac begins with the RNA polymerase II transcription in the new Mic that was formed by meiosis during conjugation or autogamy. The Mic transcripts are cleaved into the small 25 nucleotide scRNAs that are transported to the outgoing Maternal Mac for comparison scanning. This process results in the identification of RNAs for further transport to the new developing Mac from the sequences that are not found to match the Maternal Mac sequences. At this point the sequences have been identified that should remain in a new Mac and be used for gene expression in the new cell. The Mac genome sequences that should not be in the new Mac match up with these IES RNAs and are removed as IESs by the processes outlined above.

Another class of small RNAs mentioned above come from transcription of the excised IES sequences and serve in a positive feedback loop to carry out the enormous job of excising all the IES sequences in the highly amplified new Mac.

Figure 13.5c shows the third set of activities that differentiate the Mic and new Mac. Imprecise elimination of the much of the germline genome from the new Mac seems to be accomplished by chromosome fragmentation (Le Mouel et al. 2003). During Mac development the genome is fragmented into 200 linear molecules 50 kb to 1 Mg in size with telomeres added to the newly created ends. The resulting compact chromosomes with short introns and short intergenic regions are amplified to 800N. The coding density is 74–78% meaning that there is little space between the genes on each chromosome. Most of the genome is dedicated to coding for genes. It is estimated that there are 40,000 genes in the Mac genome (Aury et al. 2006).

In short, these programmed developmental steps dependent upon the new Mic and Maternal Mac create a streamlined new Mac that is designed for transcription, stripped of transposons and other repeated sequences.

Other characteristics of the *Paramecium* genome: The whole genome has been duplicated three times, providing much opportunity for repurposing genes and adaptation. However, most gene pairs remain active and situated between similar neighbors (Aury et al. 2006; Chalker and Stover 2007). The genome is AT rich (72%) which presents some challenges in cloning and sequencing but that can be overcome (Zagulski et al. 2004). There is only one stop codon for *Paramecium* translation: TGA. TAA and TAG code for glutamine. This in general poses no particular problems except in expression of *Paramecium* genes in other species. It is possible to express exogenous genes in *Paramecium*, sometimes with some editing to accommodate codon bias.

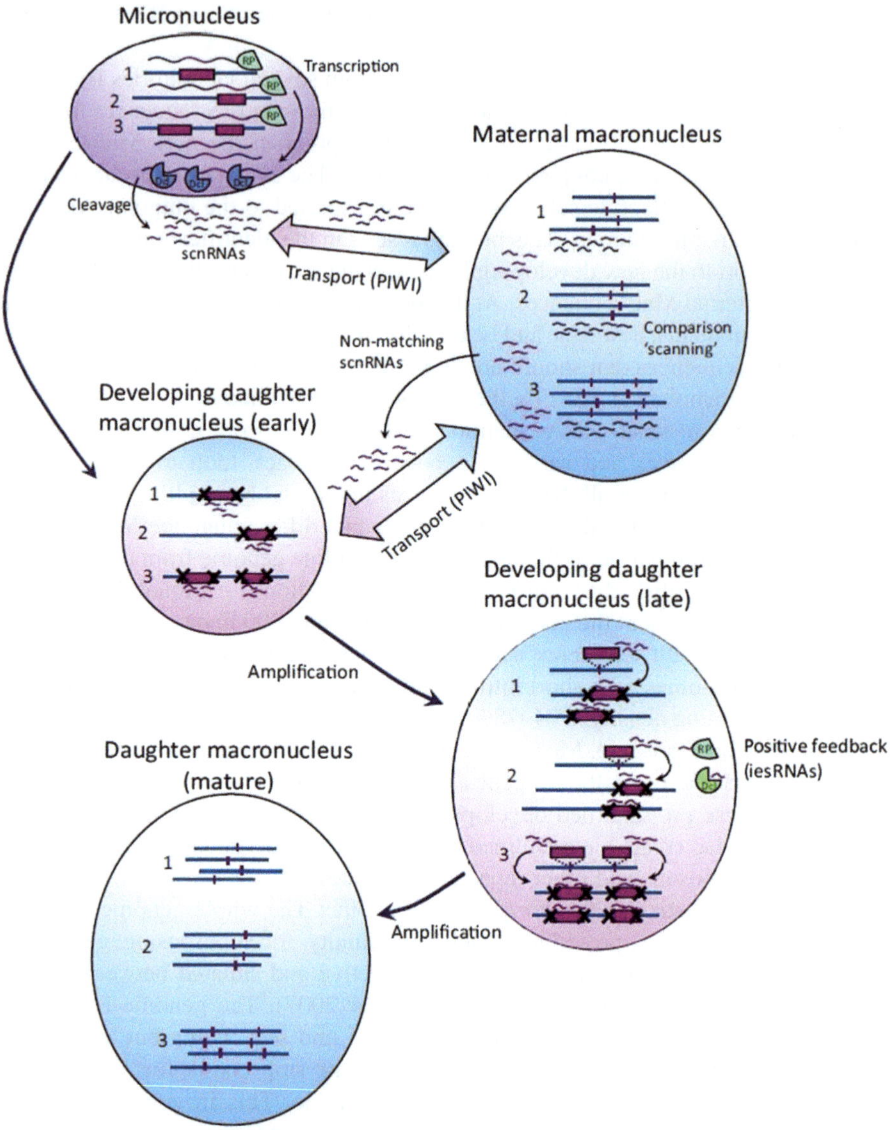

Fig. 13.6 Small RNA (sRNA)-mediated excision of internal eliminated sequences (IESs) in *Paramecium*. Steps in sexual development of the daughter Mac. The trans-nuclear comparison occurs with small single-stranded scan RNAs (scnRNA) that move from Mic to Maternal Mac where it scans transcripts. Those scnRNAs not matching with the Maternal Mac move to the developing Mac where more matching occurs to identify sequences (IESs) that should be removed from the new Mac as they are not represented in the Maternal Mac. While the new Mac is amplifying its DNA, the IES sequences are being removed, and the small RNAs from the IES are used in a positive feedback loop to ensure that all IESs are removed. Figure 2 in Allen and Nowacki (2017); with permission

13.4 Forward and Reverse Genetics

Paramecium processes have long been studied through the comparison of mutants with the wild-type phenotypes. This approach to understanding complex processes has made *Paramecium* a particularly good model for exocytosis. Trichocysts, small structures of 2–3 μ (long), sit beneath the plasma membrane. They are oblong shaped with a tip that inserts into a rosette structure at the membrane. Upon stimulation they explosively shoot out of the cell expanding their length by eightfold (Plattner 2016). This calcium-dependent regulated secretion process happens in response to stimuli, such as a synthetic aminoethyldextran, acids like picric acid, and the touch of predators (Plattner and Klauke 2001; Plattner 2016). *Didinium* and related species seem to be deterred from engulfing paramecia by the trichocysts ejected during an attack (Miyake and Haroumoto 1996; Sugibayashi and Harumoto 2000; Buonanno et al. 2013). The selection of mutants for defective trichocyst secretion has resulted in a mutant in each step (Cohen and Beisson 1980) and has facilitated the dissection of trichocyst development, trafficking through secretion (Fig. 13.7).

The calcium-dependent processes of trichocyst-regulated secretion have been documented extensively by Plattner and workers (Klauke et al. 2000; Plattner and Klauke 2001; Plattner 2013). In Fig. 13.8 there are several roles for Ca^{2+} with the source, outside or inside the cell, noted. During the stimulus-secretion coupling, first is mobilization of Ca^{2+} from alveolar sacs followed by entry of extracellular Ca^{2+} through store-operated channels (SOC) (Plattner and Klauke 2001; Plattner 2013). Alveolar sacs are ER-like calcium storage organelles with a ryanodine-sensitive Ca^{2+} release channel and a thapsigargin-sensitive SERCA pump to restore Ca^{2+} once it is released from the alveolar sac. Separately, calcium from the outside must enter the surface pore to trigger the explosive decondensation and secretion of the trichocyst. The molecular entities, such as the SOC, are not yet identified.

Figure 13.8 illustrates the physical relationship of the alveolar sacs, trichocyst, and cilia, all of which utilize $Ca^{2+}$for signaling. The right side of the figure recapitulates the steps of trichocyst secretion with Ca^{2+} sources from the SOC and alveolar sacs. The left side of the figure indicates that although the alveolar sacs are close to the base of the cilium, its Ca^{2+} from stimulated secretion does not affect the cilium, but it is possible for Ca^{2+} to leak in the opposite direction (Husser et al. 2004). We will return to this notion under the discussion of cilia (Fig. 13.9).

Plattner and his workers have pursued the identification and localization of additional pathways in *Paramecium*, specifically for the components of intracellular trafficking. They identified members of the large family of SNARE proteins (soluble *N*-ethylmaleimide-sensitive factor attachment protein receptor element) that have roles in membrane-membrane interactions in trafficking. Among the SNAREs are numerous syntaxins in the *Paramecium* genome (Kissmehl et al. 2007). The 26 members of the t-SNARE (target) syntaxins can be grouped into 15 subfamilies. Epitope tagging and immunofluorescence allowed for the localization of individual syntaxins to organelles (Fig. 13.10) and assignment of specific proteins to trafficking pathways for osmoregulation, phagocytosis, regulated exocytosis, and constitutive

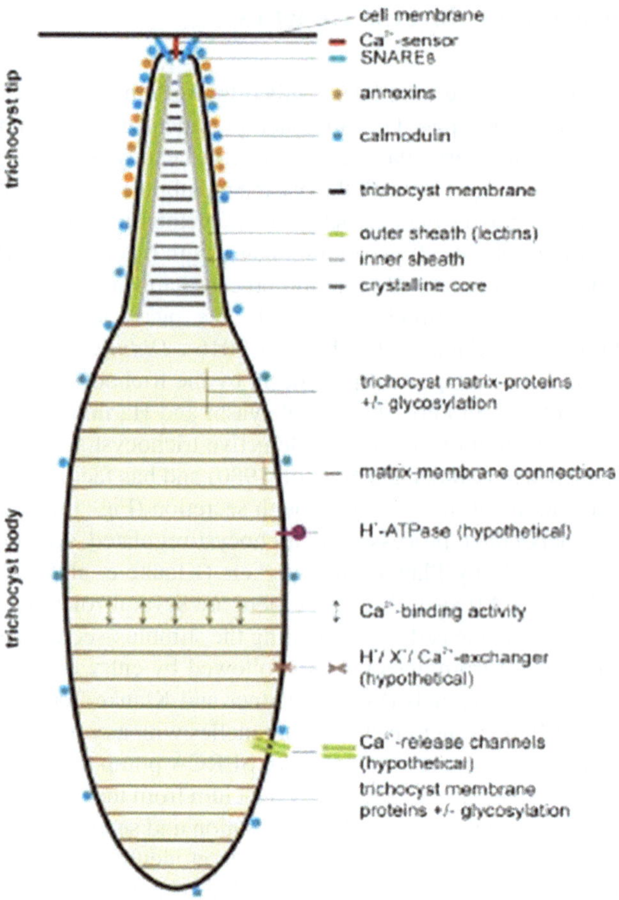

Fig. 13.7 Components of the trichocyst. Figure 13 from Plattner (2016); with permission

endocytosis through parasomal sacs (like clathrin coated pits) (Schilde et al. 2006, 2010; Kissmehl et al. 2007).

The synaptobrevin SNAREs likewise have been cloned and hunted down to their cellular locations as evidence of their roles in intracellular trafficking. Initially 12 genes in 7 gene families were located in the endoplasmic reticulum, contractive vacuole complex, parasomal sacs, cytoproct, and oral apparatus. Gene silencing showed that the synaptobrevins of the contractile vacuole are required for the function of this osmoregulatory system and cell division (Schilde et al. 2006, 2010).

As more families of synaptobrevin genes were identified, the search for the proteins in the cell broadened: various family members were found on the food vacuoles, cytostome, ciliary bases, and small vesicles. In short these synaptobrevins appear to be important in the endo-phagocytic pathways of the cell and also potentially in ciliary function (Schilde et al. 2010; Fig. 13.11).

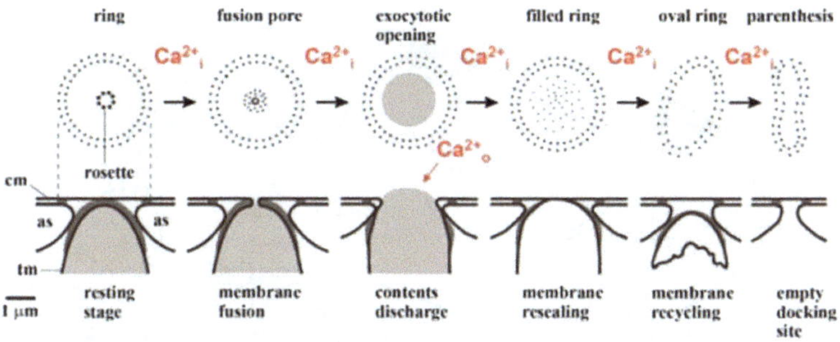

Fig. 13.8 Sequence of calcium-dependent events in stimulated trichocyst exocytosis. As alveolar sac; cm cell membrane; tm trichocyst membrane. Note the subscripts of "I" and "O" to show the origin of the calcium. Figure 3 from Plattner (2013); with permission

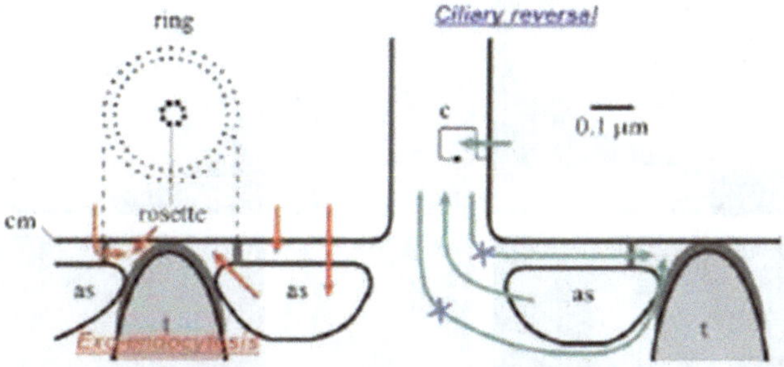

Fig. 13.9 Calcium fluxes in *P. tetraurelia* during stimulated trichocyst (t) exocytosis and during ciliary reversal. Note that during exocytosis, calcium flows from the alveolar sac (as) and the outside of the cell into the as by a store operated channel and into the rosette. During ciliary reversal on the right, calcium enters the cilium through voltage-gated calcium channels, but that calcium does not reach the trichocyst to initiate exocytosis. However, the flood of calcium from the as can enter the cilium. Figure 8 from Plattner (2013); with permission

Plattner believed that actin stood at the intersection of calcium and intracellular trafficking (Plattner 2010). Kissmehl and Plattner showed through GFP-tagged expressed proteins and RNA interference (RNAi) that actin from the nine gene families can be found throughout the cell, even at the cilia (Kissmehl et al. 2004; Sehring et al. 2007). Of the cellular trafficking pathways, cyclosis and food vacuole formation were the most clearly affected by downregulation of actin through RNAi (Fig. 13.11).

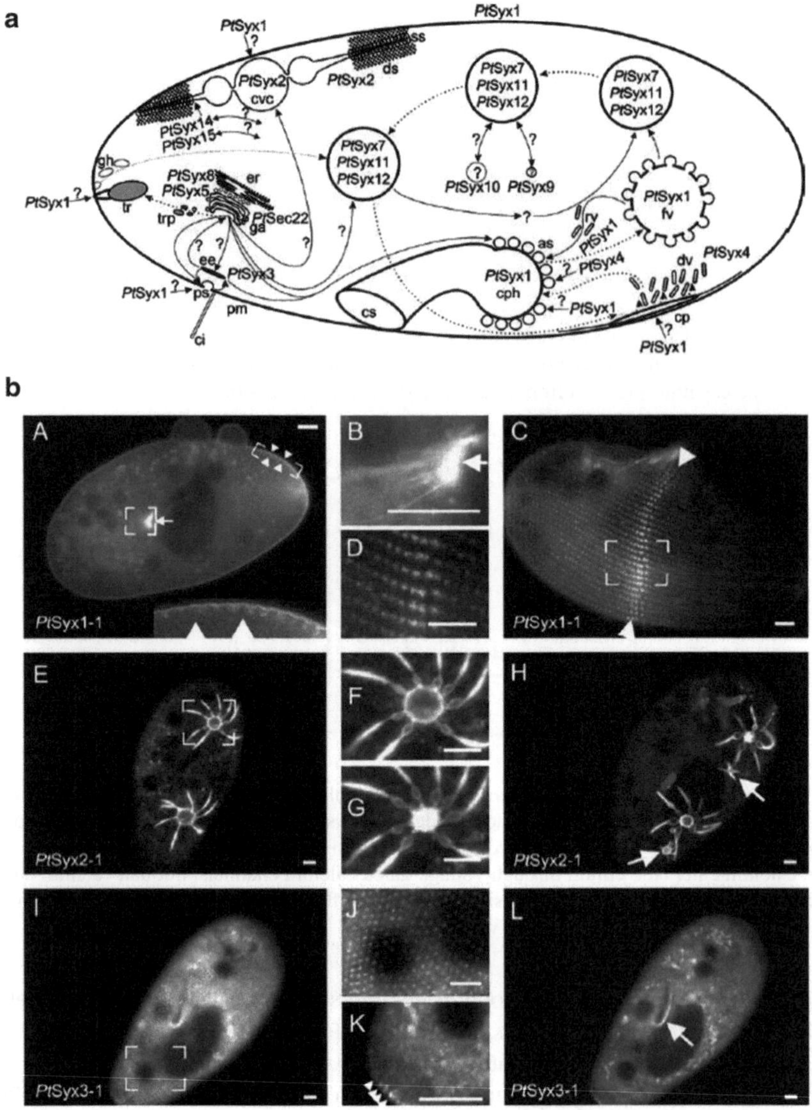

Fig. 13.10 (**a**) The localization of syntaxin isoforms throughout the *Paramecium* cell: *a* ampula, *cv* contractile vacuole, *ds* decorated spongiome, *ss* smooth spongiome. Phagosomal apparatus (*as* acidosomes, *cf* cytopharyngal fibers, *cp* cytoproct, *ci* cilia, *dv* discoidal vesicles) and other recycling vesicles, *rf*. Also in early endosomes, *er* endoplasmic reticulum, *fv* food vacuole, *ga* golgi apparatus, *gh* ghosts of released trichocysts, *oc* oral cavity, *pm* plasma membrane, *ps* parasomal sacs, *tr* trichocysts, *pof* post oral fibers. (**b**) In vivo labeling of PtSynx1-1, PtSys2-1, and Pt Syx3-1 syntaxins using GFP labels for comparison with the schematic in (**a**). Bars = 10 μm. From Figs. 16 and 5 in Kissmehl et al. (2007); with permission

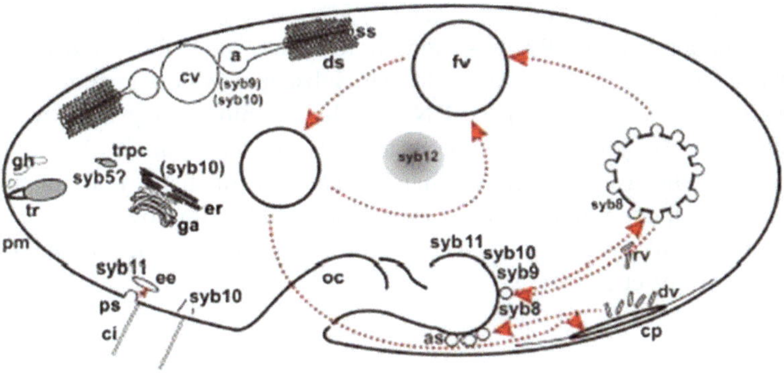

Fig. 13.11 Localization of synaptobrevin isoforms in a *Paramecium* cell. *a* ampula, *cv* contractile vacuole, *ds* decorated spongiome, *ss* smooth spongiome. Phagosomal apparatus (*as* acidosomes, *cf* cytopharyngal fibers, *cp* cytoproct, *ci* cilia, *dv* discoidal vesicles and other recycling vesicles, *rf*). Also in early endosomes, *er* endoplasmic reticulum, *fv* food vacuole, *ga* golgi apparatus, *gh* ghosts of released trichocysts, *oc* oral cavity, *pm* plasma membrane, *ps* parasomal sacs, *tr* trichocysts, *pof* post oral fibers. Figure 14 from Schilde et al. (2010); with permission

13.5 Forward and Reverse Genetics

The technique of RNA interference (RNAi) has made it possible to make null mutant phenocopies in *Paramecium* (Ruiz et al. 1998; Galvani and Sperling 2002). Since knockout gene technology is not possible in *Paramecium*, which lacks homologous recombination, RNAi allows us to make progress quickly without isolating mutants. The phenocopy lasts long enough to carry out even large-scale experiments on many liters of isolated membranes or cilia, for example (Valentine et al. 2012; Yano et al. 2013). In this overview of *Paramecium*, we have seen RNAi in action in dissecting pathways from regulated secretion to intracellular trafficking. The large number of duplicated genes in the *Paramecium* genome does not pose an impediment because an RNAi sequence for one sequence of an ohnolog pair will often downregulate the other member of the pair as well. Double and triple feeding can address the need to downregulate more than two sets of ohnologs at once. Nabi, for example, has concurrently silenced ten or more genes in the striated rootlet gene families (Nabi 2018).

Bacteria that are engineered for the purpose of producing double-stranded (ds) RNA of the desired sequence are fed to the paramecia cells. The ds RNA once inside the *Paramecium* is reduced to short 21–28 nt single-strand RNA. These in turn bind to their target sequences from transcripts of the gene of interest, which are then degraded. Proteins recognized in comparable pathways in other organisms such as Dicers, Piwis, and RdRPs play roles in *Paramecium* RNAi (Marker et al. 2014). Off-target effects can happen, and to reduce them it is possible to analyze sequences for each construct as in *Paramecium* DB (http://paramecium.cgm.cnrsgif.fr/cgi/alignment/off- target.cgi).

In order to better understand the gene-silencing process that underlies RNAi, Meyer and workers combined a forward genetic screen for RNAi factors with whole-genome sequencing to efficiently identify mutations. After UV mutagenesis, they used lethal ds RNA feeding for a gene, *NSF*, that does not a have functionally redundant paralog. If the RNAi system were intact, the cells would not survive, while those with defects in RNAi would live. Subsequent whole-genome sequencing identified two new proteins in the RNAi pathway emphasizing the power of this technology for efficient identification of mutants (Marker et al. 2014).

There have been very effective shared resources for many years in the *Paramecium* research community (Arnaiz et al. 2007, 2009). Improvements in methodology continue to evolve, most recently in transcriptomics across the stages of autogamy (Arnaiz et al. 2017).

13.6 Model for Cilia Development and Function

Cilia that cover *Paramecium* are remarkably similar to cilia across many phyla. They are slender membrane-covered organelles that protrude from the surface. Their cytoskeleton, the axoneme, with nine doublet microtubules and other components and two singlet microtubules in the center of the nine, allows the cilia to beat (Fig. 13.13a, b) (Satir et al. 2014). The dynein arms that interact with the microtubule doublets are Mg-ATPases that make the doublets slide in a coordinated way and make the cilium bend in the graceful pattern (Figs. 13.1b and 13.12a). Their motility is based on ciliary beating at about 10–20 Hz that moves the cells around in their watery environs.

As with other cilia, they grow from base to tip by intraflagellar transport (IFT) from a basal body that docks at the cell membrane. The basal body has triplets of microtubules and a transition zone between the axoneme and itself (Aubusson-Fleury et al. 2012, 2015). This zone gates the proteins and other components that move into and out of the cilium. See the review by Tassin and workers for beautiful images of the structures (Tassin et al. 2016).

The beat diagram in Fig. 13.12a shows that there is a strong (straight) power stroke toward the posterior of the cell and a lazier (curled) return stroke as it moves back toward the anterior. Forward swimming results from this beat. The cells also turn and swim backward. The turn consists of a short backward movement, twirling in place, and renewed forward swimming in a new direction. Figure 13.13 shows the cell observed in 1906 by Jennings (Jennings 1906) as it encounters a solid object, reverses, and renews swimming. In the process of this maneuver, the sweep of the cilium toward the anterior becomes the power stroke, and the stroke toward the posterior is the lazier return. When the cell is switching between these patterns, it tends to twirl in place the cilia straight out from the cell. Thus, their swimming (forward, reverse, and turn) is a read out of their ciliary function, probably observed by the first Dutch microscopists in the 1600s.

Applying electrophysiological approaches to *P. caudatum*, Naitoh, Kaneko, Eckert, and Machemer [reviewed in Machemer (1988a, b))] demonstrated that ciliary ion conductances and membrane potential control the frequency and direction of ciliary beating. These studies earned *Paramecium* its nickname of a swimming neuron. Seminal observations were that (1) ciliary frequency and speed of swimming

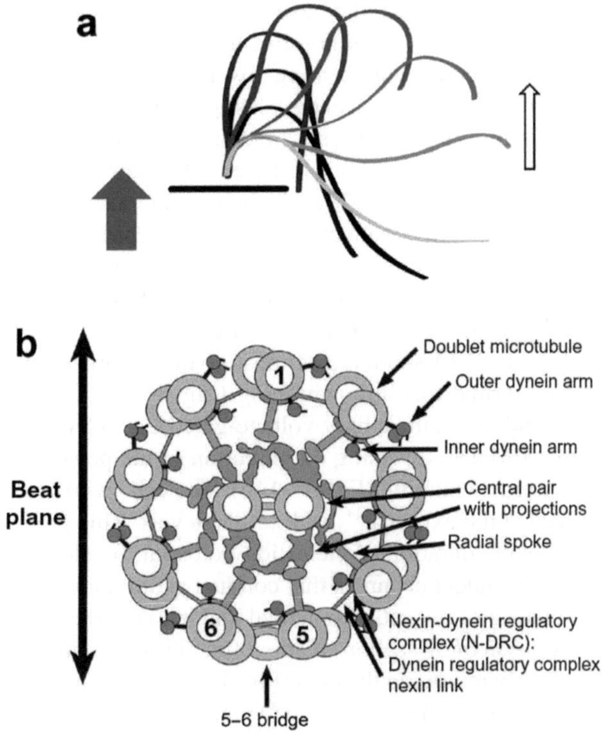

Fig. 13.12 (**a**) Bending patterns of cilia. (a) Black bar is 5 μm; the gray arrow indicates the direction of cell movement; open arrow indicates direction of the forward ciliary bend. Figure 1a from Satir et al. (2014). (**b**) Transverse section of motile cilium illustrating the main features of the 9 + 2 axoneme from motile metazoan cilia, including the outer doublet, central pair of microtubules, outer and inner dynein arms, the central pair projections, the 5–6 bridge, the radial spokes ,and the dynein regulatory complex with nexin links. Figure 2b from Satir et al. (2014); with permission

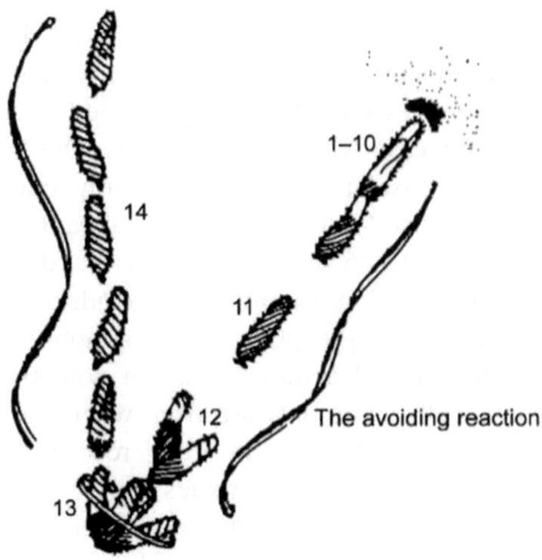

Fig. 13.13 An image based on the sketch of the stages of an avoiding reaction drawn by Jennings (1906). Anterior mechanical stimulation by a cell swimming into an object leads to depolarization, opening of the Cav channels of the cilia, movement of the cell backward for a short time, twirling in place, and forward movement of the cell in a new direction. Depolarization by ionic stimuli causes the same avoiding reaction behavior. From Fig. 5 in Eckert (1972); with permission

are dependent upon the resting membrane potential (Brehm and Eckert 1978; Machemer 1988a, b; Kutomi et al. 2012); (2) stimuli that hyperpolarize the cell slightly increase ciliary beating toward the posterior of the cell and swimming speed; (3) stimuli that depolarize the cell slightly have the opposite effect; (4) depolarization above a threshold initiates a graded Ca^{2+} action potential by opening the voltage-gated Ca^{2+} channels (Ca_v); (5) these channels are located exclusively in the ciliary membrane (Dunlap 1977; Machemer and Ogura 1979); (6) a rapidly activated voltage-gated K^+ conductance (I_{Kv}) concentrated in the cilia (Brehm and Eckert 1978); and (7) a slower Ca^{2+}-activated K^+ conductance (I_{KCa}) also acts to return the membrane potential to the resting level with the Ca^{2+} that activates this K_{Ca} channel coming from the Ca_v channels of the cilia (Satow and Kung 1980a, b).

Figure 13.14 summarizes most of this information. The cell at resting potential shows normal forward swimming (Fig. 13.14a). A sufficient depolarization above threshold opens the ciliary voltage-gated Ca^{2+} channels, increasing intraciliary Ca^{2+} above 100 nM, reversing the direction of the power stroke of the cilia, and causing backward swimming (Fig. 13.14b). The cell swims backward for as long as it takes to reduce the extra Ca^{2+} back to nM levels. This process by export or sequestering outlasts the duration of the action potential. In addition, the same Ca^{2+} can activate Ca^{2+}-dependent channels that conduct sodium ions or magnesium ions and prolong the plateau of the action potential and increase the duration of backward swimming.

A *physical* dissection of some functions of cilia can be achieved in *Paramecium* by cleanly removing the cilia (Fig. 13.15) and using the body for electrical recording (Dunlap 1977; Machemer and Ogura 1979) and the cilia for other purposes like mass spectrometry (Yano et al. 2013). As Preston observes (Preston 1990), *Paramecium* "broadcasts" the activity of ion channels through its swimming behavior. However, this connection between behavior and channel activity inspired a *genetic dissection* of swimming behavior (Kung et al. 1975). Initial studies led to the identification of many mutants such as the Pawns, which cannot reverse their swimming direction. They are devoid of the voltage-gated Ca_v conductance (Kung et al. 1975) (Fig. 13.3), as is the case with deciliated cells (Dunlap 1977). These Pawn mutants have been useful for isolating and studying ion conductances in the absence of the ciliary Ca_v conductance (Satow and Kung 1980a, b).

Figure 13.16 shows the behavioral differences between normal cells and PawnB cells in a barium solution used to induce strong action potentials and turns in normal cells. The long-exposure micrographs show tracks of the normal cells turning repeatedly and a train of action potentials from cell recordings. In contrast, the Pawn cells swim outward from where they had landed in the barium solution with no action potentials in their recording. It should be noted that Kung used *P. tetraurelia* for his studies, but work with *P. caudatum* also yielded mutants with no reversal phenotypes and insights into *Paramecium* behavior (Takahashi 1979; Haga et al. 1983; Matsuda et al. 2000; Oami and Takahashi 2003).

Unlike Pawns, other mutants called Paranoiacs and Pantophobiacs swim backward vigorously upon depolarizing stimulation and spontaneously for long periods of time. Their polar opposites, the Fast mutants, swim quickly and smoothly without turns in the same depolarizing solutions. Remarkably genetic crosses showed that these seemingly different mutants had phenotypes caused by alleles of the *same*

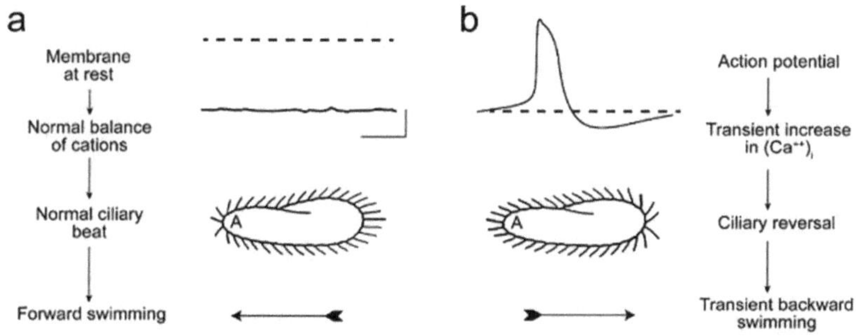

Fig. 13.14 (**a**) These images illustrate that the resting membrane potential of *Paramecium* is negative (about −25 to −40 mV); the corresponding ciliary beat is toward the posterior of the cell, and the cell swims forward. (**b**) In depolarizing solutions, such as high K^+ or Ba^{2+}, the cell's plasma membrane depolarizes and reaches threshold for the action potential. During the action potential, Ca^{2+} enters the cilia through voltage-gated channels; the high levels of Ca^{2+} change the power stroke of the cilia, which now beat most strongly toward the anterior and move the cell backward. The action potential is quickly terminated and the Ca^{2+} removed from or sequestered in the cilia, allowing ciliary beat and swimming to return to normal. From Fig. 1 in Kung et al. (1975); with permission

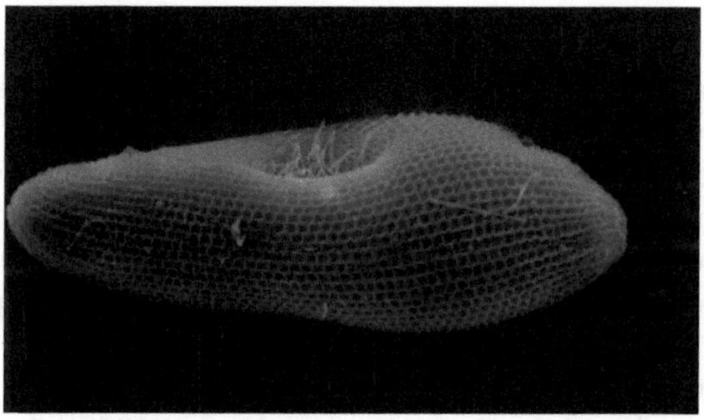

Fig. 13.15 Scanning electron micrograph image of *P. tetraurelia* cell that was deciliated by ethanol and trituration treatment. Cells survive deciliation and can be used for electrical recording or other studies. Unpublished data, J. Van Houten

gene. The gene coded for calmodulin and the analysis of these mutants resulted in a dissection of the calmodulin molecule. Amino acid substitutions in the C-terminal lobe caused a long backward swimming phenotype because the mutant calmodulin did not activate the K_{Ca} channel that is critical for repolarizing the membrane after the initiation of the action potential. Substitutions in the N-terminal domain caused a failure of backward swimming because the "Fast" mutant calmodulin failed to interact with the Ca^{2+}-activated Na^+ (Na_{Ca}) channel that, when open, should prolong the action potential and cause backward swimming (Kung et al. 1992).

Figure 13.13 illustrates the avoiding reaction, which can be initiated by ions that depolarize the cell in the response of *Paramecium* to a mechanical stimulus, i.e., bumping into an obstacle. The reaction illustrated is the bumping of the anterior,

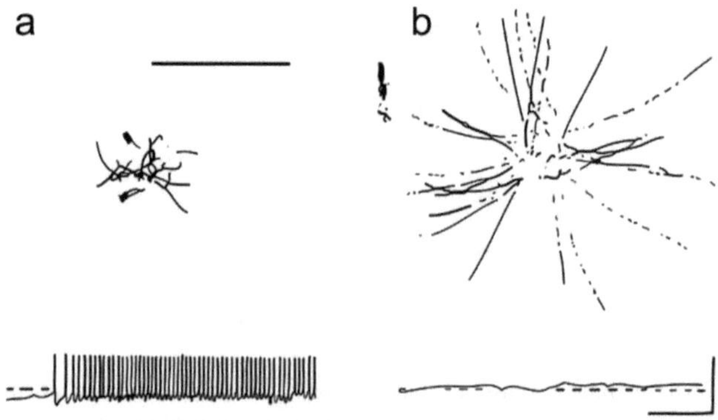

Fig. 13.16 Recordings and behavior of wild-type and pawn mutants of *Paramecium* in a depolarizing stimulus, barium chloride ($BaCl_2$). (**a**) When wild-type cells are placed in a solution of $BaCl_2$, they jerk about with rapid alternating backward and forward movements. These backward bouts correspond to the action potentials shown in the recording below. The dotted line is 0 mV. (**b**) Pawn mutants, when similarly placed in $BaCl_2$, depolarize, as shown by the potential recording above 0 mV. However, the cells never reverse their swimming direction or ciliary beating. Instead they swim smoothly away from where they were placed. From Fig. 5 in Kung et al. (1975); with permission

which sets up a depolarization by soma channels. This depolarization, in turn, activates the ciliary Ca_v channels to drive the behavioral responses of backing up away from a predator perhaps. Mechanical stimulation of the posterior end of *Paramecium* leads to hyperpolarization and fast swimming away from the presumed predator. The hyperpolarizing K^+ conductance from posterior stimulation is unchanged by deciliation. While the depolarizing Ca^{2+} receptor current from anterior stimulation is unchanged by deciliation, there are no Ca^{2+} action potentials without cilia (Machemer and Ogura 1979).

Environmental stimuli control behavior through the modulation of membrane potential and ciliary ion channel activity (Van Houten 1979). Attraction and repulsion of cells to chemical stimuli correlate with the modulation of speed and frequency of turning (i.e., frequency of action potentials) (Van Houten 1998; Valentine et al. 2010). In general, attractants hyperpolarize the cell, suppress action potentials, and cause fast swimming that leads to accumulation of a population of cells at the site of the stimulus. Repellents depolarize, triggering action potentials, and turns that cause the cells to disperse away from the site of the stimulus. Pawn mutants, which cannot turn for lack of ciliary voltage-gated Ca^{2+} channel activity, in general, fail to be attracted or repelled.

The signal information flow can go in the opposite direction. Ion channel activities of the cilia can affect signal transduction pathways, whose second messengers have secondary effects on ciliary motility. For example, hyperpolarization stimulates the adenylyl cyclase activity of the cilia and cell body (Schultz et al. 1992). Cyclic AMP produced in the cilia modifies the ciliary beat direction and frequency making the cell swim fast forward (Noguchi et al. 2004; Hamasaki et al. 1991). Schultz and workers identified a possible physical explanation (Schultz et al.

1992) by showing that an adenylyl cyclase and a K channel were domains of the same protein (Weber et al. 2004). This protein was found expressed in cilia and by proteomic analysis (Yano et al. 2013) and also in the cell membrane. This intriguing K channel does not participate in the repolarization of the action potential but may explain the long observed coupling of hyperpolarization with rapid cAMP production and rapid smooth swimming in response to the attractant glutamate (Yang et al. 1997).

The long wait for the identification of the voltage-gated Ca^{2+} channels and answer to the question of why Pawn cells do not back up: The voltage-gated Ca^{2+} channels that are not functional or missing from the Pawn cilia were physiologically characterized almost 50 years ago (Eckert 1972; Eckert and Naitoh 1972; Machemer 1988a, b). The mutants were isolated in 1969 (Kung 1971). It took until 1998–2000 to clone *PawA* and *PawB* (Haynes et al. 1998, 2000) by very heroic efforts and *PawC* still has not been identified. The elusive channels were not identified until the annotated *P. tetraurelia* genome became available. Since then we identified the genes that code for the voltage-gated Ca^{2+} channels of the cilia. There are three very large proteins (250 kd) that we tagged and expressed, which localize to the cilia. RNAi shows that they are the ones mediating the ciliary reversals. With these tools in hand, we showed that the Pawn mutants PwA and PwB lack these channels in their cilia, but once they are restored to the cilia, the Pawn mutants can swim backward (Lodh et al. 2016).

It is also possible to express tagged channel proteins and identify their location in the cilia by immunofluorescence. Figure 13.17 shows such an effort to visualize a K channel in cilia (SK1a), a folate chemoreceptor that has very different distributions from PKD2 TRP channel. Also shown are cilia and basal bodies.

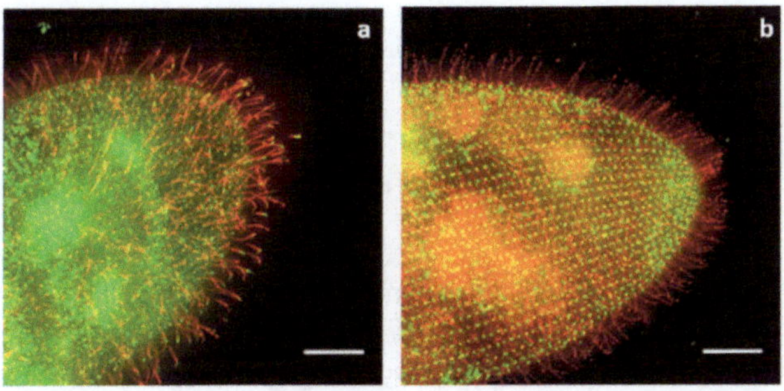

Fig. 13.17 (**a**) Wild-type *Paramecium* cell with the FLAG-epitope-tagged small-conductance Ca^{2+}-activated K^+ channel SK1a. The FLAG-tagged channel is visualized with anti-FLAG primary and red fluorescent secondary antibody. The green fluorescence of the cell body and spots in the cilia are from an antibody against the folate chemoreceptor and secondary antibody. Note the red cilia, indicating that the SK1a channel localizes primarily to cilia. (**b**) Wild-type *Paramecium* cell with the PKD2 channel FLAG-epitope tagged. The red fluorescence indicates that this channel is present in cilia and also on the general cell surface. The green punctate fluorescence on the cell body is from the primary and secondary antibodies used to identify centrin, a protein known to be present in basal bodies at the base of the cilia. The scale bar indicates 10 μm. From Fig. 6 in Valentine et al. (2012); with permission

There are many more in the zoo of mutants like eccentrics (Preston et al. 1992; Preston and Kung 1994; Haynes et al. 2002) and conductances that can be explored. The combination of electrophysiology, forward genetics, proteomics, tagged gene expression through microinjection of plasmids into the macronucleus, RNAi, and genomics/transcriptomics now makes it much more feasible to investigate these mutants and phenotypes quickly. It is routine to find that articles about *P. tetraurelia* in particular take this multidisciplinary approach to answer questions about complex interactions.

13.7 Basal Bodies and Cortical Inheritance

The surface of *Paramecium* is organized into hexagonal cortical units, each with one or two basal bodies and usually one cilium emanating from one of the basal bodies. The classic drawing by Allen in Fig. 13.18 (Allen 1971) shows the cortical units as one looks down on the surface; each one is defined by subsurface ridges with the cilia poking up through the centers of the units. Beneath the cilia are the basal bodies and their rootlets of microtubules and striated rootlet (SR) proteins. The SRs course toward the cell's anterior (top of the page) and cross through several cortical units through the ridges. Beneath all of this is a mesh called the infraciliary lattice (ICL) composed of various centrins and their binding proteins (Beisson et al. 2001). Figure 13.18 shows a cell whose cilia have been removed by gentle treatment with alcohol and pipetting (Van Houten, unpublished). This treatment reveals the cortical units with a stub of a cilium showing in most.

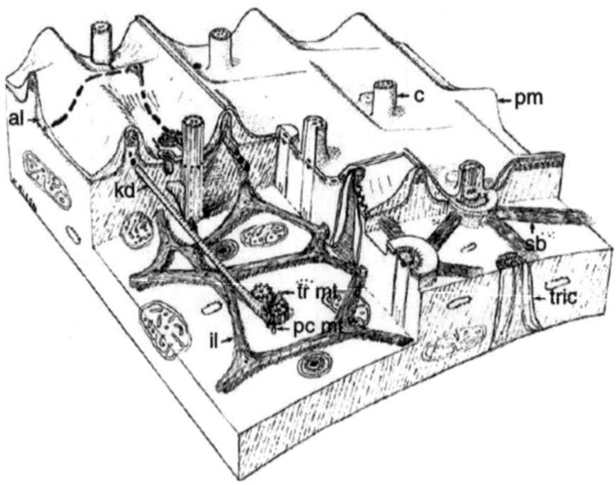

Fig. 13.18 Three D reconstruction of the cortical architecture of *P. multimicronucleatum* from ultrathin sections. *Al* alveolus, *c* cilium, *il* infraciliary lattice, *kd* kinetodesmal fiber, *pm* plasma membrane, *pc mu* post-ciliary microtubule rootlets of the basal body, *ps* parasomal sac, *sb* striated rootlet, *tric* trichocyst, *trmt* transverse microtubule rootlet. Originally from J. Cell Biol 49: 1–20 (1971); reprinted in Hausmann and Allen (2010); Allen image library, with permission

Looking back at Fig. 13.1b, the cell appears covered in cilia with the cilia emanating from the cortical units running in clear parallel rows between the posterior and anterior. As the cells divide to increase their numbers, the basal bodies must duplicate and develop more cilia while maintaining their strict pattern. If this pattern is followed, cilia will beat with the power stroke toward the posterior, and the spacing between cilia will be sufficient for metachrony. There is no random orientation of new cilia upon duplication of basal bodies because the surface pattern is dictated by the one already in place. A process called cortical inheritance. A critical benefit of this non-Mendelian inheritance is that even slight deviations from the very orderly surface pattern stand out, even of the effects are subtle. For example, in Fig. 13.19, the basal bodies are in green, and the SR that runs through several cortical units toward the anterior is in red. The control cell shows beautifully how the rows of basal bodies run straight between the posterior and anterior. The cell treated with RNAi for an SR component loses the straight orientation, and it is easy to observe that there is a change in cell surface (Nabi 2018). This is an advantage over the use of cells with primary cilia which are limited to one cilium per cell.

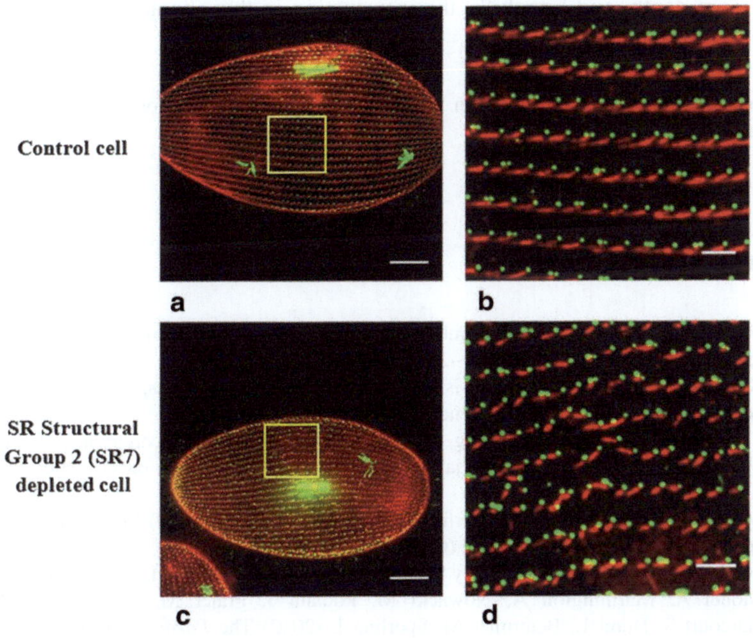

Fig. 13.19 SR gene depletion can lead to basal body row misalignment and SR abnormal appearance. Panels (**a**) and (**c**) show control and SR7-depleted cells, respectively. The yellow box in each image is enlarged (**b** and **d**) to highlight basal body rows. The basal bodies are green (anti-ID5) and SRs are red (anti-SR). Panels (**a**) and (**b**) show straight rows of basal bodies as well as SRs that extend from the posterior pole toward the anterior pole. Panels (**c**) and (**d**) show the severely misaligned basal body rows as well as SRs with abnormal appearance (shorter and not pointed toward the anterior pole). Misalignment occurs over the entire cell surface except oral groove. Scale bars are 15 mm (**a** and **c**) and 3 mm (**b** and **d**). M. Nabi PhD thesis, University of Vermont (Nabi 2018); with permission

The *Paramecium* cell is not a flat surface; it has cilia in its oral groove as well as on the curved surface. The detailed work of Iftode and others (Iftode et al. 1989) carefully documents the waves of basal body duplications and other cortical components that recapitulate the *Paramecium* surface (Beisson and Sonneborn 1965).

13.8 Conclusion

In conclusion, the highlights of this chapter are intended as an entry point into various aspects of *Paramecium* biology and research. Topics not touched on include tubulin modifications (Louka et al. 2018) and immobilization antigens, the major proteins of the surface that are tethered to the outer leaflet of the cell by a glycosyl phosphatidylinositol anchor (Paquette et al. 2001; Capdeville and Benwakrim 1996). *Paramecium* species have served for many years as model systems for endosymbiosis (Görtz 1988; Beisson et al. 2010). The topics covered of course have much more depth, and hopefully the information in this chapter helps to direct the reader to more material in depth.

Acknowledgments I thank Dr. Megan Valentine and Dr. Ashik Nabi for contributing images to this chapter.

References

Allen RD (1971) Fine structure of membranous and microfibrillar systems in the cortex of Paramecium caudatum. J Cell Biol 49:1–20

Allen S, Nowacki M (2017) Necessity is the mother of invention: ciliates, transposons, and transgenerational inheritance. Trends Genet 33:197–207

Arnaiz O, Cain S, Cohen J, Sperling L (2007) ParameciumDB: a community resource that integrates the Paramecium tetraurelia genome sequence with genetic data. Nucleic Acids Res 35:D439–D444

Arnaiz O, Malinowska A, Klotz C, Sperling L, Dadlez M, Cohen J (2009) Cildb: a knowledgebase for centrosomes and cilia. Database (Oxford) 2009:bap022

Arnaiz O, Mathy N, Baudry D, Malinsky S, Aury J, Denby W, Garnier O, Labadie K, Lauderdale B, LeMouel A, Marmington A, Nowacki M, Poulain J, Prajer M, Wincker P, Meyer E, Duharcourt S, Duret L, Betermier M, Sperling L (2012) The *Paramecium* germline genome provides a niche for intragenic parasitic DNA: evolutionary dynamics of internal eliminated sequences. PLoS Genet 8:e1002984

Arnaiz O, Van Dijk E, Bétermier M, Lhuillier-Akakpo M, de Vanssay A, Duharcourt S, Sallet E, Gouzy J, Sperling L (2017) Improved methods and resources for Paramecium genomics: transcription units, gene annotation and gene expression. BMC Genomics 18:483

Aubusson-Fleury A, Lemullois M, de Loubresse N-G, Laligné C, Cohen J, Rosnet O, Jerka-Dziadosz M, Beisson J, Koll F (2012) The conserved centrosomal protein FOR20 is required for assembly of the transition zone and basal body docking at the cell surface. J Cell Sci 125: 4395–4404

Aubusson-Fleury A, Lemullois M, Bengueddach H, Abdallah S, Shi L, Cohen J, Tassin AM, Koll F (2015) Transition zone: the sequential assembly of its components parallels its dual role in basal body anchoring and ciliary function. Cilia 4(Suppl 1):P26

Aury JM, Jaillon O, Duret L, Noel B, Jubin C, Porcel BM, Segurens B, Daubin V, Anthouard V, Aiach N, Arnaiz O, Billaut A, Beisson J, Blanc I, Bouhouche Camara F, Duharcourt S, Guigo R, Gogendeau D, Katinka M, Keller AM, Kissmehl R, Klotz C, Koll F, LeMouel A, Lepere G, Malinsky S, Nowacki M, Nowak JK, Plattner H, Poulain J, Ruiz F, Serrano V, Zagulski M, Dessen P, Betermier M, Weissenbach J, Scarpelli C, Schachter V, Sperling L, Meyer E, Cohen J, Wincker P (2006) Global trends of whole-genome duplications revealed by the ciliate *Paramecium tetraurelia*. Nature 444:171–178

Beisson J, Sonneborn TM (1965) Cytoplasmic inheritance of the organization of the cell cortex in *Paramecium aurelia*. Proc Natl Acad Sci USA 53:275–282

Beisson J, Clerot JC, Fleuro-Aubusson A, Garreau de Loubresse N, Ruiz F, Klotz C (2001) Basal body-associated nucleation center for the centrin-based cortical cytoskeletal network in Paramecium. Protist 152:339–354

Beisson J, Betermier M, Bre M-H, Cohen J, Duharcourt S, Duret L, Kung C, Malinsky S, Meyer E, Preer J, Sperling L (2010) *Paramecium tetraurelia*: renaissance of an early unicellular model. In: Crotty D, Grann A (eds) Emerging model organisms, vol 2. Cold Spring Harbor Press, New York, pp 1–30

Bischerour J, Bhullar S, Wilkes C, Regnier V, Mathy N, Dubois E, Singh A, Swart E, Arnaiz O, Sperling L, Nowacki M, Betermier M (2018) Six domesticated PiggyBac transposases together carry out programmed DNA elimination in *Paramecium*. eLife 7:e37927

Brehm P, Eckert R (1978) An electrophysiological study of the regulation of ciliary beating frequency in *Paramecium*. J Physiol 283:557–568

Buonanno F, Harumoto T, Ortenzi C (2013) The defensive function of trichocysts in *Paramecium tetraurelia* against metazoan predators compared with the chemical defense of two species of toxin-containing ciliates. Zool Sci 30:255–261

Capdeville Y, Benwakrim A (1996) The major ciliary membrane proteins in Paramecium primaurelia are all glycosylphosphatidylinositol-anchored proteins. Eur J Cell Biol 70:339–346

Chalker DL, Stover NA (2007) Genome evolution: a double take for *Paramecium*. Curr Biol 17: R97–R99

Chang S-Y, Van Houten JL, Robles LJ, Lui SS, Kung C (1976) An extensive behavioural and genetic analysis of the Pawn mutants in *Paramecium aurelia*. Gene Res 23:165–173

Cohen J, Beisson J (1980) Genetic analysis of the relationships between the cell surface and the nuclei in *Paramecium tetraurelia*. Genetics 95:797–818

Dunlap K (1977) Localization of calcium channels in *Paramecium caudatum*. J Physiol 271: 119–133

Eckert R (1972) Bioelectric control of ciliary activity. Science 176:473–481

Eckert R, Naitoh Y (1972) Bioelectric control of locomotion in the ciliates. J Protozool 19:237–243

Galvani A, Sperling L (2002) RNA interference by feeding in *Paramecium*. Trends Genet 18:11–12

Görtz H-D (1988) Endocytobiosis. In: Görtz H-D (ed) Paramecium. Springer, Berlin

Haga N, Saimi Y, Takahashi M, Kung C (1983) Intra- and interspecific complementation of membrane-inexcitable mutants of *Paramecium*. J Cell Biol 97:378–382

Hamasaki T, Barkalow K, Richmond J, Satir P (1991) cAMP-stimulated phosphorylation of an axonemal polypeptide that copurifies with the 22S dynein arm regulates microtubule translocation velocity and swimming speed in *Paramecium*. Proc Natl Acad Sci USA 88:7918–7922

Hausmann K, Allen RD (2010) Electron microscopy of *Paramecium* (Ciliata). In: Müller-Reichert T (ed) Methods in cell biology, vol 96. Academic Press, New York, pp 143–173

Haynes WJ, Vaillant B, Preston RR, Saimi Y, Kung C (1998) The cloning by complementation of the pawn-A gene in *Paramecium*. Genetics 149:947–957

Haynes WJ, Ling KY, Preston RR, Saimi Y, Kung C (2000) The cloning and molecular analysis of pawn-B in *Paramecium tetraurelia*. Genetics 155:1105–1117

Haynes WJ, Kung C, Saimi Y, Preston RR (2002) An exchanger-like protein underlies the large Mg2+ current in *Paramecium.* Proc Natl Acad Sci USA 99:15717–15722

Husser M, Hardt M, Blanchard MP, Hentschel J, Klauke N, Plattner H (2004) One-way calcium spill-over during signal transduction in *Paramecium* cells: from cortex into cilia but not the reverse. Cell Calcium 36:349–358

Iftode F, Cohen J, Ruiz F, Rueda AT, Chen-Shan L, Adoutte A (1989) Development of surface pattern during division in *Paramecium.* I. Mapping of duplication and reorganization of cortical cytoskeletal structures in the wild type. Development 105:191–211

Jennings H (1906) Behavior of the lower organisms. Indiana University Press, Bloomington

Kissmehl R, Sehring I, Wagner E, Plattner H (2004) Immunolocalization of actin in *Paramecium* cells. J Histochem Cytochem 52:1543–1559

Kissmehl R, Schilde C, Wassmer T, Danzer C, Nuehse K, Lutter K, Plattner H (2007) Molecular identification of 26 syntaxin genes and their assignment to the different trafficking pathways in *Paramecium.* Traffic 8(5):523–542

Klauke N, Blanchard MP, Plattner H (2000) Polyamine triggering of exocytosis in *Paramecium* involves and extracellular Ca2+/(polyvalent)cation-sensing receptor, subplasmalemmal Ca-store mobilization and store-operated Ca2+-influx via unspecific cation channels. J Membr Biol 174:141–156

Kung C (1971) Genic mutants with altered system of excitation in *Paramecium aurelia.* II. Mutagenesis, screening and genetic analysis of the mutants. Genetics 69:29–45

Kung C, Chang S, Satow Y, Van Houten J, Hansma H (1975) Genetic dissection of behavior in *Paramecium.* Science 188:898–904

Kung C, Preston RR, Maley ME, Ling KY, Kanabrocki JA, Seavey BR, Saimi Y (1992) *In vivo Paramecium* mutants show that calmodulin orchestrates membrane responses to stimuli. Cell Calcium 13:413–425

Kutomi O, Hori M, Ishida M, Tominaga T, Kamachi H, Koll F, Cohen J, Yamada N, Noguchi M (2012) Outer dynein arm light chain 1 is essential for controlling the ciliary response to cyclic AMP in *Paramecium tetraurelia.* Eukaryot Cell 11:645–653

Le Mouel A, Butler A, Caron F, Meyer E (2003) Developmentally regulated chromosome fragmentation linked to imprecise elimination of repeated sequences in paramecia. Eukaryot Cell 2:1076–1090

Lodh S, Yano J, Valentine MS, Van Houten JL (2016) Voltage-gated calcium channels of Paramecium. J Exp Biol 219:3028–3038

Louka P, Vasudevan KK, Guha M, Joachimiak E, Wloga D, Tomasi RFX, Baroud CN, Dupuis-Williams P, Galati DF, Pearson CG, Rice LM, Moresco JJ, Yates JR, Jiang YY, Lechtreck K, Dentler W, Gaertig J (2018) Proteins that control the geometry of microtubules at the ends of cilia. J Cell Biol 217:4298–4313

Machemer H (1988a) Electrophysiology. In: Gortz H-D (ed) Paramecium. Springer, Berlin, pp 186–215

Machemer H (1988b) Motor control of cilia. In: Görtz H-D (ed) Paramecium. Springer, Berlin, pp 216–235

Machemer H, Ogura A (1979) Ionic conductances of membranes in ciliated and deciliated *Paramecium.* J Physiol 296:49–60

Marker S, Carradec Q, Tanty V, Arnaiz O, Meyer E (2014) A forward genetic screen reveals essential and non-essential RNAi factors in *Paramecium tetraurelia.* Nucleic Acids Res 23: 1–13

Matsuda A, Saim Y, Takahashi M (2000) An unusual complementation in non-excitable mutants in Paramecium. Genet Res 76:125–133

Miyake A, Haroumoto T (1996) Defensive function of trichocysts in *Paramecium* against the predatory ciliate *Monadenium balbiani.* Eur J Protistol 32:128–133

Nabi MA (2018) Multiple functions of the striated rootlet proteins of the *Paramecium* basal body. PhD thesis, University of Vermont

Noguchi M, Kurahashi S, Kamachi H, Inoue H (2004) Control of the ciliary beat by cyclic nucleotides in intact cortical sheets from *Paramecium*. Zool Sci 21:1167–1175

Oami K, Takahashi M (2003) K+-induced Ca2+ conductance responsible for the prolonged backward swimming in K+-agitated mutant of *Paramecium caudatum*. J Membr Biol 195:85–92

Orias E, Singh DP, Meyer E (2017) Genetics and epigenetics of mating type determination in *Paramecium* and *Tetrahymena*. Annu Rev Microbiol 71:133–156

Paquette CA, Rakochy V, Bush A, Van Houten JL (2001) Glycophosphatidylinositol-anchored proteins in *Paramecium tetraurelia*: possible role in chemoresponse. J Exp Biol 204:2899–2910

Plattner H (2010) Membrane trafficking in protozoa: SNARE proteins, H+ -ATPase, actin, and other key players in ciliates. Int Rev Cell Mol Biol 280:79–184

Plattner H (2013) Calcium regulation in the protozoan model, *Paramecium tetraurelia*. J Eukaryot Microbiol 61:95–114

Plattner H (2016) Trichocysts—*Paramecium*'s projectile-like secretory organelles. J Eukaryot Microbiol 64:106–133

Plattner H, Klauke N (2001) Calcium in ciliated Protozoa: sources, regulation and calcium-regulated cell functions. Int Rev Cytol 201:115–208

Preston RR (1990) Genetic dissection of Ca2+dependent ion channel function in *Paramecium*. BioEssays 12:273–228

Preston RR, Kung C (1994) Isolation and characterization of *Paramecium* mutants defective in their response to magnesium. Genetics 137:759–769

Preston RR, Martinac Y, SC KC (1992) Genetic analysis of ion channels of prokaryotes and lower eukaryotes. Curr Opin Genet Dev 2:780–784

Ruiz F, Vayssie L, Klotz C, Sperling L, Madeddu L (1998) Homology-dependent gene silencing in *Paramecium*. Mol Biol Cell 9:931–943

Satir P, Heuser T, Sale W (2014) A structural basis for how motile cilia beat. Bioscience 64: 1073–1083

Satow Y, Kung C (1980a) Ca-induced K+-outward current in *Paramecium tetraurelia*. J Exp Biol 88:293–303

Satow Y, Kung C (1980b) Membrane currents of Pawn mutants of the pwA group in *Paramecium tetraurelia*. J Exp Biol 84:57–71

Schilde C, Wassmer T, Mansfeld J, Plattner H, Kissmehl R (2006) A multigene family encoding R-SNAREs in the ciliate *Paramecium tetraurelia*. Traffic 7:440–455

Schilde C, Schönemann B, Sehring IM, Plattner H (2010) Distinct subcellular localization of a group of synaptobrevin-like SNAREs in *Paramecium tetraurelia* and effects of silencing SNARE-specific chaperone NSF. Eukaryot Cell 9:288–305

Schultz JE, Klumpp S, Benz R, Schurhoff-Goeters WJ, Schmid A (1992) Regulation of adenylyl cyclase from *Paramecium* by an intrinsic potassium conductance. Science 255:600–603

Sehring IM, Reiner C, Mansfeld J, Plattner H, Kissmehl R (2007) A broad spectrum of actin paralogs in *Paramecium tetraurelia* cells display differential localization and function. J Cell Sci 120:177–190

Singh D, Saudemont B, Guglielmi G, Arnaiz O, Goût JF, Prajer M, Potekhin A, Przybòs E, Aubusson-Fleury A, Bhullar S, Bouhouche K, Lhuillier-Akakpo M, Tanty V, Blugeon C, Alberti A, Labadie K, Aury JM, Sperling L, Duharcourt S, Meyer E (2014) Genome-defence small RNAs exapted for epigenetic mating-type inheritance. Nature 509:447–452

Sonneborn T (1937) Sex, sex inheritance and sex determination in *Paramecium aurelia*. Proc Natl Acad Sci USA 23:378–385

Sugibayashi R, Harumoto T (2000) Defensive function of trichocysts in *Paramecium tetraurelia* against heterotrich ciliate *Climacostomum virens*. Eur J Protistol 36:415–422

Takahashi M (1979) Behavioral mutants in *Paramecium caudatum*. Genetics 91:393–408

Tassin A-M, Lemullois M, Aubusson-Fleury A (2016) *Paramecium tetraurelia* basal body structure. Cilia 5:6

Valentine M, Yano Y, Van Houten J (2010) Chemosensory transduction in *Paramecium*. Jpn J Protozool 41:1–8

Valentine MS, Rajendran A, Yano J, Weeraratne SD, Beisson J, Cohen J, Koll F, Van Houten J (2012) *Paramecium* BBS genes are key to presence of channels in cilia. Cilia 1:16

Van Houten JL (1979) Membrane potential changes during chemokinesis in *Paramecium*. Science 204:1100–1103

Van Houten J (1998) Chemosensory transduction in *Paramecium*. Eur J Protistol 34:301–307

Weber JH, Vishnyakov A, Hambach K, Schultz A, Schultz JE, Linder JU (2004) Adenylyl cyclases from *Plasmodium*, *Paramecium* and *Tetrahymena* are novel ion channel/enzyme fusion proteins. Cell Signal 16:115–125

Yang W, Braun C, Plattner H, Purvee J, Van Houten JL (1997) Cyclic nucleotides in glutamate chemosensory signal transduction of *Paramecium*. J Cell Sci 110:1567–1572

Yano J, Rajendran A, Valentine MS, Saha M, Ballif BA, Van Houten JL (2013) Proteomic analysis of the cilia membrane of *Paramecium tetraurelia*. J Proteome 78:113–122

Zagulski M, Nowak JK, LeMouel A, Nowacki M, Migdalski A, Gromadka R, Noel B, Blanc I, Dessen P, Wincker P, Keller AM, Cohen J, Meyer E, Sperling L (2004) High coding density on the largest *Paramecium tetraurelia* somatic chromosome. Curr Biol 14:1397–1404

Part III
Evo-Devo in Comparative Context

Chapter 14
Insights into Germline Development and Differentiation in Molluscs and Reptiles: The Use of Molecular Markers in the Study of Non-model Animals

Liliana Milani and Maria Gabriella Maurizii

Abstract When shifting research focus from model to non-model species, many differences in the working approach should be taken into account and usually methodological modifications are required because of the lack of genetics/genomics and developmental information for the vast majority of organisms. This lack of data accounts for the largely incomplete understanding of how the two components—genes and developmental programs—are intermingled in the process of evolution. A deeper level of knowledge was reached for a few model animals, making it possible to understand some of the processes that guide developmental changes during evolutionary time. However, it is often difficult to transfer the obtained information to other, even closely related, animals. In this chapter, we present and discuss some examples, such as the choice of molecular markers to be used to characterize differentiation and developmental processes. The chosen examples pertain to the study of germline in molluscs, reptiles, and other non-model animals.

14.1 Introduction

The evolutionary dynamics ("evo") and the fundamental basis of developmental mechanisms ("devo") are now increasingly addressed as integrated aspects of the same process. Evolutionary developmental biology (evo-devo) enables the formation of a scientific point of view about biological diversity that goes beyond a plain description (Minelli and Fusco 2008). However, to understand the processes that led to the extant biodiversity by connecting the organization plan of body structures to evolutionary aspects, it is evident that more information is needed from a wider

L. Milani (✉) · M. G. Maurizii
Department of Biological, Geological and Environmental Sciences (BiGeA), University of Bologna, Bologna, Italy
e-mail: liliana.milani@unibo.it

W. Tworzydlo, S. M. Bilinski (eds.), *Evo-Devo: Non-model Species in Cell and Developmental Biology*, Results and Problems in Cell Differentiation 68,
https://doi.org/10.1007/978-3-030-23459-1_14

range of animals, and that a restricted model-system perspective represents just the initial phase.

In the past, some model species were chosen because of specific biological features simplifying their use and study in the lab. For example, *Drosophila* and *Caenorhabditis elegans* are small animals, with a short generation time and a small chromosome number, which can be reared in the lab in big numbers allowing many kinds of studies, from population genetics to localization of molecules. Similarly, among vertebrates, zebrafish easily allows the visualization of developmental markers, given its transparency during embryonic development.

In this chapter, we discuss the state-of-the-art of development and differentiation of the germline in non-model animals that show particular features that make them good system of choice for specific biological investigations. When possible, we compare closely evolutionary-related model/non-model groups, and similarities and differences found in different cell types, or during cell differentiation and development, are highlighted.

14.2 The Study of Germline

In animals that reproduce sexually, gametes are the cells that carry the genetic information across generations (Woodland 2016) and meiosis—the process that produces gametes—is the mechanism of cell division specific for the germline. Germ cells are a perfect case study in developmental biology and evolutionary biology, being responsible for the generation of gametes, the reproduction of the individual, and the genetic continuity of the species (Extavour 2007).

The segregation of a gametogenic lineage, or of a pluripotent lineage that can also produce gametes, has been considered a necessary condition for the evolution of multicellular organisms (Extavour 2007). The separation of germ cells from somatic cells involves a set of signals and molecules (RNAs and proteins) localized in the primordial germ cells (PGCs)—the cells that will give rise to the germline of the individual. During development, PGCs can be segregated early from the other cell lineages (preformation mode) or are specified later by inductive signals (epigenesis or induction mode) (Extavour and Akam 2003) (Fig. 14.1). In both cases, members of a conserved gene set (*piwi*, *nanos*, *vasa*, *pl10*, *pumilio*, *tudor*, just to name some) are involved in germline formation (Ewen-Campen et al. 2010; Shukalyuk and Isaeva 2012). This common set of genes was initially named the "germline gene set" (Juliano and Wessel 2010). These genes are mostly involved in the transcriptional repression of somatic programs, in chromatin reorganization, in meiosis control, and in posttranscriptional regulation of mRNAs during gametogenesis and/or embryogenesis. Such molecular determinants are included into the germ plasm—a portion of the cytoplasm inherited through the female gamete—in the preformation mode, or derived by induction from surrounding tissues and de novo synthesis in the embryo following the induction mode of germline specification

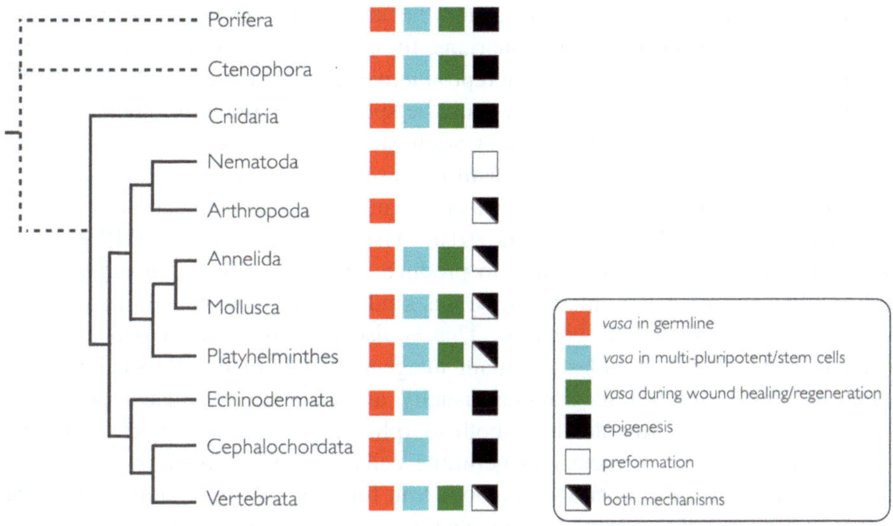

Fig. 14.1 *vasa* gene expression in different cell types and/or cell lineages, and distribution of PGC specification mechanisms across animals. No box means "not found/not expressed." Data from Extavour (2007) and Laura Fierro-Constaín et al. (2017)

(Wessel 2016). These determinants are organized in ribonucleoprotein complexes forming an electron-dense cytoplasmic structure, the so-called nuage (Eddy 1975).

14.2.1 But, Are Germline Ultrastructural and Molecular Markers Germline Specific?

In metazoans, the definition of what is a "germline" was initially based on few bilaterian model animals in which some of the genes named above appear to be germline specific. But this is not a universal feature: there are animals (e.g., Porifera and Platyhelminthes) in which multipotent cells can generate, throughout their lives, both somatic and germ cells (Funayama 2010; Rink 2013), and both germ cells and multipotent progenitor cells express at least part of the germline gene set. From an evolutionary standpoint, a recent work based on a bioinformatic analysis of a wide set of genes in sponges reinforced the hypothesis of the existence of an ancestral multipotency program (Fierro-Constain et al. 2017). Since these genes can be expressed also outside the germline, many authors have suggested to rename them as "germline multipotency program" (GMP) genes, a gene set that would operate in both pluri/multipotent stem cells and germ cells (Juliano et al. 2010) (gene example in Fig. 14.1). Thus, in some animals, GMPs are also expressed in cell types not directly involved in sexual reproduction (Juliano et al. 2010; Leclère et al. 2012; Alié et al. 2015). These cells can have different levels of self-renewal and somatic potential and would all derive from "primordial stem cells" (PriSCs) if we follow the nomenclature of Solana (2013). Planarians, for example, can regenerate germ cells from body fragments entirely lacking reproductive structures (Newmark et al. 2008), suggesting that planarian germ cells could be specified by inductive signals from a self-renewing population of totipotent adult stem cells called the neoblasts (i.e., planarian PriSCs). A similar mechanism is present at least in some cnidarians and sponges, which have populations of cells that can give rise to both germ and somatic cells (Muller et al. 2004; Funayama 2010). The expression of the GMP gene set would maintain these cells capable to produce several cell types, including germ cells (Juliano et al. 2010). Interestingly, this kind of cells show nuage-like structure in their cytoplasm as germ cells do. This is also the case of planarian neoblasts (Solana et al. 2012): neoblasts contain nuage material, in the classical form of electron-dense cytoplasmic granules usually associated with nuclear pores and mitochondria. These granules have both morphological and molecular similarities to the nuage present in metazoan germline cells; so, this cellular morphological feature cannot be used as an unambiguous trait for PGC' identification. Indeed, the nuage observed in the germ plasm of metazoans can be considered as a variant of the "P bodies," the ribonucleoprotein granules with nuage-like structure that occur in many eukaryotic cell types, and contain proteins involved in translational repression and RNA-mediated gene silencing of mRNAs present in the granules (Eulalio et al. 2007; Extavour 2007). Strikingly, many other features described for germline cells are shared by cells with maintained pluripotency, as, for example, posttranscriptional silencing that represses somatic fate (Cinalli et al. 2008; Ewen-Campen et al. 2010; van Wolfswinkel 2014).

While it is clear that many of these genes are conserved across animals, it is less clear to what extent the specific molecular interactions of these mRNAs, proteins, and cellular organelles have changed or remained the same throughout evolution (Ewen-Campen et al. 2010). Germ cells must eventually activate a specific gene regulatory network in order to go through gametogenesis and differentiate into gametes. If the core genes involved are the same, they have to interact and act differently to obtain different cell lineages, with the germ cells being directed toward meiosis and gamete production.

Molecular interactions should represent the "specificity" responsible for germ cell differentiation. However, to date, the knowledge of the molecular functions and biochemical connections between GMP gene products is vague (Ewen-Campen et al. 2010). The niche in which the cells are embedded is also very important for the specification of germ cells (Cinalli et al. 2008). In mammals, pluripotent stem cells can be derived from PGCs when cultured in the presence of fibroblast growth factor (Matsui et al. 1992; Leitch et al. 2010). Moreover, the germ cells removed from frog gonad and placed back into the embryonic blastocoel differentiate into the three primary embryonic germ layers (Wylie et al. 1985). Similar reversions can be observed in *Drosophila* (Extavour 2007 and references therein). So, PGCs can retain pluripotency and, probably, their particular environment (e.g., gonad niche) is responsible for their specific fate inside the organism (i.e., gamete production).

Thus, we have to consider the whole organism and the interactions between cell groups and tissues.

The famous metaphor of "the genetic tool-kit" traditionally used in evo-devo discussions indicates a group of interacting genes with a conserved developmental function, but it also includes transcription factors and cell-signaling and cell-adhesion factors (Wilkins 2013). An important issue in evo-devo is understanding how gene interactions at the base of animal development have evolved while retaining the elements of the genetic tool-kit (Wilkins 2013). The process must involve the evolution of the networks in which the tool-kits are embedded. Indeed, the elements of the genetic tool-kit are part of larger genetic regulatory networks (Davidson 2001). Modification in developmental processes should involve novel interplay of the genetic tool-kit genes with the upstream and downstream genes. Thus, during evolution, the core elements of the genetic tool-kit would be maintained, while the genetic regulatory networks would be reconfigured for different organs or tissues (Wilkins 2013). These "robust" modules controlling developmental processes can remain highly conserved throughout evolution, while their upstream effectors (such as inductive signals/maternal factors, etc.) and downstream targets (e.g., gametogenesis inducers) can change. Such modules appear to be conserved not only in different organisms, but also in different regions and/or times during the development of the same organism (Wagner et al. 2007; Monteiro and Podlaha 2009). So the question shifts from the genes involved in the process, which are conserved, to the complete network of interaction in which they operate.

The cell environment must have an active role in maintaining the germline after its formation—being it firstly determined by preformation or inductive mechanisms—but also in preventing somatic stem cells from producing gametes. In these processes, features of the niche in which the cells are maintained in the animal body must have a leading role.

So far, only a few molecular interactions among gene products of this genetic regulatory network have been described; thus, we know the players but not so much of how they interplay. Of course, to deeply investigate ancestral genetic tool-kit shared by animal multi/pluripotent cells, we need to improve our knowledge on these modular networks. To unveil possible specific interactions acting during germline specification, studies on non-model organisms will be extremely informative when placed into evolutionary context with model organisms.

14.2.2 Mitochondria Contribute to Germ Plasm Formation: An Evolutionary Conserved Feature?

The constant presence of mitochondria close to the nuage might be considered a germline-specific feature.

Mitochondria are associated with the somatic P bodies during the assembly of RNA-induced silencing complex (RISC; Huang et al. 2011; Ernoult-Lange et al.

2012), but, in contrast with what observed in germ cells, this interaction appears to be transitory. In such somatic context, mitochondria can provide locally high quantity of ATP or other metabolites, enhancing the assembly formation, in a manner similar to the role of mitochondria in favoring ATP-driven ion pumps and nuclear import when located close to the plasma membrane and to the nucleus (Ernoult-Lange et al. 2012).

Instead, emission of mitochondrial material merging with the nuage was observed during both male and female germline establishment, suggesting the involvement of these organelles in the process of germ plasm formation. This was documented in a few model organisms such as *Drosophila*, *Xenopus*, the sea urchin, and the mouse (Kobayashi et al. 1993, 1998; Ding et al. 1994; Iida and Kobayashi 1998; Ogawa et al. 1999; Amikura et al. 2001; Reunov and Reunova 2015). Transmission electron microscopy (TEM) analysis documented the extrusion of material and cristae from mitochondria forming the nuage both in gametes during gametogenesis (Reunov et al. 2000; Reunov 2006) and in developing embryos (Reunov et al. 2000; Amikura et al. 2001; Isaeva and Reunov 2001). In *Drosophila* early embryos, a mitochondrial-type translation is required for germ cell formation, which is disrupted by the injection of prokaryotic translation inhibitors (Amikura et al. 2005). Indeed, mitochondrial ribosomes found in the cytoplasm are required to produce proteins necessary for germ cell formation (Kobayashi et al. 1993, 1998; Ding et al. 1994; Iida and Kobayashi 1998; Ogawa et al. 1999; Amikura et al. 2001).

The germ plasm can be identified with TEM by the presence of electron-dense nuage, often positioned near the nucleus. In female germ cells, the nuage together with the associated mitochondria forms a distinctive structure known as Balbiani body (Bb) (Kloc et al. 2004). When the germ plasm segregates to the PGCs, also the associated mitochondrial population segregates to the germline precursors. In male germ cells, the nuage is called chromatoid body (Cb), and also in this case it is associated with mitochondria (Parvinen 2005). The Cb is first clearly seen in mid- and late pachytene spermatocytes as an intermitochondrial dense material. Then, during spermiogenesis, the Cb acquires a lobular structure and is surrounded by a multitude of vesicles, near the Golgi complex, and frequently connected by material continuities with the nucleus through the nuclear pores (Parvinen 2005). Recent findings indicate that microRNA and RNA-decay pathways converge to the Cb, which might function as an intracellular focal domain that organizes and controls RNA processing in male germ cells (Kotaja and Sassone-Corsi 2007). Later in spermiogenesis, at least in mammals, the Cb takes part in the formation of a ring adjacent to the mitochondrial midpiece (Parvinen 2005).

14.3 Bivalve Molluscs: A System of Choice for Studying Gonad and Germline Formation, and Mitochondrial Inheritance

The bivalve molluscs appear to have a preformation mode of germline specification and they have some characteristics that make them a suitable model to study the role of mitochondria in germline formation as well as how mitochondrial inheritance works. However, the lack of genomics data in these animals makes functional experiments very challenging.

Two main features make the bivalve molluscs particularly useful for the study on germline formation and mitochondrial inheritance. Firstly, they are a perfect system for studying the development of the gonads because they do not show an anatomically defined gonads through the majority of the year; the gonad forms every year at the beginning of the spawning season (bivalves have external fertilization), after which it is degraded. The gonadic tissue consists of a series of connected tubules organized in sack-like structures, the acini, in which germ cells differentiate centripetally from the external border to the lumen of the acini, where mature gametes accumulate during the spawning season. During the non-reproductive season, sex cannot be determined, because of the absence of secondary sexual characteristics (Devauchelle 1990; Milani et al. 2011).

Like reptiles and mammals, bivalves show all the stages of gamete differentiation inside the acinus, allowing an easy visualization of markers in cells at different stages of differentiation. Secondly, ~100 species of bivalves (Gusman et al. 2016) show doubly uniparental inheritance (DUI) of mitochondria. In DUI species, both females and males transmit mitochondria to their offspring, in contrast to what happens in the strictly maternal inheritance (SMI) of mitochondria. SMI is typical of metazoans, in which mitochondria are transmitted by the mother only because mitochondria from the spermatozoon that fertilizes the oocyte are excluded or degraded (Milani et al. 2013; Punzi et al. 2018). The high divergence between the two mitochondrial genomes (Zouros 2013) allows for the use of different approaches to investigate and describe the mechanism of DUI, which is still far from being fully understood. What is clear is that, in DUI male embryos, spermatozoon mitochondria form an aggregate (Cao et al. 2004; Cogswell et al. 2006; Milani et al. 2012), and, during development, this aggregate enters male PGCs, as proved by the exclusive presence of M-type mtDNA in spermatozoa (Venetis et al. 2006; Ghiselli et al. 2011). In DUI female embryos, the sperm mitochondria are dispersed and the mitochondrial inheritance through the female line appears to occur as in SMI species (Cao et al. 2004; Cogswell et al. 2006; Milani et al. 2011, 2012).

Thus, the bivalve molluscs are particularly interesting because:

- The gonad seasonal formation can be a useful system of choice for studying reconstitution and renewal of tissues and gametogenesis, key topics in developmental biology.

- Following fertilization, the dynamics of mitochondrial segregation in the germline can be visualized.
- DUI and SMI coexist in many bivalve families, making the arising of DUI an interesting evolutionary issue to be addressed (Milani et al. 2016).
- DUI animals have two highly divergent mitochondrial genomes, one transmitted through eggs, the other through sperm, which could be distinguished with specifically designed molecular markers.

14.3.1 Germline Identification: Using Anti-VASA as Marker

A typical component of germ plasm is the DEAD-box RNA helicase VASA, isolated in many animals (Gustafson and Wessel 2010). In these animals, VASA loss-of-function phenotypes show various germline defects at different stages of germ cell development, supporting specific roles in germline formation (Ewen-Campen et al. 2010). Induction of VASA expression can reprogram and direct chicken embryonic stem cells toward a germ cell fate (Lavial et al. 2009), supporting its function as a germ cell determinant in pluripotent cells (Ewen-Campen et al. 2010). For this reason, VASA is among the members of the GMP, and functions in RNA metabolism, as a local RNA unwinder, translational regulator, and piRNA pathway mediator (Linder and Fuller-Pace 2013; Lasko 2013) (Fig. 14.1). Products of the *vasa* gene, both mRNA and protein, are the most widely used molecular germ cell markers in Metazoa (discussed in Extavour and Akam 2003). However, there is very few information about the exact molecular function of VASA.

From an evolutionary stand point, *vasa* is among the latest innovations of GMP; instead, for example, *Pl10* and some *tudor* genes were probably present in the last eukaryotic common ancestor (LECA) (Fierro-Constain et al. 2017). However, *vasa* origin predates the evolution of the dedicated, metazoan germline (Ewen-Campen et al. 2010). Some studies suggest that *vasa* became germline specific in vertebrates (Sect. 14.5), but, as already pointed out, a different scenario comes up when considering non-vertebrates. Indeed, VASA is expressed also in multipotent somatic stem cells, at least in some invertebrate taxa, suggesting that it functions not only in the germ cells but also in broader aspects of developmental regulation (Gustafson and Wessel 2010; Lasko 2013; Yajima and Wessel 2011, 2015).

Complex regulatory network interactions have been suggested between *vasa* and other GMP genes, including in transcription, translation, and posttranslational regulation. However, very few direct VASA molecular interactors have been identified so far, and the majority of them actually act on VASA localization (Johnstone and Lasko 2004; Ewen-Campen et al. 2010), with no direct indication about VASA function. Several studies suggested that VASA RNA-binding domains are not necessary for VASA localization to the germ plasm, but instead are necessary for VASA function (Liang et al. 1994), but detailed data on that are still missing. For this reason, it is not clear whether VASA interacts differently when expressed in the germ cells or in the multipotent somatic stem cells.

14.3.1.1 VASA Antibodies from Model Organisms

When dealing with non-model organisms, the use of markers available on the market can be a first try in approaching the study. Markers are usually developed against protein homologs of model animals, and a validation in the organism under investigation is always necessary (e.g., western blot analyses, WB).

One of the DUI species utilized for VASA expression investigation was the Manila clam *Ruditapes philippinarum*, a strictly gonochoric species belonging to the family Veneridae. Given VASA sequence conservation, a first try was performed in this species with an antibody developed against the chicken VASA (anti-Cvh), previously used in birds and reptiles (Tsunekawa et al. 2000; Maurizii et al. 2009). The WB showed a single band of a plausible molecular weight deduced from the protein sequence inferred from the mRNA sequence (Milani et al. 2011; Ghiselli et al. 2012). In clam oocytes labeled with anti-Cvh, a spotted staining dispersed in the cytoplasm was found, while a strongly labeled zone was visible along the cleavage furrow in 2-bastomere embryos. This finding was similar to what was observed in chicken embryos using anti-Cvh (Tsunekawa et al. 2000), but also to what was observed in *Xenopus* embryos using other germ plasm markers (Kloc et al. 1998). The presence of VASA since the first cleavage supports a preformation mechanism for germline determination in *R. philippinarum.* However, the distribution pattern of VASA in clam gonadal tissues was not completely resolved by using this antibody.

14.3.1.2 VASA Antibodies Developed Against the VASA Protein Homolog of the Manila Clam

When no clear results are obtained with available antibodies, it can be useful, and sometimes necessary, to develop specific antibodies against the protein homolog of the species of interest. This can be achieved by the following steps: (1) Sanger sequencing of the gene of interest (starting from transcriptomic data that are increasingly present in databases, from which specific primers can be designed); (2) protein sequence inference from mRNA; (3) development of an antibody (there are companies providing such services) either against the whole protein sequence or against short peptides (chosen accordingly to their predicted performances and characteristics, such as antigenicity, solubility, epitope prediction); and (4) validation by WB of antibody specificity.

Indeed, the use of VASA antibodies developed against the VASA homolog of the Manila clam (anti-VASA *R. philippinarum*, anti-VASPH) allowed us to further investigate VASA expression in bivalve species (Milani et al. 2014, 2015a, b, 2017b, 2018). The analysis by WB of two specific antibodies developed against two different short peptides of the protein allowed the detection of two possible isoforms, with slightly different molecular weights, with different sex-specific expression level (Milani et al. 2015b). Moreover, in immunohistological preparations, the pattern of staining in the different cell types allowed the description of protein expression during gametogenesis. This pattern of expression permitted to advance hypotheses on the cell type and stage in which this protein acts. Unfortunately, as said before, when working on the non-model animals, it is not straightforward to plan experiments for protein function verification. Very few attempts are present in bibliography: for example, in vivo RNA interference in *Crassostrea gigas* showed that *vasa* is essential for germ cell development (Fabioux et al. 2009). Thus, the characterization of the pattern of protein expression is often the only feasible approach.

In *R. philippinarum*, the VASA staining was used to visually identify the germline cells. Anti-VASA stained cells might be multipotent cells, and a deep analysis of tissue morphology can help in cell-type identification. In females, as in males, VASPH-stained cells with round nuclei and compact chromatin were localized around the acinus wall; they are probably cells at an initial stage of gametogenesis (Milani et al. 2015a, b, 2017b, 2018). These cell types have to multiply considerably in animals approaching the spawning season and show a strong VASPH labeling. However, in the Manila clam, VASA-stained cells were positioned not only around the forming acini but also in other body niches, such as intestinal epithelium and in the nearby connective tissue (Fig. 14.2). From this observation, and observations on other bivalve species, it was hypothesized that these VASA-stained cells might be the PriSCs that contribute to the rebuilding of reproductive tissue (Milani et al. 2017b, 2018). According to that, the digestive epithelium might be a reservoir of PriSCs that can be deputed to multiple cell fates (Milani et al. 2017b, 2018). VASA-stained cells increase in number during gonad rebuilding and are present in large numbers both in the intestine epithelium and in the connective tissue (Milani et al. 2015a, 2017b, 2018) (Fig. 14.2); in contrast, during the spent phase of the reproductive cycle, these cells are absent from the intestine (Milani et al. 2018). Given the presence in several locations of VASA-stained cells having similar labeling pattern as well as nuclear morphology, it was hypothesized that they may actually be multipotent cells. This is in accordance with what was proposed for many animal species: namely, that VASA-positive cells can be primordial stem cells able to give rise to the germ and somatic cell lines (Juliano et al. 2010; Solana 2013).

Cells with similar rounded and compact nuclear morphology and similar VASA-staining pattern were also detected in other bivalves (Milani et al. 2017b; Cherif-Feildel et al. 2018), even if some anti-VASA antibodies did not detect the VASA-stained cells outside the gonadal tissue (*C. gigas*; Cherif-Feildel et al. 2018). The actual presence of PriSCs in bivalves is a hypothesis that should be tested in the future.

In Manila clam early oocytes, recognizable VASPH-stained spots are present, while an attenuation of the labeling is observed in oocytes during their growth and differentiation, where smaller granules are visible in the cytoplasm (Milani et al. 2015b). However, since yolk can prevent antibody entrance, the investigation of VASA distribution inside oocytes by immunohistochemistry and confocal microscopy is not straightforward. More resolutive results were obtained when TEM was used (as described in Sect. 14.3.2).

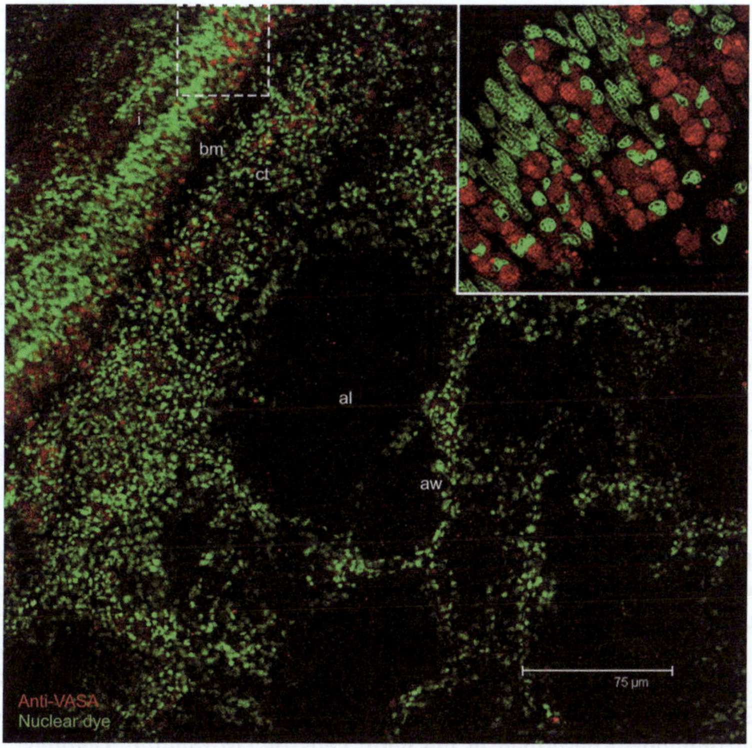

Fig. 14.2 VASA immunolocalization in the Manila clam. Confocal microscopy. Anti-VASPH highlights cells with rounded nucleus localized in the intestinal epithelium, between unstained batiprismatic cells (i), and the basal membrane (bm). Stained cells are also visible in the connective tissue (ct) and in the acinus wall (aw) that surrounds the acinus lumen (al). In the inset, a magnification of the intestinal epithelium in the box. Red: anti-VASA staining; green: nuclear dye (see Milani et al. 2015b)

14.3.2 Mitochondrial Contribution to the Formation of Germ Plasm-Related Structures: Using Anti-VASA and Antibodies Against Mitochondrial Proteins as Markers

Investigation by TEM in *R. philippinarum* allowed the identification of cytological structures related to germline formation, supporting the presence in these clams of a preformation mechanism. In *R. philippinarum*, the nuage was visible in developing oocyte in multiple Bb-like structures, with the nuage surrounded by mitochondria (Milani et al. 2011) (Fig. 14.3). As reported above, VASA is a component of the nuage; thus, the above-mentioned VASPH-stained spots visualized with immunohistochemistry in early oocytes (Sect. 14.3.1) are most likely nuage material.

In differentiating male germ cells, a clearly recognizable Cb was detected with TEM (Milani et al. 2011) (Fig. 14.3). In male acini, anti-VASPH signal decreased during male germ cell maturation showing a small labeled spot near the nucleus in spermatids; in spermatozoa the VASPH labeling, extremely reduced, was observable in the remnants of the cytoplasm after spermiohistogenesis. In this case, anti-VASPH staining could derive from the Cb—usually positioned at the base of the spermhead during spermiogenesis—but also from signal inside the mitochondria that constitute the midpiece (Milani et al. 2015b). Indeed, sperm mitochondria were reported to contain VASA also in the mouse (Reunov and Reunova 2015). That said, the analysis of clam germ cells allowed the recognition of conserved ultrastructure when compared to other animals (Kloc et al. 2004), but a few differences were also observed.

Very recently, studying the formation of germ plasm by immune-electron microscopy, some variability on the general theme was found in *R. philippinarum* (Reunov et al. 2019). The mechanisms of modification of germ plasm-related structures (GRSs) during oogenesis and spermatogenesis are similar in prezygotene/pachytene stages of males and females. During this phase, the shift from mitosis to meiosis is accompanied by (1) the dispersion of VASA-positive granules; (2) their entrance into the nucleus, possibly promoting meiotic remodeling of chromatin; and (3) their entrance into mitochondria, possibly inducing their disassembly (Reunov et al. 2019). The observed rupture of mitochondria and the presence in oocyte cytoplasm of mitochondrially encoded proteins involved in oxidative phosphorylation (Reunov et al. 2019) are in accordance with the cytoplasmic localization of mitochondrial material previously described in the germ cells of *Drosophila*, *Xenopus*, and the mouse (Sect. 14.2.2). In particular, Reunov et al. (2019) described GRS activity both for PGC-derived germ plasm—the germ plasm already present in PGC at their formation—and the additional germ plasm that is newly formed in the oocyte during its maturation; also, the involvement of mitochondria in these processes was considered. PGC-derived germ plasm dynamics are shared by both sexes, while newly formed germ plasm is stored during oogenesis and embryogenesis. Interestingly, in both the cases, there is an interplay between GRS and mitochondria. Moreover, in the Manila clam, a peculiar "germ plasm granule formation complex" (GGFC) was described, that appears morphologically different from the unique Bb described in other animals. The Bb is usually connected with the formation of multiple germ plasm granules (Kloc et al. 2002; Saffman and Lasko 1999; Cox and Spradling 2003). In *R. philippinarum*, multiple GGFCs are present (similar to multiple Bbs), each forming only one germinal granule (Reunov et al. 2019). This peculiar pattern of germ plasm formation might be a specific feature of bivalve molluscs, as a variation on the theme of Bb-like structures described in other animals (Bilinski et al. 2017). Nonetheless, a mitochondrial contribution to germ plasm formation appears to be a shared feature in animals, and, maybe, the most specific marker of germ cells.

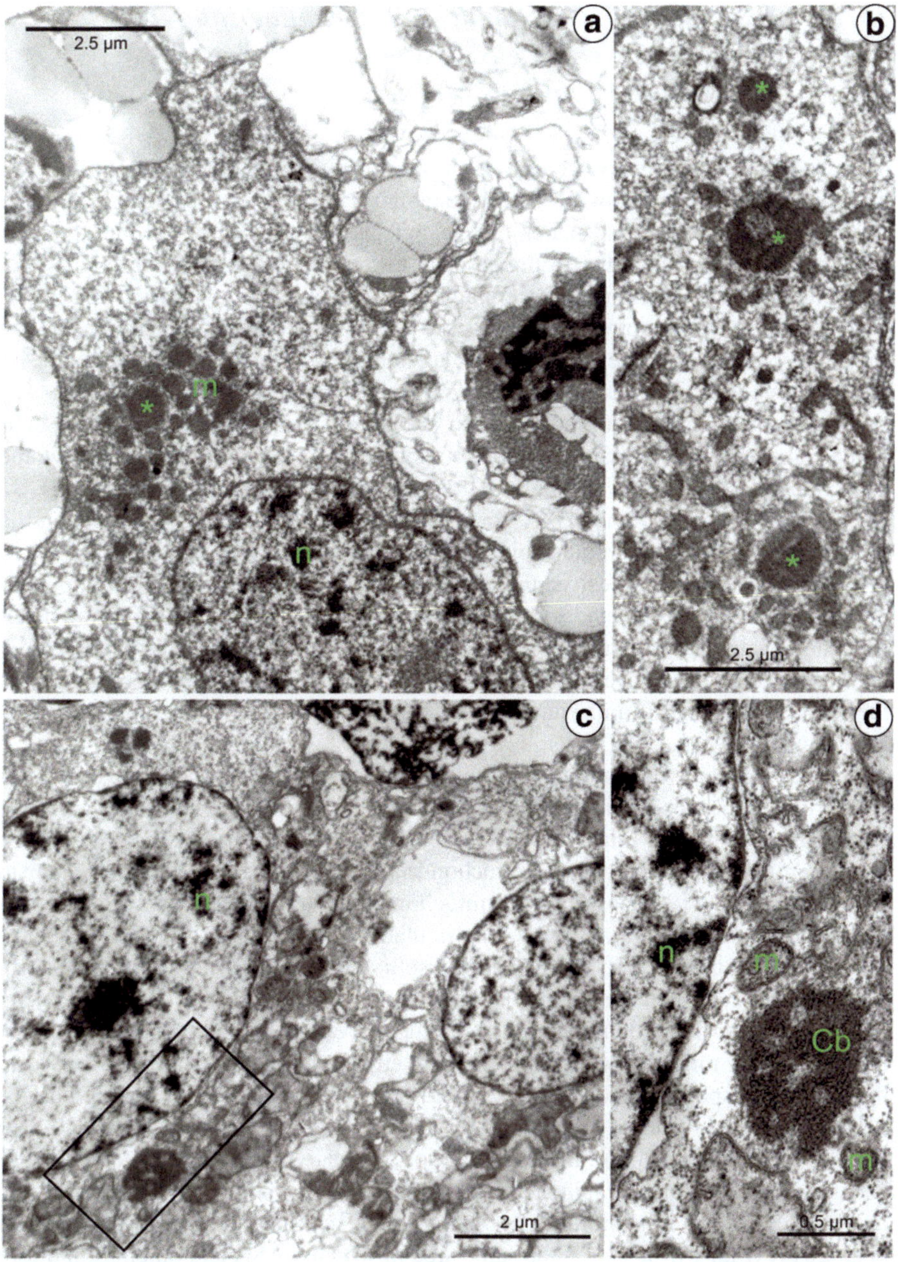

Fig. 14.3 Ultrastructural characterization of nuage structures in the Manila clam. Transmission electron microscopy. (**a**) A Balbiani body (Bb)-like structure with electro-dense, nuage material (asterisk) surrounded by mitochondria (m) close to the nucleus (n) in an oogonium. (**b**) Multiple Bb-like structures inside an oogonium. (**c**) A Chromatoid body (Cb) in a spermatogonium. (**d**) The Cb is located close to the nuclear membrane and surrounded by mitochondria (see Milani et al. 2011)

14.3.3 Dynamics of Mitochondrial Segregation into Primordial Germ Cells: Using MitoTracker and Anti-α-Tubulin as Markers

Mitochondria have an active role in germ plasm formation and consequently in the germline specification. How healthy mitochondria are segregated into the germline of the new individual is a matter of lively debate. DUI species can be a perfect system to address this issue. In these species, the embryo develops through spiral segmentation, with blastomeres receiving a specific content of cytoplasm that can be followed during development up to the formation of specific organs (Hejnol 2010). In this way, it was possible to determine that, in molluscs, the PGCs arise from the 4d mesentoblast that is formed during the sixth cell division (Lyons et al. 2012).

In metazoans, mitochondrial genomes are largely homoplasmic in a given individual, and the Bb mitochondrial population, or a part of it, may represent the mitochondrial bottleneck leading to homoplasmy (Mishra and Chan 2014). In *Drosophila* and *Xenopus*, for example, Bb-derived mitochondria were shown to be preferentially segregated to PGCs together with the nuage (Kloc et al. 2004). It was hypothesized that the organelles forming the Bb (the "mitochondrial cloud") may be selected for their higher functionality, thus assuring the better mitochondrial supply to gametes and consequently to the progeny. However, even if some interesting results were obtained (e.g., Zhou et al. 2010), more investigation on the topic is required to understand the role of selection and drift in mitochondrial inheritance.

In DUI bivalves, it is possible to follow the segregation of spermatozoon mitochondria in embryos. In males, the mitochondria that will populate the germline derive from the 4 to 5 mitochondria carried by the fertilizing spermatozoon, since eggs do not transmit sperm-type mitochondria. In fact, DUI is a stable mechanism and, for many DUI species, the two mitochondrial lineages have evolved independently for hundred million years, accumulating up to 40% of DNA sequence divergence (Zouros 2013; Milani and Ghiselli 2015; Ghiselli et al. 2018). Sperm mitochondria can be stained in vivo and used to fertilize oocytes, and it is possible to visualize sperm-derived mitochondria in the developing embryos under fluorescence/confocal microscope. During the first embryo divisions, mitochondria from the fertilizing spermatozoon are described to be (1) aggregated in male embryos, (2) separated and scattered in female embryos (Milani et al. 2012; Obata and Komaru 2005; Cogswell et al. 2006). In male embryos, it was hypothesized that the midbody derived from the mitotic spindle of the first division may act in positioning the aggregate of sperm mitochondria in the middle of the first cleavage furrow (Milani et al. 2011) (Fig. 14.4). This is the same area in which VASA and other germ plasm constituents are localized, not only in the Manila clam, but also in other animals, such as *Xenopus laevis* and the chicken (Kloc et al. 1998; Tsunekawa et al. 2000). This is an additional clue on the link between the inheritance of germ plasm material and a mitochondrial subset. The use of different type of markers, such as in vivo mitochondrial dyes (Milani et al. 2011, 2012) and microtubule staining (Milani et al.

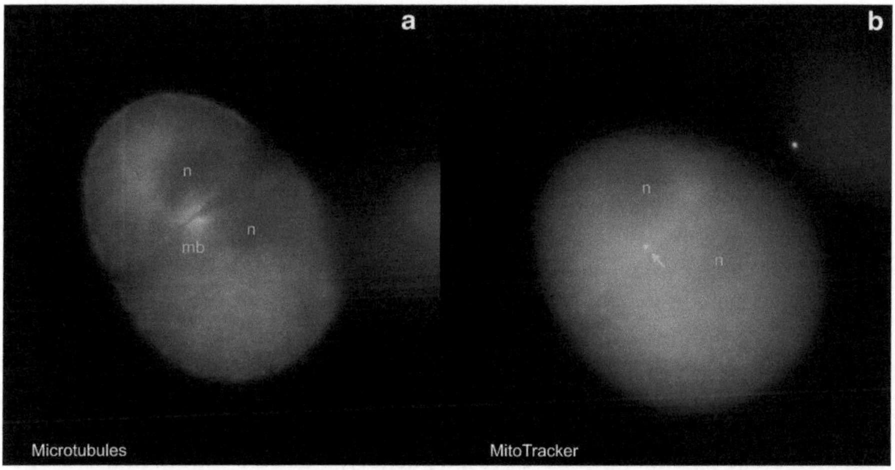

Fig. 14.4 Microtubule midbody and sperm mitochondria in Manila clam 2-blastomere embryos. Fluorescence microscopy. (**a**) Microtubule immunolocalization with anti-α-tubulin showing the midbody (mb) at the middle portion of the first cleavage furrow (n, nuclei). (**b**) MitoTracker Red staining of sperm before fertilization allows to visualize spermatozoon mitochondria (arrow) aggregated to the midbody of 2-blastomere male embryos (see Milani et al. 2011, 2012)

2011, 2012), was important to characterize the mechanisms of this unusual distribution of sperm mitochondria.

The molecular dynamics of mitochondrial segregation during development is still largely unknown. Even if DUI may look like a weird exception to SMI, molecular and phylogenetic evidence suggests that it evolved from SMI, and DUI is considered a variation of the mechanism that regulates the mitochondrial bottleneck in all metazoans (Milani and Ghiselli 2015; Ghiselli et al. 2018). The two systems likely share a basic molecular mechanism of mitochondrial inheritance, making the comparison between DUI and SMI of extreme interest from an evo-devo point of view.

14.3.4 Germline Mitochondrial Activity: Which Mitochondria Are Segregated into the Germline? Using Mitochondrial Dyes Sensitive to Mitochondrial Membrane Potential as Marker

To understand when and how mitochondria that will be transmitted across generations are segregated into the germline, it is fundamental to understand when and how germ cells originate and which mitochondria enter them.

A division of labor between male and female gamete mitochondria was proposed, with oocyte mitochondria being inactive and thus not subjected to oxidative damage by reactive oxygen species (Allen 1996; de Paula et al. 2013). However, it appears that the issue is more complicated. Data on several animals appear to support a high mitochondrial activity in inherited mitochondria, irrespectively of the gamete type carrying them (Milani and Ghiselli 2015). Investigation of mitochondrial ultrastructure, transcriptional activity, and presence of membrane potential—detected with fluorescence microscopy and mitochondrial dyes sensitive to mitochondrial membrane potential as marker—are consistent with mitochondria being active in both gamete types (Milani and Ghiselli 2015; Ghiselli et al. 2018). Indeed, mitochondrial activity may be the way to select the best performing organelles to be transmitted across generations (Milani and Ghiselli 2015). For example, mitochondria with high membrane potential are preferentially transported because of the higher level of energy produced, allowing their transport along the microtubule network (MacAskill and Kittler 2009), as could be the case for the positioning of sperm mitochondria in the embryonic first cleavage furrow in DUI species (Sect. 14.3.3 and Fig. 14.4). But also the transmission of Bb-mitochondria to embryonic PGCs may be favored by a high membrane potential (Milani 2015), a phenotype that Bb-mitochondria have in *X. laevis* (Wilding et al. 2001) and zebrafish (Zhang et al. 2008).

So, which mitochondria are segregated to PGCs? It was proposed that it would depend on two, somewhat linked, events: (1) the type of germline specification and the stage of development in which it takes place and (2) the degradation of the mitochondria from the spermatozoon and the stage of development in which it takes place (Milani 2015). In the preformation scenario, an early degradation of sperm mitochondria usually allows the segregation in the germline precursors of oocyte mitochondria with the highest membrane potential, that is, those of the Bb. However, if sperm mitochondria are not degraded, they can be segregated to the germline (possibly thanks to their higher membrane potential); this can be either accidental or the rule—so far, DUI is the only example for the latter. In the epigenesis scenario, germline specification is driven by inductive signals from surrounding tissues at a later developmental stage, and different pathways of sperm mitochondria elimination are possible, generally allowing oocyte mitochondria to be segregated into embryo germ cells.

14.4 Germline Specification in Vertebrates

Also in vertebrates, PGCs' specification occurs via two mechanisms: maternal specification (or preformation) and induction (or epigenesis) (reviewed in Extavour and Akam 2003). Independent of the mode of PGCs' formation, their precursors are the pPGCs (Swartz and Wessel 2015). The term "pPGCs" was introduced by Nieuwkoop and Sutasurya (1979), but the term has remained unused until the discovery of "blimped" Blimp1-expressing pPGCs in the mouse. "Blimped" pPGCs' fate is restricted to a PGC-fate in PGCs, that have a PGC-specific transcriptional signature, with a severe loss of DNA methylation, and an increased doubling time (Lawson and Hage 1994; Seki et al. 2005). The pPGCs divide into PGCs, but can adopt different fates if placed (or transplanted) in a different niche in the embryo.

In contrast, the PGCs retain their identity even if transplanted (de Sousa Lopes et al. 2007; Tam and Zhou 1996) and also during their migratory phase to the gonads. The preformation mechanism requires the presence of maternally generated determinants that, together with a mitochondrial cloud, constitute the germ plasm (Sects. 14.2 and 14.3). This maternal material is then inherited by the pPGCs and the PGCs. Thus, in the preformation mode, the germ cells are specified before gastrulation by germ plasm stored in the egg. In the inductive mechanism, the PGCs are formed by the inducing signals secreted by the embryonic tissues. The cells of the late blastula-early gastrula that would not normally form the germ cells can be induced to form them by exposure to an appropriate embryonic tissue or signaling molecules (Sutasurya and Nieuwkoop 1974; reviewed in Bachvarova et al. 2009a).

Vertebrate models used to investigate germ cell specification are zebrafish (*Danio rerio*) and the frog *X. laevis*, as representative of the anamniotes, and the chicken *Gallus gallus* and the mouse *Mus musculus*, as representatives of the amniotes. In zebrafish, *Xenopus*, and the chicken, the germ cell specification occurs with a preformation mechanism, while in the mouse, the germ cells (although oocytes contain a Bb in some meiotic phases: Kloc et al. 2008; Pepling et al. 2007) form by epigenesis. Although the initiation of germline development can occur through these two distinct mechanisms, many PGC-specific genes involved in the inductive mechanism are also components of the maternally synthesized germ plasm, when it is present. Interestingly, a number of these genes encode RNA-binding proteins that function in regulating zygotic genome activity, totipotency, proliferation, differentiation, and migration of PGCs (Yang et al. 2015).

14.5 Germline Specification in Vertebrate Model Systems

14.5.1 Anamniotes

Among vertebrate developmental models, the mechanisms that direct PGC commitment in *Xenopus* and the mouse are understood at the finest detail. Studies on anuran amphibian embryos have provided some of the first experimental evidence of preformation and of the role of germ plasm in vertebrate germ cell specification (Bounoure 1934). In *Xenopus*, pPGCs derive from blastomeres that inherit germ plasm localized in the vegetal cortex of the egg. Following fertilization, the vegetal plasm forms patchy aggregates in the vegetal hemisphere, which segregate unequally into blastomeres and eventually accumulate specifically in a few cells that become the PGCs. Several experiments confirmed that the vegetal plasm contains germ cell determinants; indeed, transplanted germ plasm can induce ectopic PGCs (Tada et al. 2012). Germ plasm inhibits transcription and translation in nascent PGCs (Venkatarama et al. 2010), and this prevents pPGCs from responding to signals for somatic specification. In *Xenopus*, microinjection of antibodies against *Xenopus* VASA homolog protein (XVLG) into blastomeres at the 32-cell stage caused a reduction in the number of PGCs in the tadpole stage (Ikenishi and Tanaka

1997). These findings suggest that the function of the *vasa* gene is important for germ cell differentiation and is conserved during evolution. Studies on embryos highlighted the presence of the germ plasm in several anuran species, supporting the preformation scenario as a shared mechanism for germline specification in anurans (Blackler 1958). The origin of PGCs in zebrafish was unclear until the identification of *vasa* gene homolog: *vasa* mRNA is synthesized during oogenesis, localizes near the cleavage furrows during the first embryonic cleavages, and segregates into 4 cells by the 32-cell stage of embryogenesis (Yoon et al. 1997). These four cells become the pPGCs. Thus, an origin of germ cells through the preformation mode was identified also in zebrafish.

14.5.2 Amniotes

In amniotes, early embryogenesis is more complex than in anamniotes: it includes the formation of extraembryonic tissues.

In birds, as in other vertebrates, the origin of PGCs was investigated using many approaches, but the mechanism of specification remained unclear for long time for the failure to observe the presence of pPGCs in the pregastrula phase and for the lack of reliable molecular markers. The issue about the specification mechanism of PGCs in the chicken has been deciphered by Tsunekawa and coworkers (Tsunekawa et al. 2000). They developed an antibody against chicken Ddx4 (or chicken *vasa* homolog, CVH), that resulted to be specific for the germline, and were able to localize Ddx4-positive cells in very early stages of embryonic development (inside the unlaid egg).

In 2-cell stage chicken embryos, Ddx4 was observed in the proximity of the cleavage furrow. The early localization of this germline marker prompted Tsunekawa and coworkers to suggest a maternal specification mode of PGCs formation in the chicken, in which, similar to zebrafish, the Ddx4 is part of the germ plasm (Braat et al. 1999). This idea has been accepted and is currently rooted in the field.

In mammals, the first cell fate decision distinguishes the trophoblast lineage, which participates in the formation of the placenta, from the developing epiblast, which will form the germline and somatic cells of the embryo (Rayon et al. 2014). The time and site of origin of mammalian germ cells was a controversial issue for several decades, until alkaline phosphatase activity (or Alpl), used in 1954 as a marker of germ cells in the mouse (Chiquoine 1954), allowed for localization of germ cells in the extraembryonic mesoderm. The PGCs are then incorporated into the hindgut epithelium, move into the dorsal mesentery, and from there, they reach the genital ridges on the dorsal body wall, forming the gonad primordia (Gomperts et al. 1994). Contrary to what was reported in zebrafish and the chicken, the use of the mouse *vasa* homolog (*mvh*) did not help to clarify the origin and localization of pPGCs at the earlier stages of the mouse embryonic development (Fujiwara et al. 1994; Noce et al. 2001; Toyooka et al. 2000). In the mouse, the *mvh* gene is expressed in germ cells and stem cells that are believed to possibly differentiate into oocytes (De Felici 2010), but the MVH protein is expressed solely in germ cells and it is not localized to a specific subcellular region of the oocytes and no germ plasm is present (Toyooka et al. 2000). Although the Bb is present in mouse neonatal germline cysts and oocytes of primordial follicles, it disperses as follicles begin to grow (Pepling et al. 2007; Kloc et al. 2008). In the mouse, MVH protein is an essential component of the Cb in spermatogenic germ cells (Kuramochi-Miyagawa et al. 2010; Nagamori et al. 2011) and it has been shown that the *mvh* gene plays an important role both in the specification and differentiation of male germline (Tanaka et al. 2000).

In the past 25 years, with the advent of molecular markers and transgenics, the knowledge of timing, position, and the molecular machinery involved in pPGCs' specification has been refined (Chuva de Sousa Lopes and Roelen 2010; Mikedis and Downs 2014). In fact, a true epigenetic mechanism for germline specification has been demonstrated in the mouse by both descriptive and experimental evidence (Tsang et al. 2001), identifying several Bone morphogenetic protein (Bmp) ligands as inductive signals emanating from the extraembryonic ectoderm (Lawson et al. 1999). The first Alpl-positive PGCs emerge at E7.25 and express both pluripotency and early germline markers. In particular, the nascent PGCs express the transcription factors required for PGC specification (Ohinata et al. 2005; Yamaji et al. 2008). The complex regulatory network controlling pPGC and PGC formation in the mouse is so far the only description of a molecular mechanism driving germ cell specification by induction in amniotes. At the moment, whether the mechanism described in the mouse is employed by non-rodent mammals is still unclear.

14.6 Germline Specification in Non-model Vertebrates

14.6.1 Non-model Fish

In basal vertebrates, sharks, and probably skates, the PGCs are considered to be generated by inductive mechanism (Bachvarova et al. 2009a).

In the gulf sturgeon *Acipenser oxyrhynchus*, in situ hybridization technique using *vasa* homolog probes in adult ovaries, containing growing and fully grown oocytes, did not show a regionalized localization of the marker. In contrast to zebrafish and butterfly fish oocytes, in the sturgeon, the expression of *vasa* homolog was uniformly diffused in the oocyte cytoplasm (Bachvarova et al. 2004), regardless of oocyte size. These data were consistent with an inductive mechanism of PGCs' specification. Recently, however, the use of more modern and innovative techniques allowed the visualization of germ plasm at the vegetal pole of the sturgeon eggs and documented that PGCs are generated at the vegetal pole and subsequently migrate on the yolky cell mass toward the gonadal ridge. Therefore, these results provided the evidence that PGCs are specified by inheritance of maternally supplied germ plasm in sturgeons (Saito et al. 2014). Also, these observations suggest caution before affirming that the germ plasm is not present in oocytes; in fact, failing to visualize a structure does not always mean that the structure is not present. The analysis of germ plasm localization in the oocyte cytoplasm requires the use of different molecular markers and multiple experimental approaches. The structure of the nuage in the sturgeon egg resembled that of *X. laevis* (Ikenishi et al. 1974) and is also similar to that of zebrafish, goldfish, medaka, and goby, though the structure is more abundant in the sturgeon egg than in the other teleosts (Herpin et al. 2007; Kitauchi et al. 2012).

These findings suggested that the epigenetic mode of PGCs' specification might be the first appearing pattern in the animal kingdom, with the preformation mode evolved after and independently in different lineages (Extavour 2003, 2007, 2008; Johnson et al. 2003, 2011; Bachvarova et al. 2009a). According to this hypothesis, the similarity of the patterns in the sturgeon and anurans is the result of convergent evolution. At the moment, little is known about PGCs in lungfish, the closest living relative of the tetrapod ancestor (Kemp 1982). However, some evidence suggests that their oocytes do not contain germ plasm (Johnson et al. 2003).

14.6.2 Reptilia

Among tetrapods, the study of reptiles may provide key information on the evolution of germ cell development in amniotes.

With respect to the position of early germ cells and the mode of migration in the gonad, reptiles are a diverse group. The histological evidence is difficult to interpret for some species, and, for others, it suggests that they may use preformation or epigenesis or both (Hubert 1985). The early development of germ cells has been studied in several species of turtles, identifying PGCs by their spherical shape, chromatin structure, distinct nucleolus, retention of yolk granules, distinct cell outlines, and/or periodic acid-Schiff staining (Bachvarova et al. 2009b).

With the use of molecular markers, PGCs' localization has been described in early stages of development of the turtle *Trachemys scripta* (Bachvarova et al. 2009b). Although the VASA protein was expressed only in the germ cells, *vasa* mRNA was also present in the somatic cells of the embryo (Bachvarova et al. 2009b). In addition, the expression of VASA was analyzed in turtle ovaries and neither mRNA nor protein was found in the oocytes (Bachvarova et al. 2009b). This is different from what is seen, for example, in the chicken, in which VASA protein is localized in the cortical bodies of oocytes and in a few cells of the early embryo (Tsunekawa et al. 2000), and in *Xenopus*, in which the VASA homolog XVLG1 is localized in the germ plasm of the oocyte (Bachvarova et al. 2009b). The absence of *vasa* products in turtle oocytes suggests that the germ plasm is absent in turtles, and that the PGCs of turtle arise by the inductive mode, as they do in mammals.

Recently, VASA protein has been utilized as molecular marker to trace male germ cell development in the chinese soft-shell turtle *Pelodiscus sinensis* (Li et al. 2017). In this turtle, the VASA protein is strongly expressed in the primary spermatocytes, moderately in secondary spermatocytes and round spermatids, but little in spermatogonia, elongated spermatids, spermatozoa, and somatic cells. Moreover, in *P. sinensis* male germ cells, the VASA protein is diffused in the perinuclear cytoplasm, differently from the spermatogenic cells of the mouse in which VASA was concentrated in the Cb from the primary spermatocyte to the round spermatids (Toyooka et al. 2000; Noce et al. 2001; Nagamori et al. 2011). The cellular distribution of turtle VASA protein suggests that it can be involved in meiosis entry and spermatogenesis, as documented in the mouse, while the different subcellular distribution of VASA between turtle and the mouse suggests some differences in its action during spermatogenesis. Recently, VASA expression was reported in spermatogonia, oogonia, and oocytes of the crocodile *Crocodylus moreletii* (Martínez-Juárez et al. 2018). VASA is expressed in oogonia organized in cellular niches, and in the cytoplasm of type II previtellogenic oocytes, while it was not detected in the oocytes in the following stages. The absence of localized germ cell marker in mature oocytes suggests that germ plasm is absent also in this crocodile and that its PGCs, similarly to mammals, arise by the inductive mode (Martínez-Juárez et al. 2018).

14.7 The Use of Molecular Markers in the Study of Differentiation of Germ Cells in the Lizard *Podarcis sicula*

14.7.1 Ovarian Follicles Organization

The oogenesis of *Podarcis sicula*, as that of other lizards, represents an attractive model system for the study of the process of ovarian follicle differentiation in vertebrates (Fig. 14.5). During the reproductive life span of lizards, the ovary includes the germinal bed (GB)—a small region at the dorsal surface of the ovary responsible for the continuous renewal of germ cells. This region consists of dividing oogonia, naked oocytes, and primordial follicles (Filosa 1973; Milani and Maurizii 2015; Milani et al. 2017a). Oogonia and early oocytes are restricted to the GB, and growing follicles are released from it. The released oocytes, surrounded by a single layer of follicle cells, represent the early follicle. Subsequently, some follicle cells fuse with the oocyte, enlarge, and begin differentiating into pyriform cells (Andreuccetti et al. 1978). In follicles up to 2000 μm in diameter, the oocyte is surrounded by a multilayered epithelium consisting of differentiated pyriform cells regularly distributed between small and intermediate follicle cells (precursors of pyriform cells). The "small" cells are arranged mainly at the periphery. They proliferate and migrate, making contact with the oocyte surface, where, following their fusion with the oocyte, they progressively differentiate into pyriform cells. Like nurse cells in *Drosophila*, pyriform cells transfer specific molecules and cytoplasm through intercellular bridges to the growing oocyte (Motta et al. 1995; Milani et al.

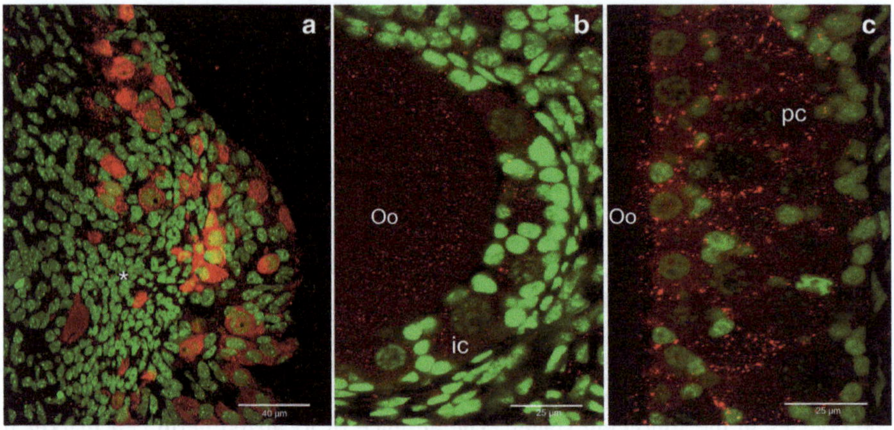

Fig. 14.5 VASA expression during *Podarcis sicula* oogenesis. Confocal microscopy. (**a**) In the germinal bed, anti-CVH stains gonia and early oocytes, while somatic cells of the stroma result unlabeled (asterisk). (**b**, **c**) The VASA protein is expressed in the cytoplasm of intermediate cells (ic), pyriform cells (pc), and in the oocyte (Oo) (see Maurizii et al. 2009). Anti-VASA: red; nuclear dye: green

2017a). However, unlike nurse cells of *Drosophila* that belong to the germline lineage, pyriform cells differentiate from small somatic follicle cells after their fusion with the oocyte (Andreuccetti et al. 1978). Intermediate cells represent the initial step in this differentiation (Filosa et al. 1979; Milani et al. 2017a). Afterward, yolk deposition rapidly enlarges the oocytes up to 9 mm in diameter, the pyriform cells are absorbed by the oocyte and disappear, and the follicular epithelium returns to its monolayer structure (Andreuccetti et al. 1978).

14.7.2 Cytoskeleton Markers

In this lizard, immunolocalization studies of cytoskeleton components, such as actin, tubulin, and intermediate filaments (IFs) (keratin and vimentin), were particularly useful to clarify the different steps of ovarian follicle differentiation and the structural and functional integration between pyriform cells and oocyte.

These kinds of studies have highlighted:

(a) The presence of vimentin in all follicle cells (small, intermediate, and pyriform cells) in accordance with their mesenchymal origin (Maurizii and Taddei 1996)
(b) That tubulin cytoskeleton is involved in the interaction between the follicle cells and the growing oocyte (Maurizii and Taddei 1996, 2012; Maurizii et al. 2004)
(c) The presence of a cytokeratin network in the cytoplasm of pyriform cells, which crosses their apex and extends through the intercellular bridges and into the oocyte cortex (Maurizii and Taddei 1996; Maurizii et al. 1997, 2000)

Immunoblotting and immunofluorescence analysis with anti-cytokeratin-specific (CKs) antibodies showed that, similarly to *X. laevis* (Franz et al. 1983), the CKs expressed in the ovarian follicle of *P. sicula* have the same immunoreactivity as those present in the monolayered epithelia, whereas they are different from those expressed in pluristratified epidermis (Maurizii et al. 1997). Furthermore, cytokeratin cytoskeleton is absent in the early oocytes located in the germinal bed, while it is present in the somatic cells of the stroma. Cytokeratin cytoskeleton first appears in the growing oocytes as a thin cortical layer, and becomes well organized in intermediate cells and in particular in fully differentiated pyriform cells. In both these cell types, a cytokeratin network connects the cytoplasm to the oocyte cortex through intercellular bridges, suggesting the peculiar function of this cytoskeleton in maintaining the cytoplasmic integration between somatic follicle cells and growing oocyte. At the end of previtellogenic oocyte growth, the intense immunolabeling of the apex in the regressing pyriform cells suggests that the cytokeratin, as other cytoplasmic components, may be transferred from these follicle cells to the oocyte. In support of this, at the end of the oocyte growth, in the larger vitellogenic oocytes surrounded by a monolayer of follicle cells, the cytokeratin constitutes a heavily immunolabeled cortical layer, thicker than in the previous stages (Maurizii et al. 1997, 2000).

14.7.3 VASA Expression During Oogenesis

To analyze the VASA distribution during *P. sicula* oogenesis, the polyclonal antibody raised against the chicken VASA homolog (anti-CVH, kindly provided by Toshiaki Noce) was used. The confocal analysis of the ovarian germinal bed showed strong anti-CVH signal in some cells that can be interpreted as gonia or oocytes in the very first stages of meiosis, while the somatic cells of the stroma in which germ cells are interspersed were unlabeled (Fig. 14.5a) (Maurizii et al. 2009). This result was subsequently confirmed using anti-Ps-VASA antibody (specific for the *P. sicula* homolog protein) (Milani and Maurizii 2015). This supported the use of VASA as a powerful tool, and a specific marker, for investigating germ cell differentiation and distinguishing them from the stromal somatic cells in this lizard. Furthermore, in ovarian follicles, the use of anti-CVH showed that the VASA protein is initially expressed by the oocytes but successively also by pyriform cells (Fig. 14.5b, c). The strong VASA signal in the cytoplasm of pyriform cells, in intermediate cells, and in the oocyte (Fig. 14.5c) suggests that these cells are either synthesizing the VASA protein or acquiring it from the oocyte: cells of somatic origin—the pyriform cells—can acquire the germ cell-specific protein following their fusion with the oocyte. This is in agreement with what observed in *Drosophila*, in which the VASA protein is expressed at the beginning by the oocyte and subsequently by the nurse cells (Lasko and Ashburner 1990). These results also highlight a functional analogy between pyriform cells and *Drosophila* nurse cells. On the other hand, in all the stages of oogenesis analyzed so far in *P. sicula,* no material that looks like the germ plasm has been observed (Maurizii et al. 2009). Thus, also in this reptile, the inductive mechanism appears to be the germline specification mode.

14.7.4 Testis Organization Using α-Tubulin Antibody

In *P. sicula* testis, like in other reptiles and differently from other amniotes (such as birds and mammals), large cohorts of germ cells develop together, each progressing through the stages of spermatogenesis as a single population, similarly to what happens in amphibians (Gribbins 2011). Therefore, a clear series of germ cell morphologies can be easily delineated during the major phases of spermatogenesis. Using TO-PRO-3 as nuclear dye and DM 1A anti-α-tubulin antibody to stain microtubule cytoskeleton on cross sections of testis during the phase of full gonadic activity, the different stages of germ cell differentiation have been identified (Milani and Maurizii 2014a, 2015) (Fig. 14.6). As in other amniotes, the epithelium of the seminiferous tubules of *P. sicula* consists of a basal compartment and of an apical region that surrounds a wide lumen. In the basal region, the spermatogonia and Sertoli cells are present in proximity of the basal membrane, and, in succession, a large population of I° and II° spermatocytes are present. Many mitotic and/or meiotic spindles were observed in the basal compartment. In the apical region, surrounding the central lumen, a large population of round spermatids and several stages of differentiating spermatids are present, while the central lumen is filled with mature spermatozoa (Fig. 14.6).

14.7.5 VASA Expression During Spermatogenesis

It is clear that the presence of germ cells in different stages of differentiation in the seminiferous tubule epithelium of *P. sicula* provides a perfect model to study VASA expression and to clarify if this protein has a role during the differentiation of male germ cells. Using anti-Ps-VASA antibody (specific for the species homolog protein) and confocal microscopy, VASA expression has been studied in male germ cells during the reproductive cycle: during full gonadal activity (spring), during regression of gonadal activity (summer), and during slow autumnal recrudescence. Furthermore, VASA expression has been analyzed in young testes when the walls of the seminiferous tubules are forming. The results showed that Ps-VASA was specifically expressed in male germ cells in all phases of the reproductive cycle from the mitotic to the meiotic stages of differentiation, but it was not detected in mature spermatozoa residing in the lumen of the tubule. In particular, spermatocytes and spermatids show cytoplasmic immunostained spots; in round spermatids, the spots converge in a notch of the nucleus. Then, in spermatids in the last stages of differentiation shortly before spermiation, a strong VASA staining is present at

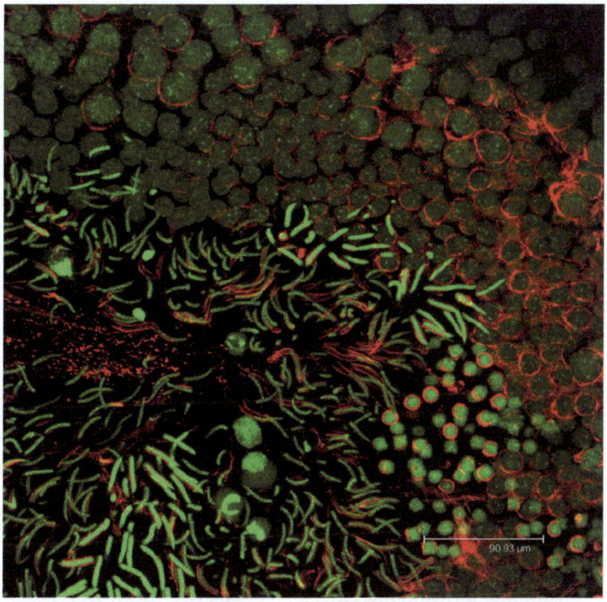

Fig. 14.6 *Podarcis sicula* testis organization with α-tubulin antibody. Confocal microscopy. Microtubule cytoskeleton staining on cross sections of testis during the phase of fully gonadic activity. Different stages of germ cell differentiation are visible (see Milani and Maurizii 2014a, b, 2015). Anti-α-tubulin: red; nuclear dye: green

one extremity of their elongated nucleus (Milani and Maurizii 2014b, 2015). This is similar to the situation in the chicken (Tsunekawa et al. 2000) where VASA-stained spots aggregate into the Cbs in spermatids (at one pole of the nucleus), and also similar to the mouse, in which VASA concentrates in the Cb from the primary spermatocyte to the round spermatids (Toyooka et al. 2000; Noce et al. 2001; Nagamori et al. 2011). Moreover, in *P. sicula* immature testis, VASA is expressed in the cytoplasm of spermatogonia and spermatocytes of forming seminiferous tubules. Thus, some differences in VASA protein distribution between turtle and lizard spermatogenesis were found, because, in the turtle, the VASA staining was strong only in primary and secondary spermatocytes, showing a perinuclear distribution. The expression pattern of VASA observed in *P. sicula* suggests that it has a role in the whole differentiation process of male germ cells, indicating a higher, apparent similarity to the mouse than to the investigated turtle.

14.8 Conclusion

The differences observed among closely related animals as well as the similarities reported for evolutionary distant species, as highlighted in the examples presented in this chapter, underline the necessity in evo-devo studies of an extensive comparative approach (see Scholtz 2008). The benefit of performing comparisons at different taxonomic levels is evident and these kinds of investigations are a fundamental complement to studies pertaining developmental biology, molecular biology, genomics, systematics, and paleontology, whose integrated contribution will increasingly allow us to understand mechanisms and processes driving evolutionary changes.

Acknowledgement We are grateful to Waclaw Tworzydlo and Szczepan Bilinski for inviting this contribution and Malgorzata Kloc for commenting on the manuscript. This work was supported by the Italian Ministry of Education, University and Research MIUR-SIR Programme no. RBSI14G0P5 funded to L.M.

References

Alié A, Hayashi T, Sugimura I, Manuel M, Sugano W et al (2015) The ancestral gene repertoire of animal stem cells. Proc Natl Acad Sci USA 112:E7093–E7100

Allen JF (1996) Separate sexes and the mitochondrial theory of ageing. J Theor Biol 180:135–140

Amikura R, Kashikawa M, Nakamura A, Kobayashi S (2001) Presence of mitochondria-type ribosomes outside mitochondria in germ plasm of Drosophila embryos. Proc Natl Acad Sci USA 98:9133–9138

Amikura R, Sato K, Kobayashi S (2005) Role of mitochondrial ribosome-dependent translation in germline formation in Drosophila embryos. Mech Dev 122:1087–1093

Andreuccetti P, Taddei C, Filosa S (1978) Intercellular bridges between follicle cells and oocyte during the differentiation of follicular epithelium in *Lacerta sicula*. J Cell Sci 33:341–350

Bachvarova RF, Masi T, Drum M, Parker N, Mason K et al (2004) Gene expression in the axolotl germ line: Axdazl, Axvh, Axoct-4, and Axkit. Dev Dyn 231:871–880

Bachvarova RF, Crother BI, Johnson AD (2009a) Evolution of germ cell development in tetrapods: comparison of urodeles and amniotes. Evol Dev 11:603–609

Bachvarova RF, Crother BI, Manova K, Chatfield J et al (2009b) Expression of Dazl and Vasa in turtle embryos and ovaries: evidence for inductive specification of germ cells. Evol Dev 11:525–534

Bilinski SM, Kloc M, Tworzydlo W (2017) Selection of mitochondria in female germline cells: is Balbiani body implicated in this process? J Assist Reprod Genet 34:1405–1412

Blackler AW (1958) Contribution to the study of germ-cells in the anura. J Embryol Exp Morphol 6:491–503

Bounoure L (1934) Recherches sur la lignée germinale chez la grenouille rousse aux premiers stades du développement. Ann Sci Nat 17:67–248

Braat AK, Speksnijder JE, Zivkovic D (1999) Germ line development in fishes. Int J Dev Biol 43:745–760

Cao L, Kenchington E, Zouros E (2004) Differential segregation patterns of sperm mitochondria in embryos of the blue mussel (*Mytilus edulis*). Genetics 166:883–894

Cherif-Feildel M, Kellner K, Goux D, Elie N, Adeline B, Lelong C, Heude Berthelin C (2018) Morphological and molecular criteria allow the identification of putative germ stem cells in a lophotrochozoan, the Pacific oyster *Crassostrea gigas*. Histochem Cell Biol 151(5):419–433. https://doi.org/10.1007/s00418-018-1740-3

Chiquoine AD (1954) The identification, origin, and migration of the primordial germ cells in the mouse embryo. Anat Rec 118:135–146

Chuva de Sousa Lopes SM, Roelen BA (2010) On the formation of germ cells: the good, the bad and the ugly. Differentiation 79:131–140

Cinalli RM, Rangan P, Lehmann R (2008) Germ cells are forever. Cell 132:559–562

Cogswell AT, Kenchington ELR, Zouros E (2006) Segregation of sperm mitochondria in two- and four-cell embryos of the blue mussel Mytilus edulis: implications for the mechanism of doubly uniparental inheritance of mitochondrial DNA. Genome 49:799–807

Cox RT, Spradling AC (2003) A Balbiani body and the fusome mediate mitochondrial inheritance during Drosophila oogenesis. Development 130:1579–1590

Davidson EH (2001) Genomic regulatory systems: development and evolution. Academic, San Diego

De Felici M (2010) Germ stem cells in the mammalian adult ovary: considerations by a fan of the primordial germ cells. Mol Hum Reprod 16:632–636

de Paula WBM, Agip A-NA, Missirlis F, Ashworth R, Vizcay-Barrena G, Lucas CH, Allen JF (2013) Female and male gamete mitochondria are distinct and complementary in transcription, structure, and genome function. Genome Biol Evol 5:196–1977

de Sousa Lopes SM, Hayashi K, Surani MA (2007) Proximal visceral endoderm and extraembryonic ectoderm regulate the formation of primordial germ cell precursors. BMC Dev Biol 7:140–148

Devauchelle N (1990) Sexual development and maturity of *Tapes philippinarum*, chapter 3. In: *Tapes philippinarum*, biologia e sperimentazione. Ente Sviluppo Agricolo Veneto (ESAV), Regione Veneto

Ding D, Whittaker KL, Lipshitz HD (1994) Mitochondrially encoded 16S large ribosomal RNA is concentrated in the posterior polar plasm of early Drosophila embryos but is not required for pole cell formation. Dev Biol 163:503–515

Eddy EM (1975) Germ plasm and the differentiation of the germ cell line. Int Rev Cytol 43:229–280

Ernoult-Lange M, Bénard M, Kress M, Weil D (2012) P-bodies and mitochondria: which place in RNA interference? Biochimie 94:1572e1577

Eulalio A, Behm-Ansmant I, Izaurralde E (2007) P bodies: at the crossroads of post-transcriptional pathways. Nat Rev Mol Cell Biol 8:9–22

Ewen-Campen B, Schwager EE, Extavour CGM (2010) The molecular machinery of germ line specification. Mol Reprod Dev 77:3–18

Extavour CGM (2003) Mechanisms of germ cell specification across the metazoans: epigenesis and preformation. Development 130:5869–5884

Extavour CGM (2007) Evolution of the bilaterian germ line: lineage origin and modulation of specification mechanisms. Integr Comp Biol 47:770–785

Extavour CGM (2008) Urbisexuality: the evolution of bilaterian germ cell specification and reproductive systems. In: Minelli A, Fusco G (eds) Evolving pathways: key themes in evolutionary developmental biology. Cambridge University Press, Cambridge, pp 321–342

Extavour C, Akam ME (2003) Mechanisms of germ cell specification across the metazoans: epigenesis and preformation. Development 130:5869–5884

Fabioux C, Corporeau C, Quillien V, Favrel P, Huvet A (2009) *In vivo* RNA interference in oyster—*vasa* silencing inhibits germ cell development. FEBS J 276:2566–2573

Fierro-Constaín L, Schenkelaars Q, Gazave E, Haguenauer A, Rocher C, Ereskovsky A, Borchiellini C, Renard E (2017) The conservation of the germline multipotency program, from sponges to vertebrates: a stepping stone to understanding the somatic and germline origins. Genome Biol Evol 9:474–488

Filosa S (1973) Biological and cytological aspect of the ovarian cycle in *Lacerta sicula* Raf. Mon Zool Ital 7:151–165

Filosa S, Taddei C, Andreuccetti P (1979) The differentation and proliferation of follicle cells during oocyte growth in *Lacerta sicula*. J Embryol Exp Morphol 54:5–15

Franz JK, Gall L, Williams MA, Picheral B, Franke WW (1983) Intermediate-size filaments in a germ cell: expression of cytokeratins in oocytes and eggs of the frog *Xenopus*. Proc Natl Acad Sci U S A 80(20):6254–6258

Fujiwara Y, Komiya T, Kawabata H, Sato M, Fujimoto H et al (1994) Isolation of a DEAD-family protein gene that encodes a murine homolog of Drosophila vasa and its specific expression in germ cell lineage. Proc Natl Acad Sci USA 91:12258–12262

Funayama N (2010) The stem cell system in demosponges: insights into the origin of somatic stem cells. Dev Growth Differ 52:1–14

Ghiselli F, Milani L, Passamonti M (2011) Strict sex-specific mtDNA segregation in the germline of the DUI species *Venerupis philippinarum* (Bivalvia Veneridae). Mol Biol Evol 28:949–961
Ghiselli F, Milani L, Chang PL, Hedgecock D, Davis JP et al (2012) De novo assembly of the Manila clam *Ruditapes philippinarum* transcriptome provides new insights into expression bias, mitochondrial doubly uniparental inheritance and sex determination. Mol Biol Evol 29:771–786
Ghiselli F, Breton S, Milani L (2018) Mitochondrial activity in gametes and uniparental inheritance—a comment on 'What can we infer about the origin of sex in early eukaryotes?'. Philos Trans R Soc Lond B Biol Sci 373:20170147
Gomperts M, Garcia-Castro M, Wylie C, Heasman J (1994) Interactions between primordial germ cells play a role in their migration in mouse embryos. Development 120:135–141
Gribbins KM (2011) Reptilian spermatogenesis. A histological and ultrastructural perspective. Spermatogenesis 1:250–269
Gusman A, Lecomte S, Stewart DT, Passamonti M, Breton S (2016) Pursuing the quest for better understanding the taxonomic distribution of the system of doubly uniparental inheritance of mtDNA. PeerJ 4:e2760
Gustafson EA, Wessel GM (2010) Vasa genes: emerging roles in the germ line and in multipotent cells. BioEssays 32:626–637
Hejnol A (2010) A twist in time—the evolution of spiral cleavage in the light of animal phylogeny. Integr Comp Biol 50:695–706
Herpin A, Rohr S, Riedel D, Kluever N, Raz E et al (2007) Specification of primordial germ cells in medaka (Oryzias latipes). BMC Dev Biol 7:3
Huang L, Mollet S, Souquere S, Le Roy F, Ernoult-Lange M et al (2011) Mitochondria associate with P-bodies and modulate MicroRNA-mediated RNA interference. J Biol Chem 286:24219–24230
Hubert J (1985) Origin and development of oocytes. In: Gans C, Billet F, Maderson PFA (eds) Biology of the Reptilia: development A, vol 14. Wiley, New York, pp 42–74
Iida T, Kobayashi S (1998) Essential role of mitochondrially encoded large rRNA for germ-line formation in Drosophila embryos. Proc Natl Acad Sci USA 95:11274–11278
Ikenishi K, Tanaka TS (1997) The cause of the decreased number of primordial germ cells in albino Xenopus resides not in the micro-environment but in the presumptive PGC. Dev Growth Differ 39:345–349
Ikenishi K, Kotani M, Tanabe K (1974) Ultrastructural changes associated with UV irradiation in the "germinal plasm" of *Xenopus laevis*. Dev Biol 36:155–168
Isaeva VV, Reunov AA (2001) Germ plasm and germ-line cell determination: the role of mitochondria. Russ J Mar Biol 27:S8–S14
Johnson AD, Drum M, Bachvarova RF, Masi T, White ME et al (2003) Evolution of predetermined germ cells in vertebrate embryos: implications for macroevolution. Evol Dev 5:414–431
Johnson AD, Richardson E, Bachvarova RF, Crother BI (2011) Evolution of the germ line-soma relationship in vertebrate embryos. Reproduction 141:291–300
Johnstone O, Lasko P (2004) Interaction with eIF5B is essential for Vasa function during development. Development 131:4167–4178
Juliano C, Wessel G (2010) Versatile germline genes. Science 329:640–641
Juliano CE, Swartz SZ, Wessel GM (2010) A conserved germline multipotency program. Development 137:4113–4126
Kemp A (1982) The embryological development of the Queensland lungfish, *Neoceratodus forsteri* (krefft). Mem Queensl Mus 20:553–598
Kitauchi T, Saito T, Motomura T, Arai K, Yamaha E (2012) Distribution and function of germ plasm in cytoplasmic fragments from centrifuged eggs of the goldfish, *Carassius auratus*. J Appl Ichthyol 28:998–1005
Kloc M, Larabell C, Pui-Yee Chan A, Etkin LD (1998) Contribution of METRO pathway localized molecules to the organization of the germ cell lineage. Mech Dev 75:81–93
Kloc M, Zearfoss NR, Etkin LD (2002) Mechanisms of subcellular mRNA localization. Cell 108:533–544

Kloc M, Bilinski S, Etkin LD (2004) The Balbiani body and germ cell determinants: 150 years later. Curr Top Dev Biol 59:1–36
Kloc M, Jaglarz M, Dougherty M, Stewart MD, Nel-Themaat L et al (2008) Mouse early oocytes are transiently polar: three-dimensional and ultrastructural analysis. Exp Cell Res 314:3245–3254
Kobayashi S, Amikura R, Okada M (1993) Presence of mitochondrial large ribosomal RNA outside mitochondria in germ plasm of Drosophila melanogaster. Science 260:1521–1524
Kobayashi S, Amikura R, Mukai M (1998) Localization of mitochondrial RNA in germ plasm of Xenopus embryos. Curr Biol 8:1117–1120
Kotaja N, Sassone-Corsi P (2007) The chromatoid body: a germ-cell-specific RNA-processing centre. Nat Rev Mol Cell Biol 8:85–90
Kuramochi-Miyagawa S, Watanabe T, Gotoh K, Takamatsu K, Chuma S et al (2010) MVH in piRNA processing and gene silencing of retrotransposons. Genes Dev 24:887–892
Lasko P (2013) The DEAD-box helicase Vasa: evidence for a multiplicity of functions in RNA processes and developmental biology. Biochim Biophys Acta 1829:810–816
Lasko PF, Ashburner M (1990) Posterior localization of Vasa protein correlates with, but is not sufficient for, pole cell development. Genes Dev 4:905–921
Lavial F, Acloque H, Bachelard E, Nieto A, Samarut J et al (2009) Ectopic expression of Cvh (Vasa, Chicken homologue) mediates the reprogramming of chicken embryonic stem cells to a germ cell fate. Dev Biol 330:73–82
Lawson KA, Hage WJ (1994) Clonal analysis of the origin of primordial germ cells in the mouse. Ciba Found Symp 182:68–84
Lawson KA, Dunn NR, Roelen BAJ, Zeinstra LM, Davis AM et al (1999) Bmp4 is required for the generation of primordial germ cells in the mouse embryo. Genes Dev 13:424–436
Leclère L, Jager M, Barreau C, Chang P, Le Guyader H et al (2012) Maternally localized germ plasm mRNAs and germ cell/stem cell formation in the cnidarian *Clytia*. Dev Biol 364:236–248
Leitch HG, Blair K, Mansfield W, Ayetey H, Humphreys P, Nichols J, Surani MA, Smith A (2010) Embryonic germ cells from mice and rats exhibit properties consistent with a generic pluripotent ground state. Development 137:2279–2287
Li W, Zhang P, Wu X, Zhu X, Xu H (2017) A novel dynamic expression of vasa in male germ cells during spermatogenesis in the Chinese soft-shell turtle (*Pelidiscus sinensis*). J Exp Zool B Mol Dev Evol 328:230–239
Liang L, Diehl-Jones W, Lasko PF (1994) Localization of vasa protein to the Drosophila pole plasm is independent of its RNA-binding and helicase activities. Development 120:1201–1211
Linder P, Fuller-Pace FV (2013) Looking back on the birth of DEAD-box RNA helicases. Biochim Biophys Acta 1829:750–755
Lyons DC, Perry KJ, Lesoway MP, Henry JQ (2012) Cleavage pattern and fate map of the mesentoblast, 4d, in the gastropod Crepidula: a hallmark of spiralian development. EvoDevo 3:21
MacAskill AF, Kittler JT (2009) Control of mitochondrial transport and localization in neurons. Trends Cell Biol 20:102–112
Martínez-Juárez A, López-Luna MA, Porras-Gómez TJ, Moreno-Mendoza N (2018) Expression of the *Sox9*, *Foxl2*, *Vasa*, and *TRPV4* genes in the ovaries and testes of the Morelet's crocodile, *Crocodylus moreletii*. J Exp Zool B Mol Dev Evol 330:148–164
Matsui Y, Zsebo K, Hogan BL (1992) Derivation of pluripotential embryonic stem cells from murine primordial germ cells in culture. Cell 70:841–847
Maurizii MG, Taddei C (1996) Immunolocalization of cytoskeletal proteins in the previtellogenic ovarian follicle of the lizard *Podarcis sicula*. Cell Tissue Res 284:489–493
Maurizii MG, Taddei C (2012) Microtubules organization and nucleation in the differentiating ovarian follicle of the lizard *Podarcis sicula*. J Morphol 273:1089–1095
Maurizii MG, Saverino O, Taddei C (1997) Cytokeratin cytoskeleton in the differentiating ovarian follicle of the lizard *Podarcis sicula* Raf. Mol Reprod Dev 48:536–542

Maurizii MG, Alibardi L, Taddei C (2000) Organization and characterization of the keratin cytoskeleton in the previtellogenic ovarian follicle of the lizard *Podarcis sicula* Raf. Mol Reprod Dev 57:159–166

Maurizii MG, Alibardi L, Taddei C (2004) α-Tubulin and acetylated α-tubulin during ovarian follicle differentiation in the lizard *Podarcis sicula* Raf. J Exp Zool A Comp Exp Biol 301:532–541

Maurizii MG, Cavaliere V, Gamberi C, Lasko P, Gargiulo G et al (2009) Vasa protein is localized in the germ cells and in the oocyte-associated pyriform follicle cells during early oogenesis in the lizard *Podarcis sicula*. Dev Genes Evol 219:361–367

Mikedis MM, Downs KM (2014) Mouse primordial germ cells: a reappraisal. Int Rev Cell Mol Biol 309:1–57

Milani L (2015) Mitochondrial membrane potential: a trait involved in organelle inheritance? Biol Lett 11:20150732

Milani L, Ghiselli F (2015) Mitochondrial activity in gametes and transmission of viable mtDNA. Biol Direct 10:22

Milani L, Maurizii MG (2014a) Differentiating male germ cells during spermatogenesis of a reptile. Mol Reprod Dev 81:777

Milani L, Maurizii MG (2014b) First evidence of Vasa expression in differentiating male germ cells of a reptile. Mol Reprod Dev 81:390

Milani L, Maurizii MG (2015) Vasa expression in spermatogenic cells during the reproductive-cycle phases of *Podarcis sicula* (Reptilia, Lacertidae). J Exp Zool B Mol Dev Evol 324:424–434

Milani L, Ghiselli F, Maurizii MG, Passamonti M (2011) Doubly uniparental inheritance of mitochondria as a model system for studying germ line formation. PLoS One 6:e28194

Milani L, Ghiselli F, Passamonti M (2012) Sex-linked mitochondrial behavior during early embryo development in *Ruditapes philippinarum* (Bivalvia Veneridae) a species with the Doubly Uniparental Inheritance (DUI) of mitochondria. J Exp Zool B Mol Dev Evol 318:182–189

Milani L, Ghiselli F, Guerra D, Breton S, Passamonti M (2013) A comparative analysis of mitochondrial ORFans: new clues on their origin and role in species with doubly uniparental inheritance of mitochondria. Genome Biol Evol 5:1408–1434

Milani L, Ghiselli F, Maurizii MG, Nuzhdin SV, Passamonti M (2014) Paternally transmitted mitochondria express a new gene of potential viral origin. Genome Biol Evol 6:391–405

Milani L, Maurizii MG, Pecci A, Ghiselli F, Passamonti M (2015a) Localization and dynamics of primordial germ cells in the bivalve species *Ruditapes philippinarum*. Mol Reprod Dev 82:406–407

Milani L, Ghiselli F, Pecci A, Maurizii MG, Passamonti M (2015b) The expression of a novel mitochondrially-encoded gene in gonadic precursors may drive paternal inheritance of mitochondria. PLoS One 10:e0137468

Milani L, Ghiselli F, Passamonti M (2016) Mitochondrial selfish elements and the evolution of biological novelties. Curr Zool 62:687–697

Milani L, Pecci A, Cifaldi C, Maurizii MG (2017a) PL10 DEAD-box protein is expressed during germ cell differentiation in the reptile *Podarcis sicula* (family Lacertidae). J Exp Zool B Mol Dev Evol 328:433–448

Milani L, Pecci A, Ghiselli F, Passamonti M, Bettini S, Franceschini V, Maurizii MG (2017b) VASA expression suggests shared germ line dynamics in bivalve molluscs. Histochem Cell Biol 148:157–171

Milani L, Pecci A, Ghiselli F, Passamonti M, Lazzari M, Franceschini V, Maurizii MG (2018) Germ cell line during the seasonal sexual rest of clams: finding niches of cells for gonad renewal. Histochem Cell Biol 149:105–110

Minelli A, Fusco G (2008) Evolving Pathways. Cambridge University Press, Cambridge

Mishra P, Chan DC (2014) Mitochondrial dynamics and inheritance during cell division, development and disease. Nat Rev Mol Cell Biol 15:634–646

Monteiro A, Podlaha O (2009) Wings, horns, and butterfly eyespots: how do complex traits evolve? PLoS Biol 7:e37

Motta CM, Castriota Scanderberg M, Filosa S, Andreuccetti P (1995) Role of pyriform cells during the growth of oocytes in the lizard *Podarcis sicula*. J Exp Zool 273:247–256

Muller WA, Teo R, Frank U (2004) Totipotent migratory stem cells in a hydroid. Dev Biol 275:215–224

Nagamori I, Cruickshank VA, Sassone-Corsi P (2011) Regulation of an RNA granule during spermatogenesis: acetylation of MVH in the chromatoid body of germ cells. J Cell Sci 124:4346–4355

Newmark PA, Wang Y, Chong T (2008) Germ cell specification and regeneration in planarians. Cold Spring Harb Symp Quant Biol 73:573–581

Nieuwkoop PD, Sutasurya LA (1979) Primordial germ cells in chordates. Cambridge University Press, Cambridge

Noce T, Okamoto-Ito S, Tsunekawa N (2001) *Vasa* homolog genes in mammalian germ cell development. Cell Struct Funct 26:131–136

Obata M, Komaru A (2005) Specific location of sperm mitochondria in mussel *Mytilus galloprovincialis* zygotes stained by MitoTracker. Develop Growth Differ 47:255–263

Ogawa M, Amikura R, Akasaka K, Kinoshita T, Kobayashi S, Shimada H (1999) Asymmetrical distribution of mitochondrial rRNA into small micromeres of sea urchin embryos. Zool Sci 16:445–451

Ohinata Y, Payer B, O'Carroll D, Ancelin K, Ono Y et al (2005) Blimp1 is a critical determinant of the germ cell lineage in mice. Nature 436:207–213

Parvinen M (2005) The chromatoid body in spermatogenesis. Int J Androl 28:189–201

Pepling ME, Wilhelm JE, O'Hara AL, Gephardt GW, Spradling AC (2007) Mouse oocytes within germ cell cysts and primordial follicles contain a Balbiani body. Proc Natl Acad Sci USA 104:187–192

Punzi E, Milani L, Ghiselli F, Passamonti M (2018) Lose it or keep it: (how bivalves can provide) insights into mitochondrial inheritance mechanisms. J Exp Zool B Mol Dev Evol 330:41–51

Rayon T, Menchero S, Nieto A, Xenopoulos P, Crespo M et al (2014) Notch and hippo converge on Cdx2 to specify the trophectoderm lineage in the mouse blastocyst. Dev Cell 30:410–422

Reunov A (2006) Structures related to the germ plasm in mouse. Zygote 14:231–238

Reunov AA, Reunova YA (2015) In mouse oocytes the mitochondrion-originated germinal body-like structures accumulate mouse Vasa homologue (MVH) protein. Zygote 23:501–506

Reunov A, Isaeva V, Au D, Wu R (2000) Nuage constituents arising from mitochondria: is it possible? Develop Growth Differ 42:139–143

Reunov A, Alexandrova Y, Reunova Y, Komkova A, Milani L (2019) Germ plasm provides clues on meiosis: the concerted action of germ plasm granules and mitochondria in gametogenesis of the clam *Ruditapes philippinarum*. Zygote 27(1):25–35. https://doi.org/10.1017/S0967199418000588

Rink JC (2013) Stem cell systems and regeneration in planaria. Dev Genes Evol 223:67–84

Saffman EE, Lasko P (1999) Germline development in vertebrates and invertebrates. CMLS 55:1141–1163

Saito T, Psenicka M, Goto R, Adachi S, Inoue K et al (2014) The origin and migration of primordial germ cells in sturgeons. PLoS One 9:e86861

Seki Y, Hayashi K, Itoh K, Mizugaki M, Saitou M et al (2005) Extensive and orderly reprogramming of genome-wide chromatin modifications associated with specification and early development of germ cells in mice. Dev Biol 278:440–458

Scholtz G (2008) On comparisons and causes in evolutionary developmental biology. In: Minelli A, Fusco G (eds) Evolving pathways: key themes in evolutionary developmental biology. Cambridge University Press, Cambridge, pp 144–159

Shukalyuk AI, Isaeva VV (2012) Molecular and sub-cellular gametogenic machinery of stem and germline cells across metazoa. In: Najman S (ed) Current frontiers and perspectives in cell biology. InTech, Rijeka. ISBN: 978-953-51-0544-2

Solana J (2013) Closing the circle of germline and stem cells: the primordial stem cell hypothesis. Evodevo 4:2

Solana J, Kao D, Mihaylova Y, Jaber-Hijazi F, Malla S et al (2012) Defining the molecular profile of planarian pluripotent stem cells using a combinatorial RNA-seq, RNA interference and irradiation approach. Genome Biol 13:R19
Sutasurya LA, Nieuwkoop PD (1974) The induction of the primordial germ cells in the urodeles. Wilhelm Roux Arch Entwickl Mech Org 175:199–220
Swartz SZ, Wessel GM (2015) Germ line versus soma in the transition from egg to embryo. Curr Top Dev Biol 113:149–190
Tada H, Mochii M, Orii H, Watanabe K (2012) Ectopic formation of primordial germ cells by transplantation of the germ plasm: direct evidence for germ cell determinant in Xenopus. Dev Biol 371:86–93
Tam PP, Zhou SX (1996) The allocation of epiblast cells to ectodermal and germ-line lineages is influenced by the position of the cells in the gastrulating mouse embryo. Dev Biol 178:124–132
Tanaka SS, Toyooka Y, Akasu R, Katoh-Fukui Y, Nakahara Y et al (2000) The mouse homolog of Drosophila Vasa is required for the development of male germ cells. Genes Dev 14:841–853
Toyooka Y, Tsunekawa N, Takahashi Y, Matsui Y, Satoh M et al (2000) Expression and intracellular localization of mouse Vasa-homologue protein during germ cell development. Mech Dev 93:139–149
Tsang TE, Khoo PL, Jamieson RV, Zhou SX, Ang SL et al (2001) The allocation and differentiation of mouse primordial germ cells. Int J Dev Biol 45:549–555
Tsunekawa N, Naito M, Sakai Y, Nishida T, Noce T (2000) Isolation of chicken *vasa* homolog gene and tracing the origin of primordial germ cells. Development 127:2741–2750
van Wolfswinkel JC (2014) Piwi and potency: PIWI proteins in animal stem cells and regeneration. Integr Comp Biol 54:700–713
Venetis C, Theologidis I, Zouros E, Rodakis GC (2006) No evidence for presence of maternal mitochondrial DNA in the sperm of *Mytilus galloprovincialis* males. Proc Biol Sci 273:2483–2489
Venkatarama T, Lai F, Luo X, Zhou Y, Newman K et al (2010) Repression of zygotic gene expression in the Xenopus germline. Development 137:651–660
Wagner G, Pavlicev M, Cheverud J (2007) The road to modularity. Nat Rev Genet 8:921–931
Wessel GM (2016) Germ line mechanics—and unfinished business. Curr Top Dev Biol 117:553–566
Wilding M, Carotenuto R, Infante V, Dale B, Marino M et al (2001) Confocal microscopy analysis of the activity of mitochondria contained within the 'mitochondrial cloud' during oogenesis in *Xenopus laevis*. Zygote 9:347–352
Wilkins AS (2013) "The genetic tool-kit": the life-history of an important metaphor, Chapter 1. In: Streelman JT (ed) Advances in evolutionary developmental biology. Wiley, Hoboken, pp 1–14
Woodland HR (2016) The birth of animal development: multicellularity and the germline. Curr Top Dev Biol 117:609–630
Wylie CC, Holwill S, O'Driscoll M, Snape A, Heasman J (1985) Germ plasm and germ cell determination in *Xenopus laevis* as studied by cell transplantation analysis. Cold Spring Harb Symp Quant Biol 50:37–43
Yajima M, Wessel GM (2011) The multiple hats of Vasa: its functions in the germline and in cell cycle progression. Mol Reprod Dev 78:861–867
Yajima M, Wessel GM (2015) Essential elements for translation: the germline factor Vasa functions broadly in somatic cells. Development 142:1960–1970
Yamaji M, Seki Y, Kurimoto K, Yabuta Y, Yuasa M et al (2008) Critical function of Prdm14 for the establishment of the germ cell lineage in mice. Nat Genet 40:1016–1022
Yang J, Aguero T, King ML (2015) The *Xenopus* maternal-to-zygotic transition from the perspective of the germline. Curr Top Dev Biol 113:271–303
Yoon C, Kawakami K, Hopkins N (1997) Zebrafish vasa homologue RNA is localized to the cleavage planes of 2- and 4-cell-stage embryos and is expressed in the primordial germ cells. Development 124:3157–3165

Zhang Y-Z, Ouyang Y-C, Hou Y, Schatten H, Chen D-Y et al (2008) Mitochondrial behavior during oogenesis in zebrafish: a confocal microscopy analysis. Develop Growth Differ 50:189–201

Zhou RR, Wang B, Wang J, Schatten H, Zhang YZ (2010) Is the mitochondrial cloud the selection machinery for preferentially transmitting wild-type mtDNA between generations? Rewinding Müller's ratchet efficiently. Curr Genet 56:101–107

Zouros E (2013) Biparental inheritance through uniparental transmission: the doubly uniparental inheritance (DUI) of mitochondrial DNA. Evol Biol 40:1–31

Chapter 15
Molecular Markers in the Study of Non-model Vertebrates: Their Significant Contributions to the Current Knowledge of Tetrapod Glial Cells and Fish Olfactory Neurons

Simone Bettini, Maurizio Lazzari, and Valeria Franceschini

Abstract The knowledge of the morphological and functional aspects of mammalian glial cells has greatly increased in the last few decades. Glial cells represent the most diffused cell type in the central nervous system, and they play a critical role in the development and function of the brain. Glial cell dysfunction has recently been shown to contribute to various neurological disorders, such as autism, schizophrenia, pain, and neurodegeneration. For this reason, glia constitutes an interesting area of research because of its clinical, diagnostic, and pharmacological relapses. In this chapter, we present and discuss the cytoarchitecture of glial cells in tetrapods from an evolutive perspective. GFAP and vimentin are main components of the intermediate filaments of glial cells and are used as cytoskeletal molecular markers because of their high degree of conservation in the various vertebrate groups. In the anamniotic tetrapods and their progenitors, Rhipidistia (Dipnoi are the only extant rhipidistian fish), the cytoskeletal markers show a model based exclusively on radial glial cells. In the transition from primitive vertebrates to successively evolved forms, the emergence of a new model has been observed which is believed to support the most complex functional aspects of the nervous system in the vertebrates. In reptiles, radial glial cells are prevalent, but star-shaped astrocytes begin to appear in the midbrain. In endothermic amniotes (birds and mammals), star-shaped astrocytes are predominant. In glial cells, vimentin is indicative of immature cells, while GFAP indicates mature ones.

Olfactory receptor neurons undergo continuous turnover, so they are an easy model for neurogenesis studies. Moreover, they are useful in neurotoxicity studies because of the exposed position of their apical pole to the external environment. Among vertebrates, fish represent a valid biological model in this field. In particular, zebrafish, already used in laboratories for embryological, neurobiological, genetic,

S. Bettini · M. Lazzari (✉) · V. Franceschini
Department of Biological, Geological and Environmental Sciences (BiGeA), University of Bologna, Bologna, Italy
e-mail: maurizio.lazzari@unibo.it

W. Tworzydlo, S. M. Bilinski (eds.), *Evo-Devo: Non-model Species in Cell and Developmental Biology*, Results and Problems in Cell Differentiation 68,
https://doi.org/10.1007/978-3-030-23459-1_15

and pathophysiological studies, is the reference organism in olfactory system research. Smell plays an important role in the reproductive behavior of fish, with direct influences also on the numerical consistency of their populations. Taking into account that a lot of species have considerable economic importance, it is necessary to verify if the model of zebrafish olfactory organ is also directly applicable to other fish. In this chapter, we focus on crypt cells, a morphological type of olfactory cells specific of fish. We describe hypothetical function (probably related with social behavior) and evolutive position of these cells (prior to the appearance of the vomeronasal organ in tetrapods). We also offer the first comparison of the molecular characteristics of these receptors between zebrafish and the guppy. Interestingly, the immunohistochemical expression patterns of known crypt cell markers are not overlapping in the two species.

15.1 Glial Cells

Early classical comparative studies brought to light the basic model of glial cytoarchitecture in the central nervous system (CNS) of vertebrates (Herrick 1948; Ramon y Cajal 1952). The studies of amphibian CNS highlighted the presence of glial cells whose cell bodies are located in the ependymal or periependymal layer. The long radially oriented processes of these cells pass through the neural wall to the meningeal surface where they form terminal expansions constituting the mambrana gliae limitans externa. These elements are found both in larvae and adults (Messenger and Warner 1989) and resemble the first glial cells appearing in the embryonic brain of amniotes (Ramon y Cajal 1952).

Radial glial cells have been identified as important elements for guidance of migrating neurons during development (see Rakic 1981; Hatten 1999). Moreover, radial glial cells are the source of neurons as well as other glial elements including astrocytes, oligodendrocytes, and ependymal cells (Campbell and Götz 2002; Weissman et al. 2003; Malatesta et al. 2008; Rowitch and Kriegstein 2010). During neurogenesis, radial glial cells descend directly from neuroepithelial cells (reviewed in Götz and Huttner 2005).

Adult amniotes show a different type of glial cell, called star-shaped astrocyte due to numerous cytoplasmic processes arising from the cell body. Star-shaped astrocytes coexist with prevailing radial glial cells in reptiles (Kalman and Pritz 2001; Lazzari and Franceschini 2001). On the contrary, star-shaped astrocytes are the main type of glial cells in the endothermic amniotes, birds and mammals (Elmquist et al. 1994; Kalman et al. 1998). The glial cytoarchitectures including star-shaped astrocytes appear functionally more complex and phylogenetically derived compared to patterns consisting of radial glial cells only.

Glial fibrillary acidic protein (GFAP), the major subunit of the gliofilaments belonging to the intermediate filament class (Eng 1985), has been recognized as a reliable marker of mature elements of the astroglial lineage (Dahl and Bignami 1985). The expression of GFAP has been reported in the glial cells of many vertebrates (Wasowicz et al. 1994; Wicht et al. 1994; Naujoks-Manteuffel and Meyer 1996). GFAP has a considerable stability in its molecular and antigenic properties throughout the vertebrate phylogenesis (see Eng et al. 2000).

Vimentin is another cytoskeletal protein of the intermediate filament group that is used in glial cell study. It was detected in not fully differentiated cells of the astroglial lineage in reptiles (Monzon-Mayor et al. 1990; Yanes et al. 1990) and mammals (Oudega and Marani 1991; Elmquist et al. 1994; Pulido-Caballero et al. 1994). It was also found in terminally differentiated glial cells of adult teleosts and amphibians (Zamora and Mutin 1988; Cardone and Roots 1990; Rubio et al. 1992).

Immunohistochemically detected GFAP abundance and vimentin scarcity can suggest a mature condition for the glial elements. Moreover, the relative abundance of radial elements in the glial pattern is morphologically indicative of an immature condition (Kalman et al. 1998; Kalman and Pritz 2001). The relationship between GFAP/vimentin expression and amount of radial elements in the glial pattern of vertebrates is still not fully clear. Exhaustive response to developmental and phylogenetic implications of specific intermediate filament molecular markers and specific glial cell types require additional studies.

The immunohistochemical detection of intermediate filament molecular markers of glial cells brought to light the cytoarchitecture of astroglial cell elements in the vertebrate CNS (Kalman 2002). The following sections describe the characteristics of astroglial cell organization in the main groups of tetrapods (Table 15.1).

15.1.1 Lungfish

The distribution of glial intermediate filament molecular markers, GFAP and vimentin, was investigated in the CNS of the African lungfish *Protopterus annectens* using immunoperoxidase technique (Lazzari and Franceschini 2004). The Dipnoi (lungfishes) are a monophyletic group of osteichthyes (Schultze and Campbell 1986) and together with their sister group *Crossopterygii* (coelacanths) represent the unique surviving lobe-finned fish (Moy-Thomas and Miles 1971). Lungfishes are considered the closest extant relatives to terrestrial vertebrates (Forey 1986). Molecular analysis, mostly DNA sequencing, supports this assumption (Amemiya et al. 2013; Biscotti et al. 2016).

GFAP immunohistochemistry revealed a glial architecture. The cell bodies of radial cells are located at the ventricular surface (the epithelial-like monostratified layer separating the neural tissue from the cavities constituting the ventricular system of the CNS). The cell body gives rise to a long radial process, which ramify into finer fibers. These fibers form the submeningeal and perivascular glial layers. In the different cerebral regions, radial glial cells show a low degree of heterogeneity in staining intensity for GFAP. In the spinal cord, the cell bodies of the glial cells are placed in the periependymal zone. GFAP immunostaining does not reveal star-shaped astrocytes in the CNS of *Protopterus*. Vimentin immunopositivity is scarce in the nervous tissue of this lungfish.

Table 15.1 GFAP- and vimentin-immunopositive glial structure in the central nervous system of tetrapods and closest, extant relatives

	Lungfish	Urodeles		Turtles	Lizards	Crocodilians	Birds	Mammals
	Protopterus annectens	*Ambystoma mexicanum*	*Triturus carnifex*	*Pelodiscus sinensis*	*Podarcis sicula* *Anolis sagrei* *Eublepharis macularius*	*Caiman crocodilus*	*Gallus domesticus*	Rat Dog
GFAP-immunopositivity	Yes	Yes	No	Yes	Yes	Yes	Yes	Yes
Brain	Ependymal radial glia (tanycytes): radial cells with cell bodies located at the ventricular surface and long positive processes	Ependymal radial glia (tanycytes): radial cells with cell bodies located at the ventricular surface and long positive processes		Predominant ependymal radial glia (tanycytes): radial cells with cell bodies located at the ventricular surface and long positive processes	Predominant ependymal radial glia (tanycytes) A few star-shaped astrocytes in the optic tectum	Predominant positive fibers from ependymal glia Non-predominant star-shaped astrocytes intermingled with radial glia processes	Restricted occurrence of radial glia Predominant star-shaped astrocytes	Radial glial fibers almost disappeared Predominant star-shaped astrocytes
Spinal cord	Radial glial cells and their processes Positive ependymocytes	Radial glia with their cell bodies displaced from the ependymal layer into a periependymal position Ependymocytes with slight positivity		Radial glial cells in periependymal position A few star-shaped astrocytes	Radial astrocytes Some star-shaped astrocytes	Predominant dense network of positive radial and nonradial fibers	Numerous intensely immunostained star-shaped astrocytes	Star-shaped astrocytes

Vimentin-immunopositivity	Yes	No	Yes	No	Yes	?	Yes	Yes
Brain	Some weakly positive fibers in the gray and white matters		Ependymal radial glia (tanycytes): radial cells with cell bodies located at the ventricular surface and long positive processes		*P. sicula* and *E. macularius*: few vimentin-positive radial glial fibers		Vimentin positive glial elements at the midline of the brain (raphe and decussations)	Only the cell bodies of some neuron-like cells
Spinal cord	Poor vimentin-positive radial processes in the white matter		Radial glial cells and their processes Positive ependymocytes		*P. sicula*: positive ependymal cells *E. macularius*: some radial processes			Ependymal cells: (tanycytes and cuboidal cells)

Note: "?": data not available

15.1.2 Urodeles

GFAP and vimentin have been used for an immunohistochemical study of glial elements in the CNS of two urodele specie, *Ambystoma mexicanum* (Ambystomatidae) and *Triturus carnifex* (Salamandridae) (Lazzari et al. 1997). In *A. mexicanum* brain, GFAP-immunopositivity appears only in the ependymal cells showing a tanycytic feature (Fig. 15.1a). GFAP-positive radial processes give rise to variously extended endfeet forming the perivascular and submeningeal glial layers (Fig. 15.1b). In the spinal cord, GFAP-immunopositive radial glial cells show their cell bodies only in periependymal position. In *A. mexicanum* CNS, no vimentin immunopositive glial structures are present. In *T. carnifex* CNS, the distribution pattern of GFAP and vimentin results quite the opposite of the *A. mexicanum* condition.

15.1.3 Turtles

Among living animals, turtles are widely accepted as the most closely related extant vertebrates to the "progenitor" amniotes that represent the common ancestor of both reptiles and mammals (Gans and Parsons 1981; Bar et al. 2000).

The glial cytoarchitecture in the CNS of the soft-shell turtle *Trionyx sinensis* was investigated by immunoperoxidase detection of GFAP and vimentin intermediate filaments (Lazzari and Franceschini 2006). In this turtle, GFAP immunostaining reveals clear positivity in both the brain and spinal cord. The stained elements consist of long radial fibers coming from cell bodies placed in the ependymal layer displaying their tanycytic nature (Fig. 15.1c, d). In the brain, star-shaped or proper astrocytes are not detected by immunostaining; on the contrary, they appear in the spinal cord. According to the more complex turtle brain compared to amphibian condition, the ependymal radial glia of the turtle show some regional specializations concerning different size and the intensity of the immunohistochemical staining, which are comparable with the results in Emydidae (Kriegstein et al. 1986; Kalman et al. 1994, 1997).

In the spinal cord, both gray and white matters appear intensely GFAP immunopositive. The ependymal layer turns out to be immunonegative, whereas evident radial astrocytes have their cell bodies in the periependymal zone. A few GFAP-positive star-shaped astrocytes are found in the lateral column of the posterior intumescence (an enlargement of the spinal cord at the level of hind limbs). Anti-vimentin immunohistochemistry reveals no positive glial structures in the CNS of *T. sinensis*.

In the turtle, GFAP immunostaining shows a certain degree of heterogeneity significantly greater than what appears in dipnoans (Lazzari and Franceschini 2004) and urodeles (Lazzari et al. 1997). The glial pattern of turtle is simpler than the condition found in lizards (Lazzari and Franceschini 2001, 2005a, b), but clearly less

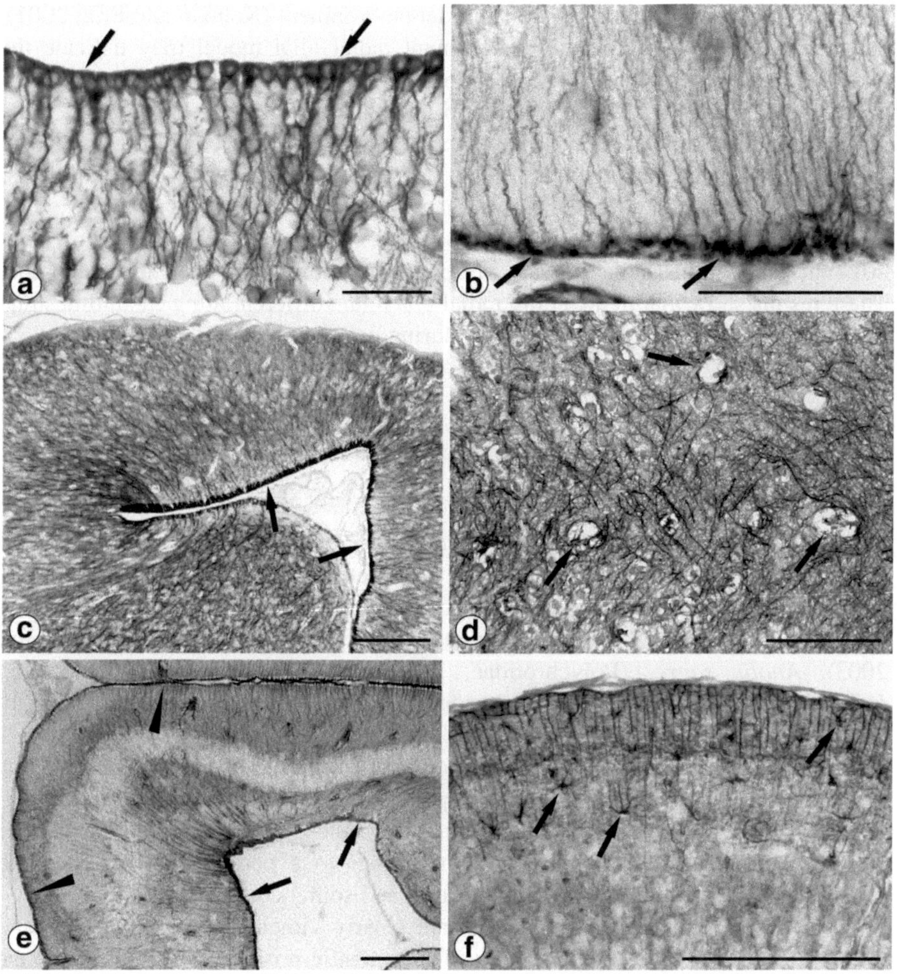

Fig. 15.1 GFAP-immunopositive structures in the brain of tetrapods. (**a**) In *Ambystoma mexicanum* telencephalon, radially oriented processes originate from ependymal radial glia (tanycytes) (arrows). (**b**) Radially oriented glial processes reach the submeningeal surface of *A. mexicanum* telencephalon where enlarged endfeet form (arrows). (**c**) In *Trionyx sinensis* telencephalon, tanycytes (arrows) mark the ventricular surface. Radial processes go through the neural wall. (**d**) In *T. sinensis* brain, glial processes give rise to perivascular glial covering (arrows). (**e**) In *Podarcis sicula* telencephalon, tanycytes (arrows) line the ventricular surface; glial endfeet (arrowheads) mark out the meningeal layer. (**f**) In *P. sicula* optic tectum, some star-shaped astrocytes (arrows) are intermingled with radial glial processes reaching the meningeal surface. Bars: 50 μm (**a**, **b**, **d**), 100 μm (**e**, **f**), 200 μm (**c**)

complex than in birds (Kalman et al. 1993a, 1998) and mammals (Elmquist et al. 1994). According to Kalman and Pritz (2001), star-shaped astrocytes, prevailing in birds and mammalian CNS, are derived cells compared to radial glia. They give rise to a glial network which permits an increase of the neural wall thickness and addresses the needs of the regional neuronal environment (Kalman and Pritz 2001). Star-shaped astrocytes inside a predominant radial glial model may indicate the local, more complex nervous areas (Simard et al. 2003; Mulligan and MacVicar 2004; Benediktsson et al. 2005; Volterra and Meldolesi 2005). In the adult turtle brain, the radial glial pattern and the absence of star-shaped astrocytes may represent an ancient character, shared with the stem reptile and present at early developmental stages in the mammalian brain (Clinton et al. 2014).

A more recent study on the Emydidae *Trachemys scripta elegans* (Clinton et al. 2014) has shown that radial glia are neuronal precursors in the turtle telencephalon. These results show that in the adult, radial glia retain differentiating capabilities that are similar to the activity they carry out during development.

15.1.4 Lizards

The glial pattern of lizards has been studied in different species, *Gallotia galloti* (Lacertidae, Monzon-Mayor et al. 1990; Yanes et al. 1990), *P. sicula* (Lacertidae, Lazzari and Franceschini 2001), *Agama impalearis* (Agamidae), *Eumeces algeriensis* (Scincidae), *Tarentola mauritanica* (Gekkonidae, Ahboucha et al. 2003), *Anolis sagrei* (Polychrotidae, Lazzari and Franceschini 2005a), and *Eublepharis macularius* (Eublepharidae, Lazzari and Franceschini 2005b; McDonald and Vickaryous 2018).

In adult lizards, GFAP-immunocytochemistry demonstrates the presence of three different astroglial lineage cell type: (1) ependymal radial glia or tanycytes throughout the brain; (2) proper radial glia or radial astrocytes, in the spinal cord with their cell bodies displaced from the ependymal layer to a periependymal zone; and (3) star-shaped astrocytes in the optic tectum and spinal cord (Fig. 15.1e, f).

In the spinal cord, ependymal cells are clearly vimentin-immunopositive and GFAP-negative. Only in the brain can a few vimentin-positive radial glial fibers be observed (Monzon-Mayor et al. 1990; Lazzari and Franceschini 2001, 2005b). During development, vimentin-GFAP transition has also been observed in reptiles (Monzon-Mayor et al. 1990; Yanes et al. 1990).

The presence and arrangement of star-shaped astrocytes (true astrocytes) is another point that deserves consideration in the study of saurian glia. The glial pattern containing star-shaped astrocytes is considered more complex and a phylogenetically derived state compared to the model with only radial elements. For these reasons, the glial organization of lizards may be considered a derived character compared to that of turtles.

In lizard, the intensity of the GFAP immunostaining of glial cells turned out to be not uniform throughout the neural wall. This condition recalls the GFAP-staining heterogeneity found in turtle (Kalman et al. 1994; Lazzari and Franceschini 2006).

According to Kalman and Pritz (2001), uneven distribution and different size of the astroglial structure are the main causes of the differences in regional GFAP immunostaining of saurian brain. These authors consider differential GFAP expression less important in reptile than in birds and mammals.

Based on GFAP abundance and vimentin scarceness, the immunohistochemical reaction of adult lizard astroglial intermediate filaments is indicative of mature astroglial cell lineage. In contrast, the glial pattern of the lizard mostly based on radial elements is morphologically more immature than those found in crocodile, birds, and mammals (Kalman et al. 1998; Kalman and Pritz 2001).

However, the connections between GFAP/vimentin and radial glia/star-shaped astrocytes during CNS evolution and development are not fully clear. In lizards belonging to different families, morphology of GFAP-positive cells and their relative proportions exhibit remarkable differences (Ahboucha et al. 2003). Additional studies on the characteristic of the vertebrate astroglial elements are necessary to provide a comprehensive response.

In a recent study on the neurogenesis in the medial cortex of *E. macularius*, immunohistochemistry identified proliferating pools of neural stem/progenitor cells within the telencephalon (McDonald and Vickaryous 2018). These cells, which express GFAP and vimentin and show a clear radial morphology, are identified as the radial glia involved in neurogenesis in adult.

15.1.5 Crocodilians

GFAP-immunopositive structures were studied in the brain of caiman, *Caiman crocodilus*, as representative of the order Crocodilia (Kalman and Pritz 2001). The caiman astroglia were clearly GFAP-immunopositive, and similar to other reptilians, the predominant elements were radial astrocytes. The glial architecture of crocodilian brain is characterized by a regional variability that is much greater than in the other reptiles studied so far, but smaller than in birds. Kalman and Pritz (2001) suggested that this condition is indicative of a more regional specialization of the glial cytoarchitecture in caiman compared to the other reptiles. The preeminent characteristic of caiman glial system compared to the other reptiles is represented by the widespread distribution of GFAP-immunopositive star-shaped astrocytes that are absent in turtles and scarce in lizards and snakes. Kalman and Pritz (2001) pointed out that, also in caiman, star-shaped astrocytes appear mixed together with radial fiber elements and do not represent the prevailing glial elements throughout the brain. Moreover, their presence appears not related to the brain wall thickness. Although Crocodilians represent the extant reptilian group most closely related to birds, Kalman and Pritz (2001) documented that the features of the caiman glial system are only partially shared by birds and mammals. This indicates the independent, parallel evolution from the diapsids or synapsids, different ancestral groups of stem reptiles.

15.1.6 Birds

The immunohistochemical detection of GFAP and vimentin was studied in the domestic chicken (*Gallus domesticus*) from hatching to adulthood (Kalman et al. 1993a, b, 1998). In the telencephalon, the day-old chicken shows a vimentin-immunopositive radial fiber system. Many of these radial elements persist until adulthood and during development undergo transition from vimentin to GFAP expression.

In the posthatch telencephalon, the distribution of star-shaped astrocytes resembles the condition in the adults, except in the visual center that shows a slow maturation rate. In the diencephalon (the rear region of the forebrain) and in the mesencephalic tegmentum (the ventral part of the mesencephalon), some GFAP-positive astrocytes are present at birth, and their number increases during the first two posthatching weeks. At hatching, these districts of the chicken brain contain only a few positive fibers. Therefore, a gradual and clear substitution of radial fibers by star-shaped astrocytes does not occur. On the contrary, this substitution is evident in mammals (Kalman et al. 1998). Whereas in reptiles the GFAP-positive radial glial system persists in optic tectum throughout life, in chicken it disappears from the tectal superficial layer that appears a derived character. At hatching, numerous GFAP-positive, vimentin-negative astrocytes accompany axons in the cerebral tracts. In posthatching birds, vimentin-positive glia are observed at the midline in decussation (medial areas of the CNS where nervous fibers pass from the right side to the left one and vice versa). Probably, these vimentin-positive elements are engaged in fiber-crossing regulation. In general, disappearance of radial fibers, dominance of astrocytes, and uneven distribution of GFAP-immunopositivity characterize glial development and organization in avian brain. Since these characters are found in phylogenetically distant mammals and only rudimentary outlined in reptiles, Kalman et al. (1998) suggested that they evolved independently along parallel paths.

15.1.7 Mammals

In mammals, radial glial cells have almost disappeared in adults. They represent the first developmental stage of astrocytes from which fibrous and protoplasmic astrocytes originate through shortening of their long processes and displacement of their cell body away from the ependymal layer (Levitt and Rakic 1980).

In adult mammals, GFAP is regarded as the principal intermediate filament protein in astrocytes (Voigt 1989), whereas vimentin is found in ependymal cells (Dahl et al. 1981; Bignami et al. 1982).

During development, vimentin predominates in early radial glia; afterward, this vimentin-like immunoreactivity decreases as radial cells mature into astrocytes (Oudega and Marani 1991; Elmquist et al. 1994; Moore and Oglesbee 2015; Luo et al. 2017). Therefore, in mammals, vimentin-immunoreactivity can be interpreted as marker of immature astrocytic cells and GFAP-immunoreactivity as marker of mature astrocytes. As the brain matures, radial glia become true astrocytes, and the change in intermediate filament protein expression from vimentin to GFAP may parallel this morphological transformation (Schnitzer et al. 1981; Elmquist et al. 1994). Embryonic astroglia of higher vertebrates resemble astrocytic cells of adult lower vertebrates.

In the adult mammalian CNS, radial glial cells are restricted to few locations: the retina (Müller cells), the cerebellum (Bergmann glia), and the hypothalamus where they are called tanycytes (Sild and Ruthazer 2011). New information derives from Mack et al. (2018) who investigated the astroglial cells in the cerebral cortex of the lesser hedgehog tenrec, *Echinops telfairi*. These animals belong to the group of Afrotheria representing a clade which is possibly a sister group of all other placental mammals (O'Leary et al. 2013; Averianov and Lopatin 2014). Lesser hedgehog tenrecs are considered to have the lowest encephalization index with relatively little neocortex compared to other placental mammals (Stephan et al. 1991; Krubitzer et al. 1997). The adult lesser hedgehog tenrec brain shows also continued cell proliferation (Alpar et al. 2010), reminiscent of the continued cell addition in anamniotes.

The authors concluded that the radial glia in the adult lesser hedgehog tenrec represents an immature form of astroglia that persists in these animals throughout life. It is noteworthy that radial glial structures occur at locations with continued neurogenesis (e.g., dentate gyrus and the paleocortex). Therefore, it is possible to imagine that the radial glial fibers of the adult are involved in the differentiation of new neurons. Radial glial fibers can lead differentiating neurons to appropriate locations. In adults, this activity is similar to the role of radial glia during development. The coincidence of radial glia and continued growth is found also in the brains of many anamniotes.

15.1.8 Ensheathing Glial Cells

GFAP and vimentin, together with other cellular markers, are useful tools also in characterizing olfactory ensheathing cells (OECs) that represent a specific type of glial cells (Pellitteri et al. 2010). Throughout life, the regeneration ability of olfactory sensory neurons (OSNs) is connected to OECs that cover olfactory axons in their route from the basal lamina of the olfactory epithelium to the glomeruli in the olfactory bulb (Pellitteri et al. 2010). In recent years, growing attention was given to OECs as they appeared able to promote and support regeneration of injured CNS (Raisman 2001; Barnett and Riddell 2004; Chung et al. 2004). In particular, studies underlined their potential role in cell-based therapy for spinal cord injury repair (Lu et al. 2002; Mackay-Sim 2005).

Recent studies are focused on OECs of anamniotes (Lazzari et al. 2013, 2014, 2016). A comparative immunohistochemical study on microsmatic (*Poecilia reticulata*) and macrosmatic fish (*Carassius auratus*, *Danio rerio*) showed some differences between the two types of olfactory organization and some similarities between fish (anamniotes) and mammals (amniotes) OSNs (Lazzari et al. 2007, 2013, 2014; Bettini et al. 2009). *D. rerio* is a good model also in olfaction-related research. In fact, its olfactory system is histologically comparable to that of other vertebrates, and the cells of its olfactory neuroepithelium can be clearly characterized by immunological approaches (Germanà et al. 2007; Gayoso et al. 2011; Braubach et al. 2012; Lazzari et al. 2017, 2019). The immunohistochemical characterization of OECs was also studied in *A. mexicanum* (Lazzari et al. 2016). Urodeles have a relatively simple olfactory system and represent a good model for studying OECs in extant representatives of basal tetrapods. The olfactory pathways of *A. mexicanum*, representative of basal terrestrial anamniotes, share characteristics with both fish (aquatic anamniotes) and mammals (amniotes). It remains speculative whether these similarities are basal features conserved during evolution or evolved independently in different vertebrate groups. So far, the functional role of the regional and interspecific differences displayed by OEC immunolocalization remains unknown.

15.2 Olfactory Crypt Neurons

In addition to ciliated (cOSNs) and microvillous (mOSNs) neurons, the olfactory organ of fish is peculiarly composed of a third sensory cell type, called crypt cell. The crypt cells were observed for the first time in the olfactory neuroepithelium of zebrafish, *Danio rerio* (Hansen and Zeiske 1998). The cell body of the crypt cell is egg-shaped, and it is located in the upper layer of the epithelium. The most striking feature of the crypt cell is an invagination, called the "crypt," located in the apical part of the neuron. The border of the crypt is surrounded by microvilli, while short cilia grow in its lumen. The large nucleus of the cell resides in the lower third of the soma. A thin axon emerges from the basal portion of the neuron. The last histological characteristic of crypt cells is that they are surrounded by specialized supporting cells which are particularly electron-lucent (Fig. 15.2). Although crypt cells are randomly distributed in all the lamellae of the olfactory organ, their number is lower than that of ciliated and microvillous cells.

15.2.1 Crypt Cell Phyletic Distribution and Evolutionary Considerations

Since their first description, crypt neurons have been observed in many actinopterygian species (Hansen et al. 2003; Hamdani and Døving 2006; Lazzari et al. 2007; Camacho et al. 2010), including freshwater and marine fish (see Hansen

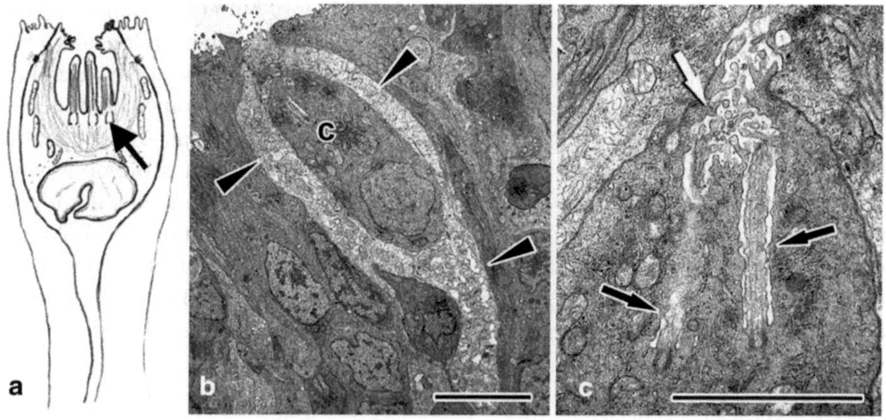

Fig. 15.2 Crypt cell morphology. (**a**) Schematic drawing of the crypt cell (modified from Camacho et al. 2010); cilia reside in the apical invagination, called "crypt" (arrow). (**b**) An ovoid crypt cell surrounded by specific electron lucent sustentacular cells (arrowheads) in the olfactory epithelium of *P. reticulata* (modified from Lazzari et al. 2007). (**c**) Magnification of the crypt showing cilia (black arrows) and microvilli (white arrow). Bars: 5 μm (**b**), 2 μm (**c**)

and Finger 2000 and Hansen and Zielinski 2005 for a review of species examined, and crypt cell distribution). Moreover, crypt cells were discovered in Chondrichthyes, both in sharks (*Scyliorhinus canicula*, Ferrando et al. 2006) and batoids (*Raja clavata*, Ferrando et al. 2007). Interestingly, no evidence exists about the presence of this cell type in lungfish, which suggests that the crypt neurons are ancestral traits evolved before actinopterygian radiation, but lost at the actinopterygian-sarcopterygian transition.

The evolution of air breathing in vertebrates could be a possible event that influenced the disappearance of this cells in tetrapods. Ultrastructural, histological, and ontogenetic studies on the olfactory organ of amphibians, the tetrapods most closely related to lungfish, did not reveal crypt cells in either terrestrial or aquatic adult species (see Hansen and Finger 2000 for a list of publications). The developmental analysis from early larval stage to post-metamorphic juveniles in different species of anurans (Benzekri and Reiss 2012; Jungblut et al. 2009, 2017; Quinzio and Reiss 2018) and urodeles (Franceschini et al. 2003; Saito et al. 2003; Nakada et al. 2014) with gill-bearing larvae did not report the presence of crypt cells. The olfactory system of the anuran model system *Xenopus laevis*, has two cavities, one of which is devoted to aquatic olfaction and the other to aerial olfaction, but no crypt cells have been detected in any of them (Hansen et al. 1998; Oikawa et al. 1998).

The most characteristic anatomical difference in the olfactory system between fish and amphibians is the presence in the latters of a vomeronasal system. The vomeronasal system is an accessory olfactory system possessed by most tetrapods, except birds, aquatic reptiles, aquatic mammals, and higher primates (including man) in which it is vestigial or lost (Bertmar 1981). It is composed by a vomeronasal organ (VNO), containing only mOSNs (while the main olfactory organ is exclusively composed by cOSNs) projecting to an accessory olfactory bulb (AOB) in the rostral telencephalon (D'Aniello et al. 2017). Interestingly, a recent study (Gonzales et al. 2010) shows that the African lungfish *Protopterus dolloi* possesses, at the base of the olfactory lamellae, the epithelial crypts expressing markers of vomeronasal receptors. Cells localized in the epithelial crypts project their axons to an AOB identified in the lateral olfactory bulb, revealed also by lectin binding in a previous work (Franceschini et al. 2000). During evolution from lungfish to tetrapods, the formation of a vomeronasal organ was accompanied by a redistribution of morphological olfactory cells subtypes, with cOSNs segregated to the main olfactory organ and mOSNs to the vomeronasal organ, and crypt cells disappeared. Possible causes of crypt cell disappearance may reside in the function of the newly developed/evolving vomeronasal organ.

15.2.2 Crypt Cells Have Similarities with the Vomeronasal Organ of Tetrapods

Crypt cells seem to be ancestral olfactory neurons, which became replaced by microvillous cells of the vomeronasal organ during the evolutionary divergence of tetrapods (Fig. 15.3). This shift was possible probably because these two cell types show common features. In fact in *D. rerio* (Oka et al. 2012), the crypt neurons express only the Gαi-coupled receptor gene *ora*4 belonging to the Vomeronasal-1-Receptor (V1R)-like family.

Moreover, in sexually mature *Oncorhynchus mykiss*, crypt cells preferentially respond to gonadal extract and pheromones from the opposite sex (Bazáes and Schmachtenberg 2012), and the vomeronasal organ is presumed to detect pheromones (Halpern and Martinez-Marcos 2003). However, the vomeronasal organ can respond to a variety of other odorants with different behavioral significance in all groups of tetrapods. Recently, zebrafish crypt cells were proposed to be involved in detecting kin odor-related signal (Biechl et al. 2016). However, which chemical compounds are detected by crypt cells remain elusive because of different results obtained in various species (Vielma et al. 2008).

Important clues supporting an evolutionary relationship between crypt cells and the vomeronasal organ come from their pathways to glomerular targets in the central nervous system.

In crucian carp (Hamdani and Døving 2006) and channel catfish (Hansen et al. 2003), olfactory bulb backtracing experiments suggested a medioventral position of the crypt neuron target region where a population of secondary neurons specific for sex pheromones are located (Lastein et al. 2006). Medial glomeruli in the bulb are connected with the medial olfactory tract, mediating reproductive behavior. Interestingly, in zebrafish, axons of crypt cells project in one single glomerulus within the mediodorsal area of the olfactory bulb (Ahuja et al. 2013), a region connected with medial amygdala (Biechl et al. 2017) that in tetrapods receives input from the

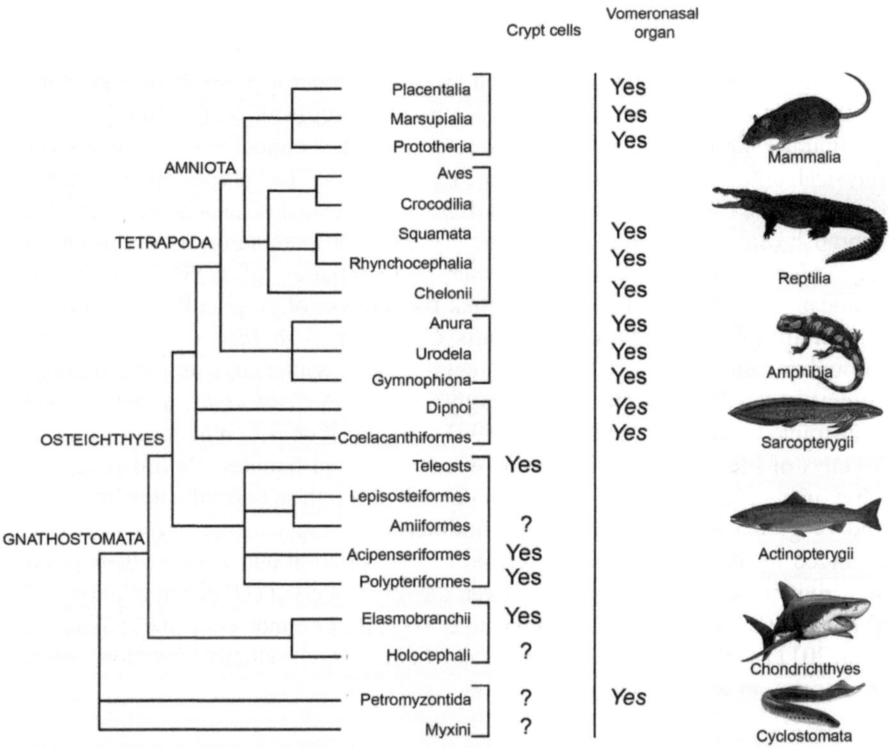

Fig. 15.3 Cladogram reporting the presence of crypt cells and vomeronasal organ in the main vertebrate groups. "Yes": the presence has been confirmed in at least one species; "*Yes*": an accessory olfactory system with segregated projections has been observed in different studies, but the organ is unique; "?": the group has never been examined

vomeronasal organ. This finding supports the hypothesis that a primordial accessory olfactory system, which originates in crypt olfactory sensory cells, may exist in teleosts.

All together, these data reinforce the hypothesis that the crypt cells may be related to the reproductive behavior and, in general, similarly to the vomeronasal organ, to the conspecific recognition (Liu et al. 2014).

15.2.3 Developmental Studies Give Useful Information in Determining a Possible Role of Crypt Cells

Camacho et al. (2010) proposed a sequence of stages in crypt cell development, characterized by morphological profiles observed with TEM: (1) at first, immature cells appear close to the epithelial basal lamina; (2) then they differentiate into cells similar to the crypt neurons (absence of knob, envelopment by supporting cells, etc.)

but with cilia still remaining on their non-invaginated apical surface; and (3) the formation of the crypt would be the final step of their differentiation.

In vertebrates, the olfactory epithelium exhibits neurogenesis throughout the life of the animal, from the embryonic/juvenile development to the adulthood. This continuous turnover is required to restore an intact epithelium after chemical or physical injuries. However, the growth dynamics of the crypt cell population is different than cOSNs and mOSNs. In *Acipenser naccarii* (Camacho et al. 2010) and zebrafish (Sandulescu et al. 2011), the crypt cell differentiation occurs in the early days of embryonic life, but later than differentiation of cOSNs and mOSNs. Hamdani et al. (2008) demonstrated that the number of crypt cells in the olfactory epithelium of the adult crucian carp *Carassius carassius* varies dramatically throughout the year, with seasonal regression in winter and renewal during the summer spawning months. Interestingly, in *Poecilia reticulata*, a species with a year-round breeding season (Houde 1997), crypt cell density remains invariant after 90 days of life, and it is different between males and females (Bettini et al. 2012). Moreover, crypt cells reach their adult density, through sex-specific dynamics, at the end of gonad development. These findings may suggest that crypt neurons are involved in mediating behavior related to reproduction and confirm their possible sensitivity to sex pheromones. However, the onset of crypt cell differentiation occurs very early (Bettini et al. 2012; Camacho et al. 2010; Ferrando et al. 2007; Sandulescu et al. 2011) so they might be involved also in the nonreproductive functions, perhaps in association with cOSNs and mOSNs.

15.2.4 Crypt Cell Markers

In all fish examined in the literature, crypt cell morphology has been fully described by TEM studies, and in various species (e.g., channel catfish, crucian carp, sturgeon, zebrafish), they have been visualized also by retrograde transport of fluorescent dye (Hansen et al. 2003; Hamdani and Døving 2006; Camacho et al. 2010; Ahuja et al. 2013). The use of specific immunohistochemical staining in the olfactory system is important for numerous fields of research, for example, environmental studies on the differential effects of pollutants on olfactory neuronal populations (Bettini et al. 2016; Lazzari et al. 2017, 2019). However, only in the zebrafish model antigenic properties of crypt neurons have been studied.

15.2.5 The Zebrafish Model

Germanà et al. (2004) reported that crypt neurons in zebrafish olfactory lamellae are specifically immunostained by antiserum against S100, a neural calcium-binding protein that stimulates cell proliferation and migration and inhibits apoptosis and differentiation (Donato et al. 2013). Thus, the anti-S100 serum was used in several subsequent studies (Gayoso et al. 2011, 2012; Sato et al. 2005). However, in the last decade, some studies (Ahuja et al. 2013; Kress et al. 2015; Oka et al. 2012) showed that the specificity of anti-S100 varies according to tissue processing: in fresh-frozen sections, the cells with crypt-neuron morphology appeared specifically stained, while, in fixed conditions, other cell types, possibly microvillous neurons, were also stained. Moreover, while anti-S100-positive microvillous cells express the *s100z* gene, anti-S100-positive crypt cells do not express *S100* genes, raising some doubts on which crypt cell antigen is recognized by the antiserum. In addition, the dorsomedial glomerular field of zebrafish olfactory bulb receives afferents from the crypt cells, as revealed by DiI retrolabeling (Gayoso et al. 2012), but their axons are S100-negative (Gayoso et al. 2011). This uncertainty precludes the use of S100-like immunoreactivity in various studies and procedures, including visualization of axonal terminal regions.

The first molecular characterization of crypt cells was made by Catania et al. (2003), who discovered that anti-TrkA, specific for a neurotrophin receptor, labeled only zebrafish crypt cells in the olfactory epithelium, but this antibody was not investigated further at the time. Ten years later, Ahuja et al. (2013) demonstrated that anti-TrkA-immunohistochemistry constitutes a robust and sensitive marker for zebrafish crypt cells, and nowadays, it is the only marker with such specificity.

15.2.6 What About Other Fish?

In scientific research, the use of only one animal model (zebrafish, in this case) allows direct comparisons of data, and this facilitates the determination of cause and effect relationship and the formulation of hypothesis. On the other hand, the characteristics described in the model could not be extended to the majority of other species. For this reason, recently, for the first time, the antigenic properties of crypt cells in another fish, *P. reticulata* (guppy) was examined (Bettini et al. 2017). *P. reticulata* possesses some anatomical and histological characteristics of the olfactory organ different from zebrafish (Lazzari et al. 2007). Crypt cell markers in the guppy appeared to overlap only partially with zebrafish markers (Fig. 15.4). While no olfactory neuron was labeled by anti-TrkA, the anti-S100 immunohistochemistry gave a pattern comparable to what has been reported in zebrafish. In addition, in the gyppy, the calretinin, a known marker of zebrafish cOSNs and mOSNs (Germanà et al. 2007), was expressed also by some S100-positive crypt neurons. However, other anti-S100-labeled crypt cells remained immunonegative. This revealed two distinct subpopulations of the same morphotype recognizable for their protein content (calretinin-positive and calretinin-negative crypt neurons). It is possible that this heterogeneity in protein expression also reflects some functional differences. All that considered, this recent study revealed that the molecular features of olfactory crypt neurons described in the fish model *D. rerio* cannot be extended to all teleosts, and the investigation of other phyletic taxa is necessary to provide more conclusive answers on the evolution of olfactory crypt cells in fish.

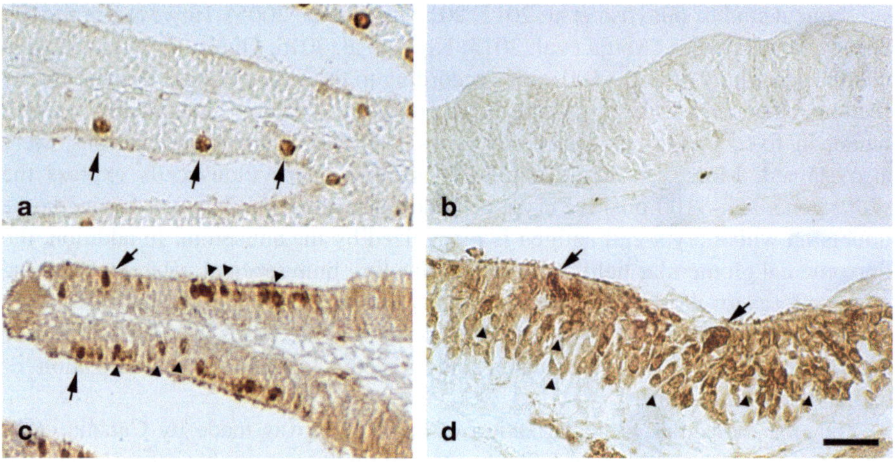

Fig. 15.4 Comparison between TrkA (**a–b**) and S100 (**c–d**) immunohistochemistry in *D. rerio* (**a**, **c**) and *P. reticulata* (**b**, **d**). TrkA is a specific marker for crypt cell (arrows) in zebrafish but not in *P. reticulata*; S100-immunohistochemistry stains crypt cells (arrows) and microvillous cells with short dendrites in the medial layer of the epithelium (arrowheads) in both species. Bar: 20 μm

References

Ahboucha S, Laalaoui A, Didier-Bazes M, Montange M, Cooper HM, Gamrani H (2003) Differential patterns of glial fibrillary acidic protein-immunolabeling in the brain of adult lizards. J Comp Neurol 464:159–171

Ahuja G, Ivandic I, Saltürk M, Oka Y, Nadler W, Korsching SI (2013) Zebrafish crypt neurons project to a single, identified mediodorsal glomerulus. Sci Rep 3:2063–2071

Alpar A, Künzle H, Gärtner U, Popkova Y, Bauer U, Grosche J, Reichenbach A, Härtig W (2010) Slow age-dependent decline of doublecortin expression and BrdU labeling in the forebrain from lesser hedgehog tenrecs. Brain Res 1330:9–19

Amemiya CT, Alfoldi J, Lee AP, Fan S, Philippe H et al (2013) The African coelacanth genome provides insights into tetrapod evolution. Nature 496:311–316

Averianov AO, Lopatin AV (2014) High-level systematics of placental mammals: current status of the problem. Biol Bull 41:801–816

Bar I, Lambert de Rouvroit C, Goffinet AM (2000) The evolution of cortical development. An hypothesis based on the role of the Reelin signaling pathway. Trends Neurosci 23:633–638

Barnett SC, Riddell JS (2004) Olfactory ensheathing cells (OECs) and the treatment of CNS injury: advantages and possible caveats. J Anat 204:57–67

Bazáes A, Schmachtenberg O (2012) Odorant tuning of olfactory crypt cells from juvenile and adult rainbow trout. J Exp Biol 215:1740–1748

Benediktsson AM, Schachtele SJ, Green SH, Dailey ME (2005) Ballistic labeling and dynamic imaging of astrocytes in organotypic hippocampal slice cultures. J Neurosci Methods 141:41–53

Benzekri NA, Reiss JO (2012) Olfactory metamorphosis in the coastal tailed frog *Ascaphus truei* (Amphibia, Anura, Leiopelmatidae). J Morphol 273:68–87

Bertmar G (1981) Evolution of vomeronasal organs in vertebrates. Evolution 35:359–366

Bettini S, Lazzari M, Ciani F, Franceschini V (2009) Immunohistochemical and histochemical characteristics of the olfactory system of the guppy, *Poecilia reticulata* (Teleostei, Poeciliidae). Anat Rec 292:1569–1576

Bettini S, Lazzari M, Franceschini V (2012) Quantitative analysis of crypt cell population during postnatal development of the olfactory organ of the guppy, *Poecilia reticulata* (Teleostei, Poeciliidae), from birth to sexual maturity. J Exp Biol 215:2711–2715
Bettini S, Lazzari M, Ferrando S, Gallus L, Franceschini V (2016) Histopathological analysis of the olfactory epithelium of zebrafish (*Danio rerio*) exposed to sublethal doses of urea. J Anat 228:59–69
Bettini S, Milani L, Lazzari M, Maurizii MG, Franceschini V (2017) Crypt cell markers in the olfactory organ of *Poecilia reticulata*: analysis and comparison with the fish model *Danio rerio*. Brain Struct Funct 222:3063–3074
Biechl D, Tietje K, Gabriele Gerlach G, Wullimann MF (2016) Crypt cells are involved in kin recognition in larval zebrafish. Sci Rep 6:24590
Biechl D, Tietje K, Ryu S, Grothe B, Gerlach G, Wullimann MF (2017) Identification of accessory olfactory system and medial amygdala in the zebrafish. Sci Rep 7:44295
Bignami A, Raju T, Dahl D (1982) Localization of vimentin, the non-specific intermediate filament protein, in embryonal glia and in early differentiating neurons. Dev Biol 91:286–295
Biscotti MA, Gerdol M, Canapa A, Forconi M, Olmo E, Pallavicini A, Barucca M, Schartl M (2016) The lungfish transcriptome: a glimpse into molecular evolution events at the transition from water to land. Sci Rep 6:21571
Braubach OR, Fine A, Croll RP (2012) Distribution and functional organization of glomeruli in the olfactory bulbs of zebrafish (*Danio rerio*). J Comp Neurol 520:2317–2339
Camacho S, Ostos-Garrido MV, Domezain A, Carmona R (2010) Study of the olfactory epithelium in the developing sturgeon. Characterization of the crypt cells. Chem Senses 35:147–156
Campbell K, Götz M (2002) Radial glia: multi-purpose cells for vertebrate brain development. Trends Neurosci 25:235–238
Cardone B, Roots BJ (1990) Comparative immunohistochemical study of glial filament proteins (glial fibrillary acidic protein and vimentin) in goldfish, octopus and snail. Glia 3:180–192
Catania S, Germanà A, Laurà R, Gonzalez-Martinez T, Ciriaco E, Vega JA (2003) The crypt neurons in the olfactory epithelium of the adult zebrafish express TrkA-like immunoreactivity. Neurosci Lett 350:5–8
Chung RS, Woodhouse A, Fung S, Dickson TC, West AK, Vickers JC, Chuah MI (2004) Olfactory ensheathing cells promote neurite sprouting of injured axons in vitro by direct cellular contact and secretion of soluble factors. Cell Mol Life Sci 61:1238–1245
Clinton BK, Cunningham CL, Kriegstein AR, Noctor SC, Martínez-Cerdeño V (2014) Radial glia in the proliferative ventricular zone of the embryonic and adult turtle, *Trachemys scripta elegans*. Neurogenesis (Austin) 1(1):e970905
D'Aniello B, Semin GR, Scandurra A, Pinelli C (2017) The vomeronasal organ: a neglected organ. Front Neuroanat 11:70
Dahl D, Bignami A (1985) Intermediate filaments in nervous tissue. In: Shay JW (ed) Cell and muscle motility, vol 6. Plenum Press, New York, pp 75–96
Dahl D, Rueger DC, Bignami A, Weber K, Osborn M (1981) Vimentin, the 57,000 molecular weight protein in fibroblast filaments in the major cytoskeletal protein in immature glia. Eur J Cell Biol 24:191–196
Donato R, Cannon BR, Sorci G, Riuzzi F, Hsu K et al (2013) Functions of S100 proteins. Curr Mol Med 13:24–57
Elmquist JK, Swanson JJ, Sakaguchi DS, Ross LR, Jacobson CD (1994) Developmental distribution of GFAP and vimentin in the Brazilian opossum brain. J Comp Neurol 344:283–296
Eng LF (1985) Glial fibrillary acidic protein (GFAP): the major protein of glial intermediate filament in differentiated astrocytes. J Neuroimmunol 8:203–214
Eng LF, Ghirnikar RS, Lee YL (2000) Glial fibrillary acidic protein: GFAP-thirty-one years (1969–2000). Neurochem Res 25:1439–1451
Ferrando S, Bottaro M, Gallus L, Girosi L, Vacchi M, Tagliafierro G (2006) Observations of crypt neuron-like cells in the olfactory epithelium of a cartilaginous fish. Neurosci Lett 403:280–282
Ferrando S, Bottaro M, Pedemonte F, De Lorenzo S, Gallus L, Tagliafierro G (2007) Appearance of crypt neurons in the olfactory epithelium of the skate *Raja clavata* during development. Anat Rec 290:1268–1272

Forey PL (1986) Relationship of lungfishes. J Morphol Suppl 1:75–91
Franceschini V, Lazzari M, Ciani F (2000) Cell surface glycoconjugates in the olfactory system of lungfish *Protopterus annectens*. Acta Zool 81:131–137
Franceschini V, Lazzari M, Ciani F (2003) Surface glycoconjugates in the olfactory system of *Ambystoma mexicanum*. J Morphol 256:301–305
Gans C, Parsons T (1981) Biology of the reptilia: morphology. Academic Press, New York
Gayoso JÁ, Castro A, Anadón R, Manso MJ (2011) Differential bulbar and extrabulbar projections of diverse olfactory receptor neuron populations in the adult zebrafish (*Danio rerio*). J Comp Neurol 519:247–276
Gayoso JÁ, Castro A, Anadón R, Manso MJ (2012) Crypt cells of the zebrafish *Danio rerio* mainly project to the dorsomedial glomerular field of the olfactory bulb. Chem Senses 37:357–369
Germanà A, Montalbano G, Laurà R, Ciriaco E, del Valle ME, Vega JA (2004) S100 protein-like immunoreactivity in the crypt olfactory neurons of the adult zebrafish. Neurosci Lett 371:196–198
Germanà A, Paruta S, Germanà GP, Ochoa-Erena FJ, Montalbano G, Vega JA (2007) Differential distribution of S100 protein and calretinin in mechanosensory and chemosensory cells of adult zebrafish (*Danio rerio*). Brain Res 1162:48–55
González A, Morona R, López JM, Moreno N, Northcutt RG (2010) Lungfishes, like tetrapods, possess a vomeronasal system. Front Neuroanat 4:1–11
Götz M, Huttner WB (2005) The cell biology of neurogenesis. Nat Rev Mol Cell Biol 6:777–788
Halpern M, Martínez-Marcos A (2003) Structure and function of the vomeronasal system: an update. Prog Neurobiol 70:245–318
Hamdani H, Døving KB (2006) Specific projection of the sensory crypt cells in the olfactory system in crucian carp, *Carassius carassius*. Chem Senses 31:63–67
Hamdani H, Lastein S, Gregersen F, Døving KB (2008) Seasonal variations in olfactory sensory neurons—fish sensitivity to sex pheromones explained? Chem Senses 33:119–123
Hansen A, Finger TE (2000) Phyletic distribution of crypt-type olfactory receptor neurons in fishes. Brain Behav Evol 55:100–110
Hansen A, Zeiske E (1998) The peripheral olfactory organ of the zebrafish, *Danio rerio*: an ultrastructural study. Chem Senses 23:39–48
Hansen A, Zielinski BS (2005) Diversity in the olfactory epithelium of bony fishes: development, lamellar arrangement, sensory neuron cell types and transduction components. J Neurocytol 34:183–208
Hansen A, Reiss JO, Gentry CL, Burd GD (1998) Ultrastructure of the olfactory organ in the clawed frog, *Xenopus laevis*, during larval development and metamorphosis. J Comp Neurol 398:273–288
Hansen A, Rolen SH, Anderson K, Morita Y, Caprio J, Finger TE (2003) Correlation between olfactory receptor cell type and function in the channel catfish. J Neurosci 23:9328–9339
Hatten ME (1999) Central nervous system neuronal migration. Annu Rev Neurosci 22:511–539
Herrick JC (1948) The brain of the tiger salamander, *Ambystoma tigrinum*. Chicago University Press, Chicago
Houde AE (1997) Reproductive biology and sexual behavior. In: Krebs JR, Clutton-Brock T (eds) Sex, color and mate choice in guppies (Monographs in behavior and ecology). Princeton University Press, Princeton
Jungblut LD, Paz DA, López-Costa JJ, Pozzi AG (2009) Heterogeneous distribution of G protein alpha subunits in the main olfactory and vomeronasal systems of *Rhinella* (*Bufo*) *arenarum* tadpoles. Zool Sci 26:722–728
Jungblut LD, Reiss JO, Paz DA, Pozzi AG (2017) Quantitative comparative analysis of the nasal chemosensory organs of anurans during larval development and metamorphosis highlights the relative importance of chemosensory subsystems in the group. J Morphol 278:1208–1219
Kalman M (2002) GFAP expression withdraws—a trend of glial evolution. Brain Res Bull 57 (3/4):509–511

Kalman M, Pritz MB (2001) Glial fibrillary acidic protein-immunopositive structures in the brain of a crocodilian, *Caiman crocodilus*, and its bearing on the evolution of astroglia. J Comp Neurol 431:460–480

Kalman M, Szekely A, Csillag A (1993a) Distribution of glial fibrillary acidic protein-immunoreactive structures in the brain of the domestic chicken (*Gallus domesticus*). J Comp Neurol 330:221–237

Kalman M, Szekely A, Csillag A (1993b) GFAP and vimentin in the developing chicken brain. Anat Anz (Suppl) 175:130–131

Kalman M, Kiss A, Majorossy K (1994) Distribution of glial fibrillary acidic protein-immunopositive structures in the brain of the red-eared freshwater turtle (*Pseudemys scripta elegans*). Anat Embryol 189:421–434

Kalman M, Martin-Partido G, Hidalgo-Sanchez M, Majorossy K (1997) Distribution of glial fibrillary acidic protein–immunopositive structures in the developing brain of the turtle *Mauremys leprosa*. Anat Embryol 196:47–65

Kalman M, Szekely AD, Csillag A (1998) Distribution of glial fibrillary acidic protein and vimentin-immunopositive elements in the developing chicken brain from hatch to adulthood. Anat Embryol 198:213–235

Kress S, Biechl D, Wullimann MF (2015) Combinatorial analysis of calcium-binding proteins in larval and adult zebrafish primary olfactory system identifies differential olfactory bulb glomerular projection fields. Brain Struct Funct 220:1951–1970

Kriegstein AR, Shen JM, Eshbar N (1986) Monoclonal antibodies to the turtle cortex reveal neuronal subsets, antigenic cross-reactivity with the mammalian neocortex, and forebrain structures sharing a pallial derivation. J Comp Neurol 254:330–340

Krubitzer L, Künzle H, Kaas J (1997) Organization of sensory cortex in a Madagascan insectivore, the tenrec (*Echinops telfairi*). J Comp Neurol 379:399–414

Lastein S, Hamdani H, Døving KB (2006) Gender distinction in neural discrimination of sex pheromones in the olfactory bulb of crucian carp, *Carassius carassius*. Chem Senses 31:69–77

Lazzari M, Franceschini V (2001) Glial fibrillary acidic protein and vimentin immunoreactivity of astroglial cells in the central nervous system of adult *Podarcis sicula* (Squamata, Lacertidae). J Anat 198:67–75

Lazzari M, Franceschini V (2004) Glial fibrillary acidic protein and vimentin immunoreactivity of astroglial cells in the central nervous system of the African lungfish, *Protopterus annectens* (Dipnoi: Lepidosirenidae). J Morphol 262:741–749

Lazzari M, Franceschini V (2005a) Astroglial cells in the central nervous system of the brown anole lizard, *Anolis sagrei*, revealed by intermediate filament immunohistochemistry. J Morphol 265:325–334

Lazzari M, Franceschini V (2005b) Intermediate filament immunohistochemistry of astroglial cells in the leopard gecko, *Eublepharis macularius*. Anat Embryol 210:275–286

Lazzari M, Franceschini V (2006) Glial cytoarchitecture in the central nervous system of the soft-shell turtle, *Trionyx sinensis*, revealed by intermediate filament immunohistochemistry. Anat Embryol 211:497–506

Lazzari M, Franceschini V, Ciani F (1997) Glial fibrillary acidic protein and vimentin in radial glia of *Ambystoma mexicanum* and *Triturus carnifex*: an immunocytochemical study. J Brain Res 38:187–194

Lazzari M, Bettini S, Ciani F, Franceschini V (2007) Light and transmission electron microscopy study of the peripheral olfactory organ of the guppy, *Poecilia reticulata* (Teleostei, Poeciliidae). Microsc Res Tech 70:782–789

Lazzari M, Bettini S, Franceschini V (2013) Immunocytochemical characterization of olfactory ensheathing cells in fish. Brain Struct Funct 218:539–549

Lazzari M, Bettini S, Franceschini V (2014) Immunocytochemical characterisation of olfactory ensheathing cells of zebrafish. J Anat 224:192–206

Lazzari M, Bettini S, Franceschini V (2016) Immunocytochemical characterisation of ensheathing glia in the olfactory and vomeronasal systems of *Ambystoma mexicanum* (Caudata: Ambystomatidae). Brain Struct Funct 221:955–967

Lazzari M, Bettini S, Milani L, Maurizii MG, Franceschini V (2017) Differential response of olfactory sensory neuron populations to copper ion exposure in zebrafish. Aquat Toxicol 183:54–62

Lazzari M, Bettini S, Milani L, Maurizii MG, Franceschini V (2019) Differential nickel-induced responses of olfactory sensory neuron populations in zebrafish. Aquat Toxicol 206:14–23

Levitt P, Rakic P (1980) Immunoperoxidase localization of glial fibrillary acidic protein in radial glial cells and astrocytes of the developing rhesus monkey brain. J Comp Neurol 193:815–840

Liu Y, Zhang J, Liu D, Zhang J (2014) Vomeronasal organ lesion disrupts social odor recognition, behaviors and fitness in golden hamsters. Integr Zool 9:255–264

Lu J, Féron F, Mackay-Sim A, Waite PM (2002) Olfactory ensheathing cells promote locomotor recovery after delayed transplantation into transacted spinal cord. Brain 125:14–21

Luo H, Wu XQ, Zhao M, Wang Q, Jiang GP et al (2017) Expression of vimentin and glial fibrillary acidic protein in central nervous system development of rats. Asian Pac J Trop Med 10:1185–1189

Mack AF, Künzle H, Lange M, Mages B, Reichenbach A, Härtig W (2018) Radial glial elements in the cerebral cortex of the lesser hedgehog Tenrec. Brain Struct Funct 223:3909–3917

Mackay-Sim A (2005) Olfactory ensheathing cells and spinal cord repair. Keio J Med 54:8–14

Malatesta P, Appolloni I, Calzolari F (2008) Radial glia and neural stem cells. Cell Tissue Res 331:165–178

McDonald RP, Vickaryous MK (2018) Evidence for neurogenesis in the medial cortex of the leopard gecko, *Eublepharis macularius*. Sci Rep 8:9648

Messenger NJ, Warner AE (1989) The appearance of neural and glial cell markers during early development of the nervous system in the amphibian embryo. Development 107:43–54

Monzon-Mayor M, Yanes C, Ghandour MS, De Barry J, Gombos G (1990) GFAP and vimentin immunohistochemistry in the adult and developing midbrain of the lizard *Gallotia galloti*. J Comp Neurol 295:569–579

Moore SA, Oglesbee MJ (2015) Spinal cord ependymal responses to naturally occurring traumatic spinal cord injury in dogs. Vet Pathol 52:1108–1117

Moy-Thomas JA, Miles RS (1971) Paleozoic fishes. WB Saunders, Philadelphia

Mulligan SJ, MacVicar BA (2004) Calcium transients in astrocyte endfeet cause cerebrovascular constrictions. Nature 431:195–199

Nakada T, Hagino-Yamagishi K, Nakanishi K, Yokosuka M, Saito TR, Toyoda F, Hasunuma I, Nakakura T, Kikuyama S (2014) Expression of G proteins in the olfactory receptor neurons of the newt *Cynops pyrrhogaster*: their unique projection into the olfactory bulbs. J Comp Neurol 522(15):3501–3519

Naujoks-Manteuffel C, Meyer DL (1996) Glial fibrillary acidic protein in the brain of the caecilian *Typhlonectes natans* (Amphibia, Gymnophiona): an immunocytochemical study. Cell Tissue Res 283:51–58

O'Leary MA, Bloch JI, Flynn JJ, Gaudin TJ, Giallombardo A, Giannini NP, Goldberg SL, Kraatz BP, Luo ZX, Meng J, Ni X, Novacek MJ, Perini FA, Randall Z, Rougier GW, Sargis EJ, Silcox MT, Simmons NB, Spaulding M, Velazco PM, Weksler M, Wible JR, Cirranello AL (2013) The placental mammal ancestor and the post-K-Pg radiation of placentals. Science 339:662–667

Oikawa T, Suzuki K, Saito TR, Takahashi KW, Taniguchi K (1998) Fine structure of three types of olfactory organs in *Xenopus laevis*. Anat Rec 252:301–310

Oka Y, Saraiva LR, Korsching SI (2012) Crypt neurons express a single V1R-related ora gene. Chem Senses 37(3):219–227

Oudega M, Marani E (1991) Expression of vimentin and glial fibrillary acidic protein in the developing rat spinal cord: an immunocytochemical study of the spinal cord glial system. J Anat 179:97–114

Pellitteri R, Spatuzza M, Stanzani S, Zaccheo D (2010) Biomarkers expression in rat olfactory ensheathing cells. Front Biosci 2:289–298

Pulido-Caballero J, Jiménez-Sampedro F, Echevarría-Aza D, Martínez-Millán L (1994) Posnatal development of vimentin-positive cells in the rabbit superior colliculus. J Comp Neurol 343:102–112

Quinzio SI, Reiss JO (2018) The ontogeny of the olfactory system in ceratophryid frogs (Anura, Ceratophryidae). J Morphol 279:37–49
Raisman G (2001) Olfactory ensheathing cells: another miracle cure for spinal cord injury. Nat Rev Neurosci 2:369–375
Rakic P (1981) Neuronal-glial interactions during development. TINS 4:184–187
Ramon y Cajal S (1952) Histologie du système nerveux de l'homme et des vertébrés, vol. 1. Consejo superior de Investigaciones Cientificas, Instituto Ramon Y Cajal, Madrid
Rowitch DH, Kriegstein AR (2010) Developmental genetics of vertebrate glial-cell specification. Nature 468:214–222
Rubio M, Suarez I, Bodega G, Fernandez B (1992) Glial fibrillary acidic protein and vimentin immunohistochemistry in the posterior rhombencephalon of the Iberian barb (*Barbus comiza*). Neurosci Lett 134:203–206
Saito S, Matsui T, Kobayashi N, Wakisaka H, Mominoki K, Matsuda S, Taniguchi K (2003) Lectin histochemical study on the olfactory organ of the newt, *Cynops pyrrhogaster*, revealed heterogeneous mucous environments in a single nasal cavity. Anat Embryol (Berl) 206:349–356
Sandulescu CM, Teow RY, Hale ME, Zhang C (2011) Onset and dynamic expression of S100 proteins in the olfactory organ and the lateral line system in zebrafish development. Brain Res 1383:120–127
Sato Y, Miyasaka N, Yoshihara Y (2005) Mutually exclusive glomerular innervation by two distinct types of olfactory sensory neurons revealed in transgenic zebrafish. J Neurosci 25 (20):4889–4897
Schnitzer J, Franke WW, Schachner M (1981) Immunocytochemical demonstration of vimentin in astrocytes and ependymal cells of developing and adult mouse nervous system. J Cell Biol 90:435–447
Schultze HP, Campbell KSW (1986) Characterization of the Dipnoi, a monophyletic group. J Morphol Suppl 1:25–37
Sild M, Ruthazer ES (2011) Radial glia: progenitor, pathway partner. Neuroscience 17:288–302
Simard M, Arcuino G, Takano T, Liu OS, Nedergaard M (2003) Signaling at the glio-vascular interface. J Neurosci 23:9254–9262
Stephan H, Baron G, Frahm H (1991) Comparative brain characteristics. In: Insectivora. Comparative brain research in mammals, vol 1. Springer, New York
Vielma A, Ardiles A, Delgado L, Schmachtenberg O (2008) The elusive crypt olfactory receptor neuron: evidence for its stimulation by amino acids and cAMP pathway agonists. J Exp Biol 211:2417–2422
Voigt T (1989) Development of glial cells in the cerebral walls of ferrets: direct tracing of their transformation from radial glia into astrocytes. J Comp Neurol 289:74–88
Volterra A, Meldolesi J (2005) Astrocytes, from brain glue to communication elements: the revolution continues. Nat Rev Neurosci 6:626–640
Wasowicz M, Pierre J, Reperant J, Ward R, Vesselkin NP, Versaux-Botteri C (1994) Immunoreactivity to glial fibrillary acidic protein (GFAP) in the brain and spinal cord of the lamprey (*Lampetra fluviatilis*). J Brain Res 35:71–78
Weissman T, Noctor SC, Clinton BK, Honig LS, Kriegstein AR (2003) Neurogenic radial glial cells in reptile, rodent and human: from mitosis to migration. Cereb Cortex 13:550–559
Wicht H, Derouiche A, Korf H-W (1994) An immunocytochemical investigation of glial morphology in the Pacific hagfish: radial and astrocyte-like glia have the same phylogenetic age. J Neurocytol 23:565–576
Yanes C, Monzon-Mayor M, Ghandour MS, De Barry J, Gombos G (1990) Radial glia and astrocytes in developing and adult telencephalon of the lizard *Gallotia galloti* as revealed by immunohistochemistry with anti-GFAP and anti-vimentin antibodies. J Comp Neurol 295:559–568
Zamora AJ, Mutin M (1988) Vimentin and glial fibrillary acidic protein filaments in radial glia of the adult urodele spinal cord. Neuroscience 27:279–288

[illegible]

Chapter 16
Embryogenesis of Marsupial Frogs (Hemiphractidae), and the Changes that Accompany Terrestrial Development in Frogs

Eugenia M. del Pino

Abstract The developmental adaptations of the marsupial frogs *Gastrotheca riobambae* and *Flectonotus pygmaeus* (Hemiphractidae) are described and compared with frogs belonging to seven additional families. Incubation of embryos by the mother in marsupial frogs is associated with changes in the anatomy and physiology of the female, modifications of oogenesis, and extraordinary changes in embryonic development. The comparison of early development reveals that gene expression is highly conserved. However, the timing of gene expression varies between frog species. There are two modes of gastrulation according to the onset of convergent extension. In gastrulation mode 1, convergent extension is an intrinsic mechanism of gastrulation. This gastrulation mode occurs in frogs with aquatic reproduction, such as *Xenopus laevis*. In gastrulation mode 2, convergent extension occurs after the completion of gastrulation movements. Gastrulation mode 2 occurs in frogs with terrestrial reproduction, such as the marsupial frog, *G. riobambae*. The two modes of frog gastrulation resemble the two transitions toward meroblastic cleavage of ray-finned fishes (Actinopterygii). The comparison indicates that a major event in the evolution of frog terrestrial development is the separation of convergent extension from gastrulation.

16.1 Introduction

Marsupial frogs (Hemiphractidae) (Fig. 16.1a) are attractive for studies of development because the incubation of embryos in the body of the mother resembles mammalian pregnancy (Fig. 16.1a) (del Pino 1989, 2018; del Pino et al. 2007; Elinson and del Pino 2012). Our comparison of development in various frogs with

E. M. del Pino (✉)
Escuela de Ciencias Biológicas, Pontificia Universidad Católica del Ecuador, Quito, Ecuador
e-mail: edelpino@puce.edu.ec

W. Tworzydlo, S. M. Bilinski (eds.), *Evo-Devo: Non-model Species in Cell and Developmental Biology*, Results and Problems in Cell Differentiation 68,
https://doi.org/10.1007/978-3-030-23459-1_16

Fig. 16.1 Marsupial frogs (Hemiphractidae). (**a**) Female of *Gastrotheca riobambae* with embryos in her pouch. The outlines of embryos appear as protuberances on the mother's back. The pouch occupies the back and the sides of the body and extends to the back of the head. The pouch aperture has an inverted V shape and is located rostrad to the cloaca. (**b**) Female of *Flectonotus pygmaeus*. The pouch resembles lateral folds of skin with a middorsal opening along the length of the body. During embryonic incubation, the lateral folds of the pouch close in the midline enclosing the embryos inside a pouch. *c* cloaca, *pa* pouch aperture. Figure 16.1b is reprinted with the permission of S. Karger AG, Basel from Schmid et al. (2012). The hemiphractid frogs: phylogeny, embryology, life history, and cytogenetics. Cytogenet Genome Res 138:69–384

X. laevis is reasonable because key cellular and developmental processes are highly conserved in vertebrates. The *Xenopus* model organism approach provides an in-depth knowledge of frog development with insights into human diseases and allows evolutionary interpretations (Harland and Grainger 2011). This chapter provides a comparative update of reproduction, oogenesis, and embryonic development in frogs of the family Hemiphractidae in comparison with other frogs and *X. laevis*.

16.2 Reproductive Modes and Development of Species Representing Various Anuran Lineages

Frogs evolved diverse solutions to reduce or eliminate aquatic dependence for reproduction, including direct development and viviparity (Duellman and Trueb 1986; Haddad and Prado 2005) (Table 16.1). Aquatic eggs, terrestrial or arboreal eggs, and eggs retained in the oviducts are the three major categories of frog reproduction (Duellman and Trueb 1986). These reproductive categories have been subdivided into 39 reproductive modes (Haddad and Prado 2005) (Table 16.1). This study reviews the development of 16 frog species that represent six modes of reproduction, according to Haddad and Prado (2005). In addition, early development of frogs of the family Hemiphractidae is reviewed including the marsupial frog *Gastrotheca riobambae* (Tables 16.1 and 16.2). Gastrulation speed of the various frogs was compared with the *X. laevis* gastrulation time (Table 16.2). Our analysis indicates that reproductive modes are associated with modifications of early development.

Table 16.1 Reproductive modes in anurans[a,b]

Type[a] Egg location	Number of reproductive modes[b]	Species analyzed in this work[c]
I. Aquatic eggs		
Eggs deposited in water	9	*Xenopus laevis*
		Ceratophrys stolzmanni
Eggs in bubble nest	1	
Eggs in foam nest (aquatic)	4	*Engystomops coloradorum*
		Engystomops randi
		Engystomops pustulosus
Eggs embedded in dorsum of aquatic female	2	
II. Terrestrial or arboreal eggs (not in water)		
Eggs on ground, on rocks, or in burrows	7	Six species of Dendrobatidae
		Eleutherodactylus coqui
		Pristimantis unistrigatus
Arboreal eggs	4	*Espadarana callistomma*
		Hyalinobatrachium fleischmanni
		Agalychnis spurrelli
Eggs in foam nest (terrestrial or arboreal)	6	
Eggs carried by adult	4	Several species of Hemiphractidae
III. Eggs retained in oviducts		
Eggs kept inside the oviducts	2	

[a]Duellman and Trueb (1986)
[b]Haddad and Prado (2005)
[c]del Pino (2018) and Schmid et al. (2018); Table 16.2

Table 16.2 Reproduction and development of frogs analyzed in this work[a]

Family and species	Egg diameter (mm)	Gastrulation mode[b] (time in hours)	Gastrocoel roof plate[c]	References
Aquatic eggs and embryos				
Pipidae				
Xenopus laevis	1.2	1 (5.5)	Yes	Nieuwkoop and Faber (1994)
Ceratophryidae				
Ceratophrys stolzmanni	2.2	1 (5)	Yes	Sáenz-Ponce et al. (2012b)
Foam nest floating in water				
Leptodactylidae				
Engystomops randi	1.1	1 (12.5)	Yes	Romero-Carvajal et al. (2009)
Engystomops coloradorum	1.3	1 (12.5)	Unknown	Romero-Carvajal et al. (2009)
Arboreal eggs, tadpoles drop in water				
Centrolenidae				
Espadarana callistomma	2.1	1 (23)	Yes	Salazar-Nicholls and del Pino (2015)
Hyalinobatrachium fleischmanni	2.1	1 (24)	Yes	Salazar-Nicholls and del Pino (2015)
Arboranae: Phyllomedusidae				
Agalychnis spurrelli	3.0	1 (19)	Unknown	Schmid et al. (2018)
Terrestrial eggs, tadpoles transported to water by an adult				
Dendrobatidae				
Epipedobates machalilla	1.6	2 (65)	Yes	del Pino et al. (2004)
Epipedobates anthonyi	2.0	2 (36)	Unknown	Moya et al. (2007)
Epipedobates tricolor	2.0	2 (36)	Unknown	Moya et al. (2007)
Hyloxalus vertebralis	2.6	2 (39)	Unknown	Hervas et al. (2015)
Ameerega bilinguis	3.0	2 (55)	Unknown	Hervas et al. (2015)
Dendrobates auratus	3.5	2 (72)	Unknown	Hervas et al. (2015)
Direct development of terrestrial eggs				
Terraranae: Eleutherodactylidae				
Eleutherodactylus coqui	3.5	2 (24)	Unknown	Townsend and Stewart (1985)

(continued)

Table 16.2 (continued)

Family and species	Egg diameter (mm)	Gastrulation mode[b] (time in hours)	Gastrocoel roof plate[c]	References
Eggs carried in the pouch of the mother, tadpoles in ponds				
Hemiphractidae				
Gastrotheca riobambae	3.0	2 (168)	Yes	del Pino (1996)

[a]del Pino et al. (2007), Hervas et al. (2015), Salazar-Nicholls and del Pino (2015), and Vargas and del Pino (2017). Reproductive modes according to Haddad and Prado (2005); Modified from del Pino (2018)

[b]Gastrulation mode 1: Notochord elongation starts in the mid-gastrula; Gastrulation mode 2: Notochord elongation starts after gastrulation (Developmental time from stage 10 to 13). Embryo culture temperatures for: *X. laevis* 23 °C, and 18–23 °C for other frogs

[c]Gastrocoel roof plate with mono-ciliated cells (Sáenz-Ponce et al. 2012b)

16.2.1 *Aquatic Eggs and Embryos of* Xenopus *and* Ceratophrys

The sub-Saharan African clawed frog, *X. laevis* (Pipidae), and the Pacific horned frog, *Ceratophrys stolzmanni* (Ceratophryidae), have aquatic development (Nieuwkoop and Faber 1994; Sáenz-Ponce et al. 2012b). Frogs with this mode of reproduction deposit hundreds to thousands of small-sized eggs in the water. This reproductive strategy ensures that a few embryos reach the adult stage in spite of predation and desiccation (Duellman and Trueb 1986).

The animal hemisphere of *C. stolzmanni* eggs has dark pigment, and the vegetal hemisphere is light-yellow, as in *X. laevis* eggs (Nieuwkoop and Faber 1994; Sáenz-Ponce et al. 2012b). However, the volume of *C. stolzmanni* eggs is six times the volume of *X. laevis* eggs (Table 16.2). In spite of this size difference, the gastrulation speed of *X. laevis* and *C. stolzmanni* is similar, an indication of the rapid development of these frogs (Table 16.2). In these two species, gastrulation movements overlap with body elongation, as detected by the elongation of the notochord (Keller et al. 1985; Keller and Winklbauer 1992; Sáenz-Ponce et al. 2012b). The overlap of notochord elongation with other gastrulation movements was defined as gastrulation mode 1 (Fig. 16.2a; Table 16.2) (del Pino et al. 2007; Vargas and del Pino 2017). The alternative is gastrulation mode 2. Convergent extension is part of the morphogenetic processes of gastrulation mode 1 (Fig. 16.2a), whereas convergent extension begins after closure of the blastopore in gastrulation mode 2 (Fig. 16.2b) (del Pino et al. 2007; Vargas and del Pino 2017). Comparative aspects of gastrulation are discussed in Sect. 16.7.

Cilia-driven fluid flow toward the left side occurs in the gastrocoel roof plate of the *X. laevis* neurula. Leftward fluid flow triggers the asymmetric placement of organs toward the left (Blum et al. 2014). A comparison of the neurulae of *C. stolzmanni*, *Engystomops randi* (Leptodactylidae), *Epipedobates machalilla*, and *Epipedobates tricolor* (Dendrobatidae) and *G. riobambae* (Hemiphractidae)

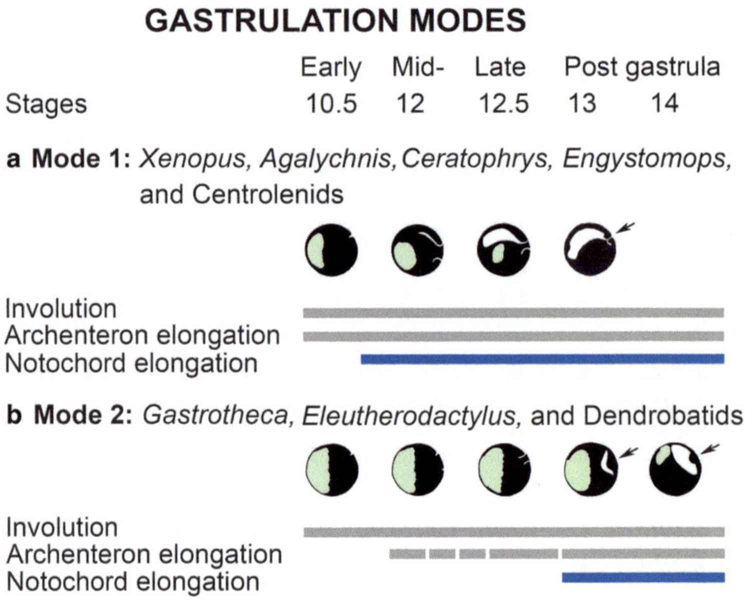

Fig. 16.2 Gastrulation modes. (**a**) Mode 1: The drawings represent the stages of gastrulation in *Xenopus laevis*. The species examined were *X. laevis*, *Agalychnis spurrelli*, *Ceratophrys stolzmanni*, *Engystomops coloradorum*, *Engystomops randi*, and the centrolenid frogs *Hyalinobatrachium fleischmanni* and *Espadarana callistomma* (Table 16.2). (**b**) Mode 2: The drawings represent the stages of gastrulation in *Gastrotheca riobambae*. The species examined were *G. riobambae*, *Eleutherodactylus coqui*, and the dendrobatid frogs *Epipedobates machalilla*, *Epipedobates anthonyi*, *Epipedobates tricolor*, *Hyloxalus vertebralis*, *Ameerega bilinguis*, and *Dendrobates auratus* (Table 16.2). Involution, archenteron, and notochord elongation are indicated by bars. Onset of archenteron elongation starts at different times during gastrulation in dendrobatid frogs, indicated by a broken bar and after blastopore closure in *G. riobambae*. The arrows indicate the circumblastoporal collar. Dorsal is oriented toward the top. Colors: light green is the blastocoel; white is the archenteron; black indicates other regions of the embryo. *Xenopus laevis* and *G. riobambae* embryonic stages are denoted according to Nieuwkoop and Faber (1994) and Moya et al. (2007) [Modified from Moya et al. (2007)]

with *X. laevis* (Pipidae) revealed the presence of monociliated cells in the gastrocoel roof plate of these frogs (Sáenz-Ponce et al. 2012b). The comparison suggests that the leftward fluid flow across the gastrocoel roof plate may trigger the placement of organs to the left side in frog embryos (Sáenz-Ponce et al. 2012b).

16.2.2 Foam Nests of Engystomops

Foam nesting in frogs provides a nonaquatic environment for early development (Duellman and Trueb 1986). Foam nests floating in water characterize the reproduction of the túgara frogs (Leptodactylidae) (Duellman and Trueb 1986)

(Table 16.1). The development of the túngara frogs, *Engystomops randi*, *Engystomops coloradorum*, and *Engystomops pustulosus* (Leptodactylidae), was examined by Romero-Carvajal et al. (2009). These frogs reproduce in captivity, and the foam nests contain about 100–240 eggs, 1.1–1.3 mm in diameter (Table 16.2) (Romero-Carvajal et al. 2009). Gastrulation of *E. randi* and *E. coloradorum* requires twice the time of *X. laevis* embryos, in spite of the similar size of their eggs (Table 16.2). The protection provided by the túngara frog foam nest may allow slow development in comparison with the aquatic development of *X. laevis* embryos. Túngara frog gastrulation is the mode 1 type as in *X. laevis* (Fig. 16.2a; Table 16.2) (del Pino et al. 2007; Romero-Carvajal et al. 2009; Vargas and del Pino 2017).

The white eggs of túngara frogs resemble albino *X. laevis* eggs. However, embryogenesis differs from *X. laevis*. Túngara frog embryos are not albino and contain dark pigment distributed around the cell nuclei (Romero-Carvajal et al. 2009). The neural plate, neural folds, and streams of cranial neural crest cells are larger than in *X. laevis* embryos (Romero-Carvajal et al. 2009). Myogenesis of the túngara frogs involves intercalation of small cells, as found in *Bombina variegata*, *G. riobambae*, and frogs of the family Dendrobatidae (Radice et al. 1989; Gatherer and del Pino 1992; del Pino et al. 2007; Romero-Carvajal et al. 2009; Hervas-Sotomayor and del Pino 2013). In contrast, myogenesis in *X. laevis* involves rotation of large cells that span the somite length (Radice et al. 1989; Gatherer and del Pino 1992; Afonin et al. 2006). The yolk sac bulges out in tailbud embryos of túngara frogs (del Pino et al. 2007; Romero-Carvajal et al. 2009). In contrast, the tailbud embryos of *X. laevis* are elongated, and the yolk sac does not protrude from the body of the embryo (Fig. 16.3a) (Hausen and Riebesell 1991; Nieuwkoop and Faber 1994).

Anurans and urodeles, including species with large eggs, have holoblastic cleavage (del Pino and Loor-Vela 1990; Ninomiya et al. 2001; Collazo and Keller 2010; Schmid et al. 2018). In contrast, the large amniote eggs of birds and reptiles have meroblastic cleavage, with cleavage occurring only in the yolk-free cytoplasm of the animal hemisphere. The cleaved area forms a disk of small cells on top of a large mass of uncleaved yolk, and the body of the embryo develops from the disk of cells (Arendt and Nübler-Jung 1999; Elinson and Beckham 2002). The yolk sac of frog embryos bulges out in the tailbud stage, resembling the meroblastic embryos of amniotes. However, the yolk is divided into cells in túngara frogs, the gliding tree frog, glass frogs, the marsupial frog *G. riobambae*, a frog without tadpoles *Eleutherodactylus coqui*, and the six species of the Dendrobatidae that were studied (del Pino and Escobar 1981; Duellman and Trueb 1986; del Pino et al. 2004, 2007; Romero-Carvajal et al. 2009; Elinson and del Pino 2012; Salazar-Nicholls and del Pino 2015).

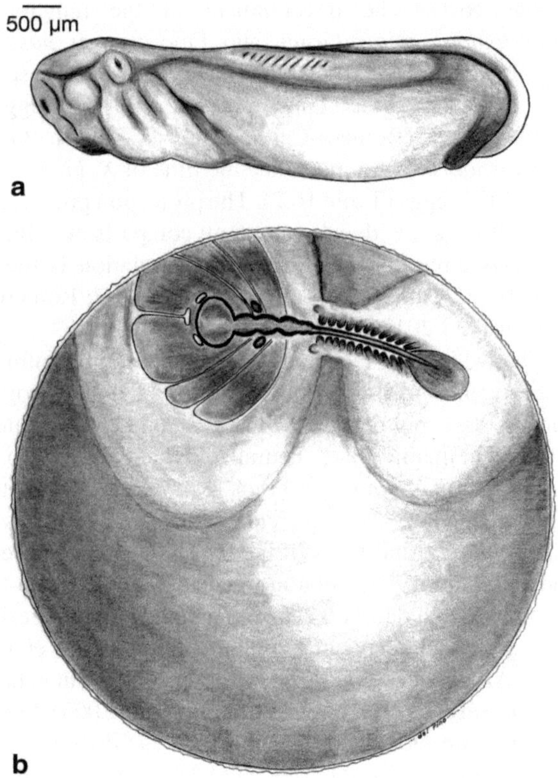

Fig. 16.3 Tailbud embryos of *X. laevis* and *Gastrotheca riobambae*. (**a**) The *X. laevis* tailbud embryo is elongated, an adaptation for early swimming; (**b**) The telolecithal embryo of *G. riobambae* has a flat orientation on the embryonic disk over a large mass of cleaved yolk. The *X. laevis* embryo has been matched with the *G. riobambae* embryo with respect to development of the gill arches, eye, and otocyst. Reproduced from del Pino and Elinson (1983)

16.2.3 Arboreal Eggs of Espadarana, Hyalinobatrachium, *and* Agalychnis

The deposition of eggs outside of the water is an important trend towards terrestrial reproduction in frogs (Duellman and Trueb 1986). The frogs with arboreal eggs analyzed in this work are the glass frogs *Espadarana callistomma*, *Hyalinobatrachium fleischmanni* (Centrolenidae), and the gliding tree frog, *Agalychnis spurrelli* (Arboranae: Phyllomedusidae) (Table 16.2) (Salazar-Nicholls and del Pino 2015; Schmid et al. 2018). These frogs deposit their eggs on the vegetation above water. At hatching, the tadpoles drop into the water. After an aquatic development period, the tadpoles undergo metamorphosis to change into four-legged terrestrial froglets (Table 16.2). Embryonic development of the glass frogs from the 32-cell stage to tadpole hatching resembles the patterns of *X. laevis* and of the túngara frogs (Salazar-Nicholls and del Pino 2015). The developmental time from cleavage to hatching, under laboratory conditions, was 6 and 12 days, and the number of eggs in the nest ranges from 23 to 35 eggs in *H. fleischmanni* and *E. callistomma*, respectively (Salazar-Nicholls and del Pino 2015). The egg volume in these glass frogs is five times that of *X. laevis* eggs (Table 16.2). The embryos of

H. fleischmanni and *E. callistomma* take four times longer than *X. laevis* embryos for the gastrulation process (Table 16.2). Accordingly, the protection provided by the arboreal nest allows development to proceed at a slower pace in comparison with the aquatic embryos of *X. laevis*. Glass frog embryos follow gastrulation mode 1 (Fig. 16.2a; Table 16.2) (Vargas and del Pino 2017; del Pino 2018).

Early development of the gliding tree frog, *A. spurrelli*, has significant deviation from the *X. laevis* developmental pattern (Schmid et al. 2018). The number of eggs in the *A. spurrelli* nest ranges from 12 to 322 eggs, and tadpoles hatch 8 days after amplexus (Vargas-Salinas and Torres 2013). The large eggs of *A. spurrelli* are uniformly light-green, and their volume is 14 times the volume of *X. laevis* eggs (Table 16.2) (Schmid et al. 2018). The gastrulation process takes three times longer than in *X. laevis* (Table 16.2). Embryos of *A. spurrelli* follow gastrulation mode 1 (Fig. 16.2a; Table 16.2) (Vargas and del Pino 2017; Schmid et al. 2018).

Cleavage of *A. spurrelli* eggs is atypical holoblastic, in contrast with the cleavage, pattern of *X. laevis* and of frogs with small eggs and aquatic reproduction (Nelsen 1953). Atypical holoblastic cleavage also occurs in *G. riobambae* and *E. coqui* (Nelsen 1953; del Pino and Loor-Vela 1990; Ninomiya et al. 2001; Elinson 2009; Schmid et al. 2018). Atypical holoblastic cleavage represents a tendency toward meroblastic cleavage. This cleavage pattern is similar to patterns observed in other amphibians with large eggs (del Pino and Loor-Vela 1990; Collazo and Keller 2010; Schmid et al. 2018) and in primitive ray-finned fishes, such as gar (*Lepisosteus*) and bowfin (*Amia*) (Actinopterygii) (Bolker 1994; Cooper and Virta 2007; Takeuchi et al. 2009). An extraordinary feature of *A. spurrelli* early development is the comparatively small blastocoel in comparison with *X. laevis* and other frogs (Schmid et al. 2018). Small blastocoel size was reported for the African frog, *Hyperolius puncticulatus* (Hyperoliidae) (Chipman et al. 1999). The significance of small blastocoel size for embryonic development is unknown.

16.2.4 Terrestrial Eggs with Tadpoles Transported to Water in Dendrobatidae

Terrestrial eggs and tadpoles transported to water by an adult characterize the reproductive modes of frogs belonging to the family Dendrobatidae (Duellman and Trueb 1986). Terrestrial reproduction is associated with tendencies toward larger egg size and fewer eggs in each reproductive event, as observed in Dendrobatidae (Duellman and Trueb 1986). Dendrobatid frogs reproduce readily in captivity, a feature that facilitates the study of their development (del Pino et al. 2004). Early development was examined in six species of dendrobatid frogs, *Epipedobates machalilla*, formerly known as *Colostethus machalilla*, *Epipedobates anthonyi*, *Epipedobates tricolor*, *Hyloxalus vertebralis*, *Ameerega bilinguis*, and *Dendrobates auratus* (Table 16.2) (Benítez and del Pino 2002; Moya et al. 2007; del Pino et al. 2007; Montenegro-Larrea and del Pino 2011; Hervas-Sotomayor and del Pino 2013; Hervas et al. 2015).

Epipedobates machalilla early development until the neurula resembles *X. laevis* development. Thereafter, development follows the large telolecithal egg pattern (del Pino et al. 2004). Early development of the additional species of dendrobatids was characterized according to the developmental table of *E. machalilla* (Table 16.2) (del Pino et al. 2004, 2007; Moya et al. 2007; Montenegro-Larrea and del Pino 2011; Hervas-Sotomayor and del Pino 2013; Hervas et al. 2015). Eggs of these dendrobatids have a darkly pigmented animal pole and a light-colored vegetal region (del Pino et al. 2004; Moya et al. 2007; Montenegro-Larrea and del Pino 2011; Hervas et al. 2015). Egg volume ranges from 2 to 25 times the volume of *X. laevis* eggs (Table 16.2). The number of eggs per egg clutch ranges from 5 to 18 eggs (Hervas et al. 2015). Gastrulation in dendrobatids ranges from 7 to 13 times the developmental time of *X. laevis* (Table 16.2). The development in the terrestrial nest of dendrobatids proceeds more slowly than in frogs with aquatic development, foam nests, or arboreal eggs (Table 16.2). Elongation of the notochord in dendrobatids begins once gastrulation is completed (gastrulation mode 2) (Fig. 16.2b; Table 16.2) (Benítez and del Pino 2002; Moya et al. 2007; Vargas and del Pino 2017).

Epiboly, internalization, and convergent extension are the main cell movements in gastrulation of vertebrates (Solnica-Krezel 2005). Convergent extension is a powerful cell movement that elongates tissues during embryonic development (Shindo 2017). Convergent extension starts in the mid-gastrula of *X. laevis* embryos and contributes to elongation of the body. Until then, the embryo maintains the spherical shape of the egg. In *X. laevis* embryos, convergent extension contributes to the closure of the blastopore (Keller et al. 1985; Keller and Danilchik 1988). Gastrulation and the onset of body elongation are coupled in the *X. laevis* embryo (Fig. 16.2a) (Keller et al. 1985). Convergence without extension, convergent thickening, occurs before convergent extension along the entire marginal zone of *X. laevis* embryos. Convergent thickening contributes to the closure of the blastopore on the ventral side of *X. laevis* embryos (Keller and Danilchik 1988; Shook et al. 2018).

The timing of convergent extension and the onset of body elongation vary in frogs according to reproductive adaptations (Elinson and del Pino 2012). In frogs with aquatic reproduction, convergent extension begins during gastrulation, as in *X. laevis* (gastrulation mode 1) (Fig. 16.2a; Table 16.2) (del Pino et al. 2007; Romero-Carvajal et al. 2009; Salazar-Nicholls and del Pino 2015; Vargas and del Pino 2017). In contrast, convergent extension begins after blastopore closure in the terrestrial embryos of frogs with gastrulation mode 2 (Fig. 16.2b; Table 16.2). Species with gastrulation mode 2 include the dendrobatid frogs, *E. coqui* and *G. riobambae* (del Pino 1996; Ninomiya et al. 2001; del Pino et al. 2007; Vargas and del Pino 2017). The blastopore lip thickens because the cells that involute during gastrulation remain in the blastopore lip. As a result, a large circumblastoporal collar develops around the closed blastopore (Moya et al. 2007). Elongation of the notochord begins after blastopore closure in frogs with gastrulation mode 2 (Fig. 16.2b). Modes of gastrulation are discussed in Sect. 16.7.

16.2.5 Direct Development of Terrestrial Eggs in Eleutherodactylus

Direct development is the advancement of embryogenesis inside the egg envelope until the formation of a miniature amphibian, without the development of a tadpole (Castroviejo-Fisher et al. 2015; Heinicke et al. 2018). Direct development is the most successful mode of amphibian reproduction as evidenced by the large number of amphibians that have eliminated the tadpole stage from their life cycle (Callery 2006). Reproduction of *Pristimantis unistrigatus* (Terraranae: Strabomantidae) in captivity was observed only once (Nina and del Pino 1977). Reproduction in captivity of other terraranan frogs was unsuccessful. The exception is *Eleutherodactylus coqui* (Terraranae: Eleutherodactylidae), a terraranan frog from Puerto Rico. *Eleutherodactylus coqui* reproduces in captivity, a condition that has allowed the extensive investigation of its development (Table 16.1) (Townsend and Stewart 1985; Callery and Elinson 2000; Elinson and del Pino 2012).

Eleutherodactylus coqui is the model organism for the study of direct development in anurans with large eggs (Elinson and del Pino 2012). The pre-hatching development of *E. coqui* until birth of a froglet requires approximately 17 days at 25 °C (Townsend and Stewart 1985). The pale-yellow eggs of *E. coqui* measure 3.5 mm in diameter, and egg clutches average 28 eggs (Elinson et al. 1990; Townsend and Stewart 1985). Egg volume is 25 times the volume of *X. laevis* eggs. Gastrulation in *E. coqui* takes four times the developmental time of *X. laevis* (Table 16.2). Embryos of *E. coqui* follow gastrulation mode 2 (Fig. 16.2b; Table 16.2) (Ninomiya et al. 2001; del Pino et al. 2007; Vargas and del Pino 2017; del Pino 2018).

Embryos of *E. coqui* develop larval features and undergo certain aspects of metamorphosis in spite of their direct developmental mode (Elinson 2013; Castroviejo-Fisher et al. 2015). Particularly noticeable, the tail is used for gas exchanges instead of swimming. The presence of rudimentary tadpole structures indicates that *E. coqui* direct development is a character derived from an ancestor with tadpoles and metamorphosis (Elinson 2001; Ziermann and Diogo 2014). The modifications of *E. coqui* early development include changes in the cleavage pattern, the blastocoel roof, and germ layer formation and the development of a special structure, the nutritional endoderm (Elinson and Fang 1998; Ninomiya et al. 2001; Elinson and Beckham 2002; Chatterjee and Elinson 2013; Karadge and Elinson 2013).

The blastula of *E. coqui* has a large blastocoel, covered by a one-cell-thick blastocoel roof (Elinson and del Pino 2012). The blastocoel roof derives from yolk-poor cells of the animal pole region, as in *X. laevis* (Elinson and del Pino 2012). Similarly, embryos of frogs and urodeles with large eggs have a single-celled blastocoel roof. For example, the blastocoel roof is a cell monolayer in *G. riobambae*, *E. machalilla*, and the urodele, *Ensatina eschscholtzii* (Elinson and del Pino 1985; del Pino et al. 2004; Collazo and Keller 2010). In embryos of *E. coqui* and *G. riobambae*, the blastocoel roof spreads, due to epiboly, to enclose the mass of cleaved yolk. The morphology resembles the thinning and spreading of the blastoderm to cover the entire yolk cell in zebrafish embryos (Bruce 2016; Elinson and del Pino 1985, 2012; Elinson and Fang 1998; Roszko et al. 2009). In embryos of *E. coqui*, the blastocoel roof develops into an extraembryonic epithelium that covers the large yolk-rich cells of the vegetal region. Later, this epithelium undergoes apoptosis and is replaced by the body wall of the frog (Elinson and Fang 1998).

Cells of the blastocoel roof of *X. laevis* and *A. mexicanum* can be experimentally induced to follow several developmental pathways (Nieuwkoop 1969; Sudarwati and Nieuwkoop 1971). As a result, the animal cap of the *X. laevis* blastula and early gastrula has become a test system for mesoderm induction and cell regulation experiments (Green 2009). Embryonic induction is the predominant mechanism of vertebrate development as it allows cells to differentiate, leading to the organization of cells into different tissues and organs (Gurdon 1987). The inducing and the responding tissues interact during embryonic induction. As a result, the responding tissue changes its direction of differentiation (Gurdon 1987). The future endoderm of the vegetal hemisphere is the inducing tissue, and the future ectoderm of the animal hemisphere is the responding tissue. This interaction allows the development of the mesoderm between the ectoderm and the endoderm in embryos of the urodele, *Ambystoma mexicanum* and *X. laevis* (Nieuwkoop 1969; Sudarwati and Nieuwkoop 1971).

One can ask how inductive signals of the vegetal hemisphere reach the responding animal cells in larger eggs (Table 16.2) (del Pino and Escobar 1981; Elinson et al. 1990; Schmid et al. 2012). Frog eggs can be larger, as indicated by the egg volume ratios in *G. riobambae* and *E. coqui* eggs, reaching 16 and 25 times the *X. laevis* egg volume, respectively (Table 16.2). However, the eggs of some hemiphractid frogs are much larger. The question of mesoderm induction was addressed in the 3.5 mm eggs of *E. coqui* (Ninomiya et al. 2001). The pluripotency of the blastocoel roof of *E. coqui* embryos differs from *X. laevis* (Elinson and Fang 1998; Elinson and del Pino 2012). The blastocoel roof of *E. coqui* does not respond to mesoderm induction (Elinson and Fang 1998). The *E. coqui* pattern indicates that large egg size is accompanied by displacement of the inducing signals more animally and superficially (Ninomiya et al. 2001; Elinson and Beckham 2002; Elinson and del Pino 2012). This change in the geometry of mesoderm induction is associated with increased egg size due to the large reserves of yolk, a requisite for frog development in the absence of tadpoles.

The *E. coqui* vegetal hemisphere gives rise to the embryonic endoderm and to the nutritional endoderm. The embryonic endoderm is displaced superficially and animally (Ninomiya et al. 2001). The vegetal region, extending from the blastocoel floor of *E. coqui* blastula, does not have mesoderm-inducing properties and develops into the nutritional endoderm. The nutritional endoderm, consisting of yolk-rich cells, provides nutrients for the direct development of a frog. The nutritional endoderm has no inductive function and does not differentiate into digestive-tract tissues (Elinson and Beckham 2002; Buchholz et al. 2007; Elinson 2009; Chatterjee and Elinson 2013; Karadge and Elinson 2013).

The prospective endoderm of *E. coqui* expresses typical endodermal genes. The analyzed genes were *Ec-vegt EcSox 17*, *activin b*, *derriere*, *Ecsmad2*, *smad4*, *vegt*, *and mix1* (Beckham et al. 2003; Buchholz et al. 2007; Chatterjee and Elinson 2013; Karadge and Elinson 2013; Sudou et al. 2016). The prospective endoderm of *E. coqui* differentiates into the embryonic endoderm and the nutritional endoderm. Differentiation of the nutritional endoderm begins in the gastrula due to transcriptional repression (Chatterjee and Elinson 2013; Karadge and Elinson 2013). The cellularized nutritional endoderm of *E. coqui* is an innovation for development on land and may represent a transitional state between holoblastic and meroblastic cleavages. The cellularized nutritional endoderm may parallel the changes that occurred in the evolution of the amniote egg (Buchholz et al. 2007).

The gene *vegt* encodes two isoforms, *mvegT* and *zvegT* (Gentsch et al. 2017). The transcription factor T-box gene *mvegt* occurs only in amphibians (Pérez et al. 2007; Gentsch et al. 2017). *mvegt* is a regulator for the specification of the germ layers and dorsoventral patterning of the mesoderm in *X. laevis* and other amphibians. Conserved orthologs of *vegt* have been found in the genomes of the frogs *E. coqui*, *R. pipiens*, *Pipa pipa*, *E. machalilla*, and *G. riobambae* and in the urodele *A. mexicanum* (Beckham et al. 2003; Nath et al. 2005; Nath and Elinson 2007; Pérez et al. 2007). The high level of conservation indicates that the function of *vegt* as meso-endodermal determinant is conserved in frogs and urodeles (Nath et al. 2005). The strategies for endoderm specification involving the protein transcription factors VegT and Mix1 were analyzed in the frogs *E. randi*, *E. machalilla*, *G. riobambae*, and *E. coqui*, in comparison with *X. laevis* embryos (Sudou et al. 2016). The conclusion of the analysis is that the mechanisms of endoderm specification are evolutionarily conserved in frogs (Sudou et al. 2016).

16.3 Reproductive Modes of the Family Hemiphractidae

Embryos of hemiphractid frogs develop in specialized epidermal pouches on the back of the mother or directly on the dorsal epidermis instead of water, vegetation, or humid soil (Fig. 16.1a; Tables 16.1 and 16.2) (Duellman 2015; Schmid et al. 2012). The protection provided by the mother allows slow development in comparison with *X. laevis* and frogs with other modes of reproduction (Table 16.2). Embryonic incubation is associated with modifications in the anatomy and physiology of both the mother and embryos (Schmid et al. 2012).

Frogs of the family Hemiphractidae are classified into 112 species and six genera, *Cryptobatrachus*, *Flectonotus*, *Fritziana*, *Gastrotheca*, *Hemiphractus*, and *Stefania*, by morphology and molecular phylogeny (Castroviejo-Fisher et al. 2015; Duellman 2015; Frost 2018). *Gastrotheca* is the largest genus, representing 63% of the total number of hemiphractid species (Frost 2018). The females of *Cryptobatrachus*, *Hemiphractus*, and *Stefania* do not have pouches, and the embryos are exposed to the external environment on the back of the mother during brooding (Fig. 16.4a) (Schmid et al. 2012; Castroviejo-Fisher et al. 2015; Duellman 2015). Females of the

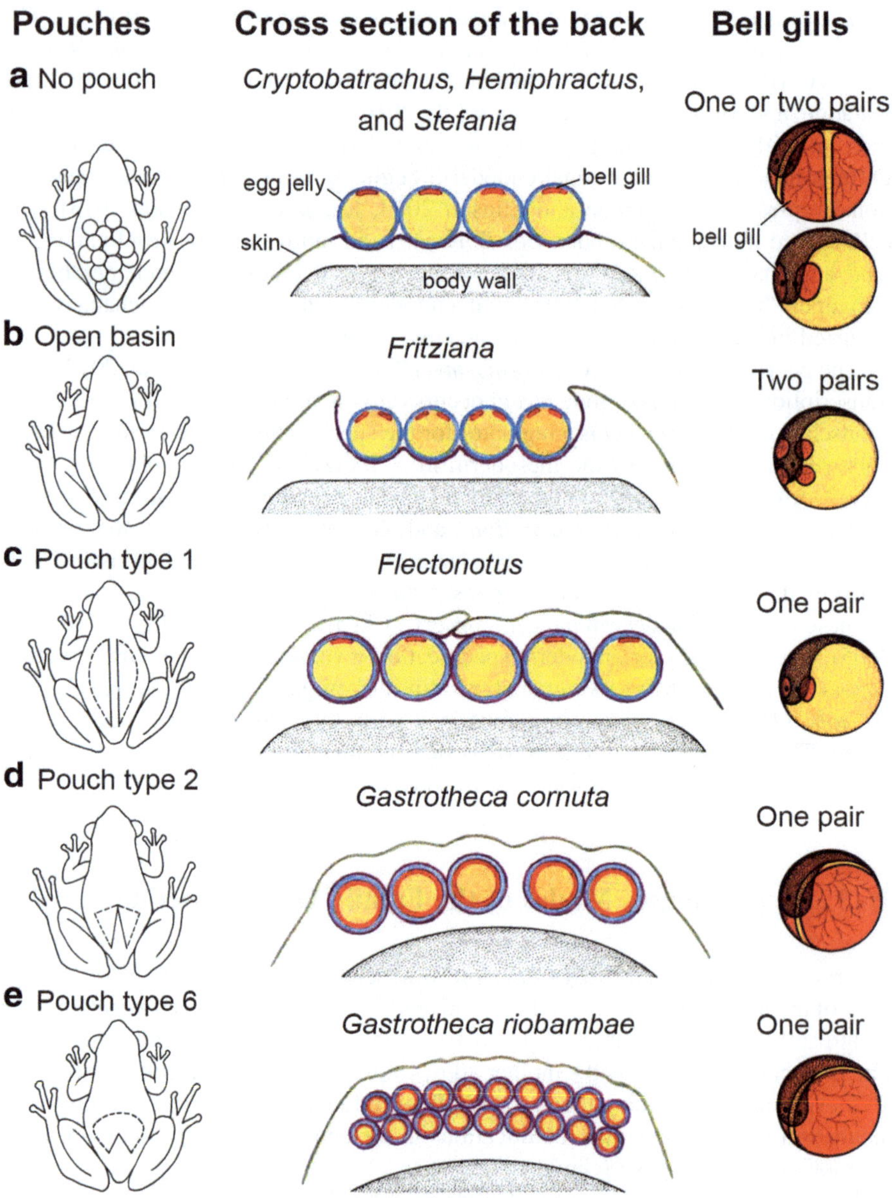

Fig. 16.4 Variation in pouches, embryos, and bell gills in hemiphractid frogs. Left images: The shape of the pouch or the area of the back to which the eggs adhere in hemiphractid frogs. Frogs are represented in dorsal view with the head toward the top. Middle images: Cross sections through the back of females reveal the relation between the mother and her embryos. The richly vascularized tissues of the back of the mother in frogs without pouches (purple) or the pouch (purple) adhere to each egg. The maternal tissues are separated by a thin layer of egg jelly (blue) from the bell gills of embryos (red). Embryos are shown in yellow. Right images: The morphology of the bell gills varies between genera of hemiphractid frogs. In some genera, the bell gills cover only the head region. In

genus *Fritziana* have an open pouch on their backs, consisting of lateral folds of skin that form an open basin for the incubation of embryos (Fig. 16.4b). The genera *Gastrotheca* and *Flectonotus*, known as marsupial frogs, incubate their embryos inside a closed pouch (Fig. 16.1a, b) (del Pino et al. 1975; Duellman 2015; Schmid et al. 2012). In consequence, the embryos of these genera are isolated from the external environment (Figs. 16.1a–b and 16.4c–e) (Duellman 2015; Schmid et al. 2012).

The development of *Gastrotheca* and *Flectonotus* embryos occurs in the absence of environmental water. In several species of *Gastrotheca*, the development inside the pouch advances until the birth of froglets. However, in *Flectonotus* and some other *Gastrotheca* species, maternal incubation lasts only until advanced tadpole stages and the tadpoles complete their development in water (del Pino and Escobar 1981; Duellman 2015; Schmid et al. 2012). The extraordinary features of marsupial frog reproduction resemble mammalian reproduction in some ways. The uterus of the mammalian female and the maternal pouch of marsupial frogs isolate the embryos from the external environment.

A small number of very large eggs characterizes the reproduction of hemiphractid frogs. The number of eggs ranges from 3 to 138 eggs in *Flectonotus fitzgeraldi* and *Gastrotheca marsupiata*, respectively. Egg diameter ranges from 2.5 mm in *G. marsupiata* to 10 mm in *G. cornuta* (del Pino and Escobar 1981). Large eggs accommodate the abundant yolk reserves required to support embryonic development, in many cases to the froglet stage (Fig. 16.3b). Frogs have an enormous diversity of egg sizes. The volume of the largest frog eggs is 4000 times the volume of the smallest eggs, with implications for the localization of morphogens (Callery 2006). The egg volume of some hemiphractid frogs is 600–1000 times the volume of *X. laevis* and *X. tropicalis* eggs (del Pino and Escobar 1981; Elinson et al. 1990). This increase in egg size likely alters the geometry of early development and the distribution of developmentally important morphogens. However, the modifications of oogenesis and embryogenesis were studied only in *E. coqui*, with 3.5 mm diameter eggs, and *G. riobambae*, with 3 mm eggs (Elinson and del Pino 2012; Schmid et al. 2012).

Three modes of reproduction have been described in hemiphractid frogs. In some hemiphractids, eggs are carried in the dorsal pouch of the female with feeding tadpoles in ponds (exotrophic tadpoles). Other hemiphractid frogs carry their eggs on the dorsum or in a dorsal pouch of the female with nonfeeding tadpoles deposited in water (endotrophic tadpoles). In other hemiphractid frogs, the female incubates

Fig. 16.4 (continued) others, the bell gills totally enclose the embryo. In hemiphractid frogs, the close communication between tissues of the mother and embryos enables the exchange of gases and fluids with resemblance to the mammalian placenta. (**a**) Hemiphractid frogs without pouches. (**b**) The open basin of frogs, genus *Fritziana*. (**c**) Pouch type one, typical of frogs, genus *Flectonotus*. (**d**) Pouch type two, found in some species of frogs, genus *Gastrotheca*. (**e**) Pouch type six, found in *Gastrotheca riobambae* and some other species of *Gastrotheca* (Published with permission © Patricia Wynne 2018. All Rights Reserved)

her embryos until the birth of froglets (direct development) (Duellman and Trueb 1986; Haddad and Prado 2005). The patterns of development in hemiphractid frogs can be explained by truncation or acceleration of the normal tadpole developmental program (Wassersug and Duellman 1984). The reproductive and developmental features of hemiphractid frogs, representing these three modes of reproduction, are reviewed here.

The free-living tadpoles of hemiphractids are the only known case of reversal from frog ancestors with direct development to the re-evolution of tadpoles (Wiens et al. 2007; Castroviejo-Fisher et al. 2015). In *Flectonotus*, *Fritziana*, and *Gastrotheca*, the tadpoles are born at different developmental stages, and the tadpoles have different morphological and trophic features (Castroviejo-Fisher et al. 2015). It is believed that free-living tadpoles of hemiphractid frogs evolved independently 2–8 times from ancestors characterized by direct development (Castroviejo-Fisher et al. 2015).

16.3.1 Development of Exotrophic Tadpoles

Females of *G. riobambae* and several other *Gastrotheca* species give birth to exotrophic tadpoles that complete aquatic development in ponds (Duellman 2015). In *G. riobambae*, fertilization is external and occurs in a terrestrial environment. During mating, the *G. riobambae* male inserts each egg inside the dorsal pouch of the female with its hind limbs. Presumably, the eggs are fertilized during their journey to the pouch. Environmental water is not required for fertilization of *G. riobambae* eggs (del Pino et al. 1975). *Gastrotheca riobambae* gametes swell and burst in low ionic saline solutions normally used for the in vitro fertilization of *X. laevis* eggs. In contrast, eggs and sperm of *G. riobambae* survive in physiological saline solutions (del Pino 1973; del Pino et al. 1975). *Gastrotheca riobambae* responds to the administration of human chorionic gonadotropin to female and male frogs with ovulation and mating, as in *X. laevis* (Elinson et al. 1990).

Gastrotheca riobambae eggs are uniform pale yellow, and the egg volume is 9–16 times the volume of *X. laevis* eggs (Fig. 16.5a, c; Table 16.2) (del Pino and Escobar 1981). The development of *G. riobambae* embryos proceeds very slowly when compared to *X. laevis* and other frogs analyzed in this study (Table 16.2). *Gastrotheca riobambae* gastrulation takes 31 times the developmental time of *X. laevis*, and its embryos follow gastrulation mode 2 (Fig. 16.2b; Table 16.2). Females of this species incubate approximately 130 eggs inside the pouch for 90–120 days (del Pino and Escobar 1981). At birth, the female releases the tadpoles in temporary pools or in ponds. The mother inserts her hind limbs into the pouch to facilitate the release of tadpoles (del Pino et al. 1975; del Pino and Escobar 1981). The tadpoles have hind limb buds at birth (del Pino and Escobar 1981). The tadpoles feed and grow in the water and transform into juvenile frogs at metamorphosis. Metamorphosis occurs after 41 days to several months, according to environmental conditions (del Pino and Escobar 1981).

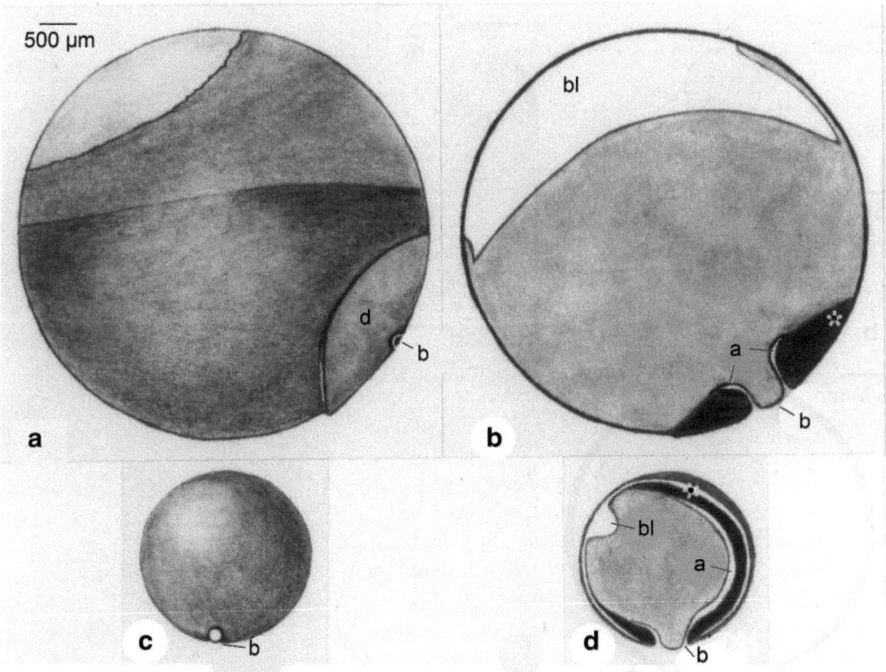

Fig. 16.5 Comparison of *Gastrotheca riobambae* and *X. laevis* gastrulae. Embryos of *G. riobambae* (**a**, **b**) and *X. laevis* (**c**, **d**) are shown at comparable stages. (**a**, **c**) External views. (**b**, **d**) Mid-sagittal sections. (**a**) A late gastrula of *G. riobambae.* The closed blastopore is surrounded by the embryonic disk. (**b**) A mid-sagittal section of a slightly younger embryo illustrates the large blastocoel cavity, the small archenteron, and the accumulation of yolk poor cells in a large circumblastoporal collar, colored in black, around the closing blastopore. The circumblastoporal collar forms the embryonic disk. (**c**) The late gastrula of *X. laevis* does not have an embryonic disk. (**d**) The archenteron in *X. laevis* at blastopore closure. The extension of the archenteron and the distribution of yolk poor cells, colored in black, differ from the *G. riobambae* gastrula. *a* archenteron, *b* blastopore, *bl* blastocoel, *d* disk. The asterisk indicates the area that will form the head of the embryo. Reproduced from del Pino and Elinson (1983)

16.3.2 Development of Endotrophic Tadpoles

Females of *Fritziana* and *Flectonotus* and some species of *Gastrotheca* give birth to non-feeding or endotrophic tadpoles (del Pino and Escobar 1981; Duellman 2015; Haddad and Prado 2005). *Flectonotus pygmaeus* females incubate a mean of seven eggs 3.0 mm in diameter for 28 days inside the pouch (del Pino and Escobar 1981). There is precocious development of the limbs (del Pino and Escobar 1981). The tadpoles lack keratinized beaks and denticles and do not swim or feed (del Pino and Escobar 1981). The tadpoles complete metamorphosis 11–15 days after birth. Gastrulation has not been studied in this frog. A major discovery was the finding of thousands of nuclei in each previtellogenic oocyte of *F. pygmaeus* (Fig. 16.6a–e). This oogenesis mode is termed multinucleated oogenesis (del Pino and Humphries

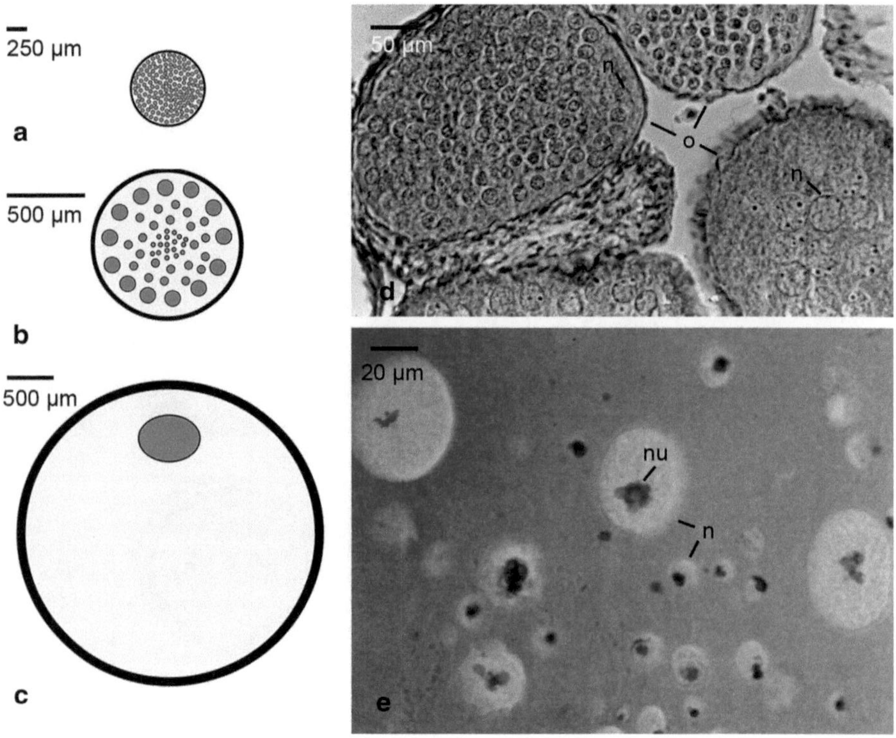

Fig. 16.6 Multinucleated oocytes of *Flectonotus pygmaeus*. (**a–c**) Diagrams of oocytes showing the decrease in nuclear numbers with the advancement of oogenesis. (**d–e**) Histological sections of previtellogenic oocytes. (**a**) Small previtellogenic oocytes with nuclei of similar diameter. Oocytes contain 2000–3000 nuclei. (**b**) In somewhat larger previtellogenic oocytes, nuclei located toward the periphery enlarge, whereas the centrally located nuclei remain small. (**c**) The fully-grown oocyte with a single nucleus. The number of nuclei decreases as the oocytes grow until only one nucleus remains in each oocyte. (**d**) Section through three previtellogenic oocytes. In each oocyte, the nuclei are of similar size. (**e**) Section through a previtellogenic oocyte with nuclei of various sizes. Nucleoli were detected in large and small nuclei. *n* nucleus, *nu* nucleolus, *o* oocyte

1978) (see Sect.16.4). In *Gastrotheca plumbea* and *Gastrotheca orophylax*, embryos are born as metamorphosing tadpoles that do not feed or swim (del Pino and Escobar 1981). The mother incubates 21 and 28 eggs, and the egg volume is 72–125 times the *X. laevis* egg volume in *G. orophylax* and *G. plumbea*, respectively (del Pino and Escobar 1981). The large egg volume indicates that embryos need considerable reserves of yolk to support development to metamorphosis stages (del Pino and Escobar 1981; Wassersug and Duellman 1984).

16.3.3 Direct Development

Females of *Cryptobatrachus*, *Hemiphractus*, and *Stefania* and numerous species of *Gastrotheca* give birth to froglets (Haddad and Prado 2005; Schmid et al. 2012; Duellman 2015). Embryos maintain vestiges of the free-living larval morphology such as the tail and oral structures. Some species develop the full complement of tadpole oral structures. Other species have partial development or do not develop tadpole oral structures (Wassersug and Duellman 1984). Wassersug and Duellman (1984) suggest that this diversity indicates that direct development evolved multiple times in hemiphractid frogs. Hemiphractid frogs with direct development incubate few eggs, ranging from 10 to 24 eggs (Schmid et al. 2012). Eggs are very large with diameters of 4–10 mm (del Pino and Escobar 1981; Elinson et al. 1990).

16.4 Mononucleated and Multinucleated Oogenesis

Oocytes of 21 frog species belonging to seven different families contain a single nucleus, termed the germinal vesicle in each oocyte (del Pino 2018; Schmid et al. 2018). A remarkable exception is the tailed frog of North America, *Ascaphus truei*, with 8 nuclei in each oocyte (Macgregor and Kezer 1970). In addition, oocytes with 1–3000 nuclei have been described in hemiphractid frogs (del Pino and Humphries 1978; Macgregor and del Pino 1982). More than one nucleus in each oocyte occurs in 12 of the 35 species of hemiphractid frogs that were studied (del Pino and Humphries 1978; Elinson et al. 1990). Oocytes have a single nucleus in the remaining 23 species (del Pino and Humphries 1978; Elinson et al. 1990). Oocytes with more than one nucleus are termed multinucleated oocytes, whereas those with a single nucleus are mononucleated oocytes (del Pino and Humphries 1978). Mononucleated oogenesis is the basic character of oogenesis in frogs. Multinucleated oogenesis is a derived character. It has been suggested that the evolution of multinucleated oogenesis occurred several times during hemiphractid frog evolution (Schmid et al. 2012).

Multinucleated oogenesis was studied in the marsupial frog, *F. pygmaeus*. Previtellogenic oocytes of *F. pygmaeus* contain between 2000 and 3000 nuclei (Fig. 16.6a, d) (del Pino and Humphries 1978; Macgregor and del Pino 1982). Each nucleus in a multinucleated oocyte is a meiotic nucleus surrounded by a nuclear envelope. The level of ribosomal gene amplification varies between nuclei (Macgregor and del Pino 1982). In small oocytes, all nuclei are active in RNA synthesis (del Pino and Humphries 1978). The number of nuclei decreases gradually with oocyte growth, and there is a diversity of nuclear sizes. Nuclei located at the periphery of the oocyte enlarge (Fig. 16.6b, e). They contain transcriptionally active lampbrush chromosomes and several nucleoli (Fig. 16.6e). In contrast, nuclei located in the central region of the oocyte remain small and disappear (Fig. 16.6e). The number of nuclei decreases until a single nucleus remains in the fully-grown

F. pygmaeus oocyte (Fig. 16.6c) (del Pino and Humphries 1978; Macgregor and del Pino 1982). The nuclei of a multinucleated oocyte may accelerate the accumulation of transcripts during the process of oogenesis in comparison with the mononucleated oocytes of *X. laevis* (del Pino and Humphries 1978; Macgregor and del Pino 1982).

Many questions remain unanswered regarding multinucleate oogenesis. It is unknown if the nuclei of a multinucleated oocyte are derived from just one oogonium. Similarly, the mechanisms that allow the localization of many nuclei inside the oocyte cytoplasm are unknown. In addition, the mechanisms of nuclear degradation and the biochemical signals protecting the final nucleus from degradation are unanswered questions (Macgregor and Kezer 1970; del Pino and Humphries 1978; Schmid et al. 2012; del Pino 2018). The study of multinucleated oogenesis has been difficult because frogs with multinucleated oocytes are rarely found in nature (del Pino and Humphries 1978; Elinson et al. 1990).

The mononucleated oocytes of *G. riobambae* have no obvious similarity with the multinucleated oocytes of *F. pygmaeus* (del Pino et al. 1986; del Pino 1989). Instead, *G. riobambae* oocytes resemble the oocytes of *X. laevis* (del Pino et al. 1986; del Pino 1989). The ovaries of *G. riobambae* juveniles comprise germline cysts in which oogonia are interconnected by cytoplasmic bridges, as in ovarian cysts of *X. laevis* (Coggins 1973; del Pino et al. 1986; Hausen and Riebesell 1991; Schmid et al. 2012). Similarly to *X. laevis*, the mitochondrial cloud or Balbiani body is present in early *G. riobambae* oocytes (del Pino et al. 1986; Hausen and Riebesell 1991; Kloc and Etkin 1995; Kloc et al. 2004; Schmid et al. 2012; Bilinski et al. 2017). Nuclei contain numerous nucleoli and lampbrush chromosomes. In addition, several nuclear proteins of *G. riobambae* oocytes are similar to oocyte nuclear proteins of *X. laevis* (del Pino 1989; del Pino et al. 1986, 2002; Schmid et al. 2012). The haploid genome of *G. riobambae* contains a low number of 5S ribosomal genes (rRNA) similar to the somatic type 5S rRNA genes of *X. laevis* (del Pino et al. 1992). A major difference between *G. riobambae* and *X. laevis* oocytes is the final diameter of the oocyte, due to the large reserves of yolk in *G. riobambae* (Fig. 16.5a, c; Table 16.2).

Germline cysts are a universal feature of the early gonads of invertebrates and vertebrates (Pepling et al. 1999). Female germline cysts of animals contain the future oocyte and sister germ cells, the nurse cells. The nurse cells transfer their cytoplasm and organelles to the future oocyte and finally are eliminated by apoptosis (Pepling et al. 1999; Lei and Spradling 2016). Mitochondria and other organelles move from the nurse cells to the oocyte to form the Balbiani body, also known as mitochondrial cloud (Kloc and Etkin 1995; Cox and Spradling 2003; Kloc et al. 2004; Bilinski et al. 2017). The Balbiani body, displaced toward the future vegetal hemisphere, is a landmark of amphibian oocytes, including the oocytes of *G. riobambae* (Coggins 1973; Kloc and Etkin 1995; Kloc et al. 2004; Hausen and Riebesell 1991; Schmid et al. 2012).

Another universal feature of early oogenesis in invertebrates and vertebrates is the massive death of oocytes due to atresia (Matova and Cooley 2001; Lei and Spradling 2016). The number of oocytes of the juvenile ovary decreases by 50% due to atresia in *R. temporaria* and *X. laevis* ovaries (Dumont 1972; Ogielska et al. 2013). The decrease in oocyte numbers in the juvenile frog ovary may result from the death of nurse cells. However, nurse cells have not been described in the frog ovary. The multinucleated oocytes of *F. pygmaeus* and other hemiphractid frogs may represent fascinating examples of the key roles of nurse cells in the building of an oocyte (del Pino 2018).

16.5 Incubation of Embryos in the Body of the Mother in Hemiphractid Frogs

The attachment of embryos to the back of the mother in hemiphractid frogs is mediated by maternal and embryonic structures. The maternal structures are the pouch of the mother in marsupial frogs, *Gastrotheca* and *Flectonotus*, the open basin in *Fritziana*, and the integument of the dorsal skin in hemiphractid frogs without pouches, *Cryptobatrachus*, *Hemiphractus*, and *Stefania* (Fig. 16.4a–e). The embryonic structures associated with incubation are the embryonic envelope and a set of bell-shaped gills (Fig. 16.7d, e) (Noble 1927).

16.5.1 The Reproductive Changes of the Maternal Pouch

The pouch and incubatory integument were examined in 36 species, representing the six genera of the family Hemiphractidae (del Pino et al. 1975; del Pino 1980). The diversity of embryonic incubation in Hemiphractidae ranges from the dorsal incubatory integument of the frogs without pouches to the complex pouches of marsupial frogs (Fig. 16.4a–e). The pouch is absent in *Cryptobatrachus*, *Hemiphractus*, and *Stefania* (Fig. 16.4a). Lateral skin folds not adherent middorsally characterize the open basin found in some species of the genus *Fritziana* (Fig. 16.4b). The two longitudinal flaps of skin become glued together middorsally over the eggs, forming an enclosed pouch in *F. pygmaeus* (Fig. 16.4c). The longitudinal pouch opening of *Flectonotus* suggests that the pouches of marsupial frogs are derived from lateral folds of dorsal skin (del Pino 1980, 2018). Pouches with a posterior opening occur in species of *Gastrotheca* and are classified in six different pouch types (Fig. 16.4d–e) (del Pino 1980).

The characteristics of the *G. riobambae* pouch are the standard for comparison with other hemiphractids (del Pino et al. 1975; del Pino 1980). The pouch is a derivative of dorsal skin in *G. riobambae*, and once it develops in juvenile females, it becomes a permanent structure of the female frog. During embryonic incubation, the pouch becomes distended to accommodate a mean of 130 eggs in the back and the sides of the body (Fig. 16.1a) (del Pino et al. 1975; del Pino and Escobar 1981). The *G. riobambae* pouch has fewer chromatophores and poison glands and less epidermal keratin in comparison with the dorsal skin (Jones et al. 1973). The mucous secretions of the pouch of several species of *Gastrotheca* as well as those of the

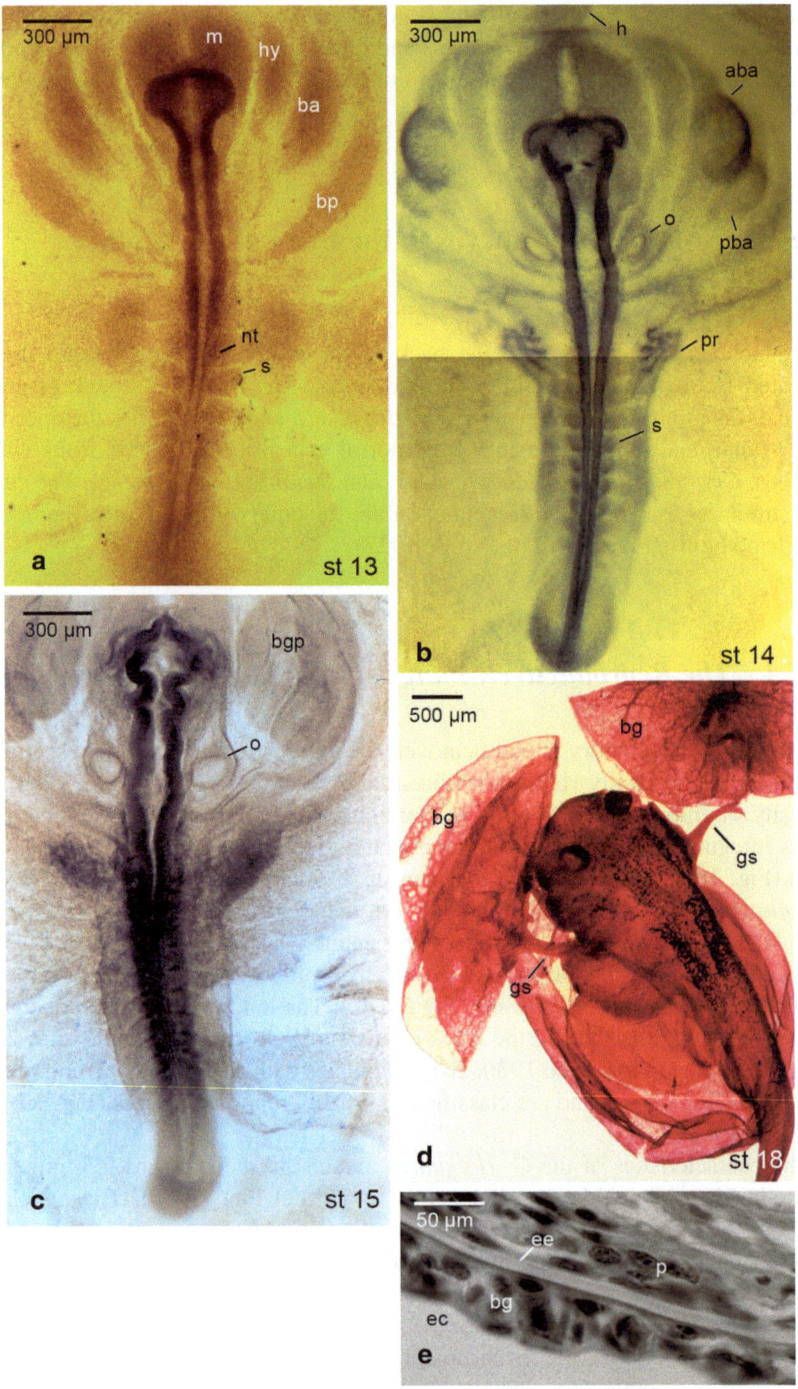

Fig. 16.7 The development of the bell gills in *Gastrotheca riobambae* embryos. Embryos in (**a–d**) are oriented with the head toward the top. The embryonic disk was dissected from the yolky vegetal hemisphere in (**a–c**) and mounted on slides. In (**d**), the yolky intestine was partially removed. Embryo in (**a**) was immunostained against vimentin. Embryos in (**b, c**) were immunostained against

incubatory integument of the hemiphractid frogs without pouches are PAS-positive mucopolysaccharides (del Pino 1980). The mucous secretions of the incubatory integument and the outer jelly envelopes of embryos cement the eggs to the incubatory integument of the mother (del Pino 1980).

The pouch and the skin merge at the level of the pouch aperture in *G. riobambae* females (Fig. 16.1a). The pouch aperture has an inverted V shape, and it is located rostrad to the cloaca (Fig. 16.1a). The pouch is open when the borders of the pouch aperture are separated (Fig. 16.4e), and it is closed when the pouch borders come together in the midline (Fig. 16.1a) (del Pino 1983). Females with open pouches have small oocytes in their ovaries. However, when the pouch is closed, the ovary contains fully grown oocytes, and the female is ready for mating. The pouch remains closed during embryonic incubation, and it reopens at the time of tadpole birth (del Pino 1983). The pouch undergoes major changes with the incubation of embryos. The walls of the pouch become attenuated, and a vascularized pouch chamber encloses each embryo (del Pino et al. 1975). After the birth of tadpoles, the pouch undergoes reorganization and returns to the non-incubatory condition, similar to normal frog skin (del Pino et al. 1975). The dorsal integument of the hemiphractid female frogs, *Cryptobatrachus*, *Hemiphractus*, and *Stefania*, resembles normal frog skin (del Pino 1980). However, during embryonic incubation, the dorsal integument becomes vascularized and forms individual shallow depressions for each embryo. The mucous secretion of the maternal integument forms an egg matrix that adheres tightly to the jelly envelope of embryos (del Pino 1980).

Gonadal steroid hormones regulate the *G. riobambae* pouch morphogenesis. Estradiol administration induces the development of the pouch in juvenile females (Jones et al. 1973). In adult females, progesterone controls pouch reproductive changes which include pouch closure and the formation of embryonic chambers. The insertion of inert beads in the pouch of non-brooding females at the time of progesterone administration triggered the formation of embryonic chambers around

Fig. 16.7 (continued) generalized neural antigens. Embryo in (**d**) was stained with borax carmine. (**a**) Morphology of a stage 13 embryo. The cranial neural crest cell streams are visible. Approximate age 21 days. (**b**) Fusion of branchial arches. The first and second branchial arches enlarge, become thicker, and fuse with each other. The heart is anterior to the head and does not beat. Approximate age 23 days. (**c**) The branchial arches have fused to originate the single pair of bell gill primordia, located lateral to the head. Blood circulation has not started. Approximate age 26 days. (**d**) Embryo with pigmented eyes. The embryo appears as a small and slightly pigmented tadpole. The bell gills have grown. In the living animal, the bell gills envelop the embryo. The two gill stalks of each bell gill have collapsed due to a preparation artifact. Approximate age 40 days. (**e**) Histological section of the pouch bell gill association from an embryo within the pouch. The pouch of the mother and the bell gills of embryos are in intimate contact. However, the pouch is separated from the bell gills by the egg envelope. A large embryonic chamber separates the embryo from the association between the pouch and bell gills. *aba* anterior branchial arch, *pba* posterior branchial arch, *bg* bell gill, *bgp* bell gill primordium, *ec* embryonic chamber, *ee* egg envelope, *h* heart, *gs* gill-stalk, *nt* neural tube, *o* otocyst, *p* pouch, *pr* pronephros, *s* somite, Streams of cranial neural crest cells: *ba* branchial anterior, *bp* branchial posterior, *hy* hyoid, *m* mandibular. Embryonic stages according to del Pino and Escobar (1981)

each bead (del Pino and Elinson 1983). The long-lived postovulatory follicles of the *G. riobambae* ovary may synthesize progesterone to support embryonic incubation, similar to the role of corpora lutea of the mammalian ovary for the maintenance of pregnancy (del Pino 1983; del Pino and Sánchez 1977).

Endogenous gonadotropins probably stimulate progesterone production in ovarian follicles of *G. riobambae*. Progesterone influences the breakdown of the nuclear envelope at oocyte maturation, the closure of the pouch, and the maintenance of embryonic incubation (de Albuja et al. 1983; del Pino 1989; del Pino and Sánchez 1977). Hormonal control of oogenesis and the pouch by sex steroids in the *G. riobambae* female parallels the hormonal control of mammalian reproduction (de Albuja et al. 1983; del Pino 1983, 1989).

16.5.2 The Egg Envelope

An egg envelope surrounds the embryos in the species of hemiphractid frogs examined, including those with aquatic and terrestrial reproductive modes. This is an indication of the conserved role of the egg envelope in frog reproduction (Table 16.2) (Elinson et al. 1990; Schmid et al. 2012). The egg envelope of *G. riobambae* (Fig. 16.7e) consists of the vitelline envelope and three layers of egg jelly, secreted by the oviducts, as in *X. laevis* (Schmid et al. 2012). The *G. riobambae* egg-jelly layers are named J-1 to J-3 (Schmid et al. 2012). The jelly layers J-1 and J-2 are dense and fibrous, whereas the outer layer, J-3, is somewhat liquid (Schmid et al. 2012). The liquid character of the egg jelly may be important to provide water to support early embryonic development. At later stages, the mucous secretion of the pouch may be the source of water for the development of *G. riobambae* embryos. The *G. riobambae* oviduct and its histology are similar to that of other frogs and urodeles (Schmid et al. 2012).

The thickness of the egg envelope varies between frogs. The egg envelope is quite thick in terraranan and dendrobatid frogs. The egg surrounded by the egg envelope of the terraranan frog, *P. unistrigatus*, measures 5.0 mm diameter, while the egg itself measures only 3.0 mm diameter (Nina and del Pino 1977). In contrast, the egg envelope of *G. riobambae* eggs is extremely thin and is only 10 μm thick (Fig. 16.7e) (del Pino et al. 1975). The egg envelope of *G. riobambae* separates the maternal pouch from the embryo during embryonic incubation and participates in the exchanges between the *G. riobambae* mother and embryos (Fig. 16.7e) (del Pino et al. 1975). The egg envelope of *G. riobambae* probably functions in sperm-egg recognition, as in *X. laevis* (Hedrick 2008). Moreover, the egg envelope may protect the embryos against infection and may maintain the appropriate humidity for embryonic development, as in other frogs. Novel functions of the *G. riobambae* egg envelope are the attachment of eggs to the maternal pouch and participation in the exchanges between each embryo with the mother (del Pino et al. 1975).

16.5.3 The Cortical Granules and Fertilization

This discussion centers on fertilization in *E. coqui* and *G. riobambae* (del Pino et al. 1975; Elinson 1987; Schmid et al. 2012). Fertilization of most anurans is monospermic, as in *X. laevis* (Elinson 1987; Hedrick 2008). However, the large *E. coqui* eggs allow the entry of numerous sperm cells in spite of the presence of egg cortical granules (Elinson 1987). Motility of *E. coqui* sperm is triggered by low-tonicity conditions, as in *X. laevis* sperm. Moreover, under experimental conditions, the terrestrial embryos of *E. coqui* develop normally in a low-tonicity aqueous solution (Elinson 1987). Therefore, the saline requirements for gamete fusion in *E. coqui* are similar to those of amphibians with external fertilization and aquatic development (Elinson 1987). Fertilization conditions of the marsupial frog *G. riobambae* differ from *E. coqui* and *X. laevis*. Eggs, sperm, and embryos of *G. riobambae* swell and die in solutions of low tonicity and survive in physiological saline solutions (del Pino et al. 1975, 1994; Elinson et al. 1990).

Oocytes of *G. riobambae* contain cortical granules, located near the plasma membrane, as in *X. laevis* and other frogs (Schmid et al. 2012). After artificial activation of *G. riobambae* eggs, by puncturing with a needle, the cortical granules undergo exocytosis, as in *X. laevis* eggs (Schmid et al. 2012). Fertilization of *G. riobambae* eggs apparently occurs during the transit of eggs from the cloaca of the female to the pouch, during amplexus (del Pino et al. 1975). In contrast, fertilization is internal in *E. coqui* (Townsend et al. 1981). It is important to determine whether internal fertilization occurs in other terraranan frogs, in the terrestrially reproducing dendrobatid frogs and in hemiphractid frogs to document the variations in fertilization associated with frog terrestrial reproduction.

16.5.4 The Cranial Neural Crest, Branchial Arches, and the Bell Gills

Embryos of frogs belonging to the family Hemiphractidae have unusual bell-shaped gills during incubation in the body of the mother (Fig. 16.7d) (Noble 1927). Embryos of *G. riobambae* develop bell gills, internal gills, and lungs during development in the maternal pouch (del Pino and Escobar 1981). The bell gills are modified external gills that, together with the lining of the maternal pouch, mediate exchanges between embryos and the mother (Fig. 16.7d, e) (del Pino and Escobar 1981). The internal gills are the respiratory organs of *G. riobambae* tadpoles, and lungs become the respiratory organs of the froglets after metamorphosis (del Pino and Escobar 1981).

The embryo of *G. riobambae* maintains an extraordinary planar orientation on the surface of an egg until the onset of organogenesis (Figs. 16.3b, 16.7a–c, and 16.8). The flat orientation may be advantageous for exchanges with the surrounding tissues of the *G. riobambae* maternal pouch. The planar condition of embryos allows the identification of the relationships of cranial neural crest cell streams, branchial

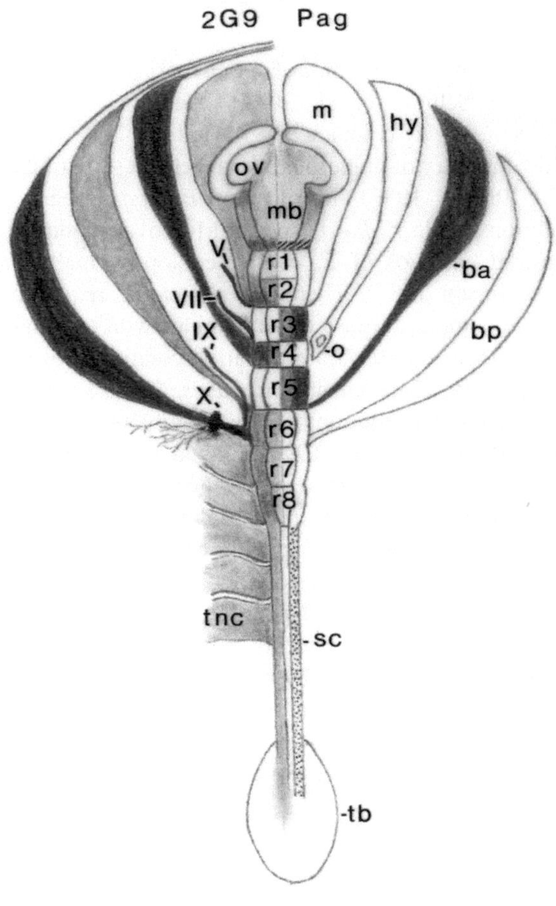

Fig. 16.8 Composite diagram of neural marker expression in the flat neurula of *Gastrotheca riobambae*. Neural structures were identified by the RNA expression of the tyrosine kinase, EPH (ephrin) receptor A4 (*epha4*) (also known as *pag, pagliaccio*). In addition, the neural antigen 2G9 and the antigens against the neural proteins Pax2 and Hoxd9 allowed the identification of neural structures (del Pino and Medina 1998). The image combines the features of stages 13 and 14 embryos. Stages are according to del Pino and Escobar (1981). The streams of cranial neural crest cells are shown as in embryos of stage 13, including thin rostral extensions, as shown on the left of the image. Marker expression in otic vesicles and pronephros is not shown. Cranial nerves are shown as in stage 14 embryos. The trunk neural crest segments of stage 13 embryos are shown on the left. The pattern of the neural antigen 2G9 is shown on the left and the pattern of *pag* (*epha4*) on the right. The intensity of marker expression is shown with different grades of shading. Cranial neural crest cell streams and rhombomeres of the hindbrain positive for *pag* RNA alternate with cranial neural crest cell streams and rhombomeres positive for antigen 2G9. The patterns of Pax2 (hatched) and Hoxd9 (stippled) are shown on the right. The expression patterns of the neural markers coincide with the patterns described for other vertebrates, suggesting that these genes and their expression patterns have been conserved in *G. riobambae* development. *mb* midbrain, r1 to r8 rhombomeres 1 to 8, *o* otocyst, *ov* optic vesicle, *sc* spinal cord, *tb* tailbud, *tnc* neural crest of the trunk, *V* trigeminal nerve, *VII* facial nerve, *IX* glossopharyngeal nerve, *X* vagus nerve, streams of

arches, and bell-gill primordia (Figs. 16.7a–c and 16.8) (del Pino and Medina 1998). Expression of the neural gene *epha4* (also known as *pag*, *Pagliaccio*) (Winning and Sargent 1994) and of the neural proteins (Pax2, Hoxd9, and antigen 2G9) (Jones and Woodland 1989) aided in the identification of the streams of cranial neural crest cells, the central nervous system, and other structures of *G. riobambae* embryos (Fig. 16.8) (del Pino and Medina 1998). Neural genes and their expression patterns are similar to other vertebrates, indicating that *G. riobambae* neural development was conserved in evolution (del Pino and Medina 1998). The heart is the anterior most structure of *G. riobambae* embryos (Fig. 16.7b). Anterior development of the heart is not unusual because the heart mesoderm of *X. laevis* has an anterior location (Warkman and Krieg 2007).

The morphogenesis of the bell gills has been analyzed in *G. riobambae* (Fig. 16.7a–c) (del Pino and Medina 1998). It has been found that the branchial anterior and branchial posterior streams of cranial neural crest cells populate the first and second branchial arches and contribute to the formation of the bell gills (Fig. 16.7a–c) (del Pino and Medina 1998). The first and second branchial arches fuse with each other on each side of the body to develop a single pair of bell gills (Fig. 16.7b–c) (del Pino and Escobar 1981). Two gill stalks connect each bell gill with the body of the embryo. The blood vessels of bell gills were traced to the first and second aortic arches (Spannhof and Spannhof 1972). As development advances, the bell gills grow and enclose the embryo in a vascularized membranous structure (Figs. 16.7d and 16.4e). The bell gills are in contact with the fertilization envelope (Fig. 16.7e). The bell gills of *F. pygmaeus* embryos are small and cover only the dorsal side of the head (del Pino and Escobar 1981). The *F. pygmaeus* bell gills derive from the first branchial arch. The origin of bell gills in other species of hemiphractid frogs was identified by observation of the blood vessels found in their gill stalks in comparison with the *G. riobambae* and *F. pygmaeus* bell gills (del Pino and Escobar 1981).

The number of bell gills and the level of embryo enclosure vary greatly in hemiphractid frogs (del Pino and Escobar 1981). A single pair of bell gills, each derived from one branchial arch, may be the simplest situation. This occurs in embryos of *Cryptobatrachus* and *Flectonotus* (Fig. 16.4a, c) (del Pino and Escobar 1981). Two pairs of bell gills, each derived from one branchial arch, characterize the embryos of *Fritziana*, *Stefania*, and most species of *Hemiphractus* (Fig. 16.4a, b). The fusion of two branchial arches on each side of the head to form a single pair of bell gills, each with two gill stalks, is a more complex morphology and occurs in species of *Gastrotheca* (Fig. 16.4d, e) and in *Hemiphractus fasciatus* (del Pino and Escobar 1981).

Fig. 16.8 (continued) cranial neural crest cells: *ba* branchial anterior, *bp* branchial posterior, *hy* hyoid, *m* mandibular. Reprinted with the permission of UPV/EHU Press from del Pino, E.M. and Medina, A. (1998). Neural development in the marsupial frog *Gastrotheca riobambae*. Int. J. Dev. Biol. 42: 723–731

16.5.5 The Incubation of Embryos

In *G. riobambae*, the vascularized bell gills of embryos and the vascularized pouch chambers of the mother are covered with villi that increase the surface area and may facilitate exchanges between the mother and embryos (del Pino et al. 1975; del Pino 2018; Schmid et al. 2012). The bell gills and the pouch are separated by the egg capsule, about 10 μm thick (Fig. 16.7e) (del Pino et al. 1975). The wet weight of embryos within the egg envelope increases threefold during incubation. In contrast, the dry weight of embryos does not change notably. Accordingly, water and gas exchanges apparently occur between the mother and her embryos (del Pino and Escobar 1981). Nutrient transfer from the mother to embryos in the pouch was reported in *Gastrotheca excubitor* (Warne and Catenazzi 2016). The relationship between the mother and embryos in marsupial frogs and the hormonal control of embryonic incubation resemble the association between the mother and embryos in placental mammals.

16.5.6 Nitrogen Waste Excretion and the In Vitro Culture of Embryos

In frogs with aquatic reproduction, the change from ammonotelism to ureotelism occurs at metamorphosis (Duellman and Trueb 1986). Embryos of *E. coqui* excrete ammonia during their intracapsular development. Nitrogen waste excretion changes to urea shortly before the hatching of froglets (Callery and Elinson 1996). Therefore, the nitrogenous waste excretion of *E. coqui* embryos resembles the pattern of frogs with aquatic tadpoles and metamorphosis. These findings indicate that the evolution of direct development in *E. coqui* is not associated with changes in the mode of nitrogenous waste excretion (Callery and Elinson 1996; Schmid et al. 2012).

A different pattern of nitrogenous waste excretion occurs in *G. riobambae*, as urea is the main excretory product of embryos, tadpoles, and adults (Alcocer et al. 1992; del Pino et al. 1994). Excretion of urea in tadpoles may be derived from the re-evolution of tadpoles in marsupial frogs from a direct developing ancestor, as already assessed by phylogenetic studies (Sect. 16.3) (Wiens et al. 2007; Castroviejo-Fisher et al. 2015). Urea concentration in the capsular fluid of embryos is significantly higher than in the blood of adults, including females during embryonic incubation (del Pino et al. 1994). The concentration of urea in the capsular fluid increases as development advances (del Pino et al. 1994). The accumulation of urea in the capsular fluid of *G. riobambae* embryos may play an osmoregulatory role for the conservation of water, needed for embryonic development. Therefore, embryonic ureotelism is an adaptation for development under the conditions of water stress of the maternal pouch (Alcocer et al. 1992; del Pino et al. 1994; del Pino 2018). Further analysis is required to determine whether ureotelism is the excretory mode of embryos, tadpoles, and adults of other hemiphractid frogs.

A major difficulty in the study of *G. riobambae* early development was the in vitro culture of embryos. Early embryos removed from the pouch do not survive in solutions of low tonicity and only survived for a couple of days in a humid chamber (del Pino et al. 1975; Elinson et al. 1990). The addition of urea to a physiological saline solution allowed survival of embryos in culture for a longer period, permitting the study of early development and gastrulation (del Pino et al. 1994; Moya et al. 2007). Culture of embryos in a urea-containing physiological saline solution became the standard method for the study of *G. riobambae* early development (del Pino et al. 1994).

16.6 *Gastrotheca riobambae* Early Development

Cleavage and gastrulation of *G. riobambae* embryos differ from *X. laevis*. *Gastrotheca riobambae* cleavage is holoblastic and becomes irregular and asynchronous after a few cell division. A blastopore with an inconspicuous dorsal blastopore lip develops during gastrulation. The blastopore closes over a tiny archenteron (Fig. 16.5a, b). In the absence of convergent extension, the blastopore lip of *G. riobambae* embryos enlarges and forms a bulky circumblastoporal collar (Figs. 16.5b and 16.9c). On the surface, an embryonic disk of small cells develops around the closed blastopore (Figs. 16.5a and 16.9a, b). The yolky vegetal region of the *G. riobambae* gastrula is cleaved into large blastomeres (Fig. 16.9d). A thin monolayer of cells, derived from the animal region, envelops the embryo (Fig. 16.9d). After blastopore closure, the archenteron extends anteriorly and the notochord elongates (Fig. 16.9e). The body of the embryo develops from the embryonic disk (Fig. 16.3b). The development of the body from an embryonic disk in *G. riobambae* (Fig. 16.3b) resembles the development from a disk of cells in birds and reptiles (del Pino and Elinson 1983). In contrast, *X. laevis* does not develop an embryonic disk, and the archenteron expands during gastrulation (Fig. 16.5c, d). The *X. laevis* blastopore lip remains thin, as the cells that involute at the blastopore move anteriorly during gastrulation (Fig. 16.5d). The development of *G. riobambae* was reviewed previously in detail (del Pino 1989, 2018; del Pino et al. 2007; Schmid et al. 2012; Elinson and del Pino 2012).

16.6.1 Cleavage and the Midblastula Transition

The embryos of *G. riobambae* develop slowly (Table 16.2). The timing of *G. riobambae* cleavage and gastrulation are more similar to the timing in mammals rather than frogs. The first cell cycle of *G. riobambae* embryos requires 16–20 h, and embryos take about 24 h from fertilization to the eight-cell stage. Cleavage is atypical holoblastic (Nelsen 1953). Cell divisions become asynchronous and irregular after the eight-cell stage (del Pino and Loor-Vela 1990). Nuclei of eight-cell embryos comprise nucleoli, indicating that transcription of ribosomal RNA has started at this early stage.

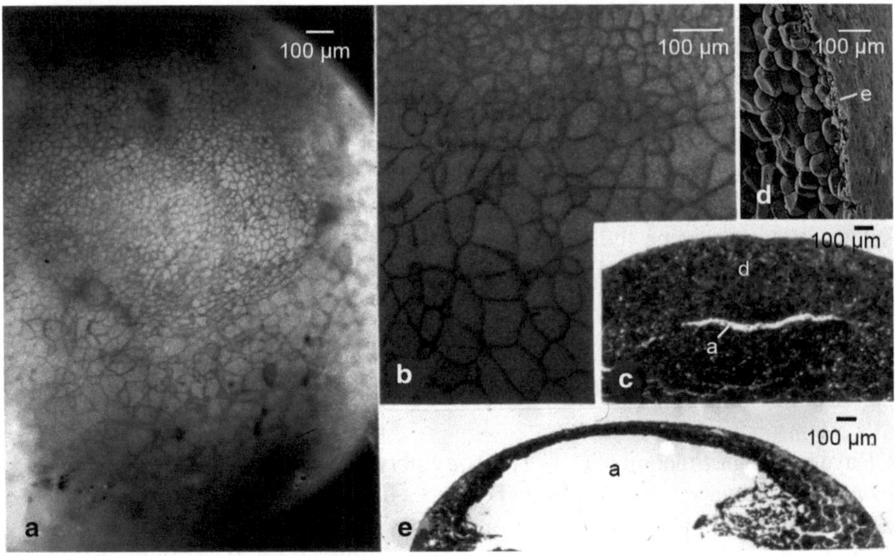

Fig. 16.9 The embryonic disk of *Gastrotheca riobambae* embryos. (**a**) Late gastrula is silver-stained to show cell borders. The embryonic disk appears as a group of smaller cells around the closed blastopore on the surface of embryos. (**b**) The image in (**a**) is shown at higher magnification. There is a clear boundary between the cells of the embryonic disk and the rest of the embryo. (**c**) The archenteron remains small during gastrulation. The slit-like archenteron separates the small cells of the embryonic disk from the large yolky cells. (**d**) The large yolky cells, outside the area of the embryonic disk, are covered by a thin, one-cell thick epithelium. (**e**) The archenteron expands after blastopore closure. The embryonic disk is stretched to form the roof of the archenteron. *a* archenteron, *d* disk, *e* epithelium. Reproduced from del Pino and Elinson (1983)

Embryos of *G. riobambae* lack a midblastula transition (del Pino and Loor-Vela 1990). Embryos of *X. laevis* have 12 rounds of synchronous cleavage, each cycle with duration of about half an hour. Thereafter, cleavage becomes asynchronous and transcription is activated. These changes characterize the midblastula transition (Newport and Kirschner 1982; Nieuwkoop and Faber 1994). It has been suggested that the rapid and synchronous early cleavage accelerates the vulnerable early development of *X. laevis* (del Pino and Loor-Vela 1990; Schmid et al. 2012). On the other hand, slow cleavage and the lack of a midblastula transition characterize the cleavage of mammalian and of *G. riobambae* embryos. Embryos of *E. coqui* develop rapidly in spite of their direct development pattern (Table 16.2).

16.6.2 The Gastrula and the Embryonic Disk

The epiboly, involution, development of the dorsal blastopore lip, formation of a blastopore around a large yolk plug, and convergent extension are typical morphogenetic processes of frog gastrulation (Solnica-Krezel 2005; Solnica-Krezel and

Sepich 2012). Convergent extension, notochord elongation, and archenteron elongation begin in the midgastrula of *X. laevis* (Keller and Winklbauer 1992). In contrast, these processes occur after blastopore closure in *G. riobambae* embryos (del Pino 1996; del Pino and Elinson 1983; Elinson and del Pino 1985; Moya et al. 2007). This modification leads to the formation of an embryonic disk in embryos of *G. riobambae*, derived from the blastopore lip (del Pino and Elinson 1983; Elinson and del Pino 1985). The closure of the blastopore of *G. riobambae* embryos, in the absence of extension, may depend on convergent thickening (Shook et al. 2018).

An increase in egg size is a common characteristic of the amphibians that undergo convergent extension after gastrulation (gastrulation mode 2) (del Pino et al. 2007; Vargas and del Pino 2017). In embryos of *G. riobambae* and dendrobatid frogs, convergent extension was identified by expression of the protein transcription factor Brachyury in the notochord. Convergent extension and notochord elongation are separated from involution and blastopore closure (Fig. 16.2b) (del Pino 1996; Benítez and del Pino 2002; Moya et al. 2007; Hervas et al. 2015, see also Sect. 16.2.4). Similarly, elongation of the notochord, as detected by the expression of *EcBra*, occurs after blastopore closure in *E. coqui*. In contrast with *G. riobambae*, embryos of dendrobatid frogs and *E. coqui* do not develop an embryonic disk (Fig. 16.2b) (Ninomiya et al. 2001; Benítez and del Pino 2002; Moya et al. 2007; Hervas et al. 2015). Retardation of convergent extension followed by the formation of an embryonic disk occurs in the urodele, *E. eschscholtzii* (Collazo and Keller 2010).

The significance of convergent extension for frog life histories was analyzed by event pairing between blastopore closure, notochord elongation, somite, and neural development (Sáenz-Ponce et al. 2012a). In frogs with overlap of convergent extension with gastrulation (gastrulation mode 1), elongation of the notochord occurs early, relative to blastopore closure, and cranial neural crest cell streams develop more rapidly relative to somite formation. Neural tube closure is temporally uncoupled from somite formation (Sáenz-Ponce et al. 2012a). This comparison indicates that rapid development is accompanied by overlap of convergent extension with gastrulation (gastrulation mode 1). In contrast, frogs that develop more slowly, convergent extension begins after gastrulation (gastrulation mode 2) (Sáenz-Ponce et al. 2012a; Vargas and del Pino 2017).

The modular nature of gastrulation allows the experimental separation of convergent extension from gastrulation in embryos of *X. laevis* and zebrafish (Ewald et al. 2004; Roszko et al. 2009; Scharf et al. 1989). During frog evolution, the modularity of gastrulation probably allowed great diversification of development processes, which was important for the evolution of terrestrial reproductive modes (Elinson and del Pino 2012). In particular, the retardation of notochord and body elongation is associated with the development from large eggs in *G. riobambae*, dendrobatid frogs, and *E. coqui*. Large egg size allows storage of sufficient yolk reserves to support direct development. The embryonic disk in *G. riobambae* embryos is a remarkable example of the modularity of frog gastrulation (Elinson and del Pino 2012; del Pino 2018).

16.6.3 *The Organizer*

Hilde Mangold and Hans Spemann (Spemann and Mangold 1924) discovered that the dorsal blastopore lip of a urodele gastrula has the capacity to induce a new dorsal axis in a host gastrula. The early dorsal blastopore lip induces head structures. The late dorsal blastopore lip induces formation of the tail (Holtfreter 1933; Mangold 1933; Nieuwkoop 1955). The dorsal blastopore lip of frog and urodele gastrulae is known as the Spemann-Mangold organizer or the organizer because of its inductive properties. The organizer is divided into head, trunk, and tail organizers, according to inductive characteristics. The prechordal plate represents the head organizer, and the notochord is the trunk organizer (Saxén 1989; Harland and Gerhart 1997; De Robertis 2006).

Gene expression in the organizer is conserved. However, the timing of gene expression varies between frogs. One of the organizer-specific genes is the highly conserved homeobox gene *lim1* (*lhx1*) (Taira et al. 1992; Karavanov et al. 1996; Kodjabachian et al. 2001). The axial mesoderm of *X. laevis* expresses *xlim1* in the gastrula and neurula (Taira et al. 1992, 1994). Similarly, the protein Lim1 is expressed in the axial mesoderm of the gastrula in *E. randi*, *E. machalilla*, *G. riobambae*, and *X. laevis* (Venegas-Ferrín et al. 2010). There were differences in the expression time among these frogs. The prechordal plate and the notochord simultaneously express Lim1 in embryos of *X. laevis* and *E. randi* (del Pino et al. 2007; Venegas-Ferrín et al. 2010). In embryos of *E. machalilla*, only the prechordal plate expresses Lim1 during gastrulation, and Lim1 expression in the notochord occurs after blastopore closure. Similarly, expression of the protein Brachyury in the notochord occurs after blastopore closure in *E. machalilla* and *G. riobambae* embryos (del Pino 1996; Benítez and del Pino 2002; Venegas-Ferrín et al. 2010).

The separation of head and trunk organizers of *G. riobambae* and other frog embryos may derive from inhibition of convergent extension during gastrulation. A candidate for this inhibition may be the gene *goosecoid* (*gsc*) according to the role of *gsc* in gastrulation (Ulmer et al. 2017). In fact, *gsc* inhibits convergent extension, in the prechordal plate of *X. laevis* and mouse embryos (Ulmer et al. 2017). The results indicate that the genes that guide early development are highly conserved in frogs. However, the timing of gene expression may vary leading to modifications of embryogenesis. In particular, the shifting of convergent extension to the post-gastrula (gastrulation mode 2) is likely important in the evolution of large eggs and the diversification of frog reproductive modes (del Pino 2018).

16.7 Comparative Analysis of Frog Gastrulation

Frog gastrulation was classified into gastrulation modes 1 and 2, according to the onset of convergent extension (Fig. 16.2a, b) (del Pino et al. 2007; Vargas and del Pino 2017). Gastrulation mode 1 occurs in frogs with aquatic reproduction and rapid development. These frogs have small eggs, 1.1–2.2 mm diameter. The exception is the gliding tree frog, *A. spurrelli*, with 2.9 mm diameter eggs (Fig. 16.2a) (Schmid et al. 2018). In contrast, gastrulation mode 2 occurs in frogs with terrestrial adaptations for reproduction (Fig. 16.2b). Egg size of the frogs with gastrulation mode 2 ranges from 1.6 to 3.5 mm diameter. The development occurs slowly in comparison with frogs with gastrulation mode 1. The exception is the direct developing frog *E. coqui* with 3.5 mm diameter eggs and rapid development (Table 16.2) (del Pino et al. 2007; Vargas and del Pino 2017).

A typical example of gastrulation mode 1 is the gastrulation pattern of *X. laevis*. Convergent extension and body elongation begin in the *X. laevis* midgastrula (Figs. 16.2a and 16.5c, d), and the process of gastrulation requires a few hours (Keller and Danilchik 1988; Wallingford et al. 2002). Other frogs with gastrulation mode 1, discussed in this work, are *A. spurrelli*, *C. stolzmanni*, *E. callistomma*, *E. coloradorum*, *E. randi*, and *H. fleischmanni* (Table 16.2). These frogs develop rapidly, as evidenced by the time required for gastrulation (Fig 16.2a; Table 16.2).

Gastrulation mode 2 occurs in dendrobatid frogs, *G. riobambae* and *E. coqui* (Fig. 16.2b). The development of dendrobatid frogs and *G. riobambae* is very slow, taking 7–31 times the time for *X. laevis* gastrulation (Table 16.2). *Eleutherodactylus coqui* develops as fast as species with gastrulation mode 1 and requires four times the *X. laevis* gastrulation time (Table 16.2). However, *E. coqui* eggs are among the largest in the frogs included in our work (del Pino et al. 2007; Vargas and del Pino 2017). Retardation of convergent extension to the late gastrula and the neurula also occurs in urodeles (Shi et al. 1987). The shift of convergent extension to the amphibian post-gastrula accompanies variation in egg size and terrestrial reproductive modes (del Pino 2018).

16.8 The Marginal Zone, the Blastopore, and the Evolution of Gastrulation

Cells of the blastocoel roof, marginal zone, and vegetal region were measured in the early gastrula of ten frog species (Vargas and del Pino 2017). Cell size does not correlate with gastrulation modes (Vargas and del Pino 2017). Strong correlation of cell and egg sizes occurs in the blastocoel roof and vegetal region of the early gastrula, an indication that larger eggs had larger cells in these two regions (Vargas and del Pino 2017). Large vegetal cells store the reserves of yolk needed for development, a requisite for the evolution of frog terrestrial reproductive modes. In contrast, the correlation of cell size with egg size was weak in the marginal zone (Vargas and del Pino 2017). Cells of the marginal zone were small irrespective of egg size. The small cells of the marginal zone seem to be important for the movements of gastrulation and the formation of the blastopore. Small cell size in the marginal zone is associated with the conservation of involution, maintaining the blastopore as a universal feature of frog gastrulation (Vargas and del Pino 2017).

Comparative studies of gastrulation in ray-finned fishes (Actinopterygii), performed in the nineteenth century, provide the best view of the evolution of vertebrate gastrulation (reviewed in Cooper and Virta 2007). Ray-finned fishes have basal, intermediate, and highly derived modes of gastrulation (Cooper and Virta 2007; Takeuchi et al. 2009). The basal vertebrate mode of early development, considered to represent the mode in the ancestors to teleosts and amphibians, includes holoblastic cleavage with gastrulation through a blastopore (Bolker 1994; Cooper and Virta 2007). Examples of this gastrulation pattern are gastrulation in the sturgeon (*Acipenser*), bichir (*Polypterus*), and frogs with gastrulation mode 1, exemplified by *X. laevis* (del Pino 2018). A tendency toward meroblastic cleavage occurs in the intermediate gastrulation mode of ray-finned fishes (Cooper and Virta 2007). Similarly, frogs with gastrulation mode 2 have a tendency toward meroblastic cleavage (del Pino 2018). The major similarities between the intermediate gastrulation mode of ray-finned fishes and frogs with gastrulation mode 2 are small blastomeres in the animal region, large size of vegetal blastomeres, and the accumulation of small cells in the blastopore lip (del Pino 2018). This pattern is exemplified by gastrulation in the bowfin (*Amia*) and in *G. riobambae* (Cooper and Virta 2007; del Pino 2018). In frogs, this pattern is associated with separation of convergent extension from gastrulation (gastrulation mode 2). The retardation of convergent extension to the post-gastrula in gastrulation mode 2 is a powerful mechanism associated with the evolution of terrestrial reproduction in frogs (del Pino 2018). The cellular and molecular mechanisms underlying this mode of gastrulation in frogs need additional investigation. Meroblastic cleavage, the most highly derived mode of fish gastrulation (Cooper and Virta 2007), does not occur in frogs.

16.9 Conclusion

The transfer of eggs from aquatic to terrestrial environments required major developmental and reproductive changes in evolution. In particular, the incubation of embryos inside the maternal pouch of marsupial frogs is accompanied by adaptations in the mother, modifications of oogenesis, and developmental changes. The comparison of gastrulation between frogs revealed an important developmental theme, the separation of convergent extension from gastrulation in frogs that evolved terrestrial reproduction. This work provides a baseline for the study of the reproductive and developmental adaptations in anurans.

Acknowledgements I acknowledge with gratitude the collaboration of former students and colleagues. I thank the late Michael Schmid for revision of the text and Jennifer Davis and Clifford Keil for language revision and valuable comments. The Pontificia Universidad Católica del Ecuador supported research in my laboratory.

References

Afonin B, Ho M, Gustin JK, Meloty-Kapella C, Domingo CR (2006) Cell behaviors associated with somite segmentation and rotation in *Xenopus laevis*. Dev Dyn 235:3268–3279

Alcocer I, Santacruz X, Steinbeisser H, Thierauch K-H, del Pino EM (1992) Ureotelism as the prevailing mode of nitrogen excretion in larvae of the marsupial frog *Gastrotheca riobambae* (Fowler) (Anura, Hylidae). Comp Biochem Physiol 101A:229–231

Arendt D, Nübler-Jung K (1999) Rearranging gastrulation in the name of yolk: evolution of gastrulation in yolk-rich amniote eggs. Mech Dev 81:3–22

Beckham YM, Nath K, Elinson RP (2003) Localization of RNAs in oocytes of *Eleutherodactylus coqui*, a direct developing frog, differs from *Xenopus laevis*. Evol Dev 5:562–571

Benítez M-S, del Pino EM (2002) Expression of Brachyury during development of the dendrobatid frog *Colostethus machalilla*. Dev Dyn 225:592–596

Bilinski SM, Kloc M, Tworzydlo W (2017) Selection of mitochondria in female germline cells: is Balbiani body implicated in this process? J Assist Reprod Genet 34:1405–1412

Blum M, Schweickert A, Vick P, Wright CVE, Danilchik MV (2014) Symmetry breakage in the vertebrate embryo: when does it happen and how does it work? Dev Biol 393:109–123

Bolker JA (1994) Comparison of gastrulation in frogs and fish. Am Zool 34:313–322

Bruce AEE (2016) Zebrafish epiboly: spreading thin over the yolk. Dev Dyn 245:244–258

Buchholz DR, Singamsetty S, Karadge U, Williamson S, Langer CE, Elinson RP (2007) Nutritional endoderm in a direct developing frog: a potential parallel to the evolution of the amniote egg. Dev Dyn 236:1259–1272

Callery EM (2006) There's more than one frog in the pond: a survey of the Amphibia and their contributions to developmental biology. Semin Cell Dev Biol 17:80–92

Callery EM, Elinson RP (1996) Developmental regulation of the urea-cycle enzyme arginase in the direct developing frog *Eleutherodactylus coqui*. J Exp Zool 275:61–66

Callery EM, Elinson RP (2000) Thyroid hormone-dependent metamorphosis in a direct developing frog. Proc Natl Acad Sci USA 97:2615–2620

Castroviejo-Fisher S, Padial JM, De la Riva I, Pombal JP Jr, Da Silva HR, Rojas-Runjaic FJM, Medina-Méndez E, Frost DR (2015) Phylogenetic systematics of egg-brooding frogs (Anura: Hemiphractidae) and the evolution of direct development. Zootaxa 4004:1–75

Chatterjee S, Elinson RP (2013) Commitment to nutritional endoderm in *Eleutherodactylus coqui* involves altered nodal signaling and global transcriptional repression. J Exp Zool Mol Dev Evol 322:27–44

Chipman AD, Haas A, Khaner O (1999) Variations in anuran embryogenesis: yolk-rich embryos of *Hyperolius puncticulatus* (Hyperoliidae). Evol Dev 1:49–61

Coggins LW (1973) An ultrastructural and radioautographic study of early oogenesis in the toad *Xenopus laevis*. J Cell Sci 12:71–93

Collazo A, Keller R (2010) Early development of *Ensatina eschscholtzii*: an amphibian with a large, yolky egg. EvoDevo 1:6

Cooper MS, Virta VC (2007) Evolution of gastrulation in the ray-finned (Actinopterygian) fishes. J Exp Zool Mol Dev Evol 308:591–608

Cox RT, Spradling AC (2003) A Balbiani body and the fusome mediate mitochondrial inheritance during *Drosophila* oogenesis. Development 130:1579–1590

de Albuja CM, Campos M, del Pino EM (1983) Role of progesterone on oocyte maturation in the egg-brooding hylid frog *Gastrotheca riobambae* (Fowler). J Exp Zool 227:271–276

De Robertis EM (2006) Spemann's organizer and self-regulation in amphibian embryos. Nat Rev Mol Cell Biol 7:296–302

del Pino EM (1973) Interactions between gametes and environment in the toad *Xenopus laevis* (Daudin) and their relationship to fertilization. J Exp Zool 185:121–132

del Pino EM (1980) Morphology of the pouch and incubatory integument in marsupial frogs (Hylidae). Copeia 1980:10–17 del Pino EM (1983) Progesterone induces incubatory changes in the brooding pouch of the frog *Gastrotheca riobambae* (Fowler). J Exp Zool 227:159–163

del Pino EM (1989) Modifications of oogenesis and development in marsupial frogs. Development 107:169–187

del Pino EM (1996) The expression of Brachyury (T) during gastrulation in the marsupial frog *Gastrotheca riobambae*. Dev Biol 177:64–72

del Pino EM (2018) The extraordinary biology and development of marsupial frogs (Hemiphractidae) in comparison with fish, mammals, birds, amphibians and other animals. Mech Dev 154:2–11

del Pino EM, Elinson RP (1983) A novel development pattern for frogs: gastrulation produces an embryonic disk. Nature 306:589–591

del Pino EM, Escobar B (1981) Embryonic stages of *Gastrotheca riobambae* (Fowler) during maternal incubation and comparison of development with that of other egg-brooding hylid frogs. J Morphol 167:277–295

del Pino E, Humphries AA (1978) Multiple nuclei during early oogenesis in *Flectonotus pygmaeus* and other marsupial frogs. Biol Bull 154:198–212

del Pino EM, Loor-Vela S (1990) The pattern of early cleavage of the marsupial frog *Gastrotheca riobambae*. Development 110:781–789

del Pino EM, Medina A (1998) Neural development in the marsupial frog *Gastrotheca riobambae*. Int J Dev Biol 42:723–731

del Pino EM, Sánchez G (1977) Ovarian structure of the marsupial frog *Gastrotheca riobambae* (Fowler). J Morphol 153:153–162

del Pino EM, Galarza ML, de Albuja CM, Humphries AA (1975) The maternal pouch and development in the marsupial frog *Gastrotheca riobambae* (Fowler). Biol Bull 149:480–491

del Pino EM, Steinbeisser H, Hofmann A, Dreyer C, Campos M, Trendelenburg MF (1986) Oogenesis in the egg-brooding frog *Gastrotheca riobambae* produces large oocytes with fewer nucleoli and low RNA content in comparison to *Xenopus laevis*. Differentiation 32:24–33

del Pino EM, Murphy C, Masson PH, Gall JG (1992) 5S rRNA-encoding genes of the marsupial frog *Gastrotheca riobambae*. Gene 111:235–238

del Pino EM, Alcocer I, Grunz H (1994) Urea is necessary for the culture of embryos of the marsupial frog *Gastrotheca riobambae*, and is tolerated by embryos of the aquatic frog *Xenopus laevis*. Develop Growth Differ 36:73–80

del Pino EM, Sáenz FE, Pérez OD, Brown FD, Ávila M-E, Barragán VA, Haddad N, Paulin-Levasseur M, Krohne G (2002) Lamina-associated polypeptide 2 (LAP2) expression in fish and amphibians. Int J Dev Biol 46:227–234

del Pino EM, Avila M-E, Pérez OD, Benítez M-S, Alarcón I, Noboa V, Moya IM (2004) Development of the dendrobatid frog *Colostethus machalilla*. Int J Dev Biol 48:663–670

del Pino EM, Venegas-Ferrín M, Romero-Carvajal A, Montenegro-Larrea P, Sáenz-Ponce N, Moya IM, Alarcón I, Sudou N, Yamamoto S, Taira M (2007) A comparative analysis of frog early development. Proc Natl Acad Sci USA 104:11882–11888

Duellman WE (2015) Marsupial frogs: *Gastrotheca* and allied genera. Johns Hopkins University Press, Baltimore

Duellman WE, Trueb L (1986) Biology of amphibians. McGraw Hill, New York

Dumont JN (1972) Oogenesis in *Xenopus laevis* (Daudin). I. Stages of oocyte development in laboratory maintained animals. J Morphol 136:153–179

Elinson RP (1987) Fertilization and aqueous development of the Puerto Rican terrestrial-breeding frog, *Eleutherodactylus coqui*. J Morphol 193:217–224

Elinson RP (2001) Direct development: an alternative way to make a frog. Genesis 29:91–95

Elinson RP (2009) Nutritional endoderm: a way to breach the holoblastic-meroblastic barrier in tetrapods. J Exp Zool Mol Dev Evol 312:526–532

Elinson RP (2013) Metamorphosis in a frog that does not have a tadpole. Curr Top Dev Biol 103:259–276

Elinson RP, Beckham Y (2002) Development in frogs with large eggs and the origin of amniotes. Zoology 105:105–117

Elinson RP, del Pino EM (1985) Cleavage and gastrulation in the egg-brooding, marsupial frog, *Gastrotheca riobambae*. J Embryol Exp Morphol 90:223–232

Elinson RP, del Pino EM (2012) Developmental diversity of amphibians. WIREs Dev Biol 1:345–369

Elinson RP, Fang H (1998) Secondary coverage of the yolk by the body wall in the direct developing frog, *Eleutherodactylus coqui*: an unusual process for amphibian embryos. Dev Genes Evol 208:457–466

Elinson RP, del Pino EM, Townsend DS, Cuesta FC, Eichhorn P (1990) A practical guide to the developmental biology of terrestrial-breeding frogs. Biol Bull 179:163–177

Ewald AJ, Peyrot SM, Tyszka JM, Fraser SE, Wallingford JB (2004) Regional requirements for Dishevelled signaling during *Xenopus* gastrulation: separable effects on blastopore closure, mesendoderm internalization and archenteron formation. Development 131:6195–6209

Frost DR (2018) Amphibian species of the world: an online reference. Version 6 (04 June, 2018). Electronic database. American Museum of Natural History, New York. http://research.amnh.org/herpetology/amphibia/index.html

Gatherer D, del Pino EM (1992) Somitogenesis in the marsupial frog *Gastrotheca riobambae*. Int J Dev Biol 36:283–291

Gentsch GE, Monteiro RS, Smith JC (2017) Cooperation between T-Box factors regulates the continuous segregation of germ layers during vertebrate embryogenesis. Curr Top Dev Biol 122:117–159

Green J (2009) The animal cap assay. Methods Mol Biol 127:1–13

Gurdon JB (1987) Embryonic induction—molecular prospects. Development 99:285–306

Haddad C, Prado C (2005) Reproductive modes in frogs and their unexpected diversity in the Atlantic Forest of Brazil. Bioscience 55:207–217

Harland R, Gerhart J (1997) Formation and function of Spemann's organizer. Annu Rev Cell Dev Biol 13:611–667

Harland RM, Grainger RM (2011) *Xenopus* research: metamorphosed by genetics and genomics. Trends Genet 27:507–515

Hausen P, Riebesell M (1991) The early development of *Xenopus laevis*. Springer, Berlin

Hedrick JL (2008) Anuran and pig egg zona pellucida glycoproteins in fertilization and early development. Int J Dev Biol 52:683–701

Heinicke MP, Lemmon AR, Lemmon EM, McGrath K, Hedges SB (2018) Phylogenomic support for evolutionary relationships of New World direct-developing frogs (Anura: Terraranae). Mol Phylogenet Evol 118:145–155

Hervas F, Torres KP, Montenegro-Larrea P, del Pino EM (2015) Development and gastrulation in *Hyloxalus vertebralis* and *Dendrobates auratus* (Anura: Dendrobatidae). Amph Reptil Conserv 8:121–135 (e90)

Hervas-Sotomayor F, del Pino EM (2013) Desarrollo neural, somitogénesis y morfología interna de los embriones de *Hyloxalus vertebralis* y *Dendrobates auratus* (Anura: Dendrobatidae). Rev Ecuat Med Cienc Biol 34:99–112

Holtfreter J (1933) Nachweis der Induktionsfähigkeit abgetöteter Keimteile: Isolations– und Transplantationsversuche. Wilhelm Roux Arch Entwickl Mech Org 128:584–633

Jones EA, Woodland HR (1989) Spatial aspects of neural induction in *Xenopus laevis*. Development 107:785–791

Jones RE, Gerrard AM, Roth JJ (1973) Estrogen and brood pouch formation in the marsupial frog *Gastrotheca riobambae*. J Exp Zool 184:177–184

Karadge U, Elinson RP (2013) Characterization of the nutritional endoderm in the direct developing frog *Eleutherodactylus coqui*. Dev Genes Evol 223:351–362

Karavanov AA, Saint-Jeannet J-P, Karavanova I, Taira M, Dawid IB (1996) The LIM homeodomain protein Lim-1 is widely expressed in neural, neural crest and mesoderm derivatives in vertebrate development. Int J Dev Biol 40:453–461

Keller R, Danilchik M (1988) Regional expression, pattern and timing of convergence and extension during gastrulation of *Xenopus laevis*. Development 103:193–209
Keller R, Winklbauer R (1992) Cellular basis of amphibian gastrulation. Curr Top Dev Biol 27:39–89
Keller RE, Danilchik M, Gimlich R, Shih J (1985) The function and mechanism of convergent extension during gastrulation of *Xenopus laevis*. J Embryol Exp Morphol 89:185–209
Kloc M, Etkin LD (1995) Two distinct pathways for the localization of RNAs at the vegetal cortex in *Xenopus* oocytes. Development 121:287–297
Kloc M, Bilinski S, Etkin LD (2004) The Balbiani body and germ cell determinants: 150 years later. Curr Top Dev Biol 59:1–36
Kodjabachian L, Karavanov AA, Hikasa H, Hukriede NA, Aoki T, Taira M, Dawid IB (2001) A study of Xlim1 function in the Spemann-Mangold organizer. Int J Dev Biol 45:209–218
Lei L, Spradling AC (2016) Mouse oocytes differentiate through organelle enrichment from sister cyst germ cells. Science 352:95–99
Macgregor HC, del Pino EM (1982) Ribosomal gene amplification in multinucleate oocytes of the egg brooding hylid frog *Flectonotus pygmaeus*. Chromosoma 85:475–488
Macgregor HC, Kezer J (1970) Gene amplification in oocytes with 8 germinal vesicles from the tailed frog *Ascaphus truei* Stejneger. Chromosoma 29:189–206
Mangold O (1933) Über die Induktionsfähigkeit der verschiedenen Bezirke der Neurula von Urodelen. Naturwissenschaften 21:761–766
Matova N, Cooley L (2001) Comparative aspects of animal oogenesis. Dev Biol 231:291–320
Montenegro-Larrea P, del Pino EM (2011) La gastrulación de *Epipedobates anthonyi* (Anura: Dendrobatidae). Rev Ecuat Med C Biol 32:24–32
Moya IM, Alarcón I, del Pino EM (2007) Gastrulation of *Gastrotheca riobambae* in comparison with other frogs. Dev Biol 304:467–478
Nath K, Elinson RP (2007) RNA of AmVegT, the axolotl orthologue of the *Xenopus* meso-endodermal determinant, is not localized in the oocyte. GEP 7:197–201
Nath K, Boorech JL, Beckham YM, Burns MM, Elinson RP (2005) Status of RNAs, localized in *Xenopus laevis* oocytes, in the frogs *Rana pipiens* and *Eleutherodactylus coqui*. J Exp Zool Mol Dev Evol 304B:28–39
Nelsen OE (1953) Comparative embryology of the vertebrates. McGraw Hill, New York
Newport J, Kirschner M (1982) A major developmental transition in early *Xenopus* embryos: I. Characterization and timing of cellular changes at the midblastula stage. Cell 30:675–686
Nieuwkoop PD (1955) Independent and dependent development in the formation of the central nervous system in amphibians; a review of experimental analysis. Exp Cell Res 3:262–273
Nieuwkoop PD (1969) The formation of the mesoderm in urodelean amphibians. I. Induction by the endoderm. Wilhelm Roux Arch Entwickl Mech Org 162:341–373
Nieuwkoop PD, Faber J (1994) Normal table of *Xenopus laevis* (Daudin). Garland Publishing, New York
Nina L, del Pino EM (1977) Estructura histológica del ovario del sapo *Eleutherodactylus unistrigatus* y observaciones sobre el desarrollo embrionario. Rev Univ Catolica Ecuador 5:31–41
Ninomiya H, Zhang Q, Elinson RP (2001) Mesoderm formation in *Eleutherodactylus coqui*: body patterning in a frog with a large egg. Dev Biol 236:109–123
Noble GK (1927) The value of life history data in the study of the evolution of the Amphibia. Ann N Y Acad Sci 30:31–128
Ogielska M, Kotusz A, Augustyńska R, Ihnatowicz J, Paśko L (2013) A stockpile of ova in the grass frog *Rana temporaria* is established once for the life span. Do ovaries in amphibians and in mammals follow the same evolutionary strategy? Anat Rec 296:638–653
Pepling ME, de Cuevas M, Spradling AC (1999) Germline cysts: a conserved phase of germ cell development? Trends Cell Biol 9:257–262

Pérez O, Benítez M-S, Nath K, Heasman J, del Pino EM, Elinson RP (2007) Comparative analysis of *Xenopus* VegT, the meso-endodermal determinant, identifies an unusual conserved sequence. Differentiation 75:559–565

Radice GP, Neff AW, Shim YH, Brustis J-J, Malacinski GM (1989) Developmental histories in amphibian myogenesis. Int J Dev Biol 33:325–343

Romero-Carvajal A, Sáenz-Ponce N, Venegas-Ferrín M, Almeida-Reinoso D, Lee C, Bond J, Ryan MJ, Wallingford JB, del Pino EM (2009) Embryogenesis and laboratory maintenance of the foam-nesting túngara frogs, genus *Engystomops* (= *Physalaemus*). Dev Dyn 238:1444–1454

Roszko I, Sawada A, Solnica-Krezel L (2009) Regulation of convergence and extension movements during vertebrate gastrulation by the Wnt/PCP pathway. Semin Cell Dev Biol 20:986–997

Sáenz-Ponce N, Mitgutsch C, del Pino EM (2012a) Variation in the schedules of somite and neural development in frogs. Proc Natl Acad Sci USA 109:20503–20507

Sáenz-Ponce N, Santillana Ortiz JD, del Pino EM (2012b) The gastrocoel roof plate in embryos of different frogs. Differentiation 83:S62–S66

Salazar-Nicholls MJ, del Pino EM (2015) Early development of the glass frogs *Hyalinobatrachium fleischmanni* and *Espadarana callistomma* (Anura: Centrolenidae) from cleavage to tadpole hatching. Amphib Reptil Conserv 8:89–106 (e88)

Saxén L (1989) Neural induction. Int J Dev Biol 33:21–48

Scharf SR, Rowning B, Wu M, Gerhart JC (1989) Hyperdorsoanterior embryos from *Xenopus* eggs treated with D_2O. Dev Biol 134:175–188

Schmid M, Steinlein C, Bogart JP, Feichtinger W, Haaf T, Nanda I, del Pino EM, Duellman WE, Hedges SB (2012) The Hemiphractid frogs phylogeny, embryology, life history, and cytogenetics. Cytogenet Genome Res 138:69–384

Schmid M, Steinlein C, Haaf T, Feichtinger W, Guttenbach M, Bogart JP, Gruber SL, Kasahara S, Kakampuy W, del Pino EM, Carrillo A-B, Romero-Carvajal A, Mahony M, King M, Duellman WE, Hedges SB (2018) The Arboranan frogs: evolution, biology, cytogenetics. Cytogenet Genome Res 155(1–4):1–326

Shi D, Delarue M, Darribere T, Riou J-F (1987) Experimental analysis of the extension of the dorsal marginal zone in *Pleurodeles waltl* gastrulae. Development 100:147–161

Shindo A (2017) Models of convergent extension during morphogenesis. WIREs Dev Biol 7:e293

Shook DR, Kasprowicz EM, Davidson LA, Keller R (2018) Large, long range tensile forces drive convergence during *Xenopus* blastopore closure and body axis elongation. Elife 7:e26944

Solnica-Krezel L (2005) Conserved patterns of cell movements during vertebrate gastrulation. Curr Biol 15:R213–R228

Solnica-Krezel L, Sepich DS (2012) Gastrulation: making and shaping germ layers. Annu Rev Cell Dev Biol 28:687–717

Spannhof I, Spannhof L (1972) Beobachtungen zur Brutbiologie und Larvenentwicklung von *Gastrotheca marsupiata*. Wiss Zeit Univ Rostock Math Naturwiss Reihe 20:97–104

Spemann H, Mangold H (1924) Über Induktion von Embryonalanlagen durch Implantation artfremder Organisatoren. Arch Mikroskop Anat Entwicklungsmech 100:599–638

Sudarwati S, Nieuwkoop PD (1971) Mesoderm formation in the anuran *Xenopus laevis* (Daudin). Wilhelm Roux Arch Entwickl Mech Org 166:189–204

Sudou N, Garcés-Vásconez A, López-Latorre MA, Taira M, del Pino EM (2016) Transcription factors Mix1 and VegT, relocalization of vegt mRNA, and conserved endoderm and dorsal specification in frogs. Proc Nat Acad Sci USA 113:5628–5633

Taira M, Jamrich M, Good PJ, Dawid IB (1992) The LIM domain-containing homeo box gene Xlim-1 is expressed specifically in the organizer region of *Xenopus* gastrula embryos. Genes Dev 6:356–366

Taira M, Otani H, Jamrich M, Dawid IB (1994) Expression of the LIM class homeobox gene Xlim-1 in pronephros and CNS cell lineages of *Xenopus* embryos is affected by retinoic acid and exogastrulation. Development 120:1525–1536

Takeuchi M, Takahashi M, Okabe M, Aizawa S (2009) Germ layer patterning in bichir and lamprey; an insight into its evolution in vertebrates. Dev Biol 332:90–102

Townsend DS, Stewart MM (1985) Direct development in *Eleutherodactylus coqui* (Anura: Leptodactylidae): a staging table. Copeia 1985:423–436

Townsend DS, Stewart MM, Harvey Pough F, Brussard PF (1981) Internal fertilization in an oviparous frog. Science 212:469–471

Ulmer B, Tingler M, Kurz S, Maerker M, Andre P, Mönch D, Campione M, Deißler K, Lewandoski M, Thumberger T, Schweickert A, Fainsod A, Steinbeißer H, Blum M (2017) A novel role of the organizer gene *Goosecoid* as an inhibitor of Wnt/PCP-mediated convergent extension in *Xenopus* and mouse. Sci Rep 7:43010

Vargas A, del Pino EM (2017) Analysis of cell size in the gastrula of ten frog species reveals a correlation of egg with cell sizes, and a conserved pattern of small cells in the marginal zone. J Exp Zool Mol Dev Evol 328B:88–96

Vargas-Salinas F, Torres OL (2013) *Agalychnis spurrelli* (Boulenger 1913) Rana arbórea planeadora. Catálogo Anfibios Reptil Colomb 1:19–25

Venegas-Ferrín M, Sudou N, Taira M, del Pino EM (2010) Comparison of Lim1 expression in embryos of frogs with different modes of reproduction. Int J Dev Biol 54:195–202

Wallingford JB, Fraser SE, Harland RM (2002) Convergent extension: the molecular control of polarized cell movement during embryonic development. Dev Cell 2:695–706

Warkman AS, Krieg PA (2007) *Xenopus* as a model system for vertebrate heart development. Semin Cell Dev Biol 18:46–53

Warne RW, Catenazzi A (2016) Pouch brooding marsupial frogs transfer nutrients to developing embryos. Biol Lett 12:20160673

Wassersug RJ, Duellman WE (1984) Oral structures and their development in egg-brooding hylid frog embryos and larvae: evolutionary and ecological implications. J Morphol 182:1–37

Wiens JJ, Kuczynski CA, Duellman WE, Reeder TW (2007) Loss and re-evolution of complex life cycles in marsupial frogs: does ancestral trait reconstruction mislead? Evolution (N Y) 61:1886–1899

Winning RS, Sargent TD (1994) *Pagliaccio*, a member of the *Eph* family of receptor tyrosine kinase genes, has localized expression in a subset of neural crest and neural tissues in *Xenopus laevis* embryos. Mech Dev 46:219–229

Ziermann JM, Diogo R (2014) Cranial muscle development in frogs with different developmental modes: direct development versus biphasic development. J Morphol 275:398–413

Chapter 17
Evolution and Regulation of Limb Regeneration in Arthropods

Yuichiro Suzuki, Jacquelyn Chou, Sarah L. Garvey, Victoria R. Wang, and Katherine O. Yanes

Abstract Regeneration has fascinated both scientists and non-scientists for centuries. Many organisms can regenerate, and arthropod limbs are no exception although their ability to regenerate is a product shaped by natural and sexual selection. Recent studies have begun to uncover cellular and molecular processes underlying limb regeneration in several arthropod species. Here we argue that an evo-devo approach to the study of arthropod limb regeneration is needed to understand aspects of limb regeneration that are conserved and divergent. In particular, we argue that limbs of different species are comprised of cells at distinct stages of differentiation at the time of limb loss and therefore provide insights into regeneration involving both stem cell-like cells/precursor cells and differentiated cells. In addition, we review recent studies that demonstrate how limb regeneration impacts the development of the whole organism and argue that studies on the link between local tissue damage and the rest of the body should provide insights into the integrative nature of development. Molecular studies on limb regeneration are only beginning to take off, but comparative studies on the mechanisms of limb regeneration across various taxa should not only yield interesting insights into development but also answer how this remarkable ability evolved across arthropods and beyond.

17.1 Introduction

For centuries, regeneration has captured the imagination of both scientists and non-scientists alike. Regeneration is featured in Ancient Greek mythology; after Prometheus stole fire to give it to the humans, the gods punished him by having his liver eaten by an eagle every day. The liver regenerated daily, condemning Prometheus to eternal punishment. In another myth, the serpentine water monster Hydra had the ability to regenerate its heads; when one was cut off, two heads would grow

Y. Suzuki (✉) · J. Chou · S. L. Garvey · V. R. Wang · K. O. Yanes
Department of Biological Sciences, Wellesley College, Wellesley, MA, USA
e-mail: ysuzuki@wellesley.edu

W. Tworzydlo, S. M. Bilinski (eds.), *Evo-Devo: Non-model Species in Cell and Developmental Biology*, Results and Problems in Cell Differentiation 68,
https://doi.org/10.1007/978-3-030-23459-1_17

back. The ability of organisms to regenerate lost body parts was noted by Aristotle, and numerous accounts of regeneration have since been documented. During the twentieth century, arthropod regeneration was extensively studied at the anatomical and cellular level (Maruzzo et al. 2005). However, in the 1980s, focus shifted away from regeneration to normal development as tools became available to probe development at the molecular level. With the emergence of molecular tools in non-model species and rapid advances in stem cell biology, there has been a renewed interest in regenerative biology in recent years. While arthropod limb regeneration has only been explored at the molecular level in a few species, such as crabs, crickets, beetles, and fruit flies (Das and Durica 2013; Hariharan and Serras 2017; Nakamura et al. 2008a; Shah et al. 2011), it is now possible to take an evolutionary developmental biology approach to study limb regeneration in a variety of arthropod species. In this chapter, we focus on some of the recent advances in our understanding of limb regeneration in crustaceans and insects where limb regeneration has been best studied. We argue that the diversity of limb types and differences in regenerative abilities provide opportunities for understanding how regenerative abilities have diverged across this group. In addition, we argue that the interface between regeneration and developmental physiology provides a framework for understanding organismal development as an integrated system that coordinates the growth of various body parts. Finally, we argue that while recent advances in *Drosophila* have begun to shed light on growth coordination mechanisms, these mechanisms are evolutionarily derived, necessitating studies in non-model systems. A summary of key questions we identify is outlined in Fig. 17.1.

17.2 Overview of the Regeneration Process

Regeneration occurs by two different mechanisms: epimorphosis or morphallaxis. Epimorphosis is characterized by cell division to recreate lost structures. A classic example of epimorphic regeneration is observed in amputated urodele amphibian limbs, where cells dedifferentiate to form a structure known as the blastema and continue replicating to generate enough cells to reform the limb. After the blastema becomes large enough, the cells redifferentiate and regrow the appendage (Brockes 1997). In contrast to epimorphosis, cell division is not involved in morphallaxis. Instead, the remaining tissue is remodeled to rebuild the missing structures, resulting in a smaller but fully functional organism (Agata et al. 2007). For example, in *Hydra*, cell division was shown to be unnecessary for regeneration as inhibition of cell division through gamma radiation did not prevent regeneration of bisected animals (Hicklin and Wolpert 1973). However, processes of regeneration cannot be neatly categorized under these two terms. Evidence of both epimorphosis and morphallaxis has been reported for flatworm regeneration as well as crustacean limb regeneration (Agata et al. 2007; Alwes et al. 2016). Nevertheless, much of arthropod limb regeneration involves the growth of additional tissues, so arthropod limb regeneration primarily undergoes epimorphosis.

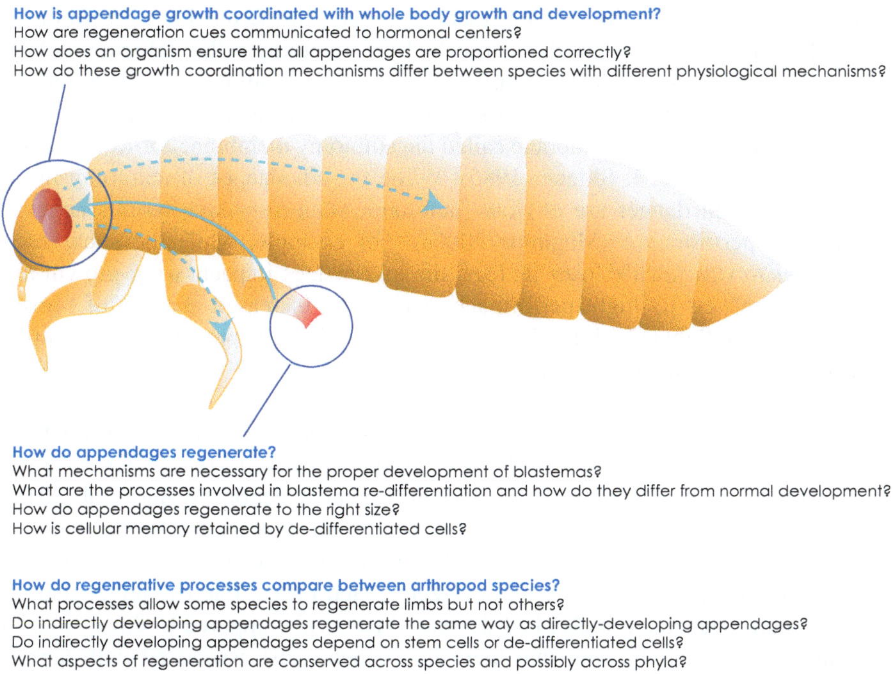

Fig. 17.1 Major questions in arthropod limb regeneration

A unique feature of arthropod limb regeneration is that they must undergo molts for external changes in regenerating structures (Das 2015). This does not mean that regeneration only occurs during the molt. On the contrary, substantial cell movement and rearrangements occur during the intermolt period. However, external changes only manifest after a molt as the cuticle cannot grow once it is established.

17.3 Adaptive Significance of Limb Regeneration in Arthropods

A survey by Maruzzo and Bortolin (2013) of arthropod species shows that not all species are capable of regenerating their limbs to the same extent. This begs the question, what is the adaptive significance of limb regeneration? An excellent review on this issue has been written previously (Maginnis 2006a), so here we provide a brief discussion focusing on arthropod limb regeneration.

In some arthropods, anatomical modifications have evolved to enable the rapid loss of appendages when trapped (Maruzzo and Bortolin 2013). These events occur at the joints, which form weak points that enable the legs to be quickly lost in the event of a predatory attack or unsuccessful shedding of the exoskeleton during a molt (Maginnis 2006b). In addition, many decapod crustaceans have the ability to cast off their legs using a mechanism called the autotomy reflex (Hopkins 1993). This is an involuntary process that involves a specialized autotomy muscle that can quickly break off the leg at a preformed weakened breakage plane. Once the leg is broken off, a specialized membrane called the autotomy membrane forms to seal the opening (Hopkins 1993; Skinner 1985). While the advantages of autotomy in the face of imminent danger are obvious, autotomies can also have serious negative impacts on performance (Maginnis 2006a). For example, while the stick insect *Sipyloidea sipylus* readily loses its legs, the loss of legs impacts wing growth and negatively impacts its flight performance leading to increased chances of "crash" landings after a flight (Maginnis 2006b). In addition, loss of appendages may impact foraging and reproductive success (Brautigam and Persons 2003; Maginnis 2006a; Uetz et al. 1996).

Regenerating an appendage may partially offset such costs incurred through the loss of an appendage. A recent study of the beetle *Harmonia axyridis* demonstrated that regeneration may have evolved through sexual selection (Wang et al. 2015). When females were presented with regenerated or non-regenerated beetles, females preferred to mate with regenerated beetles. Fecundity and egg fertility were also reduced in females that mated with non-regenerated males. Interestingly, offspring sired by non-regenerated males had decreased survival compared to those sired by regenerated males, indicating that there is a genetic cost associated with lack of regeneration (Wang et al. 2015). Taken together, these results indicate that male fitness is reduced when legs cannot regenerate and that the ability to regenerate is associated with higher reproductive success.

While there are clear advantages to being able to regenerate, certain costs have also been documented. Regeneration of limbs is likely energetically expensive, and trade-offs have been documented between regeneration and reproduction [reviewed in Maginnis (2006a)]. In addition, regenerated appendages are often imperfect and have performance costs. The size of the regenerated appendage may be smaller, reducing foraging efficiency, reproductive success, and survival. Finally, as we will discuss later, regeneration can impact developmental timing. Developmental delays can increase predation risks. Changes to the duration of juvenile stages also impact reproductive success for species that must reach sexual maturity at a specific time of the year. The ability to regenerate is thus likely to be shaped in a species-specific manner that takes into account various trade-offs underlying appendage regeneration.

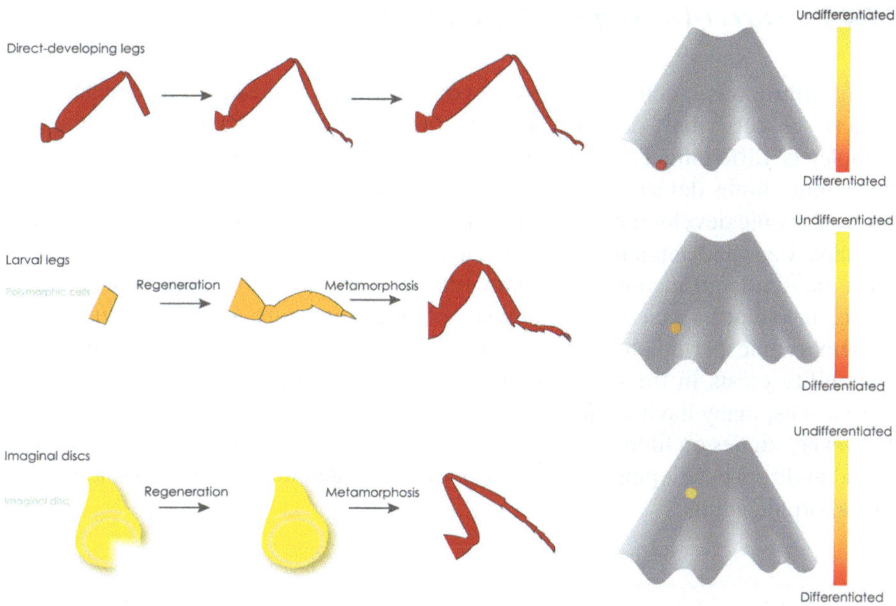

Fig. 17.2 The diversity of limb types in arthropods (left) and Waddington's epigenetic landscape model [adopted from Waddington (1957)], showing the distinct starting points of cells at the time of autotomy or limb ablation (right). (Top) Direct-developing legs in species that hatch with completely differentiated legs or differentiated legs found in adults. (Middle) Larval legs in species that undergo indirect development. (Bottom) Imaginal discs found in some holometabolous insects. These are precursors of limbs that differentiate during metamorphosis. In the epigenetic landscape, the balls represent hypothetical starting points of cells in the respective limbs. Color scale indicates the degree of differentiation (yellow = undifferentiated; red = differentiated). See text for more explanation

17.4 The Diversity of Limb Developmental Types in Arthropods

For a proper understanding of limb regeneration in arthropods, it is critical that we clarify which cell types make up the limb that is regenerating. While limbs in some arthropods fully differentiate during embryogenesis, limbs in other arthropods do not differentiate until later in postembryonic development (Fig. 17.2). Such heterochronic shifts in development likely impact the molecular processes necessary for regeneration. Thus, it is important that we clearly define the developmental stage of the limbs at the time of limb loss or amputation.

17.4.1 Direct-Developing/Differentiated Limbs

Many arthropods, such as spiders, several crustacean species, and ametabolous insects—insects that do not undergo metamorphosis—have limbs that are completely differentiated when the juveniles hatch out of the eggs. These direct-developing limbs develop adult-like morphologies during embryogenesis, and their postembryonic development primarily involves an increase in size without a change in shape. Legs and antennae of newly hatched nymphs of hemimetabolous insects—those that undergo incomplete metamorphosis—also resemble their adult counterparts. In some species, the number of leg segments increases during the postembryonic period, but much of the leg differentiates during embryogenesis. Variability exists in the degree of limb differentiation at the time of hatching in crustaceans, many having a juvenile nauplius stage with distinct limb morphologies. However, studies on limb regeneration in crustaceans have primarily focused on legs found in direct-developing crustaceans or differentiated limbs that are found in post-metamorphic adults.

17.4.2 Larval Limbs

Holometabolous insects—those that undergo complete metamorphosis with a larval, pupal, and adult stages—contain some of the most specious taxa. While important, the origin of holometaboly—in particular, the origin of the larval stage—remains a topic of contention (Erezyilmaz 2006; Konopova et al. 2011; Truman and Riddiford 1999). Over the last century, many different theories have been proposed to explain the origin of the larval form, from which two major lines of thought have emerged. One proposes that the larval and pupal stages are modified nymphal stages, that is, they are simply morphological alterations of the hemimetabolous nymphal stage. The other proposes that the larva is, in fact, an extension of the embryonic phase and that the pupal phase is equivalent to the nymphal stage. Unfortunately, without a consensus view on the origin of the larval body, it remains difficult to assess the degree of differentiation of cells found in the larval body.

Regardless of the theories, it is likely that larval limbs are neither identical to embryonic limbs (Švácha 1992) nor completely differentiated. They likely represent a partially differentiated state. Studies on the tobacco hornworm, *Manduca sexta*, have demonstrated that the larval leg of this lepidopteran species is comprised of three distinct cell populations: imaginal cells, polymorphic cells, and larval-specific cells (Tanaka and Truman 2005). Imaginal cells are precursor cells that proliferate extensively during metamorphosis and make up much of the adult leg. In contrast, polymorphic cells contribute to both the larval and adult appendages. Finally, the larval-specific cells undergo apoptosis at the end of the larval stage and do not contribute to the adult appendages. Different holometabolous insect species appear to utilize these cells to varying degrees. Tenebrionid beetle larval legs, for example, appear to be comprised primarily of polymorphic cells (Truman and Riddiford 2002). Despite the distinct terminology, both imaginal cells and polymorphic cells likely represent precursor cells that have not completed differentiation. This is because both types of cells proliferate dramatically during metamorphosis to create an adult appendage with a completely different morphology. Thus, regardless of how the larval stage evolved, larval appendages likely represent a heterochronic shift that led to the retention of stem cell-like precursor cells in the appendages.

17.4.3 Imaginal Discs

Imaginal discs are internal precursors of appendages. Imaginal discs give rise to adult appendages and do not contribute to larval structures. Typically, imaginal discs form through invagination of the epidermis late in the larval stages. However, in some species that undergo rapid development or grow large appendages—such as lepidopterans and dipterans, including *Drosophila*—imaginal discs form during embryogenesis (Švácha 1992). These early developing structures have served as the model for studying imaginal disc regeneration, and their cells have been suggested to exhibit stem cell-like properties (McClure and Schubiger 2007; Wei et al. 2000).

17.4.4 Waddington's Epigenetic Landscape

Conrad H. Waddington (1957) developed the epigenetic landscape as a metaphor to illustrate the various paths a cell might take as it differentiates. The model depicts a hill with a single valley at the top that branches into a number of valleys. A ball at the top of the hill represents a pluripotent or totipotent stem cell. Each individual valley represents a different path that can be taken by the cell as it differentiates into distinct cell types (Waddington 1957). We can visualize the various limb types as cells at different stages of differentiation. A direct-developing limb would be represented by a ball at the bottom of the hill (a differentiated cell), whereas imaginal cells and larval appendages would have cells that can be visualized as balls still rolling down the hill (precursor cells) (Fig. 17.2). The cells, therefore, have distinct starting points as they begin to undergo regeneration.

The diversity of limb development provides an interesting opportunity to explore how the cellular and molecular processes of regeneration differ between distinct types of limbs. Direct-developing appendages of crustaceans and hemimetabolous insects offer insights into how differentiated cells remaining at the site of the amputation dedifferentiate and regenerate an appendage. In contrast, the larval appendages of holometabolous insects may provide an opportunity to investigate stem cell-like/progenitor cell-based regeneration. Understanding the distinct origin of cell types that make up the regenerated appendage is critical to our understanding of the molecular nature of regeneration, especially regarding processes involved in the initial steps of regeneration when a blastema forms.

17.5 Steps Involved in Regeneration

The process of limb regeneration can be divided into three key steps, which have counterparts in vertebrate limb regeneration: wound healing, blastema formation, and repatterning. Here we provide a general overview of the movement, proliferation, and identities of cells involved in each of these steps.

17.5.1 Cellular Basis of Wound Healing

Wound healing is one of the first steps in regeneration. Very soon after limb amputation, hemolymph accumulates at the wound site and hardens to form a scab (Hopkins 1993; Truby 1983). The muscles at the site of regeneration then degenerate, and hemocytes accumulate underneath the scab and flatten (Alwes et al. 2016; Mito et al. 2002; Truby 1983). At the same time, the epidermal cells adjacent to the wound site undergo an activation phase, which is characterized by the enlargement of cells and their nuclei and the detachment of the cells from the overlying cuticle. Subsequently, the epidermal cells migrate under the flattened hemocytes to eventually create a continuous epidermal layer underneath the scab. Only a few cells adjacent to the wound site proliferate, and much of the wound healing appears to rely on cell migration (Alwes et al. 2016; Hopkins 1993; Mito et al. 2002; Truby 1983). In vertebrate limbs, wound healing has been demonstrated to provide critical signals for the next step of regeneration, the formation of the blastema (Chablais and Jazwinska 2010; Fernando et al. 2011; Kumar et al. 2007; Simkin et al. 2015; Stocum 2017).

17.5.2 Origin and Fate of Blastema Cells

A blastema is comprised of a layer of epidermal cells surrounding a mesodermal core (Alwes et al. 2016). The formation of the blastema is characterized by the thickening of the wound epidermis and the spread of epidermal cell activation signals that ultimately lead to the formation of a layer of free epidermal cells (Mito et al. 2002; Truby 1983). The blastema grows rapidly as the mesodermal cells within it proliferate to form precursor cells that will regenerate the lost appendage (Mito et al. 2002).

The formation of blastema is thought to be a critical phase of regeneration (Endo et al. 2004). Human adult amputees and other mammals are unable to regenerate their appendages as scar tissue forms in place of a blastema, preventing regeneration (Jaźwińska and Sallin 2016; McCusker et al. 2015). Thus, the development of scar-free wound healing and the blastema has attracted much interest among researchers working on regeneration in a variety of systems (Londono et al. 2018; Muneoka et al. 2008; Newmark and Alvarado 2002; Owlarn and Bartscherer 2016; Seifert et al. 2012; Simkin et al. 2015).

The nature of cells contributing to the blastema has been closely debated in the field of limb regeneration. Distinct cell types appear to contribute to the regenerated legs in different species of salamanders. In the Mexican axolotl *Ambystoma mexicanum*, Pax7-positive satellite cells give rise to regenerated muscles, whereas in the red-spotted newt *Notophthalmus viridescens*, Pax7-negative myofibers dedifferentiate to give rise to the regenerating muscle cells (Sandoval-Guzman et al. 2014). In the Japanese fire-bellied newt *Cynops pyrrhogaster*, Pax7-positive satellite cells contribute to the regeneration of larval legs, whereas after metamorphosis, differentiated muscle fiber cells give rise to regenerated legs (Tanaka et al. 2016). Thus, at least in this species, the extent to which tissues have matured determines the cell type that gives rise to the regenerated limb. Taken together, distinct cell types appear to contribute to the regenerated tissues in a species- and stage-specific manner (Kragl et al. 2009; Sandoval-Guzman et al. 2014).

Given the diversity of cell types comprising the appendages of regenerating arthropods, further studies are needed to assess whether distinct cell types contribute to blastema in different species. Thus far, studies done in the amphipod *Parhyale hawaiensis* have provided insights into the origin and fates of cells that make up its blastema (Alwes et al. 2016; Konstantinides and Averof 2014). In this species, regeneration of leg segments occurs rapidly within the same instar, although the size of the regenerated leg is somewhat reduced compared to the unablated leg. The fast regeneration time along with the ability to fluorescently tag cells has allowed the system to be visualized real time, by gluing the regenerating leg to a coverslip and directly imaging the regeneration process under a microscope (Alwes et al. 2016).

At least in the case of *Parhyale* leg regeneration, the blastema cells appear to derive from cell populations locally present in the leg stump and not from circulating cells (Konstantinides and Averof 2014). The epidermal cells and the mesodermal cells within the blastema derive from the ectoderm and mesoderm, respectively, and stay separate throughout the process of regeneration (Alwes et al. 2016; Konstantinides and Averof 2014). By tracking the fates of fluorescently labeled cells, Konstantinides and Averof (2014) demonstrated that the developmental potential of cells giving rise to the regenerated leg is restricted. They are likely to be either unipotent or at most multipotent and do not transdifferentiate from one germ layer to another.

No specialized stem cells within the remaining leg stump appear to contribute to the epidermal cells of the blastema. Rather, it appears all epidermal cells can proliferate to recreate the legs, most likely reflecting the ability of amphipods to undergo an indefinite number of molts and continuous growth. The positional information of the epidermal cells appears to be somewhat flexible; the daughter cells of epidermal progenitor cells within one leg segment may contribute to parts of another leg segment (Alwes et al. 2016).

The mesoderm cells within the blastema originate from preexisting mesoderm cells although the precise identity of these cells remains unclear (Konstantinides and Averof 2014). Konstantinides and Averof (2014) demonstrated that Pax3/7-positive cells associated with muscles contribute to regeneration. These cells resemble vertebrate satellite cells, that is, muscle precursor cells that give rise to differentiated muscle cells, which also express Pax3/7. However, other cells also appear to contribute to the regenerated muscles (Konstantinides and Averof 2014).

A few ectoderm-derived cells remain in the blastema (Alwes et al. 2016). These likely represent glial and neural cells. Although contentious, nerves in some arthropod species have been shown to play roles during regeneration. Damage to the nerve postamputation delays regeneration (Needham 1946), and studies in both insects and crustaceans have suggested that nerves may provide some signal for muscle regeneration (Goss 1969; Needham 1965; Nüesch 1968). In vertebrate leg regeneration, nerves and nerve-dependent signals also play a critical role in the initiation of blastema formation (Endo et al. 2004; Kumar et al. 2007; Singer 1952). However, these nerve-dependent factors appear to be taxon-specific (Garza-Garcia et al. 2010; Geng et al. 2015), so it remains to be seen whether nerves play similar roles during arthropod regeneration.

Whether or not all arthropods use similar mechanisms for regeneration remains to be seen. As mentioned above, some arthropod species have indirectly developing limbs, whereas others, such as *Parhyale*, have directly developing limbs. Comparisons of the cellular basis of regeneration across various species with distinct types of limbs will shed light on this issue.

17.5.3 Overview of Appendage Repatterning

Once a blastema is formed and begins to grow, the appendage undergoes the repatterning phase. During this phase, the appendage segments become reestablished through the constriction of the epidermis, and the lost structures are recreated. Distal segments are established before proximal segments, indicating that repatterning proceeds in the distoproximal direction (Truby 1983). Given that appendage morphology is unique to individual taxa, the molecular mechanisms that recreate each appendage are likely to vary between taxa and even between serially homologous appendages found within one species. It remains to be seen whether the genes that regulate the transition out of the blastema stage are conserved.

The time needed for regeneration varies across species. The amphipod *Parhyale hawaiensis* can regenerate its legs within a week, whereas leg regeneration in the cricket *Gryllus bimaculatus* can take up to 35 days (Alwes et al. 2016; Mito et al. 2002). Regeneration also varies based on the duration of an instar. In some species, repatterning is already well underway by the time the animal reaches the first molt, and a smaller limb with all the segments emerges after ecdysis, the shedding of the cuticle (Hopkins 1993; Konstantinides and Averof 2014; Truby 1983). In other species, only a blastema forms after the first molt (Nakamura et al. 2008b).

17.6 How Regeneration Actually Works: A Look into the Mechanisms Behind the Regeneration

Much of our early understanding about the regulation of insect limb regeneration derived from amputation and grafting experiments using *Drosophila* imaginal discs and cockroach legs (Bryant 1971, 1975; French 1976, 1978; French et al. 1976; Schubiger 1971). For example, grafting together cockroach leg segments cut at different locations along the proximodistal axis leads to intercalary regeneration (reviewed in French et al. 1976). Grafting together leg segments with opposing anteroposterior or dorsoventral axes leads to the formation of extra appendages. This body of work led to the development of various models of regeneration which have been previously reviewed in depth (Campbell and Tomlinson 1995; French 1978; French et al. 1976). Since the mid-1990s, studies have begun to identify key molecular regulators involved in regeneration. Here, we will summarize key findings from *Drosophila*, *Gryllus*, and *Tribolium* (Fig. 17.3).

In contrast, most studies in crustaceans have historically focused on the role of hormones during regeneration due to the lack of molecular tools (until now). We will review the key findings on the role of hormones during crustacean limb and insect imaginal disc regeneration. Although developmental genes and hormones have been studied in different systems, they likely interact to promote regeneration in all arthropod species.

17.6.1 Key Pathways of Regeneration

17.6.1.1 General Overview of JNK Signaling

One of the key signaling pathways that are activated in response to wounding or genetically induced cell death is the activation of the c-Jun N-terminal kinase (JNK) signaling pathway, a mitogen-activated protein kinase (MAPK) pathway. Multiple studies show that the JNK pathway is activated by a signal from an epidermal wound, exposure of cells to cytokines, or environmental stress (Galko and Krasnow 2004; Ip and Davis 1998; Ramet et al. 2002). In *Drosophila*, the core pathway involves four kinases, which are involved in a phosphorylation cascade: the JNKKKK Misshapen, the JNKKK Slipper, the JNKK Hemipterous (Hep), and the JNK Basket (Bsk) (Alfonso-Gonzalez and Riesgo-Escovar 2018). Bsk phosphorylates the AP-1 transcription factor Jra, which together with Dfos/Kayak (the vertebrate homolog of Fos) activates immediate early genes, such as *puckered* (*puc*).

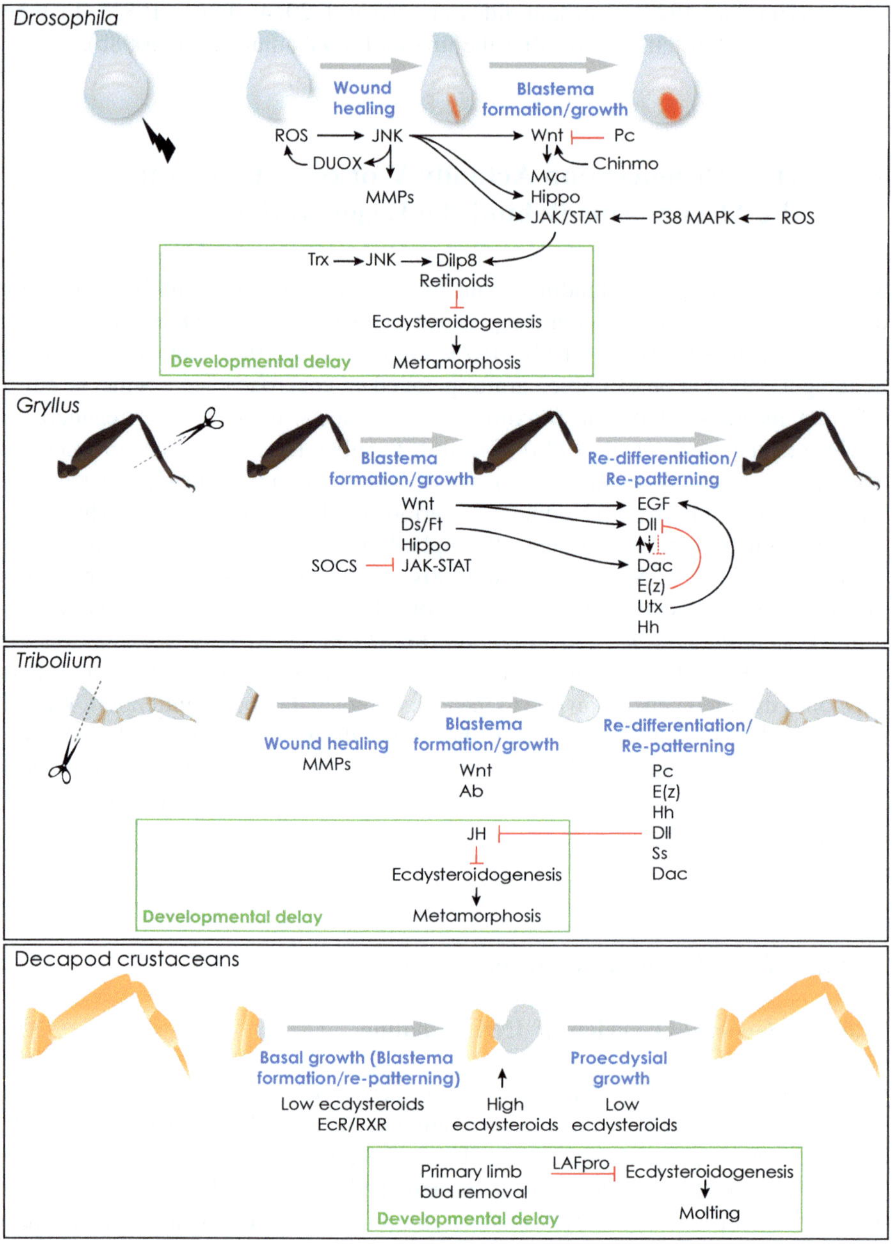

Fig. 17.3 Summary of major molecular regulators during limb regeneration and regeneration-dependent developmental delays in *Drosophila*, *Gryllus*, *Tribolium*, and decapod crustaceans. Dotted arrow represents low expression of Dll; dotted inhibitory interaction represents high expression of Dll. See text for details

17.6.1.2 General Overview of JAK-STAT Signaling

The JAK-STAT signaling pathway is activated when cytokines and growth factors bind membrane receptors associated with JAK tyrosine kinases (Rawlings et al. 2004). Upon ligand binding, the receptors dimerize, bringing together the cytoplasmic domain of the receptor subunits and activating the JAKs through transphosphorylation. The activated JAKs further activate signal-dependent transcription factors (STATs), which enter the nucleus to regulate target gene expression (Rawlings et al. 2004). In *Drosophila*, three related Unpaired (Upd) ligands [Upd1, Upd2 and Upd3] and one receptor [Domeless (Dome)], have been identified (Zeidler and Bausek 2013). Binding of Upd to Dome leads to the phosphorylation of Hopscotch (Hop), a homolog of the mammalian JAK kinase. Hop, in turn, activates the STAT92E transcription factor. As described in Sects. 17.6.2.2, 17.6.3.1 and 17.7.1, the JAK-STAT pathway plays critical roles during regeneration and regeneration-dependent developmental delays.

17.6.1.3 General Overview of Hippo Signaling

The Hippo signaling pathway has emerged as a key regulator of organ size (Pan 2010). The core components of the Hippo signaling pathway are the kinase Hippo, the WW domain-containing protein Salvador (Sav), the NDR family protein kinase Warts (Wts), the adaptor protein Mob-as-tumor-suppressor (Mats), the transcriptional coactivator Yorkie (Yki), and the transcription factor Scalloped (Sd) (Pan 2010). The first four proteins form the Hippo signaling cascade (a kinase cascade), and mutations of any one of these tumor suppressor genes lead to tissue overgrowth. The Hippo-Sav kinase complex phosphorylates and activates the Wts-Mats kinase complex. The activated Wts-Mats kinase complex then phosphorylates Yki, which acts as an oncoprotein (Huang et al. 2005). When Hippo signaling is active and Yki is phosphorylated, it becomes cytoplasmically localized and inhibited. In contrast, when Hippo signaling is inactivated, Yki accumulates in the nucleus and binds transcription factors, such as Sd, to regulate downstream target genes (Goulev et al. 2008; Wu et al. 2008; Zhang et al. 2008). The Hippo signaling pathway is modulated by several other regulators. Merlin (Mer) and Expanded (Ex) promote Wts phosphorylation, thereby acting as tumor suppressors (Hamaratoglu et al. 2006). The loss of cell adhesion proteins, Dachsous (Ds) or Fat (Ft), also leads to a reduction of Hippo signaling activity, leading to overgrowth of tissues (Bennett and Harvey 2006; Blair and McNeill 2018). In addition to its role during normal development, the Hippo signaling pathway has emerged as a pathway that contributes to the proper shape and dimension of regenerating appendages (Bando et al. 2009).

17.6.1.4 General Overview of Wnt Signaling

Wnt signaling regulates a variety of developmental processes by promoting cell proliferation and differentiation (Swarup and Verheyen 2012). In addition, the Wnt signaling pathway has emerged as a critical regulator of regeneration in many metazoan species (Whyte et al. 2012). In the canonical Wnt signaling pathway, Wnt ligand binds to the Frizzled receptor. This leads to the phosphorylation of Disheveled and membrane tethering of Axin via the coreceptor LRP5/6 or the *Drosophila* homolog Arrow (Clevers 2006). When Axin is bound to the membrane, β-catenin, a homolog of *Drosophila* Armadillo (Arm), is stabilized and becomes nuclear-localized, where it promotes transcription of target genes by interacting with Tcf/Lef transcription factor, or the *Drosophila* homolog, Pangolin (Behrens et al. 1996; Brunner et al. 1997; Clevers 2006; Molenaar et al. 1996; Willert and Nusse 1998). In the absence of Wnt, Axin initiates the formation of a destruction complex, which phosphorylates β-catenin/Arm, and ultimately leads to its destruction by proteasomes.

17.6.2 Drosophila melanogaster *Imaginal Disc Regeneration*

Much of the earlier work on *Drosophila* imaginal disc regeneration was done by fragmenting disc tissues and studying their behavior or by implanting these tissues into a host (Bryant 1971, 1975; Schubiger 1971). Advances in molecular tools have led to the development of powerful genetic tools to damage imaginal discs without having to manually manipulate them (Klebes et al. 2005; McClure et al. 2008; Smith-Bolton et al. 2009). Although *Drosophila* imaginal discs are internal precursors of the adult appendages, their regenerative mechanisms can shed light on the inner workings of appendage regeneration in other arthropods. Here we provide a brief summary of the key molecular mechanisms of imaginal disc regeneration (Fig. 17.3). Detailed reviews of imaginal disc regeneration and transdetermination are available elsewhere (Hariharan and Serras 2017; Worley et al. 2012).

17.6.2.1 Wound Healing in Imaginal Discs

The JNK signaling pathway is one of the first pathways to be activated in response to wounding or genetically induced cell death: within a few hours of disc ablation or cell death induction, the JNK effector gene *puc* is activated at the site of injury (Bergantinos et al. 2010; Bosch et al. 2005). Studies on wing imaginal discs have shown that *hep* is necessary for proper wound healing (Bergantinos et al. 2010; Bosch et al. 2005). Similarly, Bsk and Dfos/Kayak are also necessary for wound healing (Bosch et al. 2005). JNK signaling is upregulated in response to reactive oxygen species (ROS), which are produced by dying cells (Santabárbara-Ruiz et al. 2015).

In response to JNK signaling pathway activation, multiple downstream targets are activated to prepare for healing. A likely target of JNK signaling is *matrix metalloproteinase 1* (*Mmp1*), a gene coding for one of the two Matrix metalloproteinases (MMPs) found in *Drosophila* (Llano et al. 2000, 2002). MMPs have been shown to proteolytically modify signaling molecules essential for wound healing (Mott and Werb 2004). At least in the punctured epithelia, the JNK pathway upregulates *Mmp1* expression after wounding, and both Mmp1 and Mmp2 act nonredundantly to regulate reepithelialization of *Drosophila* larvae (Stevens and Page-McCaw 2012). Similarly, in imaginal discs, *Mmp1* expression is induced upon fragmentation (McClure et al. 2008) and likely contributes to wound closure.

17.6.2.2 Blastema Formation in Imaginal Discs

JNK signaling activates several key pathways, which are necessary for blastema formation and growth. The JAK-STAT pathway is a major target of the JNK signaling and p38 MAPK pathways, both of which are activated by ROS produced by the dying cells (Pastor-Pareja et al. 2008; Santabárbara-Ruiz et al. 2015). In *Drosophila*, Upd ligand and other components of the JAK-STAT pathway are activated in regenerating imaginal discs (Katsuyama et al. 2015; Pastor-Pareja et al. 2008; Santabárbara-Ruiz et al. 2015). Disruption of the JAK-STAT pathway results in failure of blastema cell proliferation: regenerative proliferation fails to occur in imaginal discs expressing a dominant negative form of Dome or discs with mutant *hop* alleles (Katsuyama et al. 2015).

Another downstream target of JNK signaling is the Hippo pathway (Grusche et al. 2011; Sun and Irvine 2011). In *Drosophila* imaginal discs, localized apoptosis induction leads to the activation of JNK signaling, which in turn leads to the nuclear localization of the Yki protein. This nuclear localization of Yki appears to be due to changes in the activity of the Hippo signaling pathway (Grusche et al. 2011; Sun and Irvine 2011). During compensatory proliferation of imaginal disc cells following the induction of apoptosis, Yki and Sd regulate the expression target cell proliferation genes, such as *Cyclin E*, which is involved in S-phase entry (Meserve and Duronio 2015). When the functions of Yki or Sd are compromised, cells of damaged imaginal discs fail to proliferate (Meserve and Duronio 2015; Sun and Irvine 2011). Thus, the Hippo signaling pathway appears to play a critical role in regulating the cell cycle during blastema formation.

JNK signaling has also been shown to activate the Wingless (Wg)/(or Wnt) signaling pathway. Wg regulates the expression of Myc, which actively promotes wing disc regeneration by upregulating the cell cycle regulator Cyclin E and promotes blastema cell proliferation (Smith-Bolton et al. 2009). Wg is upregulated quickly when imaginal discs are damaged (Katsuyama et al. 2015; Smith-Bolton et al. 2009) and cooperatively promotes blastema cell proliferation with the JAK-STAT pathway (Katsuyama et al. 2015). Wg is also associated with transdetermination events and can cause blastemas of leg imaginal discs to transdifferentiate into wing imaginal tissues (Maves and Schubiger 1995, 1998). The frequency of transdifferentiation is enhanced by the absence of the Polycomb group (PcG) proteins, which promote the trimethylation of histone H3 on lysine 27 (H3K27), silencing gene expression. Intriguingly, JNK signaling suppresses PcG proteins, indicating that PcG proteins may play critical roles in cell type memory during regeneration and perhaps more generally in the process of dedifferentiation (Lee et al. 2005).

Given the importance of JNK signaling in activating key signaling pathways involved in blastema formation and proliferation, JNK signaling needs to be maintained while the blastema grows. A recent study has demonstrated that a ROS-mediated positive feedback mechanism exists to ensure prolonged JNK signaling activity (Khan et al. 2017). Specifically, JNK signaling activates the expression of *moladietz* (*mol*), which encodes the Dual oxidase maturation factor DuoxA/NIP. DuoxA/NIP in turn upregulates NADPH dual oxidase (DUOX), which catalyzes the formation of ROS, stimulating JNK signaling. Taken together, sustained production of ROS appears to play critical roles in maintaining JNK signaling to ensure proper regeneration of imaginal discs (Khan et al. 2017).

17.6.2.3 Age-Dependent Changes in Pathways

Once *Drosophila* larvae reach the post-feeding stage, imaginal discs lose their ability to regenerate (Harris et al. 2016). Harris et al. (2016) demonstrated that although the JNK and JAK-STAT pathways are activated when cell death is induced in imaginal discs of day 9 larvae (which have ceased to feed), the expression of their target genes, such as *puc*, *MMP-1*, *wg*, and *myc*, is reduced in day 9 imaginal discs compared to day 7 imaginal discs. Detailed studies on *wg* expression regulation demonstrated that the inability to upregulate *wg* is at least in part due to changes that occur in the *wg* enhancer BRV118, which contains a damage-responsive module activated by JNK signaling, and an epigenetically controlled module, which recruits PcG proteins. Interestingly, H3K27 methylation of the BRV118 enhancer increases between days 7 and 9, indicating that the decline in regenerative ability is associated with epigenetic silencing of the *wg* gene. Ectopic expression of Myc is sufficient to override this phenomenon and restores some regenerative ability to day 9 imaginal discs. Thus, epigenetic silencing of regulatory genes may underlie age-dependent decline in the regenerative ability of imaginal discs (Harris et al. 2016).

A recent study has shown that ecdysteroids—the steroid hormones of arthropods—induce imaginal discs to switch from a regeneration program to a differentiation program during the final instar (Narbonne-Reveau and Maurange 2019). Specifically, when imaginal discs are competent to regenerate, they express the Broad-complex, Tramtrack, and Bric-à-brac/poxvirus and zinc finger (BTB/POZ) gene *chinmo*, which maintains the cells in an undifferentiated state. During the final instar, ecdysteroids are produced, activating the expression of another BTB/POZ gene *broad* (*br*), which represses *chino* expression. Precocious silencing of *chinmo* leads to loss of *wg* upregulation, precocious induction of *br*, and reduced regeneration in genetically ablated wing discs. In contrast, when *br* is silenced, *chinmo* expression is sustained even in late final instar larvae (Narbonne-Reveau and Maurange 2019). As *br* is a pupal commitment gene (Zhou and Riddiford 2002), the loss of regenerative potential may be associated with pupal commitment of imaginal discs. How this relates to the age-dependent epigenetic changes remains unknown.

17.6.3 **Gryllus bimaculatus** *Leg Regeneration*

The cricket *Gryllus bimaculatus* was one of the first non-drosophilid arthropod systems in which the mechanisms of regeneration were investigated at the molecular level, primarily using RNA interference (RNAi). Studies in this species have focused on the molecular mechanisms underlying blastema formation and leg repatterning after femur or tibia amputation (Fig. 17.3). In *Gryllus*, the first molt after regeneration results in the formation of a blastema-like bump. After the second molt, the tarsus and the claw become visible externally. However, restoration of all structures is not completed until the third molt (Nakamura et al. 2008b).

17.6.3.1 Blastema Formation/Growth in Crickets

Mito et al. (2002) reported the first molecular expression patterns of *wg*, *decapentaplegic (dpp)*, and *hedgehog (hh)* during cricket regeneration. They found that *wg*, *decapentaplegic (dpp)*, and *hedgehog(hh)* are expressed in the ventral, dorsal, and posterior compartments, respectively (Mito et al. 2002; Nakamura et al. 2007). The expression domains mirror what is seen in the leg bud of the cricket embryo. The similarity between expression patterns during embryonic development and regeneration is an indication that these genes play a role in the regulation of the proximodistal axis during regeneration.

RNAi-mediated knockdowns have been attempted for *wg*, *dpp*, and *hh*, but only *hh* knockdown resulted in a phenotype (Nakamura et al. 2008a). When *hh* is silenced, a supernumerary proximodistal axis regenerates. In contrast, knockdown of *wg* or *dpp* did not generate any effects during regeneration (Nakamura et al. 2007, 2008a). Wg is one of the many Wnt ligands of the Wnt signaling pathway (Miyawaki et al. 2004), and *dpp* encodes one of the three BMP ligands that signals via the BMP signaling pathway (Donoughe et al. 2014). Functional redundancy between these ligands could explain the lack of phenotypic effects in response to RNAi.

Besides the expression of *wg* in leg buds, *arm* knockdown phenotypes have provided additional evidence for the involvement of Wnt signaling during regeneration (Nakamura et al. 2007). During *Gryllus* regeneration, nuclear Arm is localized to the ventral side of the regenerating leg, corresponding to the region of *wg* expression (Nakamura et al. 2007). When *arm* is silenced in the *Gryllus* nymph, the amputated leg fails to regenerate after two subsequent molts when the lost tarsal segments normally would have regenerated (Nakamura et al. 2007). Thus, Wnt signaling likely plays a key role during *Gryllus* leg regeneration.

In addition to the demonstrated role of Wnt signaling, Ds/Ft signaling is also involved in blastema growth. During regeneration following tibial amputation, *ft* is activated in the proximal region of the tarsus, whereas *ds* is expressed in the distal regions of the tibia and the tarsus (Bando et al. 2009). Removal of *ds* or *ft* during regeneration results in short and thick regenerated nymphal legs due to increased cell proliferation in the blastema. In addition, removal of *ds* or *ft* leads to a reduction in *dachshund* (*dac*) expression, impacting repatterning of the legs (Ishimaru et al. 2015). Thus, Ds/Ft signaling appears to play a critical role in both cell proliferation and repatterning of the blastema.

The Hippo signaling pathway also appears to be involved in blastema growth. Knockdown of *ex* and *Mer*, which encode activators of the Hippo signaling pathway (Su et al. 2017), leads to the regeneration of an elongated tibial segment when the distal tibia is amputated (Bando et al. 2009). This elongation is accompanied by increased cell proliferation in the tibial segment, suggesting that Ex and Mer regulate blastema cell proliferation. In contrast, knockdown of Yki prevents regeneration (Bando et al. 2009).

One additional pathway involved in blastema proliferation is the JAK/STAT signaling pathway (Bando et al. 2013). *Dome*, *hop*, and *stat* are upregulated soon after amputation. Knockdown of these genes leads to the regeneration of a shortened tibial segment and either a complete loss of the tarsal segments or regeneration of a single non-segmented structure. Furthermore, knockdown of genes encoding suppressors of cytokine signaling (SOCS) proteins, negative regulators of the JAK/STAT signaling pathway, leads to the regeneration of an elongated tibia. These observations and results of cell proliferation assays indicate that the JAK/STAT signaling pathway is involved in regulating the proliferation of blastema cells (Bando et al. 2013). Thus, in both *Gryllus* and *Drosophila*, both Hippo and JAK-STAT signaling pathways appear to play critical roles during blastema development.

17.6.3.2 Repatterning in *Gryllus*

Repatterning is initiated at the intersection of *wg* and *dpp* on the distal end of the amputated leg segment (Mito et al. 2002), likely by initiating the expression of epidermal growth factor (EGF)-related ligands (Nakamura et al. 2008a). In particular, knockdown of Arm during regeneration leads to the loss of *epidermal growth factor receptor* (*Egfr*) expression (Nakamura et al. 2008b). When a tibia is amputated, *Egfr* expression is induced in the distal blastema 2 days post amputation (Nakamura et al. 2008b). Subsequently, *Egfr* expression appears in the presumptive leg segments, in a manner reminiscent of embryonic limb bud development. Removal of *Egfr* does not result in any observable effects after the first molt. However, after the second molt, the claw fails to differentiate, and the tarsus lacks one of the three tarsal segments. These patterning defects are reflected in alteration of distal expression patterns of the downstream patterning genes, *aristaless* (*al*) and *dac*, but not *Distal-less* (*Dll*) (Nakamura et al. 2008b).

Dll acts downstream of Arm but independently of EGF signaling to regulate patterning (Nakamura et al. 2008b). In the absence of Dll, tibial amputation leads to the failure of tarsal segmentation leading to the regeneration of a single tarsal segment (Ishimaru et al. 2015). The length of this tarsal segment varies depending on the amount of *Dll* removed: a low concentration of double-stranded RNA leads to the formation of a large tarsal segment, whereas a high concentration of double-stranded RNA leads to a single highly reduced segment. These phenotypic effects are reflected in *dac* expression changes. When higher amounts of *Dll* dsRNA are administered, *al* expression is disrupted and *dac* expression is eliminated (Ishimaru et al. 2015; Nakamura et al. 2008b). In contrast, when lower concentration of *Dll* dsRNA is injected, *dac* expression domain increases (Ishimaru et al. 2015). Silencing of *dac* expression leads to a shortening of the tibial segment and the elimination of the first tarsal segment. Together, these studies demonstrate that Dll is necessary for patterning the tarsal segments and the claw, whereas Dac is necessary for the patterning of the tibial and the first tarsal segment.

Finally, a recent study has demonstrated that histone modification also impacts patterning. *Enhancer of zeste* (*E*(*z*)) and *ubiquitously transcribed tetratricopeptide repeat gene on the X chromosome* (*Utx*) are upregulated in the blastema of *Gryllus* (Hamada et al. 2015). *Utx* encodes a histone H3K27 demethylase, whereas *E*(*z*) encodes a histone H3K27 methyltransferase. Knockdown of *E*(*z*) leads to the development of an extra tarsal segment with a correlated increase in *dac* expression. In contrast, the knockdown of *Utx* leads to the fusion of the first and second tarsal segments with a loss of *Egfr* expression at the presumptive boundary. Delaying the timing of leg ablation for 72 h after dsRNA administration did not affect regeneration, indicating that these genes do not play any role during dedifferentiation and only affect the repatterning phase (Hamada et al. 2015).

17.6.4 Tribolium castaneum *Limb Regeneration*

Tribolium has emerged as a useful model for developmental genetics given its sequenced genome and susceptibility to systemic RNAi (Tomoyasu and Denell 2004; Tribolium-Genome-Sequencing-Consortium 2008). Unlike crickets, these insects have immature legs that appear to be comprised primarily of polymorphic cells, which are not yet fully differentiated (Villarreal et al. 2015). This species normally undergoes seven to eight molts, and studies on regeneration have been done in sixth instar larvae (Shah et al. 2011). When limbs are removed on the second day of the sixth instar, a bump that resembles a blastema develops after one molt. After a second molt, a regenerated limb is typically observed with most of the lost segments replaced albeit with a slightly distorted morphology; proper morphology is not restored until the third molt. The distinct morphologies observed after each molt provide a means to assess the molecular mechanisms underlying wound healing, blastema formation, and repatterning (Fig. 17.3).

17.6.4.1 Wound Healing in *Tribolium*

While the key signaling pathways involved in wound healing have not yet been explored in this species, the role of MMPs has been studied. Three *Mmp* genes exist in this species: *Mmp1*, *Mmp2*, and *Mmp3*. Of the three genes, *Mmp1* and *Mmp3* are upregulated at the site of leg amputation 2 days postamputation. Knockdown of *Mmp2* alone—but neither *Mmp1* nor *Mmp3*—is sufficient to delay wound healing. However, only when another *Mmp* gene is knocked down in addition to *Mmp2* can wound healing be completely inhibited, indicating that these genes play partially redundant roles during regeneration. When the wound fails to heal, adult legs never form, indicating that MMP-mediated wound healing is required for the subsequent formation of a blastema followed by regeneration (Mitten et al. 2012).

17.6.4.2 Blastema Formation/Growth in *Tribolium*

Once the wound closes, the blastema must form and grow. Knockdown of several genes prevents the blastema formation in *Tribolium*. Specifically, the Wnt signaling pathway appears to play a critical role in blastema formation and growth. Silencing either *wnt-1* or one of the two *armadillo* paralogous genes, *arm-2*, leads to the inhibition of blastema formation during larval leg regeneration. When these larvae undergo metamorphosis, no adult legs are observed, indicating that the canonical Wnt signaling pathway is necessary for limb regeneration in this species (Shah et al. 2011).

In addition, knockdown of *abrupt* (*ab*) prevents blastema formation and leads to the development of a similar flat surface at the site of ablation (Lee et al. 2013). Ab codes for a BTB-zinc finger transcription factor and has been shown to promote a progenitor-like state in epithelia during cancer formation (Turkel et al. 2013). While the precise role of Ab during regeneration is unknown, Ab may be necessary for retention of a stem cell-like progenitor state in the blastema to promote its growth.

17.6.4.3 Redifferentiation and Repatterning in *Tribolium*

Knockdown of *Dll* and *hh* causes blastemas to grow but prevents further redifferentiation of limbs (Lee et al. 2013; Villarreal et al. 2015). Silencing *Dll* during larval leg and antenna regeneration leads to the formation of a rounded bump that never undergoes obvious patterning. Knockdown of *Dll* in larvae has been shown to cause dedifferentiation of uncut larval appendages leading to the loss of tissue identity and homeotic transformation of legs to antennae (Suzuki et al. 2009). Thus, *Dll* appears to be required for maintaining and promoting differentiation.

Knockdown of *hh* also leads to the formation of a blastema-like bump at the site of leg and antennal ablation but prevents further regeneration of limb segments even after two molts. When these larvae eventually metamorphose, the resulting pupae fail to regenerate any of the lost segments. Thus, the absence of Hh signaling prevents repatterning of the blastema (Villarreal et al. 2015).

In addition, E(z) and Polycomb (Pc), another member of the PcG proteins, appear to play a key role in beetle leg regeneration (Chou et al. 2019). In contrast to crickets, where *E(z)* knockdown generates an extra tarsal segment, *E(z)* knockdown in *Tribolium* results in a small blastema but prevents limb regeneration. Such difference in phenotypes appears to be due to species-specific differences rather than differences in the experimental procedure as the leg fails to grow back even when the femur is cut instead of the coxa. *E(z)* is also necessary for antennal regeneration in this species. Similarly, *Pc* knockdown also leads to disrupted leg regeneration although the effect is variable. In addition, *Pc* plays a critical role in the maintenance of antennal identity during antennal regeneration: when *Pc* is silenced, the antenna regenerates a leg-like structure (Chou et al. 2019). Thus, epigenetic regulators appear to play a critical role in the redifferentiation/repatterning of appendages and the maintenance of appendage identity.

Finally, several genes have been identified in this species that are necessary for proper repatterning. *Dac* is necessary for the medial leg regeneration, similar to crickets: in the absence of *dac*, the larval leg regenerates a leg with fused tibiotarus and femur segments (Lee et al. 2013). The repatterning process appears to involve the activation of genes used during embryonic limb development rather than pupal limb development. For example, the gene *spineless* (*ss*) is necessary for proper tarsal segmentation during pupation but not for the development of legs during embryogenesis. *ss* knockdown larvae regenerate their larval legs without abnormalities. In contrast, the antennae require Ss for antennal specification during embryogenesis and larval antennal regeneration (Lee et al. 2013).

17.6.5 Hormonal Regulation of Regeneration

As mentioned before, in addition to developmental genes, hormones have been shown to play critical roles during regeneration. Ecdysteroids regulate a variety of processes during arthropod development, ranging from growth, molting, metamorphosis, reproduction, to regeneration (Das 2015; Hopkins 2001; Madhavan and Schneiderman 1969; Nijhout 1998). Much of the earlier studies on the role of ecdysteroids during regeneration were performed on lepidopteran imaginal discs. By placing a ligature between the prothoracic and mesothoracic segments, the imaginal disc-bearing segments in the thorax (meso- and metathorax) can be separated from the head and the prothoracic segment, which contains the prothoracic glands. After placing such ligatures in *Galleria*, the imaginal discs failed to regenerate, especially when the timing of imaginal disc extirpation was delayed. Regeneration was restored when crystalline ecdysone was injected (Madhavan and

Schneiderman 1969). In the flesh fly, *Sarcophaga peregrina*, low levels of ecdysteroids have also been shown to be necessary for wound healing and induction of *wg* expression in the fragmented imaginal disc in vitro (Kunieda et al. 1997).

In the fiddler crab, *Uca pugilator*, two stages of regeneration have been identified following autotomy: the basal growth stage and the proecdysial growth stage (Fig. 17.3; Hopkins 1993). During the basal growth phase, the blastema cells proliferate and the regenerated limb is differentiated. During the proecdysial phase, the size of the limb increases by protein synthesis and water uptake. The two stages of regeneration are synced with the molt cycle of the animal, which is regulated by ecdysteroids (Hopkins 2001). While functional studies have not yet been performed except for a few RNAi studies, circulating levels of ecdysteroids are correlated with the timing of different phases of regenerative growth observed in crustaceans. In particular, much of the regenerative growth appears to occur when ecdysteroid titers are low, indicating that ecdysteroids play an inhibitory role (Hopkins 2001). However, low levels of ecdysteroids appear to be necessary as knockdowns of both *ecdysone receptor* (*EcR*) and *retinoid X receptor* (*RXR*), which code for ecdysone receptor heterodimeric partners, lead to growth arrest of the blastema cells through inhibition of cell division (Das and Durica 2013). Moreover, the two stages of regeneration are separated by a small ecdysteroid pulse, indicating that ecdysteroids may be necessary for shifting the molecular processes that are involved in different phases of regeneration (Hopkins 1989, 2001).

17.6.6 Summary of Molecular Interactions

Figure 17.3 summarizes the key molecular players that have been reported for flies, crickets, and beetles. It is worth noting that some of the differences between these species may arise due to differences in timing of amputation and differences in position of the amputation. Studies on *Drosophila* have primarily used genetic tools to induce cell death and manipulate gene expression in imaginal discs. In *Gryllus*, amputation has typically been performed on the same day as the dsRNA injection. In contrast, in *Tribolium*, amputation has been performed 2 days post-dsRNA injection. Given that the full effect of RNAi is unlikely to be immediate, some of the earlier functions of genes may not have been captured in the *Gryllus* studies where amputations and dsRNA injections were performed on the same day. In addition, amputations of *Gryllus* legs have typically been performed at the tibia or the femur, whereas *Tribolium* legs have mostly been amputated at the coxa. Thus, comparisons between species are not straightforward. However, there are a number of key regulators that are likely conserved across species such as Wnt signaling, which appears to be essential for the formation of blastema in many species across metazoans. Similarly, the Hippo and JAK-STAT signaling pathways are involved in both *Gryllus* and *Drosophila* regeneration, and we suspect that they also play key roles during *Tribolium* regeneration. Other factors, such as histone modifiers, appear to play distinct roles indicating that a comparative evo-devo approach to studying limb regeneration would be fruitful. Once the molecular mechanisms have been determined, we will be in the position to actively investigate how distinct limb types differ in their molecular mechanisms and why some insects can readily regenerate their limbs while others cannot.

In addition, ecdysteroids likely play key roles in a variety of arthropods species. While modulating ecdysteroid without altering the timing of molts is challenging in many systems, more studies are needed in species where tissue-specific knockouts/knockdowns of hormonal signaling pathway components are possible. In addition, while hormones likely influence developmental genes in regenerating tissues, the link between intrinsic pathways and extrinsic cues remains unexplored.

17.7 Developmental Delays Caused by Regeneration

In many species, injuries and amputations cause developmental delays, which are thought to allow the juveniles to regrow their appendages to the correct sizes before reaching maturation. Classical studies have shown that tissue damage induced in a variety of ways extends the larval phase of development by unknown mechanisms (Hussey et al. 1927; Malá et al. 1987; Poodry and Woods 1990; Russell 1974; Simpson et al. 1980). Recent studies have begun to uncover some of the molecular regulators of developmental delays, but we are far from understanding how regenerating tissues impact the development of the whole organism. Given that molting and metamorphosis are regulated by ecdysteroids, understanding the link between regeneration and developmental timing requires that we understand how regenerating structures interact with endocrine regulation of metamorphosis. Here, we review some of the recent findings that demonstrate the molecular links between regeneration and hormonal centers that coordinate the onset of metamorphosis. These studies are beginning to provide insight into how tissue growth is coordinated within the whole organism and the integrative nature of organismal development. Yet, studies have come primarily from work done in *Drosophila*, a species with a rather unique developmental physiology. We argue that more studies are needed in nonmodel systems with a more typical mode of development.

17.7.1 Developmental Delays in Drosophila

Drosophila imaginal discs have been shown to play significant roles in coupling tissue growth, maturation, and patterning during development. When disc growth is impaired, the larval period is extended, allowing tissues to regenerate or grow to their target size before entering metamorphosis (Parker and Shingleton 2011; Poodry and Woods 1990; Russell 1974; Simpson et al. 1980). In particular, work on *Drosophila* has led to the hypothesis that there might be a regeneration checkpoint that functions in a similar manner as the cell cycle checkpoint to determine when the imaginal discs have regenerated and grown to the correct size (Halme et al. 2010; Jaszczak and Halme 2016). When a Gal4-UAS system was used to induce apoptosis in imaginal discs, the amount of developmental delay correlated with the amount of regeneration of the imaginal discs (Smith-Bolton et al. 2009). Similarly, proliferating cells of imaginal discs have been shown to delay metamorphosis (Bryant and Schubiger 1971; Woods and Bryant 1989). Thus, the timing of metamorphosis is influenced by imaginal disc cell proliferation rather than the absolute size of imaginal discs.

The developmental delays are accompanied by an increase in the critical size, the size above which starvation no longer delays metamorphosis, as well as the extension of the period between attainment of critical size and metamorphosis (Stieper et al. 2008). However, once the larvae attain the mid-third instar transition—a period marked by extensive gene expression changes due to the metamorphic ecdysteroid peak—wing disc damage no longer delays the timing of metamorphosis (Hackney et al. 2012). Thus, events leading up to the production of ecdysteroids alter the timing of metamorphosis. Correspondingly, when imaginal discs are irradiated, the larval growth period is extended by delaying the transcription of the gene encoding prothoracicotropic hormone (PTTH), a neuropeptide that promotes the release of ecdysone (Halme et al. 2010).

Damaged imaginal discs appear to release at least two independent signals to the hormonal centers that coordinate the timing of metamorphosis: retinoids and *Drosophila* insulin-like peptide 8 (Dilp8) (Colombani et al. 2012; Garelli et al. 2012; Halme et al. 2010). When imaginal discs are damaged by X-irradiation, larvae delay the timing of metamorphosis, and this delay can be partially rescued when components of the retinoid biosynthesis pathway are disrupted (Halme et al. 2010). Moreover, feeding larvae a diet lacking in β-carotene significantly reduces irradiation-induced developmental delays.

In 2012, two independent research groups reported Dilp8 as a critical factor that is released from the imaginal disc to mediate developmental delays (Colombani et al. 2012; Garelli et al. 2012). *dilp8* encodes a peptide with a 6-cysteine motif characteristic of the insulin-relaxin peptide family, and elevated *dilp8* expression was identified in imaginal discs undergoing abnormal tumor-like growth (Colombani et al. 2012), apoptosis (Garelli et al. 2012), and leg-to-wing transdetermination (Klebes et al. 2005). This factor is necessary and sufficient for developmental delays and acts on the hormonal centers that coordinate the timing of metamorphosis. Ectopic expression of *dilp8* in imaginal discs delays the expression of *PTTH* in the brain, delaying the subsequent production of ecdysteroids in the ring glands, which contain prothoracic glands, the site of ecdysone production (Colombani et al. 2012; Garelli et al. 2012). When *dilp8* is downregulated in larvae with apoptotic imaginal discs, the *PTTH* delay is abrogated, suggesting that Dilp8 delays metamorphosis by inhibiting the production of PTTH (Garelli et al. 2012). Recent studies have demonstrated that Dilp8 acts via leucine-rich repeat-containing G protein-coupled receptor 3 (Lgr3), which is expressed in neurons that project to PTTH-expressing neurons in the brain (Colombani et al. 2015; Garelli et al. 2015). Taken together, Dilp8 from damaged imaginal discs appears to act on Lgr3-expressing neurons to block PTTH secretion from the brain. This Dilp8-mediated pathway appears to act independently of the retinoid pathway as feeding a diet supplemented with β-carotene does not impact the timing of metamorphosis in larvae that ectopically express Dilp8 (Garelli et al. 2012).

dilp8 expression is activated by JAK-STAT pathway in response to the JNK signaling pathway (Colombani et al. 2012; Katsuyama et al. 2015). JNK, in turn, appears to be regulated by the histone methyltransferase, Trithorax (Trx): when Trx is mutated, JNK signaling fails to be activated in damaged wing discs, leading to reduced *dilp8* expression (Skinner et al. 2015). As wound healing and blastema proliferation appear to progress normally in *Trx* mutants, the effect of Trx on JNK appears to be limited to developmental delay, with no effect on wound healing and blastema formation (Skinner et al. 2015). Thus, both epigenetic regulators and wound response signaling pathways appear to regulate the expression of *dilp8*.

While developmental delay buys time for damaged imaginal discs to grow back to the correct size, undamaged imaginal discs must delay their development in order for the adults to enclose with correctly proportioned appendages. Dilp8 restricts overgrowth of imaginal discs and acts via Lgr3 to reduce fluctuating asymmetry (Colombani et al. 2015; Garelli et al. 2015). Specifically, the growth restriction appears to be mediated by Dilp8-dependent nitric oxide production in the prothoracic glands: nitric oxide production inhibits ecdysteroid production, and ecdysteroids in turn act on the undamaged imaginal discs to delay their growth (Jaszczak et al. 2015, 2016). Thus, Dilp8 is released from damaged discs to suppress other imaginal discs from overgrowth while delaying the timing of metamorphosis.

17.7.2 Developmental Delays in Non-drosophilid Insects

When considering developmental delays in non-drosophilid insect species, two types of developmental delays must be considered. First, the timing of a molt can be delayed, such that the intermolt period is lengthened during earlier instars (intermolt delay). Secondly, the timing of metamorphosis can be delayed during the final instar either by lengthening the final instar or through the induction of extra molts called supernumerary molts (final instar delay). The latter is unique to non-drosophilid species as unlike *Drosophila*, many species readily undergo extra larval-larval molts. The two types of delays depend on distinct mechanisms and therefore must be discussed separately.

Intermolt delays have been reported in many species, ranging from cockroaches, such as *Blattella germanica*, to lepidopterans, such as *Ephestia kuehniella* and *Galleria mellonella* (Kunkel 1977; Madhavan and Schneiderman 1969; Pohley 1965). As would be expected from a delay in the timing of the molt, ecdysteroid peak is delayed in cockroaches with limb autotomy (Kunkel 1977; Stock and O'Farrell 1954). Based on this observation, Kunkel (1977) proposed that the regenerating limb sends signals to the brain, which in turn inhibits ecdysteroid production/release. In *Galleria*, imaginal disc removal also leads to delay in regeneration. Intriguingly, when the epidermal cuticle site of wing disc removal was cauterized, the delay period was substantially reduced—although not completely eliminated—indicating that the wound itself causes developmental delays (Madhavan and Schneiderman 1969). Signals controlling these developmental delays have not yet been identified. While Dilp8 appears to be unique to dipterans, relaxin receptors have been found in other species and may contribute to developmental delays in non-dipteran species (Gontijo and Garelli 2018).

Injury has been shown to delay and induce supernumerary molts in a number of species including *Galleria* and *Tribolium* (Krishnakumaran 1972; Malá et al. 1987; Suzuki et al. 2009). Final instar delays may be regulated distinctly from intermolt delays because an additional regulator called juvenile hormone (JH) plays a critical role in the timing of ecdysteroid release. JH is a status quo hormone that remains relatively high during the earlier larval instars and maintains the larval state. Ecdysteroid biosynthesis and release can occur in the presence of JH during the earlier instars, and JH ensures that ecdysteroids activate larval-specific genes during larval-larval molts. In contrast, during the final instar of many non-dipteran species, JH inhibits the secretion of PTTH from the brain/corpora cardiaca, delaying ecdysteroid biosynthesis/release. Only once JH disappears can the larva undergo metamorphosis. If JH remains high, or JH is exogenously applied, larvae of many species can often undergo an extra larval molt (Nijhout 1998; Riddiford 1996). Thus, injury-induced developmental delays in non-dipteran species likely involve JH.

Some preliminary findings indicate that this is indeed the case. In *Tribolium*, knockdown of *Dll* leads to disorganization of larval limbs and concomitant induction of supernumerary larval molts (Suzuki et al. 2009). These supernumerary molts could be terminated by inhibiting JH signaling. While additional roles of Dll in endocrine centers cannot be ruled out, ablation of larval appendages also leads to supernumerary molts, indicating that disorganized larval limbs may send an unknown signal to the hormonal centers that regulate the timing of metamorphosis (Suzuki et al. 2009). Thus, a regeneration checkpoint may also exist in other species besides *Drosophila*. Like so many developmental phenomena, *Drosophila* undergo a unique mode of development that chiefly derives from their adaptation to a rapid life cycle, and JH appears to play minimal roles in regulating the timing of metamorphosis in this species (Hatem et al. 2015; Nijhout 2015). Non-model systems will be essential for understanding the regulation of developmental delays during regeneration.

17.7.3 Developmental Delays in Crustaceans

Among crustaceans, the number of appendages removed can also affect the duration of the instar: in decapod crabs, removal of more than five legs leads to a precocious molt, which is not observed when fewer legs are removed (Skinner 1985; Skinner and Graham 1972). In decapod crustaceans, these precocious molts occur because multiple autotomies lead to a reduction in the neuropeptide, molt-inhibiting hormone

(MIH), which inhibits ecdysteroid production by the Y-organ (Skinner 1985). Intriguingly, when the regenerating limb bud (primary limb bud) is removed before a critical period, the proecdysis is delayed until the secondary limb bud differentiates and grows. This delay is caused by a reduction in ecdysteroid titers. Additionally, similar to what happens in regenerating *Drosophila* larvae, the removal of the primary limb bud suspends growth of the other primary limb buds that are unperturbed. Secondary limb bud extracts can inhibit the growth of primary limb buds. Thus, regenerating secondary limb buds secrete a factor, called Limb autotomy factor-proecdysis (LAFpro), which delays ecdysteroid production and impacts other tissues. This factor behaves like MIH but appears to be distinct from MIH (Yu et al. 2002). Whether insects and crustaceans use similar molecular regulators to promote developmental delays remains unknown. However, regeneration-dependent delays appear to converge on ecdysteroid biosynthesis, suggesting that the delay mechanism may be an ancient pathway that was present in the common ancestor of insects and crustaceans.

17.8 Future Studies

17.8.1 How Do Correctly Proportioned Appendages Regenerate and How Is Whole Body Growth Coordinated?

An important yet unexplained phenomenon is the ability of limbs to grow back to the correct size. Once an appendage is removed, the blastema must proliferate rapidly and recreate a correctly proportioned appendage relatively quickly. How the resulting appendage develops into a correctly proportioned appendage is unknown. Molecular players that affect the size of the regenerated limbs have been identified, but we are far from understanding the mechanism by which the regenerating tissue determines its size. Understanding this process will likely provide insights into how limbs attain the correct proportion during normal nonregenerative development.

In addition, as discussed above, whole-body development is influenced by regenerating structures and vice versa. We are only beginning to understand how regenerating tissues communicate with endocrine centers that coordinate developmental timing and growth of other structures. In addition, how hormones impact intrinsic pathways during limb regeneration remains unexplored. How these processes have evolved and how they are regulated in various species with distinct developmental physiology requires us to expand work in organisms beyond *Drosophila*, which has unique, derived physiology.

17.8.2 Are Cell Types and Regeneration Mechanisms Linked?

An emerging notion both within arthropods and other animals is that limb regeneration mechanisms vary between relatively closely related species. We think that these differences are at least partially due to differences in types and degrees of the differentiation of cells used during regeneration. Exploring mechanisms underlying blastema formation and redifferentiation in various species in tandem with how these processes operate during normal development will likely provide insights into this question. In addition, explorations of cell types that give rise to the blastema in various species will provide insights into the variability of the potency of these cells across species.

17.8.3 Is There a Conserved Gene Regulatory Network for Limb Regeneration Across Metazoans?

Limbs in arthropods and vertebrates are clearly not homologous, and the Urbilateria—the common ancestor of all metazoans—likely lacked limbs used for walking. However, homologous gene products, such as Dll, regulate limb development in both vertebrates and invertebrates. This may imply that the Urbilateria may have possessed some sort of appendage, perhaps sensory in nature. If these appendages regenerated in the Urbilateria, there may be some conserved gene regulatory network kernels that are used in both invertebrates and vertebrates (Davidson and Erwin 2006). Some tantalizing evidence for such partial homology exists at the level of wound healing and blastema proliferation. Proteins such as MMPs and Wnt are repeatedly used during the early stages of regeneration (Bai et al. 2005; Kawakami et al. 2006; Leontovich et al. 2000; Mitten et al. 2012; Nakamura et al. 2007; Shah et al. 2011; Stoick-Cooper et al. 2007; Vinarsky et al. 2005; Yang and Bryant 1994; Yang et al. 1999). In contrast, taxon-specific factors for regeneration have also been discovered (Garza-Garcia et al. 2009), supporting the view that regeneration is an evolutionarily labile process that has evolved numerous times during the evolution of metazoan species. One possible way to reconcile these discrepancies is if the same gene regulatory network kernels were co-opted multiple times in various taxa whenever benefits of regeneration outweighed the costs of regeneration. Such a view would imply that there are certain fundamental regulators that are always necessary for regeneration. Identifying these critical regulators may provide insights into why human limbs cannot regenerate and advance regenerative medicine.

17.9 Conclusions

In the mid-twentieth century, there was much interest in understanding limb regeneration in arthropods, but in the late twentieth and early twenty-first centuries, research was focused elsewhere. Recent advances in molecular tools in non-model organisms have revived interest in regeneration, and further studies are needed for understanding how regeneration mechanisms evolve. Novel molecular approaches, such as single-cell RNA-seq and CRISPR, should allow us to delve into the molecular and cellular underpinnings of limb regeneration across a variety of species. Studies in non-model species are essential if we hope to understand how regeneration might have evolved and make progress in regenerative medicine.

Acknowledgments We thank the anonymous reviewer, members of the Suzuki lab, and Heidi Park for their constructive feedback on this review. This work was supported by Wellesley College and by the National Science Foundation grants IOS-1027453 and IOS-1354608 to YS.

References

Agata K, Saito Y, Nakajima E (2007) Unifying principles of regeneration I: epimorphosis versus morphallaxis. Develop Growth Differ 49:73–78

Alfonso-Gonzalez C, Riesgo-Escovar JR (2018) Fos metamorphoses: lessons from mutants in model organisms. Mech Dev 154:73–81

Alwes F, Enjolras C, Averof M (2016) Live imaging reveals the progenitors and cell dynamics of limb regeneration. elife 5:e19766

Bai S, Thummel R, Godwin AR, Nagase H, Itoh Y, Li L, Evans R, McDermott J, Seiki M, Sarras MP Jr (2005) Matrix metalloproteinase expression and function during fin regeneration in zebrafish: analysis of MT1-MMP, MMP2 and TIMP2. Matrix Biol 24:247–260

Bando T, Mito T, Maeda Y, Nakamura T, Ito F, Watanabe T, Ohuchi H, Noji S (2009) Regulation of leg size and shape by the Dachsous/Fat signalling pathway during regeneration. Development 136:2235–2245

Bando T, Ishimaru Y, Kida T, Hamada Y, Matsuoka Y, Nakamura T, Ohuchi H, Noji S, Mito T (2013) Analysis of RNA-Seq data reveals involvement of JAK/STAT signalling during leg regeneration in the cricket Gryllus bimaculatus. Development 140:959–964

Behrens J, von Kries JP, Kuhl M, Bruhn L, Wedlich D, Grosschedl R, Birchmeier W (1996) Functional interaction of beta-catenin with the transcription factor LEF-1. Nature 382:638–642

Bennett FC, Harvey KF (2006) Fat cadherin modulates organ size in Drosophila via the Salvador/Warts/Hippo signaling pathway. Curr Biol 16:2101–2110

Bergantinos C, Corominas M, Serras F (2010) Cell death-induced regeneration in wing imaginal discs requires JNK signalling. Development 137:1169–1179

Blair S, McNeill H (2018) Big roles for Fat cadherins. Curr Opin Cell Biol 51:73–80

Bosch M, Serras F, Martin-Blanco E, Baguna J (2005) JNK signaling pathway required for wound healing in regenerating Drosophila wing imaginal discs. Dev Biol 280:73–86

Brautigam SE, Persons MH (2003) The effect of limb loss on the courtship and mating behavior of the wolf spider Pardosa milvina (Araneae: Lycosidae). J Insect Behav 16:571–587

Brockes JP (1997) Amphibian limb regeneration: rebuilding a complex structure. Science 276:81–87

Brunner E, Peter O, Schweizer L, Basler K (1997) pangolin encodes a Lef-1 homologue that acts downstream of Armadillo to transduce the Wingless signal in Drosophila. Nature 385:829–833
Bryant PJ (1971) Regeneration and duplication following operations in situ on the imaginal discs of Drosophila melanogaster. Dev Biol 26:637–651
Bryant PJ (1975) Pattern formation in the imaginal wing disc of Drosophila melanogaster: fate map, regeneration and duplication. J Exp Zool 193:49–77
Bryant PJ, Schubiger G (1971) Giant and duplicated imaginal discs in a new lethal mutant of Drosophila melanogaster. Dev Biol 24:233–263
Campbell G, Tomlinson A (1995) Initiation of the proximodistal axis in insect legs. Development 121:619–628
Chablais F, Jazwinska A (2010) IGF signaling between blastema and wound epidermis is required for fin regeneration. Development 137:871–879
Chou J, Ferris AC, Chen T, Seok R, Yoon D, Suzuki Y (2019) Roles of Polycomb group proteins Enhancer of zeste (E(z)) and Polycomb (Pc) during metamorphosis and larval leg regeneration in the flour beetle Tribolium castaneum. Dev Biol 450(1):34–46. https://doi.org/10.1016/j.ydbio.2019.03.002
Clevers H (2006) Wnt/beta-catenin signaling in development and disease. Cell 127:469–480
Colombani J, Andersen DS, Leopold P (2012) Secreted peptide Dilp8 coordinates Drosophila tissue growth with developmental timing. Science 336:582–585
Colombani J, Andersen DS, Boulan L, Boone E, Romero N, Virolle V, Texada M, Leopold P (2015) Drosophila Lgr3 couples organ growth with maturation and ensures developmental stability. Curr Biol 25:2723–2729
Das S (2015) Morphological, molecular, and hormonal basis of limb regeneration across Pancrustacea. Integr Comp Biol 55:869–877
Das S, Durica DS (2013) Ecdysteroid receptor signaling disruption obstructs blastemal cell proliferation during limb regeneration in the fiddler crab, Uca pugilator. Mol Cell Endocrinol 365:249–259
Davidson EH, Erwin DH (2006) Gene regulatory networks and the evolution of animal body plans. Science 311:796–800
Donoughe S, Nakamura T, Ewen-Campen B, Green DA, Henderson L, Extavour CG (2014) BMP signaling is required for the generation of primordial germ cells in an insect. Proc Natl Acad Sci USA 111:4133–4138
Endo T, Bryant SV, Gardiner DM (2004) A stepwise model system for limb regeneration. Dev Biol 270:135–145
Erezyilmaz DF (2006) Imperfect eggs and oviform nymphs: a history of ideas about the origins of insect metamorphosis. Integr Comp Biol 46:795–807
Fernando WA, Leininger E, Simkin J, Li N, Malcom CA, Sathyamoorthi S, Han M, Muneoka K (2011) Wound healing and blastema formation in regenerating digit tips of adult mice. Dev Biol 350:301–310
French V (1976) Leg regeneration in the cockroach, Blatella germanica. II. Regeneration from a non-congruent tibial graft/host junction. J Embryol Exp Morphol 35:267–301
French V (1978) Intercalary regeneration around the circumference of the cockroach leg. J Embryol Exp Morphol 47:53–84
French V, Bryant PJ, Bryant SV (1976) Pattern regulation in epimorphic fields. Science 193:969–981
Galko MJ, Krasnow MA (2004) Cellular and genetic analysis of wound healing in Drosophila larvae. PLoS Biol 2:E239
Garelli A, Gontijo AM, Miguela V, Caparros E, Dominguez M (2012) Imaginal discs secrete insulin-like peptide 8 to mediate plasticity of growth and maturation. Science 336:579–582
Garelli A, Heredia F, Casimiro AP, Macedo A, Nunes C, Garcez M, Dias AR, Volonte YA, Uhlmann T, Caparros E, Koyama T, Gontijo AM (2015) Dilp8 requires the neuronal relaxin receptor Lgr3 to couple growth to developmental timing. Nat Commun 6:8732

Garza-Garcia A, Harris R, Esposito D, Gates PB, Driscoll PC (2009) Solution structure and phylogenetics of Prod1, a member of the three-finger protein superfamily implicated in salamander limb regeneration. PLoS One 4:e7123

Garza-Garcia AA, Driscoll PC, Brockes JP (2010) Evidence for the local evolution of mechanisms underlying limb regeneration in salamanders. Integr Comp Biol 50:528–535

Geng J, Gates PB, Kumar A, Guenther S, Garza-Garcia A, Kuenne C, Zhang P, Looso M, Brockes JP (2015) Identification of the orphan gene Prod 1 in basal and other salamander families. EvoDevo 6:9

Gontijo AM, Garelli A (2018) The biology and evolution of the Dilp8-Lgr3 pathway: a relaxin-like pathway coupling tissue growth and developmental timing control. Mech Dev 154:44–50

Goss RJ (1969) Principles of regeneration. Academic Press, New York

Goulev Y, Fauny JD, Gonzalez-Marti B, Flagiello D, Silber J, Zider A (2008) SCALLOPED interacts with YORKIE, the nuclear effector of the hippo tumor-suppressor pathway in drosophila. Curr Biol 18:435–441

Grusche FA, Degoutin JL, Richardson HE, Harvey KF (2011) The Salvador/Warts/Hippo pathway controls regenerative tissue growth in Drosophila melanogaster. Dev Biol 350:255–266

Hackney JF, Zolali-Meybodi O, Cherbas P (2012) Tissue damage disrupts developmental progression and ecdysteroid biosynthesis in Drosophila. PLoS One 7:e49105

Halme A, Cheng M, Hariharan IK (2010) Retinoids regulate a developmental checkpoint for tissue regeneration in Drosophila. Curr Biol 20:458–463

Hamada Y, Bando T, Nakamura T, Ishimaru Y, Mito T, Noji S, Tomioka K, Ohuchi H (2015) Leg regeneration is epigenetically regulated by histone H3K27 methylation in the cricket Gryllus bimaculatus. Development 142:2916–2927

Hamaratoglu F, Willecke M, Kango-Singh M, Nolo R, Hyun E, Tao C, Jafar-Nejad H, Halder G (2006) The tumour-suppressor genes NF2/Merlin and Expanded act through Hippo signalling to regulate cell proliferation and apoptosis. Nat Cell Biol 8:27–36

Hariharan IK, Serras F (2017) Imaginal disc regeneration takes flight. Curr Opin Cell Biol 48:10–16

Harris RE, Setiawan L, Saul J, Hariharan IK (2016) Localized epigenetic silencing of a damage-activated WNT enhancer limits regeneration in mature Drosophila imaginal discs. elife 5: e11588

Hatem NE, Wang Z, Nave KB, Koyama T, Suzuki Y (2015) The role of juvenile hormone and insulin/TOR signaling in the growth of Manduca sexta. BMC Biol 13:44

Hicklin J, Wolpert L (1973) Positional information and pattern regulation in hydra: the effect of gamma-radiation. J Embryol Exp Morphol 30:741–752

Hopkins PM (1989) Ecdysteroids and regeneration in the fiddler crab Uca pugilator. J Exp Zool 252:293–299

Hopkins PM (1993) Regeneration of walking legs in the fiddler-crab Uca-pugilator. Am Zool 33:348–356

Hopkins PM (2001) Limb regeneration in the fiddler crab, Uca pugilator: hormonal and growth factor control. Am Zool 41:389–398

Huang J, Wu S, Barrera J, Matthews K, Pan D (2005) The Hippo signaling pathway coordinately regulates cell proliferation and apoptosis by inactivating Yorkie, the Drosophila Homolog of YAP. Cell 122:421–434

Hussey RG, Thompson WR, Calhoun ET (1927) The influence of X-rays on the development of Drosophila larvae. Science 66:65–66

Ip YT, Davis RJ (1998) Signal transduction by the c-Jun N-terminal kinase (JNK)—from inflammation to development. Curr Opin Cell Biol 10:205–219

Ishimaru Y, Nakamura T, Bando T, Matsuoka Y, Ohuchi H, Noji S, Mito T (2015) Involvement of dachshund and Distal-less in distal pattern formation of the cricket leg during regeneration. Sci Rep 5:8387

Jaszczak JS, Halme A (2016) Arrested development: coordinating regeneration with development and growth in Drosophila melanogaster. Curr Opin Genet Dev 40:87–94

Jaszczak JS, Wolpe JB, Dao AQ, Halme A (2015) Nitric oxide synthase regulates growth coordination during Drosophila melanogaster imaginal disc regeneration. Genetics 200:1219–1228

Jaszczak JS, Wolpe JB, Bhandari R, Jaszczak RG, Halme A (2016) Growth coordination during Drosophila melanogaster imaginal disc regeneration is mediated by signaling through the relaxin receptor Lgr3 in the prothoracic gland. Genetics 204:703–709

Jaźwińska A, Sallin P (2016) Regeneration versus scarring in vertebrate appendages and heart. J Pathol 238:233–246

Katsuyama T, Comoglio F, Seimiya M, Cabuy E, Paro R (2015) During Drosophila disc regeneration, JAK/STAT coordinates cell proliferation with Dilp8-mediated developmental delay. Proc Natl Acad Sci USA 112:E2327–E2336

Kawakami Y, Esteban CR, Raya M, Kawakami H, Marti M, Dubova I, Belmonte JCI (2006) Wnt/beta-catenin signaling regulates vertebrate limb regeneration. Genes Dev 20:3232–3237

Khan SJ, Abidi SNF, Skinner A, Tian Y, Smith-Bolton RK (2017) The Drosophila Duox maturation factor is a key component of a positive feedback loop that sustains regeneration signaling. PLoS Genet 13:e1006937

Klebes A, Sustar A, Kechris K, Li H, Schubiger G, Kornberg TB (2005) Regulation of cellular plasticity in Drosophila imaginal disc cells by the Polycomb group, trithorax group and lama genes. Development 132:3753–3765

Konopova B, Smykal V, Jindra M (2011) Common and distinct roles of juvenile hormone signaling genes in metamorphosis of holometabolous and hemimetabolous insects. PLoS One 6:e28728

Konstantinides N, Averof M (2014) A common cellular basis for muscle regeneration in arthropods and vertebrates. Science 343:788–791

Kragl M, Knapp D, Nacu E, Khattak S, Maden M, Epperlein HH, Tanaka EM (2009) Cells keep a memory of their tissue origin during axolotl limb regeneration. Nature 460:60–65

Krishnakumaran A (1972) Injury induced molting in Galleria mellonella larvae. Biol Bull 142:281–292

Kumar A, Godwin JW, Gates PB, Garza-Garcia AA, Brockes JP (2007) Molecular basis for the nerve dependence of limb regeneration in an adult vertebrate. Science 318:772–777

Kunieda T, Kurata S, Natori S (1997) Regeneration of Sarcophaga imaginal discs in vitro: implication of 20-hydroxyecdysone. Dev Biol 183:86–94

Kunkel JG (1977) Cockroach molting. II. The nature of regeneration-induced delay of molting hormone secretion. Biol Bull 153:145–162

Lee N, Maurange C, Ringrose L, Paro R (2005) Suppression of Polycomb group proteins by JNK signalling induces transdetermination in Drosophila imaginal discs. Nature 438:234–237

Lee AK, Sze CC, Kim ER, Suzuki Y (2013) Developmental coupling of larval and adult stages in a complex life cycle: insights from limb regeneration in the flour beetle, Tribolium castaneum. EvoDevo 4:20

Leontovich AA, Zhang J, Shimokawa K, Nagase H, Sarras MP Jr (2000) A novel hydra matrix metalloproteinase (HMMP) functions in extracellular matrix degradation, morphogenesis and the maintenance of differentiated cells in the foot process. Development 127:907–920

Llano E, Pendas AM, Aza-Blanc P, Kornberg TB, Lopez-Otin C (2000) Dm1-MMP, a matrix metalloproteinase from Drosophila with a potential role in extracellular matrix remodeling during neural development. J Biol Chem 275:35978–35985

Llano E, Adam G, Pendas AM, Quesada V, Sanchez LM, Santamaria I, Noselli S, Lopez-Otin C (2002) Structural and enzymatic characterization of Drosophila Dm2-MMP, a membrane-bound matrix metalloproteinase with tissue-specific expression. J Biol Chem 277:23321–23329

Londono R, Sun AX, Tuan RS, Lozito TP (2018) Tissue repair and epimorphic regeneration: an overview. Curr Pathobiol Rep 6:61–69

Madhavan K, Schneiderman HA (1969) Hormonal control of imaginal disc regeneration in Galleria mellonella (Lepidoptera). Biol Bull 137:321–331

Maginnis TL (2006a) The costs of autotomy and regeneration in animals: a review and framework for future research. Behav Ecol 17:857–872

Maginnis TL (2006b) Leg regeneration stunts wing growth and hinders flight performance in a stick insect (Sipyloidea sipylus). Proc Biol Sci 273:1811–1814

Malá J, Sehnal F, Kumaran AK, Granger NA (1987) Effects of starvation, chilling, and injury on endocrine gland function in Galleria mellonella. Arch Insect Biochem Physiol 4:113–128

Maruzzo D, Bortolin F (2013) Arthropod regeneration. In: Minelli A, Boxshall G, Fusco G (eds) Arthropod biology and evolution. Springer, Berlin

Maruzzo D, Bonato L, Brena C, Fusco G, Minelli A (2005) Appendage loss and regeneration in arthropods: a comparative view. In: Koenemann S, Jener R (eds) Crustacea and arthropod relationships. Crustacean issues 16. CRC Press, Boca Raton

Maves L, Schubiger G (1995) Wingless induces transdetermination in developing Drosophila imaginal discs. Development 121:1263–1272

Maves L, Schubiger G (1998) A molecular basis for transdetermination in Drosophila imaginal discs: interactions between wingless and decapentaplegic signaling. Development 125:115–124

McClure KD, Schubiger G (2007) Transdetermination: Drosophila imaginal disc cells exhibit stem cell-like potency. Int J Biochem Cell Biol 39(6):1105–1118

McClure KD, Sustar A, Schubiger G (2008) Three genes control the timing, the site and the size of blastema formation in Drosophila. Dev Biol 319:68–77

McCusker C, Bryant SV, Gardiner DM (2015) The axolotl limb blastema: Cellular and molecular mechanisms driving blastema formation and limb regeneration in tetrapods. Regeneration 2:54–71

Meserve JH, Duronio RJ (2015) Scalloped and Yorkie are required for cell cycle re-entry of quiescent cells after tissue damage. Development 142:2740–2751

Mito T, Inoue Y, Kimura S, Miyawaki K, Niwa N, Shinmyo Y, Ohuchi H, Noji S (2002) Involvement of hedgehog, wingless, and dpp in the initiation of proximodistal axis formation during the regeneration of insect legs, a verification of the modified boundary model. Mech Dev 114:27–35

Mitten EK, Jing D, Suzuki Y (2012) Matrix metalloproteinases (MMPs) are required for wound closure and healing during larval leg regeneration in the flour beetle, Tribolium castaneum. Insect Biochem Mol Biol 42:854–864

Miyawaki K, Mito T, Sarashina I, Zhang H, Shinmyo Y, Ohuchi H, Noji S (2004) Involvement of wingless/armadillo signaling in the posterior sequential segmentation in the cricket, Gryllus bimaculatus (Orthoptera), as revealed by RNAi analysis. Mech Dev 121:119–130

Molenaar M, van de Wetering M, Oosterwegel M, Peterson-Maduro J, Godsave S, Korinek V, Roose J, Destree O, Clevers H (1996) XTcf-3 transcription factor mediates beta-catenin-induced axis formation in Xenopus embryos. Cell 86:391–399

Mott JD, Werb Z (2004) Regulation of matrix biology by matrix metalloproteinases. Curr Opin Cell Biol 16:558–564

Muneoka K, Han M, Gardiner DM (2008) Regrowing human limbs. Sci Am 298:56–63

Nakamura T, Mito T, Tanaka Y, Bando T, Ohuchi H, Noji S (2007) Involvement of canonical Wnt/Wingless signaling in the determination of the positional values within the leg segment of the cricket Gryllus bimaculatus. Develop Growth Differ 49:79–88

Nakamura T, Mito T, Bando T, Ohuchi H, Noji S (2008a) Dissecting insect leg regeneration through RNA interference. Cell Mol Life Sci 65:64–72

Nakamura T, Mito T, Miyawaki K, Ohuchi H, Noji S (2008b) EGFR signaling is required for re-establishing the proximodistal axis during distal leg regeneration in the cricket Gryllus bimaculatus nymph. Dev Biol 319:46–55

Narbonne-Reveau K, Maurange C (2019) Developmental regulation of regenerative potential in Drosophila by ecdysone through a bistable loop of ZBTB transcription factors. PLoS Biol 17: e3000149

Needham AE (1946) Peripheral nerve and regeneration in Crustacea. J Exp Biol 22:107–109

Needham AE (1965) Regeneration in arthropods and its endocrine control. In: Kiortis V, Trampusch H (eds) Regeneration in Animals. North Holland, Amsterdam, pp 283–323

Newmark PA, Alvarado AS (2002) Not your father's planarian: a classic model enters the era of functional genomics. Nat Rev Genet 3:210–219
Nijhout HF (1998) Insect hormones. Princeton University Press, Princeton
Nijhout HF (2015) Big or fast: two strategies in the developmental control of body size. BMC Biol 13:57
Nüesch H (1968) The role of the nervous system in insect morphogenesis and regeneration. Annu Rev Entomol 13:27–44
Owlarn S, Bartscherer K (2016) Go ahead, grow a head! A planarian's guide to anterior regeneration. Regeneration 3:139–155
Pan D (2010) The hippo signaling pathway in development and cancer. Dev Cell 19:491–505
Parker NF, Shingleton AW (2011) The coordination of growth among Drosophila organs in response to localized growth-perturbation. Dev Biol 357:318–325
Pastor-Pareja JC, Wu M, Xu T (2008) An innate immune response of blood cells to tumors and tissue damage in Drosophila. Dis Model Mech 1:144–154
Pohley H (1965) Regeneration and the moulting cycle in *Ephestia kuehniella*. In: Kiortis V, Trampusch H (eds) Regeneration in animals. North Holland, Amsterdam, pp 324–330
Poodry CA, Woods DF (1990) Control of the developmental timer for Drosophila pupariation. Roux Arch Dev Biol 199:219–227
Ramet M, Lanot R, Zachary D, Manfruelli P (2002) JNK signaling pathway is required for efficient wound healing in Drosophila. Dev Biol 241:145–156
Rawlings JS, Rosler KM, Harrison DA (2004) The JAK/STAT signaling pathway. J Cell Sci 117:1281–1283
Riddiford LM (1996) Juvenile hormone: the status of its "status quo" action. Arch Insect Biochem Physiol 32:271–286
Russell MA (1974) Pattern formation in the imaginal discs of a temperature-sensitive cell-lethal mutant of Drosophila melanogaster. Dev Biol 40:24–39
Sandoval-Guzman T, Wang H, Khattak S, Schuez M, Roensch K, Nacu E, Tazaki A, Joven A, Tanaka EM, Simon A (2014) Fundamental differences in dedifferentiation and stem cell recruitment during skeletal muscle regeneration in two salamander species. Cell Stem Cell 14:174–187
Santabárbara-Ruiz P, López-Santillán M, Martínez-Rodríguez I, Binagui-Casas A, Pérez L, Milán M, Corominas M, Serras F (2015) ROS-induced JNK and p38 signaling is required for unpaired cytokine activation during Drosophila regeneration. PLoS Genet 11:e1005595
Schubiger G (1971) Regeneration, duplication and transdetermination in fragments of the leg disc of Drosophila melanogaster. Dev Biol 26:277–295
Seifert AW, Kiama SG, Seifert MG, Goheen JR, Palmer TM, Maden M (2012) Skin shedding and tissue regeneration in African spiny mice (Acomys). Nature 489:561–565
Shah M, Namigai E, Suzuki Y (2011) The role of canonical Wnt signaling in leg regeneration and metamorphosis in the red flour beetle Tribolium castaneum. Mech Dev 128:342–400
Simkin J, Sammarco MC, Dawson LA, Schanes PP, Yu L, Muneoka K (2015) The mammalian blastema: regeneration at our fingertips. Regeneration 2:93–105
Simpson P, Berreur P, Berreur-Bonnenfant J (1980) The initiation of pupariation in Drosophila: dependence on growth of the imaginal discs. J Embryol Exp Morphol 57:155–165
Singer M (1952) The influence of the nerve in regeneration of the amphibian extremity. Q Rev Biol 27:169–200
Skinner DM (1985) Molting and regeneration. In: Bliss DE, Mantel LH (eds) Biology of crustacea, integument, pigments, and hormonal processes. Academic Press, New York, pp 43–146
Skinner DM, Graham DE (1972) Loss of limbs as a stimulus to ecdysis in Brachyura (True Crabs). Biol Bull 143:222–233
Skinner A, Khan SJ, Smith-Bolton RK (2015) Trithorax regulates systemic signaling during Drosophila imaginal disc regeneration. Development 142:3500–3511
Smith-Bolton RK, Worley MI, Kanda H, Hariharan IK (2009) Regenerative growth in Drosophila imaginal discs is regulated by Wingless and Myc. Dev Cell 16:797–809

Stevens LJ, Page-McCaw A (2012) A secreted MMP is required for reepithelialization during wound healing. Mol Biol Cell 23:1068–1079

Stieper BC, Kupershtok M, Driscoll MV, Shingleton AW (2008) Imaginal discs regulate developmental timing in Drosophila melanogaster. Dev Biol 321:18–26

Stock A, O'Farrell AF (1954) Regeneration and the moulting cycle in Blattella germanica L. II. Simultaneous regeneration of both metathoracic legs. Aust J Biol Sci 7:302–307

Stocum DL (2017) Mechanisms of urodele limb regeneration. Regeneration 4:159–200

Stoick-Cooper CL, Weidinger G, Riehle KJ, Hubbert C, Major MB, Fausto N, Moon RT (2007) Distinct Wnt signaling pathways have opposing roles in appendage regeneration. Development 134:479–489

Su T, Ludwig MZ, Xu J, Fehon RG (2017) Kibra and Merlin activate the Hippo pathway spatially distinct from and independent of Expanded. Dev Cell 40:478–490 e473

Sun G, Irvine KD (2011) Regulation of Hippo signaling by Jun kinase signaling during compensatory cell proliferation and regeneration, and in neoplastic tumors. Dev Biol 350:139–151

Suzuki Y, Squires DC, Riddiford LM (2009) Larval leg integrity is maintained by Distal-less and is required for proper timing of metamorphosis in the flour beetle, Tribolium castaneum. Dev Biol 326:60–67

Švácha P (1992) What are and what are not imaginal discs: Reevaluation of some basic concepts (insecta, holometabola). Dev Biol 154:101–117

Swarup S, Verheyen EM (2012) Wnt/Wingless signaling in Drosophila. Cold Spring Harb Perspect Biol 4:a007930

Tanaka K, Truman JW (2005) Development of the adult leg epidermis in Manduca sexta: contribution of different larval cell populations. Dev Genes Evol 215:78–89

Tanaka HV, Ng NC, Yang Yu Z, Casco-Robles MM, Maruo F, Tsonis PA, Chiba C (2016) A developmentally regulated switch from stem cells to dedifferentiation for limb muscle regeneration in newts. Nat Commun 7:11069

Tomoyasu Y, Denell RE (2004) Larval RNAi in Tribolium (Coleoptera) for analyzing adult development. Dev Genes Evol 214:575–578

Tribolium-Genome-Sequencing-Consortium (2008) The genome of the model beetle and pest Tribolium castaneum. Nature 452:949–955

Truby PR (1983) Blastema formation and cell division during cockroach limb regeneration. J Embryol Exp Morphol 75:151–164

Truman JW, Riddiford LM (1999) The origins of insect metamorphosis. Nature 401:447–452

Truman JW, Riddiford LM (2002) Endocrine insights into the evolution of metamorphosis in insects. Annu Rev Entomol 47:467–500

Turkel N, Sahota VK, Bolden JE, Goulding KR, Doggett K, Willoughby LF, Blanco E, Martin-Blanco E, Corominas M, Ellul J, Aigaki T, Richardson HE, Brumby AM (2013) The BTB-zinc finger transcription factor abrupt acts as an epithelial oncogene in Drosophila melanogaster through maintaining a progenitor-like cell state. PLoS Genet 9:e1003627

Uetz GW, McClintock WJ, Miller D, Smith EI, Cook KK (1996) Limb regeneration and subsequent asymmetry in a male secondary sexual character influences sexual selection in wolf spiders. Behav Ecol Sociobiol 38:253–257

Villarreal CM, Darakananda K, Wang VR, Jayaprakash PM, Suzuki Y (2015) Hedgehog signaling regulates imaginal cell differentiation in a basally branching holometabolous insect. Dev Biol 404:125–135

Vinarsky V, Atkinson DL, Stevenson TJ, Keating MT, Odelberg SJ (2005) Normal newt limb regeneration requires matrix metalloproteinase function. Dev Biol 279:86–98

Waddington CH (1957) The strategy of the genes. A discussion of some aspects of theoretical biology. Allen & Unwin, London

Wang S, Tan XL, Michaud JP, Shi ZK, Zhang F (2015) Sexual selection drives the evolution of limb regeneration in Harmonia axyridis (Coleoptera: Coccinellidae). Bull Entomol Res 105:245–252

Wei G, Schubiger G, Harder F, Müller AM (2000) Stem cell plasticity in mammals and transdetermination in Drosophila: common themes? Stem Cells 18(6):409–414

Whyte JL, Smith AA, Helms JA (2012) Wnt signaling and injury repair. Cold Spring Harb Perspect Biol 4:a008078

Willert K, Nusse R (1998) Beta-catenin: a key mediator of Wnt signaling. Curr Opin Genet Dev 8:95–102

Woods DF, Bryant PJ (1989) Molecular cloning of the lethal(1)discs large-1 oncogene of Drosophila. Dev Biol 134:222–235

Worley MI, Setiawan L, Hariharan IK (2012) Regeneration and transdetermination in Drosophila imaginal discs. Annu Rev Genet 46:289–310

Wu S, Liu Y, Zheng Y, Dong J, Pan D (2008) The TEAD/TEF family protein scalloped mediates transcriptional output of the Hippo growth-regulatory pathway. Dev Cell 14:388–398

Yang EV, Bryant SV (1994) Developmental regulation of a matrix metalloproteinase during regeneration of axolotl appendages. Dev Biol 166:696–703

Yang EV, Gardiner DM, Carlson MRJ, Nugas CA, Bryant SV (1999) Expression of Mmp-9 and related matrix metalloproteinase genes during axolotl limb regeneration. Dev Dyn 216:2–9

Yu XL, Chang ES, Mykles DL (2002) Characterization of limb autotomy factor-proecdysis (LAF (pro)), isolated from limb regenerates, that suspends molting in the land crab Gecarcinus lateralis. Biol Bull 202:204–212

Zeidler MP, Bausek N (2013) The Drosophila JAK-STAT pathway. JAKSTAT 2:e25353

Zhang L, Ren F, Zhang Q, Chen Y, Wang B, Jiang J (2008) The TEAD/TEF family of transcription factor Scalloped mediates Hippo signaling in organ size control. Dev Cell 14:377–387

Zhou X, Riddiford LM (2002) Broad specifies pupal development and mediates the 'status quo' action of juvenile hormone on the pupal-adult transformation in Drosophila and Manduca. Development 129:2259–2269

Chapter 18
Viviparity in Two Closely Related Epizoic Dermapterans Relies on Disparate Modifications of Reproductive Systems and Embryogenesis

Szczepan M. Bilinski, Mariusz K. Jaglarz, and Waclaw Tworzydlo

Abstract Nutritional modes operating during embryonic/larval development of viviparous species range from "pure" lecitothrophy in which embryos rely solely on reserve materials (yolk spheres, lipid droplets, and glycogen particles) accumulated in the egg cytoplasm to matrotrophy in which embryos are continuously supplied with nutrients from a parental organism. Interestingly, a wide spectrum of diverse "mixed" modes employed in the embryo nourishment have also been described among viviparous species. Here, we summarize results of histochemical, ultrastructural, and biochemical analyses of reproductive systems as well as developing embryos of two closely related viviparous species of earwigs (Dermaptera), *Hemimerus talpoides* and *Arixenia esau*. These analyses clearly indicate that morphological as well as physiological modifications (adaptations) supporting viviparity and matrotrophy in *Hemimerus* and *Arixenia*, with the exception of a complex biphasic respiration, are markedly different. Most importantly, *Hemimerus* embryos complete their development inside terminal (largest) ovarian follicles, whereas *Arixenia* embryos, after initial developmental stages, are transferred to highly modified lateral oviducts, that is the uterus, where they develop until the release (birth) of larvae. The obtained results strongly suggest that viviparity in hemimerids and arixeniids had evolved independently and might therefore serve as an example of evolutionary parallelism as well as remarkable functional plasticity of insect reproduction and embryonic development.

S. M. Bilinski (✉) · M. K. Jaglarz · W. Tworzydlo
Department of Developmental Biology and Invertebrate Morphology, Institute of Zoology and Biomedical Research, Jagiellonian University in Krakow, Krakow, Poland
e-mail: szczepan.bilinski@uj.edu.pl

W. Tworzydlo, S. M. Bilinski (eds.), *Evo-Devo: Non-model Species in Cell and Developmental Biology*, Results and Problems in Cell Differentiation 68,
https://doi.org/10.1007/978-3-030-23459-1_18

18.1 Introduction

Two disparate reproductive strategies, oviparity (or free egg release) and viviparity (or embryonic incubation), have been recognized in animal kingdom. Oviparous species lay eggs comprising reserve materials that are used during embryonic development. In contrast, the embryos of viviparous species develop either inside reproductive organs or body cavity of the parent, which subsequently gives birth to active offspring. A wide spectrum of nutritional modes have been described among viviparous species. On one side of the spectrum, the embryos developing inside the parent rely entirely on reserve materials (previously) accumulated in eggs during vitellogenesis. This pattern is termed lecithotrophy. On the other, the developing embryos are continuously supplied with nutrients from a parental organism; this pattern is termed matrotrophy. Importantly, diverse intermediate patterns, in which lecitotrophy and matrotrophy are combined to a different degree, have been described in various taxa (for further details, see Wourms 1981; Wheeler 2003; Ostrovsky 2013a, b; Blackburn 1999, 2015; Ostrovsky et al. 2016).

Among insects, viviparous reproduction is not as rare as one may expect. It has been reported in 11 out of 44 orders, including cockroaches (Blattodea), aphids (Aphidoidea), strepsipterans (Strepsiptera), flies (Diptera), and earwigs (Dermaptera) (see Retnakaran and Percy 1985 for further discussion). In spite of this, structural and physiological aspects of viviparity and matrotrophy in insects are poorly characterized and remain partly obscure.

Earwigs (Dermaptera) is a relatively small insect order with more than 2000 species traditionally classified into three subgroups (suborders): free-living Forficulina and two epizoic taxa, Arixeniina (living on molossid bats) and Hemimerina (living on giant murid rats). Although the classification of the extant dermapterans is still under debate, recent morphological and molecular studies imply that both epizoic groups should not be ranked as separate suborders but are phylogenetically subordinate in the Forficulina (Klass 2001; Tworzydlo et al. 2013b; Kocarek et al. 2013). According to this concept, the clearly disparate morphology of hemimerids and arixeniids is interpreted as a result of the epizoic life strategy. The majority of dermapteran species is, like most insects, oviparous. Lecithotrophic viviparity has been described in two free-living species, *Marava arachidis* (Patel and Habib 1978) and *Chaetospania borneesis* (Kocarek 2009), while the matrotrophic one in both epizoic taxa. It is believed that in hemimerids and arixeniids viviparity (and matrotrophy) represents essential adaptations for epizoic life conditions providing nymphs with an immediate contact with the host and accelerating the life cycle (Nakata and Maa 1974).

The embryonic development of arixeniids and hemimerids drew the attention of entomologists more than a century ago (Heymons 1912; Burr and Jordan 1912). The results of these classical studies have been later discussed and reinterpreted by Hagan in his monumental work *Embryology of the Viviparous Insects* (Hagan 1951). In this chapter, we summarize and discuss the results of recent ultrastructural, histochemical, and biochemical analyses of structures and processes involved in the matrotrophic viviparity in these groups.

18.2 Hemimerids

Hemimerids live on giant rats (Gambian pouched rat, *Cricetomus gambianus*, and greater hamster rat, *Beamys major*) in sub-Saharan Africa. They feed on skin scales and fungi and are unable to survive outside their host (Nakata and Maa 1974; Kocarek et al. 2013). Traditionally, it had been suggested that hemimerids constitute a separate monophyletic suborder, Hemimerina. Recent morphological and molecular analyses, however, have demonstrated that this highly specialized dermapteran lineage should be placed inside the most derived dermapteran taxon, Eudermaptera (Klass 2001), as a sister group of the family Forficulidae (common earwigs) (Kocarek et al. 2013). In the studies summarized in this chapter, specimens of *Hemimerus talpoides* have been used.

18.2.1 Morphology of the Female Reproductive Organs and Oogenesis

The female reproductive system of *H. talpoides* is composed of paired ovaries, narrow lateral oviducts, and a short common oviduct. Each ovary consists of 6–11 asynchronously (sequentially) developing ovarioles, attached to one of the lateral oviducts (Fig. 18.1). Only three (rarely four) anterior-most ovarioles comprise ovarian follicles, whereas the others contain growing embryos (Fig. 18.1; see Sect. 18.2.2). All ovarioles are surrounded by a thick basal lamina and an external sheath composed of epithelial cells, muscle fibers, and several layers of branching tracheae (Fig. 18.1; Bilinski et al. 2017).

Hemimerus ovarian follicles consist of two germline cells (an oocyte and a single nurse cell) connected by a simple intercellular bridge and the surrounding follicular cells. The oocytes are completely devoid of canonical reserve materials (i.e., yolk spheres and lipid droplets) and filled with large translucent vacuoles separated by thin strands of the cytoplasm (ooplasm) (Fig. 18.1, insert; Bilinski et al. 2017). During final stages of oogenesis, elements of the endoplasmic reticulum (ER) present in these strands gather forming characteristic stacks of parallel closely adjoining cisternae (Fig. 18.1, insert). The aggregated cisternae lose ribosomes and consequently do not participate in protein syntheses. The fully grown oocytes are not protected and/or sealed by structurally defined continuous egg envelopes. They are instead surrounded by highly columnar somatic cells constituting a characteristic pseudostratified follicular epithelium (Fig. 18.1, insert; Bilinski et al. 2017). In contrast to oviparous species, this epithelium does not degenerate after fertilization; it remains functional and plays an important role in embryo nourishment (see Sect. 18.2.2.1).

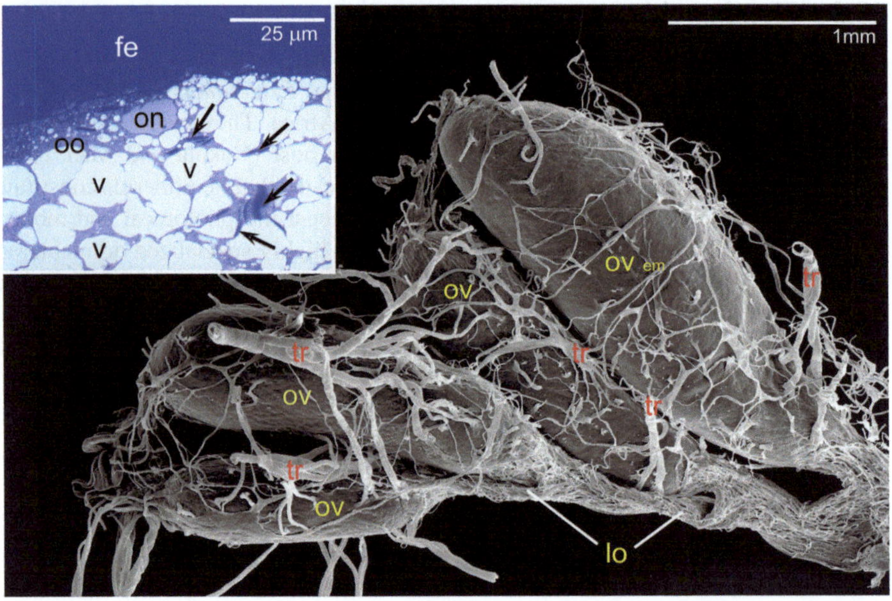

Fig. 18.1 *Hemimerus talpoides.* Ovary fragment showing the first four ovarioles (ov) attached to the lateral oviduct (lo). Three anterior-most ovarioles contain sequentially larger oocytes, whereas the last one (ov em) contains a developing embryo; trachea (tr). SEM. Insert: fragment of the fully grown oocyte (oo). The ooplasm contains large translucent vacuoles (v) separated by thin strands of the cytoplasm. Arrows indicate gathered ER cisternae. Oocyte nucleus (on), pseudostratified follicular epithelium (fe). Semithin section stained with methylene blue

18.2.2 Embryonic Development

The course of events during *Hemimerus* development had been originally described by Heymons (1912) and later reinterpreted by Hagan (1951). It transpires from Hagan's reinterpretation that the embryonic development of *Hemimerus* comprises two well-separated phases: early and late. The early phase is highly modified and reveals important adaptations to the viviparous mode of development and/or matrotrophy (see Sect. 18.2.2.1). The late phase follows, with one notable exception (see Sect. 18.2.2.2), the pattern typical for oviparous dermapterans. Importantly, both phases take place within terminal ovarian follicles and rely on nutrients transferred from mother tissues, whereas the lateral oviducts are narrow, are not transformed into uteri, and apparently do not participate in the embryo nourishment. Such a mode of viviparity has been recently referred to as intraovarian matrotrophic viviparity (Ostrovsky et al. 2016).

18.2.2.1 Early Embryogenesis: Fate and Function of the ER Stacks

A characteristic feature of the early *Hemimerus* embryogenesis is a relatively quick and distinct reduction of the embryo size (Fig. 18.2a, b). Measurements and calculations revealed that after fertilization, till the germ band formation the volume of the embryo

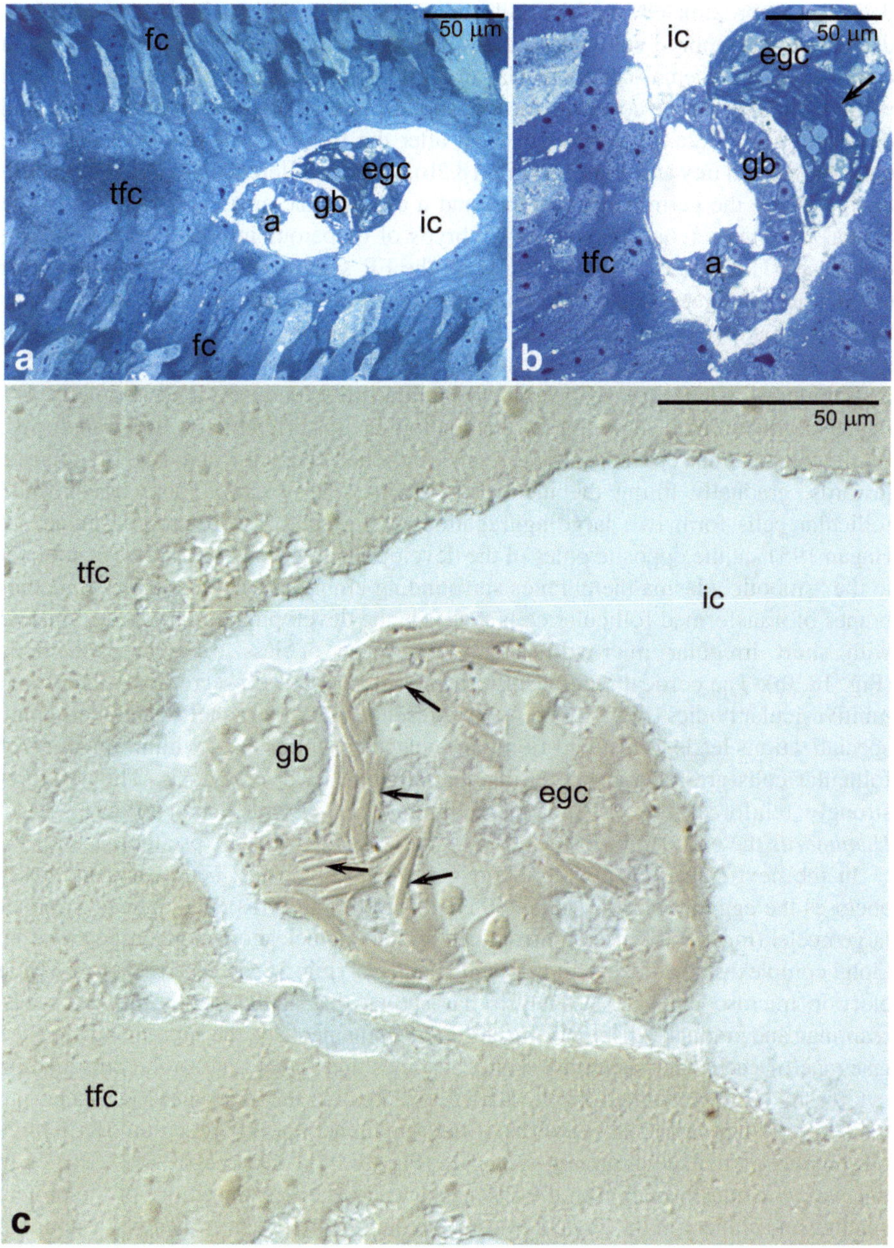

Fig. 18.2 *Hemimerus talpoides*. Early embryo developing inside the incubation cavity (ic); note a germ band (gb), amnion (a), and egg cytoplasm (egc); the latter comprises stacks of ER (arrows). Follicular cells surrounding the incubation cavity (fc), transformed follicular cells forming the placenta (tfc). (**a**, **b**) Semithin sections stained with methylene blue. (**c**) Nomarski interference contrast

decreases approximately tenfold (Bilinski et al. 2017, 2018). Such a tremendous decrease in the embryo volume coincides with the formation of a spacious incubation cavity (pseudoplacental cavity in Hagan 1951) surrounded by the pseudostratified epithelium (previously covering the oocyte). In this cavity, the embryo floats and develops (Fig. 18.2a, c). Although, even after the formation of the germ band, the embryos are still tiny and compact (Fig. 18.2b), they are differentiated into three clearly distinct parts: the germ band, amnion, and a cytoplasmic area corresponding to the central yolk mass (characteristic for embryos of oviparous insects), termed the egg cytoplasm (Hagan 1951). Interestingly, all the ER stacks present in the fully grown oocytes are segregated to the latter embryo part (Fig. 18.2c; Bilinski et al. 2018). It has been additionally shown that till the late germ band stage, the ER stacks are permanently devoid of ribosomes and apparently dormant (Fig. 18.3a; Bilinski et al. 2018).

Simultaneously to the germ band formation, the pseudostratified (formerly follicular) epithelium surrounding the incubation cavity differentiates and transforms. Some cells of this epithelium lose their epithelial character and start to migrate inwards, gradually filling the incubation cavity (Fig. 18.2a). These transformed follicular cells form two large aggregations (referred to as "pseudoplacentae" in Hagan 1951) at the opposite poles of the developing embryo (Fig. 18.2a). In contrast to the "smooth" plasma membranes surrounding embryonic cells, the plasma membranes of transformed follicular cells that face the developing embryo are furnished with short irregular microvilli and larger more or less irregular projections (Fig. 18.3b). The cortical cytoplasm of these cells comprises secretory vesicles and multivesicular bodies (Fig. 18.3b, c). The presence of such organelles and membrane specializations leads to a suggestion that molecules synthesized within transformed follicular cells are released to the incubation cavity via exocytosis. This, in turn, strongly reinforces the old idea of Heymons (1912) and Hagan (1951) that in *Hemimerus* the early embryos develop being nourished by "the pseudoplacentae".

In the next stage, the initially short germ band elongates, bends, and gradually encloses the egg cytoplasm (Hagan 1951; Bilinski et al. 2018). The latter comprises large nuclei (Fig. 18.3d, insert) surrounded by folded nuclear envelope, mitochondria, Golgi complexes, and dozens of dormant ER stacks (Fig. 18.3d, insert). Transmission electron microscopic (TEM) analysis has shown that at this stage, the ER stacks rearrange and gradually disperse (Fig. 18.3d). Simultaneously, the superficially located cisternae of each stack associate with ribosomes and translocate to the surrounding cytoplasm (Bilinski et al. 2018, 2019). Eventually, the dormant ER stacks are transformed into canonical networks of the rough endoplasmic reticulum (RER) that presumably participate in protein syntheses (Fig. 18.3d; Bilinski et al. 2018, 2019). In this way, the transformation of the ER stacks leads to the reappearance of the protein synthetic machinery in the egg cytoplasm. To our knowledge, it is the only account of such a unique process involving the transfer of inactive organelles from the oocyte to the embryo, and their activation during subsequent stages of the embryonic development in animals (Bilinski et al. 2018, 2019). Interestingly, a similar process has been described in vegetative cells of developing pollen grains in certain plant species. In this system, as in *Hemimerus*, large presumably dormant stacks of the ER present in mature vegetative cells rearrange into individual cisternae after pollination and pollen activation (Cresti et al. 1992).

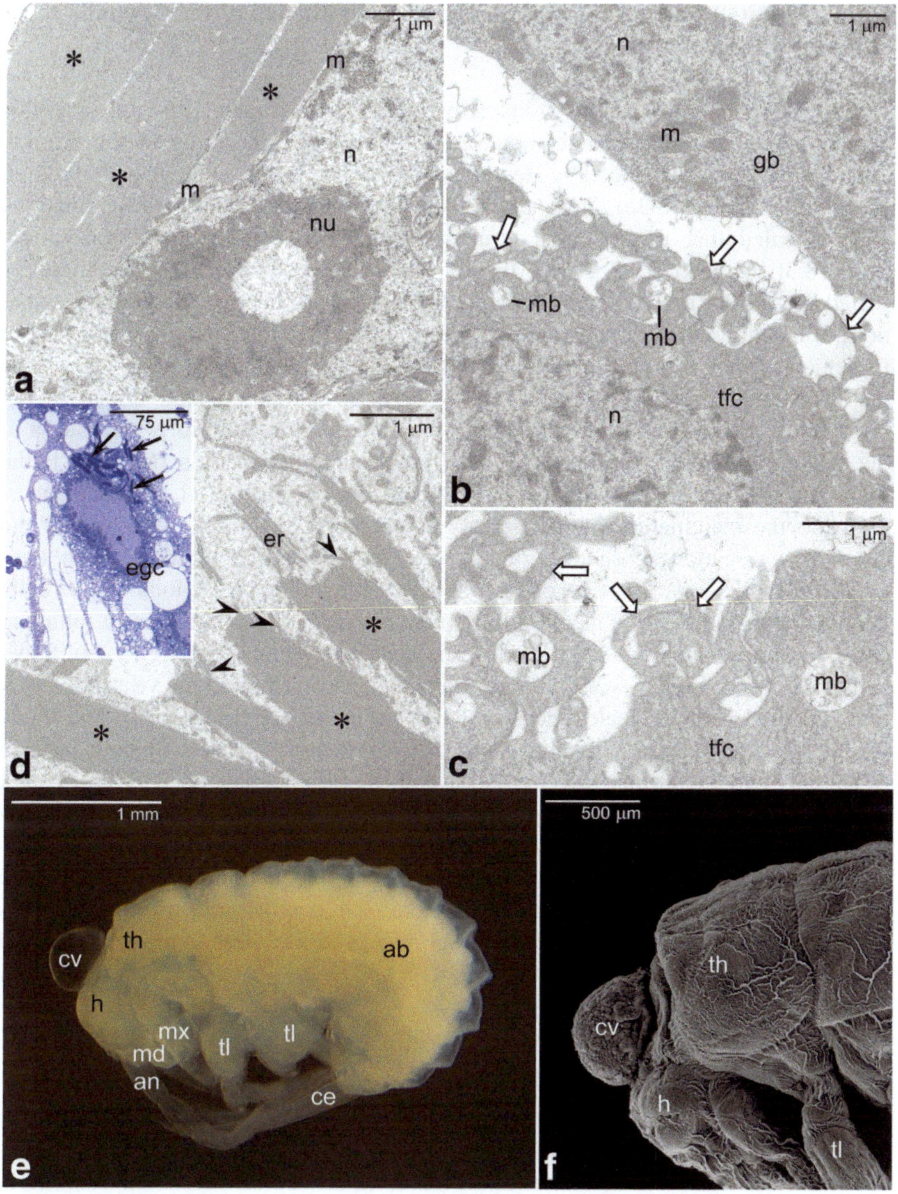

Fig. 18.3 *Hemimerus talpoides*. (**a–c**) Early embryo. (**a**) Fragment of the egg cytoplasm containing dormant stacks of endoplasmic reticulum (asterisks). TEM. (**b**, **c**) Fragment of the germ band (gb) in the vicinity of transformed follicular cells (tfc). Note irregular microvilli and irregular projections of these cells (white arrows). TEM. (**d** and insert) More advanced embryo; fragment of the egg cytoplasm (egc) with stacks of ER (asterisks; arrows in insert). Note individual cisternae (arrowheads) protruding from the stack of ER and canonical elements of endoplasmic reticulum (er). TEM, insert: semithin section stained with methylene blue. (**e**, **f**) Late embryo soon after dorsal closure and katatrepsis (the stage during which the embryo assumes its final position on the egg surface); note the cephalic vesicle (cv) located between the head (h) and the first thoracic segment

18.2.2.2 Late Embryogenesis: Formation and Function of the Cephalic Vesicle

According to classical descriptions (Heymons 1912; Hagan 1951), during this developmental phase, a characteristic cephalic vesicle [translation of the original term "Kopfblase" introduced by Heymons (1912)] is present "at the occipital region of the head" (Fig. 18.3e). Recent scanning electron microscopic (SEM) analyses have revealed that this peculiar structure forms between the head and thorax (Fig. 18.3f; Bilinski et al. 2018). In this region, the body wall is not completely sealed, leaving a single circular opening, through which the diverticulum (evagination) of the embryonic dorsal vessel extends outside the embryo participating in the formation of the vesicle (Fig. 18.4a; Bilinski et al. 2018). Interestingly, the wall of the diverticulum consists exclusively of a basal lamina of the dorsal vessel (Fig. 18.4b; Bilinski et al. 2018). Below the diverticulum (anteriorly in relation to the long embryo axis), the dorsal vessel bifurcates into two vessels: one leads to the brain and the other opens directly to the hemocoel (Bilinski et al. 2018). Concurrently to the evagination of the dorsal vessel, the somatic cells constituting the anterior aggregation reorganize once more: they lose tight contacts with their neighbors and move apart forming gradually elongating projections (Fig. 18.4c; Bilinski et al. 2018). As a result, a loose network of the distant, formerly follicular cells, arises in contact with the head of the developing embryo (Fig. 18.4d). Some cells of this network (as well as their projections) associate with the vessel diverticulum (Fig. 18.4b) and create a peculiar lattice on its surface (Fig. 18.4e). The above results indicate that the cephalic vesicle of *Hemimerus* consists of both ovarian (maternal) and embryonic layers/tissues and, therefore, may be interpreted as a placental contact. It has been suggested that this intriguing cephalic placental analog is involved in the transfer of low molecular weight compounds produced by metabolically active follicular cells and/or present in the incubation cavity, directly into the circulatory system of the embryo. Here, the contractions of muscle fibers encompassing the embryonic dorsal vessel may transport these compounds toward the brain and hemocoel (Bilinski et al. 2018).

18.3 Arixeniids

Arixeniid earwigs live in a relatively loose association with their main host, free-tailed bats (the family Molossidae) in the caves of Indonesia, Malaysia, and the Philippines. They feed on skin and gland secretions, but apparently may survive

---

Fig. 18.3 (continued) (th). (**e**) Stereomicroscope. (**f**) SEM. Abdomen (ab), antennae (an), cerci (ce), mandible (md), maxillae (mx), mitochondria (m), multivesicular bodies (mb), nucleus (n), nucleolus (nu), thoracic limbs (tl)

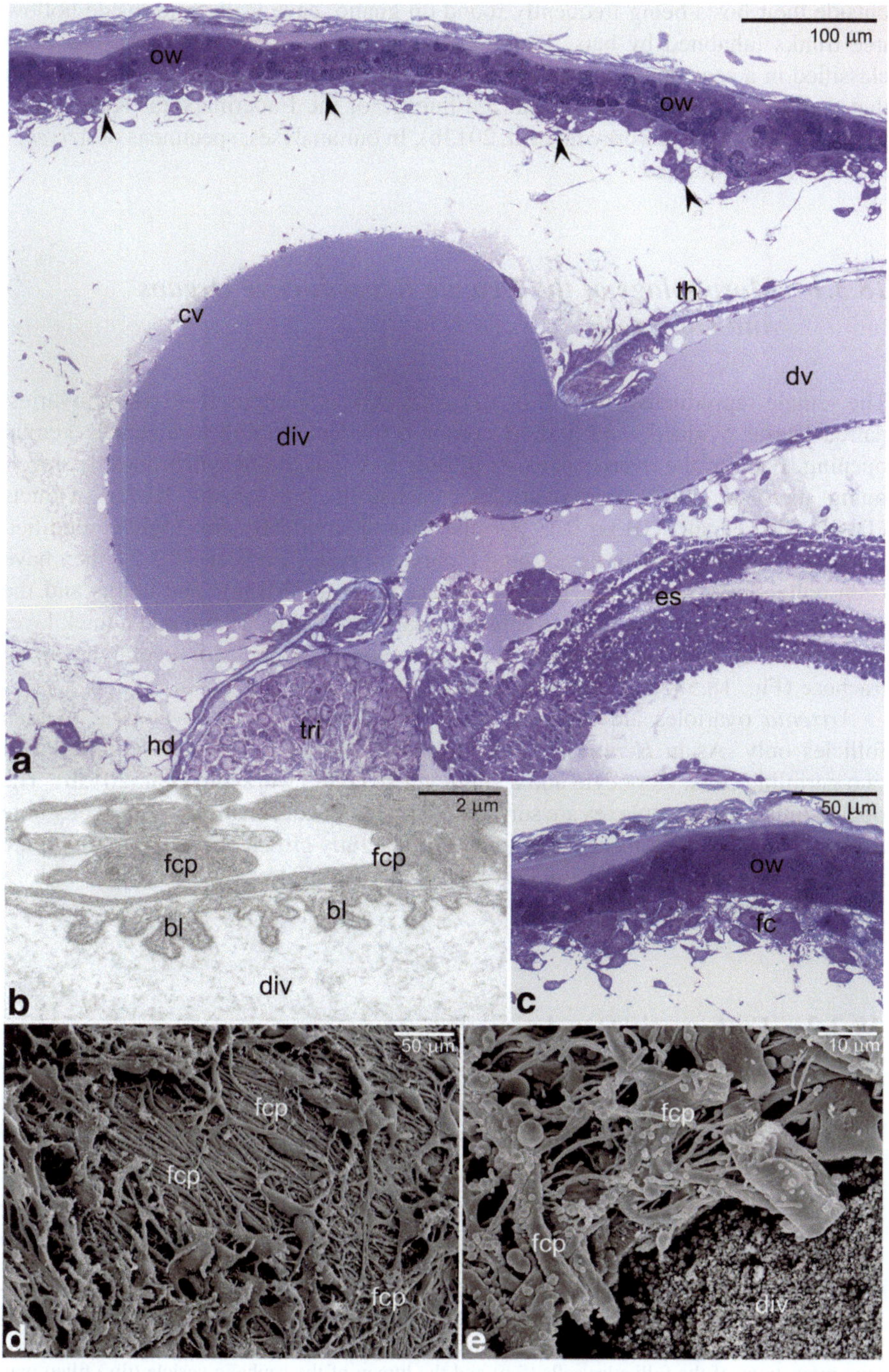

Fig. 18.4 *Hemimerus talpoides.* Morphology and ultrastructure of the cephalic vesicle. (**a**) Longitudinal section through the anterior region of an ovariole containing a late embryo; note that the cephalic vesicle (cv) comprises a diverticulum (div) of the dorsal vessel (dv); the wall of the ovariole (ow); network of follicular cells underlying the ovariole wall (arrowheads), esophagus (es),

outside their hosts being frequently found on guano, cave walls, and inside hollow tree trunks inhabited by bats. Traditionally, arixeniids (as hemimerids) had been classified in a separate suborder, Arixeniina. Recently, however, it has been shown that they represent a highly specialized lineage of the Eudermaptera (Klass 2001; Kocarek et al. 2013; Tworzydlo et al. 2013b). In our analyses, specimens of *Arixenia esau* have been studied.

18.3.1 Morphology of the Female Reproductive Organs and Oogenesis

The female reproductive system of *Arixenia esau* is composed of paired ovaries, paired lateral oviducts, and a short common oviduct leading to a female genital opening. Each of the ovaries consists of only three, as a rule synchronously developing meroistic-polytrophic ovarioles attached to one of the lateral oviducts (Tworzydlo et al. 2013a, b). As the lateral oviducts are highly modified (transformed) and contain developing embryos/larvae (see Sect. 18.3.2), they have been collectively termed the uterus (Tworzydlo et al. 2013a). The uterus and the ovarioles are surrounded by a prominent basal lamina (Fig. 18.5a) and a thick layer of external sheath composed of numerous muscle fibers and several layers of tracheae (Fig. 18.5b, d; Tworzydlo 2015).

Arixenia ovarioles are relatively short and comprise two developing ovarian follicles only. As in *Hemimerus*, the *Arixenia* ovarian follicles are composed of two germline cells: an oocyte and a single nurse cell (Tworzydlo et al. 2013b). The oocyte–nurse cell complexes are surrounded by flattened somatic cells that constitute a simple one-cell-thick follicular epithelium. The fully grown oocytes comprise quite numerous reserve materials (lipid droplets and yolk spheres) and are enclosed by a thin, two-layered egg envelope.

18.3.2 Embryonic Development

Analysis of freshly dissected material has revealed that development of the embryos in the uterus of a given female is always perfectly synchronized. This observation

---

Fig. 18.4 (continued) head (hd), first thoracic segment (th), tritocerebrum (tri). Semithin section stained with methylene blue. (**b**) Transverse section through the wall of the cephalic vesicle; note basal lamina (bl) and highly extended projections of the follicular cells (fcp) in contact with the lamina. TEM. (**c**) Fragment of the ovariole wall (ow); note follicular cells (fc) equipped with long projections. Semithin section stained with methylene blue. (**d, e**) Fractured cephalic vesicle. Note long projections of the follicular cells (fcp) and the lumen of the cephalic vesicle (div) filled with hemolymph macromolecules. SEM

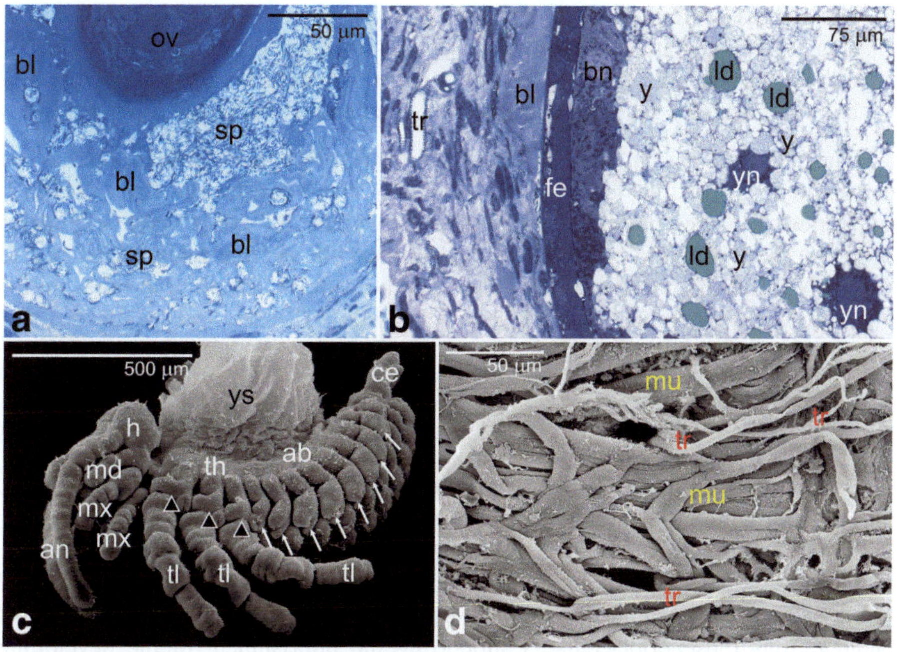

Fig. 18.5 *Arixenia esau.* (**a**) Insemination. Tangential section through the uterus wall; spermatozoa (sp) are present in contact with the basal part of an ovariole (ov). Note that uterus is covered with extensively folded basal lamina (bl). (**b**) Intraovarian phase of embryogenesis, blastoderm stage. Fragment of the blastoderm; note blastoderm nucleus (bn), follicular epithelium (fe), basal lamina (bl), trachea (tr), yolk nuclei (yn), lipid droplets (ld), yolk spheres (y). (**a**, **b**) Semithin sections stained with methylene blue. (**c**) Intrauterine phase of embryogenesis, stage 1; note that at this stage, the embryo is attached to the yolk sac (ys). (**d**) Fragment of the uterus wall covered with muscles (mu) and tracheal branches (tr). (**c**, **d**) SEM. Head (h), thorax (th), abdomen (ab), antenna (an), cerci (ce), mandible (md), maxillae (mx), thoracic limbs (tl), abdominal outgrowths (white arrows), thoracic outgrowths (triangles)

indicates that insemination (Fig. 18.5a) and the initiation of the embryonic development occur simultaneously in all six ovarioles (Bilinski and Tworzydlo 2019).

Historically, the embryonic development of *Arixenia* was considered similar if not identical to that of *Hemimerus* (Heymons 1912; Hagan 1951; Retnakaran and Percy 1985). Recent studies have indicated however that the arixeniid embryogenesis is much more complex than assumed previously and consists of two clearly separated phases that take place in different parts of the female reproductive system and involve disparate modes of nourishment (Tworzydlo et al. 2013a; Bilinski and Tworzydlo 2019):

- Early *Arixenia* embryos, as those of *Hemimerus*, develop inside the terminal ovarian follicles (the intraovarian phase) (Fig. 18.5b);
- After the germ band formation, the embryos are transferred to the uterus where they develop until the birth (the intrauterine phase). This phase is itself complex

and comprises three successive stages: young embryos before the dorsal closure (Fig. 18.5c), i.e., attached to the yolk sac and surrounded by an egg envelope (1); older embryos, after the dorsal closure, still surrounded by an egg envelope (2); and first instar larvae, after hatching, i.e., liberation from an egg envelope (3).

The first (intraovarian) developmental phase and first stage of the intrauterine one apparently rely on the lecithotrophic mode of the nourishment. This notion is supported by the presence of reserve materials in the oocyte cytoplasm (Fig. 18.5b; Tworzydlo et al. 2013b) and in the yolk sac that remains attached to the embryo till the dorsal closure. From this moment on, the embryo relies on the matrotrophic nourishment (Bilinski and Tworzydlo 2019; see Sect. 18.4).

18.3.2.1 Intraovarian Phase

This phase lasts from the fertilization till the late germ band stage (Tworzydlo et al. 2013a). Formation of a cellular blastoderm (Fig. 18.5b) and germ band proceeds as in oviparous insects. In contrast to the situation found in *Hemimerus*, the somatic (formerly follicular) cells surrounding the embryo remain flat and relatively small (Fig. 18.5b). They neither migrate toward the embryo nor transform into distinct aggregations ("pseudoplacentae") supporting the nourishment of the developing embryo.

18.3.2.2 Intrauterine Phase

The most characteristic and important feature of this developmental phase is the formation of serial multilobed outgrowths on first eight abdominal segments (Bilinski and Tworzydlo 2019). These outgrowths (termed abdominal outgrowths, AOs) are clearly recognizable as early as in stage 1 embryos (Fig. 18.5c). Interestingly, during the very same stage, morphologically similar outgrowths also develop on thoracic segments (Fig. 18.5c). These have been termed thoracic outgrowths (TOs). Both AOs and TOs are perfectly aligned and located at the same "level"—dorsally in relation to the thoracic limbs (Fig. 18.5c). However, the fates of TOs versus AOs are significantly different. Till the end of stage 2, AOs grow and ramify into lobes (Fig. 18.6a, b), whereas TOs gradually disappear. As a result, the abdomen of late stage 2 embryos is almost entirely covered with flattened multilobed and partly overlapping AOs (Fig. 18.6c). Soon after hatching from the egg envelope, AOs, or more precisely their lobes, start to rearrange and straighten out (Fig. 18.6d). Consequently, AOs of the early first instar larva protrude almost perpendicularly from the abdomen surface (Fig. 18.6d). SEM analyses revealed that each AO is ramified into four lobes (Fig. 18.7a). Morphologically, they fall into two categories that are clearly distinguishable only on sections. In each AO, one lobe appears "empty" and comprises a relatively wide hemocoelic lumen surrounded by markedly flattened epithelial cells constituting a one-cell thick epithelium (Fig. 18.7b, c). The

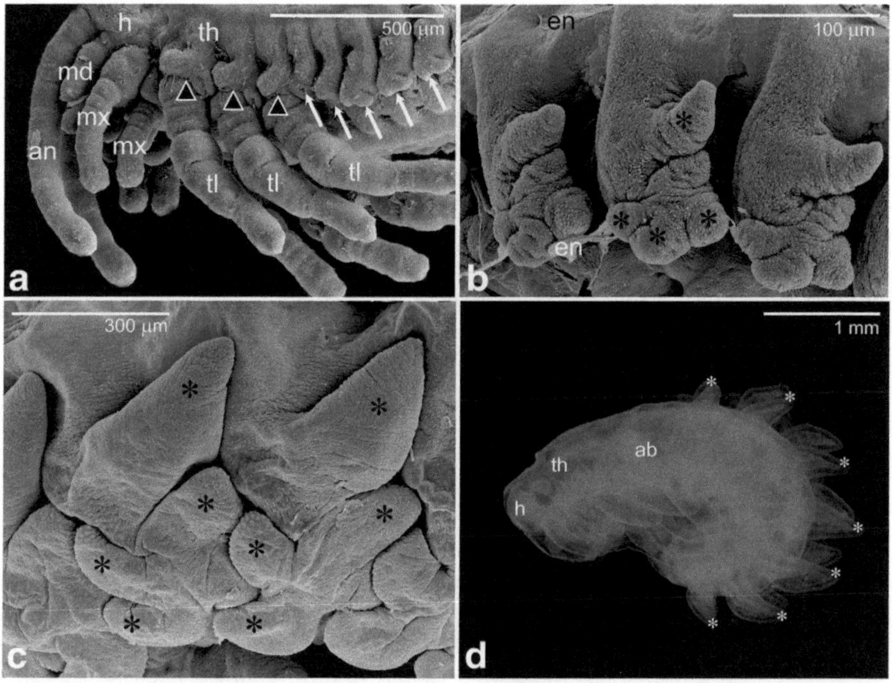

Fig. 18.6 *Arixenia esau.* (**a–c**) Intrauterine phase, stage 2. Note that as abdominal outgrowths develop, they gradually ramify into four lobes (white arrows in **a**, asterisks in **b**, **c**). (**a–c**) SEM. (**d**) Intrauterine phase, stage 3. The first instar larva after liberation from the egg envelope; note that the abdominal outgrowths (asterisks) are arisen and protrude almost perpendicularly from the abdomen. Stereomicroscope. Abdomen (ab), antennae (an), egg envelope (en), head (h), mandible (md), maxillae (mx), thoracic limbs (tl), thorax (th), thoracic outgrowths (triangles)

remaining three lobes are much more "solid" and encircled by highly prismatic epithelial cells that almost completely fill the lobe lumen (Fig. 18.7d). The lumen of both lobe types is connected with the hemocoel and contains slender strips of highly vacuolated fat body and elongated muscle fibers (Fig. 18.7b, c) attached to the epithelial cells covering the lobes (Fig. 18.7c, arrow). As the first instar larvae grow and expand, all the AO lobes come into contact with the epithelium of the uterus forming a series of discrete spot-like contact points. At these points, the two apposed epithelia, i.e., the "mother" epithelium lining the uterus lumen and the embryonic one (flat or highly prismatic) covering the given AO lobe, are directed to each other with their apical parts (i.e., microvilli) and separated only by a thin, electron-transparent (presumably at least partly permeable) larval cuticle already deposited on the surface of AOs (Bilinski and Tworzydlo 2019). Collectively, all the contact points between the mother and embryo tissues constitute a dispersed placental analog extended along the dorsal side of the larval abdomen. Although any functional analyses (for obvious reasons) have not been performed, all the accumulated data suggest that this peculiar organ could be involved in both the embryo

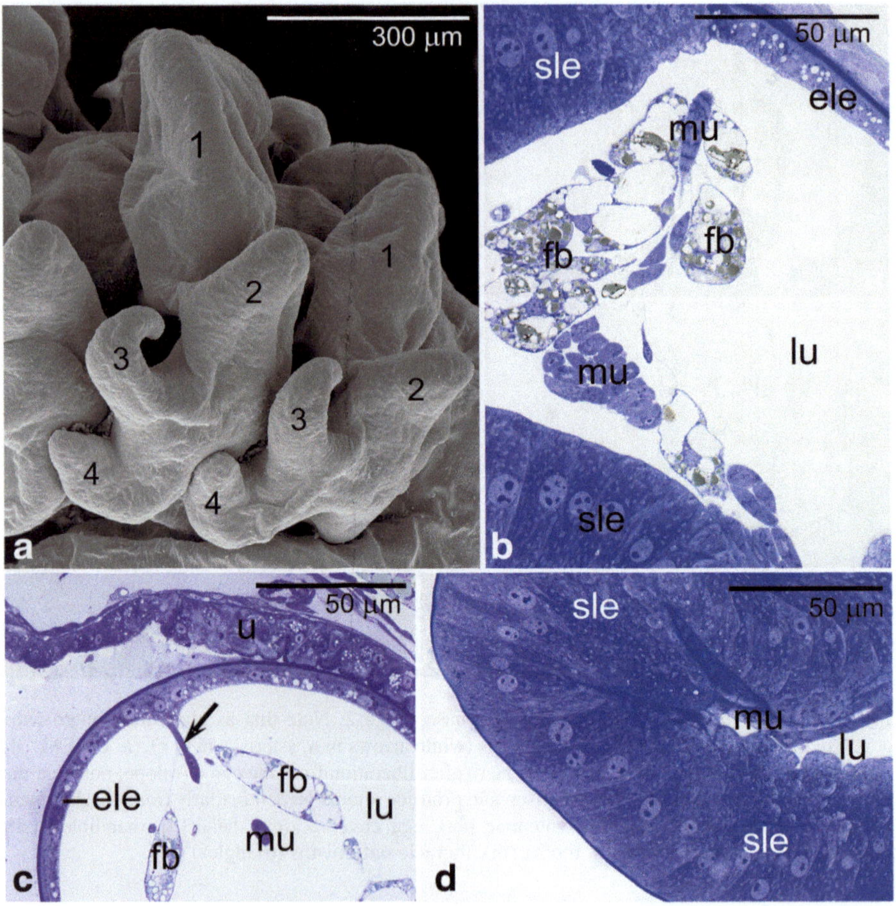

Fig. 18.7 *Arixenia esau.* Morphology of the abdominal outgrowth of the first instar larva. (**a**) Fully grown abdominal outgrowth divided into four lobes that fall into two categories: single empty lobe (1) and three solid ones (2–4). SEM. (**b**) A transitional zone between morphologically disparate epithelia covering empty and solid outgrowth lobes. (**c**) Cross section through an empty lobe. Note that the lobe is firmly surrounded by epithelium lining the uterus (u); arrow points to a muscle fiber attached to epithelium. (**d**) Cross section through a solid outgrowth lobe. (**b–d**) Semithin sections stained with methylene blue. Note that epithelial cells surrounding empty lobes (ele) are flattened, whereas those covering the solid ones (sle) are highly prismatic and resemble midgut epithelial cells of insects; fat body (fb), lumen of the lobe (lu), mucles (mu)

nourishment and the gas exchange (Bilinski and Tworzydlo 2019; Jaglarz et al. 2019; see also Sects. 18.4 and 18.5).

The discovery of the serial outgrowths on abdominal segments of *Arixenia* embryos and larvae also poses an additional, intriguing question: what is the origin (nature) of these peculiar structures? Unfortunately, at the present state of our investigations, the available data do not provide a clear answer. We hypothesize, however, that AOs might represent either outgrowths of the bases of serial wing homologs (see Clark-Hachtel et al. 2013; Clark-Hachtel and Tomoyasu 2016; Ohde et al. 2013 for the information on origin and function of wing homologs) or de novo formed, structurally complicated protrusions of the dorsal and/or lateral aspects of the abdominal segments. These two possibilities will be tested in future studies using modern techniques of 3-dimensional imaging as well as cytochemical, hybrydocytochemical, and molecular analyses.

18.4 Embryo Nourishment

As it has been stated earlier, the physiological aspects of the embryonic nutrition in epizoic dermapterans have not been analyzed to date. However, morphological and ultrastructural studies clearly indicate that the nourishment of *Hemimerus* and *Arixenia* embryos/larvae involves different processes and structures. Oocytes of *Hemimerus* are devoid of reserve materials, and consequently the development of its embryos and larvae relies purely on the matrotrophic nutrition. At least two structures take part in this process during *Hemimerus* development: aggregations of transformed "follicular cells" (pseudoplacentae in Heymons 1912; Hagan 1951) and the cephalic vesicle comprising the diverticulum of the embryonic dorsal vessel (Bilinski et al. 2018). The nourishment of *Arixenia* embryos/larvae is more complex and relies on both lecitotrophy (discussed in Sect. 18.3.2) and matrotrophy. Although the functional analyses have not been performed, it has been suggested that the solid lobes of the abdominal outgrowths (see Sect. 18.3.2.2) are a good candidate to play a role in embryo nourishment. The epithelial cells covering these lobes are equipped with microvilli and comprise numerous active organelles (Bilinski and Tworzydlo 2019). Interestingly, these cells (Fig. 18.7d) morphologically resemble the epithelial cells lining the insect midgut (compare Chapman 1985). The only essential difference lies in the "orientation" of these cell types: the apical parts of the midgut cells are directed inwards (to the lumen of this organ), whereas those of the epithelial cells covering the solid AO lobes face the mother tissue that is the uterus. Obviously, in both instances the apical cell parts face the external space (environment) from where the nutrients are obtained.

18.5 Respiration and Excretion in the Mother's Body

In addition to nourishment, viviparity poses two other physiological challenges for a mother and offspring: the removal of metabolic waste generated by embryos (excretion) and the exchange of respiratory gases (respiration).

18.5.1 Excretion

In viviparous mode of reproduction, several scenarios of excretion are possible. They include a total dependence of the progeny on the excretory system of the mother, a partial contribution of the mother, or a total excretory independence of the progeny (Blackburn 1999). The quantification of elemental composition (with energy dispersive spectrometry, EDS) and TEM analyses have shown that the latter scenario is likely to operate in *Arixenia esau* (Jaglarz et al. 2018). The larval Malpighian tubules of this species are apparently functional, and the epithelial cells that build the tubule wall accumulate dense granules (termed spheroids), characteristic for Malpighian tubules of adult insects (see Bradley 1985, 1998 for further details). Metabolic waste is subsequently transported toward the larval alimentary tract and accumulated in a cellular diaphragm occluding this tract at the midgut–hindgut junction (Jaglarz et al. 2018). This unusual process apparently prevents fouling of the mother's organism with the offspring metabolic waste, and therefore can be interpreted as an adaptation for viviparity (Jaglarz et al. 2018).

18.5.2 Respiration

The ovaries (ovarioles) and the oviducts of both *Arixenia* and *Hemimerus* are surrounded by a thick basal lamina and a prominent external sheath composed of muscle fibers and several layers of extensively branching tracheae (Figs. 18.1 and 18.5d). The latter lead to several thin-walled (not strengthened by taenidia) tubules, the tracheoles that end with tracheolar cells (Jaglarz et al. 2019). In contrast, the tracheal system of embryos and first instar larvae developing inside the ovarian follicles and/or the uterus of both epizoic dermapterans is not yet formed. The presence of extensive tracheal system (in adult females) and complete lack of the tracheae (in the first instar larvae) correspond well with the apparently disparate level of hemocyanin in the tissues of these developmental stages: both subunits (I and II) of this respiratory protein are almost entirely absent in hemocoelic fluid and tissues of adult females and abundant in homogenates of whole embryos and larvae of *Arixenia* (Jaglarz et al. 2019) and *Hemimerus* (Fig. 18.8a). These results indicate that respiration of embryos/larvae of the epizoic dermapterans is a complex process that consists of two phases. In the first, oxygen is transported via the tracheal system of the mother to the surface of the ovarian follicles/uterus and diffuses through their thin walls; in the second phase, oxygen bound to hemocyanin is distributed throughout the embryo or larva hemocoel (Jaglarz et al. 2019). Interestingly, the respiration of *Arixenia* larvae involves diffusion of oxygen through contact points between AOs and the epithelium lining the uterus (Jaglarz et al. 2019). One may imply in this context that contracting muscle fibers present inside AOs may facilitate the transfer of the hemocoelic fluid, together with hemocyanin-bound oxygen from AOs to the larval body cavity.

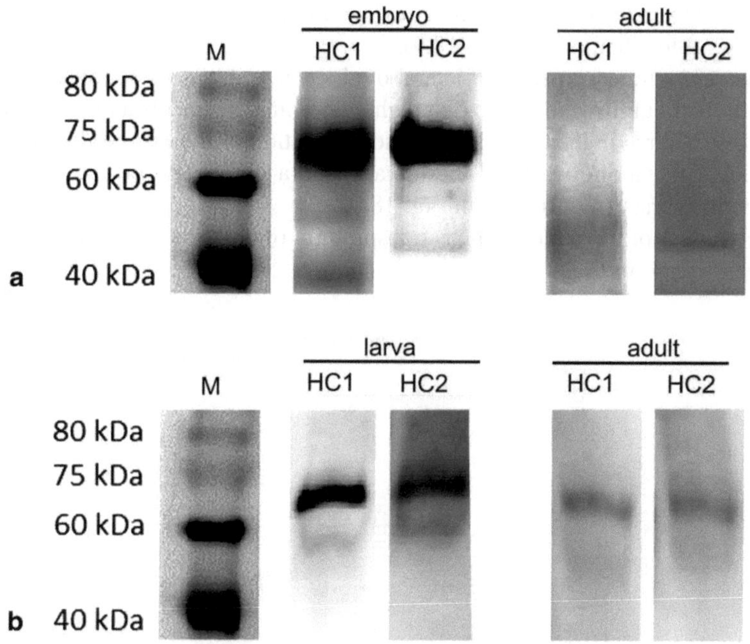

Fig. 18.8 Immunoblots demonstrating the presence (or absence) of hemocyanin subunits (HC1, HC2) in embryos/larvae and adults of *Hemimerus talpoides* (**a**) and in hemolymph of larvae and adults of *Forficula auricularia* (**b**). The hemocyanin subunits were recognized by antibodies raised against the two hemocyanin subunits of the Dubia roach, *Blaptica dubia* (the antibodies were generously provided by Prof. Burmester). The position on the immunoblots of the subunits in the studied dermapterans corresponds well with the estimated MW 75 kDa of these subunits in other arthropods (for further details, see van Holde et al. 2001; Burmester 2002)

18.6 Conclusions

Analyses of the reproductive organs and embryonic development of both *Hemimerus talpoides* and *Arixenia esau* with advanced microscopical and histochemical techniques revealed several unusual processes and/or structures (adaptations) related to specialized reproductive strategies of these species. The most important are as follows:

- In *Hemimerus*, lack of reserve materials in the egg cytoplasm; transfer of inactive (dormant) ER stacks from the ooplasm to the embryo; segregation of these organelle assemblages to the egg cytoplasm and their (re)activation at the onset of the late germ band stage; transformation of the pseudostratified follicular epithelium into two aggregations of "follicular" cells that are involved in the nourishment of the early embryo; formation of the "cephalic vesicle" during late embryogenesis; and participation of this peculiar structure in a direct transfer of nutrients from the incubation cavity to the circulatory system and hemocoel of the embryo (blue diamond in Fig. 18.9).

- In *Arixenia*, transformation of lateral oviducts into the uterus; formation of serial multilobed outgrowths on the first eight abdominal segments and involvement of these structures in the nourishment and respiration of the embryo/larva; accumulation of metabolic waste of the larva inside a diaphragm occluding the larval alimentary tract (red triangle in Fig. 18.9).
- Evolvement of a complex biphasic respiration of embryos/larvae in both species (green square in Fig. 18.9).

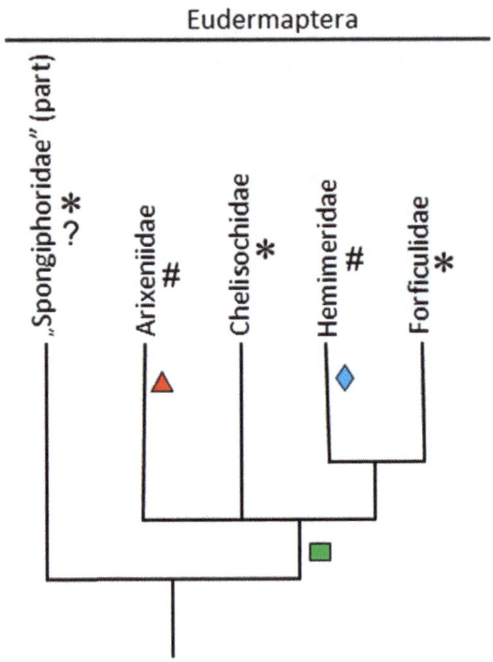

Fig. 18.9 Simplified cladogram of the Eudermaptera based on phylogenetic hypotheses of Jarvis et al. (2005), Kocarek et al. (2013), and Naegle et al. (2016). Reproductive strategies (patterns) are superimposed: oviparity is denoted with asterisks, matrotrophic viviparity with #, lecitotrophic viviparity reported in two unrelated spongiphorid species (Patel and Habib 1978; Kocarek 2009) with ?. Green square (characters 1 and 2): extensive tracheolar system in imaginal forms (1), presence of hemocyanins in larval forms (2); both characters shared by all eudermapterans (Tworzydlo and Bilinski 2008; Tworzydlo et al. 2010; Jaglarz et al. 2019; this study). Blue diamond (3–7): adaptations associated with intraovarian matrotrophy of *Hemimerus*: oocytes devoid of reserve materials and not surrounded by envelopes (3); formation of incubation cavity in which embryo develops (4); reactivation of dormant ER stacks during late germ band stage (5); aggregations of transformed "follicular" cells involved in the nourishment of early embryos (6); cephalic vesicle involved in the nourishment of older embryos (7) (Bilinski et al. 2017, 2018). Red triangle (8–10): adaptations associated with the intrauterine phase of *Arixenia* embryogenesis: oviducts transformed into the uterus (8); serial abdominal outgrowths involved in embryo nourishment and respiration (9); diaphragm occluding the alimentary tract (10) (Tworzydlo et al. 2013a; Bilinski and Tworzydlo 2019; Jaglarz et al. 2018)

We suggest that all the above listed adaptations represent apomorphic characters (evolutionary novelties) that had arisen during earwig phylogeny.

Interestingly, the structures and mechanisms underlying reproductive strategies of *Hemimerus* and *Arixenia* are significantly disparate (with one exception—see below). This, in turn, suggests that the matrotrophic viviparity had evolved in these closely related taxa (compare Fig. 18.9) independently and might therefore serve as a perfect example of evolutionary parallelism as well as functional plasticity of insect reproductive organs and embryonic development.

Finally, it should be emphasized that the reproductive strategies of the discussed dermapteran taxa share only a single character in common, i.e., a complex biphasic respiration. If so, the presence of an extensive tracheal system in imaginal forms (Figs. 18.1 and 18.5d) as well as the retention of functional hemocyanins in larvae (Fig. 18.8a, b) (both characters shared by all eudermapterans; compare Tworzydlo and Bilinski 2008; Tworzydlo et al. 2010; Jaglarz et al. 2019) might had preceded all other adaptations and therefore may represent important prerequisites for the evolvement of viviparity in both dermapteran lineages. A suggested scenario of the parallel rise of viviparity in arixeniids and hemimerids is superimposed on the cladogram (Fig. 18.9 and its legend).

Acknowledgments We would like to thank Dr. Agnieszka Rak (Jagiellonian University, Poland) for the help with western blot analyses, Prof. Thorsten Burmester (Hamburg University, Germany) for providing hemocyanin antibodies, and Dr. Ali Halajian (University of Limpopo, South Africa) for his help during collecting of *Hemimerus* specimens. We are grateful to Wladyslawa Jankowska for excellent technical support and Prof. Andrew N. Ostrovsky (University of Vienna, Austria) for valuable comments and suggestions that helped us to improve the final version of this chapter. The Hitachi S-4700 scanning electron microscope was available at the SEM Laboratory at the Institute of Geological Sciences, Jagiellonian University. The Jeol JEM 2100 transmission electron microscope was available at the Laboratory of Microscopy, Department of Cell Biology and Imaging, Institute of Zoology and Biomedical Research, Jagiellonian University. This study was funded by a research grant OPUS 11 (UMO-2016/21/B/NZ8/00560) from the National Science Centre, Poland.

References

Bilinski SM, Tworzydlo W (2019) Morphogenesis of serial abdominal outgrowths during development of viviparous dermapteran, *Arixenia esau* (Insecta, Dermaptera). Arthropod Struct Dev 49:62–69

Bilinski SM, Halajian A, Tworzydlo W (2017) Ovaries and oogenesis in an epizoic dermapteran, *Hemimerus talpoides* (Dermaptera, Hemimeridae): structural and functional adaptations to viviparity and matrotrophy. Zoology 125:32–40

Bilinski SM, Jaglarz MK, Halajian A, Tworzydlo W (2018) Unusual morphological adaptations and processes associated with viviparity in an epizoic dermapteran. PLoS One 13:e0195647

Bilinski SM, Jaglarz MK, Tworzydlo W (2019) Organelle assemblages implicated in the transfer of oocyte components to the embryo: an insect perspective. Curr Opin Insect Sci 31:1–7

Blackburn DG (1999) Viviparity and oviparity: evolution and reproductive strategies. In: Knobil E, Neill JD (eds) Encyclopedia of reproduction. Academic, London

Blackburn DG (2015) Evolution of vertebrate viviparity and specializations for fetal nutrition: a quantitative and qualitative analysis. J Morphol 276:961–990

Bradley TJ (1985) The excretory systems: structure and physiology. In: Kerkut GA, Gilbert LI (eds) Comprehensive insect physiology, biochemistry and pharmacology. Pergamon, Oxford

Bradley TJ (1998) Malpighian tubules. In: Harrison FE, Locke M (eds) Microscopic anatomy of invertebrates. Willey-Liss, New York

Burmester T (2002) Origin and evolution of arthropod hemocyanins and related proteins. J Comp Physiol B 172:95–107

Burr M, Jordan K (1912) On Arixenina Burr, a suborder of Dermaptera. In: Transactions of the second international congress of entomology, vol 2, pp 398–421

Chapman RF (1985) Structure of the digestive system. In: Kerkut GA, Gilbert LI (eds) Comprehensive insect physiology, biochemistry and pharmacology. Pergamon, Oxford

Clark-Hachtel CM, Tomoyasu Y (2016) Exploring the origin of insect wings from an evo-devo perspective. Curr Opin Insect Sci 13:77–85

Clark-Hachtel CM, Linz DM, Tomoyasu Y (2013) Insights into insect wing origin provided by functional analysis of vestigial in the red flour beetle, *Tribolium castaneum*. Proc Natl Acad Sci USA 110:16951–16956

Cresti M, Blackmore S, van Went JL (1992) Atlas of sexual reproduction in flowering plants. Springer, Berlin

Hagan HR (1951) Embryology of viviparous insects. Ronald, New York

Heymons R (1912) Uber den Genitalapparat und die Entwicklung von *Hemimerus talpoides* Walker. Zool Jahrb 15:141–184

Jaglarz MK, Tworzydlo W, Bilinski SM (2018) Excretion in the mother's body: modifications of the larval excretory system in the viviparous dermapteran, *Arixenia esau*. Protoplasma 255:1799–1809

Jaglarz MK, Tworzydlo W, Rak A, Kotula-Balak M, Sekula M, Bilinski SM (2019) Viviparity in the dermapteran *Arixenia esau:* respiration inside mother's body requires both maternal and larval contribution. Protoplasma. https://doi.org/10.1007/s00709-019-01402-1

Jarvis KJ, Haas F, Whiting MF (2005) Phylogeny of earwigs (Insecta: Dermaptera) based on molecular and morphological evidence: reconsidering the classification of Dermaptera. Syst Entomol 30:442–453

Klass KD (2001) The female abdomen of the viviparous earwig *Hemimerus vosseleri* (Insecta, Dermaptera: Hemimeridae), with a discussion of the postgenital abdomen of Insecta. Zool J Linnean Soc 131:251–307

Kocarek P (2009) A case of viviparity in a tropical non-parasiting earwig (Dermaptera, Spongiphoridae). Trop Zool 22:237–241

Kocarek P, John V, Hulva P (2013) When the body hides the ancestry: phylogeny of morphologically modified epizoic earwigs based on molecular evidence. PLoS One 8:e66900

Naegle MA, Mugleston JD, Bybee SM, Whiting MF (2016) Reassessing the phylogenetic position of the epizoic earwigs (Insecta: Dermaptera). Mol Phylogenet Evol 100:382–390

Nakata S, Maa TC (1974) A review of the parasitic earwigs. Pac Insects 16:307–374

Ohde T, Yaginuma T, Niimi T (2013) Insect morphological diversification through the modification of wing serial homologs. Science 340:495–498

Ostrovsky AN (2013a) Evolution of sexual reproduction in marine invertebrates: example of gymnolaemate bryozoans. Springer, Berlin

Ostrovsky AN (2013b) From incipient to substantial: evolution of placentotrophy in aquatic colonial invertebrates. Evolution 67:1368–1382

Ostrovsky AN, Lidgard S, Gordon DP, Schwaha T, Genikhovich G, Ereskovsky AV (2016) Matrotrophy and placentation in invertebrates: a new paradigm. Biol Rev 91:673–711

Patel PN, Habib MEM (1978) Biological and behavioral studies of an ovoviviparous earwig, *Marava arachidis* (Yersin, 1860) (Dermaptera: Forficulidae). Rev Biol Trop 26:385–389

Retnakaran A, Percy J (1985) Fertilization and special modes of reproduction. In: Kerkut GA, Gilbert LI (eds) Comprehensive insect physiology, biochemistry and pharmacology. Pergamon, Oxford

Tworzydlo W (2015) Relationship between lateral oviduct morphology and reproductive strategy in earwigs. Zool Anz 254:41–47

Tworzydlo W, Bilinski SM (2008) Structure of ovaries and oogenesis in dermapterans. I. Origin and functioning of the ovarian follicles. Arthropod Struct Dev 37:310–320

Tworzydlo W, Bilinski SM, Kocarek P, Haas F (2010) Ovaries and germline cysts and their evolution in Dermaptera (Insecta). Arthropod Struct Dev 39:360–368

Tworzydlo W, Kisiel E, Bilinski SM (2013a) Embryos of the viviparous dermapteran, *Arixenia esau* develop sequentially in two compartments: terminal ovarian follicles and the uterus. PLoS One 8:e64087

Tworzydlo W, Lechowska-Liszka A, Kocarek P, Bilinski SM (2013b) Morphology of the ovarioles and the mode of oogenesis of *Arixenia esau* support the inclusion of Arixeniina to the Eudermaptera. Zool Anz 252:410–416

van Holde KE, Miller KI, Decker H (2001) Hemocyanins and invertebrate evolution. J Biol Chem 276:15563–15566

Wheeler D (2003) Reproduction. Female. In: Resh WH, Carde RT (eds) Encyclopedia of insects. Academic, Cambridge

Wourms JP (1981) Viviparity: the maternal-fetal relationships in fishes. Am Zool 21:473–515

Chapter 19
Morphology of Ovaries and Oogenesis in Chelicerates

Izabela Jędrzejowska

Abstract The subphylum Chelicerata represents one of the oldest groups among arthropods and comprises more than a dozen orders. Representatives of particular orders differ significantly in their external morphology, reproductive biology, behavior, and structure of internal organs, e.g. of the respiratory system. However, in almost all chelicerates (excluding some mites) the female gonads show a similar architecture. In this chapter, the chelicerate-type ovary structure and the course of oogenesis are described. Structural and functional diversities of the chelicerate-type ovary in non-matrotrophic and matrotrophic arachnids are also presented.

19.1 Introduction

Chelicerata is a sister group to the remaining extant arthropods, Mandibulata, a clade which comprises Myriapoda and Pancrustacea (for more details concerning the phylogeny of arthropods, see the review by Edgecombe and Legg 2014).

Ovaries of Arthropoda are usually classified into Chelicerata and Mandibulata types according to Makioka (1988) or exogenous and endogenous types sensu Mayer and Tait (2009). However, recent studies show the occurrence of one more type of ovary in myriapods, defined as the Myriapoda type (Miyachi and Yahata 2012; Yahata et al. 2018). The major difference among the ovary types is reflected in the position of growing oocytes relative to the somatic part of the gonad. In the Chelicerata type of the ovary, oocytes protrude from the ovarian tube, and thus develop on the gonad surface (Fig. 19.1a). In the Mandibulata type, the oocytes grow inside the gonad (Fig. 19.1c). In the Myriapoda type, the oocytes develop in hemocoelic spaces of ovarian pouches. The latter are formed by the ovarian epithelium (Fig. 19.1b).

I. Jędrzejowska (✉)
Department of Animal Developmental Biology, Institute of Experimental Biology, University of Wroclaw, Poland
e-mail: izabela.jedrzejowska@uwr.edu.pl

W. Tworzydlo, S. M. Bilinski (eds.), *Evo-Devo: Non-model Species in Cell and Developmental Biology*, Results and Problems in Cell Differentiation 68,
https://doi.org/10.1007/978-3-030-23459-1_19

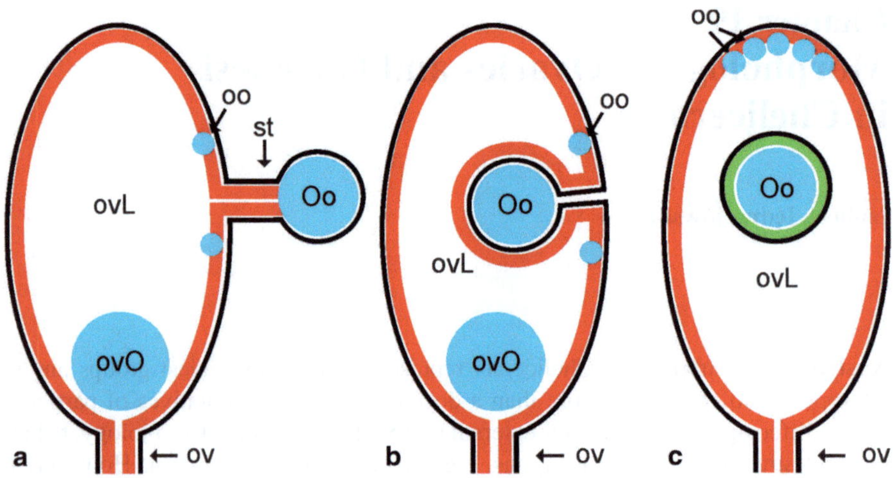

Fig. 19.1 Schematic representation of the ovary types in Arthropoda. (**a**) Chelicerata type of ovary. The oocyte is located on the gonad surface and joined to the ovarian wall by the epithelial stalk. The basal lamina is the only barrier that separates the oocyte from hemocoelic space. (**b**) Myriapoda type of ovary. The oocyte growth occurs within an ovarian pouch, the infolding of ovarian epithelium to the ovarian lumen. The basal lamina of epithelium faces the oocyte. There is no barrier between the oocyte and hemocoel. (**c**) Mandibulata (Pancrustacea) type of ovary. The oocyte grows surrounded by the follicular epithelium inside the ovary. *oo* oogonium, *Oo* oocyte, *ov* oviduct, *ovL* ovarian lumen, *ovO* ovulated oocyte, *st* oocyte stalk; red line—ovarian epithelium; black line—basal lamina; green line—follicular epithelium (modified from Yahata et al. 2018)

Extant chelicerates include Pycnogonida (sea spiders) and Euchelicerata. The latter group comprises Xiphosura (horseshoe crabs), Scorpiones (scorpions), Opiliones (harvestmen), Solifugae (solifuges, camel spiders, wind scorpions, sun spiders), Palpigradi (whip scorpions), Pseudoscorpiones (false scorpions, book scorpions), Araneae (spiders), Amblypygi (whip spiders, tailless whip scorpions), Thelyphonida (whip scorpions, vinegaroons), Ricinulei (hooded tickspiders), Schizomida (short-tailed whip scorpions), Acariformes (Actinotrichida), and Parasitiformes (Anactinotrichida). The interrelationships of Euchelicerata have yet not been established and still remain debatable (Sharma et al. 2014; Garwood and Dunlop 2014).

In chelicerates, the course of oogenesis and ovary structure are poorly known. The best investigated are diphyletic Acari, while Amblypygi, Telyphonida, and Palpigardi are largely neglected. In most chelicerates, the ovaries are panoistic, which means that all germline cells have the potential to differentiate into oocytes. In some mites, however, functional germ cell clusters diversify into oocytes and their siblings, nurse cells (trophocytes), and such ovaries are defined as meroistic or nutrimentary (Witaliński et al. 1990, 2014; Alberti and Coons 1999; Di Palma and Alberti 2001; Witaliński 2014). Panoistic ovaries of chelicerates represent the basic (plesiomorphic) condition (Alberti and Coons 1999).

In this chapter, only the structure of panoistic ovaries in chelicerates is presented.

19.2 Anatomy and Morphology of Chelicerate Ovaries

Female gonads of chelicerates exhibit anatomical variability which is based on different numbers of ovarian tubes and their branching. Branched ovaries are characteristic of basal chelicerate taxa such as Pycnogonida and Xiphosura. In pycnogonids, the gonad is a single U-shaped organ located in the trunk, and branches of the gonad extend into the legs (Miyazaki and Bilinski 2006). In Xiphosura, the ovary consists of anastomosing tubules in a reticulated network (Munson 1898; Makioka 1988). A similar "branched" structure of the female gonad is typical of scorpions. In the latter group, the gonad consists of longitudinal and transverse tubes interconnected in the form of a ladder-like structure (Matthiesen 1970; Francke 1982; Hjelle 1990; Polis and Sissom 1990). It is worth noting that the phylogenetic position of scorpions is still unresolved (Dunlop 1999; Lehmann and Melzer 2013; Sharma et al. 2015). Therefore, it remains unclear whether the "branched" state of the scorpion gonads represents a plesiomorphic character. In other non-scorpion terrestrial chelicerates, the ovarian tubes do not branch and the ovaries are either paired or unpaired. Paired and elongated ovaries that extend along the antero-posterior body axis are characteristic of e.g. spiders, solifuges, and hooded tickspiders (Foelix 1996; Klann 2009; Talarico et al. 2009), while unpaired elongated tubes are typical of schizomids (Alberti and Palacios-Vargas 2015), amblypygids (Weygoldt et al. 1972), and pseudoscorpions (Weygoldt 1969). Single and U-shaped ovaries, devoid of branches, have been found in harvestmen (Makioka 1988) and ticks (Coons and Alberti 1999).

Despite the anatomical diversity, the structure of female gonads is alike. The most characteristic feature of chelicerate ovaries is a grape-like appearance caused by the external position of oocytes. Oocytes exposed to the hemocoel cavity undergo successive stages of growth and maturation. Such localization of the growing oocytes facilitates the uptake of vitellogenins from the hemolymph by the oocyte (e.g., Fahrenbach 1999). Growth of oocytes is asynchronous, and so the ovaries contain oocytes at different developmental stages (Coons and Alberti 1999; Morishita et al. 2003; Michalik et al. 2005). The externally positioned oocytes maintain an indirect connection with the ovarian wall via oocyte stalks formed by somatic epithelial cells (Fig. 19.2a, b). Both oocytes and their stalks are covered by the basal lamina (Fig. 19.2c), which is continuous with the basal lamina of the ovarian epithelium (Fig. 19.3a).

19.3 Chelicerate Ovary in Non-matrotrophic Versus Matrotrophic Chelicerates

In chelicerates, two types of embryonic nutrition evolved: lecitrotrophy and matrotrophy. In the lecitotrophic type, nutrients for developing embryos come from the yolk that is stored in the oocyte during oogenesis. In the matrotrophic

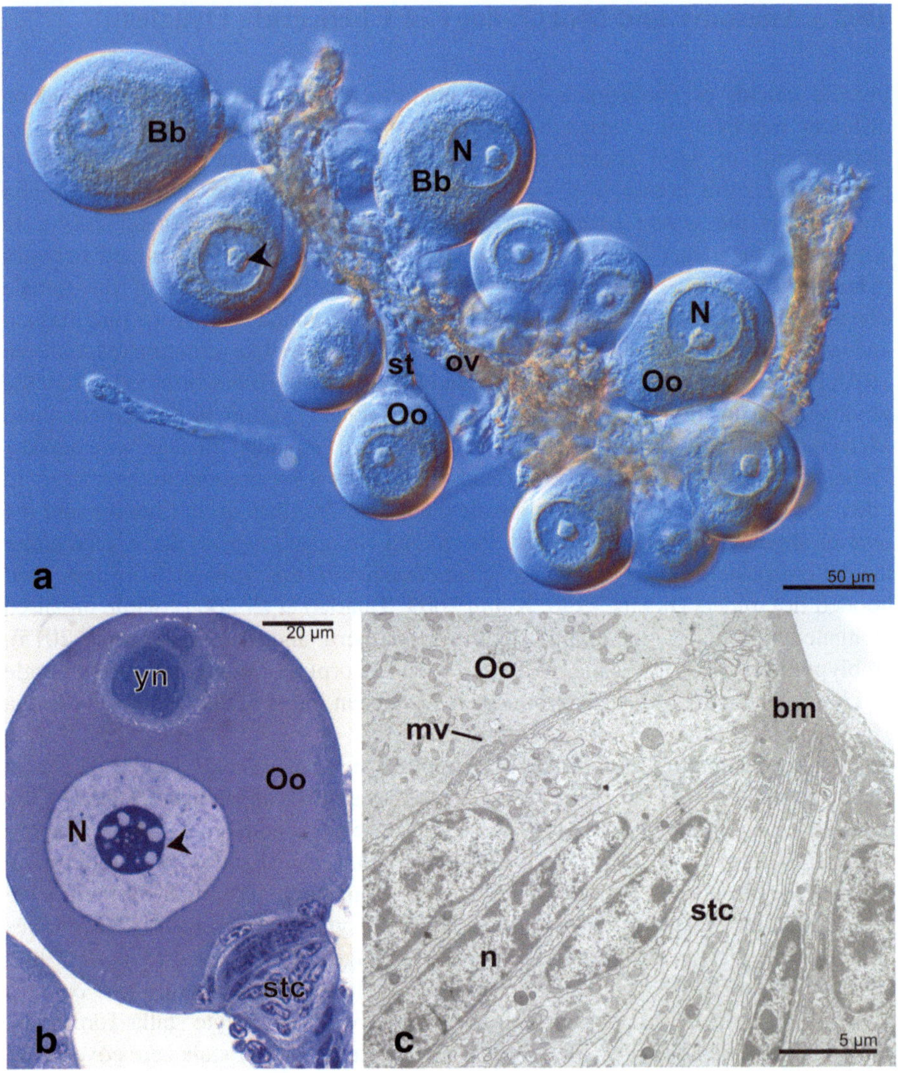

Fig. 19.2 Structure of "basic" type of chelicerate ovary in spiders. (**a**) The ovary shows a grape-like appearance. It is built of numerous oocytes at different developmental stages located on the ovary surface. Oocytes are connected to the ovarian tube by oocyte stalks. Each oocyte nucleus contains a prominent nucleolus pointed by the arrowhead. The ooplasm, in the close vicinity of the nucleus, is occupied by a dispersed Balbiani body. *Nuctenea* sp., Nomarski contrast, whole mount. (**b**) Late previtellogenic oocyte and its stalk composed of elongated stalk cells. The oocyte nucleus is located eccentrically and contains a prominent nucleolus pointed by arrowhead. A spherical and compact aggregate of organelles, the yolk nucleus, is visible in the peripheral ooplasm. *Xysticus*, LM, semithin section stained with methylene blue. (**c**) Ultrastructure of the oocyte stalk. The stalk cells are elongated and externally supported by the basal lamina. Microvilli of the stalk cells and the oocyte interdigitate. *Tegenaria ferruginea*, TEM. *Bb* Balbiani body, *bm* basal lamina; *mv* microvilli, *N* oocyte nucleus, *n* stalk cell nucleus, *Oo* oocyte, *ov* ovarian wall, *stc* stalk cells, *st* oocyte stalk, *yn* yolk nucleus

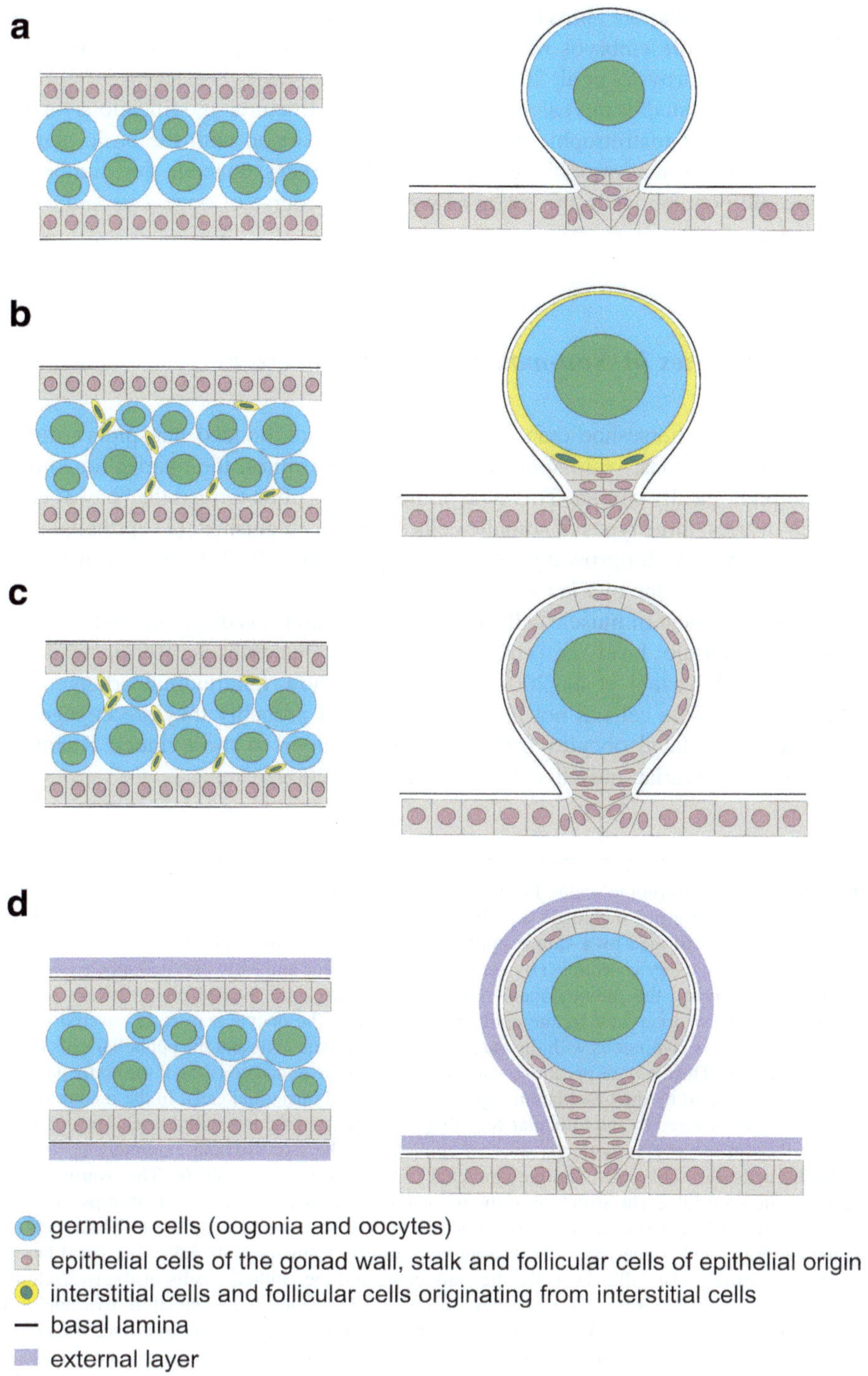

Fig. 19.3 Schematic diagram showing similarities and dissimilarities in the structure of the immature (on the left-hand side) and mature chelicerate ovaries (on the right-hand side)

type, the nutrients are provided by maternal tissues. Most chelicerates are lecitotrophic, while embryos of scorpions and pseudoscorpions are matrotrophic (Farley 1998; Ostrovsky et al. 2016).

Comparative studies reveal the occurrence of structural modifications in the female gonads of matrotrophic chelicerates in comparison with the basic plan of the ovary in non-matrotrophic groups. The differences concern the number of somatic cell populations in the ovary as well as the course of oogenesis and ovary functions.

19.3.1 Ovaries in Non-matrotrophic Chelicerates

In pycnogonids, horseshoe crabs, spiders, sun spiders, hooded tickspiders, short-tailed whip scorpions, whip spiders, ticks, and some mites, the ovaries are simple and morphologically similar. This type of the ovary is often termed the "basic" type. In such ovaries, three types of somatic cells can be distinguished: (1) epithelial stalk cells forming stalks for growing oocytes, (2) epithelial cells of the ovarian wall, and (3) muscle cells of the ovarian wall.

The arrangement of muscle cells in various chelicerates is disparate. In horseshoe crabs, the muscle layer is principally circular (Fahrenbach 1999), while in schizomids (Miyazaki et al. 2001) the muscles show longitudinal arrangement. The role of muscle cells is not fully elucidated but probably they are engaged in transport of ovulated oocytes from the ovarian lumen to the oviduct. Ricinulei, uniquely, lack ovarian muscle cells (Talarico et al. 2009).

Fig. 19.3 (continued) (not to scale). For the simplicity in mature ovaries only one oocyte is shown, and muscle cells are not included. (**a**) The "basic" type of chelicerate ovary. The wall of the immature ovary is formed by a simple epithelium externally supported by the basal lamina. The internal part of the ovary is occupied by germline cells (oogonia and early previtellogenic oocytes). In the mature ovary, the previtellogenic/vitellogenic oocytes are located on the ovary surface connected to the ovarian wall by the epithelial stalk cells. The oocyte surface is covered by the basal lamina that is continuous with the basal lamina of the stalk cells and the epithelial cells of the ovarian wall. (**b**) The ovary of pseudoscorpions. The immature gonad consists of the epithelial cells externally supported by the basal lamina, germline cells, and interstitial cells. In the mature ovary, the oocytes are covered by the follicular cells that originate from the interstitial cells. The epithelial stalk cells join the oocyte with the ovarian epithelium. The basal lamina externally covers the follicular cells, the stalk cells, and the epithelial cells of the ovarian wall. (**c**) The ovariuterus of apoikogenic scorpions. The structure of the immature gonad is comparable to that in pseudoscorpions. In the mature apoikogenic ovariuterus, the oocyte is surrounded by the follicular cells of epithelial origin and connected to the ovariuterine wall by epithelial stalk cells. The basal lamina supports externally the follicular cells, the stalk cells, and the epithelial cells of the ovariuterine tubule. (**d**) The ovariuterus of katoikogenic scorpions. The wall of the immature ovariuterus is bilayered, and the germline cells are located in the internal part of the gonad. In the mature gonad, the oocyte grows inside the diverticulum. The wall of the diverticulum is bilayered and consists of the external and internal layers. In the internal layer, the follicular and stalk cells are distinguishable

Transformation of the immature ovary anlagen to the mature gonad is a multistep process and in the "basic" type of ovary it follows a similar pattern (for comparison, see e.g. Munson 1898; Kessel and Beams 1980; Makioka 1988; Morishita et al. 2003; Michalik et al. 2005; Jędrzejowska and Kubrakiewicz 2007; Alberti and Palacios-Vargas 2015). At the initial stages of gonad maturation, the ovary is cylindrical and germline cells (oogonia and early meiotic oocytes) occupy the ovary interior. The arrangement of the germline cells within the ovary is not regular. With the onset of previtellogenesis, the first stage of oocyte growth, structure of the ovary considerably changes. Previtellogenic oocytes increase in size and gradually bulge into the hemocoel cavity. Concurrently, the volume of previtellogenic oocytes progressively grows. The basal lamina of epithelial cells constituting the ovarian wall becomes conspicuously extended over the surface of growing oocytes and forms a pouch or a sac that establishes the barrier between the growing oocyte and the hemocoel cavity. The epithelial cells remaining in contact with proximal (facing the ovarian wall) oocyte pole differentiate and become the presumptive stalk cells. The events leading to the protrusion of oocytes are not synchronized in a given ovary and terminate during advanced previtellogenesis. The oocytes remain connected to the ovarian epithelium by fully formed oocyte stalks, while the most of the oocyte surface is invested by the basal lamina (Fig. 19.3a).

The stalk cells (also called funicle/peduncle/pedicel cells) form the oocyte stalk that connects the growing oocyte and the ovarian wall. The length and shape of the stalks are not uniform in chelicerate taxa. For example, in some spiders, the stalks are short and goblet-shaped (Jędrzejowska and Kubrakiewicz 2007), in solifuges crescent-shaped (Klann 2009), and in schizomids funnel-shaped (Alberti and Palacios-Vargas 2015).

The principal role of the oocyte stalks is structural and relies on mechanical support for the externally positioned oocytes. Although the stalk cells do not form a complete epithelial cover on the oocyte surface, they exhibit some features characteristic of follicular cells, and therefore they should be considered as follicular cell counterparts. First, the stalk cells are closely apposed to the oocyte and/or the vitelline membrane that covers the oocyte. In many investigated chelicerates, including pycnogonids, horseshoe crabs, short-tailed whip scorpions, spiders, harvestmen, and ticks, the contact between oocytes and stalk cells is maintained by a microvillous meshwork formed by both cell types (Dumont and Anderson 1967; Witaliński and Żuwała 1981; Coons and Alberti 1999; Fahrenbach 1999; Michalik et al. 2005; Miyazaki and Bilinski 2006; de Oliveira et al. 2007; Bilinski et al. 2008; Alberti and Palacios-Vargas 2015) (Fig. 19.2c). Second, the stalk cells show high metabolic activity. A "nutritive" role of the stalk cells has been suggested in harvestman (Witaliński and Żuwała 1981), spiders (Morishita et al. 2003), pycnogonids (Miyazaki and Bilinski 2006; Bilinski et al. 2008), and horseshoe crabs (Munson 1898; Dumont and Anderson 1967). In the tick *Amblyomma triste*, the stalk cells, similar to insect follicular cells (Büning 1994), are involved in providing compounds for oocytes during vitellogenesis (de Oliveira et al. 2007). In *Schizomus palaciosi,* the stalk cells secrete a material that is deposited between the oolemma and the basal lamina. The authors suggest that the material is added to the basal lamina to form a thicker pouch (Alberti and Palacios-Vargas 2015).

19.3.1.1 Oocyte Growth and Ovulation

Externally positioned oocytes continue previtellogenic growth, and undergo vitellogenesis (the second step of oocyte growth) and maturation. Previtellogenic oocytes enlarge and gather an increasing number of organelles and different classes of macromolecules (RNA and proteins). Because the oocyte growth is not supported by sibling nurse cells (trophocytes), the oocyte itself synthesizes and produces all necessary cytoplasmic components. The oocyte nucleus (germinal vesicle) shows high transcriptional activity. Numerous pore complexes in the nuclear envelope of previtellogenic germinal vesicles and the presence of perinuclear granules indicate occurrence of ribonucleoprotein transport between the germinal vesicle and the ooplasm (Osaki 1972; Kessel and Beams 1980; El Shoura et al. 1989; Shatrov 1997; Michalik et al. 2005; Alberti and Palacios-Vargas 2015). Prominent nucleoli have been described in previtellogenic oocytes of harvestmen (Kessel and Beams 1980), spiders (Jędrzejowska and Kubrakiewicz 2007), schizomids (Miyazaki et al. 2001; Alberti and Palacios-Vargas 2015), and ticks (El Shoura et al. 1989; Coons and Alberti 1999) (Fig. 19.2a, b).

During vitellogenesis, the rapidly growing oocytes accumulate yolk spheres, lipid droplets, and glycogen granules (Fig. 19.5b). It is widely accepted that in chelicerates vitellogenesis is of a mixed type and relies on both auto- and heterosynthetic activity. Initially, the yolk is synthesized by the oocyte and then the precursors of the yolk are taken up in the course of endocytosis from external sources. Vitellogenic oocytes are equipped with microvilli that facilitate uptake of vitellogenins (Dumont and Anderson 1967; Kessel and Beams 1980; Shatrov 1997; Alberti and Palacios-Vargas 2015).

In some spider species, vitellogenic growth of the oocyte is supported by the occurrence of a complex assemblage of organelles termed the Balbiani body. The Balbiani body is a conservative structure present in many invertebrates and vertebrates including humans (Kloc et al. 2004). Among the main components of the Balbiani bodies are fibrogranular *nuage* material and mitochondria. What is very important, the Balbiani bodies in different animal groups exhibit substantial structural modifications, which indicates their multifunctional potential. One of the best established functions of the Balbiani body is participation in germ plasm formation (Kloc and Etkin 1995; Kloc et al. 1998, 2004, 2014). The Balbiani body is also involved in mitochondria selection (Tworzydlo et al. 2014, 2016, 2017; Bilinski et al. 2017).

In spiders, two morphological types of Balbiani bodies have been described (Jędrzejowska and Kubrakiewicz 2007, 2010; Kloc et al. 2014) and one of them, termed the yolk nucleus, markedly differs in structure, behavior, and functions from the Balbiani bodies of other animal species (Fig. 19.2a, b; Osaki 1972; Guraya 1979; Kloc et al. 2004). The functioning of the yolk nucleus has been recently described in

Clubiona sp. (Jędrzejowska and Kubrakiewicz 2010). In this species, the yolk nucleus is a spherical body that consists of a core and a cortex and occupies the fixed juxtanuclear position. The components of the yolk nucleus include an elaborate network of microfilaments, intermediate-like filaments, mitochondria, and annulate lamellae. To date, only one function of this complex organelle assemblage has been elucidated: they serve as centers of lipid droplet biogenesis (Fig. 19.5a; Jędrzejowska and Kubrakiewicz 2010).

Growth of oocytes coincides with formation of an additional cover on the oocyte surface, a vitelline membrane (envelope). In contrast to insect, the vitelline envelope is not formed during final stages of oogenesis. Its deposition in the perivitelline space (the space between the oolemma and basal lamina) starts much earlier: during previtellogenesis [in pycnogonids and harvestmen (Kessel and Beams 1980; Bilinski et al. 2008)], early vitellogenesis [horseshoe crabs, harvestmen, and spiders (Dumont and Anderson 1967; Witaliński and Żuwała 1981; Michalik et al. 2005)], or vitellogenesis [Ricinulei (Talarico et al. 2009)]. The vitelline membrane is a secretory product of the oocyte (Kessel and Beams 1980; Witaliński 1986; Coons and Alberti 1999; Bilinski et al. 2008; Talarico et al. 2009).

Mature oocytes undergo ovulation which relies on the passage of oocytes via the stalks to the ovarian lumen (Fahrenbach 1999; Morishita et al. 2003; Klann 2009). Ovulation is usually rapid and therefore often overlooked by students of chelicerate oogenesis. This process has been described in detail only in the spider, *Loxosceles intermedia* (Morishita et al. 2003). In this species, oocytes undergo dramatic changes in shape. They become squeezed between the stalk cells, which change positions and form a narrow passageway. Following ovulation the basal lamina remains on the ovary surface as an empty sac with a creased surface (Alberti and Coons 1999; Coons and Alberti 1999; Fahrenbach 1999; Morishita et al. 2003; Michalik et al. 2005; Klann 2009; Talarico et al. 2009). It is worth noting that after ovulation the stalk cells usually degenerate (Talarico et al. 2009). Mechanisms that govern ovulation in chelicerates remain unclear.

19.3.2 Ovaries in Matrotrophic Chelicerates

In pseudoscorpions, the oocytes contain mainly lipid droplets, and the yolk spheres are inconspicuous and scarce (Fig. 19.5c; Badian and Ogorzałek 1982; Jędrzejowska et al. 2013). At the postovulatory stage, the ovary is engaged in secretion of a nutritive fluid for developing embryos. The female keeps the embryos in a brood sac located on the ventral side of the abdomen (opisthosoma) (Weygoldt 1969; Makioka 1976).

In viviparous scorpions, embryos develop in the modified ovaries, where they are provided with nutrients. Hence, in scorpions the female gonad is termed "ovariuterus" (Farley 1999). In scorpions, two main developmental types are distinguished: apoikogenic and katoikogenic (Laurie 1896). In the apoikogenic type, the oocytes are either large and rich in yolk or small and yolkless and develop on the gonad surface within the follicles connected to the wall of the ovariuterus by means of stalks. Embryos develop in the lumen of the ovariuterine tubules and are nourished with yolk deposited in the oocyte cytoplasm as well as nutrients that pass from the hemolymph through the gonad wall and a placenta-like structure (Francke 1982; Farley 1996, 1998). In the more derived katoikogenic type, both the oocyte growth and the embryo development take place in the blind outpocketings of the ovariuterine tubules called diverticula. The oocytes are small and yolkless (Fig. 19.5d). The apical part of embryonic diverticulum transforms into an elongated appendix that acts as a feeding apparatus and transports nutrients from the hepatopancreas to embryos (Mathew 1956; Polis and Sissom 1990; Farley 1999).

In pseudoscorpions and scorpions, the female gonads consist of a higher number of somatic cell populations. Unlike other chelicerates, in both taxa the oocytes grow surrounded by follicular cells (Weygoldt 1969; Makioka 1979; Badian and Ogorzałek 1982; Farley 1998; Soranzo et al. 2002).

Pseudoscorpions Among pseudoscorpions, differentiation of ovarian somatic cells has been described in *Garypus japonicus* (Makioka 1976, 1979) and *Chelifer cancroides* (Badian and Ogorzałek 1982; Jędrzejowska et al. 2013). Comparative analyses have shown that in this chelicerate subgroup five types of somatic cells can be distinguished: (1) epithelial cells of the ovarian wall, (2) stalk cells, (3) interstitial cells, (4) follicular cells, and (5) muscle cells. The latter form a rudimentary ovarian musculature.

The most comprehensive data on the ovary structure in pseudoscorpions come from studies of *C. cancroides* (Jędrzejowska et al. 2013). In this species, the immature ovaries contain two types of somatic cells: the epithelial cells forming the ovarian wall and the interstitial cells that occupy the internal part of the ovary (Fig. 19.3b). The interstitial cells are equipped with long, thin cellular processes that penetrate among the germline cells. These cells invest the oocytes protruding to the hemocoel. Somewhat later, the interstitial cells differentiate into follicular cells that completely surround the oocyte surface.

The follicular cells do not exhibit apico-basal polarization typical of epithelial cells. They are elongated and show a peculiar arrangement that resembles flower petals (Fig. 19.4a). Before ovulation, the follicular cells start to degenerate (Fig. 19.5c) that probably triggers this process. In *Chelifer* the follicular cells do not actively participate in oogenesis. The role of the follicular cells is most likely mechanical and relies on preventing premature ovulation (Jędrzejowska et al. 2013).

In *C. cancroides*, during advanced stages of oogenesis, the oocyte stalks are tubular structures and their lumen is continuous with the lumen of the ovary. The stalk cells show synthetic and secretory activities at the preovulatory and postovulatory stages (Jędrzejowska et al. 2013). Before ovulation, they secrete a granular chorion deposited on the surface of the oocytes when they pass through the lumen of the stalk. After ovulation, the stalk cells undergo significant hypertrophic growth and together with epithelial cells lining the ovarian lumen secrete material which serves as a nutritive fluid for developing embryos.

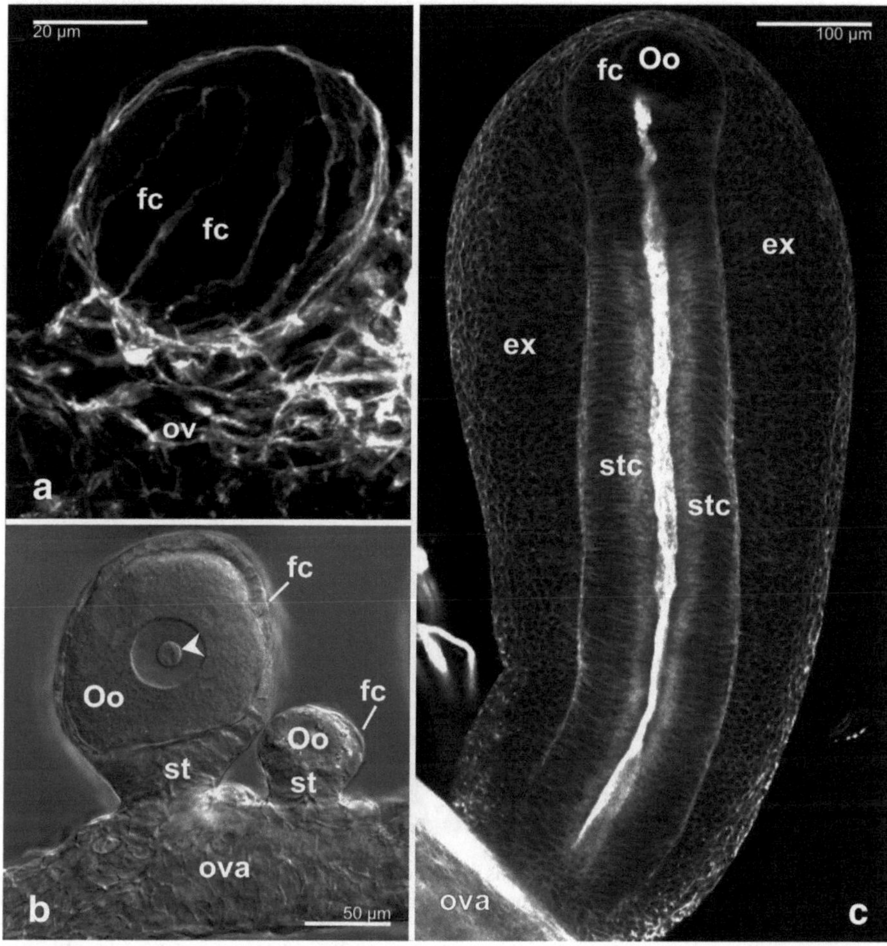

Fig. 19.4 Somatic cells that accompany growing oocytes in matrotrophic chelicerates. (**a**) Pseudoscorpion, *Chelifer cancroides*. The surface of the oocyte is covered with petal-like follicular cells. (**b**) Apoikogenic scorpion, *Euscorpius italicus*. The oocytes are covered by canonical follicular epithelium and connected to the ovariuterine tubule by epithelial stalks. Arrowhead denotes prominent nucleolus in the oocyte nucleus. (**c**) Katoikogenic scorpion, *Opistophthalmus boehmi*. The oocyte is located in the terminal part of diverticulum. The wall of diverticulum consists of the internal follicular and stalk cells and the external cell layer. (**a**, **c**) Confocal microscope; whole mounts stained with rhodamine-conjugated phalloidin, (**b**) Nomarsky optics. *ex* external layer of diverticulum, *fc* follicular cells, *Oo* oocyte, *ov* ovarian wall, *ova* ovariuterine tubule, *stc* stalk cells, *st* oocyte stalk

Apoikogenic Scorpions The apoikogenic type of development is characteristic of scorpions from the families Bothriuridae, Buthidae, Chactidae, Chaerlidae, Euscorpiidae, Iuridae, and Vaejovidae (Laurie 1896; Francke 1982; Polis and Sissom 1990). In their ovariuteri, as in pseudoscorpions five somatic cell types are distinguished: (1) epithelial wall cells, (2) interstitial cells, (3) stalk cells,

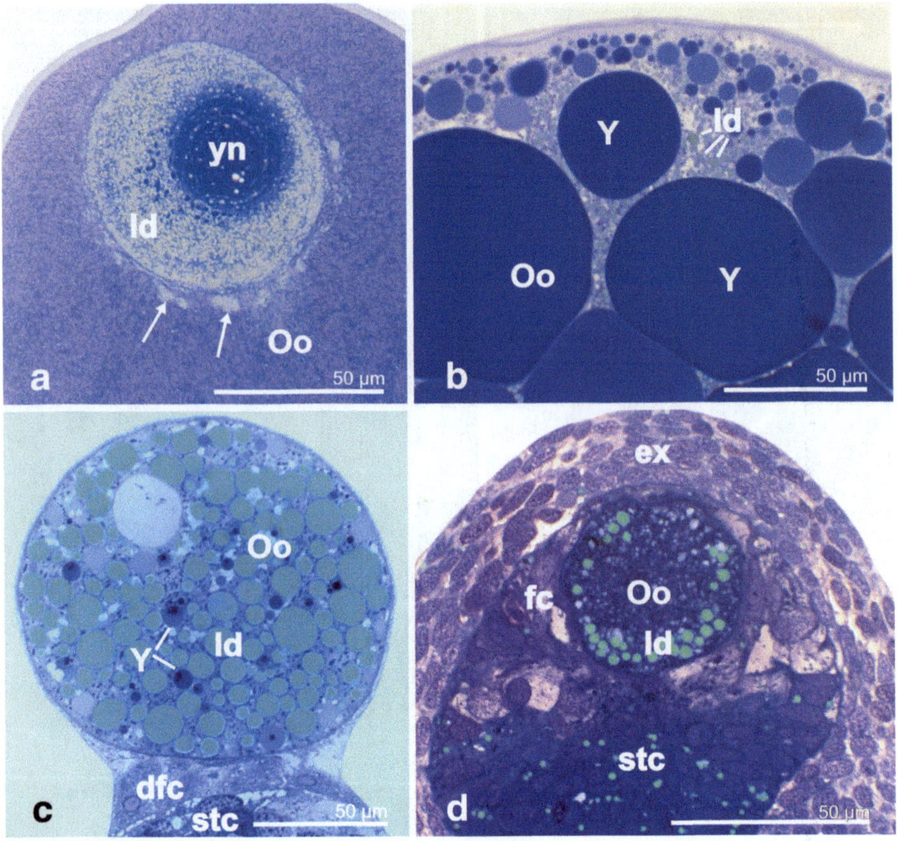

Fig. 19.5 Accumulation of reserve materials in vitellogenic oocytes. (**a**) In the spider *Clubiona* sp., the ooplasm of early vitellogenic oocyte contains the prominent yolk nucleus. Numerous lipid droplets occupy the cortex of the yolk nucleus, while some of them (denoted by arrows) are visible in the surrounding ooplasm. (**b**) A fragment of the late vitellogenic oocyte of the spider *Xysticus* sp. The ooplasm is filled with large and numerous yolk spheres and less numerous lipid droplets. (**c**) In the pseudoscorpion *Chelifer cancroides*, the ooplasm of the late vitellogenic oocyte is filled with large and numerous lipid droplets and a few small yolk spheres. (**d**) Late vitellogenic oocyte of the katoikogenic scorpion *Opistophthalmus boehmi* is small and its cytoplasm contains only a few inconspicuous lipid droplets. (**a–d**) LM, semithin sections stained with methylene blue. *ex* external layer of diverticulum, *dfc* degenerating follicular cells, *ld* lipid droplets, *Oo* oocyte, *stc* stalk cells, *Y* yolk spheres, *yn* yolk nucleus

(4) follicular cells, and (5) muscle cells. However, the fate of particular populations is different. Detailed analysis of the gonad development in the euscorpiid scorpion *Euscorpius italicus* (Jędrzejowska et al. 2014) has shown that as in pseudoscorpions, the immature ovariuterus is composed of externally located epithelium, and internally situated interstitial cells (Fig. 19.3c; for comparison, see also Fig. 19.3b). The latter are scattered among the germline cells. During protrusion of oocytes to the hemocoel cavity, the interstitial cells degenerate. In consequence, the lumen of the ovariuterine tubules appears. Bulging oocytes elevate the epithelium lining the

lumen of the ovariuterine tubules. The epithelial cells remain on the surface of growing oocytes and differentiate into follicular cells. Some other epithelial cells drawn from the epithelium stay at the base of the oocyte and form the oocyte stalk, which joins the oocyte with the wall of the ovariuterine tubule (Fig. 19.3c, 19.4b). Thus, in apoikogenic scorpions, in contrast to pseudoscorpions, the follicular cells originate from epithelial cells (Farley 1998; Soranzo et al. 2002; Jędrzejowska et al. 2014).

The role of the follicular cells is controversial. It has been suggested that in *E. carpathicus* the follicular cells are engaged in vitellogenesis synthesizing and/or transporting the oocyte reserve materials (Soranzo et al. 2002). In *E. italicus*, the apical membranes of the follicular cells facing the oocyte do not form microvilli and their cytoplasm contains only a few organelles. It indicates that they do not actively participate in vitellogenesis and formation of the vitelline membrane.

In *E. italicus*, as in other chelicerates, formation of the oocyte stalks relies on progressive differentiation of the epithelial cells of the gonad wall. In this species, the stalk cells are highly ordered and form a simple columnar epithelium. Basal parts of the stalk cells adhere to the basal lamina, whereas the apical ones form elaborate finger-like cytoplasmic processes supported by submembrane bundles of actin filaments. With these processes the stalk cells are tightly joined in a zip-like manner. In contrast to pseudoscorpions, the stalks lack a lumen. In *E. italicus*, the stalk cells are not synthetically and secretory active and play a universal mechanical role connecting the oocytes with the gonad wall (Jędrzejowska et al. 2014).

Following fertilization (in sexually reproducing apoikogenic scorpions), the structure of the ovariuterus considerably alters. Fertilized eggs pass back into the lumen of the ovariuterus where they develop until parturition. Embryos become covered with extraembryonic membranes and their arrangement within the lumen is random. Growing in size, embryos cause dilation of the tubules and the ovariuterus gains a beaded-like appearance (Polis and Sissom 1990; Warburg 2010). In the family Vaejovidae, the apoikogenic development is more complex. The embryos develop within the ovariuterine tubules attached to a trophic cell mass termed the follicular placenta. The latter apparently transfers nutrients to the embryo (Farley 1996, 1998).

Katoikogenic Scorpions The katoikogenic type of development is characteristic of scorpions from the families Scorpionidae, Hemiscorpiidae, Urodactidae, and Diplocentridae (Laurie 1896; Francke 1982; Polis and Sissom 1990). Their ovariuteri are complex organs comprising specialized structures, termed the diverticula. In the diverticula, oocyte growth, fertilization, and embryonic development take place. The wall of each diverticulum consists of two epithelia which continue into the two epithelial layers of the ovariuterine tubules (Yamazaki and Makioka 2001; Jędrzejowska et al. 2016) (Fig. 19.4c). Apart from the bilayered structure of the epithelial wall, somatic components of the female gonads are comparable to those of the apoikogenic type. Furthermore, as in other matrotrophic chelicerates, the structure of the gonad alters during consecutive stages of oogenesis and embryogenesis. Consequently, three types of diverticula have been distinguished:

(1) containing ova (rudimentary diverticula), (2) containing embryos (embryonic diverticula), and (3) empty (degenerating, post-partum) diverticula that contained the previous generation of juveniles just born (Mathew 1956; Warburg and Elias 1998; Elias et al. 1999).

Maturation of the katoikogenic gonads shows striking resemblance to the apoikogenic type (Jędrzejowska et al. 2016). In immature females, the gonad wall is built of two epithelial layers. The germline cells are distributed among the cells of the internal layer (Fig. 19.3d). Previtellogenic oocytes bulge towards the hemocoel cavity and elevate the bilayered epithelium. Each protuberance of the ovariuterine wall develops into the diverticulum, and the oocyte occupies its terminal (distal) part.

The internal epithelial cells differentiate into follicular and stalk cells (Fig. 19.4c) that show clear developmental and structural resemblances to the follicular and stalk cells in apoikogenic scorpions. In contrast, the external layer of the diverticulum and the ovariuterine wall has no counterparts in the apoikogenic type. The occurrence of the external layer in more derived katoikogenic scorpions seems to be an evolutionary novelty. The role of the external layer at the stage preceding fertilization is unknown.

As in pseudoscorpions and apoikogenic scorpions, the follicular cells in katoikogenic ovariuteri do not actively participate in oogenesis. The stalk cells play a structural role and form a mechanical barrier against premature fertilization which in katoikogenic scorpions occurs in diverticula (Jędrzejowska et al. 2016).

During embryo development, the structure of the diverticulum is significantly modified and four distinct regions of the diverticulum are distinguishable: a long stalk, a thickened collar, a conical portion, and a long appendix (Laurie 1891; Polis and Sissom 1990). Oral feeding of embryos starts very early. The appendix serves as the feeding apparatus. The maternal nutrients are absorbed by the appendix, where they are temporarily stored until passage to the embryo's mouth and digestive tract (Laurie 1891; Mathew 1948; Mathew 1956; Subburam and Gopalakrishna Reddy 1989; Polis and Sissom 1990; Farley 2011).

19.4 Conclusions

Among chelicerates, two types of embryo nourishment have been recognized: lecitotrophy and matrotrophy. The ovaries of lecitotrophic taxa have a simple or "basic" structure, whereas female reproductive systems of pseudoscorpions as well as matrotrophic scorpions are characterized by several structural and functional modifications. The major modifications include the presence of additional subpopulations of somatic cells, the site of embryo development, and the source of nutrients for developing embryos (Table 19.1). It is suggested that above modifications had evolved in pseudoscorpions and scorpions independently and that among scorpions, the katoikogenic taxa are more specialized than the apoikogenic ones.

Table 19.1 Comparison of the ovary structure and embryo development in matrotrophic chelicerates

Taxon	Follicular cells	External layer of the gonad wall	Site of embryo development	Source of nutrients
Pseudoscorpions	Present	Not present	Brood sac	Ovarian epithelial cells
Apoikogenic scorpions	Present	Not present	Ovariuterus, lumen of the ovariuterine tubule	Yolk, nutrients from hemolymph
Katoikogenic scorpions	Present	Present	Ovariuterus, diverticulum	Hepatopancreas

References

Alberti G, Coons LB (1999) Acari: Mites. In: Harrison FW (ed) Microscopic anatomy of invertebrates vol 8C. Wiley-Liss, New York, pp 515–1265

Alberti G, Palacios-Vargas JG (2015) Fine structure of the ovary of *Schizomus palaciosi* (Arachnida: Schizomida). Soil Org 87:153–168

Badian Z, Ogorzałek A (1982) Fine structure of ovary in *Chelifer cancroides* (Linnaeus, 1761) (Arachnida, Pseudosorpionidea). Zool Pol 29:137–146

Bilinski SM, Szymanska B, Miyazaki K (2008) Formation of an egg envelope in the pycnogonid, *Propallene longiceps* (Pycnogonida, Callipallenidae). Arthropod Struct Dev 37:155–162. https://doi.org/10.1016/j.asd.2007.06.001

Bilinski SM, Kloc M, Tworzydlo W (2017) Selection of mitochondria in female germline cells: is Balbiani body implicated in this process? J Assist Reprod Genet 34:1405–1412. https://doi.org/10.1007/s10815-017-1006-3

Büning J (1994) The insect ovary: Ultrastructure, previtellogenic growth and evolution. Chapman & Hall, London

Coons LB, Alberti G (1999) Acari: Ticks. In: Harisson FW, Foelix RF (eds) Microscopic anatomy of invertebrates, vol 8B. Wiley-Liss, New York, pp 267–514

de Oliveira PR, Mathias MIC, Bechara GH (2007) Vitellogenesis in the tick *Amblyomma triste* (Koch, 1844) (Acari: Ixodidae). Vet Parasitol 143:134–139. https://doi.org/10.1016/j.vetpar.2006.08.013

Di Palma A, Alberti G (2001) Fine structure of the female genital system in phytoseiid mites with remarks on egg nutrimentary development, sperm-access system, sperm transfer, and capacitation (Acari, Gamasida, Phytoseiidae). Exp Appl Acarol 25:525–591. https://doi.org/10.1023/A:1014741808835

Dumont JN, Anderson E (1967) Vitellogenesis in the horseshoe crab, *Limulus polyphemus*. J Microsc (Paris) 6:791–806

Dunlop J (1999) Fossil evidence, terrestrialization and arachnid phylogeny. J Arachnol 27:86–93. https://doi.org/10.2307/3705969

Edgecombe GD, Legg DA (2014) Origins and early evolution of arthropods. Palaeontology 57:457–468. https://doi.org/10.1111/pala.12105

El Shoura SM, Banaja AA, Roshdy MA (1989) Fine structure of the developing oocytes in adult *Argas* (*Persicargas*) *arboreus* (Ixodoidea: Argasidae). Exp Appl Acarol 6:143–156. https://doi.org/10.1007/BF01201644

Elias R, Warburg MR, Rosenberg M (1999) Seasonal changes in the ovariuterus of female *Scorpio maurus fuscus* (Scorpiones; Scorpionidae), and the effect of higher temperature. Int J Insect Morphol Embryol 28:337–348. https://doi.org/10.1016/S0020-7322(99)00035-5

Fahrenbach WH (1999) Merostomata. In: Harrison FW, Foelix RF (eds) Microscopic anatomy of invertebrates, vol 8A. Wiley-Liss, New York, pp 21–115

Farley RD (1996) Formation of maternal trophic structures for embryos of *Paruroctonus mesaensis* (Scorpionidae: Vaejovidae). Rev Suisse Zool 1:189–202

Farley RD (1998) Matrotrophic adaptations and early stages of embryogenesis in the desert scorpion *Paruroctonus mesaensis* (Vaejovidae). J Morphol 237:187–211. https://doi.org/10.1002/(SICI)1097-4687(199809)237:3<187:AID-JMOR1>3.0.CO;2-X

Farley RD (1999) Chelicerate Arthropoda. Scorpions. In: Harrison FW, Foelix R (eds) Microscopic anatomy of invertebrates, vol 8A. Wiley-Liss, New York, pp 117–222

Farley RD (2011) Pectine development in scorpion embryos and first and second instars. Euscorpius 120:1–47

Foelix RF (1996) Biology of spiders. Oxford University Press, New York

Francke OF (1982) Scorpiones. In: Parker SP (ed) Synopsis and classification of living organisms. McGraw-Hill, New York, pp 73–75

Garwood RJ, Dunlop J (2014) Three-dimensional reconstruction and the phylogeny of extinct chelicerate orders. PeerJ 2:e641. https://doi.org/10.7717/peerj.641

Guraya SS (1979) Recent advances in the morphology, cytochemistry, and function of Balbiani's vitelline body in animal oocytes. Int Rev Cytol 59:249–321

Hjelle JT (1990) Anatomy and morphology. In: Polis G (ed) The biology of scorpions. Stanford University Press, Stanford, CA, pp 9–63

Jędrzejowska I, Kubrakiewicz J (2007) The Balbiani body in the oocytes of a common cellar spider, *Pholcus phalangioides* (Araneae: Pholcidae). Arthropod Struct Dev 36:317–326. https://doi.org/10.1016/j.asd.2007.01.004

Jędrzejowska I, Kubrakiewicz J (2010) Yolk nucleus – the complex assemblage of cytoskeleton and ER is a site of lipid droplet formation in spider oocytes. Arthropod Struct Dev 39:350–359. https://doi.org/10.1016/j.asd.2010.05.001

Jędrzejowska I, Mazurkiewicz-Kania M, Garbiec A, Kubrakiewicz J (2013) Differentiation and function of the ovarian somatic cells in the pseudoscorpion, *Chelifer cancroides* (Linnaeus, 1761) (Chelicerata: Arachnida: Pseudoscorpionida). Arthropod Struct Dev 42:27–36. https://doi.org/10.1016/j.asd.2012.09.004

Jędrzejowska I, Szymusiak K, Mazurkiewicz-Kania M, Garbiec A (2014) Differentiation of somatic cells in the ovariuteri of the apoikogenic scorpion *Euscorpius italicus* (Chelicerata, Scorpiones, Euscorpiidae). Arthropod Struct Dev 43:361–370. https://doi.org/10.1016/j.asd.2013.11.004

Jędrzejowska I, Szymusiak K, Mazurkiewicz-Kania M, Garbiec A (2016) Scorpion katoikogenic ovariuterus – much more alike to apoikogenic type than it seemed to be. Arthropod Struct Dev 45:488–495. https://doi.org/10.1016/j.asd.2016.09.001

Kessel RG, Beams HW (1980) Cytodifferentiation and vitellogenesis during oogenesis in Arachnida: cytological studies on developing oocytes of a harvestman. J Morphol 163:175–190. https://doi.org/10.1002/jmor.1051630206

Klann AE (2009) Histology and ultrastructure of solifuges: comparative studies of organ systems of solifuges (Arachnida, Solifugae) with special focus on functional analyses and phylogenetic interpretations. Universitat Greifswald

Kloc M, Etkin LD (1995) Two distinct pathways for the localization of RNAs at the vegetal cortex in *Xenopus* oocytes. Development 121:287–297

Kloc M, Larabell C, Chan AP-Y, Etkin LD (1998) Contribution of METRO pathway localized molecules to the organization of the germ cell lineage. Mech Dev 75:81–93. https://doi.org/10.1016/S0925-4773(98)00086-0

Kloc M, Bilinski SM, Etkin LD (2004) The Balbiani body and germ cell determinants: 150 years later. Curr Top Dev Biol 59:1–36. https://doi.org/10.1016/S0070-2153(04)59001-4

Kloc M, Jędrzejowska I, Tworzydlo W, Bilinski SM (2014) Balbiani body, nuage and sponge bodies – the germ plasm pathway players. Arthropod Struct Dev 43:341–348. https://doi.org/10.1016/j.asd.2013.12.003

Laurie M (1891) Some points in the development of *Scorpio fulvipes*. Q J Microsc Sci 32:587–597

Laurie M (1896) XVII.—Further notes on the anatomy and development of scorpions, and their bearing on the classification of the order. J Nat Hist Ser 6(18):121–133. https://doi.org/10.1080/00222939608680422

Lehmann T, Melzer RR (2013) Looking like *Limulus*? – Retinula axons and visual neuropils of the median and lateral eyes of scorpions. Front Zool 10:40. https://doi.org/10.1186/1742-9994-10-40

Makioka T (1976) Alternative occurrence of two ovarian functions in the adult pseudoscorpion, *Garypus japonicus* Beier. Acta Arachnol 27:8–15. https://doi.org/10.2476/asjaa.27.8

Makioka T (1979) Structures of the adult ovaries in different functional phases of the pseudoscorpion, *Garypus japonicus* Beier. Acta Arachnol 28:71–81. https://doi.org/10.2476/asjaa.28.71

Makioka T (1988) Ovarian structure and oogenesis in Chelicerates and other arthropods. Proc Arthropodan Embryol Soc Jpn 23:1–11

Mathew AP (1948) Nutrition in the advanced embryos of the scorpion: *Palamnaeus scaber* Thorell. Proc Indian Acad Sci – Sect B 27:111–118. https://doi.org/10.1007/BF03049905

Mathew AP (1956) Embryology of *Heterometrus scaber*. In: Zoological memoirs. University of Travancore Research Institute, Trivandrum, pp 1–96

Matthiesen FA (1970) Reproductive system and embryos of Brazilian scorpions. An Acad Bras Cienc 42:627–632

Mayer G, Tait NN (2009) Position and development of oocytes in velvet worms shed light on the evolution of the ovary in Onychophora and Arthropoda. Zool J Linnean Soc 157:17–33. https://doi.org/10.1111/j.1096-3642.2009.00523.x

Michalik P, Reiher W, Tintelnot-Suhm M et al (2005) Female genital system of the folding-trapdoor spider *Antrodiaetus unicolor* (Hentz, 1842) (Antrodiaetidae, Araneae): ultrastructural study of form and function with notes on reproductive biology of spiders. J Morphol 263:284–309. https://doi.org/10.1002/jmor.10309

Miyachi Y, Yahata K (2012) Morphological study of ovarian structures in scolopendromorph centipedes (Myriapoda: Chilopoda) with special reference to the position of oocyte growth. Proc Arthropodan Embryol Soc Japan 47:21–28

Miyazaki K, Bilinski SM (2006) Ultrastructural investigations of the ovary and oogenesis in the pycnogonids *Cilunculus armatus* and *Ammothella biunguiculata* (Pycnogonida, Ammotheidae). Invertebr Biol 125:346–353. https://doi.org/10.1111/j.1744-7410.2006.00066.x

Miyazaki K, Ueshima R, Makioka T (2001) Structure of the female reproductive system and oogenetic mode in a schizomid, *Schizomus sawadai* (Arachnida, Schizomida). Invertebr Reprod Dev 40:1–7. https://doi.org/10.1080/07924259.2001.9652492

Morishita R, Ferreira SA, Filha AS, Faraco CD (2003) Studies on oogenesis and oviposition in the brown spider *Loxosceles intermedia* (Araneae: Sicariidae). Anat Rec 273A:575–582. https://doi.org/10.1002/ar.a.10062

Munson JP (1898) The ovarian egg of *Limulus*. A contribution to the problem of the centrosome and yolk-nucleus. J Morphol 15:11–220

Osaki H (1972) Electron microscope studies on developing oocytes of the spider, *Plexippus paykulli*. Annot Zool Jpn 45:187–200

Ostrovsky AN, Lidgard S, Gordon DP et al (2016) Matrotrophy and placentation in invertebrates: a new paradigm. Biol Rev 91:673–711. https://doi.org/10.1111/brv.12189

Polis G, Sissom W (1990) The biology of scorpions. In: Polis G (ed) Life history. Stanford University Press, Stanford, pp 81–111

Sharma PP, Kaluziak ST, Pérez-Porro AR et al (2014) Phylogenomic interrogation of arachnida reveals systemic conflicts in phylogenetic signal. Mol Biol Evol 31:2963–2984. https://doi.org/10.1093/molbev/msu235

Sharma PP, Fernandez R, Esposito LA et al (2015) Phylogenomic resolution of scorpions reveals multilevel discordance with morphological phylogenetic signal. Proc R Soc B Biol Sci 282:20142953–20142953. https://doi.org/10.1098/rspb.2014.2953

Shatrov AB (1997) Vitellogenesis and egg-shell formation in ovipositing females of the trombiculid mite *Hirsutiella zachvatkini* (Schluger) (Acariformes: Trombiculidae). Acarologia 38:140–151

Soranzo L, Stockmann R, Lautie N, Fayet C (2002) Structure of the ovariuterus of the scorpion *Euscorpius carpathicus* (L.) (Euscorpiidae) before fertilization. In: Toft S, Scharff N (eds) European arachnology 2000. Aarhus University Press, Aarhus, pp 91–96
Subburam V, Gopalakrishna Reddy T (1989) Scientific note functional morphology of the appendix of the viviparous scorpion, *Heterometrus fulvipes* (Koch) (Arachnida: Scorpionidae). Int J Insect Morphol Embryol 18:329–333. https://doi.org/10.1016/0020-7322(89)90014-7
Talarico G, Zeck-Kapp G, Palacios-Vargas JG, Alberti G (2009) Oogenesis and ovary ultrastructure in *Pseudocellus boneti* (Arachnida: Ricinulei). Contrib Nat Hist 12:1373–1386
Tworzydlo W, Kisiel E, Jankowska W, Bilinski SM (2014) Morphology and ultrastructure of the germarium in panoistic ovarioles of a basal "apterygotous" insect, *Thermobia domestica*. Zoology (Jena) 117:200–206. https://doi.org/10.1016/j.zool.2014.01.002
Tworzydlo W, Kisiel E, Jankowska W et al (2016) Exclusion of dysfunctional mitochondria from Balbiani body during early oogenesis of *Thermobia*. Cell Tissue Res 366:191–201. https://doi.org/10.1007/s00441-016-2414-x
Tworzydlo W, Marek M, Kisiel E, Bilinski SM (2017) Meiosis, Balbiani body and early asymmetry of *Thermobia oocyte*. Protoplasma 254:649–655. https://doi.org/10.1007/s00709-016-0978-7
Warburg MR (2010) Reproductive system of female scorpion: a partial review. Anat Rec 293:1738–1754. https://doi.org/10.1002/ar.21219
Warburg MR, Elias R (1998) The reproductive potential and strategy of *Scorpio maurus fuscus* (Scorpiones: Scorpionidae): anatomical clues in the ovariuterus. J Zool 246:29–37. https://doi.org/10.1017/S0952836998009042
Weygoldt P (1969) The biology of pseudoscorpions. Harvard Univerity Press, Cambridge
Weygoldt P, Weisemann A, Weisemann K (1972) Morphologisch-histologische Untersuchungen an den Geschlechtsorganen der Amblypygi unter besonderer Beracksichtigung von *Tarantula marginemaculata* C.L. Koch (Arachnida). Zeitschrift für Morphol:209–247
Witaliński W (1986) Eggshells in Mites I. A comparative ultrastructural study of vitelline envelope formation. Cell Tissue Res 244:209–214
Witaliński W (2014) Gonads and gametogenesis in astigmatic mites (Acariformes: Astigmata). Arthropod Struct Dev 43:323–340. https://doi.org/10.1016/j.asd.2014.04.003
Witaliński W, Żuwała K (1981) Ultrastructural studies of egg envelopes in harvestmen (Chelicerata, Opiliones). Int J Invertebr Reprod 4(2):95–106. https://doi.org/10.1080/01651269.1981.10553421
Witaliński W, Szlendak E, Boczek J (1990) Anatomy and ultrastructure of the reproductive systems of *Acarus siro* (Acari: Acaridae). Exp Appl Acarol 10:1–31
Witaliński W, Rożej-Pabijan E, Podkowa D (2014) Gonads in *Histiostoma mites* (Acariformes: Astigmata): structure and development. Arthropod Struct Dev 43:385–401. https://doi.org/10.1016/j.asd.2014.04.006
Yahata K, Chikami Y, Umetani E (2018) Morphological study of the ovary in *Hanseniella caldaria* (Myriapoda; Symphyla): the position of oocyte-growth and evolution of ovarian structure in Arthropoda. Arthropod Struct Dev 47:655–661. https://doi.org/10.1016/j.asd.2018.09.003
Yamazaki K, Makioka T (2001) Ovarian structural features reflecting repeated pregnancies and parturitions in a viviparous scorpion, *Liocheles australasiae*. Zool Sci 18:277–282. https://doi.org/10.2108/zsj.18.277

Chapter 20
Reproduction, Gonad Structure, and Oogenesis in Tardigrades

Izabela Poprawa and Kamil Janelt

Abstract Even though tardigrades have been known since 1772, their phylogenetic position is still controversial. Tardigrades are regarded as either the sister group of arthropods, onychophorans, or onychophorans plus arthropods. Furthermore, the knowledge about their gametogenesis, especially oogenesis, is still poor and needs further analysis. The process of oogenesis has been studied solely for several eutardigradan species. Moreover, the spatial organization of the female germ-line clusters has been described for three species only. Meroistic ovaries characterize all analyzed species. In species of the Parachela, one cell per germ-cell cluster differentiates into the oocyte, while the remaining cells become the trophocytes. In Apochela several cells in the cluster differentiate into oocytes. Vitellogenesis is of a mixed type. The eggs are covered with the egg capsule that is composed of two shells: the thin vitelline envelope that adheres to the oolemma and the thick three-layered chorion. Chorion is formed as a first followed by vitelline envelope. Several features related to the oogenesis and structure of the ovary confirm the hypothesis that tardigrades are the sister group rather for arthropods than for onychophorans.

20.1 Introduction

Tardigrades (Tardigrada) (also called water bears or moss piglets) are minute hydrophilous (from 50- to 1200-μm-long) invertebrate animals (Fig. 20.1a). They were first observed in 1773 by Goeze, but their name "Il Tradigrado" (slow stepper) was not introduced until 1776 by Spallanzani (Dewel et al. 1993; Nelson et al. 2015). Tardigrades are closely related to Onychophora and Arthropoda and are classified with them in the monophyletic clade Panarthropoda (Nielsen 2001; Telford et al. 2008; Rota-Stabelli et al. 2010; Nelson et al. 2015). Their phylogenetic position in Panarthropoda is controversial, as they are typically regarded as either the sister

I. Poprawa (✉) · K. Janelt
Department of Animal Histology and Embryology, University of Silesia in Katowice, Katowice, Poland
e-mail: izabela.poprawa@us.edu.pl

W. Tworzydlo, S. M. Bilinski (eds.), *Evo-Devo: Non-model Species in Cell and Developmental Biology*, Results and Problems in Cell Differentiation 68,
https://doi.org/10.1007/978-3-030-23459-1_20

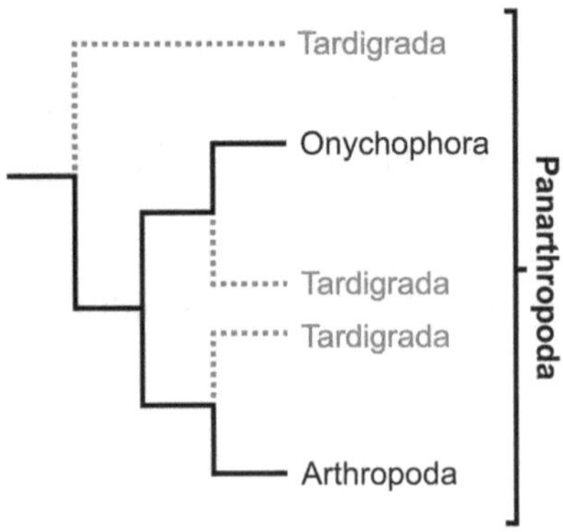

Fig. 20.1 Cladogram illustrating controversial views on the phylogenetic position of tardigrades as either the sister group of arthropods, onychophorans, or onychophorans plus arthropods (modified from Mayer et al. 2013)

group of arthropods, onychophorans, or onychophorans plus arthropods (Fig. 20.1) (Mayer et al. 2013). Tardigrades have a remarkable variability in their morphology, the environments in which they live, their reproductive modes, and their food preferences (Bertolani 1979, 1987, 2001; Ramazzotti and Maucci 1983; Dewel et al. 1993; Kinchin 1994; Nelson et al. 2015). They are widespread in marine, freshwater, and terrestrial habitats throughout the world. Thus far, about 1200 species of tardigrades have been described (Degma et al. 2009–2018). Based on their morphological characters, three classes can be distinguished in the phylum Tardigrada: Heterotardigrada, Mesotardigrada, and Eutardigrada. The class Heterotardigrada contains two orders: Arthrotardigrada (mainly marine species) and Echiniscoidea (primarily terrestrial species although some marine species have also been observed). The class Eutardigrada also contains two orders: Apochela (a limno-terrestrial species) and Parachela (a primarily limno-terrestrial species with several secondary marine taxa) (Nelson et al. 2015). The last class, Mesotardigrada, is represented by only one species *Thermozodium esakii*; however, the existence of this class is controversial because there is only one description of this species (Rahm 1937). To date, the existence of *T. esakii* has never been verified, and, what is more, the type locality was destroyed by an earthquake (Nelson 2002; Grothman et al. 2017).

Tardigrades have a bilaterally symmetrical body with four pairs of lobopodous legs. Each leg usually terminates in claws or sucking disks (Nelson 2002). The miniaturization of their body size has simplified their internal anatomy. They have a well-developed nervous system, reproductive system, and digestive system (Fig. 20.2a, b). However, they have no specialized respiratory or circulatory systems. Respiration occurs through the body wall. The body cavity is filled with the fluid that performs some functions of the circulatory system. (Dewel et al. 1993; Kinchin 1994). The storage cells (body cavity cells) freely float in this fluid. The main function of these cells is to accumulate and distribute the reserve materials as well as synthesize and secrete the vitellogenins (Szymańska 1994; Poprawa 2006; Hyra et al. 2016). Volkman and Greven (1993) observed tyrosinase activity in the cytoplasm of the storage cells. This observation indicates that the storage cells correspond to the hemocytes of insects and might participate in the immune

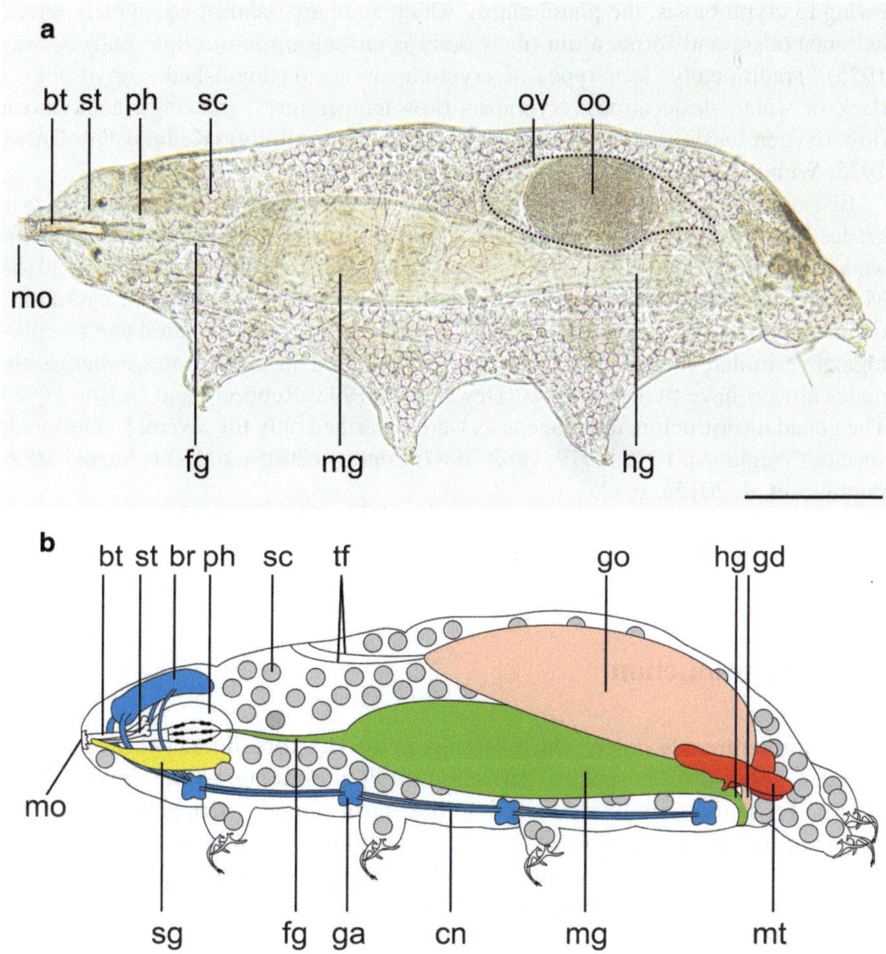

Fig. 20.2 Morphology of tardigrades. (**a**) Tardigrade *Paramacrobiotus* sp., LM, bar = 100 μm, (**b**) General morphology of eutardigrade. *Br* brain, *bt* buccal tube, *cn* connective, *fg* foregut, *ga* trunk ganglia, *gd* gonoduct, *go* gonad, *hg* hindgut, *mg* midgut, *mo* mouth opening, *mt* Malpighian tubules, *oo* oocyte, *ov* ovary, *ph* pharynx, *sc* storage cell, *sg* salivary glands, *st* stylet, *tf* terminal filament

mechanisms. Osmoregulation and excretion are performed by the cuticle or Malpighian tubules (in eutardigrades) or by specialized ventral organs (in some echiniscid heterotardigrades) (Nelson et al. 2015).

Tardigrades are a famous group of animals due to their unbelievable abilities, which allow them to survive in extreme environments such as temperatures ranging from −272 to 151 °C, pressure that is about six times greater than that found in the deepest ocean trenches, the vacuum of outer space, and under an ionizing radiation dose that is hundreds of times higher than the lethal dose for humans (Wright et al. 1992; Rebecchi et al. 2011; Horikawa 2012; Tsujimoto et al. 2015). This is possible owing to cryptobiosis, the phase during which an animal almost completely arrests its metabolism and forms a tun (their body is curling up into a little ball) (Crowe 1975). Traditionally, four types of cryptobiosis are distinguished: anhydrobiosis (lack of water, desiccation), cryobiosis (low temperatures, freezing), anoxybiosis (low oxygen tension), and osmobiosis (variations in salinity) (Keilin 1959; Crowe 1975; Wełnicz et al. 2011; Czerneková et al. 2017).

Bisexual (gonochoristic), hermaphroditic, and unisexual (parthenogenetic) tardigrades have been described (Bertolani 1983; Bertolani et al. 1983; Bertolani and Manicardi 1986; Dewel et al. 1993; Rebecchi and Nelson 1998; Nelson et al. 2015). Male and female gonochoristic tardigrades are very often distinguished based only on their primary sex characters. In both sexes, the gonads are unpaired and morphologically similar; the gonoduct (oviduct) is unpaired in the females, whereas the males always have two gonoducts (Dewel et al. 1993; Rebecchi and Nelson 1998). The gonad ultrastructure and oogenesis were described only for several Eutardigrada species (Węglarska 1975, 1979, 1982, 1987; Poprawa 2005a, b, 2011; Suzuki 2006; Poprawa et al. 2015a, b, c).

In this chapter, we will briefly review the present knowledge about the reproduction, ovary organization, and oogenesis in tardigrades.

20.2 Reproduction

The Arthrotardigrada order, which belongs to the Heterotardigrada class, primarily includes gonochoristic species (Kristensen 1977; Bertolani 2001; Nelson et al. 2015). To date, only one exception has been found—hermaphroditism was described in the population of *Orzeliscus* cf. *belopus* in Al-Ghardaqa, Egypt (Bertolani 1987). In the second order of Heterotardigrada, the Echiniscoidea of both gonochorism and parthenogenesis have been noted.

Eutardigrada species are gonochoristic, hermaphroditic, and parthenogenetic; however, hermaphroditism is rare in eutardigrades (Bertolani 1979, 1983, 2001; Altiero and Rebecchi 2001).

In the gonochoristic Eutardigrade species, the females are iteroparous. They mature and lay eggs throughout their adult lives in association with molting. The males of eutardigrades are iteroparous or semelparous (they copulate only once in their lives) (Rebecchi and Bertolani 1994). Little is known about fertilization in gonochoristic eutardigrades. Some species of *Macrobiotus*, *Ramazzottius*, and *Xerobiotus* have internal spermatheca, which is connected to the rectum. This might indicate internal fertilization (Rebecchi and Bertolani 1994; Rebecchi 1997; Bertolani 2001).

Parthenogenesis is the only way to reproduce in many Eutardigrade species that live in terrestrial, limno-terrestrial, and limnic environments (Bertolani 1987; Nelson et al. 2015). Both meiotic (automictic, in Murrayidae and *Hypsibius dujardini*) and ameiotic (apomictic) parthenogenesis have been described (Bertolani 1972, 2001; Nelson et al. 2015).

20.3 Gross Morphology of the Reproductive System

The female reproductive system (like the hermaphroditic one) of Eutardigrada is composed of a single gonad, ovary in females or ovotestis in hermaphroditic species, as well as a single gonoduct (oviduct in females) (Fig. 20.2a–b). The gonoduct, together with the hindgut and Malpighian tubules, opens into the cloaca that opens on the ventral side of the body near the fourth pair of legs (Węglarska 1979, 1987; Poprawa 2005a; Suzuki 2006; Poprawa et al. 2015a, b, c). The female reproductive system of Heterotardigrada is similar to that described for Eutardigrada with one important difference. The gonoduct of the heterotardigrades opens directly to the outside via a gonopore that is often surrounded by a rosette of cuticular folds (Dewel et al. 1993; Kinchin 1994). Only one exception to this rule has been found in *Carphania fluviatilis* (Heterotardigrada, Echiniscoidea). In this species the gonads are paired (Binda and Kristiensen 1986) and located on both sides of the alimentary tract. In some eutardigrades (e.g., *Macrobiotus hufelandi*, *Macrobiotus pallari*, *Macrobiotus polonicus*) a single seminal receptacle that opens into the hindgut near the opening of the oviduct (Marcus 1929; Bertolani 1983; Dewel et al. 1993; Poprawa et al. 2015c) has been described. This structure is used to store the sperm cells after copulation. In contrast, some marine heterotardigrades have specific paired receptacles ("annex glands") that are located laterally to the gonopore. Although their function has not yet been determined, they probably participate in reproduction or oviposition (Renaud-Debyser 1965; Pollock 1970, 1975; Grimaldi de Zio et al. 1987).

20.3.1 The Structure of the Ovary

During embryogenesis, the ovary of tardigrades develops from paired anlages; however, the adult animals, except for *Carphania fluviatilis*, have an unpaired gonad (Nelson 1982). The ovaries of tardigrades are meroistic and located dorsally or dorsolaterally above the midgut; they vary in size depending on the age and the stage of oogenesis (Fig. 20.2a–b) (Węglarska 1979; Bertolani 1983; Ramazzotti and Maucci 1983; Dewel et al. 1993; Kinchin 1994; Poprawa 2005a; Suzuki 2006; Poprawa et al. 2015b, c). The anterior part of the ovary is attached to the body wall via a terminal filament. In eutardigrades, it consists of two ligaments that are formed by protrusions of the cells of the ovary wall (Figs. 20.2b and 20.3a–b) (Węglarska 1979; Poprawa 2005a; Poprawa et al. 2015b, c), while in heterotardigrades, the terminal filament is formed by one ligament (Bertolani 1983; Ramazzotti and Maucci 1983; Dewel et al. 1993). The ovary wall is formed by a single layer of flattened somatic cells that lie on a basal lamina. The apical parts of these cells are directed toward the lumen of the gonad, whereas the basal lamina is directed to the body cavity. The epithelial cells form protrusions that penetrate between the germ cells (Węglarska 1979; Poprawa 2005a; Poprawa et al. 2015b,

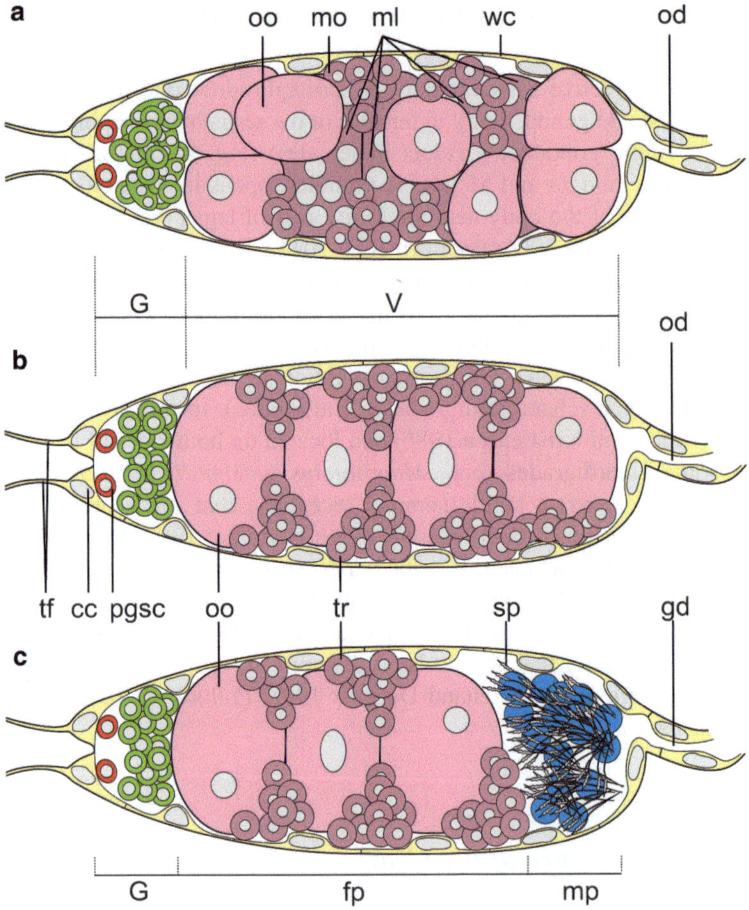

Fig. 20.3 General morphology of (**a**) the ovary of Apochela, (**b**) the ovary of Parachela, and (**c**) the ovotestis. *cc* cap cell, *fp* female part of the ovotestis, *G* germarium, *gd* gonoduct, *mo* mononuclear cell, *ml* multinuclear cell, *mp* male part of the ovotestis, *od* oviduct, *oo* oocyte, *pgsc* putative germline stem cell, *sp* sperm cell, *tf* terminal filament, *tr* trophocyte, *V* vitellarium, *wc* cell of the ovary wall

c). The main function of these cells is synthesis and secretion of the precursors of the chorion (Węglarska 1982; Poprawa 2005b, 2011; Poprawa et al. 2015b, c); therefore, they can be considered as homologous to the follicular cells of arthropods (Margaritis 1985; Büning 1994; Jędrzejowska et al. 2013). The germinal epithelium characterized for Onychophora (Mayer and Tait 2009) has not been described in tardigrades; thus, the germ cells are not intraepithelial.

The ovary of tardigrades is divided into two zones—a germarium and a vitellarium (Fig. 20.3a–b). Generally, the germarium occupies the apical part of the gonad (Fig. 20.3a, b) (Poprawa 2005a; Poprawa et al. 2015b, c), but it can also be situated on the side of the ovary as a narrow wedge or several isolated islets

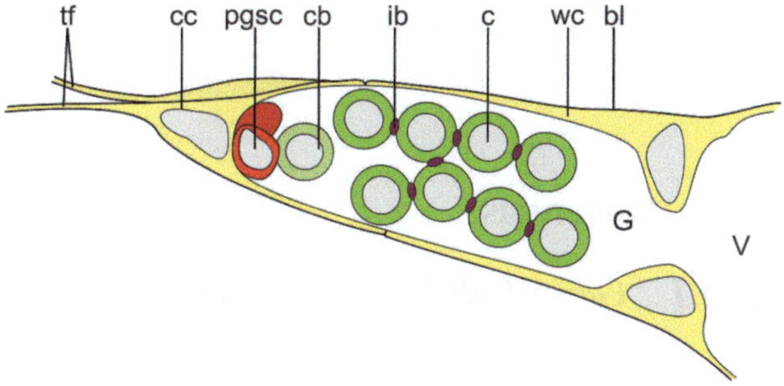

Fig. 20.4 Structure of the germarium in *Thulinius ruffoi*. *bl* basal lamina, *c* cystocyte, *cb* cystoblast, *cc* cap cell, *G* germarium, *ib*. intercellular bridge, *pgsc* putative germ-line stem cell, *tf* terminal filament, *V* vitellarium, *wc* cell of the ovary wall

(Węglarska 1979). In the germarium, the proliferation of the germ-line cells and the formation of the germ cell clusters take place. An analysis of the germarium in *Thulinius ruffoi* showed that two putative germ-line stem cells are present in the apical part of the ovary in the neighborhood of the cap cells (the cells that form the terminal filament) (Fig. 20.4). The stem cells divide completely to form cystoblasts (our unpublished studies). In contrast to stem cells, the cystoblasts divide incompletely; the resulting cells, termed the cystocytes, form clusters of interconnected sibling cells that move from the germarium to the second zone of the ovary—the vitellarium. In this part of the gonad, the cystocytes grow and differentiate into oocytes and trophocytes (Fig. 20.4a–c) (Węglarska 1979; Poprawa 2005a; Poprawa et al. 2015b, c). The number of germ-line clusters in the vitellarium depends on the species as well as the age and physiological state of the given individual (see Suzuki 2006; Poprawa et al. 2015b for further details).

20.3.2 *The Structure of the Hermaphroditic Gonad*

The saclike hermaphroditic gonad of eutardigrades is located dorsally above the midgut and attached to the body wall via a terminal filament that is composed of two ligaments (Bertolani et al. 1983; Węglarska 1987; Poprawa et al. 2015a). It is divided into three zones: a germarium, a vitellarium, and a part that is filled with male germ cells (Fig. 20.3c) (Węglarska 1987; Poprawa et al. 2015a). As in the ovary, the gonad wall is formed by a single layer of flattened somatic cells that lie on the basal lamina. In the vitellarium, the epithelial cells form protrusions that penetrate between germ-line cells, while there are no protrusions in the male part of the gonad (Poprawa et al. 2015a). The cells of the gonad wall display secretory activity and synthesize and secrete the precursors of the chorion (Poprawa 2011). In the germarium, the proliferation of the germ-line cells takes place. In the vitellarium, as in the ovary, the cells of the female germ-line clusters grow and differentiate into oocytes and trophocytes (Węglarska 1987; Poprawa et al. 2015a). The male part of the gonad is filled with clusters of male germ-line cells (Bertolani et al. 1983; Węglarska 1987; Poprawa et al. 2015a). It should be underlined that there is no boundary between the female and male parts of the hermaphroditic gonad (Poprawa et al. 2015a).

20.4 Oogenesis

20.4.1 Germ-Cell Cluster Organization

Four major types of germ-cell clusters are recognized:

(a) The two-celled clusters described in earwigs (Tworzydło et al. 2010) and the polychaete *Ophryotrocha labronica* (Brubacher and Huebner 2009, 2011)
(b) The linear germ-cell clusters or almost linear clusters that occur in some polychaetous annelids (Anderson and Huebner 1968), hexapods (Biliński 1994), and certain crustaceans (Kubrakiewicz et al. 1991)
(c) The branched clusters characteristic of insects, e.g., *Drosophila melanogaster* (Büning 1994; Lin and Spradling 1995; Kubrakiewicz 1997, 1998; Huynh and Johnston 2004; Ong and Tan 2010; Greenbaum et al. 2011; Haglund et al. 2011) and vertebrates (Kloc et al. 2004)
(d) Clusters with an anucleate mass of cytoplasm, found in flatworms, nematodes, and clitellate annelids (Foor 1968, 1983; Gibert et al. 1984; Amini et al. 2014; Świątek et al. 2009, 2018; Urbisz et al. 2017, 2018)

Although the process of oogenesis has been described in several species of tardigrades (Węglarska 1979, 1987; Poprawa 2005a, b, 2011; Poprawa et al. 2015a, b, c), the spatial organization of the female germ-cell clusters is only known for three species: *Milnesium tardigradum* (Eutardigrada, Apochela) (Dewel et al. 1993; Suzuki 2006); *Dactylobiotus parthenogeneticus* (Poprawa et al. 2015b), and *Thulinius ruffoi* (Eutardigrada, Parachela) (our unpublished studies). Residual data is also available for the *Parachela* species *Paramacrobiotus richtersi* (Węglarska 1979) and the heterotardigrade *Halechiniscus perfectus* (Dewel et al. 1993). In the parachelas the germ-cell clusters are branched (Fig. 20.5a), and only one cell per cluster differentiates into the oocyte, while the remaining cells become trophocytes that support the oocyte growth (Węglarska 1979, 1987; Poprawa 2005a; Poprawa et al. 2015a, b, c). The number of cells in a cluster varies between species (Poprawa et al. 2015b; our unpublished studies), but it can also be different in specimens of the same species or even between clusters in the same specimens (our unpublished studies). According to the literature data, the number of cells in the cluster that is formed as a result of synchronous divisions equals 2^n, where "*n*" means the number of cystocyte divisions (Biliński 1998). Węglarska described (1979) one oocyte and seven trophocytes that formed the germ-cell clusters in *Paramacrobiotus richtersi.* However, their spatial organization and formation remains obscure. In *Dactylobiotus parthenogeneticus*, the germ-cell clusters also consist of eight cells (Poprawa et al. 2015b). This suggests that these clusters were formed as a result of three synchronous divisions, while the presence of cells with

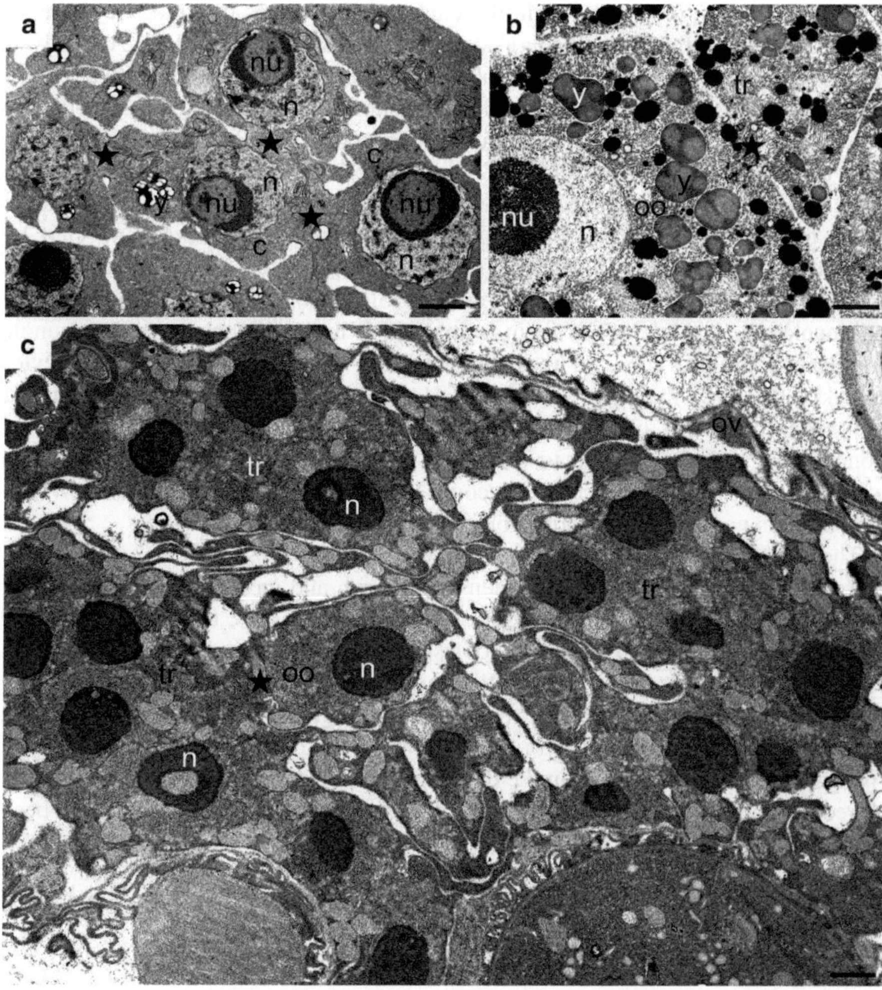

Fig. 20.5 Oogenesis of Eutardigrada, TEM. (**a**) Germ-cell cluster of *Isohypsibius* sp. during early vitellogenesis, bar = 0.9 μm, (**b**) germ-cell cluster of *Hypsibius exemplaris* during middle vitellogenesis, bar = 0.75 μm, and (**c**) germ-cell cluster of *Milnesium tardigradum* during previtellogenesis, bar = 1 μm. *c* cystocyte, *n* nucleus, *nu* nucleolus, *oo* oocyte, *ov* ovary wall, *tr* trophocyte, *y* yolk, *star* intercellular bridge

four cytoplasmic bridges indicates four divisions. It is possible that only two first divisions of the germ cells are synchronous and that synchronicity is lost in the third division. In *Dactylobiotus parthenogeneticus*, cells that have the highest number of the cytoplasmic bridges differentiate into the oocyte (Poprawa et al. 2015b). In *Thulinius ruffoi*, the situation is a little different. The germ-cell clusters of this species are huge and branched and consist of more than 30 cells that are connected by cytoplasmic bridges (our unpublished studies). Similar to *Dactylobiotus parthenogeneticus*, only one cell in the cluster differentiates into the oocyte; however, it is not the cell that has the largest number of cytoplasmic bridges. An analysis of the spatial organization and the number of cells in these clusters, as well as of cell death during oogenesis, suggests that the first four divisions during cluster formation were synchronous after which the 2^n rule began to fail (our unpublished studies). A quite different spatial organization of the germ-cell clusters was described in the apochelan species *Milnesium tardigradum* (Suzuki 2006). In this species, the center of the cluster is formed by four huge multinuclear cells that are connected by cytoplasmic bridges (Fig. 20.5c). These cells play the role of nurse cells (trophocytes). Additionally, a large number of mononuclear cells surround the multinuclear cells, and each of them is attached to the multinuclear cell by a single cytoplasmic bridge. Some of these mononuclear cells differentiate into oocytes (Suzuki 2006). In the heterotardigrade *Halechiniscus perfectus*, a multinuclear mass of cytoplasm is observed in the center of each cluster. A large number of mononuclear cells is attached to this mass. Some of these cells differentiate into oocytes, while the remaining cells become trophocytes (Dewel et al. 1993). However, there is no data on the formation of such clusters in the literature.

The cytoplasmic bridges that connect the cells in the clusters have one electron-dense rim or two rims: an electron-dense outer rim and an inner rim with a lower electron density. In *Thulinius ruffoi* and *Hypsibius exemplaris*, the presence of actin was observed in the rim of the cytoplasmic bridges (our unpublished studies). Macromolecules, ribosomes, organelles, and yolk material are transported from cell to cell via these connections (Fig. 20.5b) (Węglarska 1979; Poprawa 2005a; Poprawa et al. 2015a, b, c). The bridges become closed at the end of choriogenesis. The fusome was not observed in the developing cytoplasmic bridges of tardigrades. Moreover, the factors determining the differentiation of the oocyte has not been identified in this invertebrate taxon.

20.4.2 Previtellogenesis

Previtellogenesis is the period during which the synthesis of RNA (mRNA, rRNA) and the accumulation of ribosomes and cell organelles take place (Büning 1994; Matova and Cooley 2001; Tworzydło and Biliński 2008; Poprawa 2005a; Jaglarz et al. 2014a, b; Poprawa et al. 2015b, c).

According to the literature, during previtellogenesis of tardigrades (Węglarska 1979, 1987; Poprawa 2005a; Poprawa et al. 2015a, b, c), all of the cells in a cluster are the same size and have a similar ultrastructure. A huge nucleus with a heterogeneous nucleolus occupies the center of each cell. Very often, the presence of nucleolar vacuoles is also observed (Węglarska 1979; Poprawa 2005a; Poprawa et al. 2015a, b, c). Nucleolar vacuoles, which are also termed nucleolar cavities or interstices, are characteristic of plant cells, but they are also present in animal nucleoli and imply a high level of nucleolar activity (RNA synthesis) (Czerneková et al. 2018). Over time, the cells grow and accumulate macromolecules (mRNA, rRNA), ribosomes, and organelles such as mitochondria and short cisterns of the rough and smooth endoplasmic reticulum (Węglarska 1979, 1987; Poprawa 2005a, Poprawa et al. 2015a, b, c). Ribosomes and organelles are also observed inside the cytoplasmic bridges, which indicates the transport of these structures from cell to cell. This directional transport, which is similar to that in *Drosophila melanogaster* (Lin and Spradling 1995; Huynh and Johnston 2004), promotes the differentiation of one cell in a cluster into the oocyte.

20.4.3 Vitellogenesis

Vitellogenesis is the period during which reserve materials (mainly yolk granules) are synthesized and accumulated in the oocyte cytoplasm. These materials will be used by the developing embryo. Two major types of vitellogenesis can be distinguished: autosynthesis and heterosynthesis (Biliński 1979; Eckelbarger 1994). During autosynthesis, the yolk is synthesized by the oocyte, while in the second type of vitellogenesis, the reserve materials are synthesized outside the oocyte and/or outside the gonad (Biliński 1979; Eckelbarger 1994; Poprawa 2005a).

Three phases of vitellogenesis were distinguished in tardigrades: early, middle, and late vitellogenesis (Węglarska 1979; Poprawa 2005a; Poprawa et al. 2015a, b, c).

During early vitellogenesis, all of the cells (cystocytes) in the germ-cell clusters are morphologically similar and begin to synthesize and accumulate yolk materials (Fig. 20.5a) (Węglarska 1979; Poprawa 2005a; Poprawa et al. 2015a, b, c). The Golgi complexes as well as the rough and smooth endoplasmic reticulum are involved in the syntheses. Moreover, in some species, the mitochondria are suspected of being involved in the synthesis of lipids during vitellogenesis in tardigrades (Węglarska 1979, 1987; Poprawa 2005a).

Middle vitellogenesis is characterized by an intense growth of one (in Parachela) or several (in Apochela) cells in the clusters. These cells differentiate into the oocyte (Fig. 20.5b). The trophocytes synthesize the yolk materials and transport them to the oocyte via cytoplasmic bridges (Węglarska 1979; Poprawa 2005a; Poprawa et al. 2015a, b, c). At this stage of vitellogenesis, coated vesicles are observed at the oocyte periphery, indicating the transport of the exogenous yolk components via endocytosis (micropinocytosis) (Węglarska 1979; Poprawa 2005a). According to several authors (Poprawa 2006; Rost-Roszkowska et al. 2011; Hyra et al. 2016), in tardigrades, the yolk precursors can be synthesized by cells of the midgut epithelium and/or the storage cells. The yolk precursors can be released into the body cavity fluid and then passed from this fluid into the ovary where they are delivered to the oocyte via micropinocytosis.

During late vitellogenesis, the oocyte is filled with yolk materials. The spheres of the yolk vary in their size and structure. Many of them have a heterogeneous structure. The smaller spheres fuse to form larger structures (Węglarska 1979, 1987; Poprawa 2005a; Poprawa et al. 2015a, b, c). Histochemical analysis showed that at the end of vitellogenesis, the oocyte contains yolk materials rich in polysaccharides, proteins (mainly glycoproteins), and lipids (particularly acid lipids) (Węglarska 1979, 1987; Bertolani 1983; Poprawa 2005a; Poprawa et al. 2015a, b, c). At this stage, in some tardigrade species, regularly arranged cortical granules arise in the cortical cytoplasm of the oocyte (Węglarska 1975, 1987; Bertolani 1983; Poprawa et al. 2015c). These granules contain mucopolysaccharides (our unpublished studies).

To summarize, vitellogenesis has mixed character in tardigrades. Autosynthesis, micropinocytosis, and trophocytes all contribute to the formation of yolk.

20.4.4 Choriogenesis and the Structure of the Egg Capsule

During the choriogenesis the eggshells that build the egg capsule are formed. The egg capsule provides the optimal conditions for the developing embryo and protects the oocyte and embryo against the harmful effects of the environment (Węglarska 1982; Poprawa 2005b, 2011; Poprawa et al. 2015b, c). The most commonly used classification of eggshells is the one established by Ludwig in 1874 (Węglarska 1982). This classification is based on the method and place of synthesis of the eggshell precursors. Therefore, three types of eggshells have been distinguished:

(a) The primary type—the eggshell precursors are synthesized and secreted by the oocyte.
(b) The secondary type—the precursors of the eggshell are synthesized in the ovary, but the oocyte does not participate in this process.
(c) The tertiary type—the precursors of the eggshell are synthesized outside the ovary, e.g., in the cells of the oviduct.

The eggs of tardigrades are covered by two shells: the thin vitelline envelope that adheres to the oolemma and the thick three-layered chorion. Although the eggs of insects, onychophorans, and tardigrades are covered by similar shells, the sequence of events during the choriogenesis of these groups is different. In onychophorans and insects, the vitelline envelope is formed first, followed by the appearance of the chorion (Poprawa et al. 2002; Norman and Tait 2008). In tardigrades, at the beginning the chorion is formed, and then the formation of the vitelline envelope takes place (Węglarska 1982; Poprawa 2005b, 2011; Poprawa et al. 2015b, c).

In tardigrades, the process of choriogenesis begins during early vitellogenesis, when cystocytes in the vitellarium are undifferentiated (e.g., in *Thulinius ruffoi*), during middle vitellogenesis (e.g., *Dactylobiotus dispar*, *Dactylobiotus parthenogeneticus*, *Paramacrobiotus richtersi*, *Isohypsibius granulifer granulifer*), or during late vitellogenesis (e.g., *Macrobiotus polonicus*) (Węglarska 1982; Poprawa 2005b, 2011; Poprawa et al. 2015b, c). The first part of the chorion precursors is synthesized (and subsequently secreted to the ovary/ovotestis lumen) by the cells of the ovary/ovotestis wall. This material has flocculent or granular structure and is deposited on the surface of the oocyte, creating small isolated

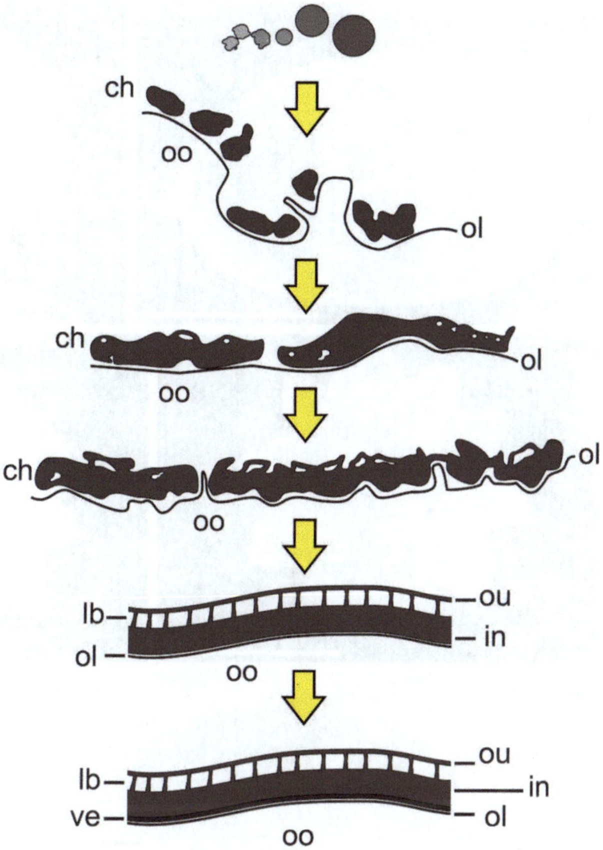

Fig. 20.6 Schematic representation of the process of choriogenesis. Initially the chorion is formed. At the beginning of choriogenesis, small parts of chorion are formed, and next they merge with each other to form larger aggregations, and as a result the continuous layer of chorion is formed. The chorion begins to differentiate into three layers: the inner, the labyrinthine, and the outer layers. After chorion formation the vitelline envelope is formed. *ch* chorion, *in* inner layer of chorion (endochorion), *lb.* labyrinthine layer, *ol* oolemma, *oo* oocyte, *ou* outer layer of chorion (exochorion), *ve* vitelline envelope

fragments of chorion. The second part of chorion precursors is synthesized and secreted by the oocyte. Then, the small fragments of chorion merge to form larger aggregations, and as a result the continuous layer of chorion is formed. Simultaneously, the chorion begins to differentiate into three layers: the inner, the labyrinthine, and the outer layers (Fig. 20.6) (Węglarska 1982; Poprawa 2005b, 2011; Poprawa et al. 2015b, c). After chorion formation, the oocyte synthesizes fibrous material that is secreted into the space between the oolemma and the chorion. Eventually, the vitelline envelope is formed from this material (Poprawa 2005b, 2011; Poprawa et al. 2015b, c).

According to the classification established by Ludwig, the vitelline envelope of tardigrades is of the primary type, while the chorion is regarded as of a mixed type—primary (secreted by the oocyte) and secondary (produced by the cells of the ovotestis/ovary wall) (Węglarska 1982; Poprawa 2005b, 2011; Poprawa et al. 2015b). The only exception has been found in the *Macrobiotus polonicus* where the chorion is purely of the secondary type—the participation of the oocyte in the synthesis of the chorion precursors was not observed in this species (Poprawa et al. 2015c).

Fig. 20.7 Oviposition and eggs. (**a**) Egg of *Milnesium tardigradum*, SEM, bar = 40 μm, courtesy of Dr. Marta Jezierska (Department of Animal Histology and Embryology, University of Silesia in Katowice, Poland); (**b**) Egg of *Macrobiotus cf. hufelandi*, SEM, bar = 30 μm, (**c**) Egg of *Paramacrobiotus cf. richtersi*, SEM, bar = 50 μm; (**d**) Egg of *Mesobiotus* sp., SEM, bar = 30 μm (**b–d**). Courtesy of Dr. Łukasz Kaczmarek (Department of Animal Taxonomy and Ecology, Faculty of Biology, Adam Mickiewicz University, Poznań, Poland), MSc Milena Roszkowska (Department of Bioenergetics, Institute of Molecular Biology and Biotechnology, Adam Mickiewicz University, Poznań, Poland), and MSc Magdalena Gawlak (Institute of Plant Protection-National Research Institute, Poznań, Poland); (**e**) Specimens of *Hypsibius exemplaris* during oviposition, LM bar = 25 μm; (**f**) *Thulinius ruffoi*—eggs inside the exuvium, LM, bar = 50 μm

A surface of chorion can be smooth (Fig. 20.7a) or ornamented (Fig. 20.7b–d) in tardigrades. The ornamented eggs have the conical processes of various numbers, shapes, and sizes. The surface of the basic chorion (the chorion between processes) is smooth, covered with mesh (Fig. 20.7b) or have areolations (Fig. 20.7c) (Węglarska 1982; Bertolani 1983; Ramazzotti and Maucci 1983; Dewel et al. 1993; Kinchin 1994; Poprawa 2005b; Nelson et al. 2015; Poprawa et al. 2015b, c). The surface of the ornamented eggs is a very useful taxonomical character.

20.5 Oviposition

Tardigrades lay their eggs freely in the environment or into the exuviae (Fig. 20.7e–f) (Kinchin 1994) although the correlation between molting and oviposition has been observed in all tardigrade species (Bertolani 1983). Generally, ornamented eggs are laid freely while smooth ones are laid into the exuvium (Bertolani et al. 1996). The number of eggs per oviposition varies widely according to the species from 1 to 60 (Bertolani 1983). The number of eggs laid can differ among specimens of one species, and it depends on an animal's age and its physiological state.

20.6 Conclusions

Despite that tardigrades have been known for almost 250 years, their exact phylogenetic position remains obscure. Although all accumulated data imply that tardigrades belong to Panarthropoda, their position within this clade is still debatable. Three hypotheses are currently considered: (1) tardigrades as the sister group of arthropods, (2) tardigrades as the sister group of onychophorans, or (3) tardigrades as the sister group of onychophorans plus arthropods. Features related to oogenesis and structure of the ovary such as (1) the presence of a germarium (germ-line proliferation zone), (2) the lack of a germinal epithelium, (3) the lack of intraepithelial germ-cells, (4) the presence of a meroistic ovary, (5) the presence of branched female germ-cell clusters, and (6) the presence of somatic cells involved in the synthesis of the precursors of eggshells confirm the hypothesis that Tardigrada are the sister group rather for Arthropoda than Onychophora.

Acknowledgments We are grateful to Dr. Łukasz Kaczmarek (Department of Animal Taxonomy and Ecology, Faculty of Biology, Adam Mickiewicz University, Poznań, Poland), MSc Milena Roszkowska (Department of Bioenergetics, Institute of Molecular Biology and Biotechnology, Adam Mickiewicz University, Poznań, Poland), MSc Magdalena Gawlak (Institute of Plant Protection-National Research Institute, Poznań, Poland), and MSc Marta Jezierska (Department of Animal Histology and Embryology, University of Silesia in Katowice, Poland) for agreeing to the use of some of their photographic documentation.

References

Altiero T, Rebecchi L (2001) Rearing tardigrades: results and problems. Zool Anz 240 (3-4):217–221

Amini R, Goupil E, Labella S, Zetka M, Maddox AS, Labbé J-C, Chartier NT (2014) *C. elegans* Anillin proteins regulate intercellular bridge stability and germline syncytial organization. J Cell Biol 206:129–143

Anderson E, Huebner E (1968) Development of the oocyte and its accessory cells of the polychaete *Diopatra cuprea* (Bosc). J Morphol 126:163–197

Bertolani R (1972) La partenogenesi nei Tardigradi. Boll Zool 39:577–581
Bertolani R (1979) Hermaphoditism in Tardigrades. Int J Invertebr Reprod 1:67–71
Bertolani R (1983) Tardigrada. In: Adiyodi KG, Adiyodi RG (eds) Reproductive biology of invertebrates, Oogenesis, oviposition and oosorption, vol I. Wiley, Chichester, pp 431–441
Bertolani R (1987) Sexuality, reproduction and propagation in tardigrades. In: Bertolani R (ed.), Biology of tardigrades. Proceedings of fourth international symposium Tardigrada, Modena. Selected Symposia and Monographs, U.Z.I. 1 Mucchi, Modena, pp 93–101
Bertolani R (2001) Evolution of the reproductive mechanisms in tardigrades – a review. Zool Anz 240(3):247–252
Bertolani R, Manicardi GC (1986) New cases of hermaphroditism in tardigrades. Int J Invert Reprod Dev 9(3):363–366
Bertolani R, Pilato G, Sabatini MA (1983) *Macrobiotus joannae*, primo macrobiotidae ermafrodito (Eutardigrada). Animal 10:327–333
Bertolani R, Rebecchi L, Claxton SK (1996) Phylogenetic significance of egg shell variation in tardigrades. Zool J Linnean Soc 116:139–148
Biliński SM (1979) Powstawanie żółtka w oogenezie bezkręgowców. Post Biol Kom 6(4):249–266
Bilinski SM (1994) The ovary of Entognatha. In: Büning J (ed) The insect ovary: ultrastructure, previtellogenic growth and evolution. Chapman and Hall, London, pp 7–30
Bilinski SM (1998) Introductory remarks. Folia Histochem Cytobiol 36:143–145
Binda MG, Kristiensen RM (1986) Notes on the genus *Oreella* (Oreellidae) and the systematic position of *Carphania fluviatilis* Binda, 1978 (Carphanidae fam. nov., Heterotardigrada). Animal 13:9–20
Brubacher JL, Huebner E (2009) Development of polarized female germline cysts in the polychaete, *Ophryotrocha labronica*. J Morphol 270:413–429
Brubacher JL, Huebner E (2011) Evolution and development of polarized germ cell cysts: new insights from a polychaete worm *Ophryotrocha labronica*. Dev Biol 357:96–107
Büning J (1994) The Insect Ovary: Ultrastructure, previtellogenic growth and evolution. Chapman and Hall, London
Crowe JH (1975) The physiology of cryptobiosis in tardigrades. Mem Ist Ital Idrobiol 32:37–59
Czerneková M, Jönsson KI, Chajec L, Student S, Poprawa I (2017) The structure of the desiccated *Richtersius coronifer* (Richters, 1903). Protoplasma 254(3):1367–1137
Czerneková M, Jönsson KI, Chajec L, Student S, Poprawa I (2018) A comparative ultrastructure study of storage cells in the eutardigrade *Richtersius coronifer* in the hydrated state and after desiccation and heating stress. PLoS One 13(8):e0201430. https://doi.org/10.1371/journal.pone.0201430
Degma P, Bertolani R, Guidetti R (2009–2018) Actual checklist of Tardigrada species (2009 2018, Ver. 34: Edition: 30-06-2018)
Dewel RA, Nelson DR, Dewel WC (1993) Tardigrada. In: Harrison FW, Rice ME (eds) Microscopic Anatomy of Invertebrates. Onychophora, Chilopoda and Lesser Protostomata, vol 12. Wiley-Liss, New York, pp 143–183
Eckelbarger KJ (1994) Diversity of metazoan ovaries and vitellogenic mechanisms – implications for life history theory. Proc Biol Soc Wash 107:193–218
Foor WE (1968) Cytoplasmic bridges in the ovary of *Ascaris lumbricoides*. Bull Tulane Univ Med Fac 27:23–30
Foor WE (1983) Nematoda. In: Adiyodi KG, Adiyodi RG (eds) Reproductive biology of invertebrates, vol I. Oogenesis, oviposition, and oosorption. Wiley, Chichester, pp 223–256
Gibert MA, Starck J, Beguet B (1984) Role of the gonad cytoplasmic core during oogenesis of the nematode *Caenorhabditis elegans*. Biol Cell 50:77–85
Greenbaum MP, Iwamori T, Buchold GM, Matzuk MM (2011) Germ cell intercellular bridges. Cold Spring Harb Perspect Biol 3:a005850
Grimaldi de Zio S, D'Addabbo Gallo M, Morone de Lucia MR, D'Addabbo L (1987) Marine Arthrotardigrade and Echiniscoidea (Tardigrada, Heterotardigrada) from the Indian Ocean. Boll Zool (4):347–357

Grothman GT, Carl Johansson C, Chilton G, Kagoshima H, Tsujimoto M, Suzuki AC (2017) Gilbert Rahm and the Status of Mesotardigrada Rahm, 1937. Zool Sci 34(1):5–10
Haglund K, Nezis IP, Stenmark H (2011) Structure and functions of stable intercellular bridges formed by incomplete cytokinesis during development. Commun Integr Biol 4:1–9
Horikawa DD (2012) Survival of tardigrades in extreme environments: a model animal for astrobiology. In: Altenbach AV, Bernhard JM, Seckbach J (eds) Anoxia. Evidence for eukaryote survival and paleontological strategies. Springer, Dordrecht, pp 205–217
Huynh JRR, St Johnston D (2004) The origin of asymmetry: early polarisation of the *Drosophila* germline cyst and oocyte. Curr Biol 14:438–449
Hyra M, Rost-Roszkowska MM, Student S, Włodarczyk A, Deperas M, Janelt K, Poprawa I (2016) Body cavity cells of Parachela during their active life. Zool J Linnean Soc 178:878–887
Jaglarz MK, Kubrakiewicz J, Jedrzejowska I, Goldyn B, Bilinski SM (2014a) Ultrastructural analysis of the ovary and oogenesis in Spinicaudata and Laevicaudata (Branchiopoda) and its phylogenetic implications. Zoology 117:207–215
Jaglarz MK, Kubrakiewicz J, Bilinski SM (2014b) The ovary structure and oogenesis in the basal crustaceans and hexapods. Possible phylogenetic significance. Arthropod Struct Dev 43:349–360
Jędrzejowska I, Mazurkiewicz-Kania M, Garbiec A, Kubrakiewicz J (2013) Differentiation and function of the ovarian somatic cells in the pseudoscorpion, *Chelifer cancroides* (Linnaeus, 1761) (Chelicerata: Arachnida: Pseudoscorpionida). Arthropod Struct Dev 42:27–36
Keilin D (1959) The Leeuwenhoek lecture. The problem of anabiosis or latent life: history and current concept. Proc R Soc London B 150:149–191
Kinchin IM (1994) The biology of tardigrades. Portland, London
Kloc M, Bilinski S, Dougherty MT, Brey EM, Etkin LD (2004) Formation, architecture and polarity of female germline cyst in Xenopus. Dev Biol 266(1):43–61
Kristensen RM (1977) On the marine genus *Styraconyx* (Tardigrada, Heterotardigrada, Halechiniscidae) with description of a new species from a warm spring on Disco Island, West Greenland. Astarte 10:87–91
Kubrakiewicz J (1997) Germ cells cluster organization in polytrophic ovaries of Neuroptera. Tissue Cell 29:221–228
Kubrakiewicz J (1998) Struktura i funkcja zespołów komórek płciowych w politroficznych owariolach sieciarek (Insecta: Neuroptera). Rozprawa habilitacyjna. Wydawnictwo Uniwersytetu Wrocławskiego, Wrocław
Kubrakiewicz J, Adamski RT, Biliński SM (1991) Ultrastructural studies on accessory nuclei in developing oocytes of the crustacean, *Siphonophanes grubei*. Tissue Cell 23:903–907
Lin H, Spradling AC (1995) Fusome asymmetry and oocyte determination in *Drosophila*. Dev Genet 16:6–12
Marcus E (1929) Tardigrada. In: Bronn HG (ed) Klassen und Ordungen des Tierreichs, vol 5, Section 4, Part 3, pp 1-608
Margaritis LH (1985) Structure and physiology of the eggshell. In: Gilbert LI, Kerkut GA (eds) Comprehensive insect biochemistry physiology and pharmacology 1, vol 6. Pergamon, Berlin, pp 151–230
Matova N, Cooley L (2001) Comparative aspects of animal oogenesis. Dev Biol 231:291–320
Mayer G, Tait NN (2009) Position and development of oocytes in velvet worms shed light on the evolution of the ovary in Onychophora and Arthropoda. Zool J Linnean Soc 157:17–33
Mayer G, Martin C, Rüdiger J, Kauschke S, Stevenson PA, Poprawa I, Hohberg K, Schill RO, Pflüger HJ, Schlegel M (2013) Selective neuronal staining in tardigrades and onychophorans provides insights into the evolution of segmental ganglia in panarthropods. BMC Evol Biol 13:230
Nelson DR (1982) Tardigrada. In: Hurlbert SH, Villalobos-Figueroa A (eds) Aquatic biota of Mexico, Central America and the West Indies. San Diego State University Press, San Diego, pp 154–158

Nelson DR (2002) Current status of Tardigrada: evolution and ecology. Integr Comp Biol 42:652–659

Nelson DR, Guidetti R, Rebecchi L (2015) Phylum Tardigrada. In: Thorp J, Rogers DC (eds) Thorp and Covich's freshwater invertebrates. Ecology and general biology, vol 1, 4th edn. Elsevier, Oxford, pp 347–380

Nielsen C (2001) Animal evolution: Interrelationships of the living phyla, 2nd edn. Oxford University Press, Oxford

Norman JM, Tait NN (2008) Ultrastructure of the eggshell and its formation in *Planipapillus mundus* (Onychophora: Peripatopsidae). J Morphol 269:1263–1275

Ong S, Tan C (2010) Germline cyst formation and incomplete cytokinesis during *Drosophila melanogaster* oogenesis. Dev Biol 337:84–98

Pollock LW (1970) Reproductive anatomy of some marine Heterotardigrada. Trans Am Microsc Soc 89:308–316

Pollock LW (1975) Tardigrada. In: Giese AC, Pearse JS (eds) Reproduction of marine invertebrates, vol 2. Academic, New York, pp 43–54

Poprawa I (2005a) The ovary structure, previtellogenic and vitellogenic stages in parthenogenetic species *Dactylobiotus dispar* (Murray, 1907) (Tardigrada: Eutardigrada). Tissue Cell 37:385–392

Poprawa I (2005b) The structure and the formation of egg shells in parthenogenetic species *Dactylobiotus dispar* Murray, 1907 (Tardigrada: Eutardigrada). Folia Biol (Krakow) 53:173–177

Poprawa I (2006) Ultrastuctural changes of the storage cells during oogenesis in *Dacylobiotus dispar* (Murray, 1907) (Tardigrada: Eutardigrada). Zool Pol 51:13–18

Poprawa I (2011) Ultrastructural studies of the formation of the egg capsule in the hermaphroditic species, *Isohypsibius granulifer granulifer* Thulin, 1928 (Eutardigrada: Hypsibiidae). Zool Sci 28(1):37–40

Poprawa I, Baran A, Rościszewska E (2002) Structure of ovaries and formation of egg envelopes in the stonefly, *Leuctra autumnalis* Aubert, 1948 (Plecoptera: Leuctridae). Ultrastructural studies. Folia Biol (Cracow) 50:29–38

Poprawa I, Hyra M, Kszuk-Jendrysik M, Rost-Roszkowska MM (2015a) Ultrastructural changes and programmed cell death of trophocytes in the gonad of *Isohypsibius granulifer granulifer* (Tardigrada, Eutardigrada, Isohypsibiidae). Micron 70:26–33

Poprawa I, Hyra M, Rost-Roszkowska MM (2015b) Germ cell cluster organization and oogenesis in the tardigrade *Dactylobiotus parthenogeneticus* Bertolani, 1982 (Eutardigrada, Murrayidae). Protoplasma 252:1019–1029

Poprawa I, Schlechte-Wełnicz W, Hyra M (2015c) Ovary organization and oogenesis in the tardigrade *Macrobiotus polonicus* Pilato, Kaczmarek, Michalczyk & Lisi, 2003 (Eutardigrada, Macrobiotidae): ultrastructural and histochemical analysis. Protoplasma 252:857–865

Rahm G (1937) Eine neue Tardigraden-ordnung aus den heissen Quellen von Unzen, Insel Kyushu, Japan. Zool Anz 121:65–71

Ramazzotti G, Maucci W (1983) Il Phylum Tardigrada. III edizione riveduta e aggiomata. Mem Ist Ital Idrobiol 41:1–1012 (an English translation, ed. by Dr. C. Beasley)

Rebecchi L (1997) Ultrastructural study of spermiogenesis and the testicular and spermathecal spermatozoon of the gonochoristic tardigrade *Xerobiotus pseudohufelandi* (Eutardigrada, Macrobiotidae). J Morphol 234(1):11–24

Rebecchi L, Bertolani R (1994) Maturative pattern of ovary and testis in eutardigrades of freshwater and terrestrial habitats. Invertebr Reprod Dev 26(2):107–117

Rebecchi L, Nelson DR (1998) Evaluation of a secondary sex character in eutardigrades. Invertebr Biol 117(3):194–198

Rebecchi L, Altiero T, Cesari M, Bertolani R, Rizzo AM, Corsetto PA, Guidetti R (2011) Resistance of the anhydrobiotic eutardigrade *Paramacrobiotus richtersi* to space flight (LIFE-TARSE mission on FOTON-M3). J Zool Syst Evol Res 49(Suppl 1):98–103

Renaud-Debyser J (1965) *Parastygarctus higginsi* n. g., n. sp., tardigrade marin interstitiel de Madagascar. C R Acad Sci 260:955–957

Rost-Roszkowska MM, Poprawa I, Wójtowicz M, Kaczmarek Ł (2011) Ultrastructural changes of the midgut epithelium in *Isohypsibius granulifer granulifer* Thulin, 1928 (Tardigrada: Eutardigrada) during oogenesis. Protoplasma 248:405–414

Rota-Stabelli O, Kayal E, Gleeson D, Daub J, Boore JL, Telford MJ, Pisani D, Blaxter M, Lavrov DV (2010) Ecdysozoan mitogenomics: evidence for a common origin of the legged invertebrates, the Panarthropoda. Genome Biol Evol 2:425–440

Suzuki A (2006) Ovarian structure in *Milnesium tardigradum* (Tardigrada: Milnesiidae) during early vitellogenesis. Hydrobiologia 558:61–66

Świątek P, Kubrakiewicz J, Klag J (2009) Formation of germ-line cysts with a central cytoplasmic core is accompanied by specific orientation of mitotic spindles and partitioning of existing intercellular bridges. Cell Tissue Res 337:137–148

Świątek P, De Wit P, Jarosz N, Chajec Ł, Urbisz AZ (2018) Micromorphology of ovaries and oogenesis in *Grania postclitellochaeta* (Clitellata: Enchytraeidae). Zoology 126:119–127

Szymańska B (1994) Interdependence between storage bodies and egg developmental stages in *Macrobiotus richtersi* Murray, 1911 (Tardigrada). Acta Biol Cracov Ser Zool 36:41–50

Telford MJ, Bourlat SJ, Economou A, Papillon D, Rota-Stabelli O (2008) The evolution of the Ecdysozoa. Philos Trans R Soc Lond Ser B Biol Sci 363(1496):1529–1537

Tsujimoto M, Imura S, Kanda H (2015) Recovery and reproduction of an Antarctic tardigrade retrieved from a moss sample frozen for over 30 years. Cryobiology 72(1):78–81

Tworzydło W, Biliński SM (2008) Structure of ovaries and oogenesis in dermapterans. I. Origin and functioning of the ovarian follicles. Arthropod Struct Dev 37:310–320

Tworzydło W, Biliński SM, Kočárek P, Haas F (2010) Ovaries and germline cysts and their evolution in Dermaptera (Insecta). Arthropod Struct Dev 39:360–368

Urbisz AZ, Chajec Ł, Brąszewska-Zalewska A, Kubrakiewicz J, Świątek P (2017) Ovaries of the white worm (*Enchytraeus albidus*, Annelida, Clitellata) are composed of 16-celled meroistic germ-line cysts. Dev Biol 426:28–42

Urbisz AZ, Chajec Ł, Ito M, Ito K (2018) The ovary organization in the marine limnodriloidin *Thalassodrilides* cf. *briani* (Annelida: Clitellata: Naididae) resembles the ovary of freshwater tubificins. Zoology 128:16–26

Volkman A, Greven H (1993) Ultrastructural localization of tyrosinase in the tardigrade cuticle. Tissue Cell 25:435–438

Węglarska B (1975) Studies on the morphology of *Macrobiotus richtersi* Murray, 1911. Mem Ist Ital Idrobiol 32:445–464

Węglarska B (1979) Electron microscope study of previtellogenesis and vitellogenesis in *Macrobiotus richtersi* J. Murr. (Eutardigrada). Zesz Nauk Uniw Jagiell Pr Zool 25:169–189

Węglarska B (1982) Ultrastructural study of the formation of egg envelopes in *Macrobiotus richtersi* (Eutardigrada). In: Nelson DR (ed) Proceedings of third international symposium Tardigrada. East Tennessee State University Press, Johnson City, pp 115–128

Węglarska B (1987) Yolk formation in *Isohypsibius* (Eutardigrada). Zoomorphology 107:287–292

Wełnicz W, Grohme MA, Kaczmarek Ł, Schill RO, Frohme M (2011) Anhydrobiosis in tardigrades: the last decade. J Insect Physiol 57(5):577–583

Wright JC, Westh P, Ramløv H (1992) Cryptobiosis in Tardigrada. Biol Rev 67:1–29

Chapter 21
Architecture and Life History of Female Germ-Line Cysts in Clitellate Annelids

Piotr Świątek and Anna Z. Urbisz

Abstract Animal female and male germ-line cells often form syncytial units termed cysts, clusters, or clones. Within these cysts, the cells remain interconnected by specific cell junctions known as intercellular bridges or ring canals, which enable cytoplasm to be shared and macromolecules and organelles to be exchanged between cells. Numerous analyses have shown that the spatial organization of cysts and their functioning may differ between the sexes and taxa. The vast majority of our knowledge about the formation and functioning of germ-line cysts comes from studies of model species (mainly *Drosophila melanogaster*); the other systems of the cyst organization and functioning are much less known and are sometimes overlooked. Here, we present the current state of the knowledge of female germ-line cysts in clitellate annelids (Clitellata), which is a monophyletic taxon of segmented worms (Annelida). The organization of germ-line cysts in clitellates differs markedly from that of the fruit fly and vertebrates. In Clitellata, germ cells are not directly connected one to another, but, as a rule, each cell has one ring canal that connects it to an anuclear central cytoplasmic core, a cytophore. Thus, this pattern of cell distribution is similar to the germ-line cysts of *Caenorhabditis elegans*. The last decade of studies has revealed that although clitellate female germ-line cysts have a strong morphological plasticity, e.g., cysts may contain from 16 to as many as 2500 cells, the oogenesis always shows a meroistic mode, i.e., the interconnected cells take on different fates; a few (sometimes only one) become oocytes, whereas the rest play the role of supporting (nurse) cells and do not continue oogenesis.

This is the first comprehensive summary of the current knowledge on the organization and functioning of female germ-line cysts in clitellate annelids.

P. Świątek (✉) · A. Z. Urbisz
Faculty of Biology and Environmental Protection, Department of Animal Histology and Embryology, University of Silesia in Katowice, Katowice, Poland
e-mail: piotr.swiatek@us.edu.pl

W. Tworzydlo, S. M. Bilinski (eds.), *Evo-Devo: Non-model Species in Cell and Developmental Biology*, Results and Problems in Cell Differentiation 68,
https://doi.org/10.1007/978-3-030-23459-1_21

21.1 Introduction

21.1.1 Germ-Line Cysts: A Conservative Trait of Animal Oogenesis

The mitotic divisions of developing female and male germ-line cells of metazoans are often not followed by complete cytokineses which leads to the formation of syncytial cysts (clusters, nests) of interconnected sibling cells (Pepling et al. 1999; Greenbaum et al. 2011; Haglund et al. 2011; Lu et al. 2017). Within germ-line cysts (GCs), the germ cells remain connected by characteristic cytoplasmic canals, termed intercellular bridges, cytoplasmic bridges, or ring canals (Pepling et al. 1999; Ventelä 2006; Greenbaum et al. 2011; Haglund et al. 2011; Lu et al. 2017; Yamashita 2018). Functioning ring canals (RCs) are short, but quite broad (from 0.25 to 10 μm in diameter), cytoplasmic channels usually stabilized by modified contractile rings, the so-called bridge rims. The rims are composed of numerous cytoskeletal proteins that lie just beneath the plasma membrane (Pepling et al. 1999; Ong and Tan 2010; Greenbaum et al. 2011; Haglund et al. 2011; Yamashita 2018). The occurrence of such wide intercellular canals between the sister cells enables the transfer of not only macromolecules (e.g., different species of ribonucleoproteins) but also cell components such as ribosomes, mitochondria, centrioles, and Golgi complexes, between the interconnected cells (Telfer 1975; Greenbaum et al. 2011; Haglund et al. 2011; Lu et al. 2017; Yamashita 2018). Although the morphology, formation, and functioning of female germ-line cysts (FGCs) have been intensively studied in the past, the majority of our knowledge comes from model organisms such as *Drosophila melanogaster* (e.g., King 1970; Lin et al. 1994; Robinson et al. 1994; Lighthouse et al. 2008; Ong and Tan 2010; recently reviewed in Bilinski et al. 2017; Yamashita 2018), *Caenorhabditis elegans* (e.g., Gumienny et al. 1999; Wolke et al. 2007; Amini et al. 2014; Seidel et al. 2018; reviewed in Amini et al. 2015), *Xenopus laevis* (Kloc et al. 2004), and *Mus musculus* (e.g., Pepling and Spradling 1998; Lei and Spradling 2013, 2016; recently reviewed by Ikami et al. 2017; Wang et al. 2017). These studies have revealed several molecular mechanisms that lead to the formation of GCs and shed light on the functioning of both male and female cysts and their role during gametogenesis (summarized in Greenbaum et al. 2011; Haglund et al. 2011; Ikami et al. 2017; Lu et al. 2017; Wang et al. 2017; Yamashita 2018). Briefly, in males, the formation of GCs is a ubiquitous phenomenon, and it seems that the interconnected male germ cells share gene products, and therefore the haploid spermatids are phenotypically diploid (Braun et al. 1989; Morales et al. 2002; Ventelä et al. 2003; for details see Guo and Zheng 2004; Greenbaum et al. 2011; Lu et al. 2017). The importance of the syncytial phase in the case of male germ cells has repeatedly been documented during experiments that have shown that disruptions in GCs formation cause infertility (e.g., Brill et al. 2000; Maddox et al. 2005; Greenbaum et al. 2006; Yamamoto et al. 2013; Lorès et al. 2014). In female germ line, the formation and functioning of cysts is more complicated. The occurrence of GCs during oogenesis is less ubiquitous than in the male line, and there are some groups of metazoans in which FGCs have never been observed, e.g., in hemimetabolous insects (Büning 1994; Tworzydło et al. 2014), arachnids (Jędrzejowska and Kubrakiewicz 2007; Jędrzejowska et al. 2013), mollusks (de Jong-Brink et al. 1983), and certain annelids (Eckelbarger 1983; Świątek et al. 2016). In these taxa, all germ-line cells (except those that are degenerating) have the potential to become functional gametes. This type of oogenesis is termed panoistic (Greek *pan* = all + *oon* = egg). In other animals as holometabolous insects (Büning 1994; Bilinski et al. 2017), basal crustaceans (Trentini and Scanabissi 1978; Jaglarz et al. 2014a), tardigrades (Poprawa et al. 2015a, b), nematodes (Wolke et al. 2007; Amini et al. 2014), several annelids (Anderson and Huebner 1968; Brubacher and Huebner 2011; Urbisz et al. 2017), cheilostome bryozoans (Dyrynda and King 1983; Moosbrugger et al. 2012), appendicularian urochordates (Ganot et al. 2007), and some vertebrates (Kloc et al. 2004; Lei and Spradling 2013, 2016; Żelazowska and Fopp-Bayat 2017), FGCs are formed, but their functioning differs in various taxa. In some invertebrates, e.g., stoneflies (Gottanka and Büning 1990), and vertebrates, e.g., amphibians (Kloc et al. 2004), the germ-line cysts are transient. As oogenesis progresses, the cyst cells separate from each other and become individual. This process leads to the rise of typical panoistic ovaries. However, in the majority of animal species, FGCs function for a longer period; simultaneously, two distinct categories of interconnected germ-line cells emerge: cells that continue oogenesis (oocytes) and become future egg cells and supporting cells (nurse cells, trophocytes) that provide the developing oocytes with macromolecules and organelles. Nurse cells are eliminated (often by apoptosis) at the end of oogenesis. The ovaries containing both categories of cells are known as meroistic (Greek *meros* = part + *oon* = egg). Unsurprisingly, *Drosophila melanogaster* became the model organism for analyses of the formation and functioning of FGCs (e.g., reviewed in de Cuevas et al. 1997; Huynh and St Johnston 2004; Lu et al. 2017; Bilinski et al. 2017; Yamashita 2018). On the other hand, the knowledge of the functioning of FGCs in other animals taxa, nematodes (especially *C. elegans*—Wolke et al. 2007; Amini et al. 2014; Seidel et al. 2018) and mammals (*M. musculus*—Pepling and Spradling 1998; Lei and Spradling 2013, 2016), is rapidly growing. It is generally believed that in meroistic ovaries, the asymmetric cell divisions that occur during cyst formation lead to the polarized (directional) transport of macromolecules and organelles along the "axis": the nurse cells—oocyte(s), which accelerates oogenesis. Some defects in this directional transfer (e.g., disrupting the microtubules as the route for transport) lead to altered oocyte growth and sterility (Huynh and St Johnston 2004).

FGCs are formed by incomplete cell divisions, but the spatiotemporal details of these divisions differ, leading to a different spatial organization (architecture) of developing and functioning cysts. There are hundreds of reports describing various patterns of FGC organization in animal taxa, but for the aims of this review, we will characterize only four of them (Fig. 21.1). In linear FGCs, sister cells form chains in which all cells, except for the terminal ones, have two RCs that connect them to their neighbors; the terminal cells have only one RC (Fig. 21.1c). The linear cysts are characteristic for certain arthropods (Biliński 1993; Büning 1994; Jaglarz et al. 2014b) and polychaetes (Anderson and Huebner 1968). Linear FGCs can be

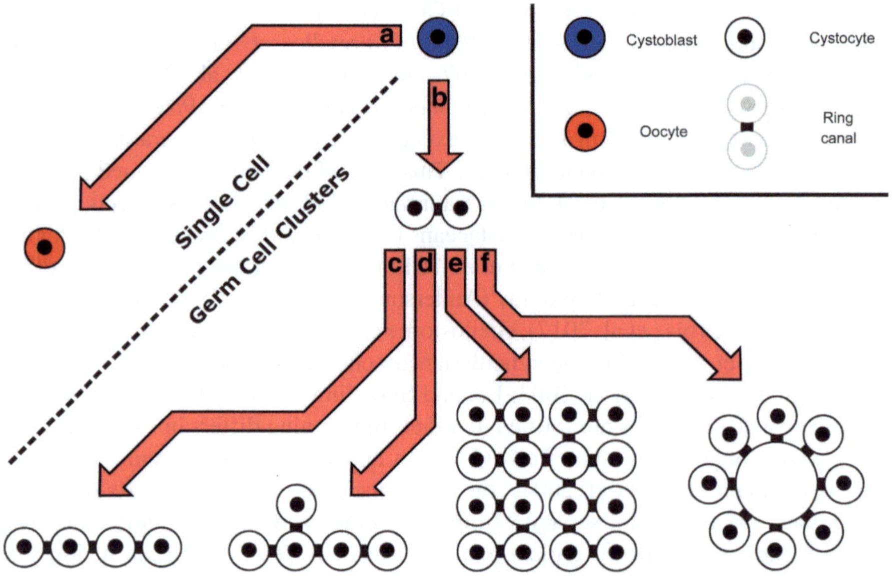

Fig. 21.1 Different patterns of female germ-line cysts (FGCs) organization. (**a**) In some animals, there is no FGC formation, and the germ cells develop individually. (**b**) In other animals the last generation of gonial cells (cystoblast) divides without full cytokinesis, and therefore FGCs uniting cells interconnected via ring canals (RCs) emerge. The spatial organization of cysts can be different: (**c**) a chain-like FGC, all of the cells, except for the terminal ones, have two RCs; (**d**) linear cyst with some branching sites, most of the cells form chains, occasionally some cells have three or more bridges; (**e**) in branched (ramified) cysts, the cystocytes have as many RCs as the number of cell division rounds that have passed; (**f**) in cysts that have a central cytoplasmic mass, irrespective of the division rounds, each cell has one RC; additionally the common cytoplasmic mass mediates the cell-to-cell contact

modified via additional mitotic divisions of the cells located somewhere in the middle of a cyst. As a result, linear clusters with branching sites arise (Fig. 21.1d). Linear FGCs with occasional branches have been found in lacewings (Neuroptera) and mice (Kubrakiewicz 1997; Lei and Spradling 2016). Female cysts can also have a more branched pattern in which some clustered cells are connected by more than two RCs. The best known are the so-called maximal branched cysts (Fig. 21.1e). Branched FGCs have been described in several insects (including *D. melanogaster*—Büning 1994) and vertebrates, e.g., in *X. laevis* (Kloc et al. 2004). The fourth system is apparently different and is also the most interesting in the context of this review. In contrast to the abovementioned modes, in this system, clustered cells do not directly connect to one another, but rather, each cell is connected to a central anuclear cytoplasmic core (mass) via only one RC. The common cytoplasmic mass occupies the center of the FGC while the cells are arranged peripherally (Figs. 21.1f and 21.2a–c). FGCs equipped with a central cytoplasmic mass have been found in nematodes (including *Caenorhabditis elegans*—Foor 1983; Gibert et al. 1984; Wolke et al. 2007; Amini et al. 2014),

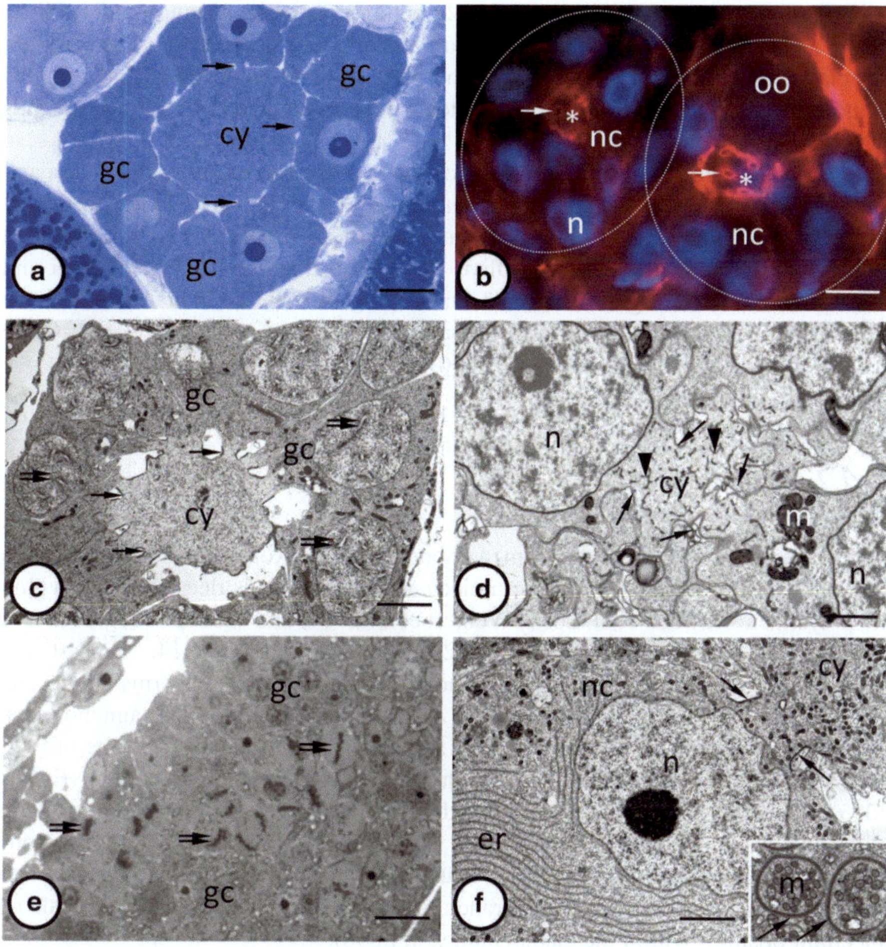

Fig. 21.2 General cysts architecture in Clitellata. (**a**) Each germ cell (gc) is connected via one RC (arrows) with a common anuclear cytoplasmic mass (cytophore—cy). *Chaetogaster diaphanus*, light microscopy (LM), epon section stained with methylene blue, bar = 20 μm. (**b**) Two cysts (encircled by a dotted line) visualized by fluorescence microscopy. Nurse cells (nc) and oocyte (oo) join to the central cytophore (asterisks) via RCs which are enriched in F-actin (arrows); *n* nurse cell nuclei. *Enchytraeus albidus*, fluorescence microscopy (FM), the whole-mounted preparation, double stained with DAPI and rhodamine-conjugated phalloidin, bar = 15 μm. (**c**) A cyst at early oogenesis. Germ cells (gc) are morphologically identical and in the same phase of meiosis (note the synaptonemal complexes in their nuclei—double arrows). Arrows mark electron-dense inner rims of RCs connecting the germ cells to the initial cytophore (cy). *Insulodrilus bifidus*, transmission electron microscopy (TEM), bar = 7 μm. (**d**) The initial cytophore (cy) is enriched in endoplasmic reticulum (arrowheads) in a cyst at the early oogenesis. Arrows mark RCs, *m* mitochondria, *n* nuclei. *Tubifex tubifex*, TEM, bar = 0.7 μm. (**e**) Partial synchronization of cystocyte divisions (double arrows) in the highly multicellular FGC of a leech. *Gc* germ cells. *Theromyzon tessulatum*, LM, epon section stained with methylene blue, bar = 35 μm. (**f**) Details of cyst organization in a medicinal leech, *Hirudo medicinalis*. Arrows point to RC, *cy* a fragment of cytophore, *n* nucleus, *nc* nurse cell, *er* endoplasmic reticulum. Note numerous mitochondria filling the RC cytoplasm and cytophore. Inset shows a transverse section through two RCs in a fish leech *Piscicola geometra*. Note electron-dense rim (arrows) and numerous mitochondria (m). TEM, bar = 2.5 μm

oribatid mites (Liana and Witaliński 2012), echiuran annelids (Leutert 1974; Pilger 1980), and clitellate annelids (Świątek et al. 2009; Urbisz et al. 2017). The common cytoplasm may attain a different form (morphology) and has been termed a rachis (in nematodes), a medulla (in oribatid mites), and a cytophore (in echiuran annelids and Clitellata).

Providing a summary of the current knowledge of the formation, architecture, and functioning of clitellate FGCs is the main aim of this review. The FGCs were also found in certain polychaetous annelids, but interestingly these cysts are not equipped with the cytophore (for details see Olive 1983b; Eckelbarger 1983; Brubacher and Huebner 2011). The organization and functioning of male GCs in Annelida have been presented in numerous original reports and have been reviewed by, e.g., Olive (1983a), Ferraguti (1999), Jamieson (2006), and Rouse (2006).

21.1.2 Clitellata

Clitellate annelids (Clitellata) form a monophyletic taxon of segmented worms (Annelida), which according to the molecular phylogenies is deeply nested within the polychaetes (Struck et al. 2011; Weigert et al. 2014) and should be regarded as a subgroup of Sedentaria (Struck et al. 2011; Weigert et al. 2014; Weigert and Bleidorn 2016). In the presented review, however, we are using the terms Polychaeta and Clitellata in the traditional sense as separate taxa. All clitellate annelids lack parapodia, are hermaphrodites, and have a specialized epithelium on some of their body segments that produce the proteins that are the components of a cocoon (Erséus 2005). Clitellate annelids were usually divided into burrowing or non-burrowing oligochaetous annelids that have chaetae (Oligochaeta), predatory or sanguivorous (blood-sucking) leeches (Hirudinida) that lack chaetae but have suckers and two ectoparasitic leechlike taxa, e.g., Branchiobdellida and Acanthobdellida, with an uncertain position within the clade (Jamieson 2006).

Traditionally, oligochaetous clitellates were classified into two ecological groups, which were known as microdriles or Microdrili (smaller organisms that are primarily associated with water, e.g., sludge worms) and megadriles or Megadrili (larger soil habitants including all earthworm species) (Beddard 1895; Jamieson 1981; Erséus 2005). True leeches (Hirudinida) are usually divided into two main groups: Rhynchobdellida—leeches that have an extensible proboscis (e.g., fish leeches) and Arhynchobdellida—leeches with a muscular pharynx with or without jaws (e.g., medicinal leeches) (Govedich and Moser 2015). Acanthobdellida includes leechlike parasites of salmonid fish that have a posterior sucker like true leeches (Hirudinida) but also have oligochaete features such as setae and intersegmental septa (Fernández et al. 1992; Bielecki et al. 2014a). Branchiobdellidans are leechlike obligate ectosymbionts of crayfish, and their position within Clitellata has been problematic (Gelder and Williams 2015).

The advent of molecular phylogeny changed traditional point of view, and it is now widely accepted that the monophyletic taxon comprising ectoparasitic, predatory, and blood-feeding Clitellata, i.e., Branchiobdellia + Acanthobdellida + Hirudinida, should be included into Oligochaeta with oligochaetous Lumbriculidae as a sister taxon, and, as a result, the terms Clitellata and Oligochaeta are regarded as synonyms (Martin 2001; Siddall et al. 2001). The phylogeny and classification of Clitellata is still a topic of intensive debate, and different systems have been proposed and used (e.g., Erséus 2005; Jamieson 2006; Marotta et al. 2008; Rousset et al. 2008; Timm and Martin 2015). For clarity, we will use the term "Oligochaeta" in its traditional sense (clitellates with chaetae) to clearly separate them from leeches and leechlike branchiobdellids and acanthobdellids. We will avoid using higher taxa names and will concentrate on species, genera, subfamilies, and families.

21.2 Germ-Line Cysts in Clitellata: General Organization and Formation

All of FGCs that have been found to date in clitellate annelids have a central cytoplasmic mass, a cytophore. As a rule, each sister cell has one RC that connects it to the cytophore. The cytophore is located in the center of the cyst, whereas the cells occupy its periphery (Figs. 21.1f and 21.2a–c) (Świątek et al. 2009; Urbisz et al. 2015, 2017). Despite this general pattern, there are significant differences in cyst morphology and functioning between taxa. They are a reflection of the quantitative differences between the main cyst components, i.e., the germ cells and cytophore. These differences are manifested by the number of interconnected cells per cyst (from 16 to as many as 2600) and the cyst shape (e.g., spherical versus conical). Additionally, the number of cysts within the entire ovary (from one to several) and the time and place where they function also vary markedly. In certain taxa, the FGCs differentiate and develop within the ovary (thus the so-called intraovarian oogenesis occurs), while in other taxa, they detach from inconspicuous ovaries quickly (at the beginning of meiosis) and function in the body cavity (extraovarian oogenesis) (Gorgoń et al. 2017; Urbisz et al. 2017; Świątek et al. 2018a). Together, all these differences affect the plasticity and functioning of FGCs in clitellate annelids and thereby the organization of the entire gonad.

21.2.1 Cyst Formation: General Rules

It is widely known that syncytial cysts are formed by modified mitotic divisions of spermatogonia or oogonia during early gametogenesis, prior to meiosis. A gonial cell that begins to divide without full cytokinesis is the mother cell of the cyst, and in the case of the female line, this cell is usually known as cystoblast, whereas the resulting interconnected sister cells are termed cystocytes (Lin and Spradling 1997; Bilinski et al. 2017; Yamashita 2018; Fig. 21.1). As was already mentioned, the morphological and molecular properties of FGCs formation are best understood in the fruit fly (recently reviewed by Bilinski et al. 2017; Yamashita 2018); however, significant progress has recently been made in vertebrates (reviewed by Elkouby 2017; Ikami et al. 2017; Wang et al. 2017). The molecular events that lead to the transformation of a cleavage furrow into a stable intercellular bridge have been intensively studied in the fruit fly and mouse (reviewed in Greenbaum et al. 2011; Haglund et al. 2011; Lu et al. 2017) but are completely unknown in annelids.

To better understand how FGCs in clitellate annelids emerge, some general rules governing cyst formation should be reminded. As was mentioned in the Introduction, in branched and linear cysts, germ cells are connected directly one to another, and clustered cells have a different number of RCs (Fig. 21.1c–e). In contrast, in the cysts found in Clitellata, an anuclear cytoplasmic mass to which each germ cell is connected by only one RC develops in the cyst center (Figs. 21.1f and 21.2a–c). In branched FGC, the number of RCs depends on the number of divisions that a given cell has passed (King et al. 1982; King and Büning 1985; de Cuevas et al. 1997; Ong and Tan 2010). In linear clusters, irrespective of the number of divisions, each cell has two RCs, while the terminal cells have one RC (Biliński 1993; Jaglarz et al. 2014b). In FGCs equipped with the cytophore, irrespective of how many division rounds a given cell has passed, each cystocyte has only one RC (Fig. 21.1f).

The specific localization of the centrioles and, as a consequence, the specific orientation of the mitotic spindles in relation to the previous bridges governs GC formation (King et al. 1982; King and Büning 1985; Büning 1994; de Cuevas et al. 1997). In branched FGCs, one centriole is "kept" in close proximity to the existing bridge (with the exception of the first division) due to the association of certain cytoskeletal proteins (e.g., cytoplasmic dynein) with a specific membranous organelle (playing a crucial role in cyst formation and oocyte specification) termed the fusomal material (Deng and Lin 1997; McGrail and Hays 1997; see below). As a result, during each round of division, only one of the daughter cells (located closer to the cyst center) receives all of the preexisting RCs, whereas a new bridge is shared with the sister cells (de Cuevas et al. 1997; Ong and Tan 2010). In linear FGCs, new cells intercalate between the older cells, and a cyst remains chain-like (de Cuevas et al. 1997; Jaglarz et al. 2014b). Another common motif that is observed during FGC formation is the synchrony of the divisions—all of the interconnected cells are in the same phase of the cell cycle, and as a consequence, the number of clustered cells equals 2^n, where n means the number of division rounds (de Cuevas et al. 1997; Biliński 1998). In *D. melanogaster* (de Cuevas et al. 1997; Ong and Tan 2010) and *X. laevis* (Kloc et al. 2004), there are 4 rounds of synchronic divisions and 16 cystocytes emerge. Exceptions to the 2^n rule, in which the number of interconnected cells is different than the power of two, have been found in several taxa, especially in insects, e.g., in scale insects (Szklarzewicz 1998) and certain coleopterans (Jaglarz 1998). This suggests that some cystocytes do not follow synchrony and fail to divide or, in contrast, that there are some extra cell divisions or some cystocytes may be eliminated via apoptosis or degeneration.

21.2.2 *Specific Alignment of the Mitotic Spindles and Preexisting Bridge Partitioning Lead to the Formation of Cysts Equipped with Central Cytoplasmic Core*

To date, there has been only one study explaining how GCs characteristic for clitellate annelids are formed (Świątek et al. 2009). The analyses of the sectioned material and whole-mounted GCs revealed that the cyst architecture results from (1) the specific positioning of the mitotic spindles during subsequent divisions and (2) modified incomplete divisions (Fig. 21.3). It was observed that although the mitotic spindles of dividing cystocytes occupy random positions in relation to their neighboring cells, the long axes of spindles are always oriented tangentially to the surface of the cytophore and that both poles of each spindle always lie far away from the cytophore and the existing RCs (Fig. 21.3d, f). This spindle orientation makes the plane of cytokinesis more or less perpendicular to the existing RCs (Fig. 21.3d, f). Then, the cleavage furrow ingresses toward the existing bridge and constricts it. As a consequence, the preexisting bridge is partitioned into two new bridges (Fig. 21.3d–f). During each round of cystocyte divisions, this mode of incomplete cytokinesis is repeated and leads to the formation of two separate daughter cells that are connected to the cytophore with only one bridge (Fig. 21.3d–g). It seems that two major mechanisms govern this process: one is responsible for the proper alignment of the mitotic spindles; the second governs the partitioning of the preexisting bridges. Because both mechanisms are unknown, only the morphological observations and literature data that describe similar mechanisms in other organisms may prompt some possible scenarios. It is known that during the branched FGC formation in *Drosophila*, the fusomal material that fills the existing bridges anchors one spindle pole of dividing cell, and, as a consequence, the division plane is more or less parallel to the existing bridge(s) (for details, see Yamashita 2018). In contrast, during the cystocyte divisions in clitellates, the centrioles are located away from the existing RCs, and therefore there is no centriole anchoring in proximity of the bridge. Why the centrioles in clitellates are located at the cystocyte pole opposite to the RC remains unknown. No specialized, morphologically detectable structure(s) that could tether and anchor the spindle has ever been found in clitellates (Świątek et al. 2009). It can only be hypothesized that an unknown mechanism causes the centrioles to be "repelled" from the cell area near an existing bridge.

The second unique character of the cell divisions in GCs equipped with a central cytoplasm is the bridge partitioning (Świątek et al. 2009; Fig. 21.3d, f). In branched and linear GCs, the bridges are not divided during cell divisions, and therefore the entire bridges are inherited or shared by the sister cells (de Cuevas et al. 1997; Ong and Tang 2010). The process of cell division found in clitellate GCs, where the leading edge ingresses and merges with a preexisting bridge, and in consequence the old bridge is divided into two "twin" bridges that eventually separate completely (Fig. 21.3d–g), should be regarded as a modified form of incomplete cytokinesis. It seems that the prerequisite for this type of RC partitioning is the division

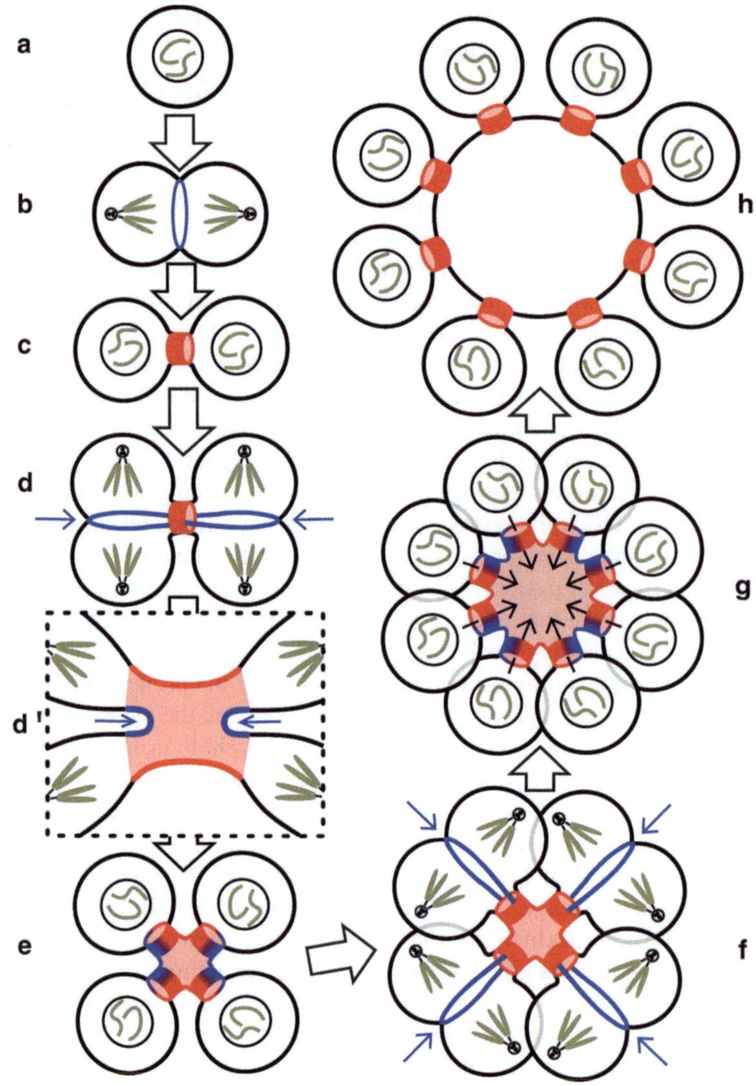

Fig. 21.3 Cyst formation. (**a**) The mother cell of a cyst, the cystoblast, starts to divide without complete cytokinesis, (**b**) and as a result of the first incomplete division, two interconnected sister cells (cystocytes) emerge (**c**). (**d**) Then, the planes of the further divisions of the cystocytes (in blue) are perpendicular (arrows) to the existing RC (in red). (**d′**) The cleavage furrows (in blue) ingress toward the existing bridge (in red) and constrict it and then merge with it; as a consequence, the already existing RC is partitioned into two new bridges that are built from the fragments of the preexisting and new bridge (red-blue ring canals marked in **e**). At the same time, the cytophore precursor (the initial cytophore—marked in pink in **d′**–**g**) begins to emerge in the cyst center. (**f** and **g**) The same mechanism is responsible for the subsequent cystocyte divisions. After the incomplete divisions cease, there is a more rapid streaming of the cytoplasm from the cystocytes towards the initial cytophore (marked by black arrows in **g**), and as a consequence, the cytophore enlarges and the RCs are completely separated as marked in (**h**)

(bifurcation) of the leading edge when it merges with a preexisting RC (Świątek et al. 2009). Such a bifurcation of the leading edge seems to generate the "dividing force" required for the division of the old bridge into two new ones (Świątek et al. 2009). A similar bifurcation of the leading edges has been described in insects during cellularization of the syncytial blastoderm (e.g., *D. melanogaster*; for details see Miller and Kiehart 1995; Mazumdar and Mazumdar 2002). It should be mentioned here that ultrastructural observations of GC formation in Clitellata suggest that the complete separation of a preexisting RC into two new RCs is postponed, at least in the female line (Fig. 21.2c). The new "twin" RCs are not completely separated at once; the dense bridge inner rims (i.e., the modified contractile rings) still connect the neighboring bridges, while the cytophore is in its initial form (see below). Ultrastructural observations suggest additionally that the bridge separation is completed when the cytophore begins to grow due to the directional streaming of the cytoplasm from the cystocytes toward cyst center. At this time, the cytophore loses its initial character, and all of RCs are completely separated (Fig. 21.3g and h).

Intuitively, it seems that the same mechanisms of cell divisions should be universal and should be responsible for the formation of all cysts that have a central cytoplasmic core. As was mentioned in the Introduction, apart from Clitellata these types of cysts have been described in the male cysts of polychaetous annelids, as well as in FGCs of echiuran annelids, oribatid mites, and nematodes. Surprisingly, although *C. elegans* has cysts with germ cells clustered around the cylindrical rachis, there have been no morphological analyses of FGC formation in this model organism. The first attempt has been made recently by Seidel et al. (2018), who provided detailed observations of the mitotic germ-cell divisions in the progenitor zone of the gonads of both male and hermaphrodite *C. elegans* specimens. Interestingly, Seidel et al. (2018) found the same pattern of cell divisions in both sexes, which is in complete accordance with the results of our analyses of FGCs formation in clitellate annelids (Świątek et al. 2009). This similarity supports the idea that the mechanism that governs the formation of GCs equipped with central cytoplasm is universal. In *C. elegans*, the mitotic spindles of the dividing cells are oriented parallel to the rachis surface, while the cleavage furrows ingress perpendicularly toward the rachis (Seidel et al. 2018). By the end of divisions, the preexisting RCs bifurcate (divide) into two new RCs. The newly formed RCs stay close to one another, and therefore the sister cells are positioned cell to cell. Compared with clitellates, one conspicuous modification of the cell divisions in *C. elegans* was observed. Upon entry into mitosis, the configuration of RCs changes from "open" to "closed" (a "closed" condition means that the cytoplasmic channel between a cell and rachis was too narrow to be seen under the confocal microscope, which means that the RC diameter is smaller than 0.3 μm). Then, at the end of cytokinesis, when the preexisting RC is already divided into two new RCs, the bridge configuration changes back from "closed" to "open" (Seidel et al. 2018). According to authors, such a synchronization of RCs opening and closing during cell division enables the neighboring germ cells to divide without synchrony (see Seidel et al. 2018 for details). The abovementioned observations of *C. elegans* germ-cell divisions seem to be a solid base for further studies, which, hopefully, will shed a light on the molecular mechanisms that are responsible for such peculiar cell divisions.

21.2.3 The Origin of the Cytophore

Another question that should be addressed in this chapter is the origin of the cytophore. Although no first cystoblast divisions in clitellates have been observed, based on older (Świątek et al. 2009) and recent observations (unpublished data), the following scenario can be presented (see Fig. 21.3). The first incomplete division of a cystoblast is not modified; the two sister cells remain connected via a cylindrical RC (Fig. 21.3a–c). During the further mitosis, two cleavage furrows move as was described above and merge with the first RC to form a small three-dimensional space that can be compared to a short cylinder-like structure with four openings that are the future separated RCs (Fig. 21.3d–d′). This central structure is filled with a small amount of cytoplasm that slowly flows from the cells toward the center of cyst and is encompassed by the inner RC rim (a modified cleavage furrow)—this structure is the precursor of the cytophore (the cytophore anlage) and has been termed the initial cytophore (Fig. 21.3e). The further cell divisions follow this pattern (Fig. 21.3f and g), and therefore after, for instance, five rounds of synchronic divisions, GC is composed of 32 cells that are arranged around the initial cytophore, which is additionally stabilized by a continuous inner bridge rim (Fig. 21.3g). An early FGC of the clitellate annelid *Insulodrilus bifidus* (Phreodrilidae) is shown in Fig. 21.2c. It is evident that the initial cytophore is in fact the limited space in the cyst center, filled with a small amount of cytoplasm. In this area the cleavage furrows of the subsequent divisions merge, and the resulting individual bridges are formed. The initial cytophore has specific ultrastructural properties. Its cytoplasm is more electron lucent than the cytoplasm of the cystocytes and contains chaotically oriented microtubules (remnants of the mitotic spindles) and scarce membranous organelles except for the short cisternae of the ER (Fig. 21.2c and d). The ER network filing the initial cytophore is especially well visible on the sections prepared using the pre-embedding contrasting method, which specifically stains the membranes (Fig. 21.2d). Initially, when the cystocytes divide, the cytoplasm flows slowly from the cells, but when the divisions cease and the cystocytes enter meiosis, the cytophore volume considerably increases, the RCs are finally completely separated, and the cytoplasm of the cytophore becomes ultrastructurally very similar to the cytoplasm of the germ cells (Urbisz et al. 2017; Świątek et al. 2018a; Figs. 21.2a and 21.3h).

Taking into account the ultrastructural properties of the initial cytophore and its behavior, it is tempting to compare it with the fusomal material. The fusome is a specialized cytoplasm enriched in ER membranes and cytoskeletal proteins as α-, β-spectrin, and Ankyrin, which fills the RCs during GCs formation in insects (Büning 1994; fusome formation, composition, and functioning in *D. melanogaster* is reviewed in, e.g., McKearin 1997; de Cuevas et al. 1997;

Huynh 2006; Yamashita 2018) and some other organisms as *X. laevis* (Kloc et al. 2004). What is more, individual portions of fusomal material may merge and form a polyfusome which stretches through all bridges in GC. In *Drosophila* it has been demonstrated that the fusome tethers the mitotic spindles of dividing cystocytes, which is crucial for forming the branched cysts (Lin and Spradling 1995; Deng and Lin 1997). To date, the molecular composition of the initial cytophore is completely unknown; however, based solely on morphological data, some similarities between fusome and the initial cytophore may be pointed out. The first is ultrastructural similarity: both structures are composed of electron-lucent granular cytoplasm devoid of membranous cell organelles except for ER cisternae (see Fig. 21.2c and d and figures in Urbisz et al. 2010, 2017; Gorgoń et al. 2017). The second similarity is the relatively fast degradation of this specific cytoplasmic area—in insect FGCs fusomal material disappears soon after completion of cystocytes divisions (Snapp et al. 2004; Haglund et al. 2010), which is similar to the transformation of the initial cytophore into definitive cytophore. The definitive cytophore is formed by directional transfer of cystocyte cytoplasm toward the cyst center. Concurrently, the cytoplasm constituting the initial cytophore seems to be dissolved or mixed by entering mass of "new" cytoplasm, and eventually the structure of the cytophore and germ-cell cytoplasm is broadly similar, as it was demonstrated in *E. albidus* (Urbisz et al. 2017; see also Fig. 21.2a). Future molecular studies should reveal the molecular composition of the initial cytophore and enable more comprehensive comparison of this structure with fusomal material. One thing seems to be clear now: if material forming the initial cytophore is really engaged in GCs formation, it acts in a different way than fusome. Fusomal material attracts centrioles, in contrast to clitellate annelids in which the mitotic spindles are always located far from the cyst center (Świątek et al. 2009), which suggests that the mechanisms of spindles positioning are different in both systems.

21.2.4 Divisions Synchrony

Usually, all germ cells in a cyst undergoing formation are in the same phase of the cell cycle. In *Drosophila*, the complete synchronization of the interconnected cells is caused by the association of cyclin A with the fusome (for more details, see, e.g., de Cuevas et al. 1997; Huynh 2006; Yamashita 2018). The interconnected germ cells enter each mitosis at the same time, and then, meiosis (not in case of *D. melanogaster*, in which only two cells—the pro-oocytes, enter meiosis) and synchronization of the divisions are responsible for the number of clustered cells. As was mentioned earlier, when the synchrony is complete and not altered by additional divisions and/or degenerations, the final number of interconnected cystocytes depends solely on the number of divisions and equals 2^n. In clitellate annelids, synchronic divisions of germ cells were reported in both male and female GCs (Świątek et al. 2009; Małota and Świątek 2016). In the female line, synchronous divisions were observed in cysts with a rather low number of cells, e.g., in fish leeches where maximally up to 60 cells are interconnected (Pérez 1907). On the other hand, there are no synchronous divisions in FGCs composed of hundreds of cells, as has been documented in some glossiphoniid leeches (own unpublished data; Fig. 21.2e). In two species from the family Enchytraeidae (*Enchytraeus albidus*—Urbisz et al. 2017 and *Grania postclitellochaeta*—Świątek et al. 2018a), FGCs are always composed of 16 cells, which strongly suggests that they undergo 4 rounds of synchronic mitosis. In another microdrile *Insulodrilus bifidus* (Phreodrilidae), FGCs are composed of 32 cells (Świątek et al. 2018b), and we assume 5 synchronic divisions. In *Chaetogaster diaphanus* and *Ripistes parasita* (Naidinae), the number of interconnected cystocytes seems to be unstable and fluctuate around 30 cells, and sometimes equals 32, while in *R. parasita* cysts with 29 and 31 cells have been recorded (Gorgoń et al. 2017). These data suggest that some of the cystocytes did not divide or degenerate. In fish leeches (Piscicolidae), the number of clustering cells can vary even in the same ovary. According to Pérez (1907) and Jörgensen (1913), there are 50–60 cells in cysts, whereas Spałek-Wołczyńska et al. (2008) recorded 24–44 germ cells in a given FGC. Most probably, in fish leeches the synchrony is lost after the first rounds of mitosis. On the other hand, in representatives of other clitellate taxa such as Tubificinae, Limnodriloidinae, Lumbriculidae, Propappidae, Branchiobdellidae, and true leeches (except for fish leeches), FGCs are composed of hundreds or even thousands of cells, and the synchrony of the divisions has never been observed. At least in some cases such as *Lumbriculus variegatus* (Lumbriculidae—Urbisz and Świątek 2013) or *Glossiphonia complanata* (Glossiphoniidae—Damas 1964), some cystocytes can still divide even in differentiated FGCs in which the oocytes and NCs are already clearly recognizable. All of the abovementioned observations suggest that the first mitotic divisions are synchronous and the number of clustering cells equal 2^n and that when the more divisions occur, the more asynchronous they are and the number of interconnected cells is significantly different from the theoretical.

21.2.5 Oocyte Specification

The formation of the polarized FGCs is connected with the differentiation of interconnected cells into two subcategories: the oocytes and the NCs. The systems that are responsible for this specification seem to be different. The molecular events that lead to the specification of the oocyte and NCs are best recognized in *D. melanogaster*. In the fruit fly, the oocyte is determined as early as during the first asymmetric division of the cystoblast where the spectrosome (the fusome precursor) is unevenly segregated between the resulting cystocytes (Lin and Spradling 1995; de Cuevas and Spradling 1998). In other insects, such as beetles, morphological observations suggest that there is a gradient of an unknown substance produced by somatic prefollicular cells that influences the oocyte determination (Matuszewski et al. 1985; Trauner and Büning 2007). In mice, the specification of the oocytes seems to be connected with the preferential transfer of cell organelles toward the perspective oocytes (Lei and Spradling 2016; reviewed in Ikami et al. 2017). In clitellate annelids, no molecular work devoted to the oocyte specification has been performed as yet, and the appropriate mechanisms are completely unknown. We can only cite a few morphological observations, the interpretation of which is difficult, and these observations are as yet inconclusive. In enchytraeids, such as *E. albidus* and *G. postclitellochaeta*, the oocyte is always located at the most distal position in the clone in relation to the rest of the ovary (Fig. 21.4c; Paschma 1962; Urbisz et al. 2017; Świątek et al. 2018a). The same situation was observed in the 32-celled cysts in *I. bifidus* (P. Świątek unpublished results). In the sludge worm *T. tubifex* where as many as 2000 cells can build a clone, several oocytes grow at the same time, and, interestingly, all of them are located on one side of the ovary (Fig. 21.4a; Urbisz et al. 2010, 2015). No markers that can suggest, e.g., asymmetric cell divisions during FGCs formation and the early specification of oocytes have been found to date.

In all clitellates, the first morphological sign of the oocyte determination is the increasing volume of cytoplasm and the formation of microvilli on the surface of perspective oocytes (Figs. 21.4a–e and 21.6a, b, d–h; Świątek 2006; Spałek-Wołczyńska et al. 2008; Urbisz et al. 2010; Gorgoń et al. 2017). Nurse cells do not form microvilli (or their microvilli are very poorly developed) and grow only slightly as has been documented in the case of three species from the subfamily Naidinae (Gorgoń et al. 2017) as well as in *T. tubifex* (Urbisz et al. 2015). While NCs also do not gather nutrients, occasionally, lipid droplets and spheres resembling proteinaceous yolk bodies have been found in these cells (Urbisz et al. 2010, 2017; Świątek et al. 2012; Urbisz and Świątek 2013; Gorgoń et al. 2017). Based on morphological observations, the NCs in clitellates have been regarded as being polyploid by some authors (e.g., Paschma 1962). However, recent analyses using the quantification of the fluorescence intensity of DAPI (a florescent dye specific for DNA) showed that the DNA level of NCs in *E. albidus* is 4C only (Urbisz et al. 2017).

21.3 The Same Content in Different Forms: Variety of Cysts in Clitellate Annelids

21.3.1 Oligochaeta and Branchiobdellida

There are numerous studies describing general aspects of oogenesis and the ovary organization in oligochaetous annelids (see Jamieson 1981 for a literature review). However, analyses devoted to the cellular processes connected with ovary functioning at the high-resolution level were rare and limited to certain species from taxa such as, e.g., Enchytraeidae (Paschma 1962; Dumont 1969), Lumbricidae (Chapron and Relexans 1971; Siekierska 2003), and Tubificinae (Casellato et al. 1987). Contemporary studies published since 2010 have delivered new data about ovary

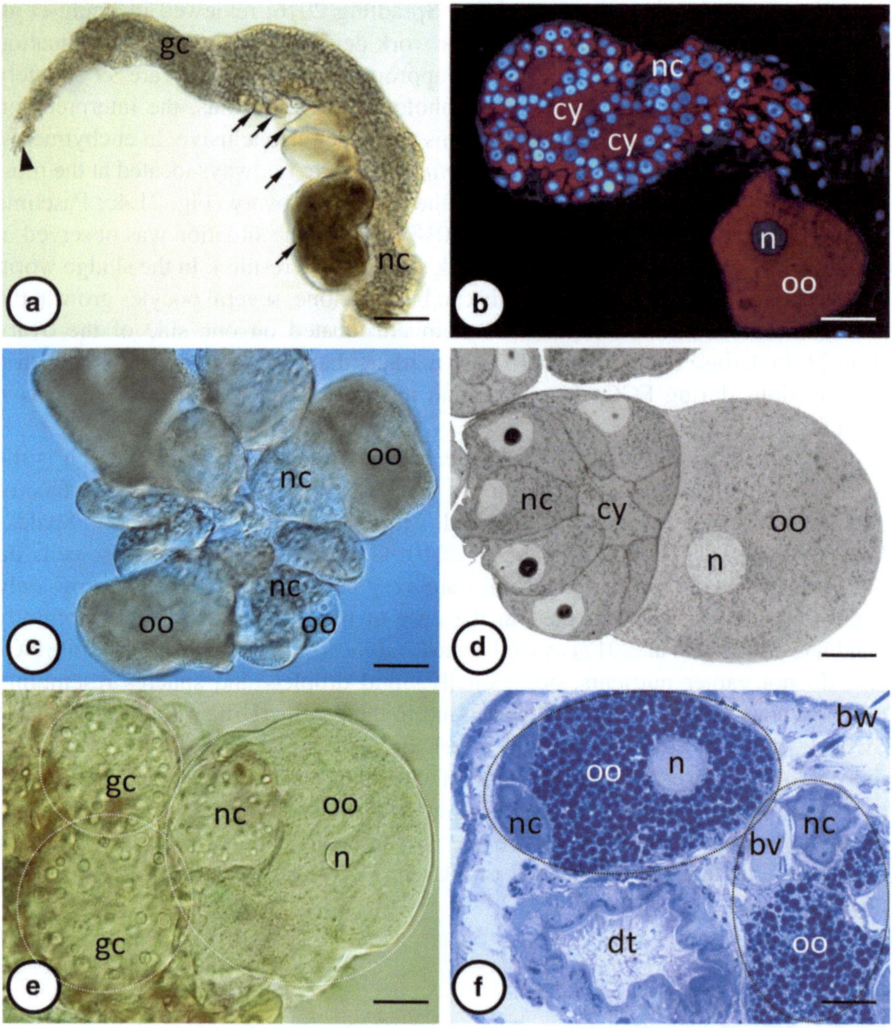

Fig. 21.4 Cysts in oligochaetous clitellates. (**a** and **b**) Multicellular and polarized cysts of "Tubifex" type. (**a**) Whole-mounted cyst/ovary of *Tubifex tubifex*. The cyst is conically shaped with narrow proximal end (arrowhead) housing young, morphologically similar germ cells (gc), and broader distal part with nurse cells (nc) and oocytes (arrows). (**b**) A transverse semi-thin section through the distal part of *T. tubifex* cyst. *Cy* cytophore, *n* oocyte nucleus, *nc* nurse cells, *oo* oocyte. (**a**) Differential interference contrast microscopy (DIC), bar = 180 μm; (**b**)FM, histocryl section double stained with DAPI and propidium iodide, bar = 40 μm. (**c** and **d**) 16-celled spherical cysts of "Enchytraeus" type. One cell differentiates into oocyte (oo), while 15 cells function as nurse cells (nc). *Cy* cytophore, *n* oocyte nucleus. *Enchytraeus albidus*, (**c**) Whole ovary visualized by DIC, bar = 40 μm; (**d**) A section through a single cyst, LM, epon section stained with methylene blue, bar = 20 μm. (**e** and **f**) Spherical cyst of "Stylaria" type floating in the body lumen visualized by DIC (**e**) and on semi-thin section (**f**). In (**e**) three cysts (encircled by a dotted line) in a consecutive developmental stages are visible; in (**f**) a section through two cysts with vitellogenic oocyte is shown (encircled). *Gc* morphologically similar germ cells, *nc* nurse cells, *oo* oocyte, *n* oocyte nucleus, *bw* body wall, *bv* blood vessel, *dt* digestive tract. (**e**) *Stylaria lacustris*, whole-mounted preparation, bar = 30 μm; (**f**) *Ripistes parasita*, LM, epon section stained with methylene blue bar = 50 μm

organization and functioning and revealed that in almost all of the oligochaetous clitellates studied to date, the germ cells are united into syncytial clusters. The only known exception is one representative of the family Capilloventridae (a family regarded as a sister clade to all other Clitellata—Erséus and Källersjö 2004; Marotta et al. 2008) in which FGCs are not formed at all, and therefore the germ cells develop individually in the gonad (Świątek et al. 2016). As was mentioned in the previous section, although clitellatous FGCs have a common pattern (each cell connected to the cytophore via a single RC), they show significant differences in their morphology and functioning. These differences are connected with number of interconnected cells, the size and shape of the cytophore, and also with the timing of cyst development in relation to their localization (ovaries versus the body cavity). Below we describe the three main functional types of FGCs that have been found in Oligochaeta to date.

In the vast majority of Oligochaeta studied to date, FGCs are highly multicellular (hundreds and more cells) and have a well-developed cytophore (Figs. 21.4a, b and 21.5a). In the sludge worm *Tubifex tubifex* (Naididae, Tubificinae) and *Thalassodrilides* cf. *briani* (Naididae, Limnodriloidinae), detailed analyses have shown that only one multicellular cyst constitutes the entire gonad (Urbisz et al. 2010, 2015, 2018a). FGC and thereby the entire ovary are clearly polarized, and the developmental gradient of the germ cells (oogonia, meiotic cells in I prophase, NCs, and the growing oocytes) occurs along the long ovary axis. During oogenesis, the germ cells differentiate and become either oocytes or NCs. The total number of interconnected cells can vary from ~1900 to ~2600 with about eight growing oocytes at a given time in *T. tubifex* (Urbisz et al. 2015) and about 300 cells in ovaries of *T.* cf. *briani* with only one growing oocyte (Urbisz et al. 2018a). Within these multicellular and polarized cysts, the long and branched cytophore extends across the whole cyst, and its smaller and smaller branches reach every germ cell via the RC. The cytophore is well developed, especially in this part of the cyst where NCs and oocytes arise (Fig. 21.4b). Here, it has the form of an extensive cytoplasmic mass rich in cell organelles and a cytoskeleton. This suggests its important role in the transfer of cell content from the NCs toward the growing oocytes (Urbisz et al. 2010, 2015, 2018a).

In this type of FGCs, several oocytes usually grow at the same time, gather cell organelles and nutritive material, and protrude from the ovary surface (Fig. 21.4a, b). Afterward, the vitellogenic oocytes lose their connections to FGC and are released into the body cavity where they remain in close contact with the blood vessels (Urbisz et al. 2010, 2015, 2018a). As almost the entire process of egg formation occurs within the gonad, the oogenesis should be regarded as intraovarian. While there is no detailed information about the final fate of the NCs after oogenesis is completed, some evidence (e.g., the morphological signs of their degeneration) strongly suggests that these cells may be eliminated from the gonad by, e.g., apoptosis.

This characteristic organization of FGC described above for *T. tubifex* and *T.* cf. *briani* was termed as "Tubifex" type (Fig. 21.5a). Cysts of "Tubifex" type were also found in certain representatives of such taxa as Tubificinae (Urbisz et al. 2010),

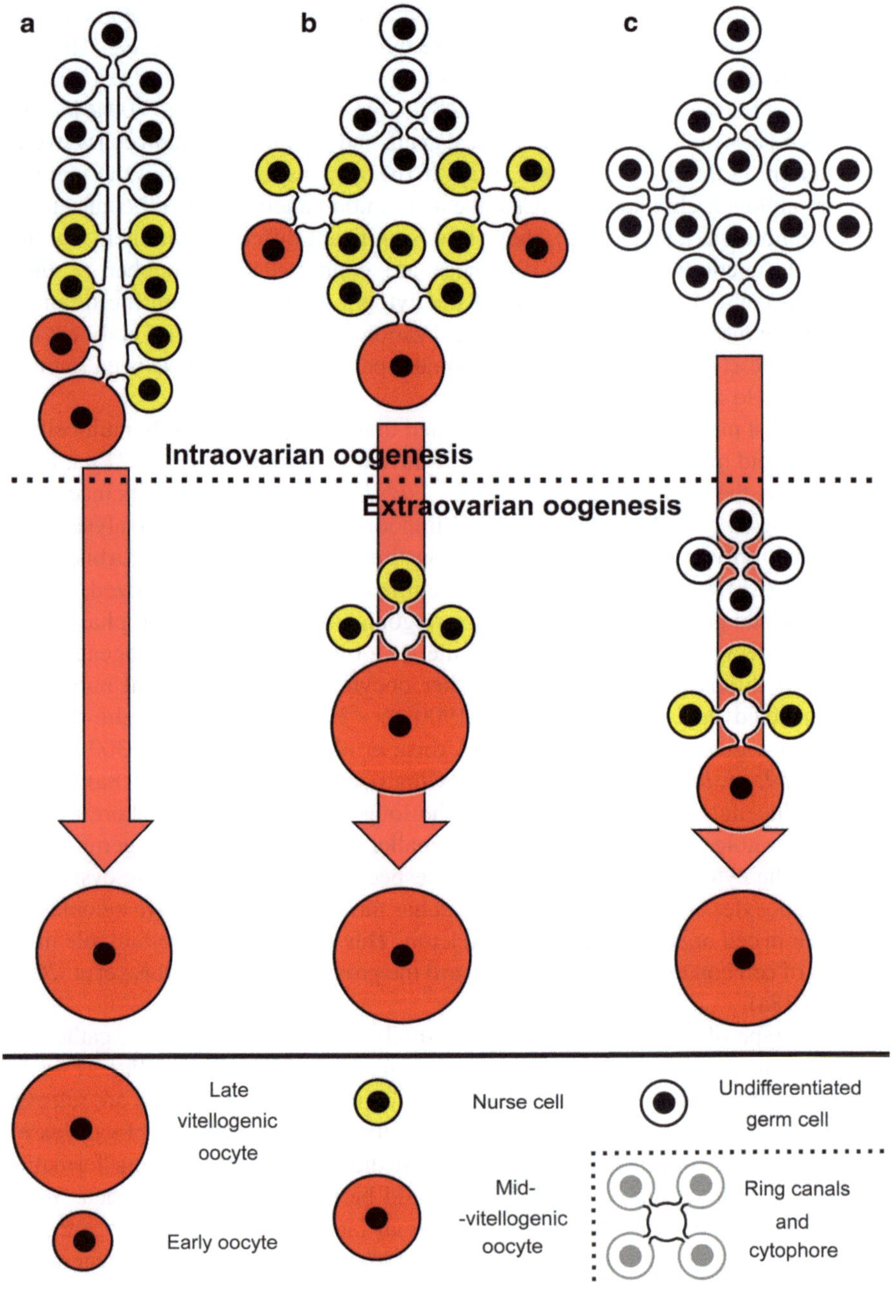

Fig. 21.5 Cysts organization in oligochaetous annelids. The germ cells are usually enveloped by the somatic cells; for clarity the latter were completely omitted. (**a**) A cyst of the "Tubifex" type. The cysts are cone-shaped and highly multicellular (even thousands of cells); the cytophore is voluminous and long. Several oocytes growing at a time are supported by numerous NCs. In some species, it was confirmed that the entire ovary is composed of one huge cyst. The oocytes accumulate the yolk when they are still connected to the cyst, and the late vitellogenic oocytes detach from the cyst/ovary. (**b**) "Enchytraeus" cyst. The cysts are composed of 16 cells (only 4 cells

Propappidae (Gorgoń et al. 2015), Lumbriculidae (Urbisz and Świątek 2013), and also in two leechlike branchiobdellids (*Branchiobdella pentodonta* and *B. parasitica*) (Świątek et al. 2012). A similar organization also occurs in one representative of the family Haplotaxidae (Urbisz et al. 2018b).

It should be mentioned that multicellular FGCs were also described in earthworms such as *Dendrobaena veneta* and *Eisenia foetida* (Lumbricidae) (Chapron and Relexans 1971; Siekierska 2003; Faron et al. 2015). It is not known how many cysts constitute a single gonad and how many cells are interconnected into a clone (Chapron and Relexans 1971; Siekierska 2003). However, it was stated that the ovaries are composed of multiple oogonia and meiotic cells, many NCs, and a number of oocytes (Siekierska 2003; Faron et al. 2015), which allows us to define cysts as being multicellular with cells differentiated into NCs and oocytes. In contrast to "Tubifex" FGCs, the cytophore found in *D. veneta* (the only earthworm species in which a cytophore was described) is extremely poorly developed and has the form of very thin, long, and anastomosing cytoplasmic strands hardly to notice on semi-thin sections (Siekierska 2003). Analyses of FGCs organization and their functioning in megadriles require much more effort in the future. We have decided not to describe a separate system of FGC organization for megadriles at this time due to the fragmentary data.

In contrast to the highly multicellular cysts with voluminous cytophore described as the "Tubifex" type, another cyst organization is known from several representatives of three other taxa that belong to the microdriles, i.e., Enchytraeidae, Naidinae, and Phreodrilidae. FGCs in these groups are small, almost spherical, and composed of dozens of cells (Figs. 21.2a–c and 21.4c–f). Each germ cell still has only one RC that connects it to the cytophore, but the cytophore is usually small and roughly spherical. The RCs have an inner rim enriched in F-actin, and they can reach up to ~4 μm or ~5 μm in diameter in the NCs and vitellogenic oocytes, respectively (Urbisz et al. 2017). The number of interconnected cells in a single cyst varies between groups. In Enchytraeidae, 16 clustering cells have always been recorded (Paschma 1962; Urbisz et al. 2017; Świątek et al. 2018a), whereas in Naidinae, this number is not constant and fluctuates around 30 (Gorgoń et al. 2017). In the only representative of Phreodrilidae that has been studied to date, *Insulodrilus bifidus*, 32 interconnected cells were found (Świątek et al. 2018b). As the number of clustered cells usually obey the 2^n rule (Paschma 1962; Urbisz et al. 2017; Gorgoń et al. 2017; Świątek et al. 2018a), it is believed that the divisions that lead to cyst formation are synchronous. Despite the fact that some time ago there was no

Fig. 21.5 (continued) are visualized). The cytophore is spherical. One cell becomes the oocyte, 15 act as the NCs. The ovaries are formed by several (dozen or so) loosely interconnected cysts. Cysts with mid- or late-vitellogenic oocytes can detach from the ovary and float in the body cavity. (**c**) "Stylaria" type. The cysts contain around 30 cells (only 4 cells are visualized). The cytophore is spherical. One oocyte emerges, and the rest of the cells are NCs. The ovaries contain only young, premeiotic, and early meiotic cysts; then, cysts leave ovaries and float in the body lumen. The reserve material is absorbed by the oocyte in the body cavity

consensus about the mode of oogenesis [in *E. albidus*, panoistic oogenesis was recognized by Dumont (1969), while Paschma (1962) found meroistic oogenesis in the same species], recent investigations have clearly shown a morphological and functional differentiation of the germ-line cells into two categories: NCs and oocytes, which confirmed ovary meroism (Urbisz et al. 2017; Świątek et al. 2016, 2018a, b). It was also found that while all of the interconnected cells initially enter meiosis, only one of them begins to grow and develops into the future egg cell; the rest withdraw meiotic prophase I and become NCs. What is interesting, the analyses of the NCs in *E. albidus* have shown that these cells do not undergo polyploidization. In contrast to the highly specialized NCs of insects, the level of ploidy in the NCs in *E. albidus* is 4C, which is the result of DNA replication before these cells begin the meiotic prophase (Urbisz et al. 2017). The next difference is connected with the mode of NCs degradation. In insects, the NCs are eliminated via apoptosis (e.g., Cavaliere et al. 1998; McCall 2004; Mpakou et al. 2011), in *E. albidus*, however, no apoptosis of the NCs is found at the end of oogenesis. Instead, the NCs are most probably engulfed (absorbed) by their sister vitellogenic oocyte (Urbisz et al. 2017).

In enchytraeid ovaries, there are several (usually more than ten) cysts that are retained within the ovary until advanced oogenesis (vitellogenesis), and therefore the entire ovary resembles a bunch of grapes, in which each unit is, in fact, a single cyst (Fig. 21.4c and d; Urbisz et al. 2017; Świątek et al. 2018a). As the ovary also shows polarization, germ-line cysts with cells during the consecutive stages of development are located along the long ovary axis. In *E. albidus*, the number of such cysts within a single ovary is about 12 (Urbisz et al. 2017). The clusters with large vitellogenic oocytes detach from the ovary, and consequently the final stages of oogenesis occur within clusters, which freely float in the coelomic fluid (Urbisz et al. 2017; Świątek et al. 2018a). Such a bunch-like system of ovary organization with 16-celled spherical meroistic FGCs is regarded as the "Enchytraeus" type (Fig. 21.5b; Urbisz et al. 2017).

In microdrile naidines (Naidinae), in contrast, older descriptions stated that the ovaries are very inconspicuous and that only the initial stages of oogenesis occur within the gonad. Then, the germ cells quickly detach from the ovary and continue oogenesis within freely floating groups of cells. Therefore, the old descriptions suggested that oogenesis in naidines is extraovarian (Dehorne 1923; Mehra 1924; Stolte 1934). This observation has been confirmed by a recent study of three common representatives of Naidinae: *Stylaria lacustris*, *Chaetogaster diaphanus*, and *Ripistes parasita* (Gorgoń et al. 2017). *Stylaria lacustris* ovaries are tiny structures composed of several cysts with oogonia and the cells that have already begun prophase I of meiosis. The majority of oogenesis occurs outside of the gonad where germ cells within the freely floating cysts differentiate into the NCs and one oocyte (meroistic oogenesis). As was mentioned earlier, in naidines, cysts are composed of *circa* 30 cells (from 28 to 32 cells have been recorded—Dehorne 1923; Gorgoń et al. 2017). The oocytes develop within the syncytial clusters until the uptake of yolk (vitellogenesis), then detach, complete oogenesis, and become mature egg cells. For comparative purposes, FGCs that form the inconspicuous ovaries and develop outside of the ovary have been termed as "Stylaria" type (Fig. 21.5c; Gorgoń et al. 2017).

To conclude the studies discussed in this section, FGCs in oligochaetous clitellates show plasticity in their architecture and functioning, despite the common rules that govern their organization. To date, three different types of FGCs have been described according to their architecture and functioning (Fig. 21.5a–c): (1) "Tubifex," multicellular and polarized cysts with an extensive cytophore found in taxa such as Tubificinae, Limnodriloidinae, Propappidae, Haplotaxidae, Lumbriculidae, and Branchiobdellidae; such FGCs can join approximately 2000 cells with several growing oocytes at a given time (in *T. tubifex*) or hundreds of cells with 1 oocyte (in *T.* cf. *briani*); (2) "Enchytraeus," 16-celled cysts, which are known from enchytraeids; at the onset of their functioning, they are roughly spherical with a small cytophore; most of the time, they function within the ovary; and (3) "Stylaria," cysts found in Naidinae which group about 30 cells; their overall morphology is similar to "Enchytraeus," but they primarily function outside the ovary. Most probably, the next different mode(s) of FGC organization and functioning occur in earthworms and in other microdriles.

It is worth noting that the spatial architecture of FGCs and the resulting diversity of the ovaries seem to be group specific and are conservative at the family/subfamily level (Gorgoń et al. 2017; Urbisz et al. 2017; Świątek et al. 2018a). In addition to the cognitive aspect, these characters may contribute to the knowledge of how the oligochaetous female gonads evolved and can also provide new data that may help to determine the relationships between the oligochaetous taxa. For instance, the ovaries and FGCs that are found in two representatives of the leechlike taxon Branchiobdellida are of the "Tubifex" type and are not organized in the same manner as FGCs found in leeches (see the next section). These findings support the idea that branchiobdellidans are not a sister taxon to true leeches (Świątek et al. 2012; Urbisz and Świątek 2013). However, the use of all characters that are connected with FGCs, oogenesis, and ovaries in phylogenic considerations (as has recently been done in leeches; see Bielecki et al. 2014b) is necessary to extend the research to other groups, which are as yet unknown in these aspects—Megadriles and also microdrile oligochaetes, e.g., Opistocystidae and Parvidrilidae.

21.3.2 True Leeches and Acanthobdellida

FGCs in leeches closely associate with somatic cells and form highly variable structures (egg follicles, ovary cords) that are enclosed within the ovarian sacs (ovisacs). As the somatic cells, similarly to the oligochaetous clitellates, play mainly mechanical role in leech cyst functioning, their descriptions and characteristics are beyond the scope of the review and are completely omitted, except for the intermediate cell and the apical cell. Moreover, as will be described below, the FGCs in leeches can be organized and can function in various ways, and therefore, we decided to identify several different systems of FGCs organization in leeches.

In the ovisacs of piscicolid leeches, there are numerous freely floating spherical structures termed egg follicles (Fig. 21.6a and b; Pérez 1907; Jörgensen 1913; Fischer and Weigelt 1975; Spałek-Wołczyńska et al. 2008). Each follicle is composed of one FGC embedded in one, most probably polyploid, somatic cell that is known as an intermediate cell. Additionally, each cyst together with the intermediate cell is surrounded by several flattened somatic cells (Fischer and Weigelt 1975; Spałek-Wołczyńska et al. 2008). The number of interconnected germ-line cells per cyst is not constant and can even differ among the follicles in the same ovary. Pérez (1907) and Jörgensen (1913) reported 50–60 cells in a follicle, whereas Spałek-Wołczyńska et al. (2008) found 24–44 cells per follicle. Although as a rule, only one germ cell per follicle develops into an oocyte (Fischer and Weigelt 1975; Spałek-Wołczyńska et al. 2008); Jörgensen (1913) observed follicles with two developing oocytes. FGC formation in fish leeches has not been analyzed using contemporary methods, the older descriptions suggest that, at least at the onset of cysts formation, the mitoses are synchronous (Pérez 1907). It seems that later on, this synchrony is lost (P. Świątek unpublished observations). However, irrespective of when the division synchrony is lost, all of the cells in a given cluster develop synchronously and have the same ultrastructure until diplotene. After this stage the clustered cells differentiate into an oocyte and numerous NCs (Spałek-Wołczyńska et al. 2008). The cytophore is more or less spherical. RCs have prominent, electron-dense inner rims enriched with F-actin (Spałek-Wołczyńska et al. 2008). While the RCs can be as wide as 2 μm, in the case of the NCs, the diameter of the oocyte RC is not known (Spałek-Wołczyńska et al. 2008). Ultrastructural analyses revealed that the cell organelles (mitochondria and ER cisternae) can pass through the bridges toward the cytophore freely (Fig. 21.2f, inset; Spałek-Wołczyńska et al. 2008). Although the final fate of the NCs is unclear, morphological observations suggest that the remnants of the cytophore and NCs are engulfed by the growing oocyte (Jörgensen 1913; Spałek-Wołczyńska et al. 2008). Functionally, the egg follicles found in piscicolids can be compared to the egg chambers that are characteristic for insects with meroistic polytrophic ovaries in which each cyst is enveloped by its own follicular epithelium (Büning 1994; Biliński 1998).

For comparative purposes, the organization of FGCs found in fish leeches is termed as cysts of the "Piscicola" type (Fig. 21.7a).

In glossiphoniids, the ovary structure and oogenesis have been analyzed several times (e.g., Aisenstadt 1964; Aisenstadt et al. 1964, 1967; Damas 1964, 1977; Fernández et al. 1992; Świątek 2005a, b, 2006), but data about cyst organization and functioning are fragmentary. The most valuable data concerning FGCs come from the ultrastructural observations presented by Fernández et al. (1992) on *Haementeria depressa*, *Helobdella triserialis*, and *Theromyzon rude*; by Świątek (2005a, b, 2006) on *Alboglossiphonia heteroclita* (previously known as *Glossiphonia heteroclita*) and from the cytochemical and autoradiographic studies on *Glossiphonia complanata* carried out by Aisenstadt (1964) and Aisenstadt et al. (1964, 1967). These analyses revealed that germ-line cysts together with the somatic cells form long and thin structures within the ovaries known as ovary strings or ovary cords (Fig. 21.6c). Despite these studies, the basic data about cyst organization, e.g.,

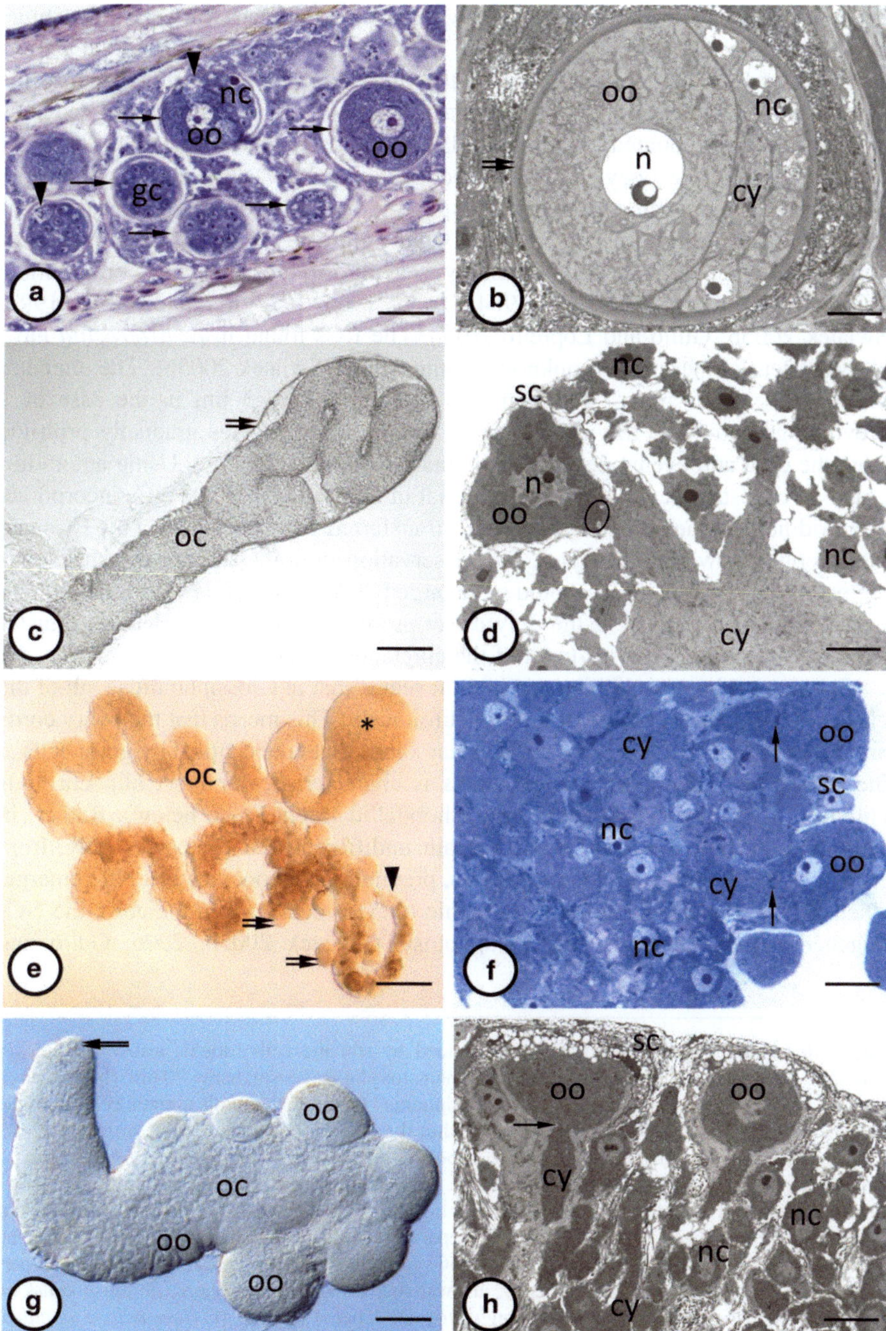

Fig. 21.6 Organization of cysts in leeches. (**a** and **b**) Cysts of "Piscicola" type. (**a**) A section through ovary, where usually dozen or so cysts can be found (arrows), (**b**) A single cyst. *Gc* morphologically similar germ cells, *nc* nurse cells, *oo* oocyte, *n* oocyte nucleus. The entire cyst is embedded within a somatic, intermediate cell hardly visible at the semi-thin sections (arrowheads)

how many cells are interconnected or how many cysts form an ovary cord (One huge cyst? Several smaller ones?) are missing. These questions will be difficult to answer as FGCs of glossiphoniids are one of the most complicated syncytial germ-line systems that have been observed in animals. Although the number of clustered cells has not yet been counted, Aisenstadt (1964) estimated that in the ovary cord of *G. complanata*, one oocyte is nourished by *circa* 2000 NCs. Besides the huge number of interconnected cells, FGCs in glossiphoniids have a spacious and irregularly branched cytophore (Fig. 21.6d), which indirectly suggests that one clone constitutes the entire cord. Most probably, several/dozen or so oocytes differentiate within a given clone (Aisenstadt 1964; Damas 1964, 1977; Fernández et al. 1992; Świątek 2005b; Gullo and Lopretto 2018). The RCs found in *A. heteroclita* have electron-dense inner rims of unknown composition (Świątek 2005b). The diameter of the RCs varies from ~2 μm in the case of NCs to 3 μm in the case of a previtellogenic oocyte (Świątek 2005b). The growing oocytes gradually protrude from the cord and eventually lose the contact with the cytophore. Using autoradiography, Aisenstadt et al. (1967) showed that in *G. complanata*, the NCs incorporate H-3 uridine, and then the labeled RNA is transferred from the NCs via the RCs and the cytophore toward the oocytes. This observation strongly suggests that the clones are polarized and oogenesis should be regarded as meroistic. It is worth mentioning that to date, this analysis is the only direct proof that the macromolecules transfer from the NCs to the oocytes in all of the clitellate annelids.

One more thing should be noted here; it seems that in Glossiphoniidae, all of the clustered cells develop more or less synchronously. This means that the ovary cords are not polarized and that there are no clear zones that are occupied by germ cells at the consecutive developmental stages as is characteristic for most oligochaetous clitellates (described above) and arhynchobdellid leeches (see below). Rather, in immature leeches, the ovary cords contain undifferentiated cystocytes only, from which, as the breeding season progresses, previtellogenic oocytes and NCs emerge. Then, at the time at which the vitellogenic oocytes detach from the cord, the NCs degenerate most probably due to apoptosis (Świątek 2005b, 2006; Gullo and

---

Fig. 21.6 (continued) and additionally surrounded by somatic cells (double arrows). *Piscicola geometra*, (**a**) LM, paraffin section stained with hematoxylin and eosin, bar = 70 μm; (**b**) LM, epon section, bar = 30 μm. (**c** and **d**) Cysts of “Glossiphonia” type together with somatic cells form the so-called ovary cord (oc) which is enclosed within the ovarian envelope (double arrow). Ellipse, RC; *cy* cytophore, *nc* nurse cells, *oo* oocyte, *n* oocyte nucleus, *sc* somatic cells. *Alboglossiphonia heteroclita*, (**c**) DIC, whole-mounted preparation, bar = 250 μm; (**d**) LM, epon section, bar = 20 μm. (**e** and **f**) Polarized and elongated cysts forming together with somatic cells the ovary cord (oc) of “Hirudo” type visualized by whole-mounted preparation (**e**) and sectioning (**f**). Asterisk shows the apical, club-shaped part of the cord; double arrows mark growing oocytes, arrowhead—distal part of the cord with degenerating germ cells; arrows, ring canals; *cy* cytophore, *nc* nurse cells, *oo* oocytes, *sc* somatic cells. (**e**) *Haemopis sanguisuga*, DIC, bar = 200 μm; (**f**) *Haemadipsa japonica*, LM, bar = 15 μm. (**g** and **h**) Cysts of “Erpobdella” type forming polarized, conically shaped and short ovary cord (oc). Double arrow marks the apical cell, *oo* growing oocytes, *cy* cytophore, *nc* nurse cells, *sc* somatic cells. (**g**) *Barbronia weberi*, DIC, whole-mounted preparation, bar = 400 μm; (**h**) *Erpobdella octoculata*, LM, epon section, bar = 15 μm

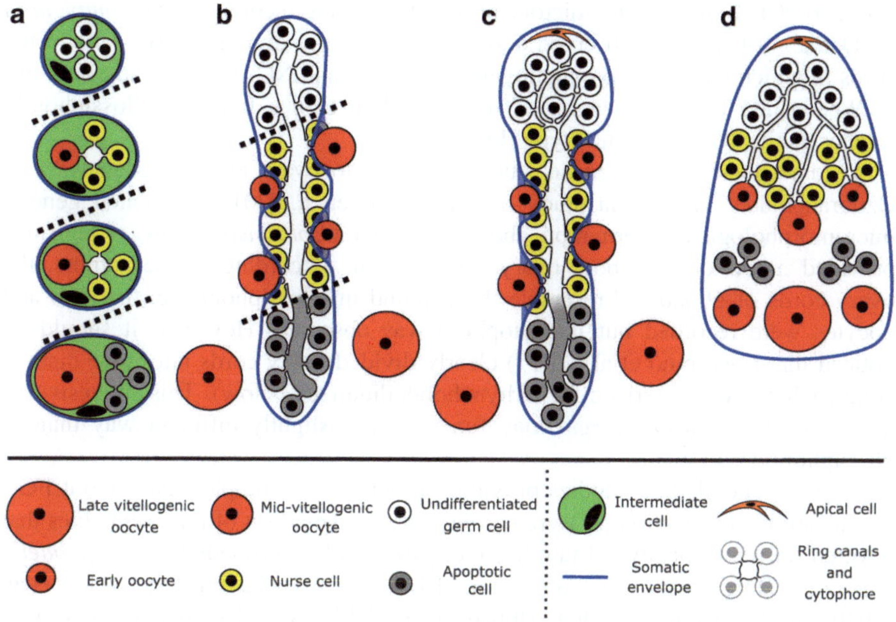

Fig. 21.7 Diversity of cyst organization in leeches. The somatic cells, except for the intermediate cell and the apical cell, were generalized as "somatic envelope" (**a**) "Piscicola" type. The cysts are spherical and embedded in a polyploid somatic intermediate cell and additionally surrounded by flat somatic cells, altogether they form egg follicles. Up to 60 cells can be interconnected; one oocyte usually develops. Most of the oogenesis cyst stays within the intermediate cell; after intermediate cell degeneration the oocyte is released into the ovary cavity (not shown). Dotted lines separate the consecutive phases of a follicle development. (**b**) "Glossiphonia" type. The cysts together with the somatic cells form elongated and generally nonpolarized structures termed as ovary cords or ovary strings. The cysts are highly multicellular. The cytophore is well developed, and several oocytes grow more or less simultaneously within the cord. The growing oocytes detach from the cord and finalize oogenesis within the ovary cavity. Because the germ cells within the cord develop synchronously, the dotted lines separate the consecutive developmental stages. (**c**) "Hirudo" type. The cyst(s) together with somatic cells form clearly polarized and elongated ovary cords. The apical part of the cord is usually club-shaped and in addition to germ-line cells contains somatic apical cell. The cysts are multicellular, and several oocytes grow at the same time within the cord. Similar to "Glossiphonia" the growing oocytes finalize oogenesis in the ovary lumen. The cytophore appears as long and convoluted cytoplasmic strands. The distal portion of the cord is occupied by the apoptotic NCs and cytophore remnants. (**d**) "Erpobdella" type. The cysts surrounded by numerous somatic cells form polarized, conically shaped, and rather short ovary cords. As in the cases of "Glossiphonia" and "Hirudo" there is no quantitative data about cyst components. Several oocytes grow simultaneously; however, they do not leave the cord until they completely absorb the reserve material. After the completion of oogenesis, they are ovulated into the ovary cavity (not shown). The cytophore is poorly developed in the form of very thin cytoplasmic strands. The apical cell is also present. The NCs undergo apoptosis at the distal part of the cord

Lopretto 2018). Some contradictory observations were recorded in *G. complanata* by Damas (1964), who observed a zone of mitotically dividing germ cells in the cords in which the young previtellogenetic oocytes and NCs were already present.

We termed the organization of the germ-line clones found in glossiphoniid leeches as the "Glossiphonia" type (Fig. 21.7b).

To date, there have been no studies devoted to ovary and FGCs analyses in Ozobranchidae. The recent study by Cascarano et al. (2017) presented general micromorphological information about the leech *Ozobranchus margoi* and also provided a few details about ovary organization. According to the authors, the ovary cords of *O. margoi* resemble those found in Glossiphoniidae; the NCs and oocytes were recorded but no cytophore was observed. However, it should be noticed that Cascarano et al. (2017) clearly divided ovary cords into three distinct zones, what is characteristic for Arhynchobdellida (see below). This suggests that the germ clones in *O. margoi* may function in a slightly different way than in glossiphoniids.

In Arhynchobdellida, studies providing more or less complete data about FGC organization and functioning were carried out on a small set of representatives that covered only four of several families recognized today: Erpobdellidae, *Erpobdella octoculata* (Jörgensen 1908; Van Damme 1974; Urbańska-Jasik 1988; Świątek et al. 2010) and *E. johanssoni* (Ben Ahmed et al. 2013); Salifidae, *Barbronia weberi* (Urbisz et al. 2014); Hirudinidae, *Hirudo medicinalis* (Świątek 2008), *H. troctina*, and *Limnatis nilotica* (Ben Ahmed et al. 2010); and Haemopidae, *Hameopis sanguisuga* (Klimaszewska-Guzik 2000; Świątek 2008). All of these analyses have revealed that cysts together with the accompanying somatic cells form long (Hirudiniformes—Fig. 21.6e) or short (Erpobdelliformes—Fig. 21.6g) ovary cords inside the ovaries of arhynchobdellids. In all of the arhynchobdellids studied to date, the ovary cords are polarized, and the germ cells being in the consecutive developmental stages are located sequentially along the long cord axis. Similarly to the situation described in Glossiphoniidae, FGCs in arhynchobdellid leeches are highly multicellular, and there is no detailed data as to how many cysts build one cord and how many cells are interconnected into one cluster. To simplify the results that were obtained in the abovementioned papers, it may be assumed that there are three zones in the ovary cords of arhynchobdellids: an apical one with the oogonia and cells entering meiosis, a middle one with cells that are clearly morphologically differentiated into numerous NCs and several growing oocytes, and a distal one with the NCs that are eliminated via apoptosis (Fig. 21.6e, g).

Although the formation of FGCs has not yet been described, some observations of the ovary cords in young leeches before the breeding season (*Hirudo verbana*—unpublished data) suggest that there are asynchronous mitotic divisions of the cystocytes in the apical zone. The RCs found in arhynchobdellids usually have an electron-dense inner rim; however, such dense rims have not been observed in *B. weberi* (Urbisz et al. 2014). In *H. medicinalis* (Świątek 2008) and *E. johanssoni* (Ben Ahmed et al. 2013), the diameter of the RCs of the previtellogenic oocytes and NCs can be estimated to be 2–3 μm. The diameters of the RCs get wider (up to at least 5 μm) when the NCs begin to degenerate (see below, Ben Ahmed et al. 2013).

The bridges have dense inner rims, which are sometimes especially thick, as in previtellogenic oocytes of *E. johanssoni* (Ben Ahmed et al. 2013) or *Haemadipsa japonica* (Fig. 21.6f). Most probably F-actin is enriched in the rims, but adequate analyses have not yet been performed. The cytophore has the form of thin cytoplasmic strands that snake between the cells (Fig. 21.6f; Świątek 2008; Ben Ahmed et al. 2010). The cytophore is especially thin in erpobdellids (Fig. 21.6h) (Świątek et al. 2010; Ben Ahmed et al. 2013). As was mentioned earlier, there is a clear zone in which the NCs and the cytophore degenerate in the ovary cords of arhynchobdellids. One apparent sign of the degradation of the NCs is the frequently recorded shift of the NC nuclei from the cells to the cytophore (Świątek 2008; Ben Ahmed et al. 2010; Urbisz et al. 2014). It seems that the RCs of these degenerating NCs widen markedly due to a decrease in the quantity and quality of the cytoskeletal proteins and that even a weak cytoplasm movement can pull the nuclei toward the cytophore. In the degeneration zone, the NCs undergo apoptosis, and the fragmented cells and the cytophore remnants are phagocytized by amoebocytes (Świątek 2008). In Hirudiniform leeches, absorbing yolk oocytes gradually protrude from the cord surface, eventually detach, and float freely in the ovisac lumen (Fig. 21.6e and f; Świątek 2008; Ben Ahmed et al. 2010). In Erpobdelliformes, the growing oocytes do not leave the cord until ovulation but rather are tightly covered by the somatic cells that constitute the cord envelope (Fig. 21.6g; Świątek et al. 2010; Ben Ahmed et al. 2013). Due to several morphological differences, the ovaries of Arhynchobdellida have been divided into two types: "Hirudo" and "Erpobdella" (Fig. 21.7c and d). In *Acanthobdella peledina* (Acanthobdellida are a sister taxon to leeches, Tessler et al. 2018), each ovary has one ovary cord with organization and functioning similar to cords found in Hirudiniformes ("Hirudo" type of ovary; Świątek et al. 2012)

One more thing is worth mentioning here. In all of the arhynchobdellid leeches studied to date, an unusual somatic cell occurs at the apical tip of the ovary cord. This cell was originally termed as "multipolaren Ganglienzelle" by Jörgensen (1908) and was recently renamed as an apical cell (Świątek 2008). It was suggested that apart from mechanical functions, it may constitute a niche for the germ-line (and somatic?) stem cells (Świątek et al. 2010). Morphologically, the apical cell resembles the somatic distal tip cell which functions as a niche that keeps the population of germ cells in a stem cell state in the *C. elegans* gonad (Kimble and White 1981; Hall et al. 1999; reviewed in, e.g., Hubbard 2007; Kimble 2011). The apical cell has peculiar ultrastructural and morphological properties, but its description is beyond the scope of this review, and for more details, see Świątek (2008), Świątek et al. (2010), and Ben Ahmed et al. (2013).

To sum up, to date four systems of ovary/FGCs organization have been distinguished in true leeches and acanthobdellids (Fig. 21.7): (A): "Piscicola" the cysts together with somatic cells form egg follicles containing a moderate number of germ cells (20–60), usually a dozen or so FGCs develop within a given ovary; (B) "Glossiphonia" a highly multicellular (number of cells not estimated, probably hundreds or even thousands) cyst(s) forms the ovary cords together with somatic cells; one, long, and convoluted cord per ovary; interconnected germ cells develop more or less synchronously within the cord; the growing oocytes detach from cord and float in the ovary lumen; (C) "Hirudo" multicellular cyst(s) together with the somatic cells form two long and convoluted cords per ovary; the cords are polarized—the club-shaped apical part is occupied by the oogonia and undifferentiated cystocytes, below NCs and oocytes differentiate; the distal end of the cord contains apoptotic NCs and the degenerating cytophore; the oocytes detach from the cord and finish oogenesis within the ovary lumen; (D) "Erpobdella" multicellular FGC (s) associate with the somatic cells and form several (7–10) cone-shaped ovary cords per ovary; the cords are polarized like in the "Hirudo" type, but the growing oocytes are retained within the cord until ovulation. In the "Hirudo" and "Erpobdella" types, the tip of the cord is occupied by the apical cell.

The formation of cysts equipped with cytophore should be regarded as a conservative aspect of leech oogenesis. Moreover, there is evolutionary tendency to build highly multicellular cysts that associate with the somatic cells and form more or less elongated ovary cords; only fish leeches have spherical FGCs composed of 20–60 cells. It seems that FGC/ovary organization is conserved at the family level, and to date, only one type of organization has been detected in a given family. These observations have led to the attempts to use the characters connected with FGC/ovary structure in phylogenetic considerations. Świątek et al. (2012) and Urbisz and Świątek (2013) pointed out that the ovary cords of *A. peledina* resemble those found in true leeches, whereas the ovaries of the two branchiobdellidans are similar to those known from oligochaetous clitellates. These results clearly strengthen the idea that acanthobdellids, but not branchibdellids, are sister to true leeches. Bielecki et al. (2014b) confirmed that the characters connected with the female reproductive systems and oogenesis are high-quality comparative data that can be useful for the phylogenetic reconstruction of Hirudinida.

21.4 Concluding Remarks

Recent years have brought new, albeit solely morphological, knowledge, which together with the older data have given us a more detailed picture of the clitellate FGC architecture and functioning. The current state of knowledge can be summarized as follows:

- The general pattern of FGCs architecture is the same in all of the clitellates studied to date: each clustered cell has one RC that connects it to a common cytoplasmic mass (the cytophore); the cells occupy FGC periphery, whereas the cytophore is positioned centrally.
- FGCs are formed via the modified incomplete mitoses of a cystoblast and cystocytes; the planes of the consecutive divisions (except for the first one) are perpendicular to the existing RCs. Each cleavage furrow constricts and merges with an existing bridge to form two new bridges. As a consequence each sister cell, irrespective of how may rounds of divisions have passed, has only one bridge. All of the bridges are gathered in a limited space in the cyst center—

around the cytophore precursor. The diameter of the fully formed RCs may reach as many as 10 μm, F-actin enriches the bridge rims.

- The cytophore emerges in the cyst center when the cytoplasm flows from the cells toward the RCs; the functional cytophore lacks a nucleus/nuclei. The cytophore mediates the exchange of cytoplasm between the cells. Ultrastructurally the early (initial) cytophore resembles the fusomal material; there are no data about the molecular composition of the cytophore.
- FGCs are polarized and oogenesis is meroistic, i.e., only one or several cells per clone (future oocyte(s)) receive(s) the macromolecules and organelles from numerous nurse cells and continue oogenesis. The NCs probably do not undergo polyploidization and are eliminated after the oocyte(s) detach from the cyst.
- Although the general architecture is conservative, FGCs show a plasticity in the context of their individual organization. For instance, the number of interconnected cells varies from 16 in enchytraeids to about 40–60 in piscicolid leeches to more than 2000 in *T. tubifex*. When the number of cells is relatively low (a dozen or so), cysts and the cytophore are spherical and contain only one oocyte. When there are hundreds or more of the clustered cells, the cytophore is voluminous and irregularly branched, cysts have conical shapes, and several oocytes develop at the same time in a given clone. Moreover, FGCs may function in the ovaries—intraovarian oogenesis or outside the ovary—extraovarian oogenesis. To systematize this knowledge, we recognized seven individual systems of FGCs/ovaries organization using the generic names of the best known representatives ("Tubifex" "Enchytraeus" "Stylaria" "Piscicola" "Glossiphonia" "Hirudo" and "Erpobdella").
- Cysts with a low number of clustered cells seem to pass synchronous cystocyte divisions and obey the 2^n rule (e.g., 16-celled cysts in enchytraeids, 32-celled cysts in *I. bifidus*). The synchrony of the divisions is lost in multicellular cysts, and the number of clustered cells is different than expected.
- FGCs with accompanying somatic cells form the ovaries; however, in some taxa, e.g., in Naidinae, cysts spent the majority of their lifetime freely floating in the coelomic fluid in the body lumen (extraovarian oogenesis).
- The individual organization of a cyst (cyst type) seems to be conservative at the family/subfamily level, and this can be used in phylogenetical considerations.

Terra incognita, which is still waiting to be explored, includes the molecular composition and mechanisms that are crucial for the cyst formation and functioning. The mechanisms that govern the cystocyte divisions, the formation of the RCs as well as the cytophore, and the specification of the oocyte and the NCs remain largely unknown. Molecular methods including mRNA sequencing and use of interfering RNAs are necessary for better understanding of the mentioned mechanisms and processes. On the other hand, the basic (morphological, ultrastructural) analyses should be still continued since organization of FGCs is still waiting to be discovered in the majority of clitellate taxa.

Acknowledgments We would like to thank Dr. Szymon Gorgoń for the drawings. We would also like to express our gratitude to all of the collaborators who have supported our study of germ cells in clitellates with their time, experience, and knowledge in recent years.

References

Aisenstadt TB (1964) Cytological studies of oogenesis. I. Morphology of the gonad of *Glossiphonia complanata* L. examined by light and electron microscopy. Citologiya 6:19–24

Aisenstadt TB, Brodskii VJ, Ivanova SN (1964) Cytological studies of oogenesis. II. A cytochemical examination of the oocyte growth in *Glossiphonia complanata* L. by UV cytophotometry and interference microscopy. Citologiya 6:77–81

Aisenstadt TB, Brodskii VJ, Gazarian KG (1967) An autoradiographic study of the RNA and protein synthesis in gonads of animals with different types of oogenesis. Citologiya 9:397–406

Amini R, Goupil E, Labella S, Zetka M, Maddox AS, Labbé J-C, Chartier NT (2014) *C. elegans* Anillin proteins regulate intercellular bridge stability and germline syncytial organization. J Cell Biol 206:129–143

Amini R, Chartier NT, Labbé JC (2015) Syncytium biogenesis: it's all about maintaining good connections. Worm 4:e992665

Anderson E, Huebner E (1968) Development of the oocyte and its accessory cells of the polychaete *Diopatra cuprea* (Bosc). J Morphol 126:163–197

Beddard FE (1895) A monograph of the order of Oligochaeta. Clarendon, Oxford

Ben Ahmed R, Fuchs AZ, Tekaya S, Harrath AH, Świątek P (2010) Ovary cords organization in *Hirudo troctina* and *Limnatis nilotica* (Clitellata, Hirudinea). Zool Anz 249:201–207

Ben Ahmed R, Tekaya S, Małota K, Świątek P (2013) An ultrastructural study of the ovary cord organization and oogenesis in *Erpobdella johanssoni* (Annelida, Clitellata: Hirudinida). Micron 44:275–286

Bielecki A, Cichocka JM, Jeleń I, Świątek P, Płachno BJ, Pikuła D (2014a) New data about the functional morphology of the chaetiferous leech-like annelids *Acanthobdella peledina* (Grube, 1851) and *Paracanthobdella livanowi* (Epshtein, 1966) (Clitellata, Acanthobdellida). J Morphol 275:528–539

Bielecki A, Świątek P, Cichocka JM, Siddall ME, Urbisz AZ, Płachno BJ (2014b) Diversity of features of female reproductive system and other morphological characters in leeches (Citellata, Hirudinida) in phylogenetic conception. Cladistics 30:540–554

Biliński S (1993) Structure of ovaries and oogenesis in entognathans (Apterygota). Int J Insect Morphol Embryol 22:255–269

Biliński SM (1998) Introductory remarks. Folia Histochem Cytobiol 36:143–145

Bilinski SM, Kubiak JZ, Kloc M (2017) Asymmetric divisions in oogenesis. In: Tassan J-P, Kubiak JZ (eds) Asymmetric cell division in development, differentiation and cancer, Results and problems in cell differentiation, vol 61. Springer, New York, pp 211–228

Braun RE, Behringer RR, Peschon JJ, Brinster RL, Palmiter RD (1989) Genetically haploid spermatids are phenotypically diploid. Nature 337:373–376

Brill JA, Hime GR, Scharer-Schuksz M, Fuller MT (2000) A phospholipid kinase regulates actin organization and intercellular bridge formation during germline cytokinesis. Development 127: 3855–3864

Brubacher JL, Huebner E (2011) Evolution and development of polarized germ cell cysts: new insights from a polychaete worm *Ophryotrocha labronica*. Dev Biol 357:96–107

Büning J (1994) The insect ovary: ultrastructure, previtellogenic growth and evolution. Chapman and Hall, London

Cascarano MC, Keklikoglou K, Arvanitidis C, Katharios P (2017) Contribution to the morphological description of the marine leech, *Ozobranchus margoi* (Apáthy) (Rhynchobdellida: Ozobranchidae) by using combined histology, micro-CT and SEM. Zootaxa 4337:91–108

Casellato S, Martinucci G, Zoja E (1987) Ultrastructural features of gametogenesis during the life cycle in *Branchiura sowerbyi* Beddard (Oligochaeta, Tubificidae). Hydrobiologia 155:145–154

Cavaliere V, Taddei C, Gargiulo G (1998) Apoptosis of nurse cells at the late stages of oogenesis of *Drosophila melanogaster*. Dev Genes Evol 208:106–112

Chapron C, Relexans J-C (1971) Connexions intercellularies et évolution nucléaire au cours de la prémémiose ovocytarie. Etude ultrastructurale chez le Lombricien *Eisenia foetida*. CR Acad Sci Paris S D 272:3307–3310

Damas D (1964) Structure et rôle du rachis ovarien chez *Glossiphonia complanata* L. (Hirudinée, Rhynchobdelle). Orgine, evolution et structure. Bull Soc Zool Fr 89:147–155

Damas D (1977) Anatomie et évolution de l'appareil génital femelle de *Glossiphonia complanata* (L.) (Hirudinée, Rhynchobdelle), au cours du cycle annuel. Étude histologique et ultrastructurale. Arch Zool Exp Gén 118:29–42

De Cuevas M, Spradling AC (1998) Morphogenesis of the *Drosophila* fusome and its implications for oocyte specification. Development 125:2781–2789

De Cuevas M, Lilly MA, Spradling AC (1997) Germline cyst formation in *Drosophila*. Annu Rev Genet 31:405–428

Dehorne A (1923) Le groupement ovocytaire et la structure de son cytophore chez *Chaetogaster diaphanus*. C R Soc Biol Paris 888–890

Deng W, Lin H (1997) Spectrosomes and fusomes anchor mitotic spindles during asymmetric germ cell divisions and facilitate the formation of a polarized microtubule array for oocyte specification in *Drosophila*. Dev Biol 189:79–94

Dumont JN (1969) Oogenesis in the annelid *Enchytraeus albidus* with special reference to the origin and cytochemistry of yolk. J Morphol 129:317–343

Dyrynda PEJ, King PE (1983) Gametogenesis in placental and non-placental ovicellate cheilostome Bryozoa. J Zool 200:471–492

Eckelbarger KJ (1983) Evolutionary radiation in polychaete ovaries and vitellogenic mechanisms: their possible role in life history patterns. Can J Zool 61:487–504

Elkouby YM (2017) All in one – integrating cell polarity, meiosis, mitosis and mechanical forces in early oocyte differentiation in vertebrates. Int J Dev Biol 61:179–193

Erséus C (2005) Phylogeny of oligochaetous Clitellata. Hydrobiologia 535/536:357–372

Erséus C, Källersjö M (2004) 18S rDNA phylogeny of Clitellata (Annelida). Zool Scr 33:187–196

Faron J, Bernaś T, Sas-Nowosielska H, Klag J (2015) Analysis of the behavior of mitochondria in the ovaries of the earthworm *Dendrobaena veneta* Rosa 1839. PLoS One 10:e0117187

Fernández J, Tellez V, Olea N (1992) Hirudinea. In: Harrison FW, Gardiner SL (eds) Microscopic anatomy of invertebrates, Annelida, vol 7. Wiley, New York, pp 323–394

Ferraguti M (1999) Euclitellata. In: Adiyodi KG, Adiyodi G (eds) Reproductive biology of invertebrates, Vol. IXB: Progress in male gamete ultrastructure and phylogeny. Oxford and IBH, New Delhi and Calcutta, pp 125–182

Fischer A, Weigelt KR (1975) Strukturelle Bezienhungen zwischen jungen Oocyten und somatischen Zellen bei den Anneliden *Platyneris* und *Piscicola*. Verh Dtsch Zool Ges 67: 319–323

Foor WE (1983) Nematoda. In: Adiyodi KG, Adiyodi RG (eds) Reproductive biology of invertebrates, Oogenesis, oviposition, and oosorption, vol I. Wiley, Chichester, pp 223–256

Ganot P, Bouquet JM, Kallesøe T, Thompson EM (2007) The *Oikopleura* coenocyst, a unique chordate germ cell permitting rapid, extensive modulation of oocyte production. Dev Biol 32:591–600

Gelder SR, Williams BW (2015) Clitellata: Branchiobdellida. In: Thorp JH, Rogers DC (eds) Thorp and Covich's freshwater invertebrates, Ecology and general biology, vol I, 4th edn. Elsevier, London, pp 551–563

Gibert MA, Starck J, Beguet B (1984) Role of the gonad cytoplasmic core during oogenesis of the nematode *Caenorhabditis elegans*. Biol Cell 50:77–85

Gorgoń S, Krodkiewska M, Świątek P (2015) Ovary ultrastructure and oogenesis in *Propappus volki* Michaelsen, 1916 (Annelida: Clitellata). Zool Anz 257:110–118

Gorgoń S, Wardas A, Krodkiewska M, Świątek P (2017) Oogenesis in the three species of Naidinae (Annelida, Clitellata) is extraovarian of the Stylaria type. Zoology 121:111–124

Gottanka J, Büning J (1990) Oocytes develop from interconnected cystocytes in the panoistic ovary of *Nemoura* sp. (Pictet) (Plecoptera: Nemouridae). Int J Insect Morphol Embryol 19:219–225

Govedich FR, Moser WE (2015) Clitellata: Hirudinida and Acanthobdellida. In: Thorp JH, Rogers DC (eds) Thorp and Covich's freshwater invertebrates, Ecology and general biology, vol I, 4th edn. Elsevier, London, pp 565–588

Greenbaum MP, Yan W, Wu MH, Lin YN, Agno JE, Sharma M, Braun RE, Rajkovic A, Matzuk MM (2006) TEX14 is essential for intercellular bridges and fertility in male mice. Proc Natl Acad Sci 103:4982–4987

Greenbaum MP, Iwamori T, Buchold GM, Matzuk MM (2011) Germ cell intercellular bridges. Cold Spring Harb Perspect Biol 3:a005850

Gullo BS, Lopretto EC (2018) Ovogénesis de *Helobdella ampullariae* (Hirudinida, Glossiphoniidae). Neotrop Biol Conserv 13:148–153

Gumienny TL, Lambie E, Hartwieg E, Horvitz HR, Hengartner MO (1999) Genetic control of programmed cell death in the *Caenorhabditis elegans* hermaphrodite germline. Development 126:1011–1022

Guo GQ, Zheng GC (2004) Hypotheses for the functions of intercellular bridges in male germ cell development and its cellular mechanisms. J Theor Biol 229:139–146

Haglund K, Nezis IP, Lemus D, Grabbe C, Wesche J, Liestol K, Dikic I, Palmer R, Stenmark H (2010) Cindr interacts with anillin to control cytokinesis in *Drosophila melanogaster*. Curr Biol 20:944–950

Haglund K, Nezis IP, Stenmark H (2011) Structure and functions of stable intercellular bridges formed by incomplete cytokinesis during development. Commun Integr Biol 4:1–9

Hall DH, Winfrey VP, Blaeuer G, Hoffman LH, Furuta T, Rose K, Hobert O, Greenstein D (1999) Ultrastructural features of the adult hermaphrodite gonad of *Caenorhabditis elegans*: relations between the germ line and soma. Dev Biol 212:101–123

Hubbard EJ (2007) *Caenorhabditis elegans* germ line: a model for stem cell biology. Dev Dynam 236:3343–3357

Huynh JR (2006) Fusome as a cell-cell communication channel of *Drosophila* ovarian cyst. In: Baluska F, Volkmann D, Barlow PW (eds) Cell-cell channels. Springer, New York, pp 217–235

Huynh JRR, St Johnston D (2004) The origin of asymmetry: early polarisation of the *Drosophila* germline cyst and oocyte. Curr Biol 14:438–449

Ikami K, Nuzhat N, Lei L (2017) Organelle transport during mouse oocyte differentiation in germline cysts. Curr Opin Cell Biol 44:14–19

Jaglarz MK (1998) The number that counts. Phylogenetic implications of the number of nurse cells in ovarian follicles of Coleoptera–Adephaga. Folia Histochem Cytobiol 36:167–178

Jaglarz MK, Kubrakiewicz J, Jedrzejowska I, Goldyn B, Bilinski SM (2014a) Ultrastructural analysis of the ovary and oogenesis in Spinicaudata and Laevicaudata (Branchiopoda) and its phylogenetic implications. Zoology 117:207–215

Jaglarz MK, Kubrakiewicz J, Bilinski SM (2014b) The ovary structure and oogenesis in the basal crustaceans and hexapods. Possible phylogenetic significance. Arthropod Struct Dev 43: 349–360

Jamieson BGM (1981) The ultrastructure of the Oligochaeta. Academic, New York, pp 356–377

Jamieson BGM (2006) Non-leech Clitellata. In: Rouse G, Pleijel F (eds) Reproductive biology and phylogeny of Annelida. Science, Enfield, pp 235–392

Jędrzejowska I, Kubrakiewicz J (2007) The Balbiani body in the oocytes of a common cellar spider, *Pholcus phalangioides* (Araneae: Pholcidae). Arthropod Struct Dev 36:317–326

Jędrzejowska I, Mazurkiewicz-Kania M, Garbiec A, Kubrakiewicz J (2013) Differentiation and function of the ovarian somatic cells in the pseudoscorpion, *Chelifer cancroides* (Linnaeus, 1761) (Chelicerata: Arachnida: Pseudoscorpionida). Arthropod Struct Dev 42:27–36

Jong-Brink de M, Boer HH, Joosse J (1983) Mollusca. In: Adiyodi KG, Adiyodi RG (eds) Reproductive biology of invertebrates, Oogenesis, oviposition, and oosorption, vol I. Wiley, Chichester, pp 397–354

Jörgensen M (1908) Untersuchungen über die Eibildung bei *Nephelis vulgaris* Mogiun Tandon (*Herpobdella atomaria* Carena). Arch Zellforsch 2:279–347

Jörgensen M (1913) Zellstudien II. Die Ei und Nahrzellen von *Piscicola*. Arch Zellforsch 10: 127–160

Kimble J (2011) Molecular regulation of the mitosis/meiosis decision in multicellular organisms. Cold Spring Harb Perspect Biol 3:a002683

Kimble J, White J (1981) On the control of germ cell development in *Caenorhabditis elegans*. Dev Biol 81:208–219

King RC (1970) Ovarian development in *Drosophila melanogaster*. Academic, New York

King RC, Büning J (1985) The origin and functioning of insect oocytes and nurse cells. In: Kerkut GA, Gilbert LI (eds) Comprehensive insect physiology, biochemistry and pharmacology, Embryogenesis and reproduction, vol I. Pergamon, Oxford, pp 37–82

King RC, Cassidy JD, Rousset A (1982) The formation of clones of interconnected cells during gametogenesis in insects. In: King RC, Akai H (eds) Insect ultrastructure, vol I. Plenum, New York, pp 3–60

Klimaszewska-Guzik R (2000) Ovary structure and some aspects of oogenesis in the leech *Haemopis sanguisuga* (L.). Acta Biol Crac s Zool 42:45–54

Kloc M, Bilinski S, Dougherty MT, Brey EM, Etkin LD (2004) Formation, architecture and polarity of female germline cyst in *Xenopus*. Dev Biol 266:43–61

Kubrakiewicz J (1997) Germ cells cluster organization in polytrophic ovaries of Neuroptera. Tiss Cell 29:221–228

Lei L, Spradling AC (2013) Mouse primordial germ cells produce cysts that partially fragment prior to meiosis. Development 140:2075–2081

Lei L, Spradling AC (2016) Mouse oocytes differentiate through organelle enrichment from sister cyst germ cells. Science 352:95–99

Leutert R (1974) Zur Geschlechtsbestimmung und Gametogenese von *Bonellia viridis* Rolando. Embryol Exp Morph 32:169–193

Liana M, Witaliński W (2012) Female and male reproductive systems in the oribatid mite *Hermannia gibba* (Koch, 1839) (Oribatida: Desmonomata). Int J Acarol 38:648–663

Lighthouse DV, Buszczak M, Spradling AC (2008) New components of the *Drosophila* fusome suggest it plays novel roles in signaling and transport. Dev Biol 317:59–71

Lin H, Spradling AC (1995) Fusome asymmetry and oocyte determination in *Drosophila*. Dev Genet 16:6–12

Lin H, Spradling AC (1997) A novel group of pumilio mutations affects the asymmetric division of germline stem cells in the *Drosophila* ovary. Development 124:2463–2476

Lin H, Yue L, Spradling AC (1994) The *Drosophila* fusome, a germline-specific organelle, contains membrane skeletal proteins and functions in cyst formation. Development 120:947–956

Lorès P, Vernet N, Kurosaki T, Van de Putte T, Huylebroeck D, Hikida M, Gacon G, Touré A (2014) Deletion of MgcRacGAP in the male germ cells impairs spermatogenesis and causes male sterility in the mouse. Dev Biol 386:419–427

Lu K, Jensen L, Lei L, Yamashita YM (2017) Stay connected: a germ cell strategy. Trends Genet 33:971–978

Maddox AS, Habermann B, Desai A, Oegema K (2005) Distinct roles for two *C. elegans* anillins in the gonad and early embryo. Development 132:2837–2848

Małota K, Świątek P (2016) Analysis of the cytoskeleton organization and its possible functions in male earthworm germ-line cysts equipped with a cytophore. Cell Tissue Res 366:175–189

Marotta R, Ferraguti M, Erséus C, Gustavsson LM (2008) Combined-data phylogenetics and character evolution of Clitellata (Annelida) using 18S rDNA and morphology. Zool J Linnean Soc 154:1–26

Martin P (2001) On the origin of the Hirudinea and the demise of the Oligochaeta. Proc Biol Sci 268:1089–1098

Matuszewski B, Ciechomski K, Nurkowska J, Kloc M (1985) The linear clusters of oogonial cells in the development of telotrophic ovarioles in polyphage Coleoptera. Dev Genes Evol 194: 462–469

Mazumdar A, Mazumdar M (2002) How one becomes many: blastoderm cellularization in *Drosophila melanogaster*. BioEssays 24:1012–1022

McCall K (2004) Eggs over easy: cell death in the Drosophila ovary. Dev Biol 274:3–14

McGrail M, Hays TS (1997) The microtubule motor cytoplasmic dynein is required for spindle orientation during germline cell divisions and oocyte differentiation in *Drosophila*. Development 124:2409–2419

McKearin D (1997) The *Drosophila* fusome, organelle biogenesis and germ cell differentiation: if you build it.... BioEssays 19:147–152

Mehra HR (1924) The genital organs of *Stylaria lacustris*, an account of their development. Q J Microsc Sci 68:147–186

Miller KG, Kiehart DP (1995) Fly division. J Cell Biol 131:1–5

Moosburgger M, Schwaha T, Walzl MG, Obst M, Ostrovsky AN (2012) The placental analogue and the pattern of sexual reproduction in the cheilostome bryozoan *Bicellariella ciliata* (Gymnolaemata). Front Zool 9:29

Morales CR, Lefrancois S, Chennathukuzhi V, El-Alfy M, Wu X, Yang J, Gerton GL, Hecht NB (2002) ATB-RBP and Ter ATPase complex accompanies specific mRNAs from nuclei through the nuclear pores and into intercellular bridges in mouse male germ cells. Dev Biol 246:480–494

Mpakou VE, Velentzas AD, Velentzas PD, Margaritis LH, Stravopodis DJ, Papassideri IS (2011) Programmed cell death of the ovarian nurse cells during oogenesis of the ladybird beetle *Adalia bipunctata* (Coleoptera: Coccinellidae). Develop Growth Differ 53:804–815

Olive PJW (1983a) Annelida—Polychaeta. In: Adiyodi KG, Adiyodi RG (eds) Reproductive biology of invertebrates, Spermatogenesis and sperm function, vol II. Wiley, Chichester, pp 321–342

Olive PJW (1983b) Annelida—Polychaeta. In: Adiyodi KG, Adiyodi RG (eds) Reproductive biology of invertebrates, Oogenesis, oviposition, and oosorption, vol I. Wiley, Chichester, pp 357–422

Ong S, Tan C (2010) Germline cyst formation and incomplete cytokinesis during *Drosophila melanogaster* oogenesis. Dev Biol 337:84–98

Paschma M (1962) The structure of the ovary and oogenesis in *Enchytraeus albidus* Henle. Zool Pol 12:145–188

Pepling ME, Spradling AC (1998) Female mouse germ cells form synchronously dividing cysts. Development 125:3323–3328

Pepling ME, De Cuevas M, Spradling AC (1999) Germline cysts: a conserved phase of germ cell development? Trends Cell Biol 9:257–262

Pérez C (1907) Notes histologique sur le Branchellion de la Torpille. II. Ovogenese. Trav Soc Sci Sta Zool Arcachon 10:307–328

Pilger JF (1980) The annual cycle of oogenesis, spawning, and larval settlement of the echiuran *Listriolobus pelodes* off Southern California. Pac Sci 34:129–142

Poprawa I, Hyra M, Rost-Roszkowska MM (2015a) Germ cell clusters organization and oogenesis in the tardigrade *Dactylobiotus parthenogeneticus* Bertolani, 1982 (Eutardigrada, Murrayidae). Protoplasma 252:1019–1029

Poprawa I, Hyra M, Kszuk-Jendrysik M, Rost-Roszkowska MM (2015b) Ultrastructural changes and programmed cell death of trophocytes in the gonad of *Isohypsibius granulifer granulifer* Thulin, 1928 (Tardigrada, Eutardigrada, Isohypsibiidae). Micron 70:26–33

Robinson DN, Cant K, Cooley L (1994) Morphogenesis of *Drosophila* ovarian ring canals. Development 120:2015–2025

Rouse GW (2006) Annelid sperm and spermiogenesis. In: Rouse G, Pleijel F (eds) Reproductive biology and phylogeny of Annelida. Science, Enfield, pp 45–76

Rousset V, Plaisance L, Erséus C, Siddall ME, Rouse GW (2008) Evolution of habitat preference in Clitellata (Annelida). Zool J Linnean Soc 95:447–464

Seidel HS, Smith TA, Evans JK, Stamper JQ, Mast TG, Kimble J (2018) *C. elegans* germ cells divide and differentiate along a folded epithelium. Dev Biol 442:173–187

Siddall ME, Apakupakul K, Burreson EM, Coates KA, Erséus C, Källersjö M, Gelder SR, Trapido-Rosenthal H (2001) Validating Livanow: molecular data agree that leeches, branchiobdellidans and *Acanthobdella peledina* are a monophyletic group of oligochaetes. Mol Phylogenet Evol 21:346–351

Siekierska E (2003) The structure of the ovary and oogenesis in the earthworm, *Dendrobaena veneta* (Annelida, Clitellata). Tissue Cell 47:37–48

Snapp EL, Iida T, Frescas D, Lippincott-Schwartz J, Lilly MA (2004) The fusome mediates intercellular endoplasmic reticulum connectivity in *Drosophila* ovarian cysts. Mol Bio Cell 15:4512–4521

Spałek-Wołczyńska A, Klag J, Bielecki A, Świątek P (2008) Oogenesis in four species of *Piscicola* (Hirudinea, Rhynchobdellida). J Morphol 269:18–28

Stolte HA (1934) Über selektive Eibildung bei *Stylaria lacustris* L. (Blastocytenstudien). Z Wiss Zool Abt A 145:79–98

Struck TH, Paul C, Hill N, Hartmann S, Hosel C, Kube M, Lieb B, Meyer A, Tiedemann R, Purschke G, Bleidorn C (2011) Phylogenomic analyses unravel annelid evolution. Nature 471:95–98

Świątek P (2005a) Structure of the germinal vesicle during oogenesis in leech *Glossiphonia heteroclita* (Annelida, Hirudinea, Rhynchobdellida). J Morphol 263:330–339

Świątek P (2005b) Oogenesis in the leech *Glossiphonia heteroclita* (Annelida, Hirudinea, Glossiphonidae). I. Ovary structure and previtellogenic growth of oocytes. J Morphol 266: 309–318

Świątek P (2006) Oogenesis in the leech *Glossiphonia heteroclita* (Annelida, Hirudinea, Glossiphonidae). II. Vitellogenesis, follicle cell structure and egg shell formation. Tissue Cell 38:263–270

Świątek P (2008) Ovary cord structure and oogenesis in *Hirudo medicinalis* and *Haemopis sanguisuga* (Clitellata, Annelida) – remarks on different ovaries organization in Hirudinea. Zoomorphology 127:213–226

Świątek P, Kubrakiewicz J, Klag J (2009) Formation of germ-line cysts with a central cytoplasmic core is accompanied by specific orientation of mitotic spindles and partitioning of existing intercellular bridges. Cell Tissue Res 337:137–148

Świątek P, Krok F, Bielecki A (2010) Germ-line cysts are formed during oogenesis in *Erpobdella octoculata* (Annelida, Clitellata, Erpobdellidae). Invertebr Reprod Dev 54:53–63

Świątek P, Urbisz AZ, Strużyński W, Płachno BJ, Bielecki A, Cios S, Salonen E, Klag J (2012) Ovary architecture of two branchiobdellid species and *Acanthobdella peledina* (Annelida, Clitellata). Zool Anz 251:71–82

Świątek P, Płachno BJ, Marchant R, Gorgoń S, Krodkiewska M, Małota K, Urbisz AZ (2016) Germ-line cells do not form syncytial cysts in the ovaries of the basal clitellate annelid *Capilloventer australis*. Zool Anz 260:63–71

Świątek P, De Wit P, Jarosz N, Chajec Ł, Urbisz AZ (2018a) Micromorphology of ovaries and oogenesis in *Grania postclitellochaeta* (Clitellata: Enchytraeidae). Zoology 126:119–127

Świątek P, Pinder A, Gajda Ł (2018b) Oogenesis and ovary organization in phreodrilid annelid – *Insulodrilus bifidus*. In: 14th International symposium on aquatic oligochaeta, Hirosaki, Japan. Organizing Committee of the ISAO 2018. Book of Abstracts, p 37

Szklarzewicz T (1998) The ovaries of scale insects (Hemiptera, Coccinea). Morphology and phylogenetic conclusions. Folia Histochem Cytobiol 36:15–165

Telfer WH (1975) Development and physiology of the oöcyte-nurse cell syncytium. Adv Insect Physiol 11:223–319

Tessler M, de Carle D, Voiklis ML, Gresham OA, Neumann JS, Cios S, Siddall ME (2018) Worms that suck: Phylogenetic analysis of Hirudinea solidifies the position of Acanthobdellida and necessitates the dissolution of Rhynchobdellida. Mol Phylogenet Evol 127:129–134

Timm T, Martin PJ (2015) Clitellata: Oligochaeta. In: Thorp JH, Rogers DC (eds) Thorp and Covich's freshwater invertebrates, Ecology and general biology, vol I, 4th edn. Elsevier, London, pp 529–549

Trauner J, Büning J (2007) Germ-cell cluster formation in the telotrophic meroistic ovary of *Tribolium castaneum* (Coleoptera, Polyphaga, Tenebrionidae) and its implication on insect phylogeny. Dev Genes Evol 217:13–27

Trentini M, Scanabissi FS (1978) Ultrastructural observations on the oogenesis of *Triops cancriformis* (Crustacea, Notostraca). I. Origin and differentiation of nurse cells. Cell Tissue Res 194:71–774

Tworzydlo W, Kisiel E, Jankowska W, Bilinski SM (2014) Morphology and ultrastructure of the germarium in panoistic ovarioles of a basal "apterygotous" insect *Thermobia domestica*. Zoology 117:200–206

Urbańska-Jasik D (1988) The ultrastructure of female reproductive cells in the ovary of *Herpobdella octooculata* (L.). Zool Poloniae 35:127–140

Urbisz AZ, Świątek P (2013) Ovary organization and oogenesis in two species of Lumbriculida (Annelida, Clitellata). Zoology 116:118–128

Urbisz AZ, Krodkiewska M, Świątek P (2010) Ovaries of Tubificinae (Clitellata, Naididae) resemble ovary cords found in Hirudinea (Clitellata). Zoomorphology 129:235–247

Urbisz AZ, Lai Y-T, Świątek P (2014) *Barbronia weberi* (Clitellata, Hirudinida, Salifidae) has ovary cords of the Erpobdella type. J Morphol 275:479–488

Urbisz AZ, Chajec Ł, Świątek P (2015) The ovary of *Tubifex tubifex* (Clitellata, Naididae, Tubificinae) is composed of one huge germ-line cyst that is enriched with cytoskeletal components. PLoS One 10:e0126173

Urbisz AZ, Chajec Ł, Brąszewska-Zalewska A, Kubrakiewicz J, Świątek P (2017) Ovaries of the white worm (*Enchytraeus albidus*, Annelida, Clitellata) are composed of 16-celled meroistic germ-line cysts. Dev Biol 426:28–42

Urbisz AZ, Chajec Ł, Ito M, Ito K (2018a) The ovary organization in the marine limnodriloidin *Thalassodrilides* cf. *briani* (Annelida: Clitellata: Naididae) resembles the ovary of freshwater tubificins. Zoology 128:16–26

Urbisz AZ, Lagnika M, Martin P, Świątek P (2018b) Ovary organization and oogenesis of *Haplotaxis* sp. (Clitellata: Haplotaxidae). In: 14th International symposium on aquatic Oligochaeta, Hirosaki, Japan. Organizing Committee of the ISAO 2018. Book of Abstracts, p 50

Van Damme N (1974) Organogénése de l'appareil génital chez la sangsue *Erpobdella octoculata* L. (Hirudinée; Pharyngobdelle). Arch Biol 85:373–397

Ventelä S (2006) Cytoplasmic bridges as cell-cell channels of germ cells. In: Baluska F, Volkmann D, Barlow PW (eds) Cell-cell channels. Springer, New York, pp 208–216

Ventelä S, Toppari J, Parvinen M (2003) Intercellular organelle traffic through cytoplasmic bridges in early spermatids of the rat: mechanisms of haploid gene product sharing. Mol Biol Cell 14:2768–2780

Wang C, Zhou B, Xia G (2017) Mechanisms controlling germline cyst breakdown and primordial follicle formation. Cell Mol Life Sci 74:2547–2566

Weigert A, Bleidorn C (2016) Current status of annelid phylogeny. Org Divers Evol 16:345–362

Weigert A, Helm C, Meyer M, Nickel B, Arendt D, Hausdorf B, Santos SR, Halanych KM, Purschke G, Bleidorn C, Struck TH (2014) Illuminating the base of the annelid tree using transcriptomics. Mol Biol Evol 31:1391–1401

Wolke U, Jezuit EA, Priess JR (2007) Actin-dependent cytoplasmic streaming in *C. elegans* oogenesis. Development 134:2227–2236

Yamamoto S, Bayat V, Bellen HJ, Tan C (2013) Protein phosphatase 1 limits ring canal constriction during *Drosophila* germline cyst formation. PLoS One 8:e70502

Yamashita YM (2018) Subcellular specialization and organelle behavior in germ cells. Genetics 208:19–51

Żelazowska M, Fopp-Bayat D (2017) Ovarian nests in cultured females of the Siberian sturgeon *Acipenser baerii* (Chondrostei, Acipenseriformes). J Morphol 278:1438–1449

MIX
Papier aus verantwortungsvollen Quellen
Paper from responsible sources
FSC® C105338
FSC
www.fsc.org

If you have any concerns about our products,
you can contact us on
ProductSafety@springernature.com

In case Publisher is established outside the EU,
the EU authorized representative is:
Springer Nature Customer Service Center GmbH
Europaplatz 3, 69115 Heidelberg, Germany

Printed by Libri Plureos GmbH
in Hamburg, Germany